Nanoparticle Assemblies and Superstructures

Nanoparticle Assemblies and Superstructures

Edited by

Nicholas A. Kotov

CRC Press
Taylor & Francis Group
Boca Raton London New York

CRC Press is an imprint of the
Taylor & Francis Group, an **informa** business

CRC Press
Taylor & Francis Group
6000 Broken Sound Parkway NW, Suite 300
Boca Raton, FL 33487-2742

First issued in paperback 2019

© 2006 by Taylor & Francis Group, LLC
CRC Press is an imprint of Taylor & Francis Group, an Informa business

No claim to original U.S. Government works

ISBN-13: 978-0-8247-2524-2 (hbk)
ISBN-13: 978-0-367-39228-4 (pbk)

Library of Congress Card Number 2005041260

Library of Congress Cataloging-in-Publication Data

Nanoparticle assemblies and superstructures / edited by Nicholas A. Kotov.
     p. cm.
  Includes bibliographical references and index.
  ISBN 0-8247-2524-7
  1. Nanoscience. 2. Nanostructures. 3. Nanostructures materials--Electric properties. 4. Nanotechnology. 5. Nanoparticles. I. Kotov, Nicholas A., 1965-

QC176.8.N35N353 2005
620'.5--dc22
                                           2005041260

Visit the Taylor & Francis Web site at
http://www.taylorandfrancis.com

and the CRC Press Web site at
http://www.crcpress.com

Dedication

Dedicated to my mother and father — one chemist and one physicist — who represent the cosmos in the miniature.

Preface

Surveying the current nanoscience literature, one can see a wide variety of shapes and morphologies of nanoscale particles that can be produced now. In addition to semiconductor and metal rods, wires, and core-shell particles that have already been extensively studied, the latest synthetic protocols demonstrate the possibility of making rings, cubes, tetrapods, triangular prisms, and many other exotic shapes. They include nanoacorns, nanocentipedes, nanoshells, nanowhiskers, and many other examples. Now the question is, What are we going to do with this nanocollection? The variety of available nanoscale objects can be considered building blocks of larger and more complex systems. Therefore, the present challenge of nanoscale science is to shift from *making* certain building blocks to *organizing* them in one-, two-, and three-dimensional structures. Such assemblies and superstructures are the next logical step in the development of nanoscience and nanotechnology. In this respect, one needs to pose the following questions:

1. What are the methods of organization of nanocolloids in more complex structures?
2. What kind of structures do we need?
3. What are the new properties appearing in the nanocolloid superstructures?

This book is the first attempt to answer these questions. It starts with two reviews assessing the current status of nanoparticle assemblies and the requirements for different applications of organized nanomaterials. The chapters in the second part of the book address changes in various properties of individual particles when they form agglomerates and simple assemblies. After that, different methods of organization of particles in the complex nanostructured superstructures are described. They include techniques involving biological ligands, force fields such as the magnetic field, layering protocols, and methods based on self-organization. The field of nanoscale assemblies and superstructures is developing very rapidly, and this list cannot be absolutely complete. There is no other way to make this book happen than to draw the line at some point and present a snapshot of the work in progress. I strongly believe that the answers to the questions mentioned above hold many interesting discoveries and surprising phenomena. This book is just the initial step toward these new scientific and technological advances that can bring a profound change to many areas of our lives.

In conclusion, I thank all the contributors of this volume. It is their hard work and excitement that give an edge to this book.

Editor

Nicholas A. Kotov was born in Moscow, Russia, in 1965. He graduated with honors from the Chemistry Department of Moscow State University in 1987. He subsequently received his Ph.D. degree in 1990 (advisor, Prof. M. Kuzmin) for research on photoinduced ion transfer processes at the liquid–liquid interface.

In 1992, he joined the group of Prof. J. Fendler in the Chemistry Department of Syracuse University as a postdoctoral associate, where he started working on the synthesis of nanoparticles and layer-by-layer assembly of nanostructured materials. Nicholas Kotov moved to Oklahoma State University to take a position as assistant professor in 1996 and was promoted to associate professor in 2001.

Currently, Nicholas Kotov is an associate professor in the Department of Chemical Engineering at the University of Michigan in Ann Arbor, sharing this appointment with the Departments of Materials Science and Biomedical Engineering. His research interests in the field of nanostructured material include synthesis of new nanocolloids, their organization in functional assemblies, layer-by-layer assembled nanocomposites, computer modeling of self-organization processes, ultrastrong materials from organized nanocolloids, nanowire-based nanodevices, biosensing applications of nanomaterials, interface of nanomaterials with living cells, and cancer treatment and diagnostics with nanoparticles.

He has received several state, national, and international awards for his research on nanomaterials, among which are the Mendeleev stipend, Humboldt fellowship, and CAREER award.

Contributors

L. Amirav
Department of Chemistry and Solid
 State Institute
Technion University
Haifa, Israel

M. Bashouti
Department of Chemistry and Solid
 State Institute
Technion University
Haifa, Israel

S. Berger
Materials Engineering
Technion University
Haifa, Israel

Shaowei Chen
Department of Chemistry and
 Biochemistry
Southern Illinois University
Carbondale, Illinois

Jinwoo Cheon
Department of Chemistry
Yonsei University
Seoul, Korea

George Chumanov
Department of Chemistry
Clemson University
Clemson, South Carolina

Jeffery L. Coffer
Department of Chemistry
Texas Christian University
Fort Worth, Texas

Helmut Cölfen
Department of Colloid Chemistry
Max Planck Institute of Colloids and
 Interfaces
MPI Research Campus Golm
Potsdam, Germany

Herwig Döllefeld
Institute of Physical Chemistry
University of Hamburg
Hamburg, Germany

Karen J. Edler
Department of Chemistry
University of Bath
Bath, United Kingdom

M. Eisen
Department of Chemistry
Technion University
Haifa, Israel

Alexander Eychmüller
Institute of Physical Chemistry
University of Hamburg
Hamburg, Germany

Latha A. Gearheart
Department of Chemistry and
 Biochemistry
University of South Carolina
Columbia, South Carolina

Sara Ghannoum
Department of Chemistry
American University of Beirut
Beirut, Lebanon

Michael Giersig
caesar research center (center of
  advanced european studies and
  research)
Bonn, Germany

Christian D. Grant
Department of Chemistry
Rutgers, the State University
  of New Jersey
Piscataway, New Jersey

Lara Halaoui
Department of Chemistry
American University of Beirut
Beirut, Lebanon

Li Han
Department of Chemistry
State University of New York at
  Binghamton
Binghamton, New York

J.W. Harrell
Center for Materials for Information
  Technology
University of Alabama
Tuscaloosa, Alabama

Gregory V. Hartland
Department of Chemistry and
  Biochemistry
University of Notre Dame
Notre Dame, Indiana

Michael Hilgendorff
caesar research center (center of
  advanced european studies and
  research)
Bonn, Germany

Min Hu
Department of Chemistry and
  Biochemistry
University of Notre Dame
Notre Dame, Indiana

Jad Jaber
Department of Chemistry
American University of Beirut
Beirut, Lebanon

Nikhil R. Jana
Department of Chemistry and
  Biochemistry
University of South Carolina
Columbia, South Carolina

Christopher J. Johnson
School of Chemistry
University of Bristol
Bristol, United Kingdom

Young-wook Jun
Department of Chemistry
Yonsei University
Seoul, Korea

Shishou Kang
Center for Materials for Information
  Technology
University of Alabama
Tuscaloosa, Alabama

Nancy Kariuki
Department of Chemistry
State University of New York at
  Binghamton
Binghamton, New York

Keisaku Kimura
Department of Material Science
Graduate School of Science
Himeji Institute of Technology
Hyogo, Japan

Nicholas Kotov
Department of Chemical Engineering
University of Michigan
Ann Arbor, Michigan

Nina I. Kovtyukhova
Department of Chemistry
Pennsylvania State University
University Park, Pennsylvania
and
Institute of Surface Chemistry
N.A.S.U.
Kiev, Ukraine

M. Krueger
Department of Physics and Solid State
  Institute
Technion University
Haifa, Israel

Seung Jin Ko
Department of Chemistry
Yonsei University
Scoul, Korca

R.D. Levine
The Fritz Haber Research Center for
  Molecular Dynamics
Hebrew University of Jerusalem
Jerusalem, Israel
and
Department of Chemistry and
  Biochemistry
University of California–Los Angeles
Los Angeles, California

E. Lifshitz
Department of Chemistry and Solid
  State Institute
Technion University
Haifa, Israel

Luis M. Liz-Marzán
Departamento de Química Física
Universidade de Vigo
Vigo, Spain

Jin Luo
Department of Chemistry
State University of New York at
  Binghamton
Binghamton, New York

Thomas E. Mallouk
Department of Chemistry
Pennsylvania State University
University Park, Pennsylvania

Serhiy Z. Malynych
Department of Chemistry
Clemson University
Clemson, South Carolina

Stephen Mann
School of Chemistry
University of Bristol
Bristol, United Kingdom

Mariezabel Markarian
Department of Chemistry
American University of Beirut
Beirut, Lebanon

Mathew M. Maye
Department of Chemistry
State University of New York at
  Binghamton
Binghamton, New York

Alf Mews
Institut für Physikalische Chemie
Universität Mainz
Mainz, Germany

Paul Mulvaney
School of Chemistry
The University of Melbourne
Melbourne, Australia

Catherine J. Murphy
Department of Chemistry and
  Biochemistry
University of South Carolina
Columbia, South Carolina

Christof M. Niemeyer
Universität Dortmund, Fachbereich
  Chemie
Biologisch-Chemische
  Mikrostrukturtechnik
Dortmund, Germany

David E. Nikles
Center for Materials for Information
  Technology
University of Alabama
Tuscaloosa, Alabama

Thaddeus J. Norman, Jr.
Lawrence Livermore National
  Laboratory
Livermore, California

Sherine O. Obare
Department of Chemistry and
  Biochemistry
University of South Carolina
Columbia, South Carolina

Isabel Pastoriza-Santos
Departamento de Química Física
Universidade de Vigo
Vigo, Spain

F. Remacle
The Fritz Haber Research Center for
  Molecular Dynamics
Hebrew University of Jerusalem
Jerusalem, Israel
and
Département de Chimie
Université de Liège
Liège, Belgium

Andrey L. Rogach
Institute of Physical Chemistry
University of Hamburg
Hamburg, Germany

Sarah K. St. Angelo
Department of Chemistry
Pennsylvania State University
University Park, Pennsylvania

A. Sashchiuk
Department of Chemistry and Solid
  State Institute
Technion University
Haifa, Israel

Seiichi Sato
Department of Material Science
Graduate School of Science
Himeji Institute of Technology
Hyogo, Japan

Elena V. Shevchenko
Institute of Physical Chemistry
University of Hamburg
Hamburg, Germany

U. Sivan
Department of Physics and Solid State
  Institute
Technion University
Haifa, Israel

Xiangcheng Sun
Center for Materials for Information
  Technology
University of Alabama
Tuscaloosa, Alabama

Dmitri V. Talapin
Institute of Physical Chemistry
University of Hamburg
Hamburg, Germany

Zhiyong Tang
Department of Chemical Engineering
University of Michigan
Ann Arbor, Michigan

Horst Weller
Institute of Physical Chemistry
University of Hamburg
Hamburg, Germany

Yan Xin
National High-Magnetic Field
  Laboratory
Florida State University
Tallahassee, Florida

Hiroshi Yao
Department of Material Science
Graduate School of Science
Himeji Institute of Technology
Hyogo, Japan

Shu-Hong Yu
Hefei National Laboratory
  for Physical Sciences at Microscale
Department of Materials Science and
  Engineering and Structural Research
  Laboratory of CAS
University of Science and Technology of
  China
Hefei, People's Republic of China

Jin Z. Zhang
Department of Chemistry
University of California
Santa Cruz, California

Chuan-Jian Zhong
Department of Chemistry
State University of New York at
  Binghamton
Binghamton, New York

Contents

Nanoscale Superstructures: Current Status

Chapter 1 Organization of Nanoparticles and Nanowires in Electronic
Devices: Challenges, Methods, and Perspectives.............................. 3

Nicholas Kotov and Zhiyong Tang

Chapter 2 Colloidal Inorganic Nanocrystal Building Blocks 75

Young-wook Jun, Seung Jin Ko, and Jinwoo Cheon

Electronic Properties of Nanoparticle Materials: From Isolated Particles to Assemblies

Chapter 3 Fluorescence Microscopy and Spectroscopy of Individual
Semiconductor Nanocrystals.. 103

Alf Mews

Chapter 4 Coherent Excitation of Vibrational Modes of Gold Nanorods 125

Gregory V. Hartland, Min Hu, and Paul Mulvaney

Chapter 5 Fabricating Nanophase Erbium-Doped Silicon into Dots,
Wires, and Extended Architectures.. 139

Jeffery L. Coffer

Chapter 6 Conductance Spectroscopy of Low-Lying Electronic States
of Arrays of Metallic Quantum Dots: A Computational Study 153

F. Remacle and R.D. Levine

Chapter 7 Spectroscopy on Semiconductor Nanoparticle Assemblies 179

Herwig Döllefeld and Alexander Eychmüller

Chapter 8 Optical and Dynamic Properties of Gold Metal Nanomaterials:
From Isolated Nanoparticles to Assemblies 193

Thaddeus J. Norman, Jr., Christian D. Grant, and Jin Z. Zhang

Chapter 9   Synthesis and Characterization of PbSe Nanocrystal Assemblies......207

M. Bashouti, A. Sashchiuk, L. Amirav, S. Berger, M. Eisen,
M. Krueger, U. Sivan, and E. Lifshitz

Biological Methods of Nanoparticle Organization

Chapter 10 Biomolecular Functionalization and Organization
of Nanoparticles 227

Christof M. Niemeyer

Chapter 11 Nature-Inspired Templated Nanoparticle Superstructures...............269

Shu-Hong Yu and Helmut Cölfen

Assembly of Magnetic Particles

Chapter 12 Magnetic Nanocrystals and Their Superstructures.........................341

Elena V. Shevchenko, Dmitri V. Talapin,
Andrey L. Rogach, and Horst Weller

Chapter 13 Synthesis, Self-Assembly, and Phase Transformation of FePt
Magnetic Nanoparticles 369

Shishou Kang, Xiangcheng Sun, J.W. Harrell, and
David E. Nikles

Chapter 14 Assemblies of Magnetic Particles: Properties and Applications 385

Michael Hilgendorff and Michael Giersig

Layered Nanoparticle Assemblies

Chapter 15 Template Synthesis and Assembly of Metal Nanowires
for Electronic Applications 413

Sarah K. St. Angelo and Thomas E. Mallouk

Chapter 16 Heteronanostructures of CdS and Pt Nanoparticles
in Polyelectrolytes: Factors Governing the Self-Assembly
and Light-Induced Charge Transfer and Transport Processes 437

Sara Ghannoum, Jad Jaber, Mariezabel Markarian, Yan Xin,
and Lara I. Halaoui

Chapter 17 Layer-by-Layer Assembly Approach to Templated Synthesis
of Functional Nanostructures..463

Nina I. Kovtyukhova

Chapter 18 Coherent Plasmon Coupling and Cooperative Interactions
in the Two-Dimensional Array of Silver Nanoparticles.................487

George Chumanov and Serhiy Z. Malynych

Self-Assembly of Nanoscale Colloids

Chapter 19 Self-Organization of Metallic Nanorods into Liquid
Crystalline Arrays 515

*Catherine J. Murphy, Nikhil R. Jana, Latha A. Gearheart,
Sherine O. Obare, Stephen Mann, Christopher J. Johnson,
and Karen J. Edler*

Chapter 20 Tailoring the Morphology and Assembly of Silver
Nanoparticles Formed in DMF..525

Isabel Pastoriza-Santos and Luis M. Liz-Marzán

Chapter 21 Interparticle Structural and Spatial Properties of Molecularly
Mediated Assembly of Nanoparticles...551

*Chuan-Jian Zhong, Li Han, Nancy Kariuki, Mathew M. Maye,
and Jin Luo*

Chapter 22 Langmuir–Blodgett Thin Films of Gold Nanoparticle
Molecules: Fabrication of Cross-Linked Networks and
Interfacial Dynamics 577

Shaowei Chen

Chapter 23 Self-Assembling of Gold Nanoparticles at an Air–Water
Interface...601

Hiroshi Yao, Seiichi Sato, and Keisaku Kimura

Index..615

Nanoscale Superstructures: Current Status

1 Organization of Nanoparticles and Nanowires in Electronic Devices: Challenges, Methods, and Perspectives

Nicholas Kotov and Zhiyong Tang

CONTENTS

1.1 Introduction 4
1.2 Prototype Devices Based on Single NPs and NWs 5
    1.2.1 Nanodevices Based on NPs 5
    1.2.2 Nanodevices Based on Semiconductor NWs 7
1.3 Prototype Devices from Thin Films of Nanocolloids 11
    1.3.1 Devices Based on Thin Films of NPs 11
        1.3.1.1 Single-Electron Charging 12
        1.3.1.2 Photoelectronic Devices from NP Thin Films 12
        1.3.1.3 Electroluminescence Devices 13
        1.3.1.4 Photovoltaic Devices 15
        1.3.1.5 Electrochromic Devices 18
        1.3.1.6 Sensors and Biosensors 21
    1.3.2 NWs as Molecular Electronic Devices...23
        1.3.2.1 Nanocircuits 24
        1.3.2.2 Electroluminescence Devices 27
        1.3.2.3 Sensors 28
1.4 Strategies of NP and NW Assembly into More Complex Structures...........30
    1.4.1 Assembly of NPs 31
        1.4.1.1 One-Dimensional Assemblies of NPs 31
        1.4.1.2 Two-Dimensional Assemblies of NPs................................37
        1.4.1.3 Three-Dimensional Assemblies of NPs 41

        1.4.2   Assemblies of Single NWs 47
                1.4.2.1   Physical Method of Aligning NWs 48
                1.4.2.2   Chemical Method of NW Alignment 50
                1.4.2.3   Biological Method of NW Alignment 52
        1.4.3   Superstructures of NWs and NPs 52
1.5   Prospects 53
References 54

1.1 INTRODUCTION

Physical dimensions of electronic devices are approaching the limit of manufacturing technologies. The elements in new integrated circuitry reached the critical size of 60 nm, which is beyond the limitation of conventional photolithography. Although some methods of lithography, such as electron beam, x-ray, and scanning probe microscopy lithography, are capable of overcoming this limitation,[1] high cost and slow processing impede their practical application for mass production.

One of the possible alternative ways to circumvent the bottleneck of low dimensions is self-organization of 60-nm and smaller structures from nanoscale building blocks. Although it is difficult to imagine an intricate electronic circuit forming from small pieces of semiconductor and metal spontaneously, the self-assembled superstructures can serve as vital elements in electronic circuits, which are interfaced to other devices by traditional lithographic means.[2-4] This approach assumes, first, the availability of nanoscale elements with dimensions from 1 to 50 nm, such as nanoparticles (NPs),[5] nanowires (NWs), and nanotubes,[6] and molecular compounds,[7] and, second, the availability of methods of their organization and self-organization into more complex structures. These methods of structural design are the primary subject of this review.

The problem of organization at the nanoscale level actually emerged long before the appearance of nanoscience as a separate discipline. Research in this area started about 30 years ago, when pioneering theoretical calculation by Aviram and Ratner showed the possibility of using single molecules as electronic elements.[8] This study opened the field of molecular electronics, which tackles very similar problems because molecules whose lengths often exceed 1 nm need to be properly positioned with respect to each other and electrodes to function as electronic devices. Molecular electronics and nanoscience also share the same problems when it comes to prospects of device manufacturing. The inevitable intrinsic defects and disorder of molecular assemblies will be a significant obstacle for reproducible molecular and nanoscale electronics.[9,10] Regardless of the many challenges lying on the way to nanocomputers, understanding the processes of self-organization of nanoscale elements will be exceptionally beneficial both for practical applications and for fundamental nanoscale science.

This article surveys current research on NPs and NWs as device elements and materials components for photonics and electronics. Since the synthesis and application of carbon nanotubes (CNTs) have been extensively reviewed before,[11-14] they are excluded from the scope of this paper. Some examples of CNTs are nonetheless mentioned here, because several experimental techniques used for organization of

CNTs are also applicable to NWs. This material is organized into three parts: (1) basic electronic properties of single NPs and NWs as functional elements of high-tech devices; (2) demonstrated applications of NP and NW assemblies in electronics, photonics, and sensors; and (3) methods of structural design of NP- and NW-based materials. We hope that such composition gives the reader a fairly complete view of the subject matter. Even though some of the important papers highly relevant to the problem of organization at the nanoscale level and superstructures from nano-colloids will come out after this book's publication, the fundamental approach taken here will still be useful for anyone interested in critical assessments of prospects of NPs and NWs for photonics and electronics.

1.2 PROTOTYPE DEVICES BASED ON SINGLE NPs AND NWs

1.2.1 NANODEVICES BASED ON NPs

Many prototype devices with NPs are related to the single-electron tunneling (SET) effect. Theoretically predicted to occur in quantized systems,[15,16] SET was first experimentally observed for microscopic systems. Fulton and Dolan observed that the charging effects in bulk metal–insulator–nanoparticle–insulator–bulk metal (MINIM) junctions could be modulated at 1 K by the gate voltage.[17] The discreteness of charge cannot be observed in conventional electronics at room temperature, because large thermal energy KT (~26 meV) overcompensates the energy of single-electron tunneling. From the equation $W = Q^2/2C$, where Q is the total charge stored on the central particle and C is the junction capacity equation, one can see that if C is small (<10 to 18 F) and the resistance is high, the work energy is large enough to resist the thermal drift at room temperature. Small capacity and large resistance of junction require the dimensions of the device to be within a few nanometers.

Colloidal synthesis of metal or semiconductor NPs made a strong impact on nanoscale electronics.[5] SET-based analogs of field effect transistors (FETs) can be smaller and faster than conventional devices but require atomic-scale precision in NP placement between the electrodes. Scanning tunneling microscopy (STM) and corresponding scanning tunneling spectroscopy (STS) are convenient experimental tools to investigate SET and to position the particles at the right distance from the source and drain electrodes.

The gaps on each side of NP (Figure 1.1A) act as double-insulator tunnel junctions with low capacity and high resistance, as required for SET to occur at room temperatures. When bias is applied between STM tips and substrates, the current discretely increases with a step of e/2C. Andres et al.[18] and, later, Taleb et al.[19] observed predicted SET phenomena at room temperature in STS spectra of Au or Ag NPs adsorbed to thiol-modified Au substrates.

The crystal structure of NPs influences the SET behavior as a result of its effect on the ladder of their electronic levels. For example, additional peaks in i-V curves were observed for crystalline, compared to amorphous, Pd NPs.[20,21] They were attributed to a certain band gap state associated with small-size metal NPs. The dependence of SET on the band gap is more prominent for semiconductor InAs NPs (Figure 1.1B).[22,23] On the side of positive bias, which corresponded to the conduction

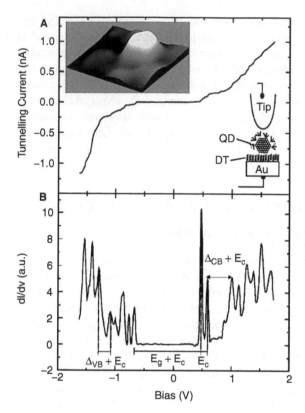

FIGURE 1.1 STM (A) and STS (B) of a single InAs nanocrystal 32 Å in radius, acquired at 4.2 K. (From Banin, U. et al., *Nature*, 400, 542–544, 1999. With permission.)

band (CB), two groups of peaks were identified. Two closely spaced peaks appeared after current onset, followed by a larger gap and a group of six peaks with equal spacing between them of about 110 mV, which arose from charging energies of SET. The first group of peaks had $1S_e$ characteristics, and thus doublets were observed. The six peaks in the second group stemmed from the degeneracy of the $1P_e$ state. It was also noticed that the energy separation was smaller for negative bias corresponding to the valence band (VB). This is consistent spin–orbit coupling and split-off[24] in the VB electronic levels. As expected, SET decreased with the increase of NP sizes, and the values of energy gaps between ground and excitation states from STS measurements were virtually identical to those from optical spectroscopy.

SET effect in single NPs was used in single-electron transistors.[25–27] Klein et al. designed a pair of source and drain electrodes with a space distance of only about 10 nm (Figure 1.2). Since nanoparticles had a much stronger Van der Waals (VDW) attraction at the place of steps or grooves than flat substrates,[27] the NPs intended to shift on the electrodes and were trapped in the nanogap between source and drain electrodes. By applying gate voltage, the number of electron transfer events can be accurately controlled.

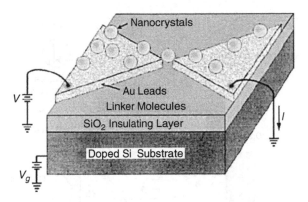

FIGURE 1.2 Scheme of single-electron transistor of CdSe NP. (From Klein, D.L. et al., *Nature*, 389, 699–701, 1997. With permission.)

Amlani and Orlov[28,29] produced an NP logic gate consisting of four quantum dots connected in a ring by tunnel junctions, and two single-dot electrometers. In such quantum dot cellular automata, the digital data are encoded in the positions of only two electrons. The device is operated by applying inputs to the gate of the cell. The logic AND and OR operations are successfully realized as the electrometer outputs.

1.2.2 NANODEVICES BASED ON SEMICONDUCTOR NWS

Compared to CNTs, semiconductor NWs represent a more recent development in the field of nanoscale electronics and have a substantially greater range of chemical properties and, thus, more practical applications and methods of organization. The chemical methods of NW synthesis were recently reviewed by Wu et al.,[6] Xia et al.,[30] and Rao et al.[31] Since the preparations strongly affect the approaches to subsequent functionalization, they will be briefly mentioned here. The NW preparations involve two main types: wet colloidal and gas phase synthesis. The first group includes template-directed,[32] solution–liquid–solid,[33–35] supercritical fluid–liquid–solid,[36] thermal decomposition of metallorganic precursors,[37] and self-assembly from single NPs.[38] Gas phase methods include vapor–liquid–solid,[39] oxide-assisted,[40] and vapor–solid.[41] Wet methods are generally more amenable to subsequent modification reactions and therefore assembly coding. At the present moment, however, the complexity of the superstructures obtained from gas phase–produced wires is somewhat higher.

Cui et al.[42] achieved an important milestone for this area, i.e., successful doping of silicon NWs by adding reaction precursors of boron (B) or phosphorus (P) to the vapor source. Analogous to the conventional bulk semiconductor doping, B-doped Si NWs exhibit characteristic p-type behavior, and the P-doped Si NWs show n-type behavior. As an alternative route, the NWs with different electrical properties can also be produced through choosing appropriate semiconductor materials.[43,44] For instance, group III-V InP and group II-VI CdS NWs are good n-type semiconductors, and IV group Si NWs exhibit p-type characteristics.

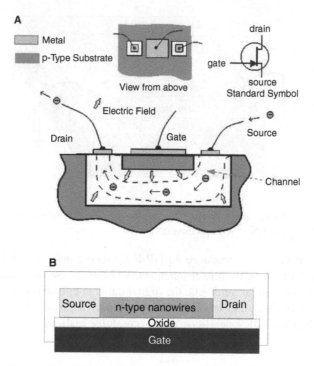

FIGURE 1.3 Scheme of conventional (A) and NW (B) FETs.

In order to fabricate NW FETs, NWs are placed on silicon substrates with a surface oxide layer, and then both conductive source and drain electrodes are sputtered on the ends of NWs (Figure 1.3). As with a conventional FET device (Figure 1.3A), only when positive potential is applied to the gate electrode is the current allowed to pass through the n-type NWs between source and drain electrodes. Although the structures and work principles are identical, NW FETs exhibit several advantages compared to traditional FETs. Besides the small size and high integration density, the transconductance of NW FETs can be as high as 3000 nAV^{-1}, which is 10 times higher than that of conventional metal oxide FETs.[45] This feature should proportionally increase the speed of the nanoscale computers. At the same time, one needs to mention that the precision of assembly of NW FET should be in the angstrom scale. Otherwise, fault-tolerant computational schemes must be used, which can operate with less reproducible behavior of elementary transistors.[46] This eventually will reduce the computational speed.

Other types of electrical characteristics of NWs can be achieved by their surface modification. Note that there are many examples of CNT superstructures with various nanoscale entities.[47–50] Reports on surface modification of semiconductor and metal NWs are significantly less abundant. Lauhon et al.[51] prepared a variety of coaxial NWs with p–n junction. The i-Ge/p-Si core-shell, i-Si/SiO$_x$/p-Si core-doubleshell, or p-Si/i-Ge/SiO$_x$/p-Ge core-multishell structures were produced by sequential chemical vapor deposition (i and p denote intrinsic and p-type conductivities, respectively). These structures can be convenient for the further minimization of NW FET by integrating all drain, source, and gate electrodes onto the single NW. Electrical measurements showed that the transconductance of coaxial structures was as high as 1500 nAV^{-1} for source–drain bias as low as 1 V. This property can be optimized depending on the device specification taking advantage of variable types of NWs.[52]

Coaxial NW structures of a different type — with an insulating layer — can be equally useful and possibly even more common than those with n–p junctions. The influence of external conditions and surrounding molecules on NW properties is beneficial for sensing[53] but must be tightly regulated when NWs are employed for other purposes. For instance, field effects and temperature-dependent surface adsorption/desorption equilibrium are likely to produce strong cross talk and operational errors. Their influence in nanocircuits is expected to be much stronger than in conventional ones. Insulation of NWs with a layer of inert material may prevent or reduce the strong noise in nanocircuits. Such a layer can protect the surface from adsorption of unwanted species, prevent charge injection, and partially screen the external fields when necessary. For sensor applications, a thin insulation layer can also be used to improve the environmental stability of the devices.[54–56] One of the materials most suitable for NW protection is silica. It has a high-voltage breakdown potential and dielectric constant,[57] considerable mechanical strength, and exceptional resilience to environmental factors. Moreover, silicon oxide processes are common for the modern electronic industry and can be transferred to the new generation of devices. Ag, SiC, Si, and other NWs have been coated with a shell of silica to meet different requirements.[58–60] All of these NWs except Ag are processed in nonaqueous solvents or solid-state materials via vapor phase or high-temperature routes. Very recently, subsequent deposition of silica and metal in the porous alumina templates was explored to prepare insulating Au nanowires.[61] Besides silica, some polymers can also be considered for making the insulating layer. As such, CdSe/poly(vinyl acetate)[62] and Au/polystyrene[63] core-shell one-dimensional structures with small aspect ratios were produced. The actual insulating capabilities of the organic or inorganic shell strongly depend on its quality and remain unknown for both types of coatings.

Liang et al.[64] used the sol-gel method to make coaxial SiO_2-on-CdTe NWs with controlled thickness of the silica layer in the range of 10 to 66 nm. Importantly, the silica layer on the ends of the NWs was consistently thinner than on its sides. This afforded the preparation of first examples of open-ended coaxial NWs with electrical accessibility of the semiconductor core (Figure 1.4A). Conducting atomic force microscopy measurements (Figure 1.4B) demonstrated that the SiO_2 coating was indeed insulating, with bias voltage reaching as high as 12 V for any thickness when the coating remained uniform (>10 nm). Taking advantage of the same technique, the current through the naked NWs was found to be mediated by mid-band-gap states. Interestingly enough, CdTe NWs stabilized with mercaptosuccinic acid produce coaxial SiO_2 composites with rather unusual morphologym, resembling nanoscale centipedes (Figure 1.5).[65] The thickness of the silica layer in the ends still remains thinner than on the NW sides. Besides the distinctive three-dimensional configuration, such geometry provides exceptionally high surface area and a structural foundation for highly branched nanocolloid superstructures. In addition, it will guarantee superb matrix connectivity in the nanowire–polymer composites for optical materials with stringent requirements to mechanical properties.[66]

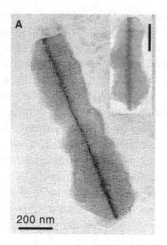

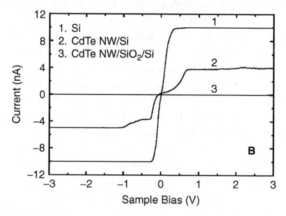

FIGURE 1.4 (A) TEM images of silica-coated NWs. Scale in the insert is 150 nm. (B) Conductive AFM *i-V* curves for As-doped silicon wafer (1), naked CdTe NW (2), and silica-coated NW samples (3). The bias in conductive AFM measurements is applied to the substrate while the probe is grounded. (From Liang, X. et al., *Langmuir*, 20, 1016–1020, 2004. With permission.)

Speaking of strength, a few words need to be said comparing devices made from NWs and CNTs because both of them share the same niche in the domain of applications. CNTs were used as field effect transistors, complementary inverters, and logic gates.[67] The advantages of nanotubes for nanoscale electronics are in exceptionally high mobility of charges, high sustainable currents, light weights, and excellent stiffness and other mechanical properties. Unfortunately, both multiwall and single-wall CNTs have a number of problems not related to the challenges of their organization. First of all, despite great success in designing methods of selective growth and separation of different CNTs,[68–72] the starting material for related devices remains a fairly complex mixture. Additionally, the modification of CNTs, which should allow for their self-assembly, also remains an unsolved issue.[73,74] Their

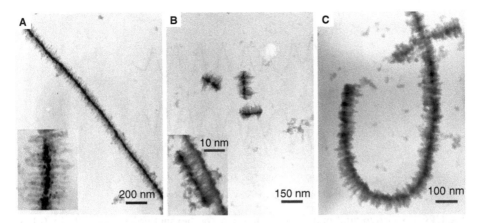

FIGURE 1.5 TEM images of mercaptosuccinic acid–stabilized CdTe/SiO$_2$ core-shell structures obtained with different amounts of sol-gel agents. (From Wang, Y. et al., *Nano Lett.*, 4, 225–231, 2004. With permission.)

exceptional properties gradually disappear with progressively higher degrees of graphene sheet functionalization. In addition, environmentally stable doping of CNTs is equally difficult.[75] Unlike CNTs,[76] the inorganic semiconductor and metal NWs provide greater flexibility in chemical and electronic properties. Simultaneously, they give rise to greater dependence of NW properties on surface passivation methods. The role of surface states in the electronic properties of NWs needs to be investigated in much greater detail. As such, the dependence of transconductance in semiconductor NWs on the surface composition and thermal activation of charge carried in them can play a significant role in their function as electronic devices.

1.3 PROTOTYPE DEVICES FROM THIN FILMS OF NANOCOLLOIDS

Taking a step up in respect to system complexity from electronic circuits based on a single NP or NW, one needs to consider devices based on thin films of nanocolloids. In addition to the control of NP/NW electrode positioning or contact, one also needs to tune the NP–NP, NP–NW, or NW–NW gaps. Despite the increase of complexity, this task might actually be easier to accomplish than it is for single-element devices because of the large number of particles involved, and these distances become averaged in nature. Consequently, lower precision in particle positioning is required, which makes it more suitable for the current experimental tools and also more practical.

1.3.1 Devices Based on Thin Films of NPs

The best progress in this field has been achieved for NP thin films largely because they have a longer history than coatings from NWs. Thus, the overview of this subject will be given first.

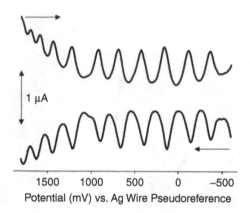

FIGURE 1.6 Differential pulse voltammogram for hexanethiolate Au NPs in 0.1 M Bu_4NPF_6/CH_2Cl_2 at Pt electrode. (From Hicks, J.F. et al., *J. Am. Chem. Soc.*, 124, 13322–13328, 2002. With permission.)

1.3.1.1 Single-Electron Charging

Similar to single NPs placed between electrodes (Figure 1.2), discrete electron transport phenomena occur in NP assemblies as long as they are separated by a tunneling gap. However, the particle and gap size distributions blur the SET peaks and make them less pronounced than those of single NPs. This problem can be alleviated by the exploration of monodispersed gold clusters protected by uniform monolayers from, say, thiols, which narrows the spread of both parameters essential for SET. Under these conditions, SET can be observed for the solution of noble metal NPs by traditional electrochemical techniques.[77–80] The discrete single-electron charging peaks, also called the ensemble Coulomb staircase, were observed in the double-layer capacitances zone of differential pulse voltammograms (DPVs) (Figure 1.6). One can observe up to 13 SET peaks in the Au NP solution by this method.[81] As expected, the amplitude and distance between individual SET peaks decreased with the increase of sizes of NPs from 1.1 to 1.9 nm (8- to 38-kDa core mass).[82] The practical application requests that the NP films register in the form of thin film, which has secondary importance as long as NPs are in a closely packed or otherwise controlled spatial arrangement.[83,84] The SET peaks of adsorbed Au NPs were also found to be very sensitive to the type of ions in the solution,[85] which indicated that the SET phenomena could be used in the thin-film ion sensors. This effect can be further enhanced by molecular design of the coating around the NPs or the matrix in which they are impeded. Selective expansion of NP gaps in response to the presence of specific analytes can be predicted.[86]

1.3.1.2 Photoelectronic Devices from NP Thin Films

Electrical and optical properties of NPs can be combined for practical purposes in three different modalities:

1. When potential is applied to semiconductor NPs, the electrons can be excited to the conduction band, and subsequent radiative recombination between hole and electron releases the stored energy as a photon, giving rise to electroluminescence.
2. Conversely, the irradiation with UV-visible light produces electron–hole pairs, which may generate photocurrent via the effective routes of charge separation.
3. Electrochromism of NP materials results from the change of optical absorption upon applying electrical potential. For instance, the electrons excited to the $1S_e$ excitation state in CB can further absorb the energy in the infrared (IR) part of the spectrum and go to a higher $1P_e$ state.[87–89]

Let us consider this combination of electrical and optical processes in NP materials and corresponding devices in the framework of NP organization.

1.3.1.3 Electroluminescence Devices

The goal of structural optimization in electroluminescent devices is to maximize radiative recombination of electrons and holes injected from both sides of an NP film sandwiched between the cathode and anode. Colvin et al.[90] and Schlamp et al.[91] were the first to report on the NP light-emitting diode (LED) made by building up binary layers of poly(p-phenylenevinylene) (PPV) polymer and CdSe or CdSe/CdS core-shell NPs onto the surface of indium tin oxide (ITO) electrodes (Figure 1.7A). The conductive polymer acts as a hole transport layer resting on an ITO hole injection electrode. The top coatings of Ag and Mg facilitated the injection of negative charge in the electron transport layer made from NPs. The radiative recombination on the PPV and NP interface gave rise to the strong luminescence near the band edge of CdSe NPs (Figure 1.7C). For the forward bias (positive bias was applied at the ITO electrode; electrons and holes are injected into the n- and p-conductive layers, respectively), the external quantum efficiency was as high as 0.1%. The reverse bias gave only little light, and the spectrum of the emission showed the additional peak from PPV emission (Figure 1.7D).

The NP layer is probably not the best electron transport layer because of the layer of stabilizer coating the semiconductor cores. Additionally, injection of charge in the NPs prevents efficient emission from them, stimulating other processes such as Auger recombination manifesting, for instance, as emission blinking.[92,93] Rational spatial organization of the layers of the device can significantly improve its performance. A recent study by Coe et al.[94] showed that when the CdSe/CdS NP monolayer was placed exactly at the interface of the electron–hole transport layer made from different polymers, the light-emitting efficiency increased 25 times, and brightness and external quantum efficiency were up to 1.6 cd A^{-1} and 0.6%, respectively.

Currently, NP and organic LEDs have approximately similar performance characteristics with some lead of the latter in longevity and commercial implementation. NP devices, however, have intrinsically higher hue purity and can easily be adjusted to emit in different parts of the spectrum from ultraviolet to visible to infrared.[95–100]

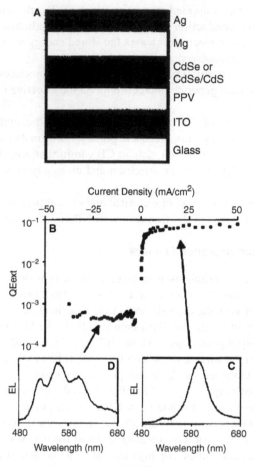

FIGURE 1.7 (A) Layer sequence in a typical light-emitting diode. (B) Dependence of external quantum efficiency vs. current density of the device on the left. Positive current values correspond to ITO positively biased with respect to Mg. Inset: Spectra of light emitted in forward (C) and reverse (D) bias. (From Schlamp, M.C. et al., *J. Appl. Phys.*, 82, 5837–5842, 1997. With permission.)

Incorporation of other nanocolloids such as NWs or nanorods, which can improve electron (or hole) transport in the films, can be expected to boost their performance.[101] Thus, optimization of the layer sequence shall remain one of the primary sources of the performance improvements; for example, Achermann et al.[102] reported the considerable enhancement of emission color of CdSe NPs placed proximal to the expitaxial quantum well via energy transfer pumping.

As with polymer LEDs, one of the future directions of further development of NP devices with electroluminescence (or with a similar process called electrogenerated chemiluminescence[103–105]) will be thin-film lasers. Observation of optical gain for semiconductor NPs, despite the strong Auger recombination process in them,[106] raises hopes that an appropriate thin-film structure acting as an optical resonator can lead to electrically excited NP lasers. Unconventional optical schemes, including photonic crystals,[107] can become quite beneficial for this purpose.

1.3.1.4 Photovoltaic Devices

The fairly broad UV-visible spectrum of NPs and the high oscillation strength of VB-to-CB transitions make them exceptional candidates for photovoltaic devices. Among various types of semiconductors, thin films from TiO_2 colloids and similar metal oxide NPs have probably been the most explored for this purpose, because of high photocorrosion stability.[108–110] They are typically used in conjunction with hole scavengers and oxidation-resistant dyes to prevent fast photobleaching. Metal and metal chalcogenides are often used with semiconducting polymers, which cannot withstand the attack of free radicals generated on the TiO_2 surface upon illumination. NPs added to the matrix of polymers cause dissociation of excitons or, in other words, physical separation of the positive and negative charges.[108] In such systems, the hole conductivity is primarily associated with a polymer component and electrical transport is typically taking place in the part containing NPs. This is true for the majority of systems, with the exception of probably CdTe and some other semiconductors with quite low redox potential.[111–114] Material-specific charge transport represents the fundamental similarity between light-emitting and photovoltaic systems. As well, the thickness of the films should be kept relatively small for both applications to avoid losses during the transport to or from the electrode. The major difference between a solar cell and LED is that the former should maximize electron–hole separation and optical density rather than light emission. The major challenges of photovoltaic devices include improving efficiency of the charge separation and charge collection, which can be answered by selecting an appropriate nanocolloid and by the intelligent design of the film structure. The numerous works on Gratzel's solar cells demonstrated that a molecular (i.e., Angstrom scale) arrangement of the photosensitizer and the charge transport layer is critical for high-output photovoltaic elements.[115] In this part, we shall focus on the effect of *nanoscale* organization of thin films.

Initial charge separation at the polymer–NP interface is typically very fast (ca. 800 fsec).[116,117] However, the back electron transfer, i.e., electron–hole recombination, is also quite fast and presents the major limitation for efficient charge separation.[118] This problem can be overcome by combining two different NPs in the photoactive centers.[119–123] Nasr et al.[120] investigated several coupled semiconductor systems, such as SnO_2/TiO_2 and $SnO_2/CdSe$, and found that fast withdrawal of the electron or hole in a separate but tightly connected NP significantly improved the element performance, resulting in the open-circuit voltage of 0.5 to 0.6 V and a power conversion efficiency of ca. 2.25%. Greenham et al.[124] and Ginger and Greenham[125] demonstrated that the mixture of CdSe or CdS NPs with the conjugated polymer poly(2-methoxy,5-(2′-ethyl)-hexyloxy-*p*-phenylenevinylene) (MEH-PPV) can effectively separate charges from the hole and electron, and the external quantum efficiency of such a device was 12%.

The combination of semiconductor NPs with metal NPs in the film was also investigated for the enhancement of light absorption due to energy/charge transfer and surface plasmon effect on the surface of Au or Ag. Strong electrical fields generated in the vicinity of illuminated noble metal NPs can further increase the transition dipole moment in the adjacent semiconductor NP. In turn, thinner, more light-absorbing films can be produced from such a composite, which will eventually increase the light/charge collection efficiency.[126–128] Of equal importance, since the energy position of the Fermi level of the noble metal is lower than that of the conduction band of the semiconductor, the electron flow toward metal NPs can effectively realize charge separation after photoexcitation of semiconductor NPs. Although this approach demonstrated some improvement in photovoltaic performance, it also left a lot of room for improving the distribution and ordering of both particles within the film.[129,130]

One of the greatest problems for all of the thin-film NP devices is the facilitation of charge transport through the film and across the electrode–polymer interface.[131,132] If there is a direct lattice contact between the crystallites, for instance, in ZnO or TiO_2 electrodes, the electron/hole hopping from NP to NP can be rather fast.[131] Studies by Noack et al.[133,134] revealed that the photocurrent in this case was a result of a fast initial charge transport, a slow transport via deep trap states, and a transport via conduction band states or shallow traps. When an NP solid is assembled from an NP bearing a coat of organic stabilizer, the necessity to tunnel through this gap makes the charge transport in nanostructured systems very slow, with resistivities more than 10^{14} ohms/cm.[135] Also, there is a strong Coulomb correlation between electron occupancy in an NP solid. The films become more resistive as more charge is injected, and the charge in them can be stored for a long time (several hundred seconds). A possible explanation for these facts is the buildup of the charge in traps near the source electrode, which reduced the field emission when a certain level was reached.[136] Morgan et al.[135] argued that the bottleneck of the charge transport was not the injecton of charges but rather the extraction of charges from the space charge region, which formed near the contact due to the slow hopping rates. The length of the space charge region is estimated to be less than 100 nm. Hikmet et al.[137] showed that the current in CdSe/ZnS composites could also be explained by space charge–limited current in the presence of defects and rationalized the role of traps. At sufficiently high voltages the traps can be filled and a trap-free space charge-limited current observed. The characteristic trap depth was estimated to be about 0.15 eV.

As can be seen, there is significant progress in the understanding of charge transport in NP films and development of new techniques for charge transport assessment such as imaging.[138] It also becomes clear that essential problems of photovoltaic devices can be addressed by proper organization of the nanocomposites.[132] As applied to photovoltaics, significant achievements in this direction were attained by using rod-like nanocolloids, which (1) can channel the charge to the electrode and (2) eliminate energy-demanding interparticle hopping.[139] Huynh et al.[140] utilized CdSe nanorods in a mixture with poly(3-hexylthiophene) (P3HT), which was spin-casted onto an ITO glass electrode coated with poly(ethylene dioxythiophene) doped with polystyrene sulfonic acid (Figure 1.8). When CdSe NPs (aspect ratio = 1) were used, the maximum external quantum efficiency was only about 20%. It increased to 55% when CdSe nanorods with an aspect ratio of 8.6 were embedded in the conductive matrix. Nanorods eliminated a large number of

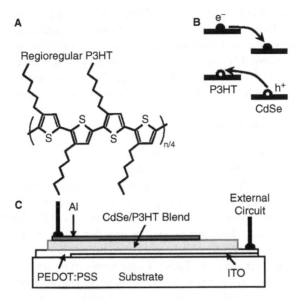

FIGURE 1.8 (A) The chemical structure of P3HT. (B) Charge separation scheme in P3HT–CdSe interface. (C) Schematics of the CdSe–P3HT photovoltaic device with a hole transport layer made from poly(ethylene dioxythiophene) doped with polystyrene sulfonic acid (PEPOT-PSS). (From Huynh, W.U. et al., *Science*, 295, 2425–2427, 2002. With permission.)

interparticle hops, which were necessary for charge transport in the film, and therefore reduced the electron–hole recombination.[141] This concept was confirmed by replacing CdSe nanorods with branched CdSe nanocrystals with a shape of tetrapods. They further enhanced the photocurrent because the tetrapods can always be pointed in the direction perpendicular to the electrode plane, which shortens the transport path.[141,142] The data for nanorods can also be compared with data obtained by Arici et al.[143] for $CuInS_2$ NPs in a very similar polymer matrix.

Other studies also indicate the potential of structural optimization for the improvement of photovoltaic performance.[144,145] Combination of two NPs of different materials not only may reduce charge carrier recombination but also can eliminate the charge traps.[146,147] Arango et al.[148] emphasized the necessity of a layered approach to the preparation of photovoltaic films rather than simple blending. In this respect, layer-by-layer (LBL) assembly, a new technique of thin-film deposition applicable to most aqueous nanocolloids, presents unique opportunities. It was initially utilized for NP photovoltaic elements from CdS and TiO_2 by Kotov et al.[119] and later was used for other systems.[149–158] This technique for photovoltaic applications is reviewed in greater detail elsewhere. For instance, LBL affords preparation of graded semiconductor films from NP and other components,[159–162] which are known to enhance the charge transport within the film due to the intrinsic gradient of the electron and hole potential.[163] Sheeney-Haj-Ichia et al.[129] used a similar principle to enhance charge transfer in hybrid NP–4,4′ bipyridinium photovoltaic element and demonstrated that thin-film architecture can be a powerful tool for controlling the direction of charge flow. It was also established that NWs in LBL films can self-assemble and have surprisingly high levels of organization.[164] Inclusion of intricately intertwined carbon nanotubes in these films[66] can be an important milestone toward improving collection efficiency of photoelements. However, the common use of insulating polyelectrolytes as LBL partners inevitably decreases the photovoltaic parameters (quantum efficiency did not exceed 7.2% for TiO_2 films, so far),[149] necessitating more active utilization of water-soluble conductive components for this purpose.[165]

1.3.1.5 Electrochromic Devices

Thin films that change color under bias probably showed the most promising results from a practical point of view.[166] The industrial prototypes of rearview mirrors and display signs have already been produced with characteristics exceeding sometimes more common liquid-crystal displays (Figure 1.9). Most commonly, NP electrochromic devices are made from metal oxide systems such as WO_3, NiO, MoO_3, IrO_2, CeO_2, Fe_2O_3, or Co_2O_3. The color change occurs due to the formation of colored intercalation compounds (WO_3), charge storage on the particle (SnO_2:Sb), or coupled redox reactions on their surface (TiO_2). Organic materials are often necessary to obtain the color change as either redox (for instance, Prussian Blue or viologens) or charge transport (polyaniline, polypyrrole, or polythiophene) agents.

The most important parameters of electrochromic devices are switching time, depth of color change, and environmental stability. zum Felde et al.[89] and Coleman et al.[167] studied the ionic and electronic processes in photochromic effect in SnO_2:Sb. They found that the switching time is controlled by the rate of ionic diffusion into the porous film of cations, which counters the negative charge of electrons injected into the NPs. The charge trapped in the Sb centers is very stable and produces up to 1 optical density unit (O.D.) change in color in the visible and IR parts of the spectrum. The response times were <10 msec,[89] which was faster than those for TiO_2–Prussian Blue films, with switching times of about 1 sec, and still significantly better than for organic polymers devices.[168,169] The diameter of the counterions was essential for maintaining the electrical and optical characteristics. When the bulky tetrabutyl ammonium cation was replaced with lithium, the photochromic effect in the visible region increased with the concomitant decrease in the response time. However, the photochromic response in the IR part of the spectrum was not affected.

Polyoxometalates (POMs) attract special interests of researchers working with electrochromic devices.[170] The sizes of POM nanocrystals can be easily controlled from several angstroms to 10 nm by simple chemical synthesis. More importantly, unlike common NPs, POMs' nanocrystals are intrinsically monodispersed and have uniform composition. Liu et al.[171,172] have demonstrated that POMs could be integrated into ultrathin films upon LBL assembly. For thin films with 20 monolayers of $Eu\text{-}(H_2O)P_5W_{30}O_{110}$, contrast changes up to 92% were observed within 7 min of turning on the bias.[173]

Metal oxide NPs are attractive for electrochromic applications due to their high environmental stability. As many as 1 million cycles of color switching were performed on SnO_2:Sb films.[89] However, metal chalcogenides became their strong competitors when narrow on/off absorption bands were required. Wang et al.[87,88]

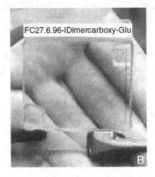

FIGURE 1.9 (A) Electrochromic sign based on Sb-doped Sn oxide produced by Monsanto. (From Coleman, J.P. et al., *Displays*, 20, 145–154, 1999. With permission.) (B, C) Prototype electrochromic window from sintered TiO_2 nanocrystalline film modified with Prussian Blue in clear (B) and colored (C) forms. (From Bonhote, P. et al., *Displays*, 20, 137–144, 1999. With permission.)

explored the electrochromic effect of CdSe NPs in solution. Upon applying negative potential, a new IR absorption peak appeared for both CdSe NP solutions with different sizes (Figure 1.10). The appearance of an IR absorption peak implied that the electrons injected at negative bias go to the 1S state of CdSe NPs. Concomitantly, the lowest interband excitation transition is bleached. The new IR peak corresponds to the transition from the 1S state in the CB to higher electronic levels in the same band. The appearance and magnitude of the IR peak were strongly dependent on

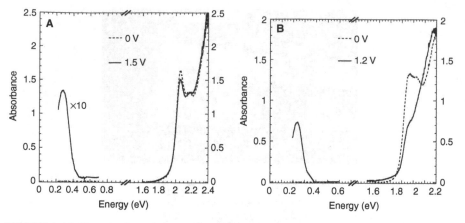

FIGURE 1.10 IR and UV-visible spectra of 5.4-nm CdSe nanocrystals (A) and 7.0-nm CdSe nanocrystals (B) at different potentials. (From Wang, C. et al., *Science*, 291, 2390–2392, 2001. With permission.)

the sizes of NPs (for example, IR peaks for 5.4-nm CdSe NPs were quite small). A decrease of NP diameter also resulted in a high-threshold voltage for electron injection, which was consistent with the wider band gap in small NPs. In addition to CdSe, identical IR peaks were observed for PbSe NPs,[174] which demonstrated the versatility of potential-induced IR absorption for semiconductor NPs.

The effect of the particle organization in the electrochromic films of NP can be as profound as for LED and photovoltaic devices.[175] The image contrast and response time are determined by the rate of charge transport between NPs, which can be controlled by the thin-film deposition process. Sintered spin-coated films appear to be the most common substrates for the studies of photochromism. Room temperature processing such as electrodeposition[176] or layer-by-layer assembly of nanoparticles[119] can lead to a more complex and sophisticated structure with excellent photochromic characteristics. For instance, the addition of Au NPs to Fe_2O_3 film greatly improves the film conductance and photochromic color depth and response time.[177–180] One such endeavor can be the improvement of the longevity of the chalcogenide NPs and acceleration of ion diffusion within the film.

Recently, the photochromic effect with NPs made directly from Prussian Blue 5 to 10 nm in diameter has been reported by DeLongchamp and Hammond.[181] The films were made by LBL assembly, which affords several performance improvements compared to other deposition methods (Figure 1.11). Poly(ethyleneimine), a weak polyelectrolyte used as an LBL partner of Prussian Blue nanoparticles, can intrinsically ionize in the vicinity of the NPs, which facilitate the switch in the redox state of the particles responsible for the color change. It also provides greater thermodynamic stability for the reduced state of Prussian Blue and, therefore, better longevity. Nanoporosity of the film supports fast diffusion of potassium counterions neutralizing the negative charge stored in the inorganic component, although with switching time in the range of seconds. A superior contrast of 77% was achieved due to eliminated reflection and scatter in the transparent state and the small size of the

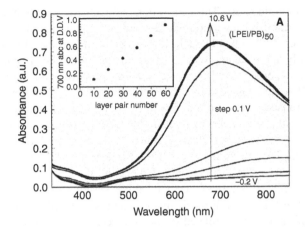

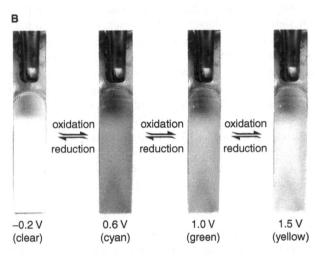

FIGURE 1.11 (A) The dependence of the absorption spectrum on the applied potential. (B) Color change of Prussian Blue films for different applied voltages. (From DeLongchamp, D.M. and Hammond, P.T., *Adv. Funct. Mater.*, 14, 224–232, 2004. With permission.)

Prussian Blue particles. Their uniform dispersion within the polymer matrix allowed for greater bleaching because the potassium ion intercalation distance is limited by the nanoparticle diameter.

1.3.1.6 Sensors and Biosensors

Metal or semiconductor NPs can also offer new transduction techniques of both an optical and electrical nature.[182] As was already mentioned, the charge transport between NPs in thin films is exceptionally sensitive to tunneling conditions, which can be altered by a variety of means, from incorporation of analytes between the NPs to their excitation by electromagnetic fields of appropriate energy. Interparticle gaps become exceptionally important for sensing applications and have to be tuned for sensitivity and selectivity. This can be done typically by variation of the chemical structure of the stabilizer layer around NPs and subsequent preparation of a closely packed structure.[183–185] This approach to the control of particle organization in thin films is relatively universal and was taken advantage of in both optical and electrical sensors.[186] The dependence of the corresponding properties on the relative degree of order in NP solids was described by Beverly et al.[187,188] The theoretical aspects of the charge transport between the NPs in two- and three-dimensional solids were extensively investigated by Remacle et al.[189] and are presented in Chapter 6. The comparison with experimental data obtained for compressed Langmuir–Blodgett (LB) films of Ag NPs strongly supports the Mott–Hubbard band model of charge transport.[190–192] This makes the system of metal NPs one of the best examples of developing collective properties in organized NP systems.[193]

Typically, the transduction mechanism in nanostructured films precipitated by the presence of an analyte is related to the change in the gap distance between the NPs due to swelling or shrinking. Some other processes making possible signaling in NP sensors include pressure-induced enhancement of photoluminescence intensity of CdS(Se) NPs,[194] reversible reduction of the crystal lattice of the nanocrystals,[195] and the magnetotunneling effect.[196] One of the most representative examples of transduction based on swelling is given by Zamborini et al.[84] and can be applied to a fairly large set of analytes. A thin film of alkanethiolate and ω-carboxylate alkanethiolate-protected Au NPs was deposited on the interdigitated array electrodes by cross-linking them with carboxylate–Cu^{2+}–carboxylate bridges (Figure 1.12A). Because a small constant bias of –0.2 V was applied to the film, an analyte-dependent current was registered. When the device was exposed to ethanol vapors, the current sharply decreased with the increase of C_2H_5OH partial pressure (Figure 1.12B). This effect stems from the high affinity of ethanol and the Au–thiol–carboxylate composite, resulting in film swelling. Correspondingly, the gap between Au NPs increases and the current displays a pronounced drop. The conductivity of such sensors will change with the chemical structure of the vapor and its concentration. The swelling effect is relatively fast and reversible, with a response time within 50 sec.

Gas sensors are probably the most abundant among the NP thin films,[183–185,195,197–199] with some examples of field/force sensors[194,196] and biosensors. The applications of NP films to biosensors were recently reviewed by Shipway and Willner[200] and Willner et al.[201] and hence will not be dwelled upon now. As a brief remark, we need to mention that the development of the latter is strongly accelerated by intense cross-fertilization between biology and nanotechnology, which can be noticed in different aspects of organized NP assemblies and superstructures presented in other chapters.

Similar to light-emitting, photovoltaic, and electrochromic devices, the major challenge for NP sensors is the accurate tuning of the two- and three-dimensional film structure to improve the sensitivity or selectivity of the device.[202,203] Quite often, the use of highly monodispersed particles is sufficient.[86] However, the preparation of more elaborate sandwich structures is necessary to fine-tune the analyte response. For instance, the work of Stich et al.[204] showed that complex layered structures are required to attain parallel screening of proteins in order to overcome some limitations of fluorescence, enzyme labels, and colloid techniques. The combination of excellent

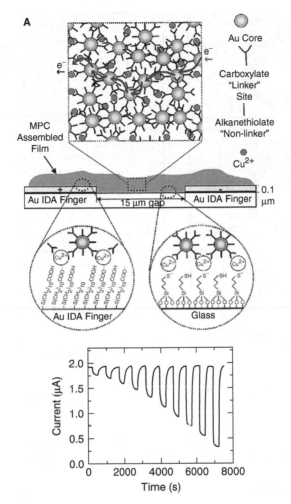

FIGURE 1.12 (A) Scheme of gas sensors of Au nanoparticles films. (B) Current changes are monitored over time. Films were alternately exposed to pure nitrogen flow and ethanol vapor in an increased sequence of fractions, 0.1, 0.2, 0.3, 0.4, 0.5, 0.6, 0.7, 0.8, 0.9, and 1.0, of the saturated ethanol vapor pressure. (From Zamborini, F.P. et al., *J. Am. Chem. Soc.*, 122, 4514–4515, 2000. With permission.)

two-dimensional ordering and tight control over the geometry of individual NPs obtained by nanosphere lithography (see Chapter 14) afforded an unprecedented level of wavelength agility of the sensor throughout the visible, near-IR, and mid-IR regions of the electromagnetic spectrum.[205,206]

1.3.2 NWs as Molecular Electronic Devices

The most common view of NWs is as interconnects and functional elements of nanocircuits, such as leads and electronic diodes. In addition, NWs also have great potential as elements for electroluminescence devices and sensors, which will be

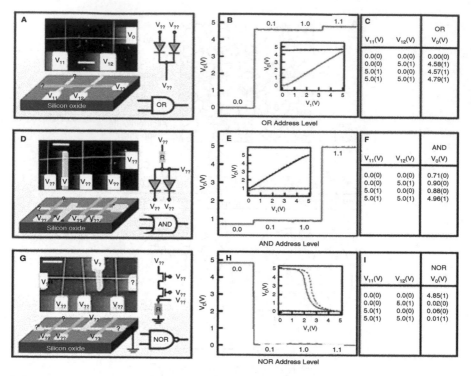

FIGURE 1.13 Schemes of logic gates from p-Si (green) and n-GaN (red) NWs for OR (A–C), AND (D–F), and NOR (G–I) operations. (From Huang, Y. et al., *Science*, 294, 1313–1317, 2001. With permission.)

discussed in this part of the review. They can also be used as voltage-driven actuators.[207] Unfortunately, the latter remains a relatively unexplored area.

1.3.2.1 Nanocircuits

Similar to conventional circuits, the integration of NWs in predesigned patterns extends their application from single electronic elements to functional devices. Prototype electronic elements on their basis have been successfully demonstrated and include p–n diodes,[43] bipolar junction transistors,[208] field effect transistors,[45,208,209] and complementary inverters.[210]

It was also shown that NWs could perform basic functions of complex logical gates, such as OR, AND, and NOR gates using p-type NWs from Si (green in Figure 1.13) and n-type NWs from GaN (red in Figure 1.13).[210] Since the understanding of potential computational schemes is important for the future development of NW assemblies, examples of gate operation are given here in some detail. Assuming that the nanoscale circuit will be operating in the binary regime similarly to the conventional circuits, one can assign logic 0 and logic 1 to low (0 V) and high (±5 V) voltage in the output, respectively. For an OR gate, two inputs were made from p-NW; one n-NW was placed over them and served as an output (Figure 1.13A). Since the NW crossings function as common diodes, logic 0 or logic 1 on the output wire appeared when at least one of the inputs had low or high voltage, respectively (Figure 1.13B and C). For a more complex AND gate, the output was assigned to a p-type NW under bias of 5 V, whereas two of n-type NWs were used as inputs (Figure 1.13D). A gate electrode was also crossing the p-type NW with a constant voltage to deplete a portion of it. Only when both input voltages were high enough to reduce the voltage drop across the constant NW resistor was the high voltage obtained at the output (logic 1). Otherwise, logic 0 appeared (Figure 1.13E and F). A NOR gate could be made when the NW crossings functioned as FETs. Two n-type NWs were connected as FET series, whereas a bias of 2.5 V was applied to the third n-NW FET to create a constant resistance of about 100 megaohms (Figure 1.13G). When either or both input voltages were high, the transistors were in the off state, i.e., had a higher resistance with respect to the constant resistor. Thus, most voltage dropped across the transistors (logic 0). Only when both input voltages were close to 0 V was the high-output voltage observed (logic 1) (Figure 1.13H and I).

The fabrication of nanoscale logic gates is an important milestone toward the practical application of NWs as nanocircuits.[211] Whereas some of the logic operations were previously demonstrated for organic molecules of rotaxane in thin films[212] and carbon nanotubes,[213,214] the inorganic NWs or carbon nanotubes are more likely to succeed as long-term, cost-effective nanodevices. Futhermore, the easy surface modification of inorganic NWs gives us numerous opportunities to produce functional electronics; for instance, recent studies showed In_2O_3 NWs can be used as data storage materials after adsoption of porphyrin or Bis(terpyridine)-Fe molecules.[215,216]

Significant advances are also being made in the manufacturing of the NW crossbar systems, grids, and other networks. Most common current approaches are based on solution chemistry or external fields[217,218] and can be almost equally well applied to carbon nanotubes.[219] Melosh et al.[220] demonstrated an elegant method of manufacturing of planar systems with aligned NWs, taking advantage of well-developed methods of thin-film deposition. The thickness of the films in molecular beam epitaxy can be controlled to the angstrom level. If the sandwich structure were turned on a side, etched, and partially filled with a different material, then parallel NWs could be stamped onto an appropriate substrate (Figure 1.14). Record junction densities as high as 10^{11} cm^{-2} were obtained with pitch (center-to-center distance) as small as 8 nm, which is substantially better than the NW arrangements obtained by electron beam lithography.

The successful demonstration of the basic devices from NWs and new advances in their manufacturing should not camouflage an important problem of a paradigm change that must happen upon transition from a conventional lithographic technology of chip production capable of organizing structures up to 100 nm in scale to the true nanometer scale devices. *One cannot and should not replicate the architecture of the computer chips from NWs or NPs by simply reducing their size.* As one can see, the accuracy of positioning of circuit elements cannot be totally mistake-proof. The variability in the device performance will also be quite high, especially for a large number of nanoscale p–n junctions or SET gaps. The same issues are to be expected for carbon nanotube electronics compounded by their intrinsic variability in properties.[12,221] Therefore, one

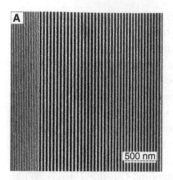

FIGURE 1.14 SEM images of NW arrays prepared by superlattice NW pattern transfer. (A) SEM image of silicon NWs deposited on SiO_2. The 19 NWs on the left are 18 nm wide and 30 nm apart (center to center). The 40 NWs on the right are 20 nm wide and 40 nm apart. (B) Model crossbar circuit made from Pt NWs with pitches from 20 to 80 nm. Scale bars, 500 nm. (From Melosh, N.A. et al., *Science*, 300, 112–115, 2003. With permission.)

needs to redesign the computing algorithm to be significantly more tolerant to defects in the circuit structure. This can be the only way to a functional and viable nanoscale circuit, which can take advantage of exceptionally high levels of integration. Although this area of research and, in particular, NW + NP superstructures supporting the defect-tolerant computing are still waiting for the active development of NW superstructure preparation, an important step in this direction was made by the creation of the Teramac computer at Hewlett-Packard.[46,222] Heath et al.[46] demonstrated that a fast and powerful computer can be made fairly inexpensively from elements prone to manufacturing and interconnect defects on the basis of a new addressing scheme that can reroute the signal around malfunctioning elements. In this respect, it is important to develop manufacturing techniques for new geometrical arrangements of the NWs topologically different from the classical crossbar scheme (Figure 1.14). Recent developments on the synthesis of branched NWs can provide more design room for devices.[223–225] One interesting example is regular dendritic NW arrays with multiple side branches from ZnO (Figure 1.15A and B), reported by Yan et al.[226] In addition to interesting electronic properties, the high refractive index of ZnO and observed waveguiding phenomena in these NWs (Figure 1.15C) make hybrid optoelectronic computing schemes potentially possible.

One can notice that the research on CNT circuits goes largely in parallel with that on NWs and NPs. Similar milestones have been achieved in both fields, such as preparation of FET,[227] logic gates,[213,214] transistor arrays,[12] photodiodes,[228] and light emitters.[229] Although there is a lot of overlap in projected applications, many performance parameters in NW devices and CNT devices are complementary; for instance, the wavelengths in light emission and different types of transistors or sensor functions. Therefore, it would be important to evaluate potential technologies to integrate the different families of the devices. Currently, this field is only in the beginning of its development, largely because of different chemical/physical approaches to device assembly.

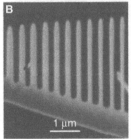

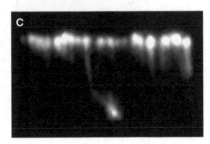

FIGURE 1.15 (A, B) SEM images of comb structures made from ZnO nanowires. (C) Far-field optical image of light emission with spatially resolved emission from individual NWs. (From Yan, H. et al., *J. Am. Chem. Soc.*, 125, 4728–4729, 2003. With permission.)

1.3.2.2 Electroluminescence Devices

The charge transport and electron–hole recombination phenomena underlying the operation of p–n junctions of electroluminescent devices from NWs are identical to those made as thin films. The challenge here is in obtaining a crossbar arrangement of the p- and n-type NWs, which was realized for p-type Zn-doped InP and n-type Te-doped InP NWs (Figure 1.16B).[44,230] Analogous to conventional bulk semiconductor p–n junctions, NW crossing exhibits characteristic current rectification in the forward bias, with a sharp current onset at 1.5 V. As expected, the NW intersection with the p–n junction serves as the light-emitting point where electrons and holes recombine upon electrical excitation, while both NWs in the device light up for the optical excitation mode (Figure 1.16A, insert). The intensity of the electroluminescent point follows the *i-V* curve of the diode (Figure 1.16B), with a threshold voltage as low as 1.7 V. The wavelength of the emitted light can be easily adjusted by using InP NWs of different diameters. As such, NW junctions with light emission peaks at 820 and 680 nm have been demonstrated (Figure 1.16C and D).

The device in Figure 1.16 is the smallest light-emitting diode currently made. Miniaturization, though, may not be the only direction for further development of

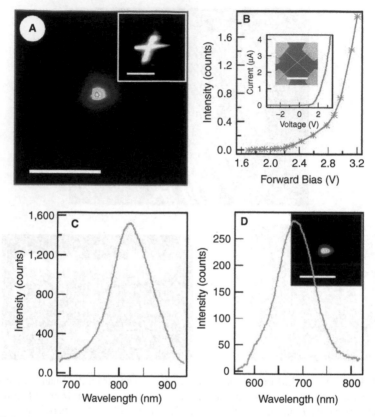

FIGURE 1.16 (A) Electroluminescence (EL) image of the light emitted from a forward-biased NW p–n junction at 2.5 V. Inset: PL image of the junction. Scale bars, 5 μm. (B) EL intensity vs. voltage. Inset: *i-V* characteristics. Inset in this inset: Field-emission SEM image of the junction itself. Scale bar, 5 μm. (C, D) The spectrum peaks at 820 and 680 nm. (From Duan, X. et al., *Nature*, 409, 66–69, 2001. With permission.)

the electroluminescent devices from single NWs. Nanoscale semiconductor rods can serve as optical resonators, which was demonstrated by Johnson et al.[231] for optically pumped single-GaN NWs. They can also be periodically doped to produce one or many p–n junctions directly on a single NW.[232] These facts indicate the prospects of electrically pumped laser on single NW, which was recently confirmed by the report by Duan et al.[233] on optical and electrical measurements made on single-crystal CdS NWs acting as Fabry–Perot optical cavities, with mode spacing inversely related to the nanowire length. Electrically driven nanowire lasers may also be assembled in arrays capable of emitting a wide range of colors.

1.3.2.3 Sensors

The optical and electronic devices described above demonstrate the state of the art of the field. At the same time, their practical applications in nanoscale electronics may be quite distant. This is not the case for various sensors that can be designed on the basis of NWs, which can significantly outperform and become commercially viable alternatives for thin-film sensors quite soon.[234] They exhibit exceptional sensitivity due to the strength of surface effects on the electrical properties in nanometer scale objects, whereas surface modification makes possible manipulations with their selectivity. The research on semiconductor and metal NW sensors was largely inspired by the success of similar applications of carbon nanotube,[235] for which chemical sensors,[236] biosensors,[237,238] and gas sensors were developed.[239]

The surface chemistry of CNTs and semiconductor NWs is quite different, which predetermines different sensor functions. Metal oxide NWs have specific interactions with many gases. Adsorption of a redox gas on the NW surface results in trapping or injection of charge carriers, whose concentration in a small volume of the wire drastically changes. The conductivity of NWs follows the charge carrier concentration and typically decreases upon exposure to the analyte, which offers a convenient transduction mechanism for gas detection.[53]

This mechanism of conductivity modulation is perceived to be advantageous to the percolation of charge in NP arrays (see above) because of better stability, signal-to-noise ratio, and faster response times.[240] NWs made of a variety of semiconducting oxides such as ZnO, SnO_2, In_2O_3, TiO_2, and Si have been explored to be a gas sensor for O_3, O_2, Cl_2, NO_2, NH_3, CO, H_2, H_2O, and EtOH.[240-246] The response time, for instance, for hydrogen sensing in mesoscopic Pd NWs was as short as 75 msec,[247] which was much faster than in sensors based on NP thin films (see above). The sensitivity for NW sensors based on ZnO nanobelts was as high as a few parts per billion, which is probably not yet the theoretical limit.[248]

Besides electrical sensors, optical transduction schemes with NWs by, for instance, surface-enhanced Raman scattering are also being explored.[249] However, they require conversion to the thin-film sensor modality, although the act of sensing still occurs on a single NW.

Preparation of NW bioconjugates for biosensors has relevance not only to life sciences but also to nanotechnology in general. Decoration with biomacromolecules can be used for organization of nanocolloids taking advantage of highly specific interactions between them. The conductivity changes in NW–protein or NW–DNA constructs are affected by the electrical field around the complementary biomolecules attaching to the protein or DNA already present on the NW surface.[250,251] When Si NWs were functionalized by a monolayer of biotin (Figure 1.17A),[252] the conductance of biotin-modified Si NWs increased rapidly to a constant value upon addition of a streptavidin solution, whereas it remained unchanged after the addition of pure buffer solution (Figure 1.17B). The plateau value linearly depended on the concentration of monoclonal antibiotin (Figure 1.17C). To some extent, the transduction mechanism in NW biosensors is similar to the operation of NW FETs. The increase of the conductance can be understood as chemical gating of charge when negatively charged streptavidin binds to the p-type Si NW surface.

Field sensors are exemplified by optical sensors from ZnO NWs (which can also be considered optical switches[253]) and magnetic field sensors with Ni NWs.[254] In the latter example, an array of highly parallel NWs defined by ion-track technology was embedded in polyimide plastic. On the top and on the bottom of the membrane, they were connected to two lithographically structured surface layers. The sensing

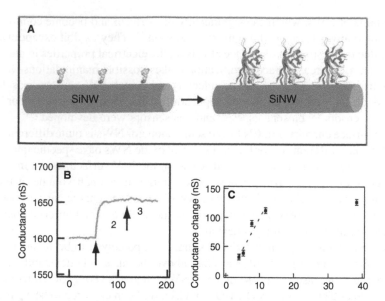

FIGURE 1.17 (A) Scheme of a biotin-modified Si NW (left) and subsequent binding of streptavidin to the Si NW surface (right). (B) Plot of conductance vs. time for a biotin-modified Si NW, where region 1 corresponds to buffer solution, region 2 corresponds to the addition of 250 n*M* streptavidin, and region 3 corresponds to pure buffer solution. (C) Plot of the conductance change of a biotin-modified Si NW vs. m-antibiotin concentration; the dashed line is a linear fit to the four low concentration data points. (From Cui, Y. et al., *Science*, 293, 1289–1292, 2001. With permission.)

mechanism was based on magnetoresistance and was critically dependent on the parallel organization of the NWs in the matrix between planar electrodes. The collaboration between Sony and the University of Hamburg resulted in a similar optical switch with aligned ZnO rods with a strong photoresponse at 366 nm.[255] The ZnO nanorods with a diameter of 15 to 30 nm and a length of 200 to 300 nm were directed into 200- to 800-nm-wide electrode gaps by using alternating electrical fields at frequencies between 1 and 10 kHz and field strengths between 10^6 and 10^7 V/m. The nanorods aligned parallel to the field lines and made contact with the Au electrodes. The *i-V* characteristics of the aligned rods were strongly nonlinear and asymmetric, showing rectifying, diode-like behavior and asymmetry factors up to 25 at 3-V bias.

1.4 STRATEGIES OF NP AND NW ASSEMBLY INTO MORE COMPLEX STRUCTURES

In order to improve the functions and simplify the manufacturing of the nanoscale devices, one needs to learn how to organize nanocolloids in the space of commensurate dimensions. Here we assume that it is the spatial organization of NPs, NWs, and CNTs — i.e., *what* one has in the end, not *how* it was made — that determines the functionality of the superstructure. This may not be entirely true in the practical sense because methods of processing affect the surface of nanocollids. Even the most advanced methods of NP/NW characterization afford only limited information about their surface, so the apparent similarity of superstructures made by different methods can be deceiving. Regardless of this caveat, it is scientifically interesting and practically important to find simpler and more efficient methods of constructing more complex structures from nanocolloids as building blocks. The strategies of organization of NPs and NWs are different in comparison to those of molecules, polymers, and biomolecules. This section will give you a fairly detailed account of current approaches to production of organized nanoscale structures.

1.4.1 ASSEMBLY OF NPS

In general, the assemblies of NPs can be divided into one-, two-, and three-dimensional systems. Methods of their preparation differ vastly, depending on the actual type of NP material. There is hardly any universal method that could be applied for most of the NPs.

1.4.1.1 One-Dimensional Assemblies of NPs

The oldest approach to the preparation of one-dimensional assemblies of NPs is the use of a linear template. This is a versatile method of preparation of all kinds of linear NPs assemblies. If necessary, after completing its role as a structural agent, the linear template may be removed. CNTs or semiconductor NWs are probably the most logical structure-directing matrix for the preparation of one-dimensional NP assemblies. For instance, multiwall CNTs were used as templates for spontaneous assembly of Au on them, which subsequently merged into complex coaxial (albeit granular) NWs.[256] In this review, we will discuss CNT-NP and NW-NP assemblies separately below, because NP NW superstructures have special relevance to nanoscale electronics and to nanocolloid constructs (discussed above). Among other templates, linear pores and channels inside polymers, alumina, and silica templates can be used to make linear agglomerates.[257] In addition to them, one-dimensional assemblies of metallic NP can be made at the edge of lattice plane terraces by chemical and electrochemical reduction of corresponding metal ions.[258] Hutchinson et al.[259] demonstrated that ditches of the corrugated carbon surface act as nucleation sites for the synthesis of aligned Au NPs. Fort et al.[260] applied the same idea to the preparation of Ag NP chains in faceted grooves of alumina.

Biomolecules, including DNAs, proteins, and sometimes even more complex structures, are rapidly becoming a very common template for one-dimensional NP assemblies. The recent popularity of this idea can be attributed to the exceptional selectivity of the biomolecular assembly, versatility of experimental methods for manipulations of their structure, and excellent intellectual and instrumental foundation for such studies developed by Life Sciences. The possibility of self-assembly of very complex structures following the amino acid or peptide code is very attractive to many scientists.

Among biotemplates, DNA is probably the most frequently used class of molecules owing to its strong electrostatic and coordination interaction with NPs. Different kinds of NPs, such as Ag,[261,262] Pd,[263] Au,[264,265] and Pt[266] have been organized

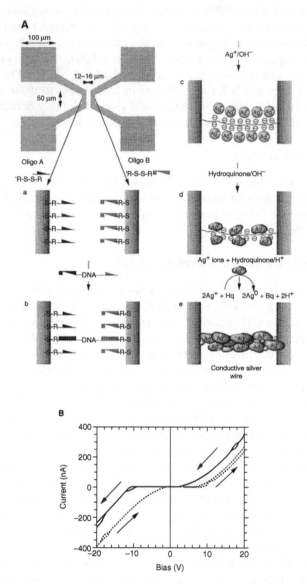

FIGURE 1.18 (A) Scheme of construction of a silver NP wire between two electrodes. (B) Experimentally observed *i-V* curves. (From Braun, E. et al., *Nature*, 391, 775–778, 1998. With permission.)

by this technique. As such, Braun et al.[261] initially stretched DNA molecules between two Au electrodes with a separation of about 15 μm. After that, Ag⁺ was ion exchanged and complexed with amino groups present on the surface of DNA. Further chemical reduction transformed Ag⁺ to Ag NPs. After several cycles of ion exchange and reduction of silver, one-dimensional Ag NP chains along the DNA molecular template were prepared (Figure 1.18A).

Most of the research in this field is done with premade NPs absorbing on the template. However, several groups took advantage of electrostatic interaction of the metal or semiconductor NPs with DNA; for instance, Torimoto et al.[267] and Jin et al.[268] made 3-nm-wide chains from CdS. Raman and x-ray photoelectron energy spectroscopy confirmed that the adsorption sites of CdS NPs were phosphate acid groups of DNA. Warner et al.[269] showed that the ribbon-like and even branched Au NP assemblies could be prepared on DNA templates. Fu et al.[270] obtained double-helical arrays by assembling Au and Pd NPs' peptide nanofibrils at different pH values.

Biotemplates of higher levels of complexity are also employed.[271] Dujardin et al.,[272] Fowler et al.,[273] and Shenton et al.[274] used the tobacco mosaic virus with a shape of a linear tube for assembly of various kinds of NPs inside or outside the tubes. Djalali et al.,[275,276] Banerjee et al.,[277] and Yu et al.[278] assembled Au NPs onto the surface of polypeptide nanotubes, and their assembly position on the biomolecules could be controlled by specific affinity of polypeptide sequences. Heterodimeric tubulin was used as a template to assemble Pd NPs.[279] Mao et al.[280,283] and Lee et al.[281,282] utilized bacteriophage to obtain oriented CdS, ZnS, CoPt, and FtPt NWs. Au and Ag NWs were also obtained by depositing Au and Ag solution onto the surface of yeast *Saccharomyces cerevisiae* Sup35P. Their results showed that after coating, the resistance of nanofibers decreased from 10^{14} to 86 ohms.[284]

There is a growing realization in the research community that the templates may not be necessary for the NP to form one-dimensional structures at all because under certain conditions they may self-assemble due to inherent anisotropy of NP–NP interactions. This statement refers not only to magnetic particles, for which chain formation and anisotropy was known a long time ago, but also to numerous non-magnetic colloids in the absence of external fields. In one of the early experiments, Korgel and Fitzmaurice demonstrated that the prolate Ag NPs self-assembled into NWs during solvent evaporation.[285] The shape anisotropy was implicated in the formation of NWs, but the details of the process were not disclosed. Pacholski et al.[286] observed a similar process in solution for ZnO NPs (Figure 1.19A). It was found that the self-orientation of the particles and aligning of the crystal lattices occurred prior to the formation of ZnO nanorods.[287] In the case of stabilizer-depleted CdTe, not only rods or NWs but also an intermediate stage, NP chains, was observed by Tang et al.[38] Analysis of the assembly process demonstrated that dipole–dipole interaction between the NPs was one of the primary forces in the self-assembly of the CdTe nanocolloid. When methanol partially stripped the TGA layer, the electrostatic repulsion preventing NP association decreased. The intrinsic dipolar moments of NPs resulted in their assembly in chains.

Later, analogous processes of the self-assembly of NP in one-dimensional structures were also observed for a variety of other colloids. By selective desorption of Trizma ligands from the (001) crystal plane, Polleux et al.[288] successfully prepared the TiO_2 NWs via the self-assembly of NPs along the (001) direction (Figure 1.19B). One-dimensional assembly was also reported by Chang et al.[289] and Liao et al.[290] for metal silver and gold NPs, respectively. Chapter 20 also provides interesting experimental data on this subject. A similar process was also observed for Ag_2S by Gao et al.[291] In some cases, linear stabilizers with two functional groups (EDTA, diamine) were believed to be the reason for the self-organization. In light of data

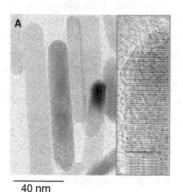
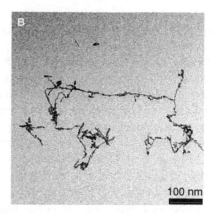

40 nm

100 nm

FIGURE 1.19 Metal oxide NWs from self-assembly of NPs. (A) ZnO. (From Pacholski, C. et al., *Angew. Chem. Int. Ed.*, 41, 1188–1191, 2002. With permission.) (B) TiO$_2$. (From Polleux, J. et al., *Adv. Mater.*, 16, 436–439, 2004. With permission.)

presented here, the actual process of chain or NW formation may be driven by other interactions.[292–295] It needs to be emphasized that a number of basic aspects of the spontaneous chaining of the NPs are not clear. For instance, it is not understood to what extent the stabilizer needs to be removed to see the effect of the dipole moment. The nonuniform distribution of the stabilizer on the nanoscale surfaces can strongly affect any transformation of NPs. There is a growing body of evidence that the conventional models of NPs as spheres with uniformly distributed molecules of layers of organic ligands are not quite accurate.[65,296] At the same time, there is some evidence that intermolecular interactions of organic stabilizers between each other or with the NP surface can play a prominent role in the self-assembly process. For instance, the anisotropic interaction of Au NPs and poly(vinyl pyrrolidone) in LB films was suspected as the reason for chain formation by Reuter et al.,[297] which was morphologically exceptionally similar to those observed by Tang et al.[38] Thus, we expect that the interplay between the NP–NP forces originating in the stabilizer layer, as well as in the NP core, can be an interesting research area for the next few years. The understanding of this interplay can yield many unique NP systems.

Other interesting one-dimensional structures are the rings (Figure 1.20), which also provide a transition to two dimensions, being often only quasi-one-dimensional. They were mainly prepared by the solvent evaporation technique from different metal colloids. Ohara et al.[298,299] reported that the rings/ribbons of a few NPs wide and only 1 NP high form on transmission electron microscopy (TEM) grids due to solvent dewetting effects in the process of evaporation. Maillard et al.[300,301] attributed the formation of rings and a variety of other closed microstructures, such as polygonal networks of nanocrystals, to thermocapillary flow during evaporation (Figure 1.20). Stowell and Korgel[302] observed spontaneous assembly of the NP layer into honeycomb structures with the hexagonal networks consisting of NP ribbons 15 nm thick and 500 nm wide. It was reported by Govor et al.[303] that rings 0.2 to 1.5 μm in diameter made from 6-nm CoPt$_3$ can also form due to phase separation. An interesting observation was recently made by Fahmi et al.[304] about the formation of

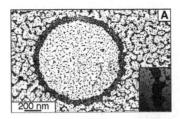

FIGURE 1.20 Exemplary ring structure obtained by solvent evaporation. (From Maillard, M., Motte, L., Pileni, M.P. *Adv. Mater.*, 13, 200–204, 2001. With permission.)

nearly perfect 150- to 200-nm-diameter rings of CdSe stabilized by an amphiphilic diblock copolymer polystyrene-block-poly(4-vinyl pyridine). They formed upon compression of the hydrophobic dispersion on the surface of water (Langmuir–Blodgett films). It was hypothesized that the use of two solvents with different vapor pressures is essential for the spontaneous formation of these superstructures. An excellent theoretic simulation was recently given to elucidate the different patterns of NPs on the substrates from the self-assembly process.[305] The only observation of ring formation without a template was very recently reported by Kong et al.[306] The perfect circular structures of ZnO formed because of the interplay of electrostatic repulsion and the mechanical deformation of ZnO ribbons in the gas phase. These rings, unlike many others described above, were monocrystalline. Their yield exceeded 40% and the synthesis was 70% reproducible.

Optical and electrical properties of one-dimensional assemblies of NPs can be a good example of collective behavior in NP assemblies, which were established both theoretically and experimentally.[307] The interactions between the aligned NPs induced strong dichroism in the absorption spectra, i.e., unequal propagation velocity in perpendicular and parallel directions to the chains. The resonance plasmon wavelengths of Ag NPs were shifted to 455 and 550 nm for transverse and longitudinal polarization, respectively.[260] Far-field polarization spectroscopy on chains of Au NPs confirmed the existence of longitudinal and transverse plasmon polariton modes.[308] Moreover, waveguiding of light in the chain of coupled metal NPs, initially proposed by Quinten et al.,[309] can also be observed experimentally. This process occurs because the oscillations of the conduction electrons are coupled to the optical excitation to form surface plasmon polaritons bound to the NPs, which are propagating from particle to particle in the chain. The actual measurements of plasmon transfer along a Ag NP chain were realized by Maier et al.[310] with the help of the near-field microscope. The characteristic propagation length of light in the chains was only several hundred nanometers. This is a rather short distance for a waveguide. Nevertheless, these data substantiate the theoretical predictions concerning the extremely strong damping of the plasmon propagation in NP assemblies.[311–313] According to theoretical calculations by Zhao et al.,[307] similar effects should be expected in two-dimensional superstructures as well.

Interestingly, a somewhat different form of waveguiding can also be realized in one-dimensional assemblies of semiconductor NPs. Experimental evidence for such a process in linear assemblies was recently obtained by Tang et al.[314] for CdTe NP chains obtained by self-assembly (Figure 1.21A). The band-edge photoluminescence

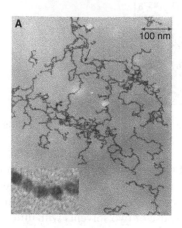
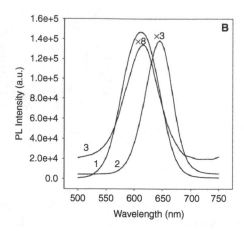

FIGURE 1.21 (A) TEM images of CdTe NP chains. (B) PL spectra for free CdTe NPs (1), NP chains (2), and ultrasonically broken chains (3). (From Tang, Z. et al., *J. Phys. Chem. B*, 108, 6927–6931, 2004. With permission.)

peak of the chains (Figure 1.21B, curve 1) showed a strong red shift of 36 nm, compared to the starting, freely suspended CdTe NPs (Figure 1.21B, curve 2). The lifetime of the luminescence in the chains was also reduced. This is consistent with the presence of Förster resonance energy transfer (FRET) arising from coupling the transition dipoles of the excited and ground states of a luminophore. The original position of the peak can be restored by physical disaggregation of the chains. In essence, the excitation is transported from one NP to another by excitation doping. Unlike metal NPs, there are limitations on the size of NPs that can act as donors and acceptors in this process.[315] At the same time, one has the possibility to direct the transport of light in these systems by controlling the band gap of CdTe. This will require the formation of NP chains in which the particles are arranged according to their diameter.

From the conductivity measurement, the electrical properties of Ag NP chains are highly nonlinear.[261] The exhibited asymmetric *i*-*V* characteristics with respect to zero bias and the shape of the curves depend on the scan direction (indicated by arrows in Figure 1.18B). In general, the zero-current plateau can be seen around zero bias with a resistance larger than 10^{13} ohms. At higher positive or negative biases, the wire becomes highly conductive. Note that the strong nonlinearity does not preclude the use of NP chains as sensors. Moreover, it improves sensitivity, as was demonstrated for Mo NPs.[247] Adsorption of H_2 gas causes the reversible expansion of the crystalline lattice of Mo. As a result, the gaps between adjacent Mo NPs decrease, improving electron transport along the chain. The gaps open up again when hydrogen is vented out. The transduction mechanism of this sensor is identical to most of the other chemical and gas sensors described above.

One should also briefly mention interesting mechanical properties of one-dimensional assemblies of NPs and related structures.[529] Organic particles in a blend with linear polystyrene macromolecules revealed a rather unusual behavior. The blend viscosity was found to decrease and scale with the change in free volume introduced by the NPs and not with the decrease in entanglement. The entanglements did not seem to be affected at all, suggesting unusual polymer dynamics.[316] Remarkable viscoelastic properties of NP composites can also be predicted on the basis of molecular dynamic simulations, as demonstrated in the pioneering works of Starr et al.,[530–532] followed by Smith et al.[317]

1.4.1.2 Two-Dimensional Assemblies of NPs

One of the methods of preparations of NP two-dimensional assemblies is STM/AFM manipulation of individual colloidal particles, as was demonstrated by different research groups.[318–321] Superstructures of a one- and three-dimensional nature can also be made in this way. There are limitations on practical feasibility, and scanning-probe microscopy (SPM) methods have some intrinsic limitations for complex three-dimensional constructs because the place where an NP should be deposited should be accessible by an SPM tip. A simple implementation of this approach is pushing NPs along the surface to create two-dimensional patterns made from cleared areas.[322] More complex nanomanipulation systems with rotation and translation of a linked two-particle structure from 5- and 15-nm Au NPs, and literal building of nanoscale pyramids resting on the silanized surface of the silicon wafer and on mica, were also reported.[323,324] SPM also affords *in situ* synthesis of NPs in a selected area of the surface when particles are made by tip-induced (predominantly electrochemical) reactions.[319,320,325–327] The same family of methods of NP organization also includes surface probe lithography, which is often realized as area-selective modification of a prospective substrate by tip-induced chemical reactions of the surface[328–330] or charge injection.[331,332] Sugimura and Nakagiri[333] demonstrated that an organosilane monolayer on a silicon wafer could be locally degraded through electrochemical decomposition of adsorbed water at the junction of the monolayer and the SPM probe. Then the probe-scanned region reacted with another organosilane, resulting in high-contrast patterns. The latter selectively adsorbed NPs or proteins.[333,334] A similar approach can be used for the modification of thiol coatings on the substrates.[335] After the tip-induced reaction of thiol replacement with dissolved R-SH compound with specific affinity to Au, noble metal NPs were adsorbed on the new surface-bound functionalities. This method yielded line widths of ca. 10-nm resolution and complex bottom-up nanostructures suitable for SET devices. Nonelectrochemical surface modification of adsorption layers was realized by Blackledge et al.[336] through selective surface modification by a catalytically active SPM tip with a layer of palladium. It induced chemical reactions in a terminal azide and carbamate functional groups of organosiloxane. The minimum measured line widths of 33 nm were reported for Au NPs subsequently adsorbed to the thiolic pattern.[336] Forty-nanometer spacing was observed for a similar SPM tip nanochemistry procedure involving chlorosilanes.[337]

One of the newer SPM methods of patterning is dip-pen lithography, pioneered by the group of Mirkin and coworkers.[338–340] The two-dimensional maps of thiols can be used to facilitate adsorption of NPs in specific areas. Demers et al.[341] functionalized thiols with oligonucleotides and observed selective adsorption of Au colloids 13 to 30 nm in diameter bearing complementary DNA strands.

Two-dimensional patterns of NP can also be created by a large variety of other lithographic techniques. One of the more popular approaches is microcontact printing (μCP), which makes possible 50- to 60-nm features. This level of spatial resolution was reproduced by several groups with both metal and semiconductor NPs.[342-347] Qin et al.[348] reported the record small spacing in NP arrays of 20 nm made by μCP. Typically, the patterned surfaces are exposed to dispersions of nanocolloids, which selectively adsorb on areas with appropriate chemical functionalities. Alternatively, the deposition of NPs can proceed as their selective synthesis, for instance, on hydrophilic areas of the pattern.[349] The use of premade NP solutions affords better size uniformity; however, the *in situ* synthesis provides additional orientation of the crystal lattice.[349]

The traditional lithography with photo- or electron beam-active resists is also widely used to create patterns on which NPs or other nanocolloids are subsequently adsorbed in the form of a monolayer.[350] The feasibility of a comprehensive all-chemical strategy for the bottom-up fabrication and spatial fixation of some of the key components of future nanoscale electronic circuits of metal NPs and metal NWs connected to addressable contact electrodes was demonstrated.[328] Vossmeyer et al.[351] developed a special photoresist based on nitroveratryloxycarbonylglycine to render the substrate photosensitive. Irradiation with 340 nm of light through a microchip mask yielded a pattern of free and protected amino groups. 12-aminododecane-capped Au particles were found to selectively bind to the surface amino groups. Importantly, the original pattern can be amplified by repeated treatment with 1,8-octanedithiol and Au NP adsorption, similar to the layer-by-layer technology, to make the pattern readily visible with the naked eye. Amplification of the pattern in the third dimension can also be achieved with a high degree of selectivity by LBL assembly.[352-355]

1.4.1.2.1 Self-Assembly of Two-Dimensional Superstructures

The geometry of the two-dimensional patterns may be self-defined rather than predetermined by a stamp or a lithographic mask, i.e., two-dimensional self-assembly. One of the prime examples of this technique is the organization of NPs by the forces during liquid evaporation[356] and other interfacial solid–liquid interactions at the nanoscale level, which we have already encountered when discussing the formation of the rings. Here, the patterns are more complex and have a definite periodicity/organization in both the X- and Y-directions. Despite the simplicity, some of the best-organized two-dimensional superstructures were produced by this method. Ordered NP mono- and multilayers in a closely packed arrangement were produced from a nanocolloid suspension upon the solvent evaporation[357-361] or adsorption of NPs on charged substrates,[362,363] provided that the NPs were highly uniform in diameter. The ordered domains can cover relatively large areas in the micron scale. For less uniform particles and somewhat lower concentrations of NP, the drying dispersion can produce circular patterns reminiscent of those considered above when discussing one-dimensional assemblies.[364] Their deliberate modulation of the fluid dynamics can make NPs assemble in surprisingly highly ordered layers[365] and networks,[366] even for fairly polydispersed colloids. A complex interplay of wetting dynamics, capillary forces, and interface instabilities, and the spontaneous formation of complex patterns, are implicated in their formation.[356,367] The predictive power of the nanoscale liquid dynamics was demonstrated by deriving the formation of characteristic bands and rings on the basis of linear stability analysis and numerical simulations, which revealed the dependence of particle distribution on equilibrium film separation distance, initial packing concentration, rate of evaporation, and NP surface activity.[356]

Highly ordered two-dimensional arrays of magnetic particles were obtained by the evaporation of aqueous solutions on octadecyltrichlorosilane-stamped surfaces.[368] The droplets of water containing metal salts prefer the hydrophilic surface of untreated silica and are confined by surface forces. The metal salt residue can be subsequently transformed into ferrites or other compounds. The resolution of 70 to 460 nm in X and Y of these arrays enables addressing each magnetic island as an individual memory cell. A similar result can be achieved by so-called nanosphere lithography, when a mono- or multilayer of uniform latex spheres serves as a mask through which a metal can be plated on the surface.[369] This method is reviewed in Chapter 14.

A two-dimensional organization method related to fluid dynamics is the assembly of NP on surfaces of copolymers, which form intricate surface patterns due to nanoscale phase separation of the polymer blocks with hydrophilic/hydrophobic balance.[370,371] Being cast from organic solution, NPs adsorb preferentially on the low-surface-tension areas[372] and often form aggregates with pronounced fractal dimensionality.[373]

One of the most powerful techniques of two-dimensional organization of NPs is Langmuir–Blodgett deposition, which combines the advantages of the self-organization of NPs and operator-controlled pattern organization. It has a very well-established reputation for nanoscale organic films and was extended to NP systems in 1994 by Kotov et al.[374] and Dabbousi et al.[375] It remains one of the most popular approaches to the preparation of remarkable two-dimensional superstructures with potential practical applications in nanoscale electronics.[374–379] Additionally, the Langmuir–Blodgett technique also affords the preparation of one-dimensional[297,304,380] and three-dimensional[379,381] systems. Unlike many of the techniques, it provides the ability to control the distance between single NPs in two-dimensional monolayers.[382] Collier et al.[383] demonstrated the transition from insulator to metal state as the distance between Ag NPs decreases. A detailed review of electrochemical and other properties of Langmuir–Blodgett NP films is given in Chapter 22. Here we should only note that insufficient mechanical strength and susceptibility to environmental effects resulted in a significant decrease of interest in Langmuir–Blodgett two-dimensional assemblies of NPs after initial enthusiasm. Although in many cases this concern is justified, there are a number of ways to make fairly robust Langmuir–Blodgett films[384,385] that can be further improved with inclusion of nanocolloids.[386] Nevertheless, a number of interesting prototype devices have been demonstrated on their basis, such as memory structures,[387,388] command surfaces,[378] SET electronics,[379,389,390] nonlinear optical elements,[391] and sensors.[379,392] The spectrum of applications is identical to other thin films from nanocolloids because the charge transfer in Langmuir–Blodgett films of NPs, as well as other properties, follow the same regularities as for other NP assemblies.[393]

In the previously considered examples of self-assembly of NP in two-dimensional superstructures, there was always a certain substrate or interface that helped NP to remain two-dimensional. Interestingly, there exists at least one NP system

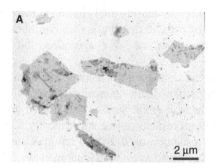
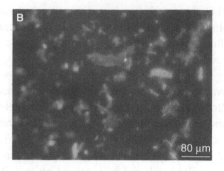

FIGURE 1.22 (A) TEM image of free-standing monolayer films of CdTe NPs. (B) Fluorescence image of monolayer films of CdTe NPs on glass substrate.

that self-assembles to become two-dimensional sheets in solution. Potentially, there can be other NP dispersions with specifically designed surface functionalities that afford the spontaneous organization of NPs in sheets based on their preferential attraction to each other.[394,395,533] 2-(dimethylamino)ethanethiol-stabilized CdTe NPs were shown to form monomolecular sheets after depletion of the stabilizer layer in a procedure similar to the formation of NP chains (Figure 1.22A).[396] The two-dimensional network of NP in the sheets provides them substantial mechanic strength compared to other monolayer films, which is likely due to partial merging of the crystal lattices of NPs. There is a substantial optical activity of CdTe remaining in such layers, which can be detected by fluorescence microscopy (Figure 1.22B). Somewhat similar self-organization of 90-nm triangular prisms from Ag in two-dimensional sheets was also observed by Chang et al.[289] The crystal lattices in the Ag platelets in the superstructure were aligned, producing an intricately interconnected network.

Self-assembly of NPs in organized two-dimensional systems can also be facilitated by proteins and other biomolecules. Considering the popularity of other combinations of biotechnology and nanotechnology (discussed above), the recent appearance of multiple studies in this area becomes quite logical. Willner and Willner consider the NP–biomolecule systems as one of the most promising techniques for programmed assemblies of nanostructures.[397] This notion is substantiated by the record performance of NP-based detection and imaging procedures, which can significantly advance the life science experimental methods.[398–401]

The biological support can be patterned by one of the lithographic methods, such as traditional photolithography,[402] dip-pen lithography,[403] or microcontact printing.[404] After that, the NP dispersions can be used as ink with specific adhesion to biomolecular surfaces. Several groups utilized the ability of proteins to self-organize in two-dimensional lattices. One of the best examples of such superstructures is proteins from bacterial surface layers, i.e., S-layers, recrystallizing *in vitro* into sheets and tube-shaped protein crystals with typical dimensions in the micrometer range.[405] Au and CdS NPs formed in the pores between the protein units repeat the square lattice with a 12.8-nm repeat distance of the S-layer.[406] The diameter of NP is typically limited to 3 to 5 nm. The NP arrays on S-layers displayed particle densities above

6×10^{11} cm^{-2}.[407] Similar film geometries were obtained with genetically engineered hollow double-ring protein chaperonins, which have either 3- or 9-nm apical pores surrounded by chemically reactive thiols.[408] The periodic solvent-exposed thiols within chaperonin templates were used to size selectively bind and organize either gold (1.4, 5, or 10 nm) or CdSe-ZnS semiconductor (4.5 nm) quantum dots into arrays. The lattices with pronounced alignment motifs in NP can be made from viral proteins.[281]

The two-dimensional crystalline layers of ferritin and apoferritin were mostly used for the preparation of two-dimensional arrays of iron-containing magnetic particles.[409] These proteins are attractive candidates for the biological approach to NP organization because the 8-nm internal cavity of apoferritin can be reconstituted with a variety of nonnative inorganic cores, such as magnetite, iron sulfide, manganese oxides, and cadmium sulfide. Encapsulation of the inorganic phase within the protein cage restricts the length scale of particle–particle interactions and prevents direct physical contact, particle fusion, and growth within the organized phase. Ferritin-based arrays could have important applications in magnetic storage and nanoelectronic devices.[410] Thus, Co/Pt NP arrays were prepared within an apoferritin S-layer. By varying the annealing conditions, the coercivities at 500 to 8000 Oe were achieved. Electrical testing of NP films shows that they are capable of sustaining recording densities greater than 12.6 Gbits/in^2.[411]

1.4.1.3 Three-Dimensional Assemblies of NPs

Three-dimensional assemblies of NPs are far less common and widespread than two-dimensional superstructures. Additionally, the level of their three-dimensional organization is not as sophisticated as for the planar arrangements. Moreover, some of them can be considered derivatives of the two-dimensional systems. For instance, electron beam and photolithography can be used to create some three-dimensional NP elements with limited complexity, inducing growth of NP islands in certain areas.[412,413] Other two-dimensional lithographic patterns transcending into three-dimensional include selective decoration of one hemisphere of colloidal particles with Au NPs and NWs.[414,415] SPM manipulation can also be expanded in another dimension: pyramidal structures were built up by pushing an NP on top of two others.[324]

Overall, the current three-dimensional NP assemblies can be classified into structures that are organized by three-dimensional supports, self-assembled systems, and stratified layers. All of them have limitations of three-dimensional organization imposed by the particular method, and none of them can be used to create a freehand three-dimensional nanostructure by design.

1.4.1.3.1 Templated Three-Dimensional Assemblies

Similar to one-dimensional and two-dimensional systems, the three-dimensional NP superstructures can be formed by adsorption on solid surfaces exhibiting three-dimensional organization. Much work in this area has been done for NP adsorption on colloidal crystals (synthetic opals). They display so-called photonic band gaps formed due to the periodic structure of the crystal. Light cannot propagate through the photonic crystal at these wavelengths, and optical effects analogous to the electronic effects in semiconductors can be observed. Thus, the optical properties of synthetic opals coated with both semiconductor and metal NP are being actively investigated.[416] Different methods of preparation of colloidal crystals with NP functionalities have been reviewed by Norris and Vlasov.[107] They include adsorption of nanocolloids on the surface,[417] incorporation of the nanoparticles in the latex and silica spheres,[418] and infiltration of interstices of colloidal crystals with NP formulations for inverted opals.[419] When luminescent quantum dots are included in the photonic crystal, their spectrum is strongly modified when coincided with the band gap.[361,417,420] Also, saturation of the excitation-to-emission conversion in the emission band and conversion efficiency improvement associated with the presence of the photonic band gap have been observed.[421]

Three-dimensional structures from metallic NPs such as Ag or Au templated by colloidal crystals are also produced.[54,422] The possibility of highly efficient surface-enhanced Raman scattering in these three-dimensional systems often serves as motivation for this work.[423] The mesoporous films from Pt/Pd NPs are considered for highly efficient catalysts.[424]

Similar assemblies from magnetic particles are less common. They are primarily made for the external control of the photonic band gap of the colloidal crystal[425] and applications in magnetic force microscopy.[426]

Considering mesoporous films, it is important to mention that NWs can also be organized with their help in three-dimensional. As such, a clean and fast process using supercritical carbon dioxide for producing ultrahigh densities, up to 1012 orders of Ge NWs per square centimeter, has been reported.[427] Uniform mesoporous thin films were also employed as templates for the nucleation and growth of unidirectional NW arrays with almost perpendicular orientation to a substrate surface.

1.4.1.3.2 Self-Assembled Three-Dimensional Superstructures

One of the most typical examples of the self-assembly of NPs in three-dimensional is also related to colloidal crystals. Now, they are made directly from the monodispersed NPs rather than serving as a micron- or submicron-scale template. Murray et al.[357] opened this field of research by making crystalline superlatices from CdSe quantum dots. As is the case with many self-assembly processes, the preparation of this type of three-dimensional assemblies is rather simple, although the synthesis of the corresponding species and substrates may not be. Most often, an aqueous or organic dispersion of uniformly sized NPs is evaporated on a solid surface.[360,428] One can also make them by controlled oversaturation in solutions,[429,430] which is somewhat reminiscent of the method of selective precipitation of NPs for size selection.[431] A nonsolvent (MeOH) was slowly introduced into the concentrated solution of the monodisperse CdSe nanocrystals in toluene. The diffusion of MeOH was carried out directly or through a buffer layer of a third component, propan-2-ol. Irregularly shaped crystals were obtained, whereas the slower diffusion through the buffer layer resulted in perfectly faceted hexagonal platelets with a size of 100 microns (Figure 1.23). After CdSe, similar three-dimensional superstructures were made from a variety of colloids, including CdS,[432,433] BaTiO$_3$,[434] PbS,[435] PbSe,[436] Ag$_2$S,[291,360] InP, InAs,[437] Pt/Pd,[438] Ag,[439] and, lately, In$_{28}$Cd$_6$S$_{54}$$^{12-}$ colloids.[440] In particular,

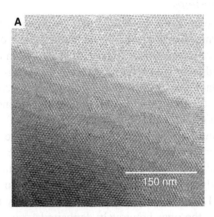

FIGURE 1.23 TEM of a three-dimensional array of the 3.5-nm CdSe NPs with an fcc superlattice. (From Tallapin, D.V. et al., *Adv. Mater.*, 13, 1868–1871, 2001. With permission.)

much work has been done for Au nanoclusters.[358,441–444] Superlattices of magnetic NPs, which were made from Co, Pt/Pd, magnetite, Ni, Fe, and alloys thereof,[359,445–450] are projected for ultra-high-density magnetic storage devices.

The NPs in the supercrystals are typically arranged in the closely packed face-centered-cubic (fcc) structure (Figure 1.23). If the interactions between the nano-clusters exhibit some anisotropy, then they assemble in different lattices, such as double diamond in the case of $Cd_{17}S_4(SCH_2CH_2OH)_{26}$ clusters,[433] cubic structure for NPs of Prussian Blue,[451] or polyhedral lattices for SnS_2/SnS spheres.[452] Superlattices from magnetic particles can also change their packing under the influence of the magnetic field.

The tendency to self-assemble in the crystalline arrays is so strong for NPs that polydispersed nanocolloids may spontaneously phase segregate according to their size. Depending on mixture composition and crystallization conditions, three types of organization can be observed: (1) different-sized particles intimately mixed, forming an ordered bimodal array; (2) size-segregated regions, with continuous hexagonal close-packed monodisperse particles; and (3) a structure in which particles of several different sizes occupy random positions in a pseudohexagonal lattice.[358] Three-dimensional binary superlattices of NP compositions are also obtained.[436,447] By selecting the sizes and ratios, different NPs, such as semiconductor and magnetic NPs, can co-crystallize into binary three-dimensional arrays with varying crystalline structures, for example, AB, AB_2, AB_5, and AB_{13} types.[436]

In addition to NPs, the three-dimensional self-assembly of Au nanorods (12 nm in diameter and 50 to 60 nm in length) has also been reported.[453] Li and Alivisatos[454] demonstrated that three-dimensional assemblies of CdSe nanorods exhibited a unique liquid-crystal state. Note the range of conditions where they form is quite narrow, and the degree of organization is not as extensive as for uniform NPs.

Electrical and optical properties and corresponding applications of the three-dimensional NP assemblies are discussed in Chapters 6 and 9 and, thus, are not considered in this chapter.

1.4.1.3.3 Layered Systems

All of the layered systems allow the researchers to organize the NP or other nano-colloids along the normal to the surface of substrates. The control over the distribution of NPs in plane remains difficult. Thus, all of the NP multilayers (sandwiches) should be characterized as pseudo-three-dimensional systems, because they have an axis of symmetry coinciding with the direction of organization. This actually is convenient for the description of, say, charge transfer phenomena, which can be treated as essentially a one-dimensional system.

Some time ago, Langmuir–Blodgett monolayers were one of the most commonly used methods of the preparation of layered systems of NPs.[374,375,382,383,455,456] Now, though, it is substantially less popular because of the low stability of the LB films. NPs in Langmuir–Blodgett films tend to self-assemble in closely packed hexagonal domains (see Section 1.4.1.2). The same packing is unlikely to remain intact when a new layer is added to the ones previously deposited.

Electrical and some optical properties of NP Langmuir–Blodgett films are described in Chapters 6 and 22. In addition to these reviews, we should mention here only a couple of interesting properties of these structures. One of them can be seen in the dependence of the multilayer capacitance on voltage. It is affected by the scan direction and exhibits strong hysteresis. This effect arises from the electron storage in the layer of NPs[387,393] combined with relatively high resistivity of the NP assemblies.[135] Note that this should be a rather common property in nanomaterials with an exceptionally high area of interfaces where the charges accumulate. Samokhvalov et al.[455] made Langmuir–Blodgett films with gradual changing of the NP diameter from the first layers to the last. Energy transfer from one layer to another should be observed in such multilayers. It is ideologically similar to the light waveguiding in semiconductor and metal NP chains (discussed before).[312–314]

Deficiency of Langmuir–Blodgett films' stability can be effectively overcome with the LBL assembly technique.[457,458] The process of LBL assembly was first described for polyelectrolytes by Decher[459] and for macroscale colloids in the some-what-forgotten early experiments of Iler and Colloid.[460] In 1995, the LBL technique was adopted for NPs by Kotov et al.[119] The principle of the LBL assembly is quite simple. It consists of repeating cycles with sequential dipping of the substrate in oppositely charged nanocolloids and polymers (Figure 1.24). This results in sequential adsorption of layers of nanocolloids and polyelectrolytes. As one builds up the film, the nature of the absorbing species may be changed periodically and thus produces an organized system of layers. Commercially available deposition systems can do this in automatic regimes and produce multilayer systems made of different NPs or other nanoscale materials, such as NWs, proteins, clay platelets, or functional polymers, with any desirable sequence of strata.

Most often, the negatively charged NPs and positively charged polymers with strong electrostatic attraction to each other are used. However, other intermolecular forces can also lead to sequential layering, such as coordination, Van der Waals forces, or hydrogen bonding. For instance, hydrophobic alkane amine-stabilized Fe and FePt NPs form multilayers paired with poly(ethylene imine) due to the coordination interaction between the imino groups of the polymer and surface atoms of the magnetic particles.[461,462] Complexation of metal ions by polyelectrolytes can also

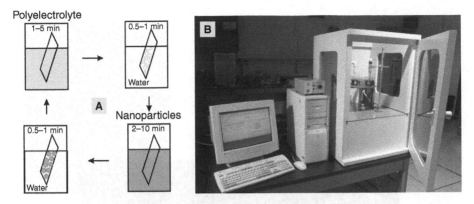

FIGURE 1.24 (A) Schematics of the layer-by-layer assembly of NPs. (B) Commercially available LBL deposition robot (NanoStrata, Inc., Tallahassee, FL).

be utilized for selective synthesis of metal and semiconductor NPs with controlled size and position inside the LBL films.[463] The accuracy of the positioning of the NP stratum made in this way approaches the interpenetration thickness of polyelectrolyte LBL films, which is two to three bilayers, i.e., equal to 5 to 7 nm.

The efficiency and robustness of the LBL process always improve with increasing of the molecular mass of the components, although successful assembly of multilayers with low-molecular-weight LBL partners has also been reported. For instance, small cations, such as Cu^{2+} and Zn^{2+}, were also used to coordinatively bridge carboxylate-functionalized Au NPs.[84,464] Shipway et al.[200] and Willner et al.[201] used the combination of oligocations of pyridine and NPs for the multilayer preparation. Shortening of the interparticle separation is important for improving electrical conductance in the NP films. This is one of the subjects highly relevant for sensors, photovoltaic cells, and luminescence collectors.

Figure 1.25A demonstrates the concept of how one can use the LBL assembly to prepare organized layered systems along the normal direction of the substrate. In this case, a multilayer composite of NPs arranged according to their sizes was obtained. Thus, the graded semiconductor material with gradual evolution of valence and conduction band energies could be prepared.[159] The model graded films were made from highly luminescent CdTe NPs stabilized by thioglycolic acid with a partner polyelectrolyte poly(diallyldimethylammonium chloride). Four different CdTe dispersions with luminescence maxima at 495 to 505, 530 to 545, 570 to 585, and 605 to 620 nm, which display green, yellow, orange, and red luminescence, respectively, were chosen. Typically, 5 to 10 CdTe NP bilayers of each of four luminescent colors were deposited onto the substrate in an order of band gap from highest to lowest. The gradual change of particle size in thin multilayer films with five NP bilayers was observed by transmission electron microscopy. For thick films, in which the individual NP strata can be partially resolved in the z-direction by confocal focusing, the gradual change of the luminescence from green to yellow, to orange, to red was observed (Figure 1.25B). The combination of size quantization and gradient nature of the material opens a possibility for new optical and electrical

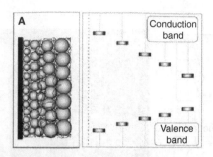

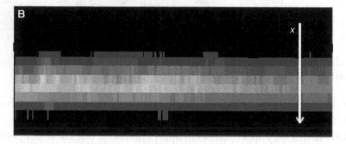

FIGURE 1.25 (A) Scheme of the graded semiconductor films as assembled from NPs. (B) Cross-sectional confocal microscopy image of the graded LBL film of CdTe NPs made of 10 bilayers of green, yellow, orange, and red NPs (40 total). Small NPs are assembled in the lower part (green and yellow luminescence), whereas bigger NPs are assembled on top of them (orange and red luminescence). (From Mamedov, A.A. et al., *J. Am. Chem. Soc.*, 123, 7738–7739, 2001. With permission.)

effects, as well as for the optimization of existing applications of NP devices based on charge transfer, with one-dimensional preferred.[465] Similar types of multilayer structures with strata of different functionalities were obtained from NPs and proteins. This approach can be taken to enhance the biocompatibility of the NP composites[466] and for the preparation of various biosensors.[467]

There also exist modifications of the multilayer technology when NPs are combined with bifunctional surfactants strongly binding to the surface of NPs. The range of organizational accuracy for them is still similar to classical LBL, because it is determined essentially by the diameter of NPs and interfacial roughness. Musick et al.[468] utilized alkane dithiol to link the layers of Au NPs and observed optical transition from semiconductor to conductor with the increase of numbers of Au NPs. Muller et al.[469] proposed a model of electron conduction in such films of Au nanopatricles. Nakanishi et al.[470] used the alkane dithiol to link the layer of CdS NPs and measured the photocurrent generated in these systems. Cassagneau et al.[471] fabricated n-type CdSe NP multilayers linked by 1,6-hexadecanethiol onto p-doped semiconducting polymers. By controlling the level of doping into the p-type junction, it was possible to prepare dissymmetric junctions and observe a rectifying behavior in the forward direction and a Zener breakdown in the reverse direction for the p–n type molecular devices. Diamines were also explored to link the layer of CdSe NPs to prepare electroluminescence devices.[472]

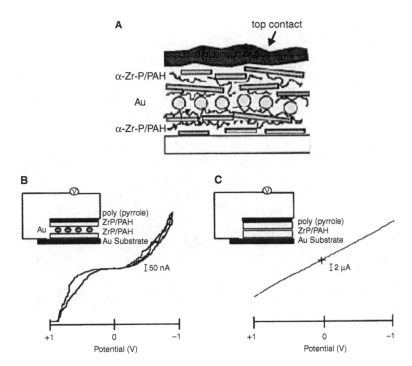

FIGURE 1.26 (A) Scheme of MINIM device. (B) Current–voltage curves for a [(α-ZrP/PAH)$_2$/Au NPs/(PAH/α-ZrP)$_2$] MINIM device (see inset). Current–voltage curve for the same device as in (C) without the Au adsorption step. Scan rate, 100 mV/sec. (From Feldheim, D.L. et al., *J. Am. Chem. Soc.*, 118, 7640–7641, 1996. With permission.)

Numerous optical and electronic prototype devices have been demonstrated on the basis of LBL materials from LEDs[97,155] to dopamine sensors[473] and single-electron charging. For instance, alternating multilayers of anionic exfoliated ziroconium phosphate (-ZrP) and poly(allylamine hydrochloride), negatively charged gold NPs were built up on mercapto-ethylamine hydrochloride-treated gold[80] (Figure 1.26A). The electrical characteristics of these multilayer systems were identical to a double-insulator tunnel junctions device (Section 1.2.1) with nanometer thickness. Due to the presence of Au NPs at a controllable distance from the electrodes, current–voltage curves displayed tunable Coulomb gaps (Figure 1.26B).[80]

Organization of the multilayer is also important when considering the mechanical properties of NPs, NWs, and similar materials. Recently, it was demonstrated that LBL films of polyelectrolytes and NPs could be prepared in free-standing form if some strong components, such as CNTs and clay, were integrated into the films.[66,474]

1.4.2 ASSEMBLIES OF SINGLE NWS

Due to their intrinsic geometrical anisotropy, virtually all current studies on NW assembly processes are aimed at their alignment, which also includes the preparation of crossbar structures. These studies address the issues of how to orient and position them in the device.[6,30] There exist examples of electronic, sensor, and optoelectronic

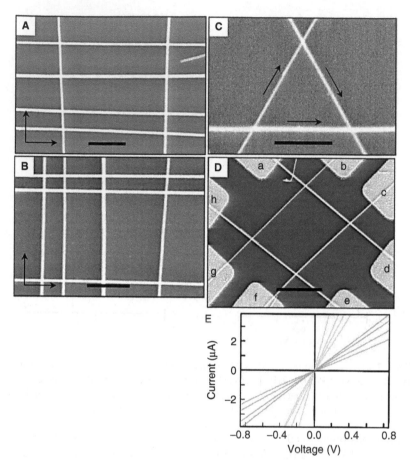

FIGURE 1.27 (A–D). Varying patterns of NWs from microfluid channel assembly. Arrow shows the direction of fluid. (E) *i-V* curves from two-terminal measurements on the 2-by-2 crossed array shown in (D). The green curves represent the *i-V* of four individual NWs (ad, bg, cf, eh), and the red curves represent *i-V* across the four n–n crossed junctions (ab, cd, ef, gh). (From Huang, Y. et al., *Science*, 291, 630–633, 2001. With permission.)

devices that require one-dimensional (chain of NWs), two-dimensional (crossbar, Figure 1.13), and three-dimensional (NW arrays perpendicular to the surface, Figure 1.28 and Figure 1.29) arrangements of NWs. Nevertheless, dimensionality of these assemblies is less important than the mutual orientation of NW. Thus, methods of NW organization will be categorized differently than for NPs. They can be ascribed to three kinds: physical, chemical, and biological.

1.4.2.1 Physical Method of Aligning NWs

One of the prevalent physical methods is the orientation of NWs with the help of external applied fields. Messer et al.[475] demonstrated the possibility of NW alignment in the microfluid channels. They succeeded in organizing $Mo_3Se_3^-$ wires into parallel

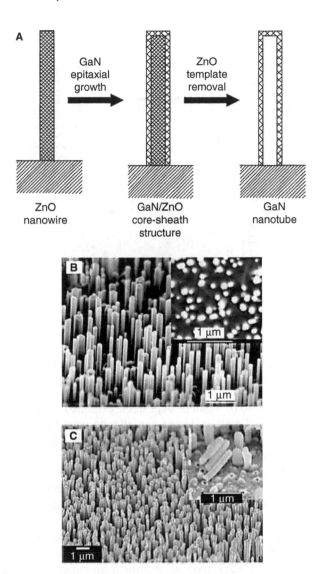

FIGURE 1.28 (A) Scheme of GaN NWs growth from ZnO NWs template. SEM images of ZnO NW (B) and GaN (C) nanotubes. (From Goldberger, J. et al., *Nature*, 422, 599–602, 2003. With permission.)

or crossbar junctions. Subsequently, Huang et al.[476] extended the applicaion of this method and assembled semiconductor InP, GaP, and Si NWs into a variety of prototype electronic devices. Complex patterns with rectangles and equilateral triangles can be easily obtained if the direction of water flow is adjusted accordingly (Figure 1.27A to D). The current–voltage measurements for each of the four crosspoints (Figure 1.27D) exhibited linear or nearly linear behavior (red curves) and are consistent with expectations for n–n type junctions (Figure 1.27E). Duan et al.[44] also utilized electric fields to assist the assembly of NWs into circuits. The p–n junctions

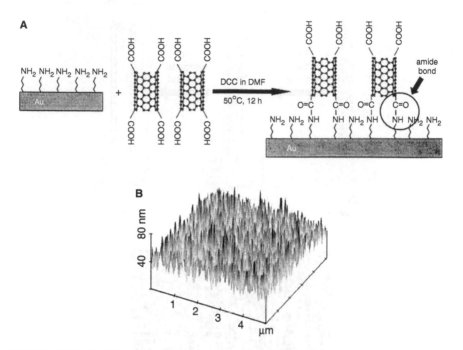

FIGURE 1.29 (A) Scheme of CNT assemblies. (B) AFM images of aligned CNTs on susb-trate. (From Diao, P. et al., *ChemPhysChem*, 3, 898–901, 2002. With permission.)

from crossed NWs formed by random deposition exhibited behavior characteristic of light-emitting diodes. Capillary forces can also be used to assist the assembly NWs or nanotubes on the surface containing microfluid channels. As such, the capillary force from the natural drying process drives NWs across the gap of chan-nels.[477] Considering magnetic force as the external stimulus for NW alignment, a high-strength magnetic field was able to direct the assembly of CNTs in the polymer matrix,[478] but so far no papers have been found on the magnetic field-induced alignment of group II-VI and similar NWs.

Another effective physical method for NW organization is the alignment on patterned templates. Several kinds of templates have been developed for this purpose. As discussed in Section 1.3.2.1, the side grooves from molecular beam epitaxy can be used as the templates to pattern NW structures.[220] Moreover, CdSe NPs could be aligned along the side edges.[479] Yang and coworkers[249,480,481] demonstrated that the LB technique is also sufficient to orient NWs in large amounts at the liquid–gas interface. An ordered LB monolayer of NWs can be further transferred onto a solid substrate.[217] Although the template methods are at the initial stage, the high output, short processing time, and low cost make them a rather competitive family of methods.

1.4.2.2 Chemical Method of NW Alignment

Oriented NWs can be obtained via chemical growth. Porous aluminum oxide, poly-mers, and silica have been used to initiate alignment of nanotubes and NWs. Applied to semiconductors, the spontaneously oriented NWs by porous template exhibit unique anisotropic photoluminescent,[427] magnetic,[482] thermoelectrical,[483] and electrical properties.[484]

The vapor–liquid–solid (VLS) method can also facilitate the preparation of aligned NW arrays when conditions of growth are properly selected. Taking Si and ZnO as examples, Wu et al.[6] demonstrated that an organized network of NWs could be obtained. NWs with specific orientation exhibit optical/electrical anisotropy and, therefore, vertical ZnO NW arrays on the substrate can be used as optical media for ultraviolet lasers[485] and a semiconductor material for Schottky diodes.[486] Additionaly, these NWs can serve as templates to prepare oriented ceramic nanotubes[487,488] (Figure 1.28A). GaN shells were coated around ZnO NWs by chemical vapor deposition. The ZnO NW templates (Figure 1.28B) were subsequently removed by thermal reduction and evaporation, leading to the ordered array of GaN nanotubes on the substrates (Figure 1.28C).

Application of external fields during the chemical synthesis can also assist the orientation of NWs and nanotubes. This method is mainly used to direct the growth of CNTs. Ural et al.[489] and Zhang et al.[490] observed alignment of CNTs grown in the electric field. Note that the external magnetic field also assisted the orientation of CNTs.[491] Huang et al.[492] describes patterns of CNTs obtained by the control of gas flow direction. Although only the alignment of CNTs has been experimentally demonstrated, it is very likely that NWs with similar anisotropy can be oriented by applying similar force fields during the NW synthesis.

Chemical methods of orientational control should also include methods based on selective reactivity of terminal points of NWs or CNTs. Liu et al.[493] found that moderate oxidation only affected the ends of CNTs. This structural feature can be taken advantage of by specific binding of CNT ends (and ends only) to Au surfaces as a result of surface condensation reaction between COOH groups of slightly oxidized CNTs and NH_2 groups of 11-amino-n-undecylmercaptan monolayers on Au substrates with dicyclohexylcarbodiimide as the condensing agent[494,495] (Figure 1.29A). AFM images (Figure 1.29B) indicated that the CNTs were vertically aligned on the modified Au substrates. The array of aligned stand-up CNTs were used as microelectrodes,[495] biosensors,[496] and field emission displays.[497]

Organized assemblies of CNTs can also be made by the LBL method by alternative adsorption of Fe^{3+} ions with the carboxy-terminated CNTs at the untethered ends.[498] Kovtyukhova et al.[499,500] fabricated two metal NWs linked by polymer heterostructures through the sequence of electrodeposition of metal, LBL assembly of polyelectrolytes, and electrodeposition of another type of metal in mesoporous templates. The i-V curves of the NW/polyelectrolyte heterostructures exhibited asymmetric hysteresis loops, which could be used as memory devices (described in Chapter 15).

Compared with physical methods, chemical methods of NW organization exhibit greater versatility and selectivity. However, in general the chemical methods also contain a danger of chemical modification of the surfaces of nanocolloids, which can be undesirable. Speaking of accuracy of positioning, both physical and chemical methods are prone to defect formation, which can be detrimental for many optoelectronic nanodevices based on single NWs but may be tolerated in optoelectronic materials.

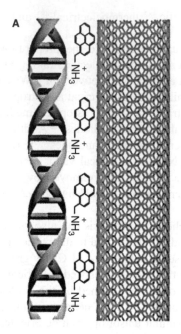
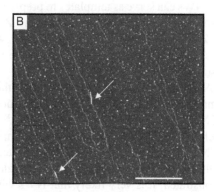

FIGURE 1.30 (A) Scheme of the interaction of CNTs and DNA template. (B) AFM images of CNTs on the aligning DAN surface. (From Xin, H. and Woolley, A.T., *J. Am. Chem. Soc.*, 125, 8710–8711, 2003. With permission.)

1.4.2.3 Biological Method of NW Alignment

Compared with NPs, the bioconjugates of metal and semiconductor NWs are more difficult to synthesize owing to their large molecular weight and, related to that, lower colloidal stability. However, the successful preparation of NW bioconjugates and their subsequent organization of biological ligands are possible. It was achieved by the groups of Nicewarner-Pena,[501] Cui,[252] Caswell,[502] and Wang.[503]

Xin and Woolley[504] showed that the biomolecules could direct the alignment of nanotubes with the help of a substrate bearing a template layer from aligned double-stranded DNA. CNTs were immobilized onto the DNA surface through electrostatic interaction (Figure 1.30A), and they revealed CNT orientation identical to that of DNA molecules (Figure 1.30B). Because of the highly aligned DNA, NWs could be easily produced by simple transfer printing[505]; it is reasonable to believe that the novel method can become a versatile way of the bottom-up construction of molecular electronic nanocircuits from both nanotubes and NWs.

1.4.3 SUPERSTRUCTURES OF NWS AND NPS

Attaching metal or semiconductor NPs to the surfaces of NWs and nanotubes is of interest for obtaining NP/NW hybrid materials with synergetic optical, electric, magnetic, and mechanical properties. This is also an imminent problem of any nanoscale electronic device.

For nanotubes, the NPs can be attached either inside the tube or on the surface. Ajayan and Iijima[506] were the first to show that a metal filled in CNTs and formed an NP–nanotube complex. This process is driven by capillary forces inside the nanotube, which are quite high due to the small diameter. Later studies demonstrated that both semiconductor and metal NPs could be synthesized in the nanotubes of SiO_2,[507] ZrO_2,[508] V_2O_3,[509] and BN.[510] The work on the assembly of NPs onto the surface of nanotubes was mainly centered on CNTs, which includes physical evaporation,[511,512] electrochemical or electroless deposition,[513,514] sol-gel,[515,516] adsorption by hydrophobic forces,[517,518] and chemical cross-linking.[519–521] Importantly, electrical and optical properties of NPs are strongly altered in these complexes.[522,523] This effect must be taken into account when designing a nanoscale device. One of the methods reducing the electronic coupling between different nanoscale species is to wrap one of them with polyelectrolytes by the LBL technique[524] or coat them with a layer of SiO_2.[64]

Unlike CNTs, the reports on complexes of NPs and semiconductor or metal NWs have been rather seldom up to now. Osterloh et al.[525] studied the assembly of CdSe and Au NPs onto the surface of $LiMo_3Se_3$ NWs upon the hydrophobic interaction and found that the optical and electronic properties of NPs were largely alerted. Metal NWs beaded with the second metal NPs were made by two-step electrodeposition.[258] Interestingly, Sun et al.[526] recently obtained the complexes of NPs and NWs for which the nanocrystals were attached only to the ends of the wires. The origin of the specificity of the attachment lies in the mechanism of Ag NW formation.[111] Faces of Ag located in the ends of NWs are not covered by the stabilizer layer, which results in a much higher rate of growth there and eventually a high apex ratio (Figure 1.31A and B). When a low concentration of 1,12-dodecanedithiol is added to the NWs' dispersion, it reacts with the uncoated Ag surface, i.e., in the NW ends. As a result, addition of Au NPs leads to their bonding with dithiol on the same spots (Figure 1.31C). Very recently, Mokari et al.[527] realized the anisotropic selective growth of gold NPs at the tips of CdSe nanorods and tetrapods. More importantly, the authors further showed that NPs at the tips of NWs provided natural contact points for self-assembly and electronic devices.

1.5 PROSPECTS

NWs and NPs will play a leading role in the technological revolution of the 21st century due to their unique electrical and optical properties and reliable preparation procedures. Bottom-up assembly of one-, two-, and three-dimensional structures is a futuristic, but still viable, technology for the next generation of optics and electronics for different purposes. More importantly, learning about the methods of their organization in space is a logical step in the development of nanotechnology, which can lead to new discoveries. On the basis of this review, one can conclude that there are many success stories in this area. Nevertheless, there are many challenges remaining, too, and the most obvious one is the ability to place NWs or NPs into desirable patterns with high efficiency and speed. The authors believe that the methods of NP and NW assembly with a precision of 0.5 to 2 nm — the most influential for electronics — will take the most effort, whereas the techniques for

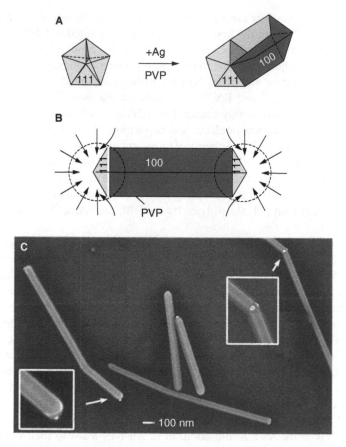

FIGURE 1.31 (A, B) Scheme of growth mechanism of Ag nanowires. Arrows show the growth direction. (C) TEM images of Ag NWs attached by Au NPs. (From Sun, Y. et al., *Nano Lett.*, 3, 955–960, 2003. With permission.)

nanoscale organization in the range of 10 to 100 nm have been developed in sufficient detail for practical applications. Biological assemblies of nanocolloids (also discussed in other chapters of this book), which cover the range between 2 and 10 nm, should play an important role in the bottom-up assembly of electronic devices not only due to high efficiency but also due to the potential complexity of the resulting organized systems. Indeed, recent research has demonstrated that single CNTs could be accurately positioned and assembled into a FET device by DNA templates,[528] which sheds light on the practical application of NPs and NWs as electronic devices.

REFERENCES

1. John, F.J. *Japanese Journal of Applied Physics, Part 1: Regular Papers, Short Notes & Review Papers* 1991, *30*, 3088–3092.
2. Service, R.F. *Science (Washington, DC)* 2001, *293*, 782–785.

3. Tseng, G.Y., Ellenbogen, J.C. *Science (Washington, DC)* 2001, *294*, 1293–1294.
4. Aviram, A., Ratner, M., Mujica, V., Editors. Molecular electronics II. *Annals of the New York Academy of Science* 2002, *960*, 249.
5. Feldheim, D.L., Keating, C.D. *Chemical Society Reviews* 1998, *27*, 1–12.
6. Wu, Y., Yan, H., Huang, M., Messer, B., Song, J.H., Yang, P. *Chemistry: A European Journal* 2002, *8*, 1260–1268.
7. Tour, J.M. *Accounts of Chemical Research* 2000, *33*, 791–804.
8. Aviram, A., Ratner, M.A. *Chemical Physics Letters* 1974, *29*, 277–283.
9. Kagan, C.R., Afzali, A., Martel, R., Gignac, L.M., Solomon, P.M., Schrott, A.G., Ek, B. *Nano Letters* 2003, *3*, 119–124.
10. Lee, J.O., Lientschnig, G., Wiertz, F., Struijk, M., Janssen, R.A.J., Egberink, R., Reinhoudt, D.N., Hadley, P., Dekker, C. *Nano Letters* 2003, *3*, 113–117.
11. Dresselhaus, M.S., Dresselhaus, G., Avouris, P., Editors. Carbon nanotubes synthesis, structure, properties, and applications. *Top. Appl. Phys.*, 2001, *80*, 447.
12. Avouris, P. *Accounts of Chemical Research* 2002, *35*, 1026–1034.
13. Dai, H. *Surface Science* 2002, *500*, 218–241.
14. Chattopadhyay, D., Galeska, I., Papadimitrakopoulos, F. *Journal of the American Chemical Society* 2003, *125*, 3370–3375.
15. Averin, D.V., Register, L.F., Likharev, K.K., Hess, K. *Applied Physics Letters* 1994, *64*, 126–128.
16. Averin, D.V., Likharev, K.K. *Springer Series in Electronics and Photonics* 1992, *31*, 3–12.
17. Fulton, T.A., Dolan, G.J. *Physical Review Letters* 1987, *59*, 109–112.
18. Andres, R.P., Bein, T., Dorogi, M., Feng, S., Henderson, J.I., Kubiak, C.P., Mahoney, W., Osifchin, R.G., Reifenberger, R. *Science (Washington, DC)* 1996, *272*, 1323–1325.
19. Taleb, A., Silly, F., Gusev, A.O., Charra, F., Pileni, M.P. *Advanced Materials (Weinheim, Germany)* 2000, *12*, 633–637.
20. Hou, J.G., Wang, B., Yang, J., Wang, K., Lu, W., Li, Z., Wang, H., Chen, D.M., Zhu, Q. *Physical Review Letters* 2003, *90*, 246803/1–246803/4.
21. Lu, W., Wang, B., Wang, K., Wang, X., Hou, J.G. *Langmuir* 2003, *19*, 5887–5891.
22. Banin, U., Cao, Y., Katz, D., Millo, O. *Nature (London)* 1999, *400*, 542–544.
23. Millo, O., Katz, D., Cao, Y., Banin, U. *Physical Review Letters* 2001, *86*, 5751–5754.
24. Norris, D.J., Bawendi, M.G. *Physical Review B: Condensed Matter* 1996, *53*, 16338–16346.
25. Klein, D.L., McEuen, P.L., Katari, J.E.B., Roth, R., Alivisatos, A.P. *Applied Physics Letters* 1996, *68*, 2574–2576.
26. Klein, D.L., Roth, R., Lim, A.K.L., Alivisatos, A.P., McEuen, P.L. *Nature (London)* 1997, *389*, 699–701.
27. Persson, S.H.M., Olofsson, L., Hedberg, L., Sutherland, D., Olsson, E. *Annals of the New York Academy of Sciences* 1998, *852*, 188–196.
28. Amlani, I., Orlov, A.O., Toth, G., Bernstein, G.H., Lent, C.S., Snider, G.L. *Science (Washington, DC)* 1999, *284*, 289–291.
29. Orlov, A.O., Amlani, I., Bernstein, G.H., Lent, C.S., Snider, G.L. *Science (Washington, DC)* 1997, *277*, 928–930.
30. Xia, Y., Yang, P., Sun, Y., Wu, Y., Mayers, B., Gates, B., Yin, Y., Kim, F., Yan, H. *Advanced Materials (Weinheim, Germany)* 2003, *15*, 353–389.
31. Rao, C.N.R., Deepak, F.L., Gundiah, G., Govindaraj, A. *Progress in Solid State Chemistry* 2003, *31*, 5–147.
32. Martin, C.R. *Chemistry of Materials* 1996, *8*, 1739–1746.

33. Trentler, T.J., Hickman, K.M., Goel, S.C., Viano, A.M., Gibbons, P.C., Buhro, W.E. *Science (Washington, DC)* 1995, *270*, 1791–1794.
34. Kan, S., Mokari, T., Rothenberg, E., Banin, U. *Nature Materials* 2003, *2*, 155–158.
35. Yu, H., Li, J., Loomis, R.A., Wang, L.W., Buhro, W.E. *Nature Materials* 2003, *2*, 517–520.
36. Holmes, J.D., Johnston, K.P., Doty, R.C., Korgel, B.A. *Science (Washington, DC)* 2000, *287*, 1471–1473.
37. Peng, X., Manna, U., Yang, W., Wickham, J., Scher, E., Kadavanich, A., Allvisatos, A.P. *Nature (London)* 2000, *404*, 59–61.
38. Tang, Z., Kotov, N.A., Giersig, M. *Science (Washington, DC)* 2002, *297*, 237–240.
39. Duan, X., Lieber, C.M. *Advanced Materials (Weinheim, Germany)* 2000, *12*, 298–302.
40. Wang, N., Tang, Y.H., Zhang, Y.F., Lee, C.S., Bello, I., Lee, S.T. *Chemical Physics Letters* 1999, *299*, 237–242.
41. Pan, Z.W., Dai, Z.R., Wang, Z.L. *Science (Washington, DC)* 2001, *291*, 1947–1949.
42. Cui, Y., Duan, X., Hu, J., Lieber, C.M. *Journal of Physical Chemistry B* 2000, *104*, 5213–5216.
43. Huang, Y., Duan, X., Cui, Y., Lieber, C.M. *Nano Letters* 2002, *2*, 101–104.
44. Duan, X., Huang, Y., Cui, Y., Wang, J., Lieber, C.M. *Nature (London)* 2001, *409*, 66–69.
45. Cui, Y., Zhong, Z., Wang, D., Wang, W.U., Lieber, C.M. *Nano Letters* 2003, *3*, 149–152.
46. Heath, J.R., Kuekes, P.J., Snider, G.S., Williams, R.S. *Science (Washington, DC)* 1998, *280*, 1716–1721.
47. Fu, Q., Lu, C., Liu, J. *Nano Letters* 2002, *2*, 329–332.
48. Shim, M., Javey, A., Kam, N.W.S., Dai, H. *Journal of the American Chemical Society* 2001, *123*, 11512–11513.
49. Carrillo, A., Swartz, J.A., Gamba, J.M., Kane, R.S., Chakrapani, N., Wei, B., Ajayan, P.M. *Nano Letters* 2003, *3*, 1437–1440.
50. Pantarotto, D., Partidos, C.D., Graff, R., Hoebeke, J., Briand, J.P., Prato, M., Bianco, A. *Journal of the American Chemical Society* 2003, *125*, 6160–6164.
51. Lauhon, L.J., Gudiksen, M.S., Wang, D., Lieber, C.M. *Nature (London)* 2002, *420*, 57–61.
52. Arnold, M.S., Avouris, P., Pan, Z.W., Wang, Z.L. *Journal of Physical Chemistry B* 2003, *107*, 659–663.
53. Maiti, A., Rodriguez, J.A., Law, M., Kung, P., McKinney, J.R., Yang, P. *Nano Letters* 2003, *3*, 1025–1028.
54. Wang, D., Salgueirino-Maceira, V., Liz-Marzan, L.M., Caruso, F. *Advanced Materials (Weinheim, Germany)* 2002, *14*, 908–912.
55. Correa-Duarte, M.A., Kobayashi, Y., Caruso, R.A., Liz-Marzan, L.M. *Journal of Nanoscience and Nanotechnology* 2001, *1*, 95–99.
56. Ung, T., Liz-Marzan, L.M., Mulvaney, P. *Journal of Physical Chemistry B* 1999, *103*, 6770–6773.
57. Dimaria, D.J. *Solid-State Electronics* 1997, *41*, 957–965.
58. Morales, A.M., Lieber, C.M. *Science (Washington, DC)* 1998, *279*, 208–211.
59. Meng, G.W., Zhang, L.D., Mo, C.M., Zhang, S.Y., Qin, Y., Feng, S.P., Li, H.J. *Journal of Materials Research* 1998, *13*, 2533–2538.
60. Yin, Y., Lu, Y., Sun, Y., Xia, Y. *Nano Letters* 2002, *2*, 427–430.
61. Kovtyukhova, N.I., Mallouk, T.E., Mayer, T.S. *Advanced Materials (Weinheim, Germany)* 2003, *15*, 780–785.
62. Xie, Y., Qiao, Z.P., Chen, M., Liu, X.M., Qian, Y.T. *Advanced Materials (Weinheim, Germany)* 1999, *11*, 1512–1515.

63. Obare, S.O., Jana, N.R., Murphy, C.J. *Nano Letters* 2001, *1*, 601–603.
64. Liang, X., Tan, S., Tang, Z., Kotov, N.A. *Langmuir* 2004, *20*, 1016–1020.
65. Wang, Y., Tang, Z., Liang, X., Liz-Marzan, L.M., Kotov, N.A. *Nano Letters* 2004, *4*, 225–231.
66. Mamedov, A.A., Kotov, N.A., Prato, M., Guldi, D.M., Wicksted, J.P., Hirsch, A. *Nature Materials* 2002, *1*, 190–194.
67. Postma, H.W., Teepen, T., Yao, Z., Grifoni, M., Dekker, C. *Science (Washington, DC)* 2001, *293*, 76–79.
68. Strano, M.S., Dyke, C.A., Usrey, M.L., Barone, P.W., Allen, M.J., Shan, H., Kittrell, C., Hauge, R.H., Tour, J.M., Smalley, R.E. *Science (Washington, DC)* 2003, *301*, 1519–1522.
69. Krupke, R., Hennrich, F., Loehneysen, H.V., Kappes, M.M. *Science (Washington, DC)* 2003, *301*, 344–347.
70. Bachilo, S.M., Balzano, L., Herrera, J.E., Pompeo, F., Resasco, D.E., Weisman, R.B. *Journal of the American Chemical Society* 2003, *125*, 11186–11187.
71. Zheng, M., Jagota, A., Semke, E.D., Diner, B.A., Mclean, R.S., Lustig, S.R., Richardson, R.E., Tassi, N.G. *Nature Materials* 2003, *2*, 338–342.
72. Zheng, M., Jagota, A., Strano, M.S., Santos, A.P., Barone, P., Chou, S.G., Diner, B.A., Dresselhaus, M.S., Mclean, R.S., Onoa, G.B., Samsonidze, G.G., Semke, E.D., Usrey, M., Walls, D.J. *Science (Washington, DC)* 2003, *302*, 1545–1548.
73. Banerjee, S., Kahn, M.G.C., Wong, S.S. *Chemistry: A European Journal* 2003, *9*, 1898–1908.
74. Lin, Y., Taylor, S., Li, H., Fernando, K.A.S., Qu, L., Wang, W., Gu, L., Zhou, B., Sun, Y.P. *Journal of Materials Chemistry* 2004, *14*, 527–541.
75. Sun, Y., Fu, K., Lin, Y., Huang, W. *Acc Chem Res* 2002, *35*, 1096–1104.
76. Appell, D. *Nature (London)* 2002, *419*, 553–555.
77. Ingram, R.S., Hostetler, M.J., Murray, R.W., Schaaff, T.G., Khoury, J., Whetten, R.L., Bigioni, T.P., Guthrie, D.K., First, P.N. *Journal of the American Chemical Society* 1997, *119*, 9279–9280.
78. Templeton, A.C., Wuelfing, W.P., Murray, R.W. *Accounts of Chemical Research* 2000, *33*, 27–36.
79. Chen, S., Ingrma, R.S., Hostetler, M.J., Pietron, J.J., Murray, R.W., Schaaff, T.G., Khoury, J.T., Alvarez, M.M., Whetten, R.L. *Science (Washington, DC)* 1998, *280*, 2098–2101.
80. Feldheim, D.L., Grabar, K.C., Natan, M.J., Mallouk, T.E. *Journal of the American Chemical Society* 1996, *118*, 7640–7641.
81. Hicks, J.F., Miles, D.T., Murray, R.W. *Journal of the American Chemical Society* 2002, *124*, 13322–13328.
82. Chen, S., Murray, R.W., Feldberg, S.W. *Journal of Physical Chemistry B* 1998, *102*, 9898–9907.
83. Chen, S. *Journal of the American Chemical Society* 2000, *122*, 7420–7421.
84. Zamborini, F.P., Hicks, J.F., Murray, R.W. *Journal of the American Chemical Society* 2000, *122*, 4514–4515.
85. Chen, S., Pei, R. *Journal of the American Chemical Society* 2001, *123*, 10607–10615.
86. Zamborini, F.P., Leopold, M.C., Hicks, J.F., Kulesza, P.J., Malik, M.A., Murray, R.W. *Journal of the American Chemical Society* 2002, *124*, 8958–8964.
87. Wang, C., Shim, M., Guyot-Sionnest, P. *Science (Washington, DC)* 2001, *291*, 2390–2392.
88. Guyot-Sionnest, P., Wang, C. *Journal of Physical Chemistry* 2003, *107*, 7355–7359.
89. zum Felde, U., Haase, M., Weller, H. *Journal of Physical Chemistry B* 2000, *104*, 9388–9395.

90. Colvin, V.L., Schlamp, M.C., Allvisatos, A.P. *Nature (London)* 1994, *370*, 354–357.
91. Schlamp, M.C., Peng, X., Alivisatos, A.P. *Journal of Applied Physics* 1997, *82*, 5837–5842.
92. Neuhauser, R.G., Shimizu, K.T., Woo, W.K., Empedocles, S.A., Bawendi, M.G. *Physical Review Letters* 2000, *85*, 3301–3304.
93. Shimizu, K.T., Woo, W.K., Fisher, B.R., Eisler, H.J., Bawendi, M.G. *Physical Review Letters* 2002, *89*, 117401/1–117401/4.
94. Coe, S., Woo, W.K., Bawendi, M., Bulovic, V. *Nature (London)* 2002, *420*, 800–803.
95. Dabbousi, B.O., Bawendi, M.G., Onitsuka, O., Rubner, M.F. *Applied Physics Letters* 1995, *66*, 1316–1318.
96. Mattoussi, H., Radzilowski, L.H., Dabbousi, B.O., Thomas, E.L., Bawendi, M.G., Rubner, M.F. *Journal of Applied Physics* 1998, *83*, 7965–7974.
97. Gao, M., Richter, B., Kirstein, S., Moehwald, H. *Journal of Physical Chemistry B* 1998, *102*, 4096–4103.
98. Mattoussi, H., Radzilowski, L.H., Dabbousi, B.O., Fogg, D.E., Schrock, R.R., Thomas, E.L., Rubner, M.F., Bawendi, M.G. *Journal of Applied Physics* 1999, *86*, 4390–4399.
99. Rogach, A.L., Koktysh, D.S., Harrison, M., Kotov, N.A. *Chemistry of Materials* 2000, *12*, 1526–1528.
100. Tesster, N., Medvedev, V., Kazes, M., Kan, S., Banin, U. *Science (Washington, DC)* 2002, *295*, 1506–1508.
101. Kim, M., Jeon, B.H., Kim, J.Y., Choi, J.H. *Synthetic Metals* 2003, *135/136*, 743–744.
102. Achermann, M., Petruska, M.A., Kos, S., Smith, D.L., Koleske, D.D., Klimov, V.I. *Nature (London)* 2004, *429*, 642–646.
103. Myung, N., Ding, Z., Bard, A.J. *Nano Letters* 2002, *2*, 1315–1319.
104. Ding, Z., Quinn, B.M., Haram, S.K., Pell, L.E., Korgel, B.A., Bard, A.J. *Science (Washington, DC)* 2002, *296*, 1293–1297.
105. Myung, N., Bae, Y., Bard, A.J. *Nano Letters* 2003, *3*, 1053–1055.
106. Klimov, V.I., Mikhailovsky, A.A., Xu, S., Malko, A., Hollingsworth, J.A., Leatherdale, C.A., Eisler, H.J., Bawendiz, M.G. *Science (Washington, DC)* 2000, *290*, 314–317.
107. Norris, D.J., Vlasov, Y.A. *Advanced Materials (Weinheim, Germany)* 2001, *13*, 371–376.
108. Hagfeldt, A., Graetzel, M. *Chemical Reviews (Washington, DC)* 1995, *95*, 49–68.
109. Hagfeldt, A., Graetzel, M. *Accounts of Chemical Research* 2000, *33*, 269–277.
110. Gratzel, M. *Nature (London)* 2001, *414*, 338–344.
111 Rincon, M.E., Hu, H., Martinez, G., Suarez, R., Banuelos, J.G. *Solar Energy Materials and Solar Cells* 2003, *77*, 239–254.
112. Wang, Y., Hang, K., Anderson, N.A., Lian, T. *Journal of Physical Chemistry B* 2003, *107*, 9434–9440.
113. Gomez-Romero, P. *Advanced Materials (Weinheim, Germany)* 2001, *13*, 163–174.
114. Godovsky, D.Y. *Advances in Polymer Science* 2000, *153*, 163–205.
115. Wang, P., Zakeeruddin, S.M., Moser, J.E., Nazeeruddin, M.K., Sekiguchi, T., Graetzel, M. *Nature Materials* 2003, *2*, 402–407.
116. Anderson, N.A., Ai, X., Lian, T. *Trends in Optics and Photonics* 2002, *72*, 368–369.
117. Anderson, N.A., Hao, E., Ai, X., Hastings, G., Lian, T. *Physica E: Low-Dimensional Systems and Nanostructures (Amsterdam)* 2002, *14*, 215–218.
118. Barazzouk, S., Lee, H., Hotchandani, S., Kamat, P.V. *Journal of Physical Chemistry B* 2000, *104*, 3616–3623.
119. Kotov, N.A., Dekany, I., Fendler, J.H. *Journal of Physical Chemistry* 1995, *99*, 13065–13069.

120. Nasr, C., Kamat, P.V., Hotchandani, S. *Journal of Physical Chemistry B* 1998, *102*, 10047–10056.
121. Nasr, C., Kamat, P.V., Hotchandani, S. *Journal of Electroanalytical Chemistry* 1997, *420*, 201–207.
122. Suarez, R., Nair, P.K., Kamat, P.V. *Langmuir* 1998, *14*, 3236–3241.
123. Vinodgopal, K., Bedja, I., Kamat, P.V. *Chemistry of Materials* 1996, *8*, 2180–2187.
124. Greenham, N.C., Peng, X., Alivisatos, A.P. *Physical Review B: Condensed Matter* 1996, *54*, 17628–17637.
125. Ginger, D.S., Greenham, N.C. *Physical Review B: Condensed Matter and Materials Physics* 1999, *59*, 10622–10629.
126. Chandrasekharan, N., Kamat, P.V. *Journal of Physical Chemistry B* 2000, *104*, 10851–10857.
127. Kamat, P.V., Shanghavi, B. *Journal of Physical Chemistry B* 1997, *101*, 7675–7679.
128. Wen, C., Ishikawa, K., Kishima, M., Yamada, K. *Solar Energy Materials and Solar Cells* 2000, *61*, 339–351.
129. Sheeney-Haj-Ichia, L., Wasserman, J., Willner, I. *Advanced Materials (Weinheim, Germany)* 2002, *14*, 1323–1326.
130. Sheeney-Haj-Ichia, L., Pogorelova, S., Gofer, Y., Willner, I. *Advanced Functional Materials* 2004, *14*, 416–424.
131. De Jongh, P.E., Meulenkamp, E.A., Vanmaekelbergh, D., Kelly, J.J. *Journal of Physical Chemistry B* 2000, *104*, 7686–7693.
132. Adams, D.M., Brus, L., Chidsey, C.E.D., Creager, S., Creutz, C., Kagan, C.R., Kamat, P.V., Lieberman, M., Lindsay, S., Marcus, R.A., Metzger, R.M., Michel-Beyerle, M.E., Miller, J.R., Newton, M.D., Rolison, D.R., Sankey, O., Schanze, K.S., Yardley, J., Zhu, X. *Journal of Physical Chemistry B* 2003, *107*, 6668–6697.
133. Noack, V., Weller, H., Eychmueller, A. *Journal of Physical Chemistry B* 2002, *106*, 8514–8523.
134. Noack, V., Weller, H., Eychmueller, A. *Physical Chemistry Chemical Physics* 2003, *5*, 384–394.
135. Morgan, N.Y., Leatherdale, C.A., Drndic, M., Jarosz, M.V., Kastner, M.A., Bawendi, M. *Physical Review B: Condensed Matter and Materials Physics* 2002, *66*, 075339/1–075339/9.
136. Ginger, D.S., Greenham, N.C. *Journal of Applied Physics* 2000, *87*, 1361–1368.
137. Hikmet, R.A.M., Talapin, D.V., Weller, H. *Journal of Applied Physics* 2003, *93*, 3509–3514.
138. Drndic, M., Markov, R., Jarosz, M.V., Bawendi, M.G., Kastner, M.A., Markovic, N., Tinkham, M. *Applied Physics Letters* 2003, *83*, 4008–4010.
139. Tian, Z.R., Voigt, J.A., Liu, J., McKenzie, B., Xu, H. *Journal of the American Chemical Society* 2003, *125*, 12384–12385.
140. Huynh, W.U., Dittmer, J.J., Alivisatos, A.P. *Science (Washington, DC)* 2002, *295*, 2425–2427.
141. Sun, B., Marx, E., Greenham, N.C. *Nano Letters* 2003, *3*, 961–963.
142. Manna, L., Milliron, D.J., Meisel, A., Scher, E.C., Alivisatos, A.P. *Nature Materials* 2003, *2*, 382–385.
143. Arici, E., Sariciftci, N.S., Meissner, D. *Advanced Functional Materials* 2003, *13*, 165–171.
144. Breeze, A.J., Schlesinger, Z., Carter, S.A., Brock, P.J. *Physical Review B: Condensed Matter and Materials Physics* 2001, *64*, 125205/1–125205/9.
145. Breeze, A.J., Schlesinger, Z., Carter, S.A., Hoerhold, H.H., Tillmann, H., Ginley, D.S., Brock, P.J. *Proceedings of SPIE: The International Society for Optical Engineering* 2001, *4108*, 57–61.

146. Qian, X., Qin, D., Bai, Y., Li, T., Tang, X., Wang, E., Dong, S. *Journal of Solid State Electrochemistry* 2001, *5*, 562–567.
147. Qian, X., Qin, D., Song, Q., Bai, Y., Li, T., Tang, X., Wang, E., Dong, S. *Thin Solid Films* 2001, *385*, 152–161.
148. Arango, A.C., Carter, S.A., Brock, P.J. *Applied Physics Letters* 1999, *74*, 1698–1700.
149. He, J.A., Mosurkal, R., Samuelson, L.A., Li, L., Kumar, J. *Langmuir* 2003, *19*, 2169–2174.
150. Hao, E., Yang, B., Ren, H., Qian, X., Xie, T., Shen, J., Li, D. *Materials Science and Engineering C: Biomimetic and Supramolecular Systems* 1999, *C10*, 119–122.
151. Maehara, Y., Takenaka, S., Shimizu, K., Yoshikawa, M., Shiratori, S. *Thin Solid Films* 2003, *438/439*, 65–69.
152. Tokuhisa, H., Hammond, P.T. *Polymer Preprints (American Chemical Society, Division of Polymer Chemistry)* 2003, *44*, 644–645.
153. Kuo, C., Kumar, J., Tripathy, S.K., Chiang, L.Y. *Journal of Macromolecular Science, Pure and Applied Chemistry* 2001, *A38*, 1481–1498.
154. Li, L.S., Li, A.D.Q. *Journal of Physical Chemistry B* 2001, *105*, 10022–10028.
155. Mattoussi, H., Rubner, M.F., Zhou, F., Kumar, J., Tripathy, S.K., Chiang, L.Y. *Applied Physics Letters* 2000, *77*, 1540–1542.
156. Li, M., Li, B., Jiang, L., Tussila, T., Tkachenko, N., Lemmetyinen, H. *Chemistry Letters* 2000, 266–267.
157. Ding, H., Ram, M.K., Nicolini, C. *J Nanosci Nanotechnol* 2001, *1*, 207–213.
158. Halaoui, L.I. *Journal of the Electrochemical Society* 2003, *150*, E455–E460.
159. Mamedov, A.A., Belov, A., Giersig, M., Mamedova, N.N., Kotov, N.A. *Journal of the American Chemical Society* 2001, *123*, 7738–7739.
160. Abdelrazzaq, F.B., Kwong, R.C., Thompson, M.E. *Journal of the American Chemical Society* 2002, *124*, 4796–4803.
161. Kaschak, D.M., Lean, J.T., Waraksa, C.C., Saupe, G.B., Usami, H., Mallouk, T.E. *Journal of the American Chemical Society* 1999, *121*, 3435–3445.
162. Guldi, D.M., Pellarini, F., Prato, M., Granito, C., Troisi, L. *Nano Letters* 2002, *2*, 965–968.
163. Boiko, B.T., Khripunov, G.S., Yurchenko, V.B., Ruda, H.E. *Solar Energy Materials and Solar Cells* 1997, *45*, 303–308.
164. Guldi, D.M., Luo, C., Koktysh, D., Kotov, N.A., Da Ros, T., Bosi, S., Prato, M. *Nano Letters* 2002, *2*, 775–780.
165. Stockton, W.B., Rubner, M.F. *Macromolecules* 1997, *30*, 2717–2725.
166. Gelinck, G.H., Huitema, H.E., van Veenendaal, E., Cantatore, E., Schrijnemakers, L., van der Putten, J.B.P.H., Geuns, T.C.T., Beenhakkers, M., Giesbers, J.B., Huisman, B.H., Meijer, E.J., Benito, E.M., Touwslager, F.J., Marsman, A.W., van Rens, B.J.E., de Leeuw, D.M. *Nature Materials* 2004, *3*, 106–110.
167. Coleman, J.P., Freeman, J.J., Madhukar, P., Wagenknecht, J.H. *Displays* 1999, *20*, 145–154.
168. Bonhote, P., Gogniat, E., Campus, F., Walder, L., Gratzel, M. *Displays* 1999, *20*, 137–144.
169. Campus, F., Bonhote, P., Gratzel, M., Heinen, S., Walder, L. *Solar Energy Materials and Solar Cells* 1999, *56*, 281–297.
170. Kurth, D.G., Liu, S., Volkmer, D. *NATO Science Series II: Mathematics, Physics and Chemistry* 2003, *98*, 441–466.
171. Liu, S., Kurth, D.G., Bredenkoetter, B., Volkmer, D. *Journal of the American Chemical Society* 2002, *124*, 12279–12287.
172. Liu, S., Volkmer, D., Kurth, D.G. *Journal of Cluster Science* 2003, *14*, 405–419.

173. Liu, S., Kurth, D.G., Mohwald, H., Volkmer, D. *Advanced Materials (Weinheim, Germany)* 2002, *14*, 225–228.
174. Wehrenberg, B.L., Guyot-Sionnest, P. *Journal of the American Chemical Society* 2003, *125*, 7806–7807.
175. Kamat, P.V. *Electrochemistry of Nanomaterials* 2001, 229–246.
176. Baeck, S.H., Jaramillo, T., Stucky, G.D., McFarland, E.W. *Nano Letters* 2002, *2*, 831–834.
177. He, T., Ma, Y., Cao, Y., Yang, W., Yao, J. *Journal of Electroanalytical Chemistry* 2001, *514*, 129–132.
178. He, T., Ma, Y., Cao, Y., Jiang, P., Zhang, X., Yang, W., Yao, J. *Langmuir* 2001, *17*, 8024–8027.
179. He, T., Ma, Y., Cao, Y., Yin, Y., Yang, W., Yao, J. *Applied Surface Science* 2001, *180*, 336–340.
180. He, T., Ma, Y., Cao, Y.A., Yang, W.S., Yao, J.N. *Physical Chemistry Chemical Physics* 2002, *4*, 1637–1639.
181. DeLongchamp, D.M., Hammond, P.T. *Advanced Functional Materials* 2004, *14*, 224–232.
182. Walt, D.R. *Nature Materials* 2002, *1*, 17–18.
183. Zhang, H.L., Evans, S.D., Henderson, J.R., Miles, R.E., Shen, T.H. *Nanotechnology* 2002, *13*, 439–444.
184. Foos, E.E., Snow, A.W., Twigg, M.E., Ancona, M.G. *Chemistry of Materials* 2002, *14*, 2401–2408.
185. Krasteva, N., Besnard, I., Guse, B., Bauer, R.E., Muellen, K., Yasuda, A., Vossmeyer, T. *Nano Letters* 2002, *2*, 551–555.
186. Viets, C., Hill, W. *Journal of Physical Chemistry B* 2001, *105*, 6330–6336.
187. Beverly, K.C., Sample, J.L., Sampaio, J.F., Remacle, F., Heath, J.R., Levine, R.D. *Proceedings of the National Academy of Sciences of the United States of America* 2002, *99*, 6456–6459.
188. Beverly, K.C., Sampaio, J.F., Heath, J.R. *Journal of Physical Chemistry B* 2002, *106*, 2131–2135.
189. Remacle, F., Beverly, K.C., Heath, J.R., Levine, R.D. *Journal of Physical Chemistry B* 2002, *106*, 4116–4126.
190. Sample, J.L., Beverly, K.C., Chaudhari, P.R., Remacle, F., Heath, J.R., Levine, R.D. *Advanced Materials (Weinheim, Germany)* 2002, *14*, 124–128.
191. Sampaio, J.F., Beverly, K.C., Heath, J.R. *Journal of Physical Chemistry B* 2001, *105*, 8797–8800.
192. Henrichs, S., Collier, C.P., Saykally, R.J., Shen, Y.R., Heath, J.R. *Journal of the American Chemical Society* 2000, *122*, 4077–4083.
193. Markovich, G., Collier, C.P., Henrichs, S.E., Remacle, F., Levine, R.D., Heath, J.R. *Accounts of Chemical Research* 1999, *32*, 415–423.
194. Filonovich, S.A., Rakovich, Y., Rolo, A.G., Artemyev, M.V., Hungerford, G., Vasilevskiy, M.I., Gomes, M.J.M., Ferreira, J.F.A. *Journal of Optoelectronics and Advanced Materials* 2000, *2*, 623–628.
195. Du, X., Wang, Y., Mu, Y., Gui, L., Wang, P., Tang, Y. *Chemistry of Materials* 2002, *14*, 3953–3957.
196. Schaadt, D.M., Yu, E.T., Sankar, S., Berkowitz, A.E. *Journal of Vacuum Science and Technology A: Vacuum, Surfaces, and Films* 2000, *18*, 1834–1837.
197. Fissan, H., Kennedy, M.K., Kruis, F.E. *Proceedings of SPIE: The International Society for Optical Engineering* 2001, *4590*, 236–242.
198. Huo, L., Gao, S., Zhao, J., Wang, H., Xi, S. *Journal of Materials Chemistry* 2002, *12*, 392–395.

199. Bochenkov, V.E., Stephan, N., Brehmer, L., Zagorskii, V.V., Sergeev, G.B. *Colloids and Surfaces A: Physicochemical and Engineering Aspects* 2002, *198/200*, 911–915.
200. Shipway, A.N., Willner, I. *Chemical Communications (Cambridge, U.K.)* 2001, *20*, 2035–2045.
201. Willner, I., Shipway, A.N., Willner, B. *ACS Symposium Series* 2003, *844*, 88–105.
202. Han, S., Lin, J., Satjapipat, M., Baca, A.J., Zhou, F. *Chemical Communications (Cambridge, U.K.)* 2001, *7*, 609–610.
203. Schmeisser, D., Appel, G., Bohme, O., Heller, T., Mikalo, R.P., Hoffmann, P., Batchelor, D. *Sensors and Actuators B: Chemical* 2000, *B70*, 131–138.
204. Stich, N., Gandhum, A., Matyushin, V., Raats, J., Mayer, C., Alguel, Y., Schalkhammer, T. *Journal of Nanoscience and Nanotechnology* 2002, *2*, 375–381.
205. Jensen, T.R., Malinsky, M.D., Haynes, C.L., Van Duyne, R.P. *Journal of Physical Chemistry B* 2000, *104*, 10549–10556.
206. McFarland, A.D., Van Duyne, R.P. *Nano Letters* 2003, *3*, 1057–1062.
207. Gu, G., Schmid, M., Chiu, P.W., Minett, A., Fraysse, J., Kim, G.T., Roth, S., Kozlov, M., Munoz, E., Baughman, R.H. *Nature Materials* 2003, *2*, 316–319.
208. Cui, Y., Lieber, C.M. *Science (Washington, DC)* 2001, *291*, 851–853.
209. Duan, X., Huang, Y., Lieber, C.M. *Nano Letters* 2002, *2*, 487–490.
210. Huang, Y., Duan, X., Cui, Y., Lauhon, L.J., Kim, K.H., Lieber, C.M. *Science (Washington, DC)* 2001, *294*, 1313–1317.
211. DeHon, A. *IEEE Transactions on Nanotechnology* 2003, *2*, 23–32.
212. Collier, C.P., Wong, E.W., Belohradsky, M., Raymo, F.M., Stoddart, J.F., Kuekes, P.J., Williams, R.S., Heath, J.R. *Science (Washington, DC)* 1999, *285*, 391–394.
213. Bachtold, A., Hadley, P., Nakanishi, T., Dekker, C. *Science (Washington, DC)* 2001, *294*, 1317–1320.
214. Derycke, V., Martel, R., Appenzeller, J., Avouris, P. *Nano Letters* 2001, *1*, 453–456.
215. Li, C., Ly, J., Lei, B., Fan, W., Zhang, D., Han, J., Meyyappan, M., Thompson, M., Zhou, C. *Journal of Physical Chemistry B* 2004, *108*, 9646–9649.
216. Li, C., Fan, W., Straus, D.A., Lei, B., Asano, S., Zhang, D., Han, J., Meyyappan, M., Zhou, C. *Journal of the American Chemical Society* 2004, *126*, 7750–7751.
217. Whang, D., Jin, S., Lieber, C.M. *Nano Letters* 2003, *3*, 951–954.
218. Smith, P.A., Nordquist, C.D., Jackson, T.N., Mayer, T.S., Martin, B.R., Mbindyo, J., Mallouk, T.E. *Applied Physics Letters* 2000, *77*, 1399–1401.
219. Jung, Y.J., Homma, Y., Ogino, T., Kobayashi, Y., Takagi, D., Wei, B., Vajtai, R., Ajayan, P.M. *Journal of Physical Chemistry B* 2003, *107*, 6859–6864.
220. Melosh, N.A., Boukai, A., Diana, F., Gerardot, B., Badolato, A., Petroff, P.M., Heath, J.R. *Science (Washington, DC)* 2003, *300*, 112–115.
221. Rao, S.G., Huang, L., Setyawan, W., Hong, S. *Nature (London)* 2003, *425*, 36–37.
222. Chandra, P., Ioffe, L.B. *Journal of Physics: Condensed Matter* 2001, *13*, L697–L704.
223. Kuang, D., Xu, A., Fang, Y., Liu, H., Frommen, C., Fenske, D. *Advanced Materials (Weinheim, Germany)* 2003, *15*, 1747–1750.
224. Dick, K.A., Deppert, K., Larsson, M.W., Martensson, T., Seifert, W., Wallenberg, L.R., Samuelson, L. *Nature Materials* 2004, *3*, 380–384.
225. Wang, D., Qian, F., Yang, C., Zhong, Z., Lieber, C.M. *Nano Letters* 2004, *4*, 871–874.
226. Yan, H., He, R., Johnson, J., Law, M., Saykally, R.J., Yang, P. *Journal of the American Chemical Society* 2003, *125*, 4728–4729.
227. Javey, A., Guo, J., Wang, Q., Lundstrom, M., Dai, H. *Nature (London)* 2003, *424*, 654–657.
228. Freitag, M., Martin, Y., Misewich, J.A., Martel, R., Avouris, P. *Nano Letters* 2003, *3*, 1067–1071.

229. Misewich, J.A., Avouris, P., Martel, R., Tsang, J.C., Heinze, S., Tersoff, J. *Science (Washington, DC)* 2003, *300*, 783–786.
230. Wang, J., Gudiksen, M.S., Duan, X., Cui, Y., Lieber, C.M. *Science (Washington, DC)* 2001, *293*, 1455–1457.
231. Johnson, J.C., Choi, H.J., Knutsen, K.P., Schaller, R.D., Yang, P., Saykally, R.J. *Nature Materials* 2002, *1*, 106–110.
232. Wu, Y., Fan, R., Yang, P. *Nano Letters* 2002, *2*, 83–86.
233. Duan, X., Huang, Y., Agarwal, R., Lieber, C.M. *Nature (London)* 2003, *421*, 241–245.
234. Mazzola, L. *Nature Biotechnology* 2003, *21*, 1137–1143.
235. Dai, H. *Accounts of Chemical Research* 2002, *35*, 1035–1044.
236. Kong, J., Franklin, N.R., Zhou, C., Chapline, M.G., Peng, S., Cho, K., Dailt, H. *Science (Washington, DC)* 2000, *287*, 622–625.
237. Wohlstadter, J.N., Wilbur, J.L., Sigal, G.B., Biebuyck, H.A., Billadeau, M.A., Dong, L., Fischer, A.B., Gudibande, S.R., Jameison, S.H., Kenten, J.H., Leginus, J., Leland, J.K., Massey, R.J., Wohlstadter, S.J. *Advanced Materials (Weinheim, Germany)* 2003, *15*, 1184–1187.
238. Besteman, K., Lee, J.O., Wiertz, F.G.M., Heering, H.A., Dekker, C. *Nano Letters* 2003, *3*, 727–730.
239. Modi, A., Koratkar, N., Lass, E., Wei, B., Ajayan, P.M. *Nature (London)* 2003, *424*, 171–174.
240. Kolmakov, A., Zhang, Y., Cheng, G., Moskovits, M. *Advanced Materials (Weinheim, Germany)* 2003, *15*, 997–1000.
241. Law, M., Kind, H., Messer, B., Kim, F., Yang, P. *Angewandte Chemie, International Edition* 2002, *41*, 2405–2408.
242. Comini, E., Faglia, G., Sberveglieri, G., Pan, Z., Wang, Z.L. *Applied Physics Letters* 2002, *81*, 1869–1871.
243. Li, C., Zhang, D., Liu, X., Han, S., Tang, T., Han, J., Zhou, C. *Applied Physics Letters* 2003, *82*, 1613–1615.
244. Zhou, X.T., Hu, J.Q., Li, C.P., Ma, D.D.D., Lee, C.S., Lee, S.T. *Chemical Physics Letters* 2003, *369*, 220–224.
245. Varghese, O.K., Grimes, C.A. *Journal of Nanoscience and Nanotechnology* 2003, *3*, 277–293.
246. Walter, E.C., Penner, R.M., Liu, H., Ng, K.H., Zach, M.P., Favier, F. *Surface and Interface Analysis* 2002, *34*, 409–412.
247. Favier, F., Walter, E.C., Zach, M.P., Benter, T., Penner, R.M. *Science (Washington, DC)* 2001, *293*, 2227–2231.
248. Wang, Z.L. *Advanced Materials (Weinheim, Germany)* 2003, *15*, 432–436.
249. Tao, A., Kim, F., Hess, C., Goldberger, J., He, R., Sun, Y., Xia, Y., Yang, P. *Nano Letters* 2003, *3*, 1229–1233.
250. Li, Z., Chen, Y., Li, X., Kamins, T.I., Nauka, K., Williams, R.S. *Nano Letters* 2004, *4*, 245–247.
251. Hahm, J.I., Lieber, C.M. *Nano Letters* 2004, *4*, 51–54.
252. Cui, Y., Wei, Q., Park, H., Lieber, C.M. *Science (Washington, DC)* 2001, *293*, 1289–1292.
253. Kind, H., Yan, H., Messer, B., Law, M., Yang, P. *Advanced Materials (Weinheim, Germany)* 2002, *14*, 158–160.
254. Lindeberg, M., Hjort, K. *Sensors and Actuators A: Physical* 2003, *A105*, 150–161.
255. Harnack, O., Pacholski, C., Weller, H., Yasuda, A., Wessels, J.M. *Nano Letters* 2003, *3*, 1097–1101.
256. Fullam, S., Cottell, D., Rensmo, H., Fitzmaurice, D. *Advanced Materials (Weinheim, Germany)* 2000, *12*, 1430–1432.

257. Martin, C.R. *Chemistry of Materials* 1996, *8*, 1739–1746.
258. Penner, R.M. *Journal of Physical Chemistry B* 2002, *106*, 3339–3353.
259. Hutchinson, T.O., Liu, Y.P., Kiely, C., Kiely, C.J., Brust, M. *Advanced Materials (Weinheim, Germany)* 2001, *13*, 1800–1803.
260. Fort, E., Ricolleau, C., Sau-Pueyo, J. *Nano Letters* 2003, *3*, 65–67.
261. Braun, E., Eichen, Y., Sivan, U., Ben Yoseph, G. *Nature (London)* 1998, *391*, 775–778.
262. Keren, K., Krueger, M., Gilad, R., Ben Yoseph, G., Sivan, U., Braun, E. *Science (Washington, DC)* 2002, *297*, 72–75.
263. Richter, J., Seidel, R., Kirsch, R., Mertig, M., Pompe, W., Plaschke, J., Schackert, H.K. *Advanced Materials (Weinheim, Germany)* 2000, *12*, 507–510.
264. Kumar, A., Pattarkine, M., Bhadbhade, M., Mandale, A.B., Ganesh, K.N., Datar, S.S., Dharmadhikari, C.V., Sastry, M. *Advanced Materials (Weinheim, Germany)* 2001, *13*, 341–344.
265. Harnack, O., Ford, W.E., Yasuda, A., Wessels, J.M. *Nano Letters* 2002, *2*, 919–923.
266. Ford, W.E., Harnack, O., Yasuda, A., Wessels, J.M. *Advanced Materials (Weinheim, Germany)* 2001, *13*, 1793–1797.
267. Torimoto, T., Yamashita, M., Kuwabata, S., Sakata, T., Mori, H., Yoneyama, H. *Journal of Physical Chemistry B* 1999, *103*, 8799–8803.
268. Jin, J., Jiang, L., Chen, X., Yang, W.S., Li, T.J. *Chinese Journal of Chemistry* 2003, *21*, 208–210.
269. Warner, M.G., Hutchison, J.E. *Nature Materials* 2003, *2*, 272–277.
270. Fu, X., Wang, Y., Huang, L., Sha, Y., Gui, L., Lai, L., Tang, Y. *Advanced Materials (Weinheim, Germany)* 2003, *15*, 902–906.
271. Coelfen, H., Mann, S. *Angewandte Chemie, International Edition* 2003, *42*, 2350–2365.
272. Dujardin, E., Peet, C., Stubbs, G., Culver, J.N., Mann, S. *Nano Letters* 2003, *3*, 413–417.
273. Fowler, C.E., Shenton, W., Stubbs, G., Mann, S. *Advanced Materials (Weinheim, Germany)* 2001, *13*, 1266–1269.
274. Shenton, W., Douglas, T., Young, M., Stubbs, G., Mann, S. *Advanced Materials (Weinheim, Germany)* 1999, *11*, 253–256.
275. Djalali, R., Chen, Y.F., Matsui, H. *Journal of the American Chemical Society* 2002, *124*, 13660–13661.
276. Djalali, R., Chen, Y.F., Matsui, H. *Journal of the American Chemical Society* 2003, *125*, 5873–5879.
277. Banerjee, I.A., Yu, L., Matsui, H. *Nano Letters* 2003, *3*, 283–287.
278. Yu, L., Banerjee, I.A., Matsui, H. *Journal of the American Chemical Society* 2003, *125*, 14837–14840.
279. Behrens, S., Rahn, K., Habicht, W., Bohm, K.J., Rosner, H., Dinjus, E., Unger, E. *Advanced Materials (Weinheim, Germany)* 2002, *14*, 1621–1625.
280. Mao, C., Flynn, C.E., Hayhurst, A., Sweeney, R., Qi, J., Georgiou, G., Iverson, B., Belcher, A.M. *Proceedings of the National Academy of Sciences of the United States of America* 2003, *100*, 6946–6951.
281. Lee, S.W., Lee, S.K., Belcher, A.M. *Advanced Materials (Weinheim, Germany)* 2003, *15*, 689–692.
282. Lee, S.W., Mao, C., Flynn, C.E., Belcher, A.M. *Science (Washington, DC)* 2002, *296*, 892–895.
283. Mao, C., Solis, D.J., Reiss, B.D., Kottmann, S.T., Sweeney, R.Y., Hayhurst, A., Georgiou, G., Iverson, B., Belcher, A.M. *Science (Washington, DC)* 2004, *303*, 213–217.

284. Scheibel, T., Parthasarathy, R., Sawicki, G., Lin, X.M., Jaeger, H., Lindquist, S.L. *Proceedings of the National Academy of Sciences of the United States of America* 2003, *100*, 4527–4532.

285. Korgel, B.A., Fitzmaurice, D. *Advanced Materials (Weinheim, Germany)* 1998, *10*, 661–665.

286. Pacholski, C., Kornowski, A., Weller, H. *Angewandte Chemie, International Edition* 2002, *41*, 1188–1191.

287. Shah, P.S., Sigman, M.B., Jr., Stowell, C.A., Lim, K.T., Johnston, K.P., Korgel, B.A. *Advanced Materials (Weinheim, Germany)* 2003, *15*, 971–974.

288. Polleux, J., Pinna, N., Antonietti, M., Niederberger, M. *Advanced Materials (Weinheim, Germany)* 2004, *16*, 436–439.

289. Chang, J.Y., Chang, J.J., Lo, B., Tzing, S.H., Ling, Y.C. *Chemical Physics Letters* 2003, *379*, 261–267.

290. Liao, J., Zhang, Y., Yu, W., Xu, L., Ge, C., Liu, J., Gu, N. *Colloids and Surfaces A: Physicochemical and Engineering Aspects* 2003, *223*, 177–183.

291. Gao, F., Lu, Q., Zhao, D. *Nano Letters* 2003, *3*, 85–88.

292. Deng, Y., Nan, C.W., Wei, G.D., Guo, L., Lin, Y.H. *Chemical Physics Letters* 2003, *374*, 410–415.

293. Wang, D., Yu, D., Mo, M., Liu, X., Qian, Y. *Solid State Communications* 2003, *125*, 475–479.

294. Liao, J.H., Chen, K.J., Xu, L.N., Ge, C.W., Wang, J., Huang, L., Gu, N. *Applied Physics A: Materials Science and Processing* 2003, *76*, 541–543.

295. Yang, J., Xue, C., Yu, S.H., Zeng, J.H., Qian, Y.T. *Angewandte Chemie, International Edition* 2002, *41*, 4697–4700.

296. Jackson, A.M., Myerson, J.W., Stellacci, F. *Nature Materials* 2004, *3*, 330–336.

297. Reuter, T., Vidoni, O., Torma, V., Schmid, G., Nan, L., Gleiche, M., Chi, L., Fuchs, H. *Nano Letters* 2002, *2*, 709–711.

298. Ohara, P., Heath, J.R., Gelbart, W.M. *Angewandte Chemie, International Edition in English* 1997, *36*, 1078–1080.

299. Ohara, P.C., Gelbart, W.M. *Langmuir* 1998, *14*, 3418–3421.

300. Maillard, M., Motte, L., Ngo, A.T., Pileni, M.P. *Journal of Physical Chemistry B* 2000, *104*, 11871–11877.

301. Maillard, M., Motte, L., Pileni, M.P. *Advanced Materials (Weinheim, Germany)* 2001, *13*, 200–204.

302. Stowell, C., Korgel, B.A. *Nano Letters* 2001, *1*, 595–600.

303. Govor, L.V., Bauer, G.H., Reiter, G., Shevchenko, E., Weller, H., Parisi, J. *Langmuir* 2003, *19*, 9573–9576.

304. Fahmi, A.W., Oertel, U., Steinert, V., Froeck, C., Stamm, M. *Macromolecular Rapid Communications* 2003, *24*, 625–629.

305. Rabani, E., Reichman, D.R., Geissler, P.L., Brus, L.E. *Nature (London)* 2003, *426*, 271–274.

306. Kong, X.Y., Ding, Y., Yang, R., Wang, Z.L. *Science (Washington, DC)* 2004, *303*, 1348–1352.

307. Zhao, L., Kelly, K.L., Schatz, G.C. *Journal of Physical Chemistry B* 2003, *107*, 7343–7350.

308. Maier, S.A., Brongersma, M.L., Kik, P.G., Atwater, H.A. *Physical Review B: Condensed Matter and Materials Physics* 2002, *65*, 193408/1–193408/4.

309. Quinten, M., Leitner, A., Krenn, J.R., Aussenegg, F.R. *Optics Letters* 1998, *23*, 1331–1333.

310. Maier, S.A., Kik, P.G., Atwater, H.A. *Physical Review B: Condensed Matter and Materials Physics* 2003, *67*, 205402/1–205402/5.
311. Krenn, J.R. *Nature Materials* 2003, *2*, 210–211.
312. Maier, S.A., Kik, P.G., Atwater, H.A., Meltzer, S., Harel, E., Koel, B.E., Requicha, A.A.G. *Nature Materials* 2003, *2*, 229–232.
313. Maier, S.A., Kik, P.G., Atwater, H.A. *Applied Physics Letters* 2002, *81*, 1714–1716.
314. Tang, Z., Ozturk, B., Wang, Y., Kotov, N.A. *Journal of Physical Chemistry B* 2004, *108*, 6927–6931.
315. Crooker, S.A., Hollingsworth, J.A., Tretiak, S., Klimov, V.I. *Physical Review Letters* 2002, *89*, 186802/1–186802/4.
316. Mackay, M.E., Dao, T.T., Tuteja, A., Ho, D.L., Van Horn, B., Kim, H.C., Hawker, C.J. *Nature Materials* 2003, *2*, 762–766.
317. Smith, G.D., Bedrov, D., Li, L., Byutner, O. *Journal of Chemical Physics* 2002, *117*, 9478–9489.
318. Hens, Z., Tallapin, D.V., Weller, H., Vanmaekelbergh, D. *Applied Physics Letters* 2002, *81*, 4245–4247.
319. Durston, P.J., Palmer, R.E., Wilcoxon, J.P. *Applied Physics Letters* 1998, *72*, 176–178.
320. Silly, F., Gusev, A.O., Taleb, A., Pileni, M.P., Charra, F. *Materials Science and Engineering C: Biomimetic and Supramolecular Systems* 2002, *C19*, 193–195.
321. Radojkovic, P., Schwartzkopff, M., Gabriel, T., Hartmann, E. *Solid-State Electronics* 1998, *42*, 1287–1292.
322. Baur, C., Gazen, B.C., Koel, B., Ramachandran, T.R., Requicha, A.A.G., Zini, L. *Journal of Vacuum Science and Technology B: Microelectronics and Nanometer Structures* 1997, *15*, 1577–1580.
323. Ramachandran, T.R., Baur, C., Bugacov, A., Madhukar, A., Koel, B.E., Requicha, A., Gazen, C. *Nanotechnology* 1998, *9*, 237–245.
324. Resch, R., Baur, C., Bugacov, A., Koel, B.E., Madhukar, A., Requicha, A.A.G., Will, P. *Langmuir* 1998, *14*, 6613–6616.
325. Li, W., Hsiao, G.S., Harris, D., Nyffenegger, R., Virtanen, J.A., Penner, R.M. *Journal of Physical Chemistry* 1996, *100*, 20103–20113.
326. Kolb, D.M., Ullmann, R., Will, T. *Science (Washington, DC)* 1997, *275*, 1097–1099.
327. Zamborini, F.P., Crooks, R.M. *Journal of the American Chemical Society* 1998, *120*, 9700–9701.
328. Hoeppener, S., Maoz, R., Cohen, S.R., Chi, L., Fuchs, H., Sagiv, J. *Advanced Materials (Weinheim, Germany)* 2002, *14*, 1036–1041.
329. Jung, Y.M., Ahn, S.J., Kim, E.R., Lee, H. *Journal of the Korean Physical Society* 2002, *40*, 712–715.
330. Yang, W., Chen, M., Knoll, W., Deng, H. *Langmuir* 2002, *18*, 4124–4130.
331. Mesquida, P., Stemmer, A. *Advanced Materials (Weinheim, Germany)* 2001, *13*, 1395–1398.
332. Mesquida, P., Stemmer, A. *Microelectronic Engineering* 2002, *61/62*, 671–674.
333. Sugimura, H., Nakagiri, N. *Journal of the American Chemical Society* 1997, *119*, 9226–9229.
334. Jung, Y.M., Ahn, S.J., Chae, H.J., Lee, H., Kim, E.R., Lee, H. *Molecular Crystals and Liquid Crystals Science and Technology, Section A: Molecular Crystals and Liquid Crystals* 2001, *370*, 231–234.
335. Kraemer, S., Fuierer, R.R., Gorman, C.B. *Chemical Reviews (Washington, DC)* 2003, *103*, 4367–4418.
336. Blackledge, C., Engebretson, D.A., McDonald, J.D. *Langmuir* 2000, *16*, 8317–8323.
337. Wouters, D., Schubert, U.S. *Langmuir* 2003, *19*, 9033–9038.

338. Piner, R.D., Zhu, J., Xu, F., Hong, S., Mirkin, C.A. *Science (Washington, DC)* 1999, *283*, 661–663.
339. Hong, S., Zhu, J., Mirkin, C.A. *Science (Washington, DC)* 1999, *286*, 523–525.
340. Bullen, D.A., Wang, X., Zou, J., Chung, S.W., Liu, C., Mirkin, C.A. *Materials Research Society Symposium Proceedings* 2003, *758*, 141–150.
341. Demers, L.M., Park, S.J., Taton, T.A., Li, Z., Mirkin, C.A. *Angewandte Chemie, International Edition* 2001, *40*, 3071–3073.
342. Li, H.W., Muir, B.V.O., Fichet, G., Huck, W.T.S. *Langmuir* 2003, *19*, 1963–1965.
343. Guo, Q., Teng, X., Rahman, S., Yang, H. *Journal of the American Chemical Society* 2003, *125*, 630–631.
344. Li, X.M., Paraschiv, V., Huskens, J., Reinhoudt, D.N. *Journal of the American Chemical Society* 2003, *125*, 4279–4284.
345. Porter, L.A., Jr., Choi, H.C., Schmeltzer, J.M., Ribbe, A.E., Elliott, L.C.C., Buriak, J.M. *Nano Letters* 2002, *2*, 1369–1372.
346. Cherniavskaya, O., Adzic, A., Knutson, C., Gross, B.J., Zang, L., Liu, R., Adams, D.M. *Langmuir* 2002, *18*, 7029–7034.
347. Toprak, M., Kim, D.K., Mikhailova, M., Muhammed, M. *Materials Research Society Symposium Proceedings* 2002, *705*, 139–144.
348. Qin, D., Xia, Y., Xu, B., Yang, H., Zhu, C., Whitesides, G.M. *Advanced Materials (Weinheim, Germany)* 1999, *11*, 1433–1437.
349. Chen, C.C., Lin, J.J. *Advanced Materials (Weinheim, Germany)* 2001, *13*, 136–139.
350. Werts, M.H.V., Lambert, M., Bourgoin, J.P., Brust, M. *Nano Letters* 2002, *2*, 43–47.
351. Vossmeyer, T., DeIonno, E., Heath, J.R. *Angewandte Chemie, International Edition in English* 1997, *36*, 1080–1083.
352. Hua, F., Shi, J., Lvov, Y., Cui, T. *Nano Letters* 2002, *2*, 1219–1222.
353. Clark, S.L., Montague, M.M., Hammond, P.T. *Polymeric Materials Science and Engineering* 1997, *77*, 400–401.
354. Jiang, X., Chen, K.M., Kimerling, L.C., Hammond, P.T. *Abstracts of Papers: American Chemical Society* 2000, *220*, COLL-301.
355. Hammond, P.T. *Current Opinion in Colloid and Interface Science* 2000, *4*, 430–442.
356. Warner, M.R.E., Craster, R.V., Matar, O.K. *Journal of Colloid and Interface Science* 2003, *267*, 92–110.
357. Murray, C.B., Kagan, C.R., Bawendi, M.G. *Science (Washington, DC)* 1995, *270*, 1335–1338.
358. Kiely, C.J., Fink, J., Brust, M., Bethell, D., Schiffrin, D.J. *Nature (London)* 1998, *396*, 444–446.
359. Sun, S., Murray, C.B., Weller, D., Folks, L., Moser, A. *Science (Washington, DC)* 2000, *287*, 1989–1992.
360. Pileni, M.P. *Journal of Physical Chemistry B* 2001, *105*, 3358–3371.
361. Rogach, A.L., Talapin, D.V., Shevchenko, E.V., Kornowski, A., Haase, M., Weller, H. *Advanced Functional Materials* 2002, *12*, 653–664.
362. Colvin, V.L., Goldstein, A.N., Alivisatos, A.P. *Journal of the American Chemical Society* 1992, *114*, 5221–5230.
363. Tang, Z., Wang, Y., Kotov, N.A. *Langmuir* 2002, *18*, 7035–7040.
364. Zhang, Z., Wei, B.Q., Ajayan, P.M. *Applied Physics Letters* 2001, *79*, 4207–4209.
365. Giraldo, O., Durand, J.P., Ramanan, H., Laubernds, K., Suib, S.L., Tsapatsis, M., Brock, S.L., Marquez, M. *Angewandte Chemie, International Edition* 2003, *42*, 2905–2909.
366. Haidara, H., Mougin, K., Schultz, J. *Langmuir* 2001, *17*, 1432–1436.
367. Haidara, H., Mougin, K., Schultz, J. *Langmuir* 2001, *17*, 659–663.

368. Zhong, Z., Gates, B., Xia, Y., Qin, D. *Langmuir* 2000, *16*, 10369–10375.
369. Li, S.P., Lew, W.S., Xu, Y.B., Hirohata, A., Samad, A., Baker, F., Bland, J.A.C. *Applied Physics Letters* 2000, *76*, 748–750.
370. Ansari, I.A., Hamley, I.W. *Journal of Materials Chemistry* 2003, *13*, 2412–2413.
371. Shenhar, R., Norsten, T.B., Sanyal, A., Uzun, O., Rotello, V.M. *Polymer Preprints (American Chemical Society, Division of Polymer Chemistry)* 2003, *44*, 515–516.
372. Kim, G., Libera, M. *Eylasutoma* 1999, *34*, 45–52.
373. Li, L.S., Jin, J., Yu, S., Zhao, Y., Zhang, C., Li, T.J. *Journal of Physical Chemistry B* 1998, *102*, 5648–5652.
374. Kotov, N.A., Meldrum, F.C., Wu, C., Fendler, J.H. *Journal of Physical Chemistry* 1994, *98*, 2735–2738.
375. Dabbousi, B.O., Murray, C.B., Rubner, M.F., Bawendi, M.G. *Chemistry of Materials* 1994, *6*, 216–219.
376. Collier, C.P., Vossmeyer, T., Heath, J.R. *Annual Review of Physical Chemistry* 1998, *49*, 371–404.
377. Greene, I.A., Wu, F., Zhang, J.Z., Chen, S. *Journal of Physical Chemistry B* 2003, *107*, 5733–5739.
378. Motschmann, H., Mohwald, H. *Handbook of Applied Surface and Colloid Chemistry* 2002, *2*, 79–98.
379. Schmid, G. *Advanced Engineering Materials* 2001, *3*, 737–743.
380. Iakovenko, S.A., Trifonov, A.S., Giersig, M., Mamedov, A., Nagesha, D.K., Hanin, V.V., Soldatov, E.C., Kotov, N.A. *Advanced Materials (Weinheim, Germany)* 1999, *11*, 388–392.
381. Takahagi, T., Huang, S., Tsutsui, G., Sakaue, H., Shingubara, S. *Materials Research Society Symposium Proceedings* 2002, *704*, 47–52.
382. Kotov, N.A., Meldrum, F.C., Fendler, J.H. *Journal of Physical Chemistry* 1994, *98*, 8827–8830.
383. Collier, C.P., Saykally, R.J., Shiang, J.J., Henrichs, S.E., Heath, J.R. *Science (Washington, DC)* 1997, *277*, 1978–1981.
384. Chen, S. *Langmuir* 2001, *17*, 2878–2884.
385. Mallwitz, F., Goedel, W.A. *Angewandte Chemie, International Edition* 2001, *40*, 2645–2647.
386. Sastry, M., Mayya, K.S., Patil, V. *Langmuir* 1998, *14*, 5921–5928.
387. Paul, S., Pearson, C., Molloy, A., Cousins, M.A., Green, M., Kolliopoulou, S., Dimitrakis, P., Normand, P., Tsoukalas, D., Petty, M.C. *Nano Letters* 2003, *3*, 533–536.
388. Poddar, P., Telem-Shafir, T., Fried, T., Markovich, G. *Physical Review B: Condensed Matter and Materials Physics* 2002, *66*, 060403/1–060403/4.
389. Erokhin, V., Facci, P., Carrara, S., Nicolini, C. *Biosensors and Bioelectronics* 1997, *12*, 601–606.
390. Takahagi, T. *Kuki Seijo* 2000, *38*, 238–244.
391. Bernard, S., Felidj, N., Truong, S., Peretti, P., Levi, G., Aubard, J. *Biopolymers* 2002, *61*, 314–318.
392. Huo, L.H., Li, X.L., Li, W., Xi, S.Q. *Sensors and Actuators B: Chemical* 2000, *B71*, 77–81.
393. Nabok, A.V., Iwantono, B., Hassan, A.K., Ray, A.K., Wilkop, T. *Materials Science and Engineering C: Biomimetic and Supramolecular Systems* 2002, *C22*, 355–358.
394. Zhang, Z., Horsch, M.A., Lamm, M.H., Glotzer, S.C. *Nano Letters* 2003, *3*, 1341–1346.
395. Lamm, M.H., Chen, T., Glotzer, S.C. *Nano Letters* 2003, *3*, 989–994.
396. Tang, Z., Wang, Y., Wang, Z.L., Kotov, N.A. Submitted for publication.

397. Willner, I., Willner, B. *Pure and Applied Chemistry* 2001, *73*, 535–542.
398. Service, R.F. *Science (Washington, DC)* 2003, *301*, 1827.
399. Nam, J.M., Thaxton, C.S., Mirkin, C.A. *Science (Washington, DC)* 2003, *301*, 1884–1886.
400. Larson, D.R., Zipfel, W.R., Williams, R.M., Clark, S.W., Bruchez, M.P., Wise, F.W., Webb, W.W. *Science (Washington, DC)* 2003, *300*, 1434–1437.
401. Wu, X., Liu, H., Liu, J., Haley, K.N., Treadway, J.A., Larson, J.P., Ge, N., Peale, F., Bruchez, M.P. *Nature Biotechnology* 2003, *21*, 41–46.
402. Hua, F., Lvov, Y., Cui, T. *Journal of Nanoscience and Nanotechnology* 2002, *2*, 357–361.
403. Lee, K.B., Lim, J.H., Mirkin, C.A. *Journal of the American Chemical Society* 2003, *125*, 5588–5589.
404. Doyle, P.S., Bibette, J., Bancaud, A., Viovy, J.L. *Science (Washington, DC)* 2002, *295*, 2237.
405. Shenton, W., Pum, D., Sleytr, U.B., Mann, S. *Nature (London)* 1997, *389*, 585–587.
406. Dieluweit, S., Pum, D., Sleytr, U.B. *Supramolecular Science* 1998, *5*, 15–19.
407. Mertig, M., Wahl, R., Lehmann, M., Simon, P., Pompe, W. *European Physical Journal D: Atomic, Molecular and Optical Physics* 2001, *16*, 317–320.
408. McMillan, R.A., Paavola, C.D., Howard, J., Chan, S.L., Zaluzec, N.J., Trent, J.D. *Nature Materials* 2002, *1*, 247–252.
409. Yamashita, I. *Thin Solid Films* 2001, *393*, 12–18.
410. Li, M., Wong, K.K.W., Mann, S. *Chemistry of Materials* 1999, *11*, 23–26.
411. Hoinville, J., Bewick, A., Gleeson, D., Jones, R., Kasyutich, O., Mayes, E., Nartowski, A., Warne, B., Wiggins, J., Wong, K. *Journal of Applied Physics* 2003, *93*, 7187–7189.
412. Stellacci, F., Bauer, C.A., Meyer-Friedrichsen, T., Wenseleers, W., Alain, V., Kuebler, S.M., Pond, S.J.K., Zhang, Y., Marder, S.R., Perry, J.W. *Advanced Materials (Weinheim, Germany)* 2002, *14*, 194–198.
413. Zhang, J., Li, X., Liu, K., Cui, Z., Zhang, G., Zhao, B., Yang, B. *Journal of Colloid and Interface Science* 2002, *255*, 115–118.
414. Nagle, L., Ryan, D., Cobbe, S., Fitzmaurice, D. *Nano Letters* 2003, *3*, 51–53.
415. Nagle, L., Fitzmaurice, D. *Advanced Materials (Weinheim, Germany)* 2003, *15*, 933–935.
416. Van Blaaderen, A. *MRS Bulletin* 1998, *23*, 39–43.
417. Gaponenko, S.V., Bogomolov, V.N., Petrov, E.P., Kapitonov, A.M., Eychmueller, A.A., Rogach, A.L., Kalosha, I.I., Gindele, F., Woggon, U. *Journal of Luminescence* 2000, *87/89*, 152–156.
418. Rogach, A.L., Nagesha, D., Ostrander, J.W., Giersig, M., Kotov, N.A. *Chemistry of Materials* 2000, *12*, 2676–2685.
419. Vlasov, Y.A., Yao, N., Norris, D.J. *Advanced Materials (Weinheim, Germany)* 1999, *11*, 165–169.
420. Valenta, J., Linnros, J., Juhasz, R., Rehspringer, J.L., Huber, F., Hirlimann, C., Cheylan, S., Elliman, R.G. *Journal of Applied Physics* 2003, *93*, 4471–4474.
421. Solovyev, V.G., Romanov, S.G., Sotomayor Torres, C.M., Muller, M., Zentel, R., Gaponik, N., Eychmuller, A., Rogach, A.L. *Journal of Applied Physics* 2003, *94*, 1205–1210.
422. Velev, O.D., Tessier, P.M., Lenhoff, A.M., Kaler, E.W. *Nature (London)* 1999, *401*, 548.
423. Wang, W., Asher, S.A. *Journal of the American Chemical Society* 2001, *123*, 12528–12535.
424. Kang, H., Jun, Y.W., Park, J.I., Lee, K.B., Cheon, J. *Chemistry of Materials* 2000, *12*, 3530–3532.

425. Lindlar, B., Boldt, M., Eiden-Assmann, S., Maret, G. *Advanced Materials (Weinheim, Germany)* 2002, *14*, 1656–1658.
426. Ramesh, S., Cohen, Y., Prozorov, R., Shafi, K.V.P.M., Aurbach, D., Gedanken, A. *Journal of Physical Chemistry B* 1998, *102*, 10234–10242.
427. Ryan, K.M., Erts, D., Olin, H., Morris, M.A., Holmes, J.D. *Journal of the American Chemical Society* 2003, *125*, 6284–6288.
428. Murray, C.B., Kagan, C.R., Bawendi, M.G. *Annual Review of Materials Science* 2000, *30*, 545–610.
429. Tallapin, D.V., Shevchenko, E.V., Kornowski, A., Gaponik, N., Haase, M., Rogach, A.L., Weller, H. *Advanced Materials (Weinheim, Germany)* 2001, *13*, 1868–1871.
430. Weller, H. *Philosophical Transactions of the Royal Society of London, Series A: Mathematical, Physical and Engineering Sciences* 2003, *361*, 229–240.
431. Murray, C.B., Norris, D.J., Bawendi, M.G. *Journal of the American Chemical Society* 1993, *115*, 8706–8715.
432. Guo, S., Konopny, L., Popovitz-Biro, R., Cohen, H., Porteanu, H., Lifshitz, E., Lahav, M. *Journal of the American Chemical Society* 1999, *121*, 9589–9598.
433. Vossmeyer, T., Reck, G., Katsikas, L., Haupt, E.T.K., Schulz, B., Weller, H. *Science (Washington, DC)* 1995, *267*, 1476–1479.
434. O'Brien, S., Brus, L., Murray, C.B. *Journal of the American Chemical Society* 2001, *123*, 12085–12086.
435. Guo, S., Popowitz-Biro, R., Arad, T., Hodes, G., Leiserowitz, L., Lahav, M. *Advanced Materials (Weinheim, Germany)* 1998, *10*, 657–661.
436. Redl, F.X., Cho, K.S., Murray, C.B., O'Brien, S. *Nature (London)* 2003, *423*, 968–971.
437. Micic, O.I., Jones, K.M., Cahill, A., Nozik, A.J. *Journal of Physical Chemistry B* 1998, *102*, 9791–9796.
438. Martin, J.E., Wilcoxon, J.P., Odinek, J., Provencio, P. *Journal of Physical Chemistry B* 2002, *106*, 971–978.
439. Courty, A., Fermon, C., Pileni, M.P. *Advanced Materials (Weinheim, Germany)* 2001, *13*, 254–258.
440. Su, W., Huang, X., Li, J., Fu, H. *Journal of the American Chemical Society* 2002, *124*, 12944–12945.
441. Wang, S., Sato, S., Kimura, K. *Chemistry of Materials* 2003, *15*, 2445–2448.
442. Stoeva, S.I., Prasad, B.L.V., Uma, S., Stoimenov, P.K., Zaikovski, V., Sorensen, C.M., Klabunde, K.J. *Journal of Physical Chemistry B* 2003, *107*, 7441–7448.
443. Brown, L.O., Hutchison, J.E. *Journal of Physical Chemistry B* 2001, *105*, 8911–8916.
444. Martin, J.E., Wilcoxon, J.P., Odinek, J., Provencio, P. *Journal of Physical Chemistry B* 2000, *104*, 9475–9486.
445. Murray, C.B., Sun, S., Doyle, H., Betley, T. *MRS Bulletin* 2001, *26*, 985–991.
446. Yin, J.S., Wang, Z.L. *Physical Review Letters* 1997, *79*, 2570–2573.
447. Shevchenko, E.V., Talapin, D.V., Rogach, A.L., Kornowski, A., Haase, M., Weller, H. *Journal of the American Chemical Society* 2002, *124*, 11480–11485.
448. Shevchenko, E., Talapin, D., Kornowski, A., Wiekhorst, F., Kotzler, J., Haase, M., Rogach, A., Weller, H. *Advanced Materials (Weinheim, Germany)* 2002, *14*, 287–290.
449. Petit, C., Legrand, J., Russier, V., Pileni, M.P. *Journal of Applied Physics* 2002, *91*, 1502–1508.
450. Legrand, J., Ngo, A.T., Petit, C., Pileni, M.P. *Advanced Materials (Weinheim, Germany)* 2001, *13*, 58–62.
451. Vaucher, S., Li, M., Mann, S. *Angewandte Chemie, International Edition* 2000, *39*, 1793–1796.

452. Hong, S.Y., Popovitz-Biro, R., Prior, Y., Tenne, R. *Journal of the American Chemical Society* 2003, *125*, 10470–10474.
453. Nikoobakht, B., Wang, Z.L., El Sayed, M.A. *Journal of Physical Chemistry B* 2000, *104*, 8635–8640.
454. Li, L.S., Alivisatos, A.P. *Advanced Materials (Weinheim, Germany)* 2003, *15*, 408–411.
455. Samokhvalov, A., Gurney, R.W., Lahav, M., Naaman, R. *Journal of Physical Chemistry B* 2002, *106*, 9070–9078.
456. Sirota, M., Fradkin, L., Buller, R., Henzel, V., Lahav, M., Lifshitz, E. *ChemPhysChem* 2002, *3*, 343–349.
457. Kotov, N.A. *MRS Bulletin* 2001, *26*, 992–997.
458. Kotov, N.A. Layer-by-layer assembly of nanoparticles and nanocolloids: intermolecular interactions, structure and materials perspectives. In *Thin Films: Polyelectrolyte Multilayers and Related Multicomposites*, edited by G. Decher and J.B. Schlenoff. Wiley-VCH Verlag GmbH & Co. KGaA, Weinheim, Germany, 2003, 207–243.
459. Decher, G. *Science (Washington, DC)* 1997, *277*, 1232–1237.
460. Iler, R.K., Colloid, J. *Interface Science* 1966, *21*, 569–594.
461. Sun, S., Anders, S., Hamann, H.F., Thiele, J.U., Baglin, J.E.E., Thomson, T., Fullerton, E.E., Murray, C.B., Terris, B.D. *Journal of the American Chemical Society* 2002, *124*, 2884–2885.
462. Sun, S., Murray, C.B., Weller, D., Folks, L., Moser, A. *Science (Washington, DC)* 2000, *287*, 1989–1992.
463. Joly, S., Kane, R., Radzilowski, L., Wang, T., Wu, A., Cohen, R.E., Thomas, E.L., Rubner, M.F. *Langmuir* 2000, *16*, 1354–1359.
464. Hicks, J.F., Zamborini, F.P., Osisek, A.J., Murray, R.W. *Journal of the American Chemical Society* 2001, *123*, 7048–7053.
465. Wang, Y., Tang, Z., Correa-Duarte, M.A., Liz-Marzan, L.M., Kotov, N. A. *Journal of the American Chemical Society* 2003, *125*, 2830–2831.
466. Sinani, V.A., Koktysh, D.S., Yun, B.G., Matts, R.L., Pappas, T.C., Motamedi, M., Thomas, S.N., Kotov, N.A. *Nano Letters* 2003, *3*, 1177–1182.
467. Sastry, M., Rao, M., Ganesh, K.N. *Accounts of Chemical Research* 2002, *35*, 847–855.
468. Musick, M.D., Keating, C.D., Keefe, M.H., Natan, M.J. *Chemistry of Materials* 1997, *9*, 1499–1501.
469. Muller, K.H., Herrmann, J., Raguse, B., Baxter, G., Reda, T. *Physical Review B: Condensed Matter and Materials Physics* 2002, *66*, 075417/1–075417/8.
470. Nakanishi, T., Ohtani, B., Uosaki, K. *Journal of Physical Chemistry B* 1998, *102*, 1571–1577.
471. Cassagneau, T., Mallouk, T.E., Fendler, J.H. *Journal of the American Chemical Society* 1998, *120*, 7848–7859.
472. Lee, J., Mathai, M., Jain, F., Papadimitrakopoulos, F. *Journal of Nanoscience and Nanotechnology* 2001, *1*, 59–64.
473. Koktysh, D.S., Liang, X., Yun, B.G., Pastoriza-Santos, I., Matts, R.L., Giersig, M., Serra-Rodriguez, C., Liz-Marzan, L.M., Kotov, N.A. *Advanced Functional Materials* 2002, *12*, 255–265.
474. Tang, Z., Kotov, N.A., Magonov, S., Ozturk, B. *Nature Materials* 2003, *2*, 413–418.
475. Messer, B., Song, J.H., Yang, P. *Journal of the American Chemical Society* 2000, *122*, 10232–10233.
476. Huang, Y., Duan, X., Wei, Q., Lieber, C.M. *Science (Washington, DC)* 2001, *291*, 630–633.
477. Chen, J., Weimer, W.A. *Journal of the American Chemical Society* 2002, *124*, 758–759.

478. Kimura, T., Ago, H., Tobita, M., Ohshima, S., Kyotani, M., Yumura, M. *Advanced Materials (Weinheim, Germany)* 2002, *14*, 1380–1383.
479. Artemyev, M., Moeller, B., Woggon, U. *Nano Letters* 2003, *3*, 509–512.
480. Kim, F., Kwan, S., Akana, J., Yang, P. *Journal of the American Chemical Society* 2001, *123*, 4360–4361.
481. Yang, P., Kim, F. *ChemPhysChem* 2002, *3*, 503–506.
482. Henry, Y., Iovan, A., George, J.M., Piraux, L. *Physical Review B: Condensed Matter and Materials Physics* 2002, *66*, 184430/1–184430/14.
483. Hillhouse, H.W., Tuominen, M.T. *Microporous and Mesoporous Materials* 2001, *47*, 39–50.
484. Wirtz, M., Martin, C.R. *Advanced Materials (Weinheim, Germany)* 2003, *15*, 455–458.
485. Huang, M.H., Mao, S., Feick, H., Yan, H., Wu, Y., Kind, H., Weber, E., Russo, R., Yang, P. *Science (Washington, DC)* 2001, *292*, 1897–1899.
486. Park, W.I., Yi, G.C., Kim, J.W., Park, S.M. *Applied Physics Letters* 2003, *82*, 4358–4360.
487. Goldberger, J., He, R., Zhang, Y., Lee, S., Yan, H., Choi, H.J., Yang, P. *Nature (London)* 2003, *422*, 599–602.
488. Fan, R., Wu, Y., Li, D., Yue, M., Majumdar, A., Yang, P. *Journal of the American Chemical Society* 2003, *125*, 5254–5255.
489. Ural, A., Li, Y., Dai, H. *Applied Physics Letters* 2002, *81*, 3464–3466.
490. Zhang, Y., Chang, A., Cao, J., Wang, Q., Kim, W., Li, Y., Morris, N., Yenilmez, E., Kong, J., Dai, H. *Applied Physics Letters* 2001, *79*, 3155–3157.
491. Lee, K.H., Cho, J.M., Sigmund, W. *Applied Physics Letters* 2003, *82*, 448–450.
492. Huang, S., Cai, X., Liu, J. *Journal of the American Chemical Society* 2003, *125*, 5636–5637.
493. Liu, J., Rinzler, A.G., Dai, H., Hafner, J.H., Bradley, R.K., Boul, P.J., Lu, A., Iverson, T., Shelimov, K., Huffman, C.B., Rodriguez-Macias, F., Shon, Y.S., Lee, T.R., Colbert, D.T., Smalley, R.E. *Science (Washington, DC)* 1998, *280*, 1253–1256.
494. Liu, Z., Shen, Z., Zhu, T., Hou, S., Ying, L., Shi, Z., Gu, Z. *Langmuir* 2000, *16*, 3569–3573.
495. Diao, P., Liu, Z., Wu, B., Nan, X., Zhang, J., Wei, Z. *ChemPhysChem* 2002, *3*, 898–901.
496. Gooding, J.J., Wibowo, R., Liu, J., Yang, W., Losic, D., Orbons, S., Mearns, F.J., Shapter, J.G., Hibbert, D.B. *Journal of the American Chemical Society* 2003, *125*, 9006–9007.
497. Lee, O.J., Jeong, S.H., Lee, K.H. *Applied Physics A: Materials Science and Processing* 2003, *76*, 599–602.
498. Chattopadhyay, D., Galeska, I., Papadimitrakopoulos, F. *Journal of the American Chemical Society* 2001, *123*, 9451–9452.
499. Kovtyukhova, N.I., Mallouk, T.E. *Chemistry: A European Journal* 2002, *8*, 4354–4363.
500. Kovtyukhova, N.I., Martin, B.R., Mbindyo, J.K.N., Smith, P.A., Razavi, B., Mayer, T.S., Mallouk, T.E. *Journal of Physical Chemistry B* 2001, *105*, 8762–8769.
501. Nicewarner-Pena, S.R., Griffith Freeman, R., Reiss, B.D., He, L., Pena, D.J., Walton, I.D., Cromer, R., Keating, C.D., Natan, M.J. *Science (Washington, DC)* 2001, *294*, 137–141.
502. Caswell, K.K., Wilson, J.N., Bunz, U.H.F., Murphy, C.J. *Journal of the American Chemical Society* 2003, *125*, 13914–13915.
503. Wang, Y., Tang, Z., Kotov, N.A. Submitted for publication.

504. Xin, H., Woolley, A.T. *Journal of the American Chemical Society* 2003, *125*, 8710–8711.
505. Nakao, H., Gad, M., Sugiyama, S., Otobe, K., Ohtani, T. *Journal of the American Chemical Society* 2003, *125*, 7162–7163.
506. Ajayan, P.M., Iijima, S. *Nature (London)* 1993, *361*, 333–334.
507. Chae, W.S., Ko, J.H., Hwang, I.W., Kim, Y.R. *Chemical Physics Letters* 2002, *365*, 49–56.
508. Bao, J., Xu, D., Zhou, Q., Xu, Z., Feng, Y., Zhou, Y. *Chemistry of Materials* 2002, *14*, 4709–4713.
509. Azambre, B., Hudson, M.J., Heintz, O. *Journal of Materials Chemistry* 2003, *13*, 385–393.
510. Golberg, D., Xu, F.F., Bando, Y. *Applied Physics A: Materials Science and Processing* 2003, *76*, 479–485.
511. Xue, B., Chen, P., Hong, Q., Lin, J., Tan, K.L. *Journal of Materials Chemistry* 2001, *11*, 2378–2381.
512. Kong, J., Chapline, M.G., Dai, H. *Advanced Materials (Weinheim, Germany)* 2001, *13*, 1384–1386.
513. Choi, H.C., Shim, M., Bangsaruntip, S., Dai, H. *Journal of the American Chemical Society* 2002, *124*, 9058–9059.
514. Ang, L.M., Hor, T.S.A., Xu, G.Q., Tung, C.H., Zhao, S., Wang, J.L.S. *Chemistry of Materials* 1999, *11*, 2115–2118.
515. Lee, S.W., Sigmund, W.M. *Chemical Communications (Cambridge, U.K.)* 2003, 780–781.
516. Han, W.Q., Zettl, A. *Nano Letters* 2003, *3*, 681–683.
517. Ellis, A.V., Vijayamohanan, K., Goswami, R., Chakrapani, N., Ramanathan, L.S., Ajayan, P.M., Ramanath, G. *Nano Letters* 2003, *3*, 279–282.
518. Zhang, J., Wang, G., Shon, Y.S., Zhou, O., Superfine, R., Murray, R.W. *Journal of Physical Chemistry B* 2003, *107*, 3726–3732.
519. Azamian, B.R., Coleman, K.S., Davis, J.J., Hanson, N., Green, M.L.H. *Chemical Communications (Cambridge, U.K.)* 2002, 366–367.
520. Liu, L., Wang, T., Li, J., Guo, Z.X., Dai, L., Zhang, D., Zhu, D. *Chemical Physics Letters* 2002, *367*, 747–752.
521. Li, X., Niu, J., Zhang, J., Li, H., Liu, Z. *Journal of Physical Chemistry B* 2003, *107*, 2453–2458.
522. Banerjee, S., Wong, S.S. *Nano Letters* 2002, *2*, 195–200.
523. Haremza, J.M., Hahn, M.A., Krauss, T.D., Chen, S., Calcines, J. *Nano Letters* 2002, *2*, 1253–1258.
524. Jiang, K., Eitan, A., Schadler, L.S., Ajayan, P.M., Siegel, R.W., Grobert, N., Mayne, M., Reyes-Reyes, M., Terrones, H., Terrones, M. *Nano Letters* 2003, *3*, 275–277.
525. Osterloh, F.E., Martino, J.S., Hiramatsu, H., Hewitt, D.P. *Nano Letters* 2003, *3*, 125–129.
526. Sun, Y., Mayers, B., Herricks, T., Xia, Y. *Nano Letters* 2003, *3*, 955–960.
527. Mokari, T., Rothenberg, E., Popov, I., Costi, R., Banin, U. *Science (Washington, DC)* 2004, *304*, 1787–1790.
528. Keren, K., Berman, R.S., Buchstab, E., Sivan, U., Braun, E. *Science (Washington, DC)* 2003, *302*, 1380–1382.
529. Glotzer, S.C., *Nature Materials* 2003, *2*, 713–714.
530. Starr, F.W., Schroder, T.B., Glotzer, S.C., *Phys. Rev. E* 2001, *64*, 021802/1–021802/5.
531. Starr, F.W., Schroder, T.B., Glotzer, S.C., *Macromolecules* 2002, *35*, 4481–4492.
532. Starr, F.W., Douglas, J.F., Glotzer, S.C., *J. Chem. Phys.* 2003, *119*, 1777–1788.
533. Zhang, Z,., Glotzer, S.C., *Nano Letters* 2004, in press.

2 Colloidal Inorganic Nanocrystal Building Blocks

Young-wook Jun, Seung Jin Ko, and Jinwoo Cheon

CONTENTS

2.1 Aspects of Nanoscale Building Blocks 75
2.2 Current Achievements and Perspectives of Nanobuilding Blocks 77
    2.2.1 Isotropic Basic Elements 77
        2.2.1.1 Spheres 77
        2.2.1.2 Cubes 79
    2.2.2 Anisotropic Basic Elements 81
        2.2.2.1 One-Dimensional Rods and Wires 81
        2.2.2.2 Two-Dimensional Discs and Plates 90
    2.2.3 Novel Nanobuilding Structures 91
        2.2.3.1 Building Blocks with High Symmetry 91
        2.2.3.2 Supercrystals: Assemblies of Nanobuilding Blocks 96
2.3 Conclusions 96
Acknowledgment 97
References 97

2.1 ASPECTS OF NANOSCALE BUILDING BLOCKS

Nanocrystals that are usually regarded as artificial atoms are the basic units for nanoscale devices,[1] and the design and operation of these devices will be easily accomplished by using nanocrystals with distinct sizes and shapes. By assembling and patterning nanobuilding blocks in a manner similar to assembling Lego blocks, the fabrication of future devices that are extremely fast, efficient, and small will be possible.[2] As with Lego blocks, various shapes and sizes of nanobuilding blocks will be required for the various components of the nanostructure.

It has been found that size and shape of nanocrystals are key factors for the determination of their chemical and physical properties. Bulk materials have peculiar

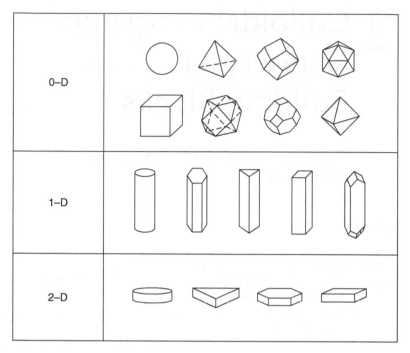

FIGURE 2.1 Basic nanoscale building blocks.

characteristics such as color, phase transition temperature, and band gap. For example, gold has a golden color and silver has metallic properties. In the nanoscale regime, however, the color of gold can be red, deep violet, and blue,[3] and silver can be an insulator, depending on their sizes, shapes, or assemblies.[2] In other words, the size and shape of nanocrystals act as critical parameters for determining materials properties. If the precise control of size and shape of nanocrystals is possible, their chemical and physical properties can be manipulated as desired.[4] Then these nanocrystals can be used as the building blocks for assembling and patterning future nanodevices. Therefore, the control of nanobuilding blocks is crucial for the success of future nanodevices.

Basic nanobuilding blocks can be classified by crystal symmetry and dimensionality (Figure 2.1). Spheres, icosahedrons, and cubes can be classified as zero-dimensional isotropic basic elements. These are the most familiar shapes in the nanoworld. Semiconductor and metal nanospheres and C_{60} are in this group. Rods, cylinders with a polygon shape, and wires are examples of one-dimensional anisotropic basic elements. These building blocks can be applicable for two-terminal circuit devices.[5,6] Carbon nanotubes are another good example of one-dimensional building blocks. Using cross-junctions of carbon nanotubes, nonvolatile random access memory devices are possible.[7] Discs and plates with polygon shape belong to two-dimensional anisotropic basic elements.[3,8,9] Examples of rather complex structured building blocks are star-shaped and multipod structures such as highly symmetric crosses.[10]

There are several approaches to preparing nanobuilding blocks, including gas phase syntheses and liquid phase syntheses. In the past decade, although efforts for controlling shapes of nanocrystals using liquid phase synthesis have been extensively carried out, these have been limited to certain systems, such as II-VI semiconductors. To obtain various shapes of nanobuilding blocks, the presentation of a general and more comprehensive synthetic scheme and the unraveling of shape-guiding mechanisms are necessary. In the next section, current achievements and perspectives in shape-controlled synthesis of nanobuilding blocks using liquid phase synthesis will be discussed.

2.2 CURRENT ACHIEVEMENTS AND PERSPECTIVES OF NANOBUILDING BLOCKS

2.2.1 ISOTROPIC BASIC ELEMENTS

2.2.1.1 Spheres

Spheres are the most basic and symmetric components among various-shaped nanocrystals and, therefore, they have been extensively studied during the past decade. Metal and semiconductor nanospheres are good examples. In an earlier period of research trends, nanocrystals were usually grown in hydrolytic media in the presence of structured micelles.[11] Several metallic and semiconducting nanocrystals have been grown from aging processes of ionic precursors inside organic micelles. However, nanocrystals obtained by this method have relatively poor crystallinity or polydispersity in their size.

As an alternative way to solve these problems, a thermal decomposition method of organometallic precursors under hot organic solution was adopted. At first, Chestnoy et al.[12] synthesized various II-VI semiconductor nanospheres with high colloidal stability with coordinating solvents (e.g., 4-ethylpyridine), but size tunability and monodispersity of nanocrystals obtained were still poor. Murray et al.[13] successfully developed a more advanced methodology to prepare various-sized CdSe nanocrystals via the method of injecting a precursor solution containing dimethyl cadmium and trioctylphosphine selenide into a hot trioctylphosphine oxide solution. The size of nanocrystals varied from 1.2 to 12 nm, with high monodispersity and crystallinity, and the nanocrystals obtained were highly soluble in various organic solvents. Optical spectra clearly exhibit size-dependent quantum confinement effects, indicating high monodispersity and high crystallinity of nanocrystals (Figure 2.2).

Researchers also utilized the same methodology of the above CdSe nanocrystal synthesis to prepare magnetic metals and metal alloys. Magnetic nanocrystals are of great interest in ultra-high-density storages and biological applications for drug delivery and diagnostics. Ely et al.,[14] Murray et al.,[15] Dinega et al.,[16] Puntes et al.,[17] and Park et al.[18] have reported various sizes of high-quality spherical cobalt nanocrystals obtained by thermal decomposition of cyclooctadienyl or octacarbonyl derivatives of cobalt complexes in the presence of various capping materials, including TOPO (trioctylphosphine oxide), oleic acid, and NaAOT (AOT = bis(2-ethylhexyl)sulfosuccinate).

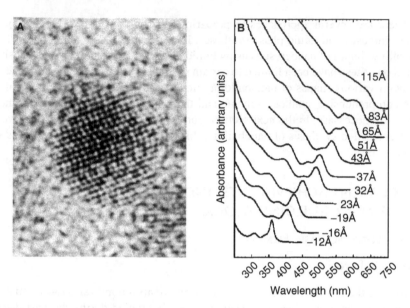

FIGURE 2.2 Trioctylphophine oxide-coated CdSe nanocrystals. (A) HRTEM image. (B) Size-dependent optical spectra. (From Murray, C.B. et al., *J. Am. Chem. Soc.*, 115, 8706–8715, 1993. With permission.)

Magnetic alloys have enhanced magnetic properties compared to those of pure magnetic metals. FePt nanoalloys were synthesized by the polyol reduction process of platinum bisacetylacetonate and thermal decomposition of iron pentacarbonyl in the presence of oleic acid and oleylamine stabilizers by Sun et al.[19] The chemical composition and size of the nanocrystals were easily controlled by changing reaction parameters. Moreover, nanocrystals obtained were self-assembled into three-dimensional superstructures, and interparticle spacing was also controllable. Annealing of as-synthesized $Fe_{48}Pt_{52}$ nanocrystals with face-centered-cubic (fcc) structure accompanied phase transition to face-centered tetragonal $Fe_{48}Pt_{52}$ nanocrystals and transformed them into room-temperature nanoscale ferromagnets. Based on the same synthetic method, CoPt analogs have been thoroughly studied by Weller et al.[20]

Another approach to synthesize magnetic alloys is via redox transmetallation reaction between two metals.[21] Park and Cheon[21] reported synthesis of both solid-solution and core-shell types of CoPt nanoalloys using this approach (Figure 2.3). When cobalt ocatacarbonyl was injected into hot toluene solution containing platinum bis-1,1,1,5,5,5-hexafluoroacetylacetonate (hfac) and oleic acid, stoichiometric fractions of cobalt atoms in solution were replaced by platinum atoms via transmetallation reaction, which resulted in solid-solution CoPt nanoalloys (Figure 2.4A). However, when cobalt nanocrystal colloids and $Pt(hfac)_2$ were refluxed in nonane solution containing dodecylisocyanide stabilizers, transmetallation reactions were initiated and proceeded on the cobalt nanocrystal surface, which produced core-shell type CoPt nanoalloys (Figure 2.4B). This approach has several advantages for producing nanoalloys:

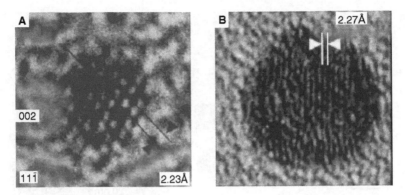

FIGURE 2.3 HRTEM images of CoPt nanoalloys. (A) Solid-solution type. (B) Core-shell type. (From Park, J.-I. and Cheon, J., *J. Am. Chem. Soc.*, 123, 5743–5746, 2001. With permission.)

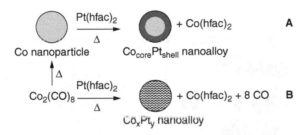

FIGURE 2.4 Schemes of CoPt nanoalloy formation via redox transmetallation processes. (A) Core-shell type. (B) Solid-solution type. (From Park, J.-I. and Cheon, J., *J. Am. Chem. Soc.*, 123, 5743–5746, 2001. With permission.)

1. There is no additional reducing agent.
2. It can produce both solid-solution and core-shell types of nanoalloys.
3. Chemical compositions and outer-shell thickness are also controllable by simply changing the Co/Pt ratio.
4. Selective growth onto a specific metal substrate is possible. As an example, when Pt(hfac)$_2$ is added to the solution containing both cobalt and gold nanocrystal colloids, Pt(hfac)$_2$ grows only onto the cobalt nanocrystal surface to make CoPt alloys, not onto the gold nanocrystal surface.

2.2.1.2 Cubes

Isotropic zero-dimensional nanocrystals with faceted faces are also obtainable by careful controlling crystal surface energy. El-Sayed[22] demonstrated the shape control of Pt nanocrystals with fcc structure. The Pt precursor is reduced by hydrogen gas bubbling in a reaction solution that includes a capping polymer. By controlling the concentration of the capping polymer, the resulting shapes are changed. At high

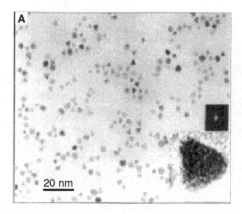
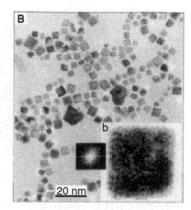

FIGURE 2.5 Various shapes of isotropic Pt nanocrystals. (A) Cubes. (B) Tetrahedrons. The nanocrystal shapes were strongly dependent on the polymer concentration. (From El-Sayed, M.A., *Acc. Chem. Res.*, 34, 257–264, 2001. With permission.)

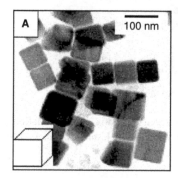
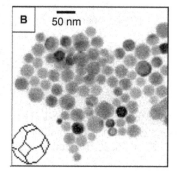

FIGURE 2.6 TEM images of PbS nanocrystals and their polygonal models. (A) Dodecy-lamine capping molecular system and corresponding crystal model (inset). (B) Dodecanethiol capping molecular system and corresponding crystal model (inset). (From Lee, S.-M. et al., *J. Am. Chem. Soc.*, 124, 11244–11245, 2002. With permission.)

polymer concentration, nanocrystals are prevented from rapid growth and the result-ing shape is a small tetrahedron with only {111} faces (Figure 2.5A). At low polymer concentrations, the deposition rate of a Pt atom on the crystal surfaces is faster on the {111} faces and the final shape becomes a more rectangular cubic shape (Figure 2.5B).

Zero-dimensional polygonal shapes were also observed in semiconductor nano-crystals. Lee et al.[10] demonstrated shape-controllable synthesis of PbS nanocrystals with an isotropic rock salt model system. Due to the high surface energy of the {111} faces in a rock salt system, a cubic shape was generally obtained in the presence of weakly binding capping molecules (Figure 2.6A). However, the presence of tightly binding capping molecules, such as dodecanethiol, mainly affected the shapes of the nanocrystals. Since dodecanethiol is much more strongly binding on the {111} faces, tetradecahedron shapes highly truncated on the {111} faces were obtained at high concentrations of dodecanethiol (Figure 2.6B). Similarly, Trindade et al.[23] also synthesized cubic-shaped PbS nanocrystals via a single-source precursor.

As described above, polygonal shape control in zero dimensions is relatively feasible by regulating the relative growth rates of different crystallographic surfaces of nanocrystals.

2.2.2 ANISOTROPIC BASIC ELEMENTS

2.2.2.1 One-Dimensional Rods and Wires

Recent studies on the shape control of nanocrystals have been highly focused on one-dimensional systems. Since one-dimensional rod growth is the fundamental step of anisotropic shape control, it may be possible to have further control of nanobuilding blocks with highly complex structures once we understand the shape-guiding mechanisms of nanorods.

To generate one-dimensional nanocrystals, researchers have explored one-step *in situ* synthesis of one-dimensional nanorods utilizing methods similar to those for the well-studied spherical nanocrystals. For example, the one-dimensional colloidal rod-based system of CdSe has been successfully demonstrated by Peng et al.[24] The use of binary capping molecules such as trioctylphosphine oxide (TOPO) and hexylphosphonic acid (HPA) was effective for the generation of shape anisotropy in CdSe, along with the nature of its intrinsic hexagonal structure. Recent studies on *in situ* anisotropic growth of the nanocrystals propose a growth process that is referred to as oriented attachment. Park et al.[25] reported iron nanorods that were formed by a stabilizer-induced aggregation process of iron nanospheres, and Tang et al.[26] demonstrated the formation of CdTe nanowires via crystal dipole-induced self-assembly of CdTe nanospheres.

Another approach for liquid phase preparation of nanorods is via a two-step seed-mediated synthesis. In the first step, spherical seed nanocrystals are synthesized and isolated. Precursors of the desired one-dimensional structure are then chemically reduced in the presence of capping molecules, where seed nanocrystals serve as nucleation points. Gold nanosphere direct growth of Si nanowires by seed catalytic growth in supercritical fluid[27] and metallic rod systems, including Au or Ag rods obtained by seed-mediated growth in solution, have been presented.[28] Although interface issues between the seed and growth materials have not been well resolved, this method has great potential for further advances of research on one-dimensional rod-structured materials. The general applicability of this method to other materials needs to be further explored.

2.2.2.1.1 Group II-VI Semiconductors

The nonhydrolytic high-temperature injection method can be heavily utilized for high-quality nanorod synthesis. Peng et al.[24] first reported CdSe nanorods via thermal decomposition of dimethylcadmium and trioctylphosphineselenide in a hot surfactant mixture of TOPO and HPA. In this synthesis, one-dimensional rod-shaped structures result from preferential growth along the [001] direction of wurtzite CdSe, which is promoted by a selective adhesion of HPA molecules on specific faces. With

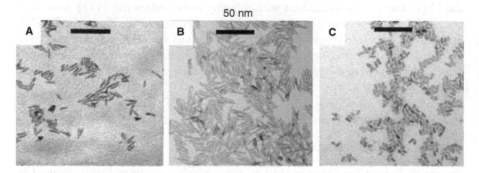

FIGURE 2.7 CdSe nanorods grown using thermal decomposition of precursors in the presence of binary surfactants. (From Peng, X. et al., *Nature*, 404, 59–61, 2000. With permission.)

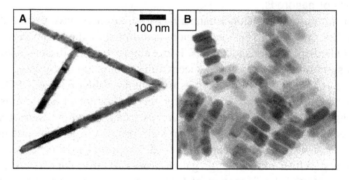

FIGURE 2.8 Various one-dimensional II-VI semiconductor nanorods. (A) ZnTe. (From Jun, Y. et al., *Chem. Commun., 1,* 101–102, 2001. With permission.) (B) CdS. (From Jun, Y. et al., *J. Am. Chem. Soc., 123,* 5150–5151, 2001. With permission.)

increasing HPA concentration, the nanocrystal shape evolves from spheres and short rods to long rods with high aspect ratios (Figure 2.7).

On the other hand, Jun et al.[29] reported a simpler synthetic method using a single-source precursor instead of dual sources. For example, zinc telluride nanorods were formed when a single-source precursor, Zn(TePh)2-TMEDA, decomposed in a hot amine surfactant mixture (Figure 2.8). Anisotropic one-dimensional growth of metal sulfide nanocrystals, including CdS, MnS, and PbS, was also possible by thermal decomposition of stable single-molecule precursors in a monosurfactant system.[10,30,31]

Hydrolytic synthesis of II-VI semiconductors also produces one-dimensional rod-shaped nanocrystals by shape transformations involving oriented attachment processes. Pacholski et al.[32] have proposed an alignment and reorganization of zinc oxide nanocrystals via an oriented attachment process (Figure 2.9). In this synthesis, zinc acetate produces zinc oxide nanospheres through hydrolysis and aging processes. The nanospheres are then aligned and fused together in order to remove high-energy surfaces. Finally, reconstruction processes of the fused nanocrystal surfaces result in rod-shaped nanocrystals with flat surfaces (Figure 2.10). Similar shape

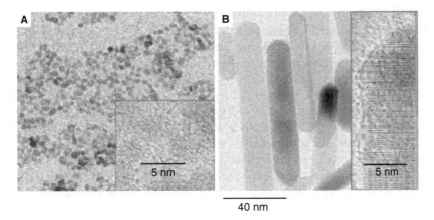

FIGURE 2.9 ZnO nanorod formation via oriented attachment processes. (A) Nanosphere aggregates. (B) Nanorods. (From Pacholski, C. et al., *Angew Chem. Int. Ed.*, 41, 1188–1191, 2002. With permission.)

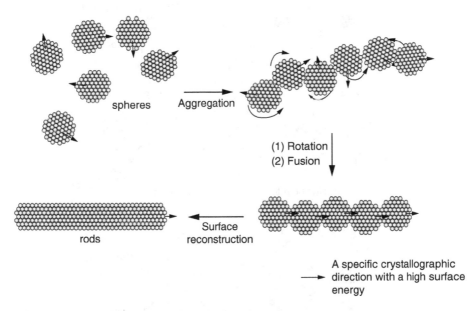

FIGURE 2.10 Schematic illustration of oriented attachment processes.

transformation processes were also observed in other II-VI semiconductor nanocrystals. Tang et al.[26] reported a shape transformation from sphere to rod by dipole-induced fusion of CdTe individual nanospheres. Zinc blende CdTe spheres were first formed by reaction between Cd and Te precursors in aqueous solution. Then CdTe spheres obtained were aligned through dipole interactions between nanocrystals. Finally, the nanospheres were fused together with a simultaneous crystal phase transition to wurtzite.

2.2.2.1.2 Group III-V Semiconductors

Unlike II-VI semiconductor systems that have been thoroughly studied, anisotropic shape control of III-V semiconductor nanocrystals has been very limited. This is most likely due to a greater degree of their covalent bonding nature and the absence of suitable precursors. Moreover, except for metal nitrides, III-V semiconductors favor isotropic zinc blende structures and, thus, zero-dimensional nanocrystal growth is commonly preferred rather than rod growth.[33] Kim et al.[34] have shown that the crystalline phase of gallium phosphide nanocrystals can be chemically controlled by adopting suitable surfactants. GaP semiconductors have two different crystalline phases that are in rotational isomerism: the zinc blende phase is a staggered conformation with [111] directions, whereas the wurtzite phase is an eclipsed conformation with [001] directions (Figure 2.11A and B).[33] Zinc blende GaP is a thermodynamically

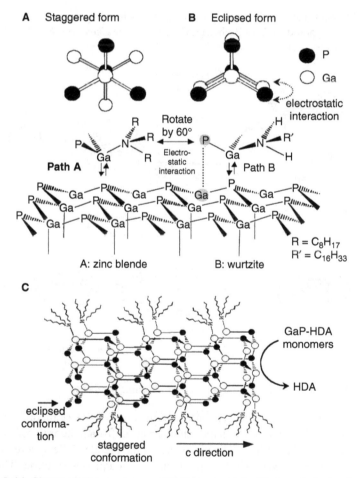

FIGURE 2.11 Shape-guiding processes of GaP nanocrystals. Steric and electronic effects on the crystalline phase (A, B) and rod growth (C) of GaP nanocrystals. (From Kim, Y.-H. et al., *J. Am. Chem. Soc.*, 124, 13656–13657, 2002. With permission.)

stable phase that is sterically favored. Wurtzite GaP is a kinetically stable phase that is electronically favored due to the electrostatic interaction between geminal gallium and phosphine atoms when wurtzite GaP monomers approach the crystal surface. When sterically bulky tertiary amines are used as capping molecules, the staggered conformation is highly favored due to the large steric hindrance between capping molecules and the crystal surface. Therefore, formation of zinc blende GaP nanospheres is favored (Figure 2.11A). However, when sterically less hindered primary amines are present, the steric energy difference between the eclipsed conformation and the staggered conformation is diminished. The steric effects on determining the binding geometry of incoming monomers are now small, and electronic effects play a major role under a kinetically driven growth regime with a high monomer concentration. Therefore, formation of the wurtzite GaP phase is facilitated (Figure 2.11B). In short, the formation of a zinc blende structure is favored with highly steric ligands such as trioctylamine (TOA), whereas wurtzite is dominant under less bulky linear alkylamines (e.g., hexadecylamine [HDA]).

Once the crystalline phase is determined, subsequent kinetic growth yields different types of shapes, which are also affected by the stereochemistry of ligands. When wurtzite seeds are formed, sterically bulky TOA selectively binds to the other faces (e.g., {100} and {110} faces) with a staggered conformation, rather than to the {002} faces, and blocks growth on these faces. On the other hand, GaP-HDA complexes continuously supply monomers on the {001} faces with high surface energy and, therefore, promote the growth in the c-direction of a rod structure (Figure 2.11C).

III-V semiconductor one-dimensional nanocrystals can also be synthesized via solution–liquid–solid (SLS) processes (Figure 2.12).[35] When InP semiconductor precursors decompose, precursor molecules generate InP monomers in solution and indium metallic seeds via two cycles of In-P bond cleavage and reformation. The monomers diffuse into liquid indium seeds and eventually precipitate and crystallize along a specific direction when the liquids seeds are supersaturated, inducing the formation of wire-shaped InP nanocrystals.

2.2.2.1.3 Group IV Semiconductors

In the case of group IV semiconductor systems, it is extremely difficult to obtain nanorods by typical solution-based precursor injection methods due to their highly covalent character. In contrast, under gas phase–based syntheses such as chemical vapor deposition, one-dimensional silicon and germanium wires can be easily obtained on a substrate using vapor–liquid–solid (VLS) growth mechanisms.[36] In VLS growth mechanisms, monomers can form an alloy in equilibrium with their pure solid in a catalytic seed and a one-dimensional group semiconductor nanowire is expelled from the seed as it reaches supersaturation. A similar approach can be used for generating colloidal nanowires. Monomers in solution diffuse onto the catalytic seeds and form an alloy with them. As the dissolution of monomers into the seed increases, the supersaturated pure solid nanowire grows out of the catalytic seed. For example, in the supercritical hexane fluid environment, Ge nanowires have been obtained via gold catalysts by Hanrath and Korgel[37] (Figure 2.13). Above 300°C, Ge and Au can form an alloy in equilibrium with pure solid Ge. Monomers

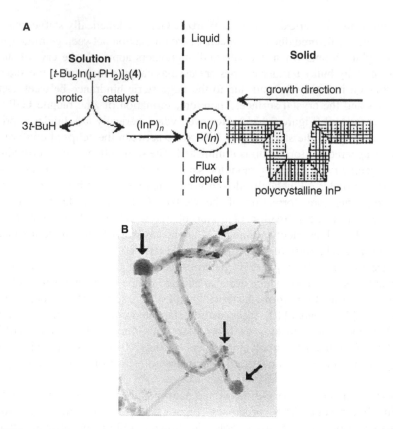

FIGURE 2.12 SLS growth of InP nanofibers. (A) Schematic SLS process. (B) A TEM image. (From Trentler, T.J. et al., *J. Am. Chem. Soc.*, 119, 2172–2181, 1997. With permission.)

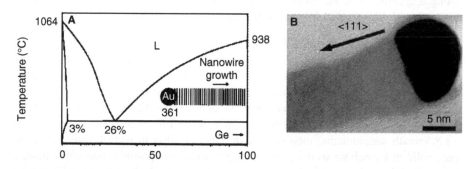

FIGURE 2.13 SLS growth of Ge nanowires. (A) Phase diagram between germanium and gold. (B) HRTEM image. (From Hanrath, T. and Korgel, B.A., *J. Am. Chem. Soc.*, 124, 1424–1429, 2002. With permission.)

made by thermal degradation of the Ge precursor dissolve into Au nanocrystals until reaching a Ge:Au alloy supersaturation and pure solid Ge is expelled from the Au seeds. Size-monodispersed Au nanocrystals are necessary to achieve nanowire

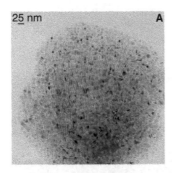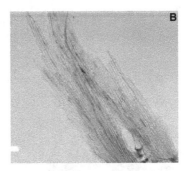

FIGURE 2.14 Alkylamine-coated magnetic nanorods and wires. (A) Nickel nanorods. (From Cordente, N. et al., *Nano Lett.*, 1, 565–568, 2001. With permission.) (B) Cobalt nanowires. (From Dumestre, F. et al., *Angew Chem. Int. Ed.*, 41, 4286–4289, 2002. With permission.)

growth. By changing the reaction pressure, it is possible to control the growth direction of nanowires.

2.2.2.1.4 Metals

Preparation of metallic rods is also possible via a high-temperature precursor injection method. Park et al.[25] reported that iron nanorods were produced by thermal decomposition of iron pentacarbonyl in a hot surfactant solution. Cordente et al.[38] and Dumestre et al.[39] reported that nickel and cobalt nanorods were formed by thermal decomposition of organometallic precursors, $Ni(COD)_2$ (COD = cyclooctadiene) and $Co(C_8H_{13})(C_8H_{12})$, in the presence of alkylamine surfactants (Figure 2.14).

Electrochemical reaction in a structured micelle template was also possible to produce gold nanorods reported by El-Sayed[22] and Feldheim and Foss.[40] Gold ions were dissolute from a gold anode, and the shape distribution of the rods was determined by the current and ratio of the concentration of the surfactant to that of the cosurfactant. Short Au nanorods with aspect ratios of ~3 to 7 mainly dominated by {100} and {110} facets, with growth directions parallel to the [001] direction.

The two-step seed-mediated growth method is also very effective for synthesizing one-dimensional nanocrystals with distinct sizes. In colloidal systems, there are two stages in nanocrystal growth — the nucleation stage and the crystal growth stage.[41] In the nucleating stage, the concentration of monomers increases rapidly and the initial seeds of the crystals are formed upon decomposition of precursors. During the crystal growth stage, monomers diffuse onto the seed and deposit on its surface, resulting in crystal growth. A prerequisite to obtaining the monodispersed nanocrystals is the short nucleating stage, a relatively slow growth stage, and the separation between these two stages. Otherwise, bimodal nucleation may occur, and the resulting sizes and shapes of the nanocrystals become polydispersed. The two-step seed-mediated growth method is to add preciously formed seeds to separate the nucleation stage from the crystal growth stage. Murphy and Jana[28] utilized nanoscale metallic seeds as nuclei for crystal growth inside a structured micelle template (Figure 2.15). At first, spherical metallic monodispersed gold or silver seeds are made by chemical reaction of metal salt with a strong reducing agent in the presence

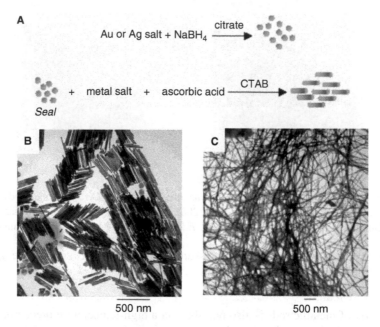

FIGURE 2.15 (A) Schematic illustration of seed-mediated Ag nanorod growth. TEM images of Ag nanorod (B) and Ag nanowire (C). (From Murphy, C.J. and Jana, N.R., *Adv. Mater.*, 14, 80–82, 2000. With permission.)

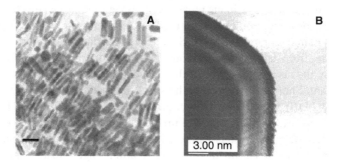

FIGURE 2.16 Photochemically induced gold nanorod growth. (A) Large area TEM image. (B) HRTEM image. (From Kim, F. et al., *J. Am. Chem. Soc.*, 124, 14316–14317, 2002. With permission.)

of capping agents such as citrate. Then these seeds are added to a solution containing metal salt, a weak reducing agent, and a rod-like micellar soft template. In this method, the aspect ratio is controlled by the ratio of metal seed to metal salt concentration. At lower concentrations of metal seeds, there is relatively higher monomer flux into the seed, and longer rods can be produced. Furthermore, delicate control of solution pH can make it possible to obtain aspect ratios as large as ~350.

Photochemical reduction of metal ions also induces formation of rod-shaped nanocrystals (Figure 2.16A).[42] When gold and silver ions are dissolved in a

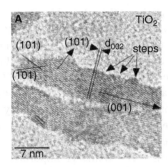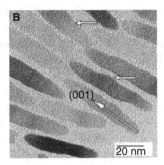

FIGURE 2.17 One-dimensional titanium dioxide nanorods. (A) Large-area TEM image. (B) HRTEM image. (From Chemseddine, A. and Moritz, T., *Eur. J. Inorg. Chem.*, 1999, 235–245. With permission.)

FIGURE 2.18 Naturally aligned titanium dioxide nanocrystals. (From Penn, R.L. and Banfield, J.F., *Geochim. Cosmochim. Acta*, 63, 1549–1557, 1999. With permission.)

hexadecyltrimethylammonium bromide (CTAB) solution, UV irradiation induces reduction of silver ions and silver clusters are formed. Then, gold ions reduce by sacrificial oxidation of silver clusters and therefore promote formation of gold nanorods. The shape-guiding mechanism of this process is not clear, but a rod-like micellar templating is unlikely in this process. In the absence of silver ions, direct UV photo reduction of gold ions results in spheres (Figure 2.16B). This result means the role of silver ions rather than CTAB molecules is crucial for determining their shapes.

2.2.2.1.5 Transition Metal Oxides

Transition metal oxides consist of an important group of materials used in white pigment, electronic ceramics, cosmetics, catalysis AS supports, and as photocatalysts. Nanostructured titania are of particular interest, with potential applications as solar cell materials. Chemseddine and Moritz[43] demonstrated elongated TiO_2 nanocrystals synthesized by hydrolysis and polycondensation of titanium alkoxide $[Ti(OR)_4]$ in the presence of tetramethyl ammonium hydroxide (Me_4NOH) as a stabilizer and reaction catalyst (Figure 2.17). Organic cations stabilized the anatase polyanionic core, and increased monomer flux induced one-dimensional growth of the titania nanocrystals.

Penn and Banfield[44] also reported naturally aligned titania nanocrystals under hydrothermal conditions, by adapting an oriented attachment mechanism into the nanocrystal development (Figure 2.18). Hydrothermal treatment of titanium alkoxide precursors produced diamond-shaped anatase titania nanocrystals. The nanocrystals were truncated with three different kinds of faces: {001}, {121}, and {101}. Because the {001} face had the largest number of dangling bonds and the {101} face had the lowest number of dangling bonds, the surface energy of the {001} face was higher than that of the {101} face. When a sufficient thermal energy was supplied to the system, the removal of high-energy surfaces was thermodynamically favorable. Thus, fusion between diamond-shaped nanocrystals along the [001] direction was preferred, resulting in a necklace-shaped nanocrystal.

A cubic perovskite-structured transition metal oxide is important due to its unique electronic, magnetic, and optical properties. The one-dimensional ternary perovskite oxide was first demonstrated in solution-based synthesis by Urban et al.[45] They synthesized single-crystalline cubic perovskite $BaTiO_3$ and $SrTiO_3$ by solution phase decomposition of bimetallic precursors in the presence of stabilizing ligands.

2.2.2.2 Two-Dimensional Discs and Plates

Unlike one-dimensional rod systems, formation of disc-shaped nanocrystals is very rare in colloidal systems. In a kinetically driven growth regime, one-dimensional nanorod growth is promoted when preferential growth along a specific direction is induced. Likewise, when growth along a specific axis is prevented, the formation of disc-shaped nanocrystals is induced. Very recently, Puntes et al.[9] reported the formation of disc-shaped cobalt nanocrystals (Figure 2.19A). Alkylamine ligands strongly bind to the {001} faces of hexagonally close-packed cobalt and prevent growth along the [001] directions. The nanodiscs obtained assemble into long ribbons where discs are stacked face-to-face via a magnetic interaction between individual nanodiscs.

Seed-mediated growth in a micelle solution also promotes disc-shaped nanocrystals (Figure 2.19B). Silver nanocrystals with a disc shape were formed when silver nitrate solution was added to concentrated acetylammonium bromide solution containing silver seeds.[46] The disc shape of the silver nanocrystals results from plane growth from silver seeds inside the CTAB micelle structure. Silver nanocrystals obtained also assemble into a necklace-like structure similar to that observed in cobalt nanodiscs. In this case, silver nanodiscs are assembled via Van der Waals interactions between monolayers on the basal plane of the discs. Polymer mesospheres can also induce the formation of nanodiscs. Hao et al.[47] reported formation of silver discs in the presence of polystyrene spheres (Figure 2.19C). When silver ions adsorb onto the carboxylate-functionalized polystyrene spheres, the mesospheres block growth along a specific direction and enhance growth along the lateral direction, which results in disc-shaped silver nanocrystals.

In addition to the disc shapes, two-dimensional prismatic shapes have been widely observed. These nanoprisms predominantly appear in fcc-structured transition metals. Photoinduced silver nanoprisms were demonstrated by Rin et al.[3] Ag nanospheres synthesized by the reduction of $AgNO_3$ were irradiated with a conventional 40-W fluorescent lamp. As the irradiation time was increased, an unexpected color change was observed from yellow (which is a characteristic surface plasmon

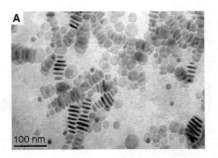

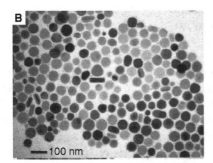

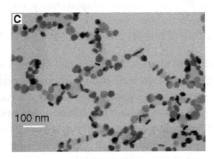

FIGURE 2.19 Various metal nanodiscs. (A) Cobalt nanodiscs. (From Puntes, V.F. et al., *J. Am. Chem. Soc.*, 124, 12874–12880, 2002. With permission.) (B) Silver nanodiscs. (From Chen, S. et al., *J. Phys. Chem. B*, 106, 10777–10781, 2002. With permission.) (C) Polymer mesosphere–silver nanodisc composites. (From Hao, E. et al., *J. Am. Chem. Soc.*, 124, 15182–15183, 2002. With permission.)

band of the spherical particle) to green, and finally blue. Simultaneously, the shape change of nanospheres into nanoprisms was also observed by transmission electron microscopy (TEM) measurements. Electron energy loss spectroscopy (EELS) analysis showed that the formed triangular-shaped Ag nanocrystals were actually flat-top prismatic shapes.

Other fcc-structured transition metal nanoprisms, such as Pd and Ni, were also demonstrated by Bradley et al.[8] Tetra-N-octylammonium carboxylates were used as reducing and stabilizing agents. Scanning tunneling microscopy (STM) measurements showed that the trigonal particles turned out to have an extended flat character in their shapes. Pinna et al.[48] also observed prismatic semiconductor nanocrystals. Triangular CdS nanocrystals were synthesized. The triangular nanocrystals turned out to be flat, and the crystalline phase was proved as a hexagonal wurtzite structure.

2.2.3 NOVEL NANOBUILDING STRUCTURES

2.2.3.1 Building Blocks with High Symmetry

As described above, shape controls of nanocrystals have been widely studied between zero-dimensional spheres, icosahedrons, one-dimensional rods, and two-dimensional plates. These basic building blocks have unique geometry and properties

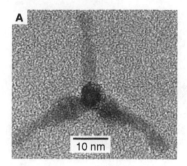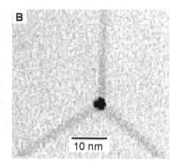

FIGURE 2.20 Various tetrapod shapes. (A) CdSe rod-based tetrapods. (From Manna, L. et al., *J. Am. Chem. Soc.*, 122, 12700–12706, 2000. With permission.) (B) MnS wire-based tetrapods. (From Jun, Y. et al., *J. Am. Chem. Soc.*, 124, 615–619, 2002. With permission.)

with their shapes, and they can be used as components of nanobuilding structures. If the assemblies of each component with different geometries can be controlled, construction of novel nanobuilding structures may be possible. For example, L-shaped structures can be obtained using two rods and a cube and V-shaped structures can be achieved by assembling two rods and a tetrahedron. CdSe tetrapods obtained by Manna[49] and MnS tetrapods by Jun[31] are good examples of such a nanosystem (Figure 2.20). Under certain conditions, zinc blende tetrahedron seeds with four {111} faces, and subsequent rod growth on these four surfaces, resulted in tetrapod-shaped nanocrystals with tetrahedral (T_d) symmetry.[30] Wire formation of MnS instead of rod-shaped nanocrystals arises from a larger surface energy difference between the {001} faces of wurtzite due to the unoccupied d orbitals of the surface manganese metals.[31]

Formation of novel multipod structures using a tetrahedron and one-dimensional rods were thoroughly studied in the case of CdS nanocrystals. Kinetic-controlled growth resulted in a systematic variation of novel structures from spheres to one-dimensional rod, bipod, tripod, and tetrapod, with corresponding point symmetry of D_h, C_{2v}, T_d, and C_{3v}, respectively (Figure 2.21 and Figure 2.22).

Geometrical control of these novel nanobuilding blocks is accomplished by carefully controlling nanocrystal growth parameters such as growth temperature, monomer concentration, and functional group of stabilizers. Geometrical shape evolution of isotropic rock salt-structured PbS nanocrystals is a good example.[10] Rapid injection of a PbS molecular precursor induces the formation of tetradecahedron seeds that are terminated by {100} and {111} faces, and subsequent competitive growth on these two different types of crystalline faces determines the final shape. Typically, in the rock salt structure, the {111} surface has a higher surface energy than the {100} surface. When excess thermal energy is supplied by utilizing high growth temperatures (e.g., E = kT, T ~250°C), the thermodynamic regime governs the growth process, and due to faster growth on the {111} faces, the formation of cube-shaped PbS nanocrystals is favored (Figure 2.23C). However, in the presence of a strongly binding capping molecule (e.g., dodecanethiol), the thiol molecule binds to the {111} surface via a^3-Pb_3-SR (R = $C_{12}H_{23}$) bridging mode while weakly

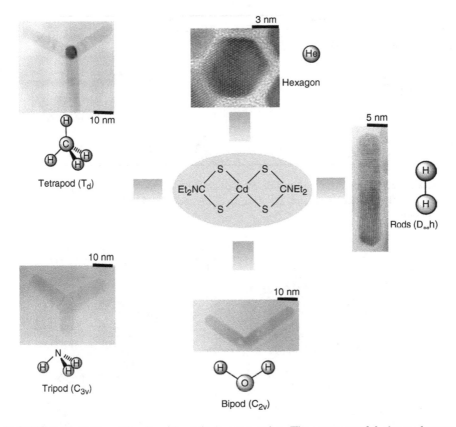

FIGURE 2.21 CdS multipods with various symmetries. The nanocrystals' shape changes from an isotropic sphere to rod, bipod, tripod, and tetrapod. (From Jun, Y. et al., *J. Am. Chem. Soc.*, *123*, 5150–5151, 2001. With permission.)

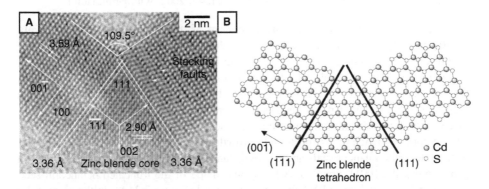

FIGURE 2.22 Atomic structure of bipod-shaped nanocrystals. (A) HRTEM image of CdS tetrapods. (B) Two-dimensional projected model of bipod-shaped CdS nanocrystals. (From Jun, Y. et al., *J. Am. Chem. Soc.*, *123*, 5150–5151, 2001. With permission.)

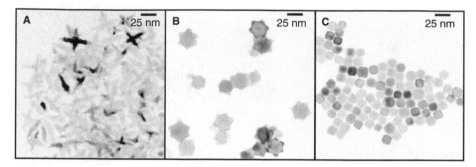

FIGURE 2.23 TEM images of rod-based multipods (A), star-shaped nanocrystals (B), and cubes (C). (From Lee, S.-M. et al., *J. Am. Chem. Soc.*, 124, 11244–11245, 2002. With permission.)

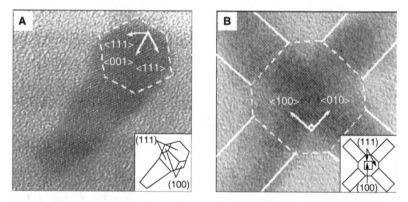

FIGURE 2.24 (A) HRTEM image of tadpole-shaped monopod with zone axis of [110]. (B) HRTEM image of cross-shaped tetrapod of PbS nanocrystals with zone axis of [001]. (From Lee, S.-M. et al., *J. Am. Chem. Soc.*, 124, 11244–11245, 2002. With permission.)

binding to the {100} faces. The surface energy of the {111} faces can selectively be lowered significantly relative to that of the {100} faces. Therefore, under the conditions of low temperature (~120°C) and in the presence of a strongly binding dodecanethiol capping molecule, the growth process shifts into the kinetic growth regime, and growth on the {100} faces with high surface energy is preferred. This results in one-dimensional rod-based single- and multipod structures (Figure 2.23A). High-resolution transmission electron microscopy (HRTEM) images of multipods can be good evidence for the preferable growth along the [100] directions. The image of tadpole-shaped nanocrystals and lattice distance measurements indicates that the rod grows along the [001] direction. A hexagonal shape of the truncated octahedral seed with four {111} and two {100} faces was observed (Figure 2.24A). In the case of the cross-shaped nanocrystals, [100] directional growth was also observed. The separation angle between the pods of the cross is 90°, and an octagonal shape consistent with the [001] projection of a tetradecahedron is observed at the central junction of the cross (Figure 2.24B). Moreover, at intermediate temperatures (e.g., 180°C), as a transient

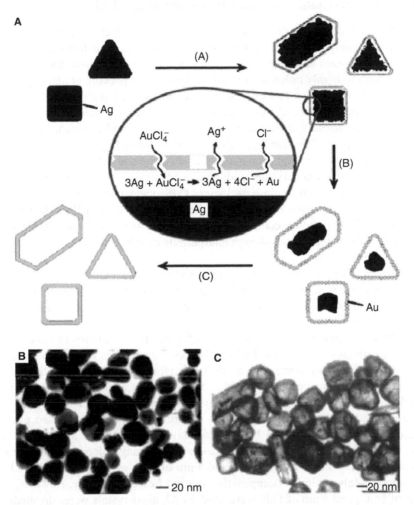

FIGURE 2.25 Hollow nanocrystals synthesized via a nanocrystal templating method. (A) Schematic illustration of the reaction procedure. (B) Silver nanocrystal templates. (C) Gold hollow nanostructures formed by reacting these silver nanoparticles with an aqueous $HAuCl_4$ solution. (From Sun, Y. et al., *Nano Lett*, 2, 481–485, 2002. With permission.)

species, star-shaped nanocrystals possessing both the characteristics of zero- and one-dimensional structures are formed (Figure 2.23B).

Various hollow nonocrystals were also synthesized via template-engaged replacement reaction by Sun et al.[50] In the fist step, silver nanocrystals with various shapes (such as cubes, triangles, wires) were synthesized by a simple reduction method. In the second step, novel metal ions with high reduction potential (e.g., Au^{3+}, Pd^{3+}) were deposited onto the surface of silver nanocrystals. In the final step, novel metals were reduced by sacrificial oxidation of silver nanocrystals, which resulted in hollow-shaped nanocrystals (Figure 2.25).

2.2.3.2 Supercrystals: Assemblies of Nanobuilding Blocks

Superstructures that are built up with basic nanobuilding blocks can also be achieved by chemical interparticle interactions. When nanocrystals are formed with highly monodispersed sizes and shapes, slow evaporation of solvent from the nanocrsytal solution leads to highly ordered three-dimensional close packing of nanocrystals via a Van der Waals interaction between nanocrystals coated with long-chain alkane molecules. The symmetry of a close-packed array of nanocrystals can be modulated by changing parameters, including particle sizes, shape, and interparticle distances.

Superlattices of CdSe nanocrystals were first reported by Murray et al.[51] using selective solvent evaporation from a mixture of octane and octanol solution containing spherical CdSe nanocrystals under reduced pressures. Superlattices obtained were in face-centered-cubic close packing of CdSe nanocrystals and exhibited novel optical characteristics that were different from those of diluted CdSe nanospheres in solution. The emission spectra of the superlattices were red shifted from those of individual CdSe nanocrystals through interparticle coupling, although the absorption spectra were identical.

The symmetry of superlattices constructed with spheres is strongly dependent on interparticle distances. Short-chain capping groups lead to formation of cubic packed superlattices, whereas long-chain capping groups induce hexagonal close packing of nanospheres.

Shapes of nanocrystals also strongly influence the symmetry of superlattices. When nanospheres were used as building blocks, superlattices with fcc or hexagonal close-packed (hcp) symmetry were observed. However, when prismatic particles with tetragonal geometry were assembled, thin flake-like aggregates of a two-dimensional superlattice with pseudorectangular symmetry were obtained.[52]

Binary assemblies consisting of two different nanocrystals are also possible. Figure 2.26 shows Fe_3O_4:FePt binary assemblies with different sizes and a fixed mass ratio of 1:10.[53] The assembly structures were strongly dependent on the sizes of nanocrystals. When 4 nm of Fe_3O_4 and 4 nm of FePt were assembled, Fe_3O_4 and FePt nanocrystals randomly occupied the sites in a hexagonal lattice. However, when 8 nm of Fe_3O_4 and 4 nm of FePt were used, Fe_3O_4 nanocrystals were surrounded by six to eight FePt nanocrystals. At a large size difference between them (12:4 nm), phase separation between two different kinds of nanocrystals was clearly observed.

2.3 CONCLUSIONS

In the past decade, various synthetic methodologies have been successfully developed for building novel nanostructures ranging from spheres and one-dimensional rods to highly sophisticated novel architectures. Although recent studies have demonstrated that the ability to control complex nanostructures, such as hollow cubes, stars, and multipods, is possible, a more comprehensive understanding of their growth mechanisms and the development of advanced synthetic schemes for the desired nanobuilding blocks are of critical importance for the continuous success and advancement of this field. Once there are enough nanobuilding blocks, or nano-Legos, and different building components available to us, the rational assembly and

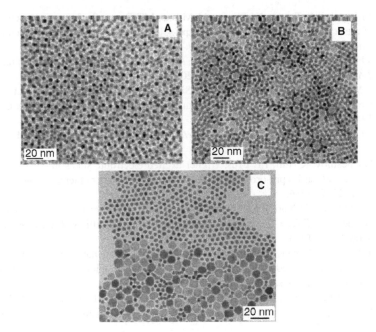

FIGURE 2.26 Various binary nanocrystal (Fe_3O_4 and $FePt_3$) superstructures: 4:4 nm (A), 8:4 nm (B), and 12:4 nm (C). (From Zeng, H. et al., *Nature*, 420, 395–398, 2002. With permission.)

manipulation of them into specific systems will follow. Futuristic nanodevices with novel applications will then be closer to realization.

ACKNOWLEDGMENT

This work was supported by the Korea Research Foundation (KRF-2003-041-C00172).

REFERENCES

1. AP Alivisatos, Semiconductor clusters, nanocrystals, and quantum dots, *Science,* 271: 933–937, 1996.
2. G Markovich, CP Collier, SE Henrichs, F Remacle, RD Levine, JR Heath, Architectonic quantum dot solids, *Acc Chem Res* 32: 415–423, 1999.
3. R Jin, Y Cao, CA Mirkin, KL Kelly, GC Schatz, JG Zheng, Photoinduced conversion of silver nanospheres to nanoprisms, *Science* 294: 1901–1903, 2001.
4. AN Goldstein, CM Echer, AP Alivisatos, *Science* 256: 1425–1427, 1996.
5. HW Ch Postma, T Teepen, Z Yao, M Grifoni, C Dekker, *Science* 293: 76–79, 2001.
6. Y Huang, XF Duan, Y Cui, LJ Lauhon, KH Kim, CM Lieber, *Science* 294: 313–1316, 2001.
7. T Eckes, K Kim, E Joselevich, GY Tseng, CL Cheung, CM Lieber, Carbon nanotube-based nonvolatile random access memory for molecular computing, *Science* 289: 94–97, 2000.

8. JS Bradley, B Tesche, W Busser, M Maase, MT Reetz, Surface spectroscopic study of the stabilization mechanism for shape-selectively synthesized nanostructured transition metal colloids, *J Am Chem Soc* 122: 4631–4636, 2000.

9. VF Puntes, D Zanchet, CK Erdonmez, AP Alivisatos, Synthesis of hcp-Co nanodisks, *J Am Chem Soc* 124: 12874–12880, 2002.

10. S-M Lee, Y Jun, S-N Cho, J Cheon, Single-crystalline star-shaped nanocrystals and their evolution: programming the geometry of nano-building blocks, *J Am Chem Soc* 124: 11244–11245, 2002.

11. AT Hubbard, in *Surfactant Science Series*, Vol. 92, 190–216, T Sugimoto, Ed., Marcel Dekker, New York, 2000.

12. N Chestnoy, R Hull, LE Brus, Higher excited electronic states in clusters of ZnSe, CdSe, and ZnS: spin-orbit, vibronic, and relaxation phenomena, *J Chem Phys* 85: 2237–2242, 1986.

13. CB Murray, DJ Norris, MG Bawendi, Synthesis and characterization of nearly monodisperse CdE (E = S, Se, Te) semiconductor nanocrystallites, *J Am Chem Soc* 115: 8706–8715, 1993.

14. TO Ely, C Amiens, B Chaudret, Synthesis of nickel nanoparticles. Influence of aggregation induced by modification of poly(vinylpyrrolidone) chain length on their magnetic properties, *Chem Mater* 11: 526–529, 1999.

15. CB Murray, S Sun, H Doyle, T Betley, Monodisperse 3D transition-metal (Co, Ni, Fe) nanoparticles and their assembly into nanoparticle superlattices, *MRS Bull* 26: 985– 991, 2001.

16. DP Dinega, MG Bawendi, A solution-phase chemical approach to a new crystal structure of cobalt, *Angew Chem Int Ed* 38: 1788–1791, 1999.

17. VF Puntes, KM Krishnan, AP Alivisatos, Colloidal nanocrystal shape and size control: the case of cobalt, *Science* 291: 2115–2117, 2001.

18. J-I Park, N-J Kang, Y-W Jun, SJ Oh, H-C Ri, J Cheon, Superlattice and magnetism directed by the size and shape of nanocrystals, *ChemPhysChem* 6: 543–547, 2002.

19. S Sun, CB Murray, D Weller, L Folks, A Moser, Monodisperse FePt nanoparticles and ferromagnetic FePt nanocrystal superlattices, *Science* 287: 1989–1992, 2000.

20. EV Shevchenko, DV Talapin, AL Rogach, A Kornowski, M Haase, H Weller, Colloidal synthesis and self-assembly of CoPt$_3$ nanocrystals, *J Am Chem Soc* 124: 11480–11485, 2002.

21. J-I Park, J Cheon, Synthesis of "solid solution" and "core-shell" type cobalt-platinum magnetic nanoparticles via transmetalation reactions, *J Am Chem Soc* 123: 5743–5746, 2001.

22. MA El-Sayed, Some interesting properties of metals confined in time and nanometer space of different shapes, *Acc Chem Res* 34: 257–264, 2001.

23. T Trindade, P O'Brien, X Zhang, M Motevalli, Synthesis of PbS nanocrystallites using a novel single molecule precursors approach: x-ray single-crystal structure of Pb(S$_2$CNEt^{i}Pr)$_2$, *J Mater Chem* 7: 1011–1016, 1997.

24. X Peng, L Manna, W Yang, J Wickham, E Scher, A Kadavanich, AP Alivisatos, Shape control of CdSe nanocrystals, *Nature* 404: 59–61, 2000.

25. S-J Park, S Kim, S Lee, ZG Khim, K Char, T Hyeon, Synthesis and magnetic studies of uniform iron nanorods and nanospheres, *J Am Chem Soc* 122: 8581–8582, 2000.

26. Z Tang, NA Kotov, M Giersig, Spontaneous organization of single CdTe nanoparticles into luminescent nanowires, *Science* 297: 237–240, 2000.

27. JD Holmes, KP Johnston, RC Doty, BA Korgel, Control of thickness and orientation of solution-grown silicon nanowires, *Science* 287: 1471–1473, 2000.

28. CJ Murphy, NR Jana, Controlling the aspect ratio of inorganic nanorods and nanowires, *Adv Mater* 14: 80–82, 2000.
29. Y Jun, C-S Choi, J Cheon, Size and shape controlled ZnTe nanocrystals with quantum confinement effect, *Chem Commun* 1: 101–102, 2001.
30. Y Jun, S-M Lee, N-J Kang, J Cheon, Controlled synthesis of multi-armed CdS nanorod architectures using monosurfactant system, *J Am Chem Soc* 123: 5150–5151, 2001.
31. Y Jun, Y Jung, J Cheon, Architectural control of magnetic semiconductor nanocrystals, *J Am Chem Soc* 124: 615–619, 2002.
32. C Pacholski, A Kornowski, H Weller, Self-assembly of ZnO: from nanodots to nanorods, *Angew Chem Int Ed* 41: 1188–1191, 2002.
33. C-Y Yeh, ZW Lu, S Froyen, A Zunger, Zinc-blende–wurtzite polytypism in semiconductors, *Phys Rev B* 46: 10086–10097, 1992.
34. Y-H Kim, Y Jun, B-H Jun, S-M Lee, J Cheon, Sterically induced shape and crystalline phase control of GaP nanocrystals, *J Am Chem Soc* 124: 13656–13657, 2002.
35. TJ Trentler, SC Goel, KM Hickman, AM Viano, MY Chiang, AM Beatty, PC Gibbons, WE Buhro, Solution–liquid–solid growth of indium phosphide fibers from organometallic precursors: elucidation of molecular and nonmolecular components of the pathway, *J Am Chem Soc* 119: 2172–2181, 1997.
36. AM Morales, CM Lieber, A laser ablation method for the synthesis of crystalline semiconductor nanowires, *Science* 279: 208–211, 1998.
37. T Hanrath, BA Korgel, Nucleation and growth of germanium nanowires seeded by organic monolayer-coated gold nanocrystals, *J Am Chem Soc* 124: 1424–1429, 2002.
38. N Cordente, M Respaud, F Senocq, M-J Casanove, C Amiens, B Chaudret, Synthesis and magnetic properties of nickel nanorods, *Nano Lett* 1: 565–568, 2001.
39. F Dumestre, B Chaudret, C Amiens, M-C Fromen, M-J Casanove, P Renaud, P Zurcher, Shape control of thermodynamically stable cobalt nanorods through organometallic chemistry, *Angew Chem Int Ed* 41: 4286–4289, 2002.
40. DL Feldheim, Jr, CA Foss, *Metal Nanoparticles: Synthesis, Charaterization, and Applications*, Marcel Dekker, New York, 2002, pp. 163–182.
41. T Sugimoto, *Monodispersed Particles*, Elsevier, Amsterdam, 2001.
42. F Kim, JH Song, P Yang, Photochemical synthesis of gold nanorods, *J Am Chem Soc* 124: 14316–14317, 2002.
43. A Chemseddine, T Moritz, Nanostructuring titania: control over nanocrystal structure, size, shape, and organization, *Eur J Inorg Chem* 2: 235–245, 1999.
44. RL Penn, JF Banfield, Morphology development and crystal growth in nanocrystalline aggregates under hydrothermal conditions: insights from titania, *Geochim Cosmochim Acta* 63: 1549–1557, 1999.
45. JJ Urban, WS Yun, Q Gu, H Park, Synthesis of single-crystalline perovskite nanorods composed of barium titanate and strontium titanate, *J Am Chem Soc* 124: 1186–1187, 2002.
46. S Chen, Z Fan, DL Carroll, Silver nanodisks, Synthesis, characterization, and self-assembly, *J Phys Chem B* 106: 10777–10781, 2002.
47. E Hao, KL Kelly, JT Hupp, George C. Schatz, Synthesis of silver nanodisks using polystyrene mesospheres as templates, *J Am Chem Soc* 124: 15182–15183, 2002.
48. N Pinna, K Weiss, J Urban, M-P Pileni, Triangular CdS nanocrystals: structural and optical studies, *Adv Mater* 13: 261–264, 2001.
49. L Manna, EC Scher, AP Alivisatos, Synthesis of soluble and processable rod-, arrow-, teardrop-, and tetrapod-shaped CdSe nanocrystals, *J Am Chem Soc* 122: 12700–12706, 2000.

50. Y Sun, BT Mayers, Y Xia, Template-engaged replacement reaction: a one-step approach to the large-scale synthesis of metal nanostructures with hollow interiors, *Nano Lett* 2: 481–485, 2002.
51. CB Murray, CR Kagan, MG Bawendi, Self-organization of CdSe nanocrystallites into three-dimensional quantum dot superlattices, *Science* 270: 1335–1338, 1995.
52. M Li, H Schnablegger, S Mann, Coupled synthesis and self-assembly of nanoparticles to give structures with controlled organization, *Nature* 402: 393–395, 1999.
53. H Zeng, J Li, JP Liu, ZL Wang, Shouheng sun exchange-coupled nanocomposite magnets by nanoparticle self-assembly, *Nature* 420: 395–398, 2002.

Electronic Properties
of Nanoparticle Materials:
From Isolated Particles
to Assemblies

3 Fluorescence Microscopy and Spectroscopy of Individual Semiconductor Nanocrystals

Alf Mews

CONTENTS

3.1   Introduction 103
3.2   Experimental 105
3.3   Fluorescence Blinking in CdSe Nanocrystals 110
3.4   Time-Resolved Spectroscopy 115
3.5   Conclusion 119
Acknowledgments 120
References 120

3.1 INTRODUCTION

One of the main driving forces for the research of nanoscale materials is the size dependence of their physical properties.[1] Consequently, many applications take advantage of these phenomena, as has been described in several contributions throughout this book. On the other hand, this strong size dependence will necessarily hamper a detailed investigation of nanostructures, because the preparation of, e.g., nanocrystals does not lead to particles that are absolutely identical. Even though several synthesis methods have been developed to prepare very different kinds of nanocrystals with size distributions of only several percent, the particles are still slightly different in diameter, shape, surface structure, and crystallinity.[2–4] As a consequence, the physical properties will be slightly different among the particles within an ensemble. However, the development of devices from nanoparticles depends strongly on the understanding of their individual properties and, therefore, the investigation of single particles is a precondition for the buildup of nanocrystals assemblies and the understanding of their collective behavior.

In this chapter we will focus on the fluorescence properties of individual semiconductor nanocrystals. In particular, we will deal with CdSe particles, which have become a prototype structure in nanocrystals research over the last 10 years.[5,6] This is mainly due to the fact that the absorption onset and the fluorescence color can be tuned over almost the whole visible range due to quantum confinement effects. Naturally, this has led to many applications that are based on color multiplexing, for example, in multicolor fluorescence labeling,[7,8] in polymer light-emitting diodes (LEDs),[9,10] as local fluorescence microscopy probes,[11] or even for active laser media.[12,13]

Whereas the absorption onset, and hence the emission wavelength, is governed by the confinement of the photogenerated charge carriers, and therefore by the volume of the particles,[14] the fluorescence quantum yield is strongly dependent on surface modifications.[15–17] In general, emission takes place a few nanoseconds after the excitation and the charge carriers can relax during this timescale into trap sites, which are believed to be on the particle surface.[18] A reason for the strong surface influence is that the specific surface of particles in the size regime between 1 and 10 nm is very high. For example, a simple calculation indicates that a particle that is 3 nm in diameter has about 50% and a 6-nm particle has 30% of its atoms at the surface. All of these surface atoms are not in an ideal crystallographic environment and can comprise, e.g., unsaturated dangling bonds, which might lead to electronic levels within the band gap.[19] Also, the molecular surface ligands can possibly donate or accept charge carriers.[20] Therefore, the photogenerated electrons or holes can be transferred to these so-called surface traps, which can lead to fluorescence quenching.

One way to minimize the surface effects is the development of spherically layered particles of different materials.[21,22] For example, it was shown that the coverage of CdSe nanocrystals with a few monolayers of CdS[23] or ZnS[24,25] can increase the fluorescence quantum yield up to more than 80%. Also, the effect of different ligands on the fluorescence quantum yield has been demonstrated.[26] In most cases, however, it was shown that the influence of the particle environment cannot be fully suppressed, so that the interaction between the confined charge carriers and electronic surface traps is still a subject of ongoing research.[20]

In this contribution we will summarize our own experiments, where the influence of the environment of semiconductor nanocrystals is investigated in detail by the studying of the fluorescence properties of individual particles.[15,27–30] In general, this technique has revealed many novel photophysical properties that could never have been seen in ensemble experiments.[31] A very intriguing phenomenon that has already been observed in the very first observations of single nanocrystals[32,33] is the blinking behavior of the particles, i.e., the temporary fluorescence intermittency under continuous excitation. As a physical reason, it is discussed that charges in and around the particles are responsible for the dark periods.[32,34] Since these charge fluctuations are at least partially caused by trapping and detrapping of the photogenerated charge carriers,[15] the method of single-nanocrystal fluorescence spectroscopy is an excellent tool to study the charge carrier dynamics within semiconductor nanocrystals.

The outline of this contribution is as follows: In the experimental section, we will describe the method of confocal microscopy to study the fluorescence of single nanocrystals. In the following, we will focus on the blinking behavior under continuous wave (cw) excitation and summarize different experiments that have been performed in order to understand the underlying mechanism. Finally, we will show time-resolved experiments with pulsed laser excitation to get a deeper insight into the charge carrier dynamics on a nanosecond timescale.

3.2 EXPERIMENTAL

The development of selective nanocrystals spectroscopy closely follows the evolution of techniques in molecular spectroscopy, where it was shown that the investigation of the optical properties of individual entities can be performed in the time,[35] frequency,[36,37] and space domains.[38] For example, in the time domain all individual molecules or particles can be coherently excited and the decay of the electronic and vibrational coherence can be directly measured.[39,40] From this, the line width and vibrational frequency can be calculated in principle, but the results still give only the average properties of many individual particles, which are investigated simultaneously.

The methods used to study subsets of molecules and nanocrystals in the frequency domain are the techniques of transient hole burning (or transient differential absorption (TDA))[41] and fluorescence line narrowing (FLN).[42] In a typical TDA experiment, a narrow nanosecond pump laser bleaches a subset out of the broad absorption band, whose electronic transition is in resonance with the pump energy. This bleach leads to a spectral hole in the absorption spectrum, which is recorded with a spectrally broad probe pulse a few nanoseconds later. From the shape of this spectral hole, the homogeneous absorption spectrum can be deduced by deconvolution techniques, in principle.[43] Possible complications are due to the fact that even with this technique, several particles are excited at the same time and the calculated absorption spectrum again shows only their average properties. Also, the bleached subset of particles might absorb in another spectral region instead. This induced absorption, which is also called an antihole, might interfere with the bleach spectrum to some degree, which further complicates the interpretation.

In an FLN experiment, the homogeneous fluorescence properties are investigated by tuning the excitation laser to the edge of the absorption spectrum, thus exciting only the biggest particles within an ensemble. However, since there are still many different nanocrystals excited, a multiparameter deconvolution procedure is needed to extract the homogeneous fluorescence spectrum with simultaneous consideration of the absorption spectrum and the inhomogeneous broadening. Also, since for spectral selection only the particles at the absorption wings are excited, the extracted fluorescence spectra reflect only the most unrepresentative entities within the sample.[44]

A way out of those complications has been the direct investigation of isolated single molecules or nanocrystals by spatial selection. The first experiments toward this goal used a near-field scanning optical microscope (NSOM), in which the sample was illuminated with a metal-coated optical fiber that has a very small aperture at the tip (<100 nm).[45,46] If the distance between the tip and the surface is in the range of the near field, i.e., smaller than the illumination wavelength, the illumination spot size, and therefore the spatial resolution, is in the range of the dimension of the aperture. This high spatial resolution is of major importance if the fluorescing objects are very close together, because the spatial resolution of conventional far-field optics is given by the natural diffraction limit; i.e., the dimension of the focus is about half of the light wavelength. However, this is usually not an issue for colloidal semiconductor nanocrystals because they can easily be diluted such that the separation of the particles is sufficient to be investigated by far-field optics.[47] The advantage of far-field experiments is that the experiment becomes much simpler because there is

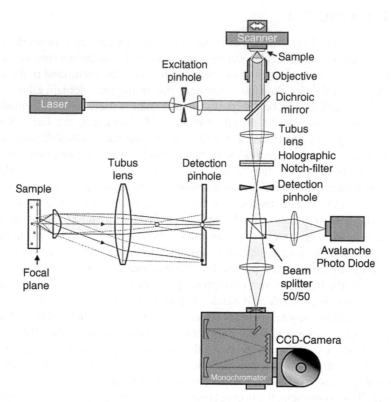

FIGURE 3.1 Scheme of a confocal optical microscope to investigate the optical properties of matter with submicrometer resolution (for details, see text). The inset shows a scheme to explain the confocal geometry, which is used to acquire the signal only from a small volume and therefore to increase the signal/noise ratio.

no need for a complicated procedure of tip preparation and manipulation. Also, the light power, which is limited in NSOM by the destruction threshold of the tip, is basically not constrained, and possible interactions between the tip and the particles are not present.

In conventional far-field microscopy, a wide area of the sample is illuminated and the luminescence or the reflection is projected on a charge coupled device (CCD) camera.[48] Numerous experiments have been performed with this technique, including the investigation of the blinking,[49] the Stark effect,[50] and the fluorescence polarization[51] of single nanocrystals. The results described in this contribution were achieved by using a so-called confocal optical microscope, where the sample is in the focus of the diffraction-limited excitation spot.[52–54]

Figure 3.1 shows the principal setup of such a microscope. The laser beam is focused on an excitation pinhole and recollimated such that the diameter of the parallel excitation beam is slightly larger than the entrance aperture of the microscope objective, to ensure the highest spatial resolution. Then the excitation light is guided into the microscope objective by a dichroic mirror, which is reflective for the short-wavelength excitation light and transparent for the longer-wavelength emission light coming back from the sample. The sample, usually a thin polymer film with embedded nanocrystals on a planar substrate, is sitting in the focus of the objective. The fluorescence is collected with the same microscope objective and passes the dichroic mirror and a holographic notch filter, which absorbs the backscattered excitation light. Then the fluorescence light is focused onto the detection pinhole and further guided simultaneously to a point detector (avalanche photo diode (APD)) and a spectrometer equipped with a liquid nitrogen-cooled CCD camera. To record an image, the sample is laterally scanned through the focal plane and the luminescence intensity is recorded at each pixel and plotted with respect to the scanner position.

The basic principle of confocal microscopy is that the excitation and detection volume are identical, due to the use of the described pinholes. This is shown in the inset of Figure 3.1 for the detection beam: if the luminescence comes from a nanocrystal, which is lying in the central axis of the microscope objective and at the same time in the focal plane, then the tubus lens focuses the light onto the detection pinhole and, hence, onto the APD. Fluorescence light that comes from particles that are not on axis will not pass this pinhole. Also, the fluorescence from particles that are on axis but not in the focal plane will not be directly focused onto the pinhole and is, therefore, strongly suppressed. Hence, confocal geometry also allows for fluorescence imaging of vertical sections through a three-dimensional sample, in principle.

Figure 3.2A shows a typical scanning confocal fluorescence image of CdSe/ZnS core-shell nanocrystals. The CdSe particles ($\lambda_{fluo.}$= 570 nm) were prepared by standard methods[6] and covered with four monolayers on ZnS.[24] The fluorescence quantum yield was of the order of 50%, compared to a solution of the dye Rhodamine 6G. All of the investigations were performed with nanocrystals embedded in a polymer matrix. For sample preparation, two drops of a solution containing 12 g/l polystyrene and 10^{-9} mol/l nanocrystals (NCs) were placed on a cover slide and spin cast at 3000 rpm, resulting in a film of about 50-nm thickness. For the $10 \times 10 \ \mu m^2$ image shown in Figure 3.2, this sample was scanned line by line (256×256 pixel^2) through the excitation spot while excited with an excitation intensity of 350 W/cm^2 ($\lambda = 488$ nm) in the center of the microscope objective. It can be seen that several spots show horizontal dark lines, e.g., the spot enlarged in Figure 3.2B, which is due to the fluorescence intermittency of the particle during the imaging. The line scan of Figure 3.2D shows that the measured fluorescence intensity is of the order of 100 cts/10 msec and also allows determination of the spatial resolution. For example, if the enlarged spot shown in Figure 3.2C is fitted with a Gaussian function, the width of the luminescent spot can be determined to be 320 nm.

In order to investigate the fluorescence blinking behavior, a single luminescent nanocrystal is positioned in the focus of the microscope objective and the fluorescence intensity is recorded in time. Figure 3.3A shows such a fluorescence transient over a period of half an hour. The overall count rate distribution is shown in the corresponding histogram of Figure 3.3B. It can be seen that this particular nanocrystal shows an almost binary blinking behavior of the fluorescence, switching between an *on* and *off state*. For analysis, an intensity threshold (dotted line) is defined where the NC is either *on* or *off* if the fluorescence count rate is above or below this threshold, respectively. From this, the length of the different *on* and *off*

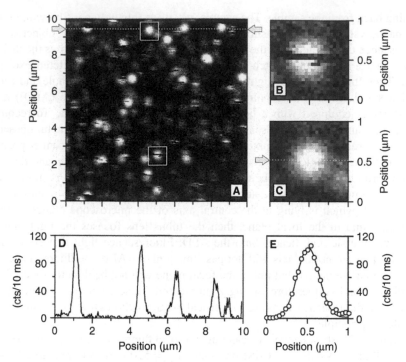

FIGURE 3.2 Confocal fluorescence image of individual semiconductor particles achieved with a setup as shown in Figure 3.1. The spots in (A) show the fluorescence of individual CdSe/ZnS particles on a glass substrate achieved by scanning the substrate line-wise through the excitation spot and collecting the fluorescence at each position. The horizontal stripes are due to fluorescence blinking, which happens occasionally during the scan, as can be seen by comparison in the enlarged parts in (B) and (C). The spatial resolution is given by the diffraction limit and can be extracted from line scans of the intensity profiles (gray arrows) in (D) and (E).

periods can be determined; the *on* periods are shaded in gray. Histograms of the *on* and *off* distributions are plotted in Figure 3.3C and D, respectively.

Even though the *on–off* statistics are very poor for one single nanocrystal, the logarithmic plots of those histograms already show that the decay of the distributions is not single exponential. However, in order to compare the blinking behavior of particles under different experimental conditions, one has to extract a value from the fluorescence transients. The simplest way is to determine the mean *on* and *off* times,[32] which are of comparable length for the distributions shown in Figure 3.3 (t_{on} = 6.5 sec and t_{off} = 6.2 sec).

Another way of characterizing the *on* and *off* characteristics is to consider the whole distribution, i.e., to calculate the probability density $P(t)$ to observe an *on* or *off* event over a time period of t.[55] Since the short events occur more often than the longer ones, the time windows do not have to be equidistant but can be increased in time. However, then the number of events has to be normalized by the length of the respective time window.

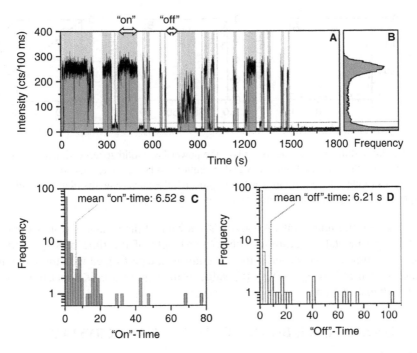

FIGURE 3.3 Analysis of the fluorescence blinking. (A) Transient fluorescence intensity of a single emitter that was positioned in the focus of the microscope objective. If the fluorescence intensity is above an arbitrarily defined threshold (dotted line), the particle is considered to be in an *on* state (gray areas) and otherwise *off*. (B) Count rate distribution used to define the threshold level. (C, D) *On* and *off* histograms as calculated from the transient in (A).

In general, one can distinguish between different probability densities $P(t)$, the most prominent of which are the exponential (Equation 3.1) and power law (Equation 3.2) statistics

$$P(t) \sim e^{-\frac{t}{\tau}} \tag{3.1}$$

$$P(t) \sim t^{-m} \tag{3.2}$$

For demonstration, Figure 3.4 shows two hypothetical distributions $P(t)$ that obey an exponential distribution (hollow spheres) and the power law (solid spheres) plotted on a linear scale (A) and single- (B) and double-logarithmic (C) scales, respectively. In case of an exponential distribution, the lifetime of the *on* or *off* state would be directly given by the value of τ in Equation 3.1. However, this would only be the case if the plot of $P(t)$ on a single logarithmic scale would result in a straight line. Since Figure 3.3 clearly shows that this is not the case, it was argued by Kuno et al.[55] that the distribution of events could instead be explained by power law statistics, where the plot on a double-logarithmic scale should result in a straight

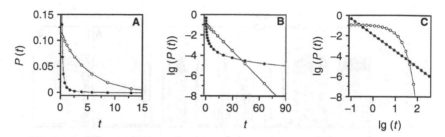

FIGURE 3.4 Arbitrary data sets to explain the power law (solid spheres, Equation 3.1) and exponential (hollow circles, Equation 3.2) distributions. Whereas an exponential decay should result in a straight line in a semilogarithmic plot (B), the power law distribution results in a straight line when plotted in a double-logarithmic manner (C).

line, as shown in Figure 3.4C. Also, it was shown that the values of m and τ depend on the experimental conditions, such as the length of the time bins t_{min} and the observation time t_{max}. However, if the experiments are performed with the same time windows, or if at least $t_{min} \ll t_{max}$, the value of the exponent m is a measure for the *on* and *off* distributions.

3.3 FLUORESCENCE BLINKING IN CdSe NANOCRYSTALS

The first investigations of the blinking behavior of CdSe and CdSe/ZnS core-shell particles have been performed by Nirmal et al.,[32] who observed that the covering of CdSe particles with ZnS leads to an increase of the *on* times but leaves the *off* times almost unaffected. Also, they investigated the blinking dynamics as a function of excitation intensity and found that the *on* times are considerably shorter at higher excitation intensity.

As a physical reason for the dark state, they proposed that a particle is not fluorescing when it is charged. After subsequent absorption, the deposited energy promotes the excess charge into an excited state, which in turn undergoes radiationless relaxation, a process called Auger quenching. The charging mechanism itself was explained by a light-induced Auger ionization, in which one charge carrier (electron or hole) is ejected into the surrounding matrix due to Coulomb repulsion if there are more than one electron–hole pair excited within the individual particle.[56] Therefore, the *on* times become shorter at higher excitation intensity because the probability of exciting multiple electron–hole pairs increases. Also, the *on* times become longer when the surface is covered with ZnS because the ZnS shell acts as a barrier for the Auger ionization.

Although it is commonly accepted that charging of a particle is the reason for the dark state, the mechanism for the transition from the particle being either *on* or *off* is still controversial.[57,58] For example, it was shown that the blinking dynamics also depend on the temperature, which was explained by a thermally activated surface-trapping process.[59] If one of the charge carriers is localized in such a surface trap, the net charge in the interior of the nanocrystals might still lead to Auger quenching processes and, therefore, to a dark particle.

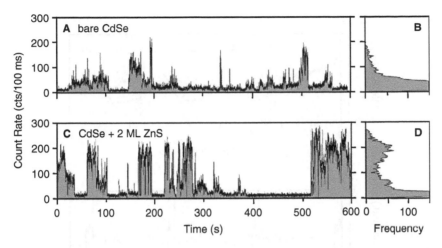

FIGURE 3.5 Influence of the ZnS coverage on the fluorescence of CdSe nanocrystals. The fluorescence transient of the bare particle (A) shows strong intensity fluctuations at any fluorescence count rate (B). After coverage, a distinct value for the particles being *off* can be defined from the transient (C) and the count rate histogram (D).

Here we show results where we chemically manipulated the surface of the nanocrystals and investigated the blinking behavior in order to examine the influence of the surface effect. Figure 3.5 shows the fluorescence intensity of single nano-crystals as a function of time and count rate distributions for bare CdSe particles (A, B) and particles that are covered with two monolayers of ZnS (C, D), respectively.[60]

In general, we reproduced the results of Nirmal et al.,[32] but we found that especially for the bare CdSe particles, it is very difficult to distinguish between *on* and *off* states. This becomes obvious when looking at the count rate distribution in Figure 3.5B because there is no discrete threshold level to distinguish between particles being either *on* or *off*. The situation changes to some extent after covering the particles with ZnS. Now one can clearly distinguish between particles in an *on* or *off* state, but still, the particles do not show a strictly binary blinking; i.e., sometimes the particles seem to emit with just a fraction of their maximal intensity. To explain this behavior, Kuno et al.[57,61] performed fluorescence blinking investigations of CdSe/ZnS particles with different time bin widths. They showed that basically the fluorescence intermittency happens on any timescale investigated from 100 μsec to 100 sec. Along these lines, it might well be that the count rate distribution below the maximum count rate level is due to *on* periods that are shorter than the bin width, which was 100 msec in our experiments. Consequently, an *off* or *on* time can only be defined in combination with the bin width; i.e., if the threshold level is set right above the noise level of the experiment, the particle must be *off* for more than 100 msec.

In any case, the difference of the blinking statistics clearly shows that the inorganic ZnS surface passivation changes the fluorescence behavior of single particles dramatically. This might be due to either the buildup of a barrier for the charge carriers to escape into the surrounding matrix or the passivation of surface traps. In

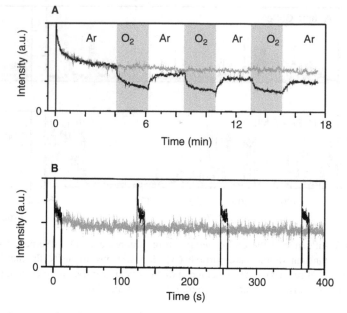

FIGURE 3.6 Influence of a different gas environment and illumination on the fluorescence intensity of a film of ZnS-covered CdSe nanocrystals. (A) The fluorescence intensity is reversibly decreased when the film is temporarily exposed to oxygen (black line) compared to argon (gray line). The intensity decay in pure argon is also reversible, as shown in (B), where the fluorescence under continuous illumination (gray line) is compared to temporarily illumination (black line).

order to distinguish between these two processes, we have performed experiments in which we investigated the influence of the gas environment of oxygen vs. argon on the blinking statistics, because adsorbed oxygen molecules might also act as traps for the charge carriers.[15] To compare the results with ensemble measurements, we first investigated the gas dependence of a solid film of ZnS-covered CdSe particles. The films for ensemble measurements were prepared by spin casting a concentrated solution of CdSe/ZnS particles onto a glass coverslip, resulting in a heterogeneous film consisting of large fluorescent aggregates of nanocrystals (1 to 10 μm in diameter).

The influence of alternating argon and oxygen exposure upon the film fluorescence is shown in Figure 3.6A (black line), compared to the fluorescence intensity in argon only (gray line). It can be seen that the fluorescence intensity decreases under constant illumination within the first minutes, even though the film is exposed to argon. However, this decrease is not due to a permanent bleach of the particles but recovers if the illumination is turned off for a certain amount of time. This is demonstrated in Figure 3.6B, in which the gray line shows the fluorescence intensity under continuous illumination, whereas the black line demonstrates that the fluorescence recovers up to its initial intensity if the laser is turned off for 2 min. Therefore, the temporary bleach might be due to the reversible charging of some of the particles, due to either light-induced Auger ionization or surface-trapping processes.

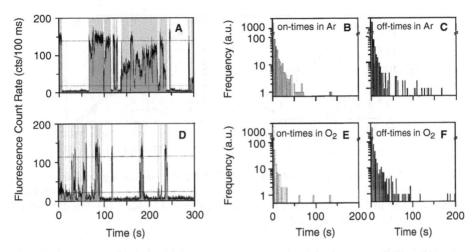

FIGURE 3.7 Influence of a different gas environment on the fluorescence of individual ZnS-covered CdSe nanocrystals. Representative fluorescence transient and *on–off* distributions for particles in the argon atmosphere (A–C) and oxygen (D–F), respectively. While the length of the *on* times is considerably shorter for particles in oxygen than argon, the *off*-time distribution is almost not affected by the gas environment.

As a second effect, the fluorescence is also decreased by changing the gas environment from argon to oxygen. It can be seen that the oxygen exposure (shaded in gray) leads to a significant decrease in the fluorescence intensity, which is nearly reversible when the surrounding gas atmosphere is changed back to argon. However, there is also an irreversible bleaching process, possibly due to a process called photocorrosion, of some of the particles, which alters the fluorescence of the nanocrystal ensemble with time. This is of great relevance for the single-particle measurements, where the data statistics can in principle be enhanced by increasing either the measurement time of each nanocrystal or the number of investigated particles. To reduce the influence of this irreversible process on the data analysis, we concentrate, for the single-particle investigations, only on the first 10 min of the fluorescence time traces.

The influence of different gas atmospheres on the fluorescence of single CdSe/ZnS particles is shown in Figure 3.7. The graphs in the upper row show a representative time trace of a single particle and the *on* and *off* times for about 20 individual ZnS-covered CdSe particles in Argon, and the lower graphs show the same for these particles in oxygen. As mentioned above, the threshold to distinguish between *on* and *off* was set right above the background level. Therefore, once the nanocrystal is at least 100 msec in its *off* state, but certainly longer than 200 msec, this event will be accounted for as an *off* event. The uncertainty can be understood by the fixed time bins being only partially filled with photons. For example, if the particle turns *off* for, say, 100 msec, 50 msec of which might be detected in two consecutive time bins, then this *off* event would not be detected at all.

By comparing the histograms, it can be seen that the *off*-time distribution seems to be almost not affected by the presence of oxygen, whereas the *on* times in oxygen

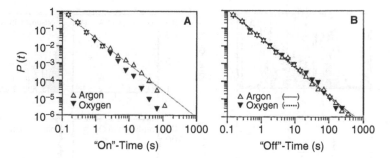

FIGURE 3.8 Normalized probability densities $P(t)$ for the *on*- and *off*-time distributions shown in Figure 3.7. Whereas all *off* times can be described with the power law distribution with $-m = 1.65$ to 1.63, the *on* times deviate from this plot at longer times. For the short *off* times, $m_{Ar} = -1.52$ and $m_{O_2} = -1.82$.

are considerably shorter than single dots in argon. Therefore, the mean *off* time in argon (3.4 sec) and in oxygen (3.3 sec) is almost the same, whereas the value for the *on* time is larger for particles in argon (2.4 sec) than oxygen (1 sec).

It has been mentioned in the experimental part already that the determination of mean *on* and *off* times is strictly valid for exponential distributions only. However, since the histograms in Figure 3.7 clearly indicate a nonexponential spreading of the measured time distributions, we also analyzed the data with the described power law statistics. This procedure has also been used to characterize the blinking behavior when the size of the particles, temperature, and excitation energy were varied.[62] As a main result, it could be seen that the probability density for the *off* times is almost the same, because the slope of the log–log plot changes only slightly (from 1.4 to 1.6) for the different-sized particles investigated under different conditions. The slope of the *on*-time probability, however, deviates considerably from a power law behavior, especially for longer *on* times. Figure 3.8 shows the power law analysis for the fluorescence blinking in different gas atmospheres, i.e., for the histograms of Figure 3.7.

In general, the power law statistics show the same trend as has been observed from the mean *on* and *off* times. As the mean *off* time did not change upon alternating gas environments, the negative slope of the power law analysis for the different conditions also remains the same, i.e., m = −1.65 in argon compared to m = −1.63 in oxygen. Additional information can be gained from the fact that all *off*-time intervals strongly follow this power law, which is indicative for a spontaneous process for the particles to turn back from its *off* state to the *on* state.

For the *on* times, the values of the slopes in argon (m = −1.52) and oxygen (m = −1.82) are different, in agreement with the different mean *on* times. Also, it can be seen that only the time intervals of fewer than 10 sec seem to follow the power law behavior. Above this value, the deviation from this linear behavior becomes very obvious. One might argue that this deviation is only due to the fact that the statistics become very poor for time intervals above 10 sec for this particular set of data. However, there is no deviation from the linear behavior on a log–log plot for the *off* times that were calculated from the same fluorescence time traces. Therefore, the deviation of the data from the power law behavior is an indication that the switching of the particles from *on* to *off* is not a spontaneous process but induced by external effects. As this deviation is more obvious when the particles are exposed to air, this is another qualitative measure of the quenching process of the particles induced by oxygen traps.

Obviously, the presence of oxygen at the surface of the particles changes the electronic properties of the nanocrystals, which in turn changes their fluorescence behavior. In the past, it has been shown that adsorption of oxygen also changes the photoconductivity of thin CdS and CdSe films.[63,64] The authors could explain this effect by assuming that the adsorbed oxygen can act as a scavenger for the charge carriers and, therefore, reduces the conductivity by more than one order of magnitude. Thus, it is reasonable that oxygen molecules at the surface of the semiconductor particles indeed might act as trapping sites for the photogenerated charge carriers. If, for example, the electron would temporarily be transferred to the adsorbed O_2, this would leave a positively charged particle behind, which represents the dark state of the nanocrystal.

3.4 TIME-RESOLVED SPECTROSCOPY

The examples mentioned above clearly show that the fluorescence blinking behavior of single semiconductor nanocrystals can be influenced by inorganic and molecular surface modifications. Therefore, it is very likely that electronic fluctuations at the surface are responsible for this effect. A main drawback of the continuous wave experiments, however, is the limited time resolution, which is given by the length of the time bins, as has been mentioned already.[61] Especially when the fluorescence intensity is strongly fluctuating, it is basically not known whether the temporarily low fluorescence intensities are due to very short *off* times that cannot be resolved or competing deactivation processes of the excited state, which happen on the timescale of the fluorescence lifetime.

In general, one can distinguish between two photophysically different processes: static and dynamic fluorescence quenching.[65] In the case of short *off* times, this would be a static fluorescence quenching process that in general occurs when there is ground-state interaction between the fluorophor and the quencher.[66] This would be given if the particle is *off* for a short time because a dark particle is already charged in its ground state and might absorb but not emit a photon. In dynamic quenching, the excited state of the particle can be deactivated by competing processes, i.e., via a radiative or nonradiative channel, as sketched in Figure 3.9.

When a particle is excited from the ground state to the excited state, the radiative relaxation is characterized by the radiative rate k_{rad} and the nonradiative relaxation by k_{nr}. A way to determine these relaxation processes is the measurement of the fluorescence decay time. The results presented in this chapter are taken by the method of time-correlated single-photon counting (TCSPC), where the single nanocrystals are excited with a ps Laser pulse ($\lambda_{ex} = 458$ nm) and the time delay between the excitation and the respective fluorescence photon is measured.[30] A histogram of thousands of those events is a direct measure of the relaxation time of the excited state. However, the observed decay rate k_{obs} is different from the radiative decay rate

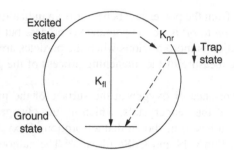

FIGURE 3.9 Sketch to explain the fluorescence dynamics in semiconductor particles. If a particle is excited from the ground state to the excited state, the relaxation can happen via a radiative channel with a rate k_{fl} or a nonradiative channel with a rate k_{nr} into trap states.

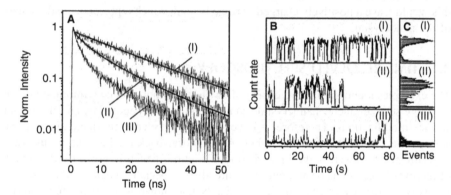

FIGURE 3.10 Comparison between fluorescence decay curves (A), fluorescence transients (B), and count rate distributions (C) for three different particles (I), (II), and (III) from the same sample of ZnS-covered CdSe particles. The solid curves in (A) show fitting curves with a stretched exponential function. Particle (I) shows an almost monoexponential decay and, at the same time, a distinct *on* level of the fluorescence transient and the respective count rate distribution. The stronger intensity fluctuations of particles (II) and (III) are correlated with the stronger bending of the fluorescence decay curves, suggesting a distribution of decay rates.

k_{rad}. In principle, the observed decay process is just a sum of all possible decay channels that can lead to depopulation of the excited state: $k_{obs} = k_{rad} + k_{nr}$. In other words, if a changing electronic or molecular environment will lead to a fluctuating nonradiative decay rate k_{nr}, this will necessarily lead to a varying observed decay rate k_{obs} and, therefore, to a multiexponential fluorescence time decay.

Figure 3.10A shows three different fluorescence decay curves that were measured from individual nanocrystals (I), (II), and (III) and plotted in a single-logarithmic manor; i.e., a monoexponential decay would result in a straight line. This behavior, however, can only be observed for nanocrystals (I), whereas the bending of curves (II) and (III) directly shows that the underlying relaxation process cannot be described with a single rate. Figure 3.10B shows the fluorescence transients that were measured at the same time as the fluorescence decay curves. Here it can clearly

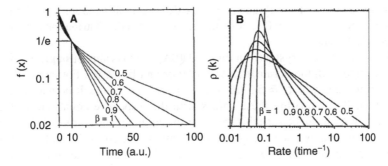

FIGURE 3.11 Arbitrary curves to explain the Kohlrausch–Williams–Watts (or stretched exponential) function and underlying decay rate distribution. The curves in (A) show KWW functions with different values of the stretching exponent β. The decay rate distributions shown in (B) are calculated from the curves in (A) by an inverse Laplace transformation. Obviously, a stronger bending of the decay curves goes along with a smaller β broader distribution of the decay rates.

be seen that the monoexponential fluorescence decay of nanocrystal (I) is correlated with an almost binary blinking behavior; i.e., this particular nanocrystal is either *on* or *off*. In contrast, the count rate distribution of particle (II) is much broader, and at the same time, the corresponding decay curve is not single exponential. Finally, for particle (III), with the strongest intensity fluctuations, the fluorescence decay is very fast and multiexponential.

For a quantitative analysis, the underlying distribution of decay rates can be extracted from the decay curves, in principle. However, it is commonly known that this is an ill-conditioned problem because there will be many different rate distributions that could describe the corresponding decay curve equally well.[67] For example, any decay curve can be reasonably fitted with a sum of four discrete exponentials. For the fluorescence decay curves of the nanocrystals, we found that they can be fitted reasonably with a so-called Kohlrausch–Williams–Watts (KWW) function, commonly known as stretched exponential function (Equation 3.3)[68]:

$$I(t) = I_0 \cdot e^{-\left(\frac{t}{\tau}\right)^{\beta}} \tag{3.3}$$

This function describes the exponential decaying fluorescence intensity $I(t)$, which drops to $1/e$ of its initial value $I(0)$ after the decay time τ for any value of the stretching exponent β, as shown in Figure 3.11A.

This stretching exponent β is basically a measure of the relaxation rates involved in the fluorescence decay, where a smaller β means a broader rate distribution. The underlying distribution of rates can be calculated in principle by a direct inverse Laplace transformation of a given decay curve or its fit function. Since the inverse Laplace transform has only analytical solutions for discrete values of β, we employed the technique of series expansion and saddle point calculation.[68] This is demonstrated in Figure 3.11B for the hypothetical decay curves of Figure 3.11A. For a decay with an arbitrary lifetime of $\tau = 10$ and $\beta = 1$, the decay rate distribution would be a delta function at the rate $k = 1/\tau = 0.1$. For smaller values of β the distribution $\rho(k)$ becomes increasingly broader.

The decay of nanocrystal (I) in Figure 3.10A, which is almost single exponential, could be fitted with the parameters $\tau_I = 19.9\ ns$ and $\beta_I = 0.97$. For nanocrystal (II), we observed values of $\tau_{II} = 6.5\ ns$ and $\beta_{II} = 0.66$, i.e., a much broader decay rate distribution. By comparing these values with the respective count rate histograms of Figure 3.10C, it can be seen that the a narrow count rate histogram of (I) correlates with a narrow fluorescence decay rate distribution. Moreover, it can be argued that a high fluorescence count rate goes along with a long fluorescence lifetime and vice versa.

To verify these assumptions, we performed experiments in which several fluorescence decay curves were measured from a single nanocrystal at different count rate thresholds. Figure 3.12A to C show fluorescence transients of one individual nanocrystal (V) on different timescales. In addition, Figure 3.12D shows the ensemble absorption and fluorescence spectra of the colloidal solution nanocrystals (dotted lines) and two fluorescence spectra of individual particles (IV and V) for comparison. The fluorescence spectrum of particle (V) was recorded at the same time as the transients in (A) to (C). It can be seen that the fluorescence bands of the single-particle fluorescence spectra are considerably narrower (50 meV) than the ensemble spectrum (100 meV).

Obviously, nanocrystals (V) showed very strong intensity fluctuations but could be investigated for more than 1000 sec, enabling us to detect decay curves at several arbitrary count rate levels. In the time periods in which the nanocrystals show very low fluorescence intensity, we recorded the decay curve labeled (i) and at medium and high fluorescence count rates the decay curves (ii) and (iii). The resulting decay curves, together with the fitted stretched exponential functions, are shown in Figure 3.12E. It can clearly be seen that the fluorescence decay changes considerably depending on the temporary fluorescence intensity, even for a single nanocrystal. The inset in Figure 3.12E shows the corresponding lifetime distributions, which were calculated as described above. Here it can clearly be seen that a lower count rate is indeed associated with a shorter fluorescence decay time.

The fact that even a single nanocrystal changes its excited-state lifetime during the measurement clearly demonstrates that indeed fluctuating dynamic quenching processes can occur. In the light of the experiments described above, this is most likely due to acceptor levels on the surface of the particle, such as temporarily adsorbed molecules or dislocations in the surrounding matrix, which could lead to nonradiative decay channels. These acceptor levels will open up new nonradiative decay channels, which will lead to a decrease of the fluorescence intensity and, at the same time, to a decrease of the excited-state lifetime. Therefore, a fluctuation of the nonradiative decay channels will necessarily lead to a broadening of the observed fluorescence decay time distribution. Further measurements in a defined environment, e.g., at low temperature in a helium atmosphere,[69] are needed to explore whether an intrinsically charged dark particle is indeed totally dark or if the quenching rate can also be resolved by time-resolved fluorescence measurements.

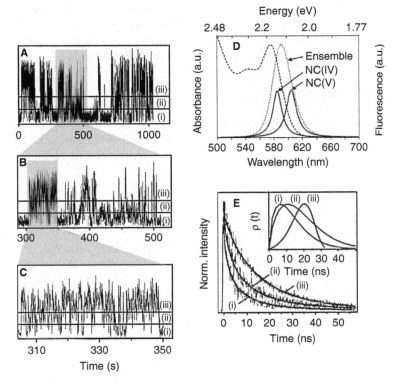

FIGURE 3.12 Fluorescence transients (A–C), fluorescence spectra (D), and fluorescence decay curves at different count rates (E) of a single ZnS-covered CdSe nanocrystal. The transients for this particular nanocrystal (V) show very strong intensity fluctuations. The fluorescence spectrum of this particle is shown in (D), together with another single-particle fluorescence spectrum (IV) and the ensemble fluorescence (dotted line) and absorption (dashed line) spectrums. The three decay curves (i, ii, and iii) in graph (E) were taken at different count rate levels as marked in the fluorescence transients and fitted with KWW functions. The calculated decay rate distributions in the inset of (E) clearly show that a lower transient fluorescence intensity is correlated with a shorter decay time.

3.5 CONCLUSION

In conclusion, we have shown that the investigation of single semiconductor particles reveals many phenomena that are difficult to observe in ensemble measurements but that are essential for the understanding of nanocrystal assemblies. We have shown that the fluorescence behavior of particles can be directly related to the charge carrier dynamics on different timescales, which is of major importance for, e.g., the conductivity or collective electronic behavior of particle assemblies. The next steps might be the buildup and investigation of small and defined particle aggregates to directly connect the single-particle research with the properties of assemblies. Another direction will be the local investigation of assemblies with the tools of single-particle spectroscopy to address the problem of sample inhomogeneities or local manipulations within the assemblies. In any case, the investigation of nano-structures and their assemblies is an ongoing field of research that requires strong interdisciplinary collaboration of researchers from different fields.

ACKNOWLEDGMENTS

I thank all of my colleagues and coworkers. In particular, I acknowledge Prof. Thomas Basché for experimental support and fruitful discussion and Dr. Felix Koberling, as well as Germar Schlegel, because the project described in this chapter is part of their Ph.D. thesis.

REFERENCES

1. K Klabunde, Ed., *Nanoscale Materials in Chemistry*, Wiley, New York, 2001.
2. H Weller, Collioidal semiconductor Q-particles: chemistry in the transition region between solid state and molecules, *Angew. Chem. Int. Ed. Engl.* 41: 41–57, 1993.
3. AP Alivisatos, Semiconductor clusters, nanocrystals, and quantum dots, *Science* 271: 933–937, 1996.
4. AD Yoffe, Semiconductor quantum dots and related systems: electronic, optical, luminescence and related properties of low dimensional systems, *Adv. Phys.* 50: 1–208, 2001.
5. MG Bawendi, AR Kortan, ML Steigerwald, LE Brus, X-ray structural characterization of larger CdSe semiconductor clusters, *J. Phys. Chem.* 91: 7282–7290, 1989.
6. CB Murray, DJ Norris, MG Bawendi, Synthesis and characterization of nearly mon-odisperse CdE (E = S, Se, Te) semiconductor nanocrystallites, *J. Am. Chem. Soc.* 115: 8706–8715, 1993.
7. M Bruchez, M Moronne, P Gin, S Weiss, AP Alivisatos, Semiconductor nanocrystals as fluorescent biological labels, *Science* 281: 2013–2016, 1998.
8. WCW Chan, SM Nie, Quantum dot bioconjugates for ultrasensitive nonisotopic detection, *Science* 281: 2016–2018, 1998.
9. VL Colvin, MC Schlamp, AP Alivisatos, Light-emitting diodes made from cadmium selenide nanocrystals and a semiconducting polymer, *Nature* 370: 354–357, 1994.
10. S Coe, WK Woo, M Bawendi, V Bulovic, Electroluminescence from single mono-layers of nanocrystals in molecular organic devices, *Nature* 420: 800–803, 2002.
11. GT Shubeita, SK Sekatskii, G Dietler, I Potapova, A Mews, T Basche, Scanning near-field optical microscopy using semiconductor nanocrystals as a local fluorescence and fluorescence resonance energy transfer source, *J. Microsc. Oxford* 210: 274–278, 2003.
12. VI Klimov, MG Bawendi, Ultrafast carrier dynamics, optical amplification, and lasing in nanocrystal quantum dots, *MRS Bull.* 26: 998–1004, 2001.
13. M Kazes, DY Lewis, Y Ebstein, T Mokari, U Banin, Lasing from semiconductor quantum rods in a cylindrical microcavity, *Adv. Mater.* 14: 317, 2002.
14. AL Efros, M Rosen, The electronic structure of semiconductor nanocrystals, *Annu. Rev. Mater. Sci.* 30: 475–521, 2000.
15. F Koberling, A Mews, T Basche, Oxygen-induced blinking of single CdSe nanoc-rystals, *Adv. Mater.* 13: 672–676, 2001.

16. CF Landes, M Braun, MA El-Sayed, On the nanoparticle to molecular size transition: fluorescence quenching studies, *J. Phys. Chem. B* 105: 10554–10558, 2001.

17. WGJHM v Sark, PLTM Frederix, AA Bol, HC Gerritsen, A Meijerink, Blueing, bleaching, and blinking of single CdSe/ZnS quantum dots, *ChemPhysChem* 3: 871–879, 2002.

18. DF Underwood, T Kippeny, SJ Rosenthal, Ultrafast carrier dynamics in CdSe nanocrystals determined by femtosecond fluorescence upconversion spectroscopy, *J. Phys. Chem. B* 105: 436–443, 2001.

19. NA Hill, KB Whaley, A theoretical study of the influence of the surface on the electronic structure of CdSe nanoclusters, *J. Phys. Chem.* 100: 2831–2837, 1994.

20. O Schmelz, A Mews, T Basche, A Herrmann, K Mullen, Supramolecular complexes from CdSe nanocrystals and organic fluorophors, *Langmuir* 17: 2861–2865, 2001.

21. AR Kortan, R Hull, RL Opila, MG Bawendi, ML Steigerwald, RJ Caroll, LE Brus, Nucleation and growth of cadmium selendie on zinc sulfide quantum crystallite seeds, and vice versa, in inverse micelle media, *J. Am. Chem. Soc.* 112: 1327–1332, 1990.

22. A Mews, A Eychmüller, M Giersig, D Schooss, H Weller, Preparation, characterization, and photophysics of the quantum dot quantum well system CdS/HgS/CdS, *J. Phys. Chem.* 98: 934–941, 1994.

23. XG Peng, MC Schlamp, AV Kadavanich, AP Alivisatos, Epitaxial growth of highly luminescent CdSe/CdS core/shell nanocrystals with photostability and electronic accessibility, *J. Am. Chem. Soc.* 119: 7019–7029, 1997.

24. MA Hines, P Guyot-Sionnest, Synthesis and charaterization of strongly luminescing ZnS-capped CdSc nanocrystals, *J. Phys. Chem.* 100: 468–471, 1996.

25. BO Dabbousi, J Rodriguez Viejo, FV Mikulec, JR Heine, H Mattoussi, R Ober, KF Jensen, MG Bawendi, (CdSe)ZnS core-shell quantum dots: synthesis and characterization of a size series of highly luminescent nanocrystallites, *J. Phys. Chem. B* 101: 9463–9475, 1997.

26. DV Talapin, AL Rogach, I Mekis, S Haubold, A Kornowski, M Haase, H Weller, Synthesis and surface modification of amino-stabilized CdSe, CdTe and InP nanocrystals, *Coll. Surf A* 202: 145–154, 2002.

27. J Tittel, W Gohde, F Koberling, A Mews, A Kornowski, H Weller, A Eychmuller, T Basche, Investigations of the emission properties of single CdS-nanocrystallites, *Ber. Bunsenges. Phys. Chem.* 101: 1626–1630, 1997.

28. F Koberling, A Mews, T Basche, Single-dot spectroscopy of CdS nanocrystals and CdS/HgS heterostructures, *Phys. Rev. B* 60: 1921–1927, 1999.

29. F Koberling, A Mews, G Philipp, U Kolb, I Potapova, M Burghard, T Basche, Fluorescence spectroscopy and transmission electron microscopy of the same isolated semiconductor nanocrystals, *Appl. Phys. Lett.* 81: 1116–1118, 2002.

30. G Schlegel, J Bohnenberger, I Potapova, A Mews, Fluorescence decay time of single semiconductor nanocrystals, *Phys. Rev. Lett.* 88: article 137401, 2002.

31. S Empedocles, M Bawendi, Spectroscopy of single CdSe nanocrystallites, *Acc. Chem. Res.* 32: 389–396, 1999.

32. M Nirmal, BO Dabbousi, MG Bawendi, JJ Macklin, JK Trautman, TD Harris, LE Brus, Fluorescence intermittency in single cadmium selenide nanocrystals, *Nature* 383: 802–804, 1996.

33. J Tittel, W Göhde, F Koberling, T Basché, A Kornowski, H Weller, A Eychmüller, Fluorescence spectroscopy on single CdS nanocrystals, *J. Phys. Chem. B* 101: 3013–3016, 1997.

34. O Cherniavskaya, LW Chen, MA Islam, L Brus, Photoionization of individual CdSe/CdS core/shell nanocrystals on silicon with 2-nm oxide depends on surface band bending, *Nano Lett.* 3: 497–501, 2003.

35. WS Warren, AH Zewail, Multiple phase-coherent laser pulses in optical spectroscopy. I. The technique and experimental applications, *J. Phys. Chem.* 78: 2279–2297, 1983.
36. J Friedrich, D Haarer, Photochemical hole burning: a spectroscopic study of relaxation processes in polymers and glasses in the optical domain, *Angew. Chem. Int. Ed. Engl.* 23: 113, 1984.
37. WE Moerner, T Basche, Optical spectroscopy of single impurity molecules in solids, *Angew. Chem. Int. Ed. Engl.* 32: 457, 1993.
38. T Basché, WE Moerner, M Orrit, *Single-Molecule Detection, Imaging and Spectroscopy*, 1st ed., Verlag Chemie, Weinheim, Germany, 1997.
39. DM Mittleman, RW Schoenlein, JJ Shiang, VL Colvin, AP Alivisatos, CV Shank, Quantum size dependence of femtosecond electronic dephasing and vibrational dynamics in CdSe nanocrystals, *Phys. Rev. B* 49: 14435–14447, 1994.
40. A Mews, F Koberling, T Basche, G Philipp, GS Duesberg, S Roth, M Burghard, Raman imaging of single carbon nanotubes, *Adv. Mater.* 12: 1210–1214, 2000.
41. DJ Norris, A Sacra, CB Murray, MG Bawendi, Measurement of the size-dependent hole spectrum in CdSe quantum dots, *Phys. Rev. Lett.* 72: 2612–2615, 1994.
42. M Nirmal, CB Murray, MG Bawendi, Fluorescence-line narrowing in CdSe quantum dots: surface localization of the photogenerated exciton, *Phys. Rev. B* 50: 2293–2300, 1994.
43. A Mews, U Banin, AV Kadavanich, AP Alivisatos, Structure determination and homogeneous optical properties of CdS/HgS quantum dots, *Ber. Bunsenges. Phys. Chem.* 101: 1621–1625, 1997.
44. A Mews, A Eychmuller, Quantum wells within quantum dots, a CdS/HgS nanoheterostructure with global and local confinement, *Ber. Bunsenges. Phys. Chem.* 102: 1343–1357, 1998.
45. JK Trautman, JJ Macklin, LE Brus, E Betzig, Near-field spectroscopy of single molecules at room-temperature, *Nature* 369: 40–42, 1994.
46. XS Xie, RC Dunn, Probing single-molecule dynamics, *Science* 265: 361–364, 1994.
47. W Gohde, J Tittel, T Basche, C Brauchle, UC Fischer, H Fuchs, A low-temperature scanning confocal and near-field optical microscope, *Rev. Sci. Instrum.* 68: 2466–2474, 1997.
48. SA Empedocles, R Neuhauser, K Shimizu, MG Bawendi, Photoluminescence from single semiconductor nanostructures, *Adv. Mater.* 11: 1243–1256, 1999.
49. RG Neuhauser, KT Shimizu, WK Woo, SA Empedocles, MG Bawendi, Correlation between fluorescence intermittency and spectral diffusion in single semiconductor quantum dots, *Phys. Rev. Lett.* 85: 3301–3304, 2000.
50. SA Empedocles, MG Bawendi, Quantum-confined stark effect in single CdSe nanocrystallite quantum dots, *Science* 278: 2114–2117, 1997.
51. SA Empedocles, R Neuhauser, MG Bawendi, Three-dimensional orientation measurements of symmetric single chromophores using polarization microscopy, *Nature* 399: 126–130, 1999.
52. SA Blanton, A Dehestani, PC Lin, P Guyot-Sionnest, Photoluminescence of single semiconductor nanocrystallites by two-photon excitation spectroscopy, *Chem. Phys. Lett.* 229: 317–322, 1994.
53. F Koberling, U Kolb, G Philipp, I Potapova, T Basche, A Mews, Fluorescence anisotropy and crystal structure of individual semiconductor nanocrystals, *J. Phys. Chem. B*, 2003, 107(30): 7463–7471.
54. JB Pawley (Hrsg.), *Handbook of Biological Confocal Microscopy*, 2nd ed., Plenum Press, New York, 1995.

55. M Kuno, DP Fromm, HF Hamann, A Gallagher, DJ Nesbitt, Nonexponential "blinking" kinetics of single CdSe quantum dots: a universal power law behavior, *J. Chem. Phys.* 112: 3117–3120, 2000.

56. AL Efros, M Rosen, Random telegraph signal in the photoluminescence intensity of a single quantum dot, *Phys. Rev. Lett.* 78: 1110–1113, 1997.

57. M Kuno, DP Fromm, ST Johnson, A Gallagher, DJ Nesbitt, Modeling distributed kinetics in isolated semiconductor quantum dots, *Phys. Rev. B* 67: article 125304, 2003.

58. R Verberk, AM van Oijen, M Orrit, Simple model for the power-law blinking of single semiconductor nanocrystals, *Phys. Rev. B* 66: article 233202, 2002.

59. U Banin, M Bruchez, AP Alivisatos, T Ha, S Weiss, DS Chemla, Evidence for a thermal contribution to emission intermittency in single CdSe/CdS core/shell nanocrystals, *J. Chem. Phys.* 110: 1195–1201, 1999.

60. F Koberling, A Mews, I Potapova, T Basché, Probing the interaction of single CdSe quantum dots with their local environment, *Proc. SPIE* 4456: 31, 2001.

61. M Kuno, DP Fromm, HF Hamann, A Gallagher, DJ Nesbitt, "On"/"off" fluorescence intermittency of single semiconductor quantum dots, *J. Chem. Phys.* 115: 1028–1040, 2001.

62. KT Shimizu, RG Neuhauser, CA Leatherdale, SA Empedocles, WK Woo, MG Bawendi, Blinking statistics in single semiconductor nanocrystal quantum dots, *Phys. Rev. B* 6320: article 205316, 2001.

63. GA Somorjai, Charge transfer controlled surface interactions between oxygen and CdSe films, *J. Phys. Chem. Solids* 24: 175–186, 1963.

64. RH Bube, The basic significance of oxygen chemisorption on the photoelectric properties of CdS and CdSe, *J. Electrochem. Soc.* 113: 793–798, 1966.

65. A Hässelbarth, A Eychmüller, H Weller, Detection of shallow electron traps in quantum sized CdS by fluorescence quenching experiments, *Chem. Phys. Lett.* 203: 271–276, 1993.

66. MC Ko, GJ Meyer, Dynamic quenching of porous silicon excited states, *Chem. Mater.* 8: 2686–2692, 1996.

67. WR Ware, *Photochemistry in Organized and Constrained Media*, VCH Publishers, New York, 1991.

68. CP Lindsey, GD Patterson, Detailed comparison of the Williams–Watts and Cole–Davidson functions, *J. Phys. Chem.* 73: 3348–3357, 1980.

69. SA Crooker, T Barrick, JA Hollingsworth, VI Klimov, Multiple temperature regimes of radiative decay in CdSe nanocrystal quantum dots: intrinsic limits to the dark-exciton lifetime, *Appl. Phys. Lett.* 82: 2793–2795, 2003.

4 Coherent Excitation of Vibrational Modes of Gold Nanorods

Gregory V. Hartland, Min Hu, and Paul Mulvaney

CONTENTS

4.1 Introduction 125
4.2 Experimental 127
4.3 Results and Discussion 128
    4.3.1 Laser-Induced Heating and Coherent Excitation
    of Vibrational Modes 128
    4.3.2 Gold Nanorods: Vibrational Period vs. Probe Laser
    Wavelength and Rod Length.. 130
    4.3.3 Hole-Burning Experiments in Nanorods 133
4.4 Summary and Conclusions 135
Acknowledgments.. 136
References.. 136

4.1 INTRODUCTION

The optical properties of gold particles have played an integral role in the development of physical chemistry and nanotechnology since Faraday's experiments in 1857.[1] A theory to explain the color of colloidal gold solutions was derived by Mie in 1908.[2] Mie's theory allows the extinction spectra of spherical particles to be calculated from the dielectric constants of the metal and the surrounding medium and forms the basis for our understanding of light scattering by small particles (the Tyndall effect). These calculations show that the distinctive colors of metal particles arise from a collective dipolar oscillation of the conduction electrons, which is called the surface plasmon band.[3] The position of the plasmon band depends on the identity of the metal, the shape of the particles, and their surroundings and is also sensitive to the distance between particles. Specifically, the characteristic purple color of flocculated Au sols arises from dipolar coupling between the plasmon oscillations of neighboring particles.[3]

For Au, Ag, and Cu, the plasmon band falls in the visible region of the spectrum and is responsible for the brilliant colors of solutions of these particles. For most other metals, the plasmon band occurs in the UV, yielding solutions with a drab gray or brown color.[3] Mie's theory is so successful that deviations from it are automatically assigned to changes in the dielectric constant of the material. For example, the broadening of the plasmon band for small metal particles is attributed to electronic scattering at the particle surface, which becomes significant when the particle diameter is less than the mean free path of the electrons in the metal.[4-6]

A topic of current interest in this field is investigating the properties — both material and optical — of nonspherical particles. Synthetically, it is difficult to control the shape of metal or semiconductor nanoparticles, although considerable progress has been made in recent years. For example, cubic Pt nanoparticles can be synthesized by H_2 reduction of $PtCl_6$,[2-7] triangular Ag particles have been produced by aging a citrate-stabilized Ag sol in light,[8] and elliptical metal particles can be made by deformation of spherical particles embedded in a glass.[9] However, the most dramatic advances have been made in the synthesis of rod-shaped particles. Metallic rods of different widths and almost any length can be made by electrochemical deposition in a template, such as the pores of a membrane.[10,11] This method has the advantage that a variety of different metals can be deposited; however, significant postsynthesis processing has to be performed to remove the rods from the template. A more straightforward technique is to use a micelle solution to direct the growth of the rods in combination with either chemical or electrochemical reduction of a metal salt.[12-15] Using this technique, Au or Ag rods can be made with controllable lengths up to 50 μm.[14,15]

Nanorods are of particular interest because of their unique optical properties. Specifically, rod-shaped particles show two plasmon bands corresponding to electron oscillations along the long or the short axis of the rod. These bands are termed the longitudinal and transverse plasmon bands, respectively.[3,12,13] For Au, the transverse plasmon band occurs at ~520 nm (which is close to the plasmon resonance for spherical Au particles), and the longitudinal band falls in the visible to near-infrared (IR) spectral region. Nanorods also show dramatically increased fluorescence quantum yields compared to spherical particles, with enhancement factors of more than a million times compared to bulk metal.[16] This makes them attractive candidates for studies using ultrafast fluorescence detection techniques.[17]

Our contribution to nanorod research has been to investigate nanorods' phonon modes by time-resolved spectroscopy.[18] In the past several years a number of groups have shown that the quantized acoustic vibrational modes of nanoparticles can be observed in ultrafast experiments.[19-27] These studies have been performed on spherical metal and semiconductor particles (both in solution and with solid supports),[19-25] core-shell bimetallic particles,[26] and Ag ellipses in a glass matrix.[27] In these experiments, an ultrafast laser pulse excites the system and induces rapid expansion. This impulsively excites the phonon mode that correlates with the expansion coordinate of the particles. For spherical particles, the measured vibrational frequencies are in excellent agreement with continuum mechanics calculations for the symmetric breathing mode.[23,24] However, for nonspherical particles, the situation is much more complex. In particular, it is not clear which vibrational modes will be excited and how the frequencies of these modes depend on the size, shape, and elastic properties of the material.

In a recent study of aligned ellipsoidal Ag particles in a glass matrix, Perner and coworkers[27] observed a beat signal that had a period of either 52 or 22 psec, depending on whether the probe laser was resonant to the longitudinal or transverse plasmon band of the particles. They attributed this signal to expansion and contraction along either the major or minor axis of the ellipse. The observed periods are approximately given by $2d/c_l$, where c_l is the longitudinal speed of sound and d is either the length or width of the ellipse. In this paper results from our recent ultrafast laser studies with Au nanorods are discussed.[18] The rods were grown by an electrochemical method and had average aspect ratios between 2.1 and 5.5 (which correspond to lengths between 20 and 70 nm). Modulations with a period on the order of 50 to 70 ps can be clearly seen in our transient absorption data. This period is much longer than that expected from Perner et al.'s [27] results. This difference could arise because (1) the vibrational mode excited in the nanorods is fundamentally different from the mode excited in the ellipsoidal particles or (2) the elastic properties of the Au nanorods cannot be estimated from values for bulk (polycrystalline) Au, due to their inherent crystallinity or small size. The exact value of the period also depends on the probe wavelength; this occurs because different wavelengths interrogate different-length rods in the sample.

4.2 EXPERIMENTAL

The experiments described in this paper were performed with a regeneratively amplified Ti:Sapphire laser (Clark-MXR, CPA-1000) pumped by a Spectra Physics Millenia Vs solid-state diode-pumped Nd:YAG laser. The output of the regenerative amplifier ($\lambda = 800$ nm, 0.5 mJ/pulse, 120-fsec full-width-at-half-maximum (fwhm) sech^2 deconvolution, 1-kHz repetition rate) is split by a 90:10 beam splitter. The 90% portion either is used directly to excite the sample or is frequency doubled in a 1-mm β-barium borate (BBO) crystal to generate 400-nm pump pulses. The intensity of the pump laser pulses is controlled with a $\lambda/2$-waveplate/polarizer combination. The 10% portion is used to generate a white-light continuum in a 3-mm sapphire window, which is used as the probe. The pump and probe are overlapped at the sample by focusing with a 10-cm lens, and the polarizations are typically set to parallel. Specific probe wavelengths are selected using a Jobin-Yvon Spex H-10 monochromator (10-nm spectral resolution) placed after the sample. Fluctuations in the probe laser intensity are normalized by splitting off and monitoring a small portion of the probe beam before the sample. The timing between the pump and probe pulses was controlled by a stepper motor-driven translation stage (Newport UTM150PP.1). The pump laser intensity at the sample was measured by a Molectron J3-02 energy detector, and the pump laser spot size was (5 to 10) $\times$ 10^{-4} cm^2 for the experiments described below. Typical pump laser powers used in these experiments were 0.5 to 2 μJ/pulse. These powers produce temperature increases on the order of 50 to 200 K. Low powers are needed to avoid laser-induced transformation of the rods into spheres.[28,29]

The gold nanorods were synthesized using the electrochemical technique first described by Yu and coworkers[12] and Link.[13] The nanorod samples used in our experiments had average aspect ratios between 2.1 and 5.5 and widths between 12 and 28 nm. The size distributions for the different samples were examined by transmission electron microscopy (TEM).[18] The specific aspect ratios (ξ) and widths (*w*) for the samples used were ξ = 2.1 ± 0.3, *w* = 26 ± 2 nm; ξ = 2.7 ± 0.2, *w* = 13 ± 1 nm; ξ = 3.0 ± 0.3, *w* = 12 ± 1 nm; ξ = 3.5 ± 0.4, *w* = 12 ± 1 nm; and ξ = 5.5 ± 1.6, *w* = 14 ± 1 nm. Detailed TEM analysis performed by others has shown that the rods produced by this technique are single crystals and that they grow along the [001] direction.[30] Despite extensive purification, these samples are polydisperse; i.e., each sample contains a wide distribution of different-length rods.

The UV-visible absorption spectra of the rod samples show the characteristic longitudinal and transverse plasmon resonances that correspond to oscillation of the conduction electrons along the length or width of the rod.[12,13] The transverse plasmon band appeared at ca. 520 nm for all of the samples, whereas the longitudinal resonance appeared between 600 and 840 nm, depending on the aspect ratio. It is important to note that the near-UV absorbance does not strongly depend on aspect ratio. Thus, in the 400-nm pump experiments all of the rods in the sample were equally excited by the pump laser pulses. In contrast, for the 800-nm pump experiments, only rods with an aspect ratio of ~4 were expected to be excited (*vida infra*). The samples were held in 2-mm cuvettes for the transient absorption experiments, and the data were collected without flowing or stirring. However, the relatively low pump laser powers and repetition rates used in these experiments mean that thermal effects that arise from heat accumulation in the solvent are not a problem.

4.3 RESULTS AND DISCUSSION

4.3.1 LASER-INDUCED HEATING AND COHERENT EXCITATION OF VIBRATIONAL MODES

In our experiments the ultrafast pump laser pulse primarily excites electrons through the $5d \rightarrow 6sp$ interband transition. The excited electrons subsequently rapidly redistribute their energy over the entire electron distribution, causing an increase in the electronic temperature.[31,32] The initial temperature of the electrons can be very large (thousands of Kelvin), because metal particles can (and do) absorb many photons (their absorption transitions do not bleach in the same way that molecules do) and electrons have a very small heat capacity.[33,34] The hot electron distribution then equilibrates with the phonon modes via electron–phonon (*e-ph*) coupling. Figure 4.1 shows transient absorption data for ca. 15-nm-diameter spherical gold particles in aqueous solution. In these experiments the probe laser was tuned to the maximum of the plasmon band, which essentially monitors the temperature of the sample.[31,32] The data show a rapid decay due to the *e-ph* coupling process and a slower decay that corresponds to heat dissipation to the environment. These results show that the heat deposited by the laser remains in the particles for hundreds of picoseconds.[35–38]

The heat deposited into the lattice causes the particles to expand. For Au particles larger than ca. 6 nm, the timescale for lattice heating is faster than the period of the phonon mode that correlates with the expansion coordinate. This means that this mode can be impulsively excited.[23,24] For spherical particles, the vibrational motion excited is the symmetric breathing mode.[39–41] The coherently excited vibrational

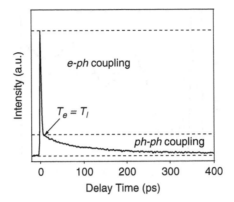

FIGURE 4.1 Transient bleach data for 15-nm-diameter Au particles recorded with 530-nm probe pulses and a pump laser power of 0.2 µJ/pulse. The different regions of the decay (coupling between the electrons and phonons within the particles, and coupling between the particles and their environment) are labeled in the figure.

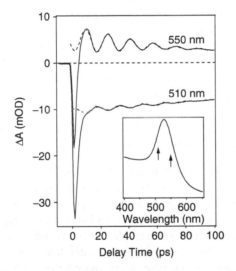

FIGURE 4.2 Transient absorption data for ~50-nm-diameter Au particles recorded with probe wavelengths of 510 and 550 nm. The dashed lines show fits to the modulated portion of the data using a damped cosine function. The frequency and damping time of the modulations are identical in the two traces, but the phases differ by ca. 180°. The insert shows the UV-visible extinction spectra of the particles, with the arrows indicating the two probe laser wavelengths.

motion shows up as a modulation in the transient absorption traces because it changes the electron density and, therefore, the position of the plasmon resonance of the particles.[22–24] Experimental data for ca. 50-nm-diameter spherical gold particles is presented in Figure 4.2. The two traces correspond to probe laser wavelengths on the red and blue sides of the plasmon band (see the insert in Figure 4.2). Note that the modulations have the same frequency, but they are ~180° out of phase. The

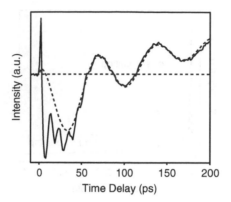

FIGURE 4.3 Transient absorption data for Au nanorods with an average aspect ratio of 2.1 ± 0.3. The pump and probe wavelengths were 400 and 640 nm, respectively. The dashed line shows a fit to the slow modulation in the data.

phase difference proves that the signal originates from a periodic change in the position of the plasmon band. The observed period depends on the size and shape of the particles and their elastic properties. For spherical particles, the frequencies can be exactly predicted by continuum mechanics calculations, which are explicitly derived for elastically isotropic (i.e., polycrystalline) particles.[23,24,39–41] An interesting question, and the main focus of this article, is to consider what happens for particles that have different shapes or are single crystals.

4.3.2 GOLD NANORODS: VIBRATIONAL PERIOD VS. PROBE LASER WAVELENGTH AND ROD LENGTH

Figure 4.3 shows representative transient absorption data for a gold nanorod sample with an aspect ratio of 2.1 ± 0.3. The transient absorption experiments were performed with 400-nm pump laser pulses and a probe laser wavelength of 640 nm, which is resonant to the longitudinal plasmon band of the sample. Note that the rod samples also contain spherical particles; however, the vibrational modes of spheres cannot be observed at probe wavelengths greater than ~600 nm.[42] Two types of modulations can be seen in our data: a fast modulation with a period of ca. 11 ps and a slower (more pronounced) modulation with a period of ca. 70 ps. The period of the faster modulation is approximately equal to the value expected for the breathing vibrational mode of the rod.[48] On the other hand, the period of the slower modulation is close to the value expected for expansion and contraction along the long axis of the rod — the extensional mode. Neither period matches the value expected for the spherical particles in the sample.

We have performed extensive measurements on the rod samples produced electrochemically.[12] Samples with different aspect ratios have been examined and we have observed that for a given sample, the exact value of the period depends on the probe wavelength. On the other hand, for spherical particles, the measured periods are independent of wavelength to within the accuracy of our experiments.[42] An example of the probe wavelength dependence is shown in Figure 4.4, where results for an

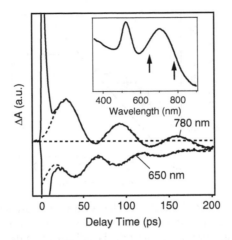

FIGURE 4.4 Transient absorption data recorded with 400-nm pump pulses and probe wavelengths of 650 and 780 nm for a nanorod sample with an average aspect ratio of 3.5. The dashed lines are fits to the data using a damped cosine function.

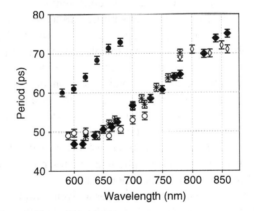

FIGURE 4.5 Period vs. probe wavelength for Au nanorod samples with aspect ratios of 2.1 (•), 2.7 (o), 3.0 (*), 3.5 (♦), and 5.5 (◊).

Au nanorod sample with an average aspect ratio of 3.5 are presented.[18] The two probe wavelengths used in these experiments lie on either side of the longitudinal plasmon band of the sample. There are several points to note. First, both the frequency and the phase are different for these two probe wavelengths. Second, the fast modulations are not present; i.e., only the slower vibrational motion is observed. Indeed, we have seen only the fast modulation, which corresponds to the breathing mode in a few experiments.

The way the period of the vibrational motion changes with probe wavelength is shown in Figure 4.5. This figure presents data collected from five different samples with a variety of probe laser wavelengths. The samples were all produced using the electrochemical method,[12] and are fairly polydisperse. The average aspect ratios for the samples are given in the figure legend. The increase in the period with increasing

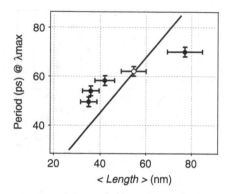

FIGURE 4.6 Period vs. average length for the Au nanorods examined in this work. The straight line is a fit to the data assuming that the period is linearly proportional to the length of the rod. The horizontal error bars reflect the polydispersity in the samples.

probe laser wavelength can be qualitatively explained as follows. The nanorod samples are polydisperse; i.e., each sample contains rods with a variety of different lengths. The different-length rods have different longitudinal plasmon resonances, with longer rods (larger aspect ratio) absorbing further into the near IR. For Au nanorods in aqueous solution, $\lambda_{max} \approx 86 \times \xi + 419$, where the aspect ratio $\xi = L/w$.[12,13,18] As the probe laser is tuned from the blue to the red, it probes (on average) longer rods in the sample. Because the vibrational period is proportional to the length of the rod, longer vibrational periods are obtained for longer probe laser wavelengths.

The period vs. probe wavelength results appear to fall onto two curves: the rods with an average aspect ratio of $\xi = 2.1$ are on one curve, and the rods with aspect ratios of $\xi = 2.7$, 3.0, 3.5, and 5.5 are on another. This occurs because the rods in the sample with $\xi = 2.1$ are approximately two times wider than the rods in the other samples. A given probe wavelength interrogates rods with a certain aspect ratio, but the actual length of the rods depends on their width.

Note that the period vs. wavelength data for a given sample is not a straight line — the period "flattens off" at high and low probe wavelengths. This occurs because of the finite size distributions of the samples. Simulations of the transient absorption experiments show that at probe laser wavelengths near the plasmon band maximum, the effect of the sample polydispersity is minimized; i.e., the measured periods are close to the values expected from the average length of the rods (this is a fairly obvious conclusion).[18] Thus, to compare different samples, the probe laser should be tuned as closely to the maximum of the longitudinal plasmon band as possible. Figure 4.6 shows the period vs. the average length for the different rod samples, where the period was determined from experiments with the probe laser wavelength $\lambda_p \approx \lambda_{max}$. The open symbol represents the average period for the 2.1 ± 0.3 aspect ratio sample, i.e., the sample that corresponds to the displaced curve in Figure 4.5. Clearly, accounting for the different width of the rods in this sample brings these results into agreement with the other samples.

The straight line in Figure 4.6 is a fit to the data, assuming that the period is linearly proportional to the length of the rod, i.e., $\tau = 2L/\alpha$. The value of α obtained is 1800 msec^{-1}. This is close to the speed expected for the extensional mode of $C = \sqrt{E/\rho}$, where E is Young's modulus and ρ is the density of the rod.[48] Using published values of E and ρ, we calculate $C = 2000$ msec^{-1}. This is essentially the same as the number listed in the *CRC Handbook* for the velocity of an extensional wave in a thin gold rod (2030 msec^{-1}).[44]

At this point, it is not clear why our Au nanorod samples give a different period from the ellipsoidal Ag particles.[27] Ther are two possibilities. First, the vibrational modes excited in the Au nanorods are fundamentally different from those for the Ag ellipses (i.e., we are exciting a motion that is much more complicated than simple stretching and compression along a single axis). Second, using the material properties for bulk (polycrystalline) gold is not appropriate for these small crystalline nanorods. In comparison, the Ag ellipses of reference 27 are polycrystalline and have a period close to the expected value of $2L/c_l$. However, the gold nanorods in these experiments grow with a specific structure.[48] The elastic properties of the single crystal nanorods may be different from those of polycrystalline gold. However, differences in the longitudinal speed of sound for propagation along different crystal directions amount to <20%.[45] Thus, the fact that the rods are not polycrystalline does not seem to be a likely explanation for the unusual frequencies.

4.3.3 HOLE-BURNING EXPERIMENTS IN NANORODS

In the experiments discussed above, the 400-nm pump laser excites all the particles in the sample. However, the probe laser only interrogates nanorods with a certain aspect ratio. Thus, different-length rods contribute in different ways for a given experiment, leading to the change in the measured vibrational period with probe wavelength. The exact way the vibrational period changes with wavelength depends on the bandwidth of the probe laser, the size distribution of the sample, and the transient absorption spectrum of a single rod.[18] Thus, the period vs. wavelength data potentially provide information about the contributions from homogeneous and inhomogeneous broadening to the spectra of nanorods.[46]

Our initial report on coherently excited vibrational modes in nanorods presented an analysis of the period vs. probe wavelength data using Gaussian functions for the distribution of lengths, and the spectra for the rods and the laser. Note that the size distribution of the sample can be determined by TEM or by modeling the steady-state UV-visible absorption spectra of the rods, and the bandwidth of the probe laser is determined by the spectral response of our detection system. Thus, in principle, the only unknown is the homogeneous line width of the longitudinal plasmon band for the rods. However, it is difficult to unambiguously fit the period vs. probe wavelength data to obtain this information, as the homogeneous and inhomogeneous (size distribution) contributions are strongly correlated in this analysis.

Another way of investigating the vibrational modes and homogeneous line width in the nanorod samples is to perform experiments with the pump laser tuned to the longitudinal plasmon band.[28,29,47] In this case, the pump laser should excite only a subpopulation of the rods, and the analysis of the results will be more straightforward. Transient absorption data recorded with 800-nm pump laser pulses and probe laser wavelengths of 700 and 740 nm are presented in Figure 4.7 for the

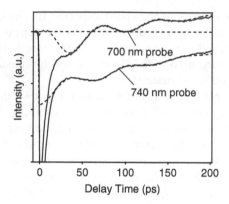

FIGURE 4.7 Transient absorption data recorded with 800-nm pump pulses and probe wavelengths of 700 and 740 nm for a nanorod sample with an average aspect ratio of 3.5. The dashed lines are fits to the data using a damped cosine function. Note that the modulations are approximately 180° out of phase.

sample with an average aspect ratio of 3.5. These two probe laser wavelengths lie on either side of the longitudinal plasmon band of this sample. The two traces have similar periods (as expected), and they are ~180° out of phase with each other.

Our previous work with spherical particles showed that the phase of the modulations differs by 180° for experiments performed on the red or blue sides of the plasmon band (see Figure 4.2). This is because the signal arises from a periodic change in the plasmon band position. The signal in the nanorod experiments is expected to arise from a similar effect; i.e., we expect a phase difference for experiments performed on the red or blue sides of the plasmon band. In Figure 4.8 the period and phase extracted from the transient absorption traces are plotted vs. probe wavelength for the 800- and 400-nm pump laser excitation experiments for the 3.5 aspect ratio sample. It is important to note two points from this data. First, the change in period with wavelength is much more dramatic for the 400-nm pump laser experiments than for the near-IR pump experiments. This is expected, as fewer rods contribute to the signal in the 800-nm experiments. Second, the modulations due to the coherently excited vibrational motion change phase at 720 nm (i.e., at the longitudinal plasmon band maximum) for *both* the 400- and 800-nm pump experiments. This result is not as easy to understand.

In the 400-nm pump experiments we expect the change in phase at the maximum of the plasmon band for the sample, since all of the nanorods are excited at this wavelength. However, in the near-IR pump experiments, the pump laser should select only rods that have their longitudinal plasmon resonance at 800 nm, i.e., rods with an aspect ratio of ~4. Thus, the change in phase for the modulations should occur at 800 nm rather than at 720 nm for the near-IR pump experiments.

A qualitative explanation for the results in Figure 4.8 is that the homogeneous line width for the longitudinal plasmon resonance must be comparable to the overall line width for the sample. In this case, nanorods that have aspect ratios significantly different from ~4 will be excited by the 800-nm pump laser pulses. Thus, the near-IR

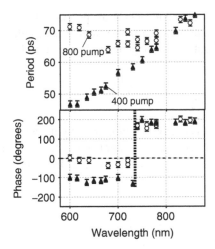

FIGURE 4.8 Period and phase vs. probe laser wavelength for the nanorod sample with an average aspect ratio of 3.5. The circles represent experiments performed with 800-nm pump laser pulses, and the triangles represent experiments with 400-nm pump pulses. The vertical line in the lower panel indicates the maximum of the longitudinal plasmon band for the sample.

pump experiments still average over all the rods in the sample, albeit with a different weighting than the 400-nm pump experiments. We are currently analyzing these results to obtain quantitative information about the homogeneous line width for the longitudinal plasmon band of gold nanorods.[48] The results from this analysis will be compared with the line widths observed in Rayleigh scattering spectra from single gold nanorods.[49] The results from these two measurements may be different because our measurements probe absorption, whereas the single-rod experiments probe scattering. These two contributions are not necessarily the same.[50]

4.4 SUMMARY AND CONCLUSIONS

Excitation of metal nanoparticles with an ultrafast laser pulse causes a rapid increase in the electronic temperature. The hot electrons subsequently equilibrate with the phonon modes on a several-picosecond timescale. These rapid heating processes can impulsively excite the phonon mode that correlates with the expansion coordinate. For spherical particles, the symmetric breathing mode is excited, and the frequency of this mode can be exactly calculated using continuum mechanics.[23,24] The situation is not as straightforward for nonspherical particles. Specifically for rods, both the breathing and extensional vibrational modes are excited.

For the Au nanorods examined in our recent experiments, we found that the period of the coherently excited phonon mode depends on the length of the rod, but the values obtained do not match those previously observed for Ag ellipses.[27] It is not clear whether this difference is due to differences in the symmetry of the particles (cylindrical compared to ellipsoidal) or their material properties (the gold nanorods in our experiments grow in a specific way, whereas the Ag particles in reference 27 are polycrystalline). Measurements of how the period and (more importantly) the phase of the modulations change with the probe laser wavelength for a given sample also provide information about the homogeneous line width of the longitudinal plasmon band.

ACKNOWLEDGMENTS

The work described in this paper was supported by the National Science Foundation by grant CHE98-16164. The authors thank Prof. John Sader for his contributions during the course of these experiments and Mr. Bill Archer for his help with the TEM measurements.

REFERENCES

1. M Faraday. Experimental relations of gold (and other metals) to light. *Philos Trans R Soc London* 147: 145–181, 1857.
2. G Mie. Beiträge zur optik trüber medien, speziell kolloidaler metallösungen. *Ann Physik* 25: 377–445, 1908.
3. U Kreibig, M Vollmer. *Optical Properties of Metal Clusters*. Berlin: Springer, 1995.
4. RH Doremus. Optical properties of small silver particles. *J Chem Phys* 42: 414–417, 1965.
5. WA Kraus, GC Schatz. Plasmon resonance broadening in small metal particles. *J Chem Phys* 79: 6130–6139, 1983.
6. H Hovel, S Fritz, A Hilger, U Kreibig, M Vollmer. Width of cluster plasmon resonances: bulk dielectric functions and chemical interface damping. *Phys Rev B* 48: 18178–18188, 1993.
7. TS Ahmadi, ZL Wang, TC Green, A Henglein, MA El-Sayed. Shape-controlled synthesis of colloidal Pt nanoparticles. *Science* 272: 1924–1926, 1996.
8. RC Jin, YW Cao, CA Mirkin, KL Kelly, GC Schatz, JG Zheng. Photoinduced conversion of silver nanospheres to nanoprisms. *Science* 294: 1901–1903, 2001.
9. R Borek, KJ Berg, G Berg. Low-temperature tensile deformation of flat glass containing metal particles to generate dichroism. *Glastech Ber-Glass Sci Technol* 71: 352–359, 1998.
10. BR Martin, DJ Dermody, BD Reiss, MM Fang, LA Lyon, MJ Natan, TE Mallouk. Orthogonal self-assembly on colloidal gold-platinum nanorods. *Adv Mater* 11: 1021–1025, 1999.
11. SR Nicewarner-Pena, RG Freeman, BD Reiss, L He, DJ Pena, ID Walton, R Cromer, CD Keating, MJ Natan. Submicrometer metallic barcodes. *Science* 294: 137–141, 2001.
12. YY Yu, SS Chang, CL Lee, CRC Wang. Gold nanorods: electrochemical synthesis and optical properties. *J Phys Chem B* 101: 6661–6664, 1997.
13. S Link, MB Mohamed, MA El-Sayed. Simulation of the optical absorption spectra of gold nanorods as a function of their aspect ratio and the effect of the medium dielectric constant. *J Phys Chem B* 103: 3073–3077, 1999.
14. NR Jana, L Gearheart, CJ Murphy. Wet chemical synthesis of high aspect ratio cylindrical gold nanorods. *J Phys Chem B* 105: 4065–4067, 2001.
15. YG Sun, YN Xia. Large-scale synthesis of uniform silver nanowires through a soft, self-seeding, polyol process. *Adv Mater* 14: 833–837, 2002.

16. MB Mohamed, V Volkov, S Link, MA El-Sayed. The "lightning" gold nanorods: fluorescence enhancement of over a million compared to the gold metal. *Chem Phys Lett* 317: 517–523, 2000.
17. O Varnavski, RG Ispasoiu, L Balogh, D Tomalia, T Goodson. Ultrafast time-resolved photoluminescence from novel metal-dendrimer nanocomposites. *J Chem Phys* 114: 1962–1965, 2001.
18. GV Hartland, M Hu, O Wilson, P Mulvaney, JE Sader. Coherent excitation of vibrational modes in gold nanorods. *J Phys Chem B* 106: 743–747, 2002.
19. M Nisoli, S De Silvestri, A Cavalleri, AM Malvezzi, A Stella, G Lanzani, P Cheyssac, R Kofman. Coherent acoustic oscillations in metallic nanoparticles generated with femtosecond optical pulses. *Phys Rev B* 55: 13424–13427, 1997.
20. TD Krauss, FW Wise. Coherent acoustic phonons in a semiconductor quantum dot. *Phys Rev Lett* 79: 5102–5105, 1997.
21. ER Thoen, G Steinmeyer, P Langlois, EP Ippen, GE Tudury, CHB Cruz, LC Barbosa, CL Cesar. Coherent acoustic phonons in PbTe quantum dots. *Appl Phys Lett* 73: 2149–2151, 1998.
22. JH Hodak, IB Martini, GV Hartland. Observation of acoustic quantum beats in nanometer sized Au particles. *J Chem Phys* 108: 9210–9213, 1998.
23. N Del Fatti, C Voisin, F Chevy, F Vallee, C Flytzanis. Coherent acoustic mode oscillation and damping in silver nanoparticles. *J Chem Phys* 110: 11484–11487, 1999.
24. JH Hodak, A Henglein, GV Hartland. Size dependent properties of Au particles: coherent excitation and dephasing of acoustic vibrational modes. *J Chem Phys* 111: 8613–8621, 1999.
25. G Cerullo, S De Silvestri, U Banin. Size dependent dynamics of coherent acoustic phonons in nanocrystal quantum dots. *Phys Rev B* 60: 1928–1932, 1999.
26. JH Hodak, A Henglein, GV Hartland. Coherent excitation of acoustic breathing modes in bimetallic core-shell nanoparticles. *J Phys Chem B* 104: 5053–5055, 2000.
27. M Perner, S Gresillon, J Marz, G Von Plessen, J Feldmann, J Porstendorfer, KJ Berg, G Berg. Observation of hot-electron pressure in the vibration dynamics of metal nanoparticles. *Phys Rev Lett* 85: 792–795, 2000.
28. SS Chang, CW Shih, CD Chen, WC Lai, CRC Wang. The shape transition of gold nanorods. *Langmuir* 15: 701–709, 1999.
29. S Link, C Burda, B Nikoobakht, MA El-Sayed. Laser-induced shape changes of colloidal gold nanorods using femtosecond and nanosecond laser pulses. *J Phys Chem B* 104: 6152–6163. 2000.
30. ZL Wang, MB Mohamed, S Link, MA El-Sayed. Crystallographic facets and shapes of gold nanorods of different aspect ratios. *Surf Sci* 440: L809–L814, 1999.
31. JH Hodak, IB Martini, GV Hartland. Spectroscopy and dynamics of nanometer-sized noble metal particles. *J Phys Chem B* 102: 6958–6967, 1998.
32. N Del Fatti, F Vallee. Ultrafast optical nonlinear properties of metal nanoparticles. *Appl Phys B* 73: 383–390, 2001.
33. G Tas, HJ Maris. Electron-diffusion in metals studied by picosecond ultrasonics. *Phys Rev B* 49: 15046–15054, 1994.
34. NW Ashcroft, ND Mermin. *Solid State Physics*. Orlando: Harcourt Brace, 1976.
35. J Hodak, IB Martini, GV Hartland. Ultrafast study of electron-phonon coupling in colloidal gold particles. *Chem Phys Lett* 284: 135–141, 1998.
36. M Hu, GV Hartland. Heat dissipation for Au particles in aqueous solution: relaxation time versus size. *J Phys Chem B* 106: 7029–7033, 2002.
37. TW Roberti, BA Smith, JZ Zhang. Ultrafast electron dynamics at the liquid–metal interface: femtosecond studies using surface-plasmons in aqueous silver colloid. *J Chem Phys* 102: 3860–3866, 1995.

38. TS Ahmadi, SL Logunov, MA El-Sayed. Picosecond dynamics of colloidal gold nanoparticles. *J Phys Chem* 100: 8053–8056, 1996.
39. H Lamb. On the vibrations of an elastic sphere. *Proc London Math Soc* 13: 189–212, 1882.
40. VA Dubrovskiy, VS Morochnik. Natural vibrations of a spherical inhomogeneity in an elastic medium. *Izv Earth Phys* 17: 494–504, 1981.
41. KE Bullen, BA Bolt. *An Introduction to Seismology.* 4th ed. Cambridge, U.K.: Cambridge University Press, 1985.
42. GV Hartland. Coherent vibrational motion in metal particles: determination of the vibrational amplitude and excitation mechanism. *J Chem Phys* 116: 8048–8055, 2002.
43. G Simmons, H Wang. *Single Crystal Elastic Constants and Calculated Aggregate Properties: A Handbook.* Cambridge, MA: The MIT Press, 1971.
44. *CRC Handbook of Chemistry and Physics.* 80th ed. Boca Raton, FL: CRC Press, 1999.
45. HF Pollard. *Sound Waves in Solids.* London: Pion, 1977.
46. S Volker. Hole-burning spectroscopy. *Ann Rev Phys Chem* 40: 499–530, 1989.
47. S Link, MA El-Sayed. Spectroscopic determination of the melting energy of a gold nanorod. *J Chem Phys* 114: 2362–2368, 2001.
48. M Hu, X Wang, GV Hartland, J Perez-Juste, P Mulvaney, JE Sadev. Vibrational response of nanorods to ultrafast laser-induced heating: theoretical and experimental analysis. *J. Am. Chem. Soc.* 125: 14925–14933, 2003.
49. C Sonnichsen, T Franzl, T Wilk, G von Plessen, J Feldmann, O Wilson, P Mulvaney. Drastic reduction of plasmon damping in gold nanorods. *Phys Rev Lett* 88: article 077402, 2002.
50. HC van de Hulst. *Light Scattering by Small Particles.* New York: Dover Publications, 1981.

5 Fabricating Nanophase Erbium-Doped Silicon into Dots, Wires, and Extended Architectures

Jeffery L. Coffer

CONTENTS

5.1 Introduction 139
5.2 Nanocrystal Fabrication 141
5.3 Structural Characterization of Erbium-Doped Silicon Dots 141
5.4 Photophysical Measurements of Erbium-Doped Silicon Dots 145
5.5 Erbium-Doped Silicon Nanowires 145
5.6 Photophysical Measurements of Erbium-Doped Silicon Wires 149
5.7 Conclusions 150
Acknowledgments 150
References 151

5.1 INTRODUCTION

Light emission from nanophase silicon is being seriously investigated for practical reasons such as amplification[1] (and possible lasing), as well as fundamental phenomena such as a proposed possible origin for extended red emission detected from the galaxy.[2] Three distinct strategies are currently prevalent for achieving light emission from this formally indirect band gap material: quantum confinement,[3] band gap engineering,[4] and rare earth doping.[5] The first category, common to materials such as porous silicon,[6] demonstrates an interesting type of quantum size effect whereby the fundamental indirect band gap remains intact but visible luminescence can be detected as a consequence of kinetic suppression of nonradiative Auger type processes in electron–hole recombination.[7] The second approach, band gap engineering,[4] entails the use of nanometer-sized compositional alloys of Si sandwiched with materials such as germanium (Ge); its properties are strongly influenced by interfacial strain between heterogeneous phases in the layered film.

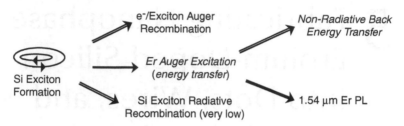

FIGURE 5.1 Possible mechanistic pathways for Si exciton energy transfer in erbium-doped silicon materials.

The third approach is through the use of optically active impurities in silicon, most commonly with rare earth centers such as erbium.[8] Erbium is of particular interest because of the $(^4I_{13/2}) \rightarrow (^4I_{15/2})$ transition and the resulting luminescence band at 1.54 µm, which lies at an absorption minimum for silica-based optical fibers and glasses. A silicon-based nanoscale light emitter, when coupled to a suitable waveguide, would prove vital to the construction of a completely Si-based optoelectronic device. In general, rare earth ions are useful extrinsic luminescent dopants in semiconductors as a consequence of the relatively narrow line width of the rare earth f–f transitions and their photophysical stability in a variety of host matrices. Most importantly, rare earth centers such as Er^{3+} commonly participate in intracenter transitions stimulated by energy transfer with host excitons of the crystalline semiconductor (Figure 5.1).[8] However, the radiative efficiency of this process is strongly affected by a number of issues, including (1) the rate of back energy transfer (as a consequence of the shallowness of the trapped e⁻ level of the Si), (2) the relatively low solubility of erbium in crystalline Si (unless a monolayer of oxygen is present), and (3) self-quenching brought about as a result of clustering of erbium centers in the lattice.

In light of the above discussion, a key issue to pursue entails an evaluation of the effect of size confinement of the Si-based exciton on the structure and resultant photophysics, in the form of either zero-dimensional dots or one-dimensional wires. Although some limited success in fabricating dot composites has been encountered in top-down approaches typically involving the high-energy ion co-implantation of erbium ions and Si into oxide matrices,[9–11] alternative bottom-up approaches from the assembly of smaller structures pose considerable synthetic challenges. Ideally, the dependence of dimensional confinement of the Si host on impurity or dopant properties should address some or all of the following features:

- Size of Si host crystal: diameter in the dots, width in the case of wires
- Average Er^{3+} concentration in a given Si nanostructure
- Local Er^{3+} coordination environment in a given nanostructure
- Average Er^{3+} spatial location in a given dot or wire

We have successfully synthesized small emissive Si dots of various sizes, with erbium (Er^{3+}) randomly dispersed throughout the nanocrystal at a fixed dopant concentration (1 or 2%).[12] Larger Si dots (20-nm mean diameter) have also been prepared and characterized; in these larger nanoparticles, the erbium centers are preferentially enriched at the surface.[13] These materials have been analyzed in further detail, both structurally and photophysically. Routes to a more directed control of Er^{3+} location warrant attention for both fundamental and applied perspectives, as spatial distribution of erbium centers should impact the recombination of carriers generated either electrically or with light. Furthermore, we have recently encountered success with the synthesis of one-dimensional Si nanowires, which contain erbium ions deliberately concentrated in a distinct shell along the surface of the silicon core structure.[14]

5.2 NANOCRYSTAL FABRICATION

Our strategy for the preparation of rare earth-doped nanoparticles involves a modification of a high-temperature aerosol reaction involving the combustion of disilane (Si_2H_6, diluted in He carrier gas) at 1000°C, followed by isolation as a colloidal solution.[12,13] Our apparatus not only is capable of producing homogeneous oxide-capped Si nanoclusters but also possesses the ability to introduce an additional vapor phase reactant concomitantly into the reactant stream. We employed the known erbium chemical vapor deposition (CVD) precursor $Er(tmhd)_3$ (tmhd = 2,2,6,6-tetramethyl-3,5-heptanedionato) as the Er source in this reactor (Figure 5.2A). Initial experiments focused on the effects of varying the pyrolysis oven length (from 1.5 to 6 cm) on the structure and spectroscopic properties of the nanoparticles isolated from these reaction products.

For the synthesis of silicon nanocrystals with deliberate erbium-rich surfaces, the pyrolysis reactor design was altered from our original configuration to decouple the initial nucleation and growth steps of the silicon nanocrystals from the erbium incorporation step (Figure 5.2B). The initial pyrolysis of dilute disilane results in the nucleation and growth of kinetically trapped Si nanocrystals; the second pyrolysis oven achieves decomposition of the $Er(tmhd)_3$ precursor followed by growth of the shell. As with the above strategy, experiments focused on the effects of varying the length of the pyrolysis oven and its temperature, the disilane flow rate, and the $Er(tmhd)_3$/helium carrier gas flow rate on the mean nanoparticle feature size.

The solubility of these erbium-doped Si nanocrystals in various organic solvents has been improved by surface modification with $-SiMe_3$, $-Si(CH_2)_3CN$, and $-Si(CH_2)_3NH_2$ moieties, while retaining the desired luminescence at 1540 nm associated with the erbium centers.[15] We anticipate this solubility enhancement to effect easier size-selective precipitation and processing for additional applications.

5.3 STRUCTURAL CHARACTERIZATION OF ERBIUM-DOPED SILICON DOTS

A typical high-resolution transmission electron micrograph for Er^{3+}-doped silicon nanocrystals prepared from the pyrolysis of disilane and a suitable erbium precursor (such as $Er(tmhd)_3$) is illustrated in Figure 5.3. Lattice fringes associated with the cubic silicon unit cell are clearly evident, with an observed spacing of 3.1 Å for the <111> direction. Based on a working hypothesis that an increased length of the

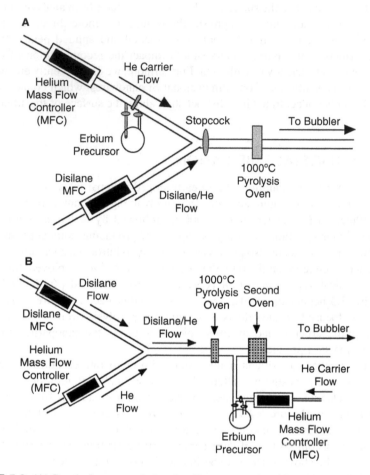

FIGURE 5.2 (A) Pyrolysis reactor design for Si nanocrystals with randomly dispersed Er^{3+} site environments. (B) Pyrolysis reactor for Si nanocrystals with surface-enriched Er^{3+}.

pyrolysis oven will increase the time for nucleation and growth of the doped nanocrystals, we have compared the quantitative size distributions of Er-doped Si nanocrystals obtained from pyrolysis in a 3.0-cm oven vs. those obtained using a 6.0-cm oven.[16] Although both sets of reaction conditions yield nanocrystals with ~2% Er, use of a 3.0-cm oven produces individual Si nanocrystals with a mean diameter of 3.3 nm. This is in clear contrast to the Er-doped Si structures synthesized from a longer 6.0-cm oven, where the mean particle size has effectively doubled to 5.9 nm.

For the Si nanocrystals with the erbium preferentially enriched at the surface, the sequential layered growth results in a slightly larger nanocrystal on average for the same disilane pyrolysis conditions, in the range of 22 to 26 nm. This lack of erbium during Si nanocrystal nucleation produces a very differently sized material, in stark contrast to the originally reported Er^{3+} randomly dispersed nanocrystals, where the presence of the Er^{3+} centers during Si nanoparticle nucleation and growth catalyzes the formation of smaller Si nanostructures.

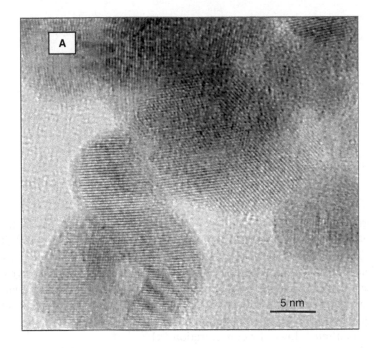

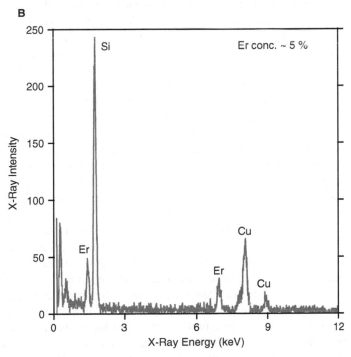

FIGURE 5.3 (A) High-resolution transmission electron microscopy (HREM) image of Er-doped Si nanocrystals. Scale bar, 5 nm. (B) Typical EDX spectrum for Er-doped Si nanocrystals.

TABLE 5.1
Erbium EXAFS Structural Data
for Randomly-Doped Si Nanocrystals

Sample	Er Coord N (±0.7)	d_{Er-O} [±0.01 (Å)]
Er_2O_3	6	2.26
3.2 nm	3.6	2.19
5.9 nm	4.7	2.27
6.4 nm	6.4	2.26

Regardless of size, the as-formed Er-doped Si nanocrystals possess the cubic unit cell structure, as evidenced by selected area electron diffraction (SAED) patterns.[12,13] There is no evidence of elemental Er, Er silicide, or Er oxide phases. For the randomly dispersed Er-doped Si nanocrystals produced from the co-pyrolysis of disilane and $Er(tmhd)_3$, an Er concentration of 2% is present, based on quantitative analyses of energy-dispersive x-ray (EDX) spectra. We have also been successful in producing Si nanocrystals of comparable mean particle diameter (6.0 nm, for example) but with a much lower Er concentration (<1%).

In order to examine the local coordination environment of the erbium centers in these doped nanocrystals, we turned to extended x-ray absorption fine-structure spectroscopy (EXAFS).[17] EXAFS measurements at the Er LIII edge (done in collaboration with Leandro Tessler of UNICAMP/LLNS (State University of Campinas and the Brazilian National Synchroton)) confirm that the erbium coordination environment in these silicon nanocrystals is oxygen rich, but interestingly, for the randomly dispersed Er-doped Si nanostructures, the erbium coordination number demonstrates some size-dependent behavior (Table 5.1). It increases from 3.6 in 3.2-nm nanocrystals to 6.4 (±0.7) in 6.2-nm nanocrystals, with a concomitant increase in Er-O bond length from 2.19 to 2.26Å (±0.01Å). It appears that in very small (3.2-nm) nanocrystals the lattice environment for the Er dopants is quite constrained, while as the particle size increases, the coordination environment becomes more Er_2O_3-like, with Er coordinated to six oxygen atoms. For the larger Er surface-enriched material (Table 5.2), samples with 26- and 22.5-nm average particle sizes were measured; a coordination number $N = 6.6 \pm 0.7$ and Er-O atomic separation $r = 2.30 \pm 0.01$ Å were determined for a 26-nm sample and $N = 6.9 \pm 0.7$ and $r = 2.31 \pm 0.01$ Å for the 22.5-nm sample. These parameters indicate an Er_2O_3-like environment, reminiscent of the oxygen octahedra that constitute the Er first coordination shell in $Er(tmhd)_3$. The Er-O bond lengths are markedly greater than those of Er_2O_3 ($r = 2.26$ Å), as the different growth mechanisms of the nanoparticles containing Er-rich surface layers manifest themselves in a different local erbium environment. In these samples, the Er sites are not constrained by the Si lattice, allowing a less constrained Er_2O_3-like Er environment. The longer bond lengths are interpreted as a consequence of the fact that virtually all of the Er is located in the first atomic layers of the nanoparticle surface.

TABLE 5.2
Erbium EXAFS Structural Data
for Erbium Surface Enriched
Si Nanocrystals

Sample	Er Coord N (±0.7)	d_{Er-O} [±0.01 (Å)]
Er_2O_3	6	2.26
26 nm	6.6	2.30
22 nm	6.9	2.31

5.4 PHOTOPHYSICAL MEASUREMENTS OF ERBIUM-DOPED SILICON DOTS

For Si dots with erbium (Er^{3+}) randomly dispersed throughout the nanocrystal, the anticipated Er^{3+} emission maximum near 1.54 μm, associated with the $(^4I_{13/2}) \rightarrow (^4I_{15/2})$ transition, is observed upon excitation at 488 nm (Figure 5.4A).[12] It is important to note that for these Er-doped Si nanoparticles, no visible luminescence associated with the Si nanoparticles is detected. Detailed measurements of the excitation wavelength and power dependence of the observed 1540-nm photoluminescence have been carried out. For an excitation wavelength range of 475 to 514 nm, we find that there is a monotonic decrease in 1540-nm emission intensity as the excitation wavelength increases (Figure 5.4B). No distinct increase in the Er^{3+} photoluminescence (PL) is observed upon excitation at either 488 or 514 nm; thus, the steady decrease in near-IR PL intensity with increasing excitation wavelength is consistent with a carrier-mediated process involving energy transfer from a silicon exciton to erbium centers.

However, the as-formed Si nanocrystals possessing Er-rich surface layers, with a coordination number near 6, do not demonstrate the 1540-nm luminescence associated with the Er^{3+} centers. This is presumably a consequence of the fact that the local Er concentration in the outer atomic shells is so high that the photoluminescence is quenched by Er-Er cross-relaxation. Preliminary thermal annealing experiments of the Si nanocrystals with Er-rich surfaces find that a brief 800°C vacuum treatment results in strong near-IR photoluminescence at 1540 nm in this material[13]; correspondingly, excitation wavelength studies suggest that this emission is direct excitation, rather than carrier mediated (Figure 5.4C). It is also fascinating to note, from a structural perspective assessed by EXAFS, that the conditions required in the annealing experiment for optical activity result in a diminution of the Er-O bond length closer to that of Er_2O_3 (observed d = 2.23 Å) and a reduction in the Er coordination number of <6 (observed, 5.7).[18]

5.5 ERBIUM-DOPED SILICON NANOWIRES

In addition to these zero-dimensional dots, it is important to fabricate one-dimensional nanoscale wires (NWs) composed of these elements for their potential value

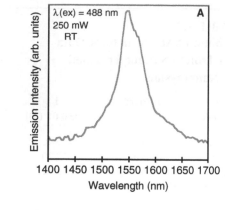

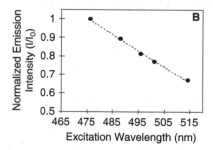

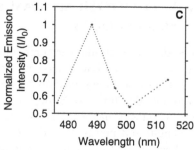

FIGURE 5.4 (A) Room-temperature photoluminescence spectrum of Er-doped Si nanocrystals, 6-nm average diameter. The excitation wavelength used for this measurement is 488 nm, at a power of 250 mW. Excitation wavelength dependence of the integrated near-infrared PL intensity for randomly dispersed Er-doped Si nanoparticles (B) and erbium surface-enriched Si nanocrystals (C).

in both electrical and optical applications. The former is of merit, given the known ability of erbium-rich silicon phases to act as efficient ohmic contacts on Si with the low-energy-available Schottky barrier height(s).[19] For the latter, one can envision (with the proper wire width) the propagation of light through a continuous Si network possessing the proper refractive index gradient as a consequence of the erbium preferentially enriched at the wire surface. Hence, two complementary approaches to the fabrication of Si nanowires possessing distinctive layered structures with Si-rich cores and Er-rich surface layers have been pursued initially.

In each case, the base strategy focuses on a combination of our previous approaches to Er-doped Si nanocrystals (with the rare earth centers in either randomly dispersed or surface-enriched varieties) coupled with the demonstrated utility of vapor–liquid–solid (VLS) type routes for Si nanowire nucleation and growth.[20,21] One straightforward combination is to initially grow the silicon nanowires on Au islands (deposited on a Si substrate) in the presence of dilute silane in He, followed by pyrolysis of a volatile $Er(tmhd)_3$ complex. Exploitation of a VLS-type process permits the desired reaction to proceed at relatively lower temperatures for shorter time intervals, when compared to that employed for the Si dots (600°C for ca. 3 h for the wires vs. 1000°C for 24 h for the dots).

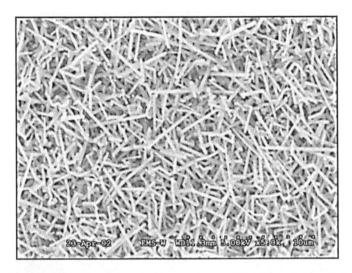

FIGURE 5.5 SEM image of an array of erbium surface-enriched Si nanowires. Scale bar, 1 μm.

Scanning electron microscopy (SEM) images (Figure 5.5) of the reaction products reveal numerous interwoven nanowires with diameters of 150 to 300 nm and lengths up to 10 μm. Energy-dispersive x-ray measurements confirmed the presence of erbium in these structures, and subsequent analysis by field emission SEM suggested appreciable roughness on the surface of the wires. In order to understand the structure in more detail, wires were detached from the substrate by dispersion in isopropanol and analyzed by transmission electron microscopy (TEM). Such studies reveal that the wires have a core-shell structure with erbium enriched at the surface. Figure 5.6 shows a typical wire with a diameter of ~250 nm. In terms of quantitative elemental analyses, three distinct areas are found from the EDX spectra. The center of the wire is clearly Si rich (96%) (area 2), and the wire clearly has a dark rim visible, which is high in Er concentration. On the surface of the wire, however, there are two different regions: a relatively smoother area closer to the core with an appreciably larger erbium concentration (~12%) (area 3), along with nodules emanating from the stalk that are nearly balanced in terms of amounts of erbium (53%) and silicon (47%) (area 1). Crystallographically SAED measurements recorded on the nanowire exhibit a (110) diffraction pattern and clearly indicate that the doped Si NW is single crystal in nature. The diffraction patterns taken from the Er-rich surface layer reveal that the surface layer is also crystalline. Unfortunately, this specific wire is apparently too thick to obtain lattice images at atomic resolution from the Si. We have subsequently synthesized smaller-diameter wires whereby the single-crystal Si nanowire core is surrounded by a polycrystalline Er/Si sheath (Figure 5.7); the lattice images associated with each type of material are evident in this case.

The second strategy to fabricate Si wires with Er-rich wires entails the concomitant presence of silane and the Er(tmhd)$_3$ precursor during the pyrolysis event under conditions that will exude the erbium from wire (as a consequence of the limited solubility of erbium in Si).[22] For the second strategy, the same reactor design is

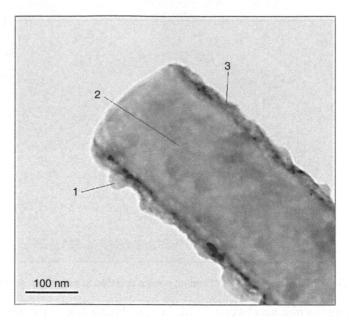

FIGURE 5.6 A typical TEM image of a surface Er-enriched Si wire. EDX analysis for the marked three areas (1–3) is presented in the text.

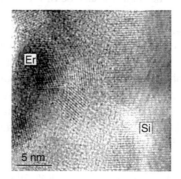

FIGURE 5.7 A high-resolution transmission electron microscopy (HREM) image of an Er-doped Si nanowire. Scale bar, 5 nm.

employed as in the first approach, but now in a three-step reaction sequence. After a brief silane preexposure (20 sccm, 5 min), the SiH_4 and $Er(tmhd)_3$ are co-pyrolyzed at 600°C for 25 min, followed by an extended Er^{3+} exposure for an additional 60 min. SEM studies of this reaction product, particularly from Z-contrast imaging in the backscattered mode, also show a network of uniform interwoven wires with a distinct Er-rich surface (as high as 50%) and a Si-rich core (Figure 5.8). The presence of the Er^{3+} centers during SiH_4 pyrolysis clearly impacts wire growth kinetics, as these wires are much larger in terms of diameter, with widths on the order of 700 to 800 nm. SAED measurements and micro-Raman spectroscopy indicate that the

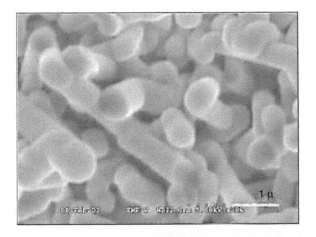

FIGURE 5.8 SEM images of Er-doped Si NWs grown by concomitant co-pyrolysis of silane and Er(tmhd)$_3$. Top: Plane view image. Bottom: Backscattered electron image illustrating the presence of the erbium preferentially at the surface (light shading) vs. the Si-rich core (dark).

nanostructures are poorly crystalline in the Si core and amorphous in the Er-enriched shell. Preliminary micro-Raman analyses (Figure 5.9) suggest that the crystallinity of such structures can be improved by high-temperature annealing at 800°C, as evidenced by a loss of the low-frequency tail associated with the amorphous Si component near 480 cm^{-1}.

5.6 PHOTOPHYSICAL MEASUREMENTS OF ERBIUM-DOPED SILICON WIRES

The room-temperature photoluminescence of these Er-doped Si nanowires has also been measured. Upon excitation at 488 nm, the anticipated Er^{3+} emission maximum

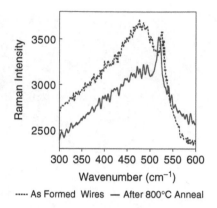

FIGURE 5.9 Effect of thermal anneal on the Raman spectra of Er-doped Si NWs grown by concomitant co-pyrolysis of silane and Er(tmhd)$_3$.

near 1.54 μm, associated with ($^4I_{13/2}$) → ($^4I_{15/2}$) transition, is observed. As noted above, detectable erbium emission at room temperature in a crystalline semiconductor is often difficult to achieve, and it is likely that dimensional suppression of Auger processes play a role. Excitation wavelength dependence measurements (in the range of 470 to 514 nm) indicate a rather insensitive response in terms of near-IR emission intensity. This would suggest that an ensemble of these wires emits with a combination of both carrier-mediated and direct-excitation pathways, perhaps a consequence of the two different average structural environments detected in their microstructure. We can only speculate at this point that the erbium present within the core is accessible to energy transfer from the Si, whereas those erbium centers at the shell of the wire are subject to direct excitation of the relevant ligand field states. This effect is currently under further investigation.

5.7 CONCLUSIONS

From the studies described here, it is apparent that the ability to produce doped nanoscale Si dots and wires with luminescent centers effectively opens the door for the fabrication of a wide variety of rare earth-doped Si nanostructures, whose emission maximum can be tuned by the selection of a particular rare earth ion (ranging from blue emission from Tb^{3+} to near infrared from Er^{3+}). Layered core-shell nanowire structures, as exemplified by the fabrication of Si-rich cores and Er-rich surface layers via a simple vapor transport process, suggest that a diverse range of longer, easier-to-manipulate wire-like materials with variable electrical and dielectric properties are also possible with similar chemistry. Overall, such materials are of enormous potential value for optical and optoelectronic applications, and a number of related experiments are in progress.

ACKNOWLEDGMENTS

Much of the research described herein was carried out by several different students and postdoctoral fellows: John St. John, Robert Senter, Junmin Ji, Yandong Chen, and Zhaoyu Wang. Their valuable contributions are gratefully acknowledged. The author also sincerely thanks the National Science Foundation and the Robert A. Welch Foundation for financial support of this research. EXAFS measurements were carried out in collaboration with Prof. Leandro Tessler of the Universidade Estadual de Campinas and LNLS — National Synchrotron Light Laboratory, Brazil. The expertise of Dr. Alan Nichols of the Electron Microscopy Facility of the Research Resources Center of the University of Illinois–Chicago is also gratefully acknowledged, along with the earlier assistance of Dr. Russell F. Pinizzotto, currently at the Missouri Academy of Math and Science at Northwest Missouri State University.

REFERENCES

1. L. Pavesi, L. DalNegro, C. Mazzoleni, G. Franzo, L. Pavesi. Optical gain in silicon nanocrystals. *Nature* 408: 440, 2000.
2. A.N. Witt, K.D. Gordon, D.G. Furton. Silicon nanoparticles: source of extended red emission? *Astrophys. J.* 501: L111, 1998.
3. L. Brus. Luminescence of silicon materials: chains, sheets, nanocrystals, nanowires, microcrystals, and porous silicon. *J. Phys. Chem.* 98: 3575, 1994.
4. P. Vogl, M. Rieger, J. Majewski, G. Arbstreiter. How to convert group IV semiconductors into light emitters. *Phys. Scripta* T49: 476, 1993.
5. A. Polman. Erbium-implanted thin film photonic materials. *J. Appl. Phys.* 82: R1, 1997.
6. L.T. Canham. Silicon quantum wire array fabrication by electrochemical dissolution of wafers. *Appl. Phys. Lett.* 57: 1046, 1990.
7. L.E. Brus, P. Szajowski, W. Wilson, T. Harris, S. Schupler, P. Citrin. Electronic spectroscopy and photophysics of Si nanocrystals: relationship to bulk c-Si and porous Si. *J. Am. Chem. Soc.* 117: 2915, 1995.
8. L. Kimerling, K. Kolenbrander, J. Michel, J. Palm. Light emission from silicon. *Solid State Phys.* 50. 353, 1997.
9. G. Franzò, V. Vinciguerra, F. Priolo. The excitation mechanism of rare-earth ions in silicon nanocrystals. *Appl. Phys. A* 69: 3, 1999.
10. P.G. Kik, A. Polman, Exciton–erbium interactions in Si nanocrystal-doped SiO_2. *J. Appl. Phys.* 88: 1992, 2000.
11. C.E. Chryssou, A.J. Kenyon, T.S. Iwayama, C.W. Pitt, D.E. Hole. Evidence of energy coupling between Si nanocrystals and Er^{3+} in ion-implanted silica thin films. *Appl. Phys. Lett.* 75: 2011, 1999.
12. J.V. St. John, J. Coffer, Y. Chen, R. Pinizzotto. Synthesis and characterization of discrete luminescent erbium-doped silicon nanocrystals. *J. Am. Chem. Soc.* 1211: 888, 1999.
13. R.A. Senter, Y. Chen, J.L. Coffer, L. Tessler. Synthesis of silicon nanocrystals with erbium-rich surface layers. *Nano Lett.* 1: 383, 2001.
14. Z. Wang, J.L. Coffer. Erbium surface enriched silicon nanowires. *Nano Lett.* 2: 1303, 2002.
15. J. Ji, R.A. Senter, J.L. Coffer. Surface modification of erbium-doped silicon nanocrystals. *Chem. Mater.* 13: 4783, 2001.
16. J.V. St. John, J. Coffer, Y. Chen, R. Pinizzotto. Size control of erbium-doped silicon nanocrystals. *Appl. Phys. Lett.* 77: 1635, 2000.
17. L. Tessler, J.L. Coffer, J. Ji, R.A. Senter. Erbium environments in silicon nanocrystals. *J. Non-Cryst. Solids* 299/302P1: 673, 2002.

18. R. Senter, L. Tessler, J.L. Coffer, Unpublished observations.
19. N. Norde, J. deSousa Pires, F. d'Heurle, F. Pessavento, S. Peterson, P.A. Tove. The Schottky-barrier height of the contacts between some rare-earth metals (and silicides) and p-type silicon. *Appl. Phys. Lett.* 38: 865, 1981.
20. J. Westwater, D.P. Gosain, S. Tomiya, S. Usui, H. Ruda. Growth of silicon nanowires via gold/silane vapor–liquid–solid reaction. *J. Vac. Sci. Technol. B* 15: 554, 1997.
21. J. Hu, T.W. Odom, C.M. Lieber. Chemistry and physics in one dimension: synthesis and properties of nanowires and nanotubes. *Acc. Chem. Res.* 32: 435, 1999.
22. A. Polman, G.N. van den Hoven, J.S. Custer, J.H. Shin, R. Serna. Erbium in crystal silicon: optical activation, excitation, and concentration limits. *J. Appl. Phys.* 77: 1256, 1995.

6 Conductance Spectroscopy of Low-Lying Electronic States of Arrays of Metallic Quantum Dots: A Computational Study

F. Remacle and R.D. Levine

CONTENTS

6.1 Introduction 153
6.2 Hamiltonian 157
6.3 The Transmission Function 158
6.4 Mechanisms for Conduction 160
6.5 Working Expression for the Matrix Elements of the Transmission Function 161
6.6 Computational Results for the Transmission Function 164
6.7 Expression for the Current: The Fermi Window Function 169
6.8 Computed Current–Voltage Curves 172
6.9 Concluding Remarks 174
Acknowledgments 174
References 174

6.1 INTRODUCTION

Spectroscopic probing of low-lying excited states of small arrays of quantum dots is discussed with special reference to the role of fluctuations of the size of the dots, the compression of the array, the temperature, and the external voltage. The measured response is the current through the array, and it is shown how, at low applied voltages and low temperatures, it faithfully characterizes the density of states of the array.

Special reference is made to conduction via domain-localized states. The chapter concludes with the study of level-crossing spectroscopy.

Electronic states of arrays of coupled metallic quantum dots have unusual properties for three reasons:

1. Inevitable variations in size occur from one dot to another. Because the size determines the energies of the highest occupied orbitals on each dot, the fluctuation in sizes means that adjacent dots are not necessarily resonantly coupled.
2. The strength of exchange coupling between dots can be tuned over a wide range by moderate variations in the compression of the array.
3. Even small dots have a charging energy that is quite low compared to that of atoms or small molecules. Ionic states can therefore be of low energy and mix with covalent states.

There is theoretical and computational evidence[1,2] that the electronic states of such an array can qualitatively change, undergoing so-called quantum phase transitions, as a function of its compression and the amount of disorder. There can be a number of qualitatively different such transitions, particularly when the disorder is high.[2] There is then the question of how to experimentally probe such transitions. Static measurements of the surface potential[3] offer one possibility.[4] Here we discuss conductance spectroscopy as a rather sensitive probe of the nature of the electronic polymorphism of such systems. Low-temperature transport experiments, e.g., references 5 to 7 (for a perspective, see reference 8), have indeed reported measurements that are rather encouraging for the potential of such an approach and suggest that the inevitable blurring due to a finite observational resolution of a current measurement that requires a finite voltage gradient and a finite temperature can be sufficiently minimized.

The arrays that we discuss are geometrically well ordered. Experimentally, quantum dots prepared[6,9–11] with a narrow size can self-assemble into a two-dimensional hexagonal lattice. For such arrays, probes of the nanostructure such as atomic force microscopy (AFM)[3] verify that the domain is without packing defects. Measurements[4] of the surface potential for geometrically well-ordered hexagonal arrays of metallic quantum dots result in contours that have a systematic structure. The structure of the surface potential responds, in a reversible fashion, to changes in the compression of the array and in the voltage gradient.[4] Experiments to measure the surface potential at low temperatures are planned but not yet carried out.

The disorder that we speak about is due to the inevitable variations of the size of the individual dots. This has two primary effects. The first is a diagonal one or an inhomogeneous broadening. The dots do not all have quite the same ionization potential or the same electron affinity. The dispersion in sizes also affects the coupling between dots. To measure a current, the array needs to be quite compressed. When dots of a few nanometers in diameter are almost touching the classical long range, dipole–dipole coupling is no longer realistic.[12] At the shorter distance, the dots are exchange coupled due to the overlap of tails of wave functions centered on adjacent dots. The magnitude of the coupling depends on the diameter of the dots because the higher-most-occupied orbitals are essentially those of a particle in a well. (Both the width and depth of the well are changing with the size of the dot.)

Ordinarily there can be a Coulomb blockade[5,13–16] to charge transport. For compressed arrays the magnitude of the exchange coupling between the dots exceeds the charging energy. By the Mott criterion, the system is then expected to be a conductor because the covalent band has effectively merged with the ionic band. However, the fluctuations in the sizes of the dots result in states of the array that are not necessarily delocalized.[17,18] Exchange coupling varies exponentially with the distance between the dots. Therefore, compression can significantly increase the coupling strength and a strong-enough coupling can bridge the local variations in the energies of the higher-most orbitals of the dots. The central point of our results is that upon compression, the transition is not simply from a localized to a delocalized state but, rather, low-lying excited states, states that are in the middle of the band and are most important for conductivity at low temperatures, will be domain localized. The domains are due to the coupling being more effective between nonneighboring dots that are nearly comparable in size, rather than between two adjacent dots that are not. The dots intermediate between the two effectively coupled ones are known to chemists as a bridge, and the longer range coupling is called super exchange.[19]

Ideally, to probe the electronic structure of the array, one would be able to directly measure its transmission function, which is a weighted density of conducting states, as a function of energy.[20,21] The advantages are that such measurements are not subject to optical selection rules and that the transmission function is a sensitive function of the size distribution of the dots in the array, of the compression, and of the external bias. Operationally, what one can measure is the transmission function as seen through the finite-width Fermi window of a conductivity measurement. This is discussed in more detail in Section 6.6 because the range of states within the observation window varies with the external bias and temperature.

Experimentally, the density of conducting states is accessed through the differential conductivity, dI/dV, vs. voltage curves. Different experimental configurations are possible, and depending on the one that is adopted, it is not necessarily the same states that fall within the Fermi window, so that different setups can lead to different I-V curves, even if nominally, the same source–drain voltage drop is applied. Measurements can be done using an asymmetric double junction. Such conditions are typical, for example, for STM measurements,[22–27] where the potential drop at one electrode is much smaller than the potential drop at the other one. In such a configuration, by changing the sign of the applied bias, it is possible to probe different kinds of levels: empty levels for positive bias and occupied levels for negative bias, so that the I-V curves are not symmetric with respect to a change of the bias sign, particularly for molecules. When the coupling to the two electrodes is more symmetric, such as, for example, in break junction single-molecule experiments in which the molecule is tethered between two electrodes of the same metal, the potential drop is evenly distributed between the two electrodes. Changing the sign of the bias no longer allows probing different kinds of states. In that case, it is always the same states that fall within the Fermi window, irrespective of the sign of the applied bias. Such I-V plots are therefore symmetric with respect to a change of the sign of the

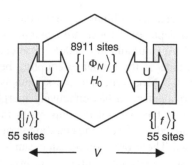

FIGURE 6.1 Schematic drawing of the geometry of the array and the electrodes. In this chapter we consider the case of the coupling U to the two electrodes being the same. This is physically realistic for a compressed array, but it need not be the case if a single dot or a single molecule is placed in the gap between the electrodes.

applied bias. This is rigorously true if the transmission function of the system remains exactly the same for positive and negative bias. Because of disorder effects, we will see (in Section 6.8) that for quantum dot (QD) arrays, the transmission function may slightly differ for the same strength of positive and negative bias, leading to slightly asymmetric I-V curves. In order to probe different kinds of states in the case of a symmetric double junction, one has to ground one of the electrodes[28,29] or apply a gate voltage.[30]

In the configuration used in the experiments of the Heath group for QD arrays,[7] which we discuss below, neither of the electrodes is grounded and the coupling to the electrodes is assumed to be symmetric. Therefore, the bias is distributed evenly between the two electrodes (see Figure 6.1 and Figure 6.2). Due to the finite width of the Fermi window, individual peaks can be resolved only for sparse systems.[22–25] In the array of quantum dots that we are discussing, even for submillielectronvolts applied bias and low temperatures, there are still many excited electronic states of the array in the Fermi window that contribute to the current, and they cannot be fully resolved.[26,27] To probe the fine details of this transition, it is necessary to be in the low T, low V regime. The temperature needs to be low so that the Fermi window is not significantly broadened. The voltage needs to be low both so that the Fermi window is as narrow as possible and so that the Stark effect of the voltage does not distort the (quite close in energy) states of the array. Of course, as a practical aspect, the voltage-induced changes in state that can result in enhanced conductivity[31] may be a desirable aspect. The change in the delocalization of the low-lying electronic states as a function of energy can be probed by moving the narrow Fermi window along the energy levels of the array, either by changing the Fermi level of the electrodes or by applying a gate voltage to the array.[30]

The computations of the transmission function and current are based on using the generalized Ehrenfest theorem.[32] The paper starts with a summary of the formalism. A full account can be found in reference 21. One can also find references there of other formalisms[33–35] and discussions of conductivity of other nanosystems.[36–42] We compute the transmission function and current for hexagonal two-dimensional arrays that have 55 Ag QDs per side, and these dots are directly coupled

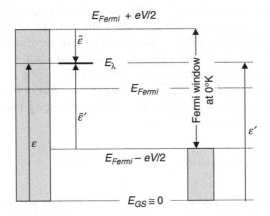

FIGURE 6.2 Schematic drawing of the energy levels. The configuration of the electrode is such that the applied bias is distributed symmetrically between the left and right sides. Other configurations are possible, however. One obvious alternative is where one of the electrodes is grounded, and this will lead to different *I-V* curves than the ones shown later in this chapter. As can be seen from Figure 6.8, an MO λ whose energy is inside the Fermi window contributes to hole or electron conduction, depending on whether it is below or above the HOMO.

to the electrodes. In total, the array is made of 8911 dots (see Figure 6.1). The size distribution of the dots is narrow, as in the best experiments,[7] and the array is compressed to the same dot–dot distance that is estimated from transmission electron microscopy (TEM) images. The heart of the formalism is the computation of the transmission function (Sections 6.3 and 6.5). As a preliminary, the Hamiltonian of the array is briefly discussed in Section 6.2. The role of conduction by electrons or holes is discussed in Section 6.4. The computed results for the transmission function, showing the evolution of the density of states as the array is compressed, are discussed in Section 6.6. The averaging of the transmission function that is required to compute the current through the array is discussed in Section 6.7, with special reference to the blurring of details by the finite Fermi window function. Computational results for current vs. voltage are shown in Section 6.8.

6.2 HAMILTONIAN

The Hamiltonian of the system in the absence of interaction of the array with the electrodes is given by

$$H_0 = \sum_{i,\,left\,electrode} E_i |i\rangle\langle i| + \sum_{f,\,right\,electrode} E_f |f\rangle\langle f| + \sum_{k,\,array} \alpha_k |k\rangle\langle k| + \sum_{l,k} \beta_{lk} |k\rangle\langle l| \quad (6.1)$$

H_0 is a Hückel or tight-binding-type Hamiltonian, and all the indices label sites (see Figure 6.1). $|i\rangle$ and $|f\rangle$ are electron states of the left and right electrodes, respectively, with Fermi levels E_i and E_f. α_k is the energy of the valence orbital of the kth dot. It carries a site index because it is determined by the size of the dot, as shown in Equation 6.2. For an isolated dot, α is the ionization potential, but in the array the value of α is renormalized due to the mean field effects of the other dots. There are two additional effects, one random and one, the Stark shift, that is systematic in the location of the dot in the array. First, α is roughly inverse to the cross section of the dot, so variations in size imply variations in the value of α. Second, applying a voltage gradient across the array means that α is changing monotonically along the gradient. We use the representation

$$\alpha_i = \alpha_0 \left(1 + \delta\alpha_i\right) + eV_i = \alpha_0 \left(1 + \delta\alpha \, \mathrm{ran}_i\right) + eV_i \tag{6.2}$$

where α_0 is the mean site energy, ran_i is a random number within the range (-0.5, 0.5), and $\delta\alpha$ is the range (in percent) of fluctuation in the site energies induced by the fluctuation in sizes, $\Delta\alpha = \alpha_0\delta\alpha$ and $\Delta\alpha/\alpha \propto \Delta R/R$. In addition, the site energies are Stark shifted due to the presence of the external bias by an amount eV_i, where V_i is the applied external field at the position of the dot i in the array. Note that the voltage V is included in the Hamiltonian, H_0, so that its effects are treated at all orders.

Neighboring dots are exchange coupled (last term of Equation 6.1). The transfer integral β is determined by the overlap of the tails of the wave function of adjacent dots. Thus, β depends on both the sizes and the separation between the dots, and we take it to decrease exponentially with increasing distance D between the dots[43,44]:

$$\beta = \left(\beta_0/2\right)\left(1 + \tanh\left(\left(D_0 - D\right)/4RL\right)\right) \xrightarrow{D>2R} \beta_0 \exp\left(D_0/2RL\right)\exp\left(-D/2RL\right) \tag{6.3}$$

For compressed arrays, an independent electron approximation for describing the electronic structure is adequate because we have shown[20] that the effect of electron correlation is negligible when the exchange coupling β is larger than the charging energy. At lower compressions, the role of electron correlation cannot be neglected. In particular, the effect of the Coulomb blockade is essential in the region of the Mott insulator-delocalized electrons' transition.[1] In Equation 6.3 the interdot distance D is scaled in units of the diameter, $2R$ of the dots, and the onset of the decline of β is governed by the parameter $D_0/2R$. L is the range parameter of the exchange coupling β. The values $\beta_0 = 0.5$ eV, $D_0/2R = 1.1$, and $1/2L = 11$ used here are the same as in our earlier studies[20,45] for Ag nanodots of 7 nm in diameter.

6.3 THE TRANSMISSION FUNCTION

To compute the current, we begin with the transmission function for an electron propagating through the array. It is computed between an initial state where the extra electron is on the electrode on the left and a final state where the electron is on the right:

$$\left|T_{IF}\right|^2 = \left\langle F\left|T(E)\right|I\right\rangle \delta\left(E_I - E_F\right)\left\langle I\left|T(E)\right|F\right\rangle \tag{6.4}$$

The transition operator $T(E)$ is given by

$$T(E) = U + UG(E)U \tag{6.5}$$

U is the coupling between the array and the electrodes, and $G(E)$ is Green's function of the full Hamiltonian:

$$H = H_0 + U \tag{6.6}$$

In Equation 6.4, the initial and final states $|I\rangle$ and $|F\rangle$ are eigenstates of the Hamiltonian H_0. They are both $N + 1$ electron states. The state $|I\rangle$ is the product of the state of the left electrode $|i\rangle$ and an N electron state $|\Phi_N\rangle$ of the array of N dots:

$$|I\rangle = |i\rangle|\Phi_N\rangle \tag{6.7}$$

On the other hand, the final state $|F\rangle$ is the product of the state of the right electrode $|f\rangle$ and the final state of the array $|\Phi'_N\rangle$. The electron transfer can be an inelastic process and leave the array in a different state. Thus, in principle, $|\Phi'_N\rangle$ need not be the same as $|\Phi_N\rangle$. Our formalism allows for inelastic processes, but we will not use them in the computations reported below. We consider the transfer to be elastic so that $|\Phi_N\rangle \equiv |\Phi'_N\rangle$. Explicitly,

$$|\Phi_N\rangle = \prod_N (\phi_\mu)^{f_\mu} \tag{6.8}$$

where f_μ is the occupancy of the orbital μ, and in a one-electron approximation, it is zero or unity. The MO μ values are obtained by diagonalization of the Hamiltonian for the array, including the effect of the external bias.

The position of the different energy levels is schematically shown in Figure 6.2. The energy of a state $|I\rangle$ is the sum of the energy of the state of the left electrode, ε, and of the energy of the state $|\Phi_N\rangle$ of the array:

$$E_I = \varepsilon + E_{\Phi_N} = E_{Fermi} + eV/2 + \tilde{\varepsilon} + E_{\Phi_N} \tag{6.9}$$

E_{Fermi} is the energy of the Fermi level of the electrode, and $eV/2$ is the shift of the Fermi level due to the applied bias, V. e is the charge of the electron. It is convenient to refer the energy of the state of the electrode to the Fermi level, shifted by the voltage, as shown in Figure 6.1, $\varepsilon = E_{Fermi} + eV/2 + \tilde{\varepsilon}$. In the computational results below, the Fermi level is taken to be slightly above the ground state of the array. We show how varying its position allows probing of the electronic structure. In a real measurement, it is not the energy of the Fermi level that is varied, but a gate voltage is applied to the array. From the computational point of view, for the configuration of the electrodes for which the effect of the bias is symmetrically distributed between the left and right electrodes, both methods are equivalent. The energy of the states $\left|F\right\rangle$ for which an electron has been transferred to the right electrode is

$$E_F = \varepsilon' + E_{\Phi'_N} = E_{Fermi} - eV/2 + \tilde{\varepsilon}' + E_{\Phi'_N} \qquad (6.10)$$

where ε' is the energy of the state $\left|f\right\rangle$ on the right electrode and $E_{\Phi'_N}$ is the energy of the state of the array $\left|\Phi'_N\right\rangle$ after the electron has been transferred to the right electrode.

In the physical electrode, the one-electron states form a continuum. We ensure that the electron does not reflect from the electrode by using outgoing boundary conditions on Green's function in the scattering operator, Equation 6.5. Another option is to mimic the continuum of the one-electron states of the electrodes by replacing the one-electron coupling operator U between the array and the electrode by an effective coupling to a continuum,[32] sometimes called a self-energy term.[34,35,39] We have used this technique in studies of electron propagation along peptide chains.[46] We also take it that the rate of charge transfer is low enough that the broadening of the levels can be neglected.

6.4 MECHANISMS FOR CONDUCTION

For an elastic transmission, there are two mechanisms for conduction between the initial and final states $\left|I\right\rangle$ and $\left|F\right\rangle$ and the states of the array. In the first mechanism, starting with the initial state $\left|I\right\rangle$ that has one electron on the left electrode, N electron on the array, and 0 electron on the right electrode,

$$\left|I\right\rangle = \left|i\right\rangle^1 \left|\Phi_N\right\rangle^N \left|f\right\rangle^0 \qquad (6.11)$$

the action of the coupling U is first to transfer an electron to the array, and then the extra electron on the array is transferred to the right electrode:

$$\text{LAR} \xrightarrow{U_L} \text{L}^+ \text{A}^- \text{ R} \xrightarrow{U_R} \text{L}^+ \text{AR}^- \qquad (6.12)$$

In terms of $N + 1$ electron states, this process is written

$$\left|i\right\rangle^1 \left|\Phi_N\right\rangle^N \left|f\right\rangle^0 \xrightarrow{U_L} \left|i\right\rangle^0 \left|\Phi_N\right\rangle^{N+1} \left|f\right\rangle^0 \xrightarrow{U_R} \left|i\right\rangle^0 \left|\Phi_N\right\rangle^N \left|f\right\rangle^1 \qquad (6.13)$$

and the final state is

$$\left|F\right\rangle = \left|i\right\rangle^0 \left|\Phi_N\right\rangle^N \left|f\right\rangle^1 \qquad (6.14)$$

In the intermediate state, the array has one extra electron, whereas the two electrodes are empty. We call this process electron conduction.

The second route leading from the same initial to the same final state is that the coupling U first transfers an electron from the array to the right electrode, and then an electron is transferred from the left electrode to the array:

$$LAR \xrightarrow{\ U_R\ } L\ A^+R^- \xrightarrow{\ U_L\ } L^+AR^- \tag{6.15}$$

which, in terms of the $N + 1$ electron states, corresponds to

$$\left|i\right\rangle^1 \left|\Phi_N\right\rangle^N \left|f\right\rangle^0 \xrightarrow{\ U_R\ } \left|i\right\rangle^1 \left|\Phi_N\right\rangle^{N-1} \left|f\right\rangle^1 \xrightarrow{\ U_L\ } \left|i\right\rangle^0 \left|\Phi_N\right\rangle^N \left|f\right\rangle^1 \tag{6.16}$$

For this route, in the intermediate state, the array has one electron fewer than in the initial state. We call this process hole conduction.

The difference between the two mechanisms is that for hole conduction, the electrodes couple to a lower-energy-occupied MO of the system, whereas for electron conduction, the coupling is via an empty, unoccupied orbital. In the computational results below, we take the coupling strength to be the same, independent of the mechanism. This need not be the case, and more experimental work is needed to allow the theory to be refined concerning this point.

In principle, the coupling to the electrodes can induce the transfer of more than one excess electron to the array; this is particularly important if the coupling to the two electrodes is of very different strengths. We here deal with a symmetric situation and assume that the current is low enough that the array carries no more than one excess electron or no more than one hole (for hole conduction). We show in the next section that because of these two mechanisms, in a symmetric junction geometry in which the bias drop is symmetrically distributed between the two electrodes, one needs to consider the transmission function of electronic states both above and below the ground state of the array.

6.5 WORKING EXPRESSION FOR THE MATRIX ELEMENTS OF THE TRANSMISSION FUNCTION

The $N + 1$ zero-order electron states are separable and complete:

$$1 = \sum_i^{55} \sum_{\{|\Phi_N\rangle\}} \left|i\right\rangle \left|\Phi_N\right\rangle \left\langle\Phi_N\right| \left\langle i\right| \tag{6.17}$$

where $\left\{\left|\Phi_N\right\rangle\right\}$ is the set of the N electron states of the array alone, and there are many such states of the order of 2^N. The possible one-electron states of the electrode are denoted by $\left\{\left|i\right\rangle\right\}$.

In the weak coupling limit we neglect the higher-order terms in U and replace the full Green's function in Equation 6.5 by the zero-order Green's function:

$$T(E) = U + UG_0(E)U \qquad (6.18)$$

We describe the states of the array at the one-electron level (Equation 6.8). For the purpose of evaluating the matrix elements of the transmission function, Equation 6.18, we need to include in the spectral resolution of the zero-order Green's function both the states that correspond to conduction by electrons and the states that correspond to conduction by holes. For the electron route (Equation 6. 13), the subset of possible $N + 1$ intermediate electron states has $N + 1$ electrons in the array (Equation 6.13) and zero electrons on the electrodes. They are of the form

$$|\Phi_{N+1}\rangle = |i\rangle^0 \left(\prod_N |\phi_k\rangle^{f_k} \right) |\phi_\lambda\rangle^1 |f\rangle^0 = \underbrace{|\Phi_N\rangle}_{N \text{ electrons}} \cdot \underbrace{|\phi_\lambda\rangle}_{\text{one electron}} \qquad (6.19)$$

The extra electron has to go in an unoccupied MO of the array, which we denote by ϕ_λ. On the other hand, in the case of the hole route, the subset of possible $N + 1$ electron states has $N - 1$ electrons in the array; they are of the form

$$|\Phi_{N+1}\rangle = |i\rangle^1 \left(\prod_{N-1} |\phi_k\rangle^{f_k} \right) |f\rangle^1 = \underbrace{|i\rangle}_{\text{one electron}} \cdot \underbrace{|\Phi_{N-1}\rangle}_{N-1 \text{ electrons}} \cdot \underbrace{|f\rangle}_{\text{one electron}} \qquad (6.20)$$

The electron that is transferred to the right electrode has to be removed from an unoccupied MO.

The expression of the matrix elements of the transmission function is, therefore,

$$\langle F|T(E)|I\rangle = \langle f|\langle \Phi_N| U \sum_{\Phi_{N+1}, \Phi_{N-1}} \frac{|\Phi_{N\pm1}\rangle\langle\Phi_{N\pm1}|}{E - E_{\Phi_{N\pm1}}} U |\Phi_N\rangle|i\rangle$$

$$= \sum_\lambda \frac{\langle f|U|\lambda\rangle\langle\lambda|U|i\rangle}{E - E_\Lambda} \qquad (6.21)$$

where the sum is on both occupied and unoccupied MOs. E_Λ is the energy of the intermediate $N + 1$ electron states. In terms of the site basis states, we have

$$|\lambda\rangle = \sum_k^N c_{k\lambda} |k\rangle \qquad (6.22)$$

We, therefore, get

$$\langle F|T(E)|I\rangle = \sum_\lambda \sum_{k,k'} \frac{\langle f|U|k\rangle c_{k\lambda}\langle k'|U|i\rangle c_{k'\lambda}}{E - E_\Lambda} = U^2 \sum_\lambda \sum_{k,k'} \frac{c_{k\lambda}c_{k'\lambda}}{E - E_\Lambda} \quad (6.23)$$

where k and k' are the sites on the left and right of the array that are coupled to sites i and f of the left and right electrodes, respectively. To obtain the final expression of the transmission function, we further make the approximation that the on-shell contribution of Green's function is the dominant contribution

$$\frac{1}{E^+ - E_\Lambda} = \begin{pmatrix} \text{a real part} \\ \text{that we negelct} \end{pmatrix} - i\pi\delta(E - E_\Lambda)$$

where the superscript + is a reminder that we use outgoing boundary conditions. This is justified because of the high density of electronic states. The expression of the computed transmission function that we discuss in Section 6.6 is, therefore,[21]

$$\langle F|T(E)|I\rangle \delta(E_I - E_F)\langle I|T(E)|F\rangle$$

$$= U^4 \sum_\lambda \sum_{k,k'} \frac{c_{k\lambda}c_{k'\lambda}}{E - E_\Lambda} \sum_\sigma \sum_{l,l'} \frac{c_{l\sigma}c_{l'\sigma}}{E - E_\Sigma} \delta(E_I - E_F)$$

$$\approx -\pi^2 U^4 \sum_\lambda \sum_{k,k'} \sum_\sigma \sum_{l,l'} c_{k\lambda}c_{k'\lambda}c_{l\sigma}c_{l'\sigma}\delta(E - E_\Sigma)\delta(E - E_\Lambda)\delta(E_I - F_F)$$

$$= -\pi^2 U^4 \sum_\lambda \left| \sum_{k,k'} c_{k\lambda}c_{k'\lambda} \right|^2 \delta(E_I - E_\Lambda)$$

$$(6.24)$$

where $E_I = \varepsilon + E_{\Phi_N}$. $E_\Lambda = E_{\Phi_N} + E_\lambda$ for electron conduction and $E_\Lambda = \varepsilon + E_{\Phi_N} - E_\lambda + \varepsilon'$ for hole conduction, so that in both cases, the resonance condition implies $\varepsilon = E_\lambda$ (the resonance condition $E_I = E_F$ implies, for elastic transmission, $E_{\Phi_N} = E_{\Phi_{N'}}$, that $\varepsilon = \varepsilon'$). The final expression for the transmission function becomes

$$|T_{if}|^2 = \langle F|T(E)|I\rangle \delta(E_I - E_F)\langle I|T(E)|F\rangle = -\pi^2 U^4 \sum_\lambda \left| \sum_{k,k'} c_{k\lambda}c_{k'\lambda} \right|^2 \delta(\varepsilon - E_\lambda) \quad (6.25)$$

Equation 6.25 shows that each molecular orbital λ of the array contributes to the transmission function at its energy E_λ and with a weight

$$T_\lambda = \left| \sum_{k,k'} c_{k\lambda} c_{k'\lambda} \right|^2 \tag{6.26}$$

It is essential to note that for the weight to be large, the molecular orbital λ has to have a component on both sides of the array. This is because the states k and k' are sites of the array that are coupled to the electrodes on the left and right, respectively. Only if both amplitudes are large for the same λ will that MO contribute to the current. It is this property that most readily distinguishes between states that are delocalized over both edges of the array and states that are not.

6.6 COMPUTATIONAL RESULTS FOR THE TRANSMISSION FUNCTION

The transmission function is computed for Ag nanodots of 7-nm diameter and 5% size fluctuations. This variation is about the best that is currently possible to achieve for Ag dots. Dots were arranged in a hexagonal array, 55 dots per side (Figure 6.1). The compression of the array was chosen to be in the narrow range, where we see the evidence for a transition from the domain-localized phase to a localized regime. At a higher compression, $D/2R \leq 1.2$, the electronic states are fully delocalized for all values of the voltage. By $D/2R > 1.24$, the states are localized and the conductivity is too low to be measured.

Figure 6.3 shows the effect of compression on the transmission function, computed for a low external bias of 1.1 meV. The values of compression are chosen in the region of the exponential decay of the transfer integral β $(D/2R > 1.1)$. The upper panel is computed for $D/2R = 1.2$ ($\beta = 0.050$ eV) and the middle panel for $D/2R = 1.22$ ($\beta = 0.033$ eV). The lower panel is for $D/2R = 1.23$ ($\beta = 0.027$ eV). The most striking effect of the decrease of the compression on the transmission function is the decrease in intensity. This is due to the decreasing of the delocalization of the MOs as the array is expanded. By $D/2R = 1.24$ (not shown), the intensity is almost zero and the array is nonconducting.

The overall shape of the transmission function with a gap of about 4.9 eV can be understood from Equation 6.25. The transmission function is a measure of probability for a MO to be localized on the edges of the array. The MOs localized on these sites must be higher in energy because these sites, which are edge sites, have a lower connectivity (= 3) than the sites that are in the middle of the array (the latter have a connectivity of 6). The lowest MOs are therefore localized in the middle of the array, as shown in the top panels of Figure 6.4 and Figure 6.5. As their energy increases, the MOs become more localized on the edges. Therefore, as can be seen by comparing the extent of the band of conducting states in Figure 6.3 to the full bandwidth (Figure 6.4), the MOs on the edges of the band are not conducting, meaning that they have small weights on the edges' sites. Figure 6.5 shows the weights on a line of 110 sites, parallel to the direction of the voltage gradient. The weights are computed for a low bias strength (1.1 meV) and averaged over a band of 100 MOs centered at the bottom of the band (top panel), in the middle of the

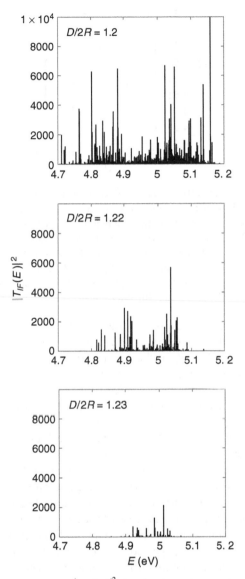

FIGURE 6.3 Tranmission function, $\left|T_{IF}\left(E\right)\right|^2$ (Equation 6.25), computed for a small external bias of 1.1 meV and 5% of disorder in site energies at different compressions $D/2R$. For $D/2R$ = 1.2, many MOs are conducting and the current is almost ohmic (see Figure 6.9). When the compression is decreased to that appropriate for an array on the verge of the domain localized to insulator transition ($D/2R$ = 1.22, middle panel), the band of conducting states is narrower (because β decreases) and the states are less conducting. For $D/2R$ = 1.23, lower panel, the array is almost not conducting.

band (middle panel), and at the top of the band (lower panel). The corresponding figures for the whole array (8911 sites) are shown in Figure 6.6. One can see that as their energy increases and gets closer to the middle of the band, the MOs have

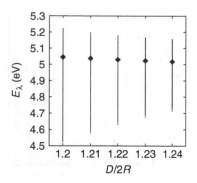

FIGURE 6.4 Energies of the 8911 MOs of the band for different values of the compression, computed for 5% of disorder in the site energies ($\Delta\alpha = 0.25$eV) and $V = 1.1$ meV. When $\beta > \Delta\alpha$, the width of the band is 12 β. As the compression decreases, the band shrinks because β decreases. When $\beta \ll \Delta\alpha$ (not shown), the width of the band is governed by $\Delta\alpha$ and is centered around α_0 ($\alpha_0 = 5$ eV in the computation). Also shown as a diamond is the position of the HOMO, which is located slightly above α_0 for high compression and tends to α_0 as β becomes negligible compared to α_0. This asymmetry in the location of the HOMO means that the density of excited states is significantly higher.

a larger weight on the edge sites, which translates into a maximum of the transmission function. Inside the conducting band, the intensity of the conducting states exhibits a gap at about 4.9 eV. In the absence of disorder, for a one-dimensional array, this gap is centered at the value of α_0 and can be analytically expressed using the weights of the MOs on the site coupled to the electrodes.[20] As the ratio $\Delta\alpha / \beta$ increases when the lattice is expanded, this gap is progressively filled because β is no longer strong enough to couple neighboring sites and the distribution of the site energies is uniform at about α_0 (Equation 6.2). The MOs around the gap are the ones that are localized closest to the edges, and because their connectivity is smaller, they are most sensitive to disorder: they are in the domain-localized regime. In this regime, the conduction is of a variable-range-hopping type and shows an $\exp\left(-T_{VRH}/T\right)^{1/2}$ temperature dependence.[20] At slightly higher energies, the transmission of the conducting states increases because of the better delocalization of the MOs. The conduction becomes activated and follows an $\exp\left(-T_{act}/T\right)$ temperature dependence. Toward the upper edge of the band (lower panels of Figure 6.5 and Figure 6.6), the transmission function decreases again because the wave function is localized and this decrease occurs for lower and lower energies as the strength of the external bias increases. There is a further cutoff of the band of conducting states due to the voltage; we discuss it below.

As can be seen from Figure 6.4, β also governs the width of the band of MOs, which is about 12 β. (Most of the 8911 sites are inner sites with a connectivity of 6; there are $6 \times 55 = 330$ edge sites with a connectivity of 3.) Because of the differences in connectivity and the disorder in the site energies, the density of states within the band is not uniform and the band is not symmetric about the mean zero-order site energy, α_0 ($= 5$ eV in the computations). As β decreases, the bandwidth decreases (Figure 6.4). When the array is expanded further, we reach the $\beta \ll \Delta\alpha$

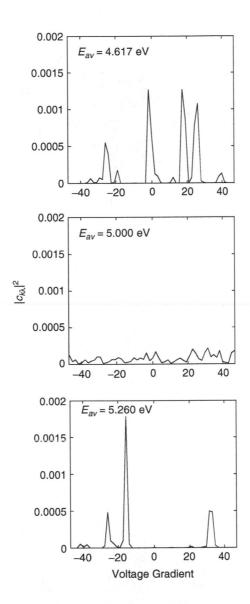

FIGURE 6.5 Plot demonstrating the localization of the MOs as a function of energy along a line of sites parallel to the voltage gradient. The weights $|c_{k\lambda}|^2$ are averaged over a band of 100 MOs at about 4.617 eV (upper panel), 5.000 eV (middle panel), and 5.260 eV (upper panel). They are computed at a compression $D/2R = 1.2$ for 5% of disorder in site energies and an external strength of the bias of 1.1 meV. At energies lower and higher than the edges of the conduction band (see Figure 6.3, upper panel), the wave function is localized in the middle of the array and has no weights on the edge sites. On the other hand, within the conduction band (middle panel), the weights of the MOs are more uniform across the array, with significant weights on the edge site, which is the condition for an MO to be conducting.

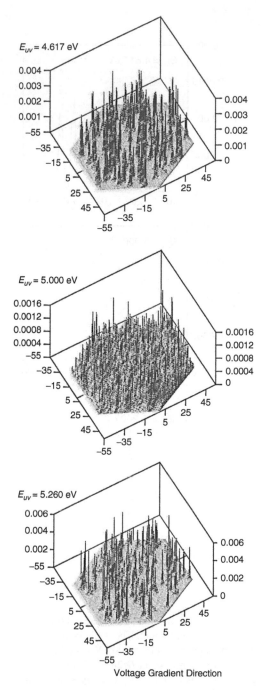

FIGURE 6.6 The localization of the MOs as a function of energy shown over the entire array (same parameters as in Figure 6.5). Both the localization into unconnected islands for energies lower (top panel) and higher (bottom panel) than the conduction band and the delocalization into touching domains for the conducting states are seen to exist across the entire array.

regime, and the bandwidth is then governed by the value of $\Delta\alpha$ and is symmetric around α_0. The position of the highest MO also shifts with β. It is slightly higher than α_0 for large compression, and it decreases to α_0 when the exchange coupling is negligible with respect to $\Delta\alpha$.

The cutoff of the transmission function induced by the external bias can be understood[45] from Equation 6.25. Consider a positive bias (Figure 6.7) and say that k is on the positive bias side and k' on the negative bias side, $\alpha_k > \alpha_{k'}$. When V becomes larger than the interdot coupling strength β and comparable to $\Delta\alpha$, the amplitude $c_{k\lambda}$ of an MO λ of higher energy will be large on site k but very small on site k', and vice versa for an MO of low energy. Since it is the product of the amplitudes that weighs the density of conducting states in Equation 6.25, there will be a cutoff of the transmission at high and low energy when $V > \beta$. Because the band of MOs is not centered around α_0, the cutoff on the low energy side (Figure 6.7) is more important than on the high energy side. When the bias is negative (not shown), the overall effect remains the same.

As can be seen from Figure 6.4 through Figure 6.7, the transmission function is a sensitive probe of the delocalized character of the MOs. It governs the value of the current across the array, but it is not directly experimentally accessible because the current is the result of a double averaging: a thermal averaging over the low-lying excited states of the array and the Fermi averaging over the states of the electrodes. The transmission function is also affected by the strength of the bias, even in the millielectronvolt regime (Figure 6.7). We have shown that this nonlinear effect of the bias can be regarded as a voltage-induced phase transition.[31] In the next section, we review the computation of the current, and we discuss in Section 6.8 how, for low voltage and low temperatures, the transition in the delocalized character of the MOs is manifested as a break in the I-V curves.

6.7 EXPRESSION FOR THE CURRENT: THE FERMI WINDOW FUNCTION

The rate of charge transport across the array is computed using the Ehrenfest theorem.[32] What we actually compute is the rate of change of the charge on the right electrode.[21] For a two-dimensional array, the stationary rate of change is given by

$$\frac{d\langle P\rangle}{dt} = \frac{2\pi}{h} \sum_{I,F} \left(p_F - p_I\right)\langle F|T(E)|I\rangle \delta\left(E_I - E_F\right)\langle I|T(E)|F\rangle \quad (6.27)$$

and the current I across the array is

$$I = 2e\frac{d\langle P\rangle}{dt} \quad (6.28)$$

where the factor 2 takes into account the two possible spin directions and e is the charge of the electron.

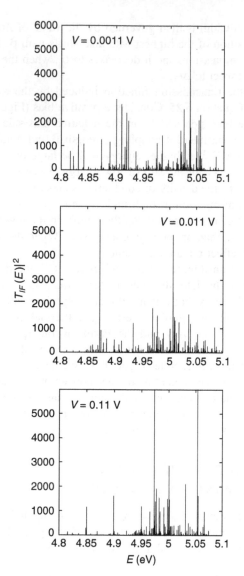

FIGURE 6.7 Effect of the voltage on the transmission function, $\left|T_{IF}(E)\right|^{2}$ (Equation 6.25), computed for 5% of disorder in site energies and a compression $D/2R$ = 1.22. The upper panel is for a small external bias of 1.1 meV. As the bias increases (middle panel, V = 11 meV; upper panel, V = 0.11 V), the band of conducting states shrinks due to the cutoff effect of the voltage. Also note that although the overall shape of the spectrum remains the same, the details of the level structure are affected by the bias.

The weights p_I and p_F correspond to a double thermal averaging. The state of the electrode has a Fermi thermal weight, $f_i(\tilde{\varepsilon}) = f_i\left(\varepsilon - E_{Fermi} - (ev/2)\right)$. The state of the array $\left|\Phi_N\right\rangle$ is a many-electron state and has a Boltzmann factor as its thermal weight. The energies of the excited states of the array are measured with respect to the ground state (GS) of the array (Figure 6.2). Since we work in a one-electron picture and use Koopman's theorem (same set of MOs for the GS as for the excited states), for a singly excited state corresponding to a transition from an MO i below the highest occupied molecular orbital (HOMO) to an MO j above the HOMO, the energy that appears in the Boltzmann factor is

$$E_{\Phi_N} = E_j - E_i \tag{6.29}$$

where E_j and E_i are the energies of the two MOs involved in the transition.

Therefore, in Equation 6.27,

$$\sum_I p_I \longrightarrow \sum_{\Phi_N} \exp\left(-E_{\Phi_N} / kT\right) \int d\varepsilon \, f_i\left(\varepsilon - E_{Fermi} - \frac{eV}{2}\right) \tag{6.30}$$

$$\sum_F p_F \longrightarrow \sum_{\Phi_N} \exp\left(-E_{\Phi_N} / kT\right) \int d\varepsilon' \, f_f\left(\varepsilon' - E_{Fermi} + \frac{eV}{2}\right) \tag{6.31}$$

The current is defined as the rate of charge transport, and it is not normalized per state of the array. Therefore, we do not divide Equations 6.30 or 6.31 by the electronic partition function of the array.

Taking into account the expression of the transmission function (Equation 6.25) and the different resonance conditions, we get for the final expression of the current

$$I = 2e\frac{2\pi}{h}\left(-\pi^2 U^4\right)\sum_{\Phi_N} \exp\left(-E_{\Phi_N} / kT\right)$$

$$\times \sum_\lambda \left(f_f\left(E_\lambda - E_{Fermi} + \frac{eV}{2}\right) - f_i\left(E_\lambda - E_{Fermi} - \frac{eV}{2}\right)\right)\left|\sum_{k,k'} c_{k\lambda} c_{k'\lambda}\right|^2 \tag{6.32}$$

It takes the form of a double average of the transmission function. At low temperatures, the width of the Fermi window function,

$$g(\varepsilon, V) \equiv \left(f_f\left(\varepsilon - E_{Fermi} + \frac{eV}{2}\right) - f_i\left(\varepsilon - E_{Fermi} - \frac{eV}{2}\right)\right) = \frac{\sinh v}{\cosh v + \cosh x} \tag{6.33}$$

with $x = \left(\varepsilon - E_{Fermi}\right) / kT$ and $v = eV / 2kT$, is governed by the bias and is only slightly broadened by temperature. The Fermi window determines the range of MOs that contribute to the conduction. Depending on the position of the Fermi level, E_{Fermi}, the window can be centered across the gap in the transmission function or well above it. It is for this reason that varying the position of the Fermi level for a small bias can be used to probe the change in the delocalization of the array. Note that the total amount of current depends on the number of excited states of the array that have a significant Boltzmann weight. We have shown that for low temperatures, the *I-V* curve's low-lying excited states do not differ significantly from that of the ground state.[21] We therefore report in Section 6.8 computed *I-V* curves for the ground electronic state of the array. To get the total current, one can in good approximation multiply the *I-V* curve of the array by the electronic partition function,

$$\sum_{\Phi_N} \exp\left(-E_{\Phi_N}/kT\right).$$

6.8 COMPUTED CURRENT–VOLTAGE CURVES

The conductance spectrum is plotted in Figure 6.8 as a function of the energy of the Fermi level. The top panel is computed for a very small bias (1.1 meV) and at low temperature (5 K) so that the Fermi function is quite narrow and the level structure of the transmission function is almost fully resolved. The density of states is two 10^5 states per electronvolt, and even for a bias 1 meV, there are still hundreds of states in the Fermi window. At $T = 20$ K and for the same small bias (middle panel), the Fermi window is broadened by the temperature and the peaks of the spectrum overlap. The lower panel is computed for a still-low temperature but a stronger bias ($V = 11$ meV). Two effects can be observed. One is trivial: the lines are wider and the intensity of the current is higher because there are more states in the Fermi window. The other effect is that the spectrum looks qualitatively different from that computed for a lower bias because of the nonlinear effects of the voltage on the electronic structure.

In Figure 6.8, the energy of the Fermi level is varied across the HOMO. Below the energy of the HOMO, conduction occurs through hole conduction, whereas above the energy of the HOMO, conduction occurs through electron conduction. It is only when the HOMO level falls within the Fermi window that both mechanisms for conduction contribute to the current.

The domain-localized phase is manifested as a break in the *I-V* curves at low voltages, as shown in Figure 6.9. The current shown in Figure 6.9 is plotted for decreasing compression of the array. The value of the energy of the Fermi level is 5 eV and $T = 25$ K. For the highest level of compression ($D/2R = 1.2$), the conduction is almost ohmic, even for a small applied bias. Due to disorder effects, the transmission functions for applied bias of the same strengths, but of opposite sign, can slightly differ, leading to slightly asymmetric *I-V* curves, more so at lower compression, where disorder effects are more important. When the compression decreases, at very small bias, the current is very small. This is because the Fermi window includes only low conducting MOs around the gap (see Figure 6.3). For this regime, the conduction is in the variable-range hopping (VRH) regime. As the value of the bias increases, there is a change of slope in the *I-V* curve and the conduction becomes activated. The critical value of the external bias to enter the ohmic regime increases with decreasing compression. The inset in Figure 6.9 shows that for bias strong enough so that the Fermi window includes the better delocalized MOs, the conduction becomes ohmic for all values of $D/2R$ except at 1.24, for which the array is an insulator.

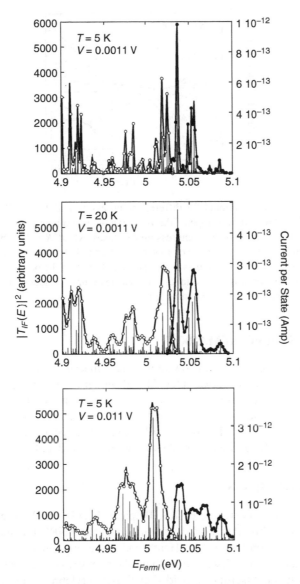

FIGURE 6.8 Conductance spectrum vs. transmission function for different temperatures and bias strengths. The spectrum is obtained by varying the position of the Fermi level across the conduction band. The current corresponding to hole conduction is plotted in open circles, whereas that corresponding to electron conduction is plotted with full circles. The two mechanisms contribute to the total current only when the Fermi window includes the HOMO. Even when computed for low bias ($V = 1.1$ meV) and low temperatures ($T = 5$ K) (top panel), the current spectrum does not correspond to the underlying fully resolved transmission function because of the high density of states (there are hundreds of states within the Fermi function). When the temperature (middle panel) or the voltage (lower panel) increases, the current spectrum further broadens. Note that for higher bias (lower panel), the overall shape of the spectrum is modified by the effects of the voltage on the electronic structure.

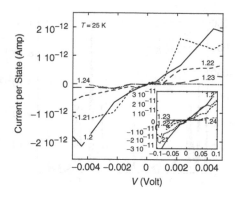

FIGURE 6.9 *I-V* curves computed for different compression and $T = 25$ K for $V < 5$ meV ($E_{Fermi} = 5$ eV). The break in the slope of the *I-V* curve as the voltage is increasing is seen to occur for higher values of V as the compression decreases. At $D/2R = 1.2$, the conduction is ohmic throughout the voltage range. For lower values, it is first domain localized and then becomes ohmic when the Fermi window is wide enough to access the better delocalized MOs. The insert shows that for stronger voltages, one reaches ohmic conduction except for $D/2R = 1.24$, which corresponds to an insulating array.

6.9 CONCLUDING REMARKS

The spectroscopic capabilities of low-temperature measurements of electrical conductance were discussed with examples. The advantages of using the lowest possible voltage were noted. Although for certain applications higher currents may be desirable, the required higher source–drain voltage means that it is not the unperturbed device that is being probed. The resulting level-crossing spectroscopy may be interesting in its own right, and an example was discussed in detail.

ACKNOWLEDGMENTS

F.R. is a Maître de Recherches, FNRS, Belgium. R.D.L. thanks Liège University for the hospitality while this work was started and the FNRS for making this stay possible. We thank Professor J.R. Heath and his group (in particular, Dr. K.C. Beverly) for many useful discussions about their experimental results. This work used the computational facilities provided by SFB 377 (Hebrew University) and NIC (University of Liège). F.R. thanks ARC (University of Liège) and FRFC (2.4562.03) for their support. The final stages of this work were supported by the United States–Israel BiNational Science Foundation.

REFERENCES

1. F Remacle and RD Levine, Electronic response of assemblies of designer atoms: the metal-insulator transition and the role of disorder, *J. Am. Chem. Soc.* 122: 4084–4091, 2000.

2. F Remacle and RD Levine, Architecture with designer atoms: simple theoretical considerations, *Proc. Natl. Acad. Sci. U.S.A.* 97: 553–558, 2000.

3. JN Barisci, R Stella, GM Spinks, and GG Wallace, Characterisation of the topography and surface potential of electrodeposited conducting polymer films using atomic force and electric force microscopies, *Electrochim. Acta* 46: 519–531, 2000.

4. JL Sample, KC Beverly, PR Chaudhari, F Remacle, JR Heath, and RD Levine, Imaging transport disorder in conducting arrays of metallic quantum dots: an experimental and computational study, *Adv. Mater.* 114: 124–128, 2002.

5. RP Andres, T Bein, M Dorogi, S Feng, JI Henderson, CP Kubiak, W Mahoney, RG Osifchin, and R Reifenberger, "Coulomb staircase" at room temperature in a self-assembled molecular nanostructure, *Science* 272: 1323–1325, 1996.

6. XM Lin, HM Jaeger, CM Sorensen, and KJ Klabunde, Formation of long range ordered nanocrystal superlattices on silicon nitride substrates, *J. Phys. Chem. B* 105: 3353, 2001.

7. KC Beverly, JF Sampaio, and JR Heath, Effects of size dispersion disorder on charge transport in self assembled 2D Ag nanoparticle arrays, *J. Phys. Chem. B* 106: 2131–2135, 2002.

8. KC Beverly, JL Sample, JF Sampaio, F Remacle, JR Heath, and RD Levine, Quantum dot artificial solids: understanding the static and dynamic role of size and packing disorder, *Proc. Natl. Acad. Sci. U.S.A.* 99: 6456–6459, 2002.

9. AP Alivisatos, Semiconductors clusters, nanocrystals, and quantum dots, *Science* 271: 933–937, 1996.

10. CP Collier, T Vossmeyer, and JR Heath, Nanocrystal superlattices, *Annu. Rev. Phys. Chem.* 49: 371–404, 1998.

11. CB Murray, CR Kagan, and MG Bawendi, Synthesis and characterization of monodisperse monocrystal and closed packed nanocrystal assemblies, *Annu. Rev. Mater. Sci.* 30, 2000.

12. JJ Shiang, JR Heath, CP Collier, and RJ Saykally, Cooperative phenomena in artificial solids made from silver quantum dots: the importance of classical coupling, *J. Phys. Chem. B* 102: 3425–3430, 1997.

13. NF Mott, *Metal-Insulator Transitions*, London, Taylor & Francis, 1990.

14. S Datta, WD Tian, SH Hong, R Reifenberger, JI Henderson, and CP Kubiak, Current-voltage characteristics of self-assembled monolayers by scanning tunneling microscopy, *Phys. Rev. Lett.* 79: 2530–2533, 1997.

15. RC Ashoori, Electrons in artificial atoms, *Nature* 379: 413–419, 1996.

16. DL Klein, PL McEuen, JE Bowen-Katari, R Roth, and AP Alivisatos, An approach to electrical studies of single nanocrystals, *Appl. Phys. Lett.* 68: 2574–2576, 1996.

17. PW Anderson, New approach to the theory of superexchange interactions, *Phys. Rev.* 115: 2–13, 1959.

18. P Phillips, Anderson localization and the exceptions, *Annu. Rev. Phys. Chem.* 44: 115–144, 1993.

19. HM McConnell, Intramolecular charge transfer in aromatic free radicals, *J. Chem. Phys.* 35: 508–515, 1961.

20. F Remacle, KC Beverly, JR Heath, and RD Levine, Conductivity of 2D Ag quantum dots arrays: computational study of the role of size and packing disorder at low temperature, *J. Phys. Chem. B* 116: 4116–4122, 2002.

21. F Remacle and RD Levine, Current–voltage–temperature characteristics of 2D arrays of metallic quantum dots, *Isr. J. Chem.* 42: 269–280, 2003.

22. D Porath, Y Levi, M Tarabiah, and O Millo, Tunneling spectroscopy of isolated C-60 molecules in the presence of charging effects, *Phys. Rev. B* 56: 9829–9833, 1997.

23. B Alperson, I Rubinstein, G Hodes, D Porath, and O Millo, Energy level tunneling spectroscopy and single electron charging in individual CdSe quantum dots, *Appl. Phys. Lett.* 75: 1751–1753, 1999.

24. U Banin, YW Cao, D Katz, and O Millo, Identification of atomic-like electronic states in indium arsenide nanocrystal quantum dots, *Nature* 400: 542–544, 1999.

25. O Millo, D Katz, YW Cao, and U Banin, Scanning tunneling spectroscopy of InAs nanocrystal quantum dots, *Phys. Rev. B* 61: 16773–16777, 2000.

26. SH Kim, G Medeiros-Ribeiro, DAA Ohlberg, RS Williams, and JR Heath, Individual and collective electronic properties of Ag nanocrystals, *J. Phys. Chem B* 103: 10341–10347, 1999.

27. C Medeiros-Ribeiro, DAA Ohlberg, RS Williams, and JR Heath, Rehybridization of electronic struture in compressed 2D quantum dots superlattices, *Phys. Rev. B* 59: 1633–1636, 1999.

28. H Park, J Park, AKL Lim, EH Anderson, AP Alivisatos, and PL McEuen, Nanomechanical oscillations in a single-C_{60} transistor, *Nature* 407: 57–60, 2000.

29. J Park, AN Pasupathy, JI Goldsmith, C Chang, Y Yaish, JR Petta, M Rinkoski, JP Sethna, HD Abruna, PL McEuen, and DC Ralph, Coulomb blockade and the Kondo effect in single-atom transistors, *Nature* 417: 722–725, 2002.

30. F Remacle, KC Beverly, JR Heath, and RD Levine, Gating the conductivity of arrays of metallic quantum dots, *J. Phys. Chem. B* 107: (50) 13892–13901, 2003.

31. F Remacle and RD Levine, Voltage-induced phase transition in arrays of metallic nano dots in computed transport and surface potential structure, *Appl. Phys. Lett.* 82: 4543–4545, 2003.

32. RD Levine, *Quantum Mechanics of Molecular Rate Processes*, New York, Dover, 1999.

33. R Laundauer, *IBM J. Res. Dev.* 1: 223, 1957.

34. Y Imry, *Introduction to Mesoscopic Physics*, Oxford, Oxford University Press, 2002.

35. S Datta, *Electronic Transport in Mesoscopic Systems*, Cambridge, U.K., Cambridge University Press, 1995.

36. DV Averin and YV Nazarov, Virtual electron diffusion during quantum tunneling of electric charge, *Phys. Rev. Lett.* 65: 2446–2449, 1990.

37. Y Alhassid, The statistical theory of quantum dots, *Rev. Mod. Phys.* 72: 895–968, 2000.

38. A Nitzan, Electron transmission through molecules and molecular interfaces, *Annu. Rev. Phys. Chem.* 52: 681–750, 2001.

39. V Mujica, M Kemp, and MA Ratner, Electron conduction in molecular wires, *J. Chem. Phys.* 101: 6849–6855, 1994.

40. V Mujica, MA Ratner, and A Nitzan, Molecular rectification: why is it so rare? *Chem. Phys.* 281: 147–150, 2002.

41. SN Yaliraki and MA Ratner, Molecule-interface coupling effects on electronic transport in molecular wires, *J. Chem. Phys.* 109: 5036–5043, 1998.

42. YQ Xue, S Datta, and MA Ratner, First-principles based matrix Green's function approach to molecular electronic devices: general formalism, *Chem. Phys.* 281: 151–170, 2002.

43. F Remacle, CP Collier, JR Heath, and RD Levine, The transition from localized to collective electronic states in a silver quantum dots monolayer examined by nonlinear optical response, *Chem. Phys. Lett.* 291: 453–458, 1998.

44. F Remacle, CP Collier, G Markovitch, JR Heath, U Banin, and RD Levine, Networks of quantum nano dots: the role of disorder in modifying electronic and optical properties, *J. Phys. Chem. B* 102: 7727–7734, 1998.

45. F Remacle and RD Levine, Voltage-induced non-linear characteristics of arrays of metallic quantum dots, *Nano Lett.* 2: 697–701, 2002.

46. F Remacle, RD Levine, EW Schlag, and R Weinkauf, Electronic control of site selective reactivity: a model combining charge migration and dissociation, *J. Phys. Chem. A* 103: 10149–10158, 1999.

7 Spectroscopy on Semiconductor Nanoparticle Assemblies

Herwig Döllefeld and Alexander Eychmüller

CONTENTS

7.1 Closely Packed Layers 180
7.2 Quantitative Description: Dipole–Dipole Interaction 183
7.3 Discussion: Closely Packed Layers 185
7.4 Crystalline Superstructures 186
7.5 Quantitative Description: Electronic Interaction 188
7.6 Summary 190
Acknowledgment 190
References 191

In the scientific community involved in research on nanoparticles, for more than 20 years there has been a call for a technology that closes the gap between the macroscopic and the nanoscopic world. Building a technology chain from the nearly molecular building blocks all the way up to the human interface is in the focus of contemporary research. The formation of competence centers and gaining funds supporting these investigations are noticeable all over the world. One approach for handling is embedding the nanoparticles in a matrix. Another approach is the buildup of superstructures forming new materials made out of nanoparticles comparable to the iron block made out of single iron atoms. Just like in the analogy, the materials properties of the bulk material, of both the iron block and the nanoparticle super-structure, consequently arise from the properties of the monomers (atom or nano-particle), from the internal structure of the arrangement in the means of distance between the building blocks and the crystallographic structure, and finally from the nature of the interaction between them.[1]

Because such an immense amount of insight has already been gained concerning both the chemical and physical properties of the isolated particles (or at least embedded in a solvent matrix, but separated from each other), in this chapter we focus on the investigations performed to determine the nature of the interactions that take place between the particles in the moment they come close to each other.

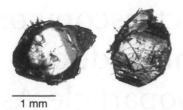

FIGURE 7.1 Microscopic image of two single-crystalline superstructures of $Cd_{17}S_4(SCH_2CH_2OH)_{26}$ nanoparticles.

Besides the brilliant work that has been published on the electronic interactions taking place between metal nanoparticles[2,3] and on Foerster energy transfer in semiconductor quantum dot solids consisting of particles of different sizes,[4,5] little is known about the interactions that take place when superstructures from semiconductor particles of the same type and size are formed. In this chapter we focus on the investigations performed on ordered (crystalline materials) and nonordered (amorphous layers) superstructures of CdS nanoparticles.

Already in 1995 Vossmeyer[6,7] succeeded in building up crystalline superstructures from CdS particles by a simple wet chemical route. Crystals from nanoparticles of different constitutions and sizes were grown. Figure 7.1 shows a microscopic image of two crystals of about 2 mm in diameter consisting of $Cd_{17}S_4(SCH_2CH_2OH)_{26}$ nanoparticles. The internal structure is clearly single crystalline, and we will refer to the investigation of these crystalline structures later in the chapter.

7.1 CLOSELY PACKED LAYERS

For investigating possible interactions between particles in nonordered superstructures, amorphous layers were built up by spin coating of the corresponding solutions.

In 1994, Vossmeyer et al.[8] noticed a red shift of the first electronic transition when comparing the absorption spectra of compact layers of very small CdS nanoparticles with the spectra of the cluster solutions. This effect appeared to be completely reversible, and consequently it was called reversible absorbance shift. Thus, 1994 marked the starting point of interest in the investigation of phenomena that take place in semiconductor particle arrangements.

Recently, Artemyev et al.[9,10] prepared unstructured solids from CdSe nanoparticles with polymers acting as spacers in order to control the distance between the particles. The pure solid built up by drop casting solutions of semiconductor nanoparticles showed very broad transitions bearing resemblance to a bulk spectrum rather than to isolated absorption bands. Micic et al.[11] obtained similar results with InP nanoparticles.

Leatherdale and Bawendi[12] published a paper on CdSe particles in different dielectrics, and a red shift of the first electronic transition was found exhibiting no significant broadening. Kim et al.[13] presented pressure-dependent studies on solutions and compact layers of semiconductor nanoparticles. Different stabilizing agents were used, and in contrast to the results from Leatherdale and Bawendi, no red shift

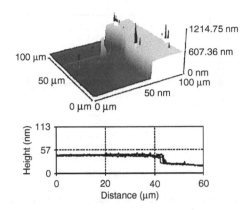

FIGURE 7.2 Top: AFM image of a layer of $Cd_{17}S_4(SCH_2CH_2OH)_{26}$ particles with a thickness of about 500 nm. The obtained surface roughness is 9 nm (1.8%, root-mean-square [RMS]). Bottom: AFM traces taken across a spin-coated film of $Cd_{17}S_4(SCH_2CH_2OH)_{26}$ particles. The traces yield a thickness of this layer of 18.9, 21.0, and 17.2 nm, respectively. The slightly bent background underneath is explained by a drifting during the measuring process. (From Döllefeld, H. et al., *J. Phys. Chem. B*, 106, 5604, 2002. With permission.)

was found for the compact layers of CdSe particles stabilized with trioctylphosphine/trioctylphosphineoxid (TOP/TOPO). Only for particles stabilized with pyridine was a red shift observed. Again, no significant broadening of the electronic transitions was found.

The compact layers in our investigations were built up by spin coating the concentrated solutions of the clusters in dimethylformamide (DMF) onto quartz substrates, whereas the preparation of the particles followed the literature.[6,7] The characterization of the particles by different methods such as UV-visible spectroscopy, powder x-ray diffraction (XRD), and single-crystal XRD yields the formulae $Cd_{17}S_4(SCH_2CH_2OH)_{26}$ and $Cd_{32}S_{14}(SCH_2CH(CH_3)OH)_{36}$ for the two tetrahedral clusters with sizes of 1.4 and 1.8 nm, respectively. A thin transparent and homogenous film remains on the substrate after the spin coating and drying of the sample. Any dielectric influence of possibly remaining solvent could be excluded spectroscopically.

Sample inhomogeneities that are obtained by simply evaporating the solvent (see below) could be avoided. The homogeneity of the films can be documented either by comparing absorption spectra recorded at different spots of the sample or by atomic force microscopy (AFM). Figure 7.2 shows AFM images of edges prepared in the closely packed layers.[14] In the upper image, the topography of a thick layer, with a thickness of about 500 nm, can be seen. The spikes are attributed to recording artifacts. In the lower image, three traces across a rather thin layer of about 19 nm are shown. Even for such thin layers (about 10 to 15 clusters in height), very uniform layers have been obtained. The thicknesses of the resulting layers depend on the concentration of the clusters in the solution. Probably, the viscosity determines the amount of cluster material remaining on the substrate. Comparing results from transmission spectroscopy and AFM, the optical density scaled linearly with the physical thickness of the layers. The films were stable in air for at least 6

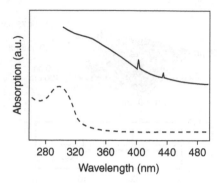

FIGURE 7.3 Different absorption spectra of one sample of $Cd_{17}S_4(SCH_2CH_2OH)_{26}$ nanoparticles prepared by drop casting. The dashed line illustrates a spectrum that corresponds to the expected result, whereas the spectrum shown by the solid line might lead to incorrect conclusions. (From Döllefeld, H. et al., *J. Phys. Chem. B*, 106, 5604, 2002. With permission.)

months. Only in very thin samples was slow degradation observed, probably at the surface of the layers.

As mentioned above, absorption spectra recorded at different spots of the sample were absolutely identical, proving the homogeneity of the obtained layers, as opposed to samples that were built by drop casting of a colloidal solution. As an illustration of this, in Figure 7.3 we show some results from an erroneous experiment: a drop of a colloidal solution of $Cd_{17}S_4(SCH_2CH_2OH)_{26}$ nanocrystals in DMF has been dried on a quartz substrate under ambient conditions.[14] The two spectra have been taken from different spots of this sample, proving the importance of correct sample preparation. The dashed line resembles a rather proper spectrum, whereas the solid line might lead to false conclusions. Namely, a bulk-like absorption spectrum might be identified by mistake by neglecting scattering effects.

After irritations resulting from insufficient sample preparation could be excluded, absorption spectra of the compact layers were recorded. Figure 7.4 shows these spectra in comparison with the spectra of the corresponding cluster solutions for $Cd_{17}S_4(SCH_2CH_2OH)_{26}$ nanocrystals.[14] The spectrum of the compact layer recorded in transmission geometry (dashed line) was found to be shifted to the red, in comparison with the spectrum of the cluster solution (solid line), without showing any significant broadening of the first electronic transition. Surprisingly, the absolute value of the red shift varied with the thickness of the compact layer. The maximum of the red shift in comparison with the spectra of the clusters in solution was about 70 meV at an optical density of about 0.1 at the absorption maximum.

Paying attention to the baseline in the spectrum of the compact layer (dashed line), one might notice a scattering background underneath the actual spectrum, although the investigated samples appear totally transparent in the visible region. For evaluating the possible influence of the scattering background to the resulting spectrum, an integrating sphere was used, allowing the suppression of any influence caused by scattering effects. The absorption measurement of the same layer in this different setup resulted in significantly different signals, i.e., in the removal of the scattering background (dotted line in Figure 7.4).[14] In addition, although still shifted to the red, the effect of shifting

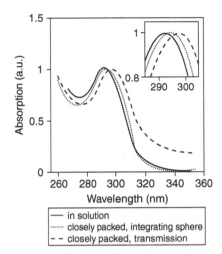

FIGURE 7.4 Absorption spectra of a compact layer of $Cd_{17}S_4(SCH_2CH_2OH)_{26}$ nanoparticles in comparison to the spectrum of the same particles in solution (solid line). Significant scattering is observed when the layer is measured in transmission geometry (dashed line) instead of in the integrating sphere (dotted line). The inset provides a closer look at the resulting changes in the position of the maximum of the transition band depending on the experimental setup. (Solution: cell thickness, 1 cm, OD = 1.15; calculated concentration, 0.7 μmol/l; integrating sphere, OD = 0.14; calculated film thickness, 30 nm; in transmission, optical density (OD) = 0.26; calculation of film thickness is difficult because of scattering background, estimated to be 25 nm.) (From Döllefeld, H. et al., *J. Phys. Chem. B*, 106, 5604, 2002. With permission.)

seems to be reduced. Furthermore, when recorded in the integrating sphere, the energy of the first electronic transition did not show a dependence on the thickness of the layers anymore. Thus, regarding the discussion of the energetic position of an absorption band, it is extremely important to suppress artifacts caused by scattering. Doing this, a red shift of the first electronic transition in the compact layer of about 29 meV, in comparison to the isolated cluster in solution, was observed.[14]

Taking into account the possible pitfalls in the preparation of the compact layers, together with the experimental setup for recording the spectra, compact layers of different cluster sizes were investigated. In all cases, the spectra of the compact layers appeared shifted to the red, in comparison with the corresponding cluster solutions, whereas the amount of the shift depended on the cluster size: the smaller the clusters, the larger the shift. The exact numbers will be given below, together with a quantitative description.

7.2 QUANTITATIVE DESCRIPTION: DIPOLE–DIPOLE INTERACTION

As mentioned above, other groups discuss the energy shift of the first electronic transition in the absorption spectra by polarization effects due to changes in the

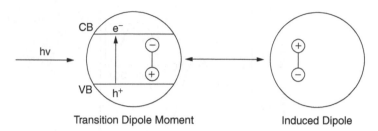

Transition Dipole Moment   Induced Dipole

FIGURE 7.5 Schematic view of the interaction of the transition dipole moment of a particle appearing with the absorption process and the induced dipole moment of a neighboring particle. (From Döllefeld, H. et al., *J. Phys. Chem. B*, 106, 5604, 2002. With permission.)

dielectric properties of the surrounding environment, sometimes called solvato-chromism.[12,13,15,16] This effect, being relevant for semiconductor particles, was first discussed by Brus.[17] Recently, Rabani and coworkers[15] presented a theoretical study on CdSe nanoparticles based on a modified pseudopotential method. The calculations yielded large dipole moments of the nanocrystals, resulting from the anisotropy of the crystal structure of hexagonal cadmium selenide particles. Those moments are affected by a polar environment. However, the authors state that their results "also imply that other nanocrystals (such as GaP, CdS, and Si) with crystal structures that belong to a higher symmetry point group will not interact strongly with their environment." Thus, we do not take into account these effects mentioned.

Leatherdale and Bawendi[12] (and references therein) explained the observed small red shift (4 meV) of the absorption band in their experiment on CdSe particles stabilized by TOP/TOPO by the polarization energy, including screening within a core-shell model. Opposed to these results, Kim et al.[13] excluded an influence of the dielectric surrounding on the first electronic transition of their CdSe nanocrystals when they also were stabilized with TOP/TOPO. A shift of about 25 meV has been observed by these authors for the system CdSe capped with pyridine. Thus, whether or not in CdSe stabilized by TOP/TOPO a significant red shift is observed may depend on the individual viewpoint.

For a quantitative discussion of the experimental results, we present a simple model again based on the influence of the polar properties of the surrounding medium on the transition energy. In contrast to other authors, we treat the surrounding medium not as a continuum, but configured from isolated units, i.e., the neighboring particles. In this model the transition dipole moment of the light-absorbing particle may induce dipole moments in the neighboring particles in their ground state, stabilizing the original transition dipole moment.[14]

In Figure 7.5 the interaction between the transition dipole moment of the absorbing particle and the induced dipole moments in the neighboring particles is sketched schematically. An analog to the additional Coulomb term for the determination of the transition energy by the Brus formula,[18] this interaction lowers the original transition energy of the absorbing particle.

For a determination of the transition dipole moment of the original absorbing particle, the oscillator strength f is evaluated from spectroscopic data:

$$f = 4.3 \cdot 10^{-9} \int \varepsilon d\tilde{v} \approx 4.3 \cdot 10^{-9} \cdot \varepsilon_{max} \cdot \text{FWHM}$$

The transition dipole moment μ_{tr} of the absorption process can be obtained by

$$\mu_{tr} = \sqrt{\frac{3 \, h \, e^2}{8 \, \pi^2 \, m_e \, v_{max}} \cdot f} = \sqrt{2.15 \cdot 10^6 \cdot \frac{f}{\tilde{v}}}$$

with the absorption coefficient ε, the wave number $\tilde{v}$ (cm^{-1}), the full width at half maximum (FWHM) (cm^{-1}), Planck's constant h, the elementary charge e, and the effective mass of the electron m_e. With the assumption that the formula of Clausius and Mossotti is appropriate, the polarizability volume of the clusters can be calculated from the bulk high-frequency dielectric constant ($\varepsilon_{hf} = 5.5$) by

$$\alpha' = \left(\frac{\varepsilon_{hf} - 1}{\varepsilon_{hf} + 2} \right) \cdot \frac{3 \, \varepsilon_0}{N} \cdot \frac{1}{4 \, \pi \, \varepsilon_0} \approx 0.6 \cdot V \cdot \frac{3}{4 \, \pi} = 0.6 \cdot r^3$$

with the particle density N, the volume V, and the radius r. Although the value of ε_{hf} could in general depend on the particle size, it is quite common to use the bulk value for a first rough estimate (see, e.g., Brus[18]). The potential energy of the interaction between a dipole moment and an induced dipole moment is given by[19]

$$E_{pot} = -\frac{\mu_1^2 \, \alpha_2'}{\pi \, \varepsilon_0} \cdot \frac{1}{r^6}$$

We obtained a value close to 1 for the oscillator strength of the 1.4-nm particles, which is expected for this kind of molecular species, having nearly perfect overlap of the wave functions of the charge carriers. The resulting transition dipole moment amounts to 7.9 D, which fits nicely in the linear dependence of the dipole moments on the cluster sizes found by others.[20-26] The resulting polarizabilities strongly depend on the assumed particle shapes and volumes. Assuming a spherical shape with a diameter of 1.4 nm, we obtained a polarizability volume of 205.8 Å^3.

7.3 DISCUSSION: CLOSELY PACKED LAYERS

For a distance between the particles (center to center) of 1.4 nm, this yields a potential energy of −51.2 meV for the interaction between the dipole of the absorbing cluster and the induced dipoles in adjacent clusters (cf. Table 7.1). Taking into account the tetrahedrally shaped particles, the polarizability volume decreases (57.0 Å^3), leading to a reduced shift of −14.2 meV to lower energies. Comparing these results with the value of the experiment (−29 meV), these simple calculations explain quite well the red shift of the absorption band of the clusters in the compact layer in comparison to the isolated clusters in solution.

TABLE 7.1
Red Shift of the First Electronic Transition of Compact Layers of CdS Nanocrystals in Comparison to the Corresponding Transition of the Same Clusters in Solution

	Experimental	Calculated Spherical	Calculated Tetrahedral
Cluster sizes	ΔE (meV)	ΔE (meV)	ΔE (meV)
d = 1.8 nm	12	27.2	6.7
d = 1.4 nm	29	51.2	14.2

Note: Shown are the experimental values for the different cluster sizes as well as the calculated results. The theoretical results for both spherical- and tetrahedral-shaped particles are given. Note that for both particle sizes, the experimental value fits between the values of the assumed particle shapes.

Source: Döllefeld, H. et al., *J. Phys. Chem. B*, 106, 5604, 2002. With permission.

Table 7.1 shows the exact numbers of the red shifts obtained for the two cluster compounds. The red shift for the larger cluster (ø = 1.8 nm) amounts to 12 meV, which transforms to 1 nm in the energy range of 330 nm.

In addition, we applied our simple model to the CdSe particles (ø = 4 nm) of Leatherdale and coworkers, mentioned above. With $\varepsilon = 6.2$, FWHM = 936 cm^{-1}, $\tilde{v}$ = 17,344 cm^{-1},[12] and ε_{max} = 200,000 l mol^{-1} cm^{-1},[27] we obtained an oscillator strength of f = 0.81, a transition dipole moment of μ_{tr} = 9.86 D, and a polarizability volume of α' = 5073 Å^3. With these numbers, the calculated red shift amounts to 3.6 meV, which fits very nicely with the experimental result from Leatherdale et al. (4 meV).

Both approaches, our model of dipole–dipole interaction and that focusing on the polarization energy, are based on the influence of the dielectric surrounding the transition energy of an absorbing cluster. The formalism introduced by Brus and cited by many other groups focuses on the dielectric properties of a continuous medium around the particles, whereas our model describes the surrounding medium as distinct particles with corresponding discrete dipoles. An advantage of our model seems to be its simplicity. Only textbook equations are used, with very little calculational expenditure. The results are in fair agreement with the obtained experimental values for the red shift of the first electronic transition.

7.4 CRYSTALLINE SUPERSTRUCTURES

As mentioned above, we will now focus on the investigation of crystalline super-structures. The monodisperse particles become useful as building blocks of big

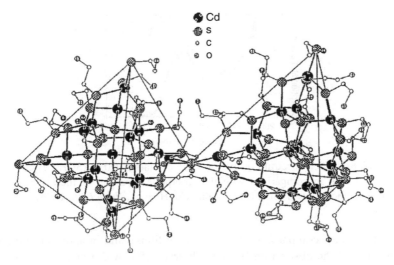

Cd
S
C
O

FIGURE 7.6 Crystal structure of a superstructure of $Cd_{17}S_4(SCH_2CH_2OH)_{26}$ nanocrystals. Two particles are shown that are bound covalently at the tip of each tetrahedron via a bridging sulfur atom from the ligands. For a better representation of the cores, the hydrogen atoms are not shown in this figure. (From Vossmeyer, T. et al., *Science*, 267, 1476, 1995. With permission.)

supramolecular units.[28–30] Reports on the self-organization into quantum dot solids built from semiconductor[1,31,32] and metal particles[33–37] already exist.

To the best of our knowledge, the results presented in our previous paper[38] are the only published work thus far on the energetic position of the first electronic transition in a crystalline superstructure made from nanocrystals of the same size.

A very fine crystalline superstructure of CdSe nanoparticles was presented by Murray et al.[31] in 1995. A comparatively small red shift could be measured in the emission spectra of the solids that was explained by small size deviations of the particles within the sample, resulting in energy transfer from the smaller to the larger particles.[4,5]

Figure 7.1 shows cluster crystals grown to sizes even in the macroscopic range (2 mm). These cluster crystals were formed from thiol-stabilized CdS nanoparticles $Cd_{17}S_4(SCH_2CH_2OH)_{26}$, which have been prepared according to the literature.[6] Single-crystal XRD showed the crystalline structure and the cubic diamond-like superstructure formed by covalently bridging between the tips of neighboring clusters via thiol stabilizer molecules (Figure 7.6).

Because of absorption coefficients of the clusters as high as 84,000 l/mol·cm, it is impossible to perform absorption spectroscopy on this kind of solids. In order to study the optical properties of this material, reflexion spectroscopy has been performed. For this, we milled the crystals down to grain sizes of about 2 μm and measured the diffuse reflection in the integrating sphere. In Figure 7.7, the UV-visible absorption spectrum of this compound in solution is shown together with the reflexion spectrum of the crystalline material.[38] It is clearly seen that in the spectrum of the crystalline material, the shift of the first electronic transition to lower energies

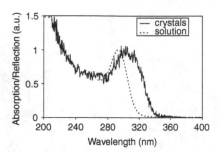

FIGURE 7.7 Reflection spectrum of micron-sized crystals of $Cd_{17}S_4(SCH_2CH_2OH)_{26}$ nano-crystals (solid line) in comparison with the absorption spectrum of the same compound dispersed in solution (dotted line). (From Döllefeld, H. et al., *Nano Lett.*, 1, 267, 2001. With permission.)

is larger (150 meV compared to 29 meV in the films), and in addition to the shift, a broadening of the absorption band is observed (full width at half maximum of about 390 to 520 meV), being absent in the nanocrystal films. For the larger red shift, a more densely packed structure of the crystals, compared to the films, is held responsible, whereas for the broadening, we propose a model of quantum mechanical coupling of the respective states of the individual particles (described below).

7.5 QUANTITATIVE DESCRIPTION: ELECTRONIC INTERACTION

Heath and coworkers published results on the behavior of metal nanoparticles arranged in a two-dimensional monolayer using the Langmuir–Blodgett technique.[1-3] Small silver nanocrystals showed a shift to the red of the low-energy electronic feature in a reflectance measurement when the distance between the particles was lowered. This was found valid down to a particular distance, whereas below this distance, the plasmon band disappeared due to quantum coupling.

Depending on the interparticle distance, those solids reflect dipole–dipole inter-action followed by pure electronic coupling when the particles come into closest contact. Recently, Remacle and Levine[39] studied theoretical electron transfer pro-cesses in arrays of quantum dots.

Quantum coupling of the electronic states of semiconductor quantum dots was found by Schedelbeck et al.[40] GaAs quantum dots were prepared in a GaAlAs environment by epitaxial growth and lithographic techniques. The authors were able to prepare these dots at different distances, and quantum coupling has been proven in the photoluminescence by the observed splitting of the energy states.

For an estimation of the strength of electronic interaction between adjacent clusters, a periodical box potential, shown in Figure 7.8 according to Kronig and Penney,[41] was used. For simulating both the crystalline arrangement of the particles with very small distances between them and the situation in solution with large distances between the particles, this setup could be used simply by varying the distance between the coupling systems. In this model, the boxes representing the particles (size = 1.4 nm) are separated by potential walls. The height of these energy

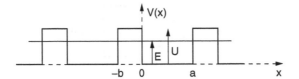

FIGURE 7.8 One-dimensional periodic box potential according to Kronig and Penney, with the width of the boxes a corresponding to the diameter of the particles and the width b and height U of the energy barrier. The width of the barrier represents the distance between the particles. The energy level E for the resulting delocalized electronic system is plotted schematically as well. (From Döllefeld, H. et al., *J. Phys. Chem. B*, 106, 5604, 2002. With permission.)

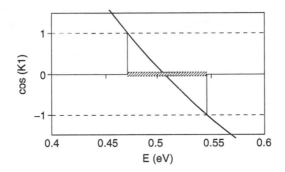

FIGURE 7.9 Graphic solution from the model of Kronig and Penney for the energy states of the electron in a nanoparticle with a small distance between the particles (for details of the calculation, see text). The boundary conditions for the wave function yield an expression including a cosine function with the wave vector K and the period length l ($l = a + b$; cf. Figure 7.8). The cosine function values (between 1 and –1) yield the permitted energy region for the electronic states. In this case, this is a subband spreading over 75 meV. (From Döllefeld, H. et al., *J. Phys. Chem. B*, 106, 5604, 2002. With permission.)

barriers was chosen to be 3 eV, and the separation of the boxes (i.e., the interparticle distance) was varied between 20 nm, reflecting large distances, and 0.7 nm, modeling the situation with neighboring clusters. The effective masses of the charge carriers were taken from the literature (0.2 and 0.7 for electrons and holes, respectively[42]). The three-dimensional arrangement of the particles was taken into account by multiplying the one-dimensional results by 3. According to Brus,[18] the transition energy was then calculated by addition of the bulk band gap and the Coulomb interaction. For large distances between the clusters, a transition energy of 3.92 eV was obtained, which is in fair agreement with the experimental value (4.24 eV). For the clusters with small distances between the inorganic cores, we obtained the formation of a subband for the electron states, which spans about 75 meV (for one-dimensional) for a separation of 0.7 nm (Figure 7.9). Due to the higher effective mass of the holes, a narrower subband (1.5 meV) was observed for this charge carrier. The centers of the subbands remained exactly at the same energetic positions as the discrete levels

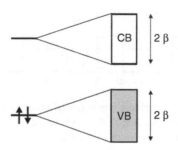

FIGURE 7.10 Schematic view of the development of bands, e.g., valence band and conduction band, from the molecular energy levels by quantum mechanical coupling, including the resonance integral β.

in the solution case. Thus, it turns out that for the electronic interaction of semiconductor particles in close contact, a broadening of the transition band would be expected without significant energetic shifting. This electronic coupling of the molecular states in the clusters forming bands in the superstructures is sketched in Figure 7.10, explaining both the unchanged center of gravity of the energetic transition and the broadening coming along with it. The additional shift of the transition energy in the crystalline material is then due only to dipole–dipole interactions of adjacent particles.

7.6 SUMMARY

The chemical and physical properties of semiconductor nanoparticles are broadly understood. For access from the macroscopic world to use these particles as new materials, they must be embedded in a matrix that could be either some appropriate material or, again, cluster material but assembled in a sort of superstructure. In both cases, the properties of the particles will be influenced by the changed surrounding medium, affecting the new materials' properties. In this article investigations were presented on the interactions between particles in ordered and nonordered superstructures. Possible pitfalls were highlighted in sample preparation and the absorption spectroscopy setup that lead to problematic or even wrong experimental results and, therefore, to false conclusions. After the spectroscopic results for the closely packed layers, an alternative model was presented for the dipole–dipole interactions that describes the experimental results quite well. In addition, a spectroscopy for crystalline superstructures was introduced. The results obviously point in the direction of an electronic coupling between the semiconductor nanoparticles. Clearly, the models presented are rough estimates for our experimental results, and detailed theoretical studies would help us to gain a more quantitative picture.

ACKNOWLEDGMENT

We thank Dr. Andreas Richter, University of Hamburg, for performing the AFM measurements and Prof. Nick Kotov, Oklahoma State University, for valuable discussions over the years. This work was supported by the German Science Foundation, SFB 508.

REFERENCES

1. Collier, C.P., Vossmeyer, T., Heath, J.R. *Annu. Rev. Phys. Chem.* 1998, *49*, 371.
2. Collier, C.P., Saykally, R.J., Shiang, J.J., Heinrichs, S.E., Heath, J.R. *Science* 1997, *277*, 1978.
3. Shiang, J.J., Heath, J.R., Collier, C.P., Saykally, R.J. *J. Phys. Chem. B* 1998, *102*, 3425.
4. Kagan, C.R., Murray, C.B., Nirmal, M., Bawendi, M.G. *Phys. Rev. Lett.* 1996, *76*, 1517.
5. Kagan, C.R., Murray, C.B., Bawendi, M.G. *Phys. Rev. B* 1996, *54*, 8633.
6. Vossmeyer, T., Reck, G., Katsikas, L., Haupt, E.T.K., Schulz, B., Weller, H. *Science* 1995, *267*, 1476.
7. Vossmeyer, T., Reck, G., Schulz, B., Katsikas, L., Weller, H. *J. Am. Chem. Soc.* 1995, *117*, 12881.
8. Vossmeyer, T., Katsikas, L., Giersig, M., Popovic, I.G., Diesner, K., Chemseddine, A., Eychmüller, A., Weller, H. *J. Phys. Chem.* 1994, *98*, 7665.
9. Artemyev, M.V., Bibik, A.I., Gurinovich, L.I., Gaponenko, S.V., Woggon, U. *Phys. Rev. B* 1999, *60*, 1504.
10. Artemyev, M.V., Woggon, U., Jaschinski, H., Gurinovich, L.I., Gaponenko, S.V. *J. Phys. Chem.* 2000, *104*, 11617.
11. Micic, O.I., Ahrenkiel, S.P., Nozik, A. *Appl. Phys. Lett.* 2001, *78*, 4022.
12. Leatherdale, C.A., Bawendi, M.G. *Phys. Rev. B* 2001, *63*, 165315.
13. Kim, B.S., Islam, M.A., Brus, L.E., Herman, I.P. *J. Appl. Phys.* 2001, *89*, 8127.
14. Döllefeld, H., Weller, H., Eychmüller, A. *J. Phys. Chem. B* 2002, *106*, 5604.
15. Rabani, E., Hetenyi, B., Berne, B.J., Brus, L.E. *J. Chem. Phys.* 1999, *110*, 5355.
16. Banyai, L., Gilliot, P. *Phys. Rev. B* 1992, *45*, 14136.
17. Brus, L.E. *J. Chem. Phys.* 1983, *79*, 5566.
18. Brus, L.E. *J. Chem. Phys.* 1984, *80*, 4403.
19. Atkins, P.A. *Physikalische Chemie*, Vol. 2, *Auflage*, German ed. VCH Verlagsgesellschaft: Weinheim, 1996.
20. Colvin, V.L., Alivisatos, A.P. *J. Chem. Phys.* 1992, *97*, 730.
21. Colvin, V.L., Cunningham, K.L., Alivisatos, A.P. *J. Chem. Phys.* 1994, *101*, 7122.
22. Blanton, S.A., Leheny, R.L., Hines, M.A., Guyot-Sionnest, P. *Phys. Rev. Lett.* 1997, *79*, 865.
23. Empedocles, S.A., Bawendi, M.G. *Science* 1997, *278*, 2114.
24. Empedocles, S.A., Neuhauser, R., Bawendi, M.G. *Nature* 1999, *399*, 126.
25. Empedocles, S.A., Neuhauser, R., Shimizu, K., Bawendi, M.G. *Adv. Mater.* 1999, *11*, 1243.
26. Shim, M., Guyot-Sionnest, P. *J. Chem. Phys.* 1999, *111*, 6955.
27. Schmelz, O., Mews, A., Basché, T., Herrmann, A., Müllen, K. *Langmuir* 2001, *17*, 2861.
28. Alivisatos, A.P. *J. Phys. Chem.* 1996, *100*, 13226.
29. Alivisatos, A.P. *Science* 1996, *271*, 933.
30. Alivisatos, A.P. *Endeavour* 1997, *21*, 56.
31. Murray, C.B., Kagan, C.R., Bawendi, M.G. *Science* 1995, *270*, 1335.
32. Rizza, R., Fitzmaurice, D., Hearne, S., Hughes, G., Spoto, G., Ciliberto, E., Kerp, H., Schropp, R. *Chem. Mater.* 1997, *9*, 2969.

33. G. Schmid, Ed. *Clusters and Colloids*. VCH: Weinheim, Germany, 1994.
34. Vossmeyer, T., DeIonno, E., Heath, J.R. *Angew. Chem. Int. Ed.* 1997, *36*, 1080.
35. Wang, Z.L., Harfenist, S.A., Vezmar, I., Whetten, R.L., Bentley, J., Evans, N.D., Alexander, K.B. *Adv. Mater.* 1998, *10*, 808.
36. Pileni, M.P. *J. Phys. Chem. B* 2001, *105,* 3358.
37. Rogach, A.L., Talapin, D.V., Shevchenko, E.V., Kornowski, A., Haase, M., Weller, H. *Adv. Funct. Mater.* 2002, *12*, 653.
38. Döllefeld, H., Weller, H., Eychmüller, A. *Nano Lett.* 2001, *1*, 267.
39. Remacle, F., Levine, R.D. *J. Phys. Chem. B* 2001, *105*, 2153.
40. Schedelbeck, G., Wegschneider, W., Bichler, M., Abstreiter, G. *Science* 1997, *278*, 1792.
41. Flügge, S. *Rechenmethoden der Quantentheorie*, Vol. 5, verbesserte Auflage ed. Springer-Verlag: Berlin, 1993.
42. *Landolt-Börnstein Numerical Data and Functional Relationship in Science and Technology*, New Series, Group III, 22a, Sec. 3.6. Springer-Verlag: Berlin, 1982.

8 Optical and Dynamic Properties of Gold Metal Nanomaterials: From Isolated Nanoparticles to Assemblies

Thaddeus J. Norman, Jr., Christian D. Grant, and Jin Z. Zhang

CONTENTS

8.1 Introduction 193
8.2 Origins of the Surface Plasma Resonance 194
8.3 The SPR and Hot Electron Relaxation Dynamics in Spherical
 Au Nanoparticles 195
8.4 Optical Emission of Metal Nanoparticles and Nanoclusters 198
8.5 The SPR and Hot Electron Relaxation Dynamics in Spheroidal,
 Rod-Shaped, and Au Nanoshell Particles 199
8.6 Optical Absorption and Hot Electron Dynamics
 in Au Nanoparticle Superlattices and Aggregates 200
8.7 Conclusion 203
References 204

8.1 INTRODUCTION

The properties of small metal particles have intrigued people for centuries (Figure 8.1). In antiquity, colloidal gold and silver were used to color glass, and in the late 19th and early 20th centuries, theories were developed to explain the color that these particles gave to glass and solutions.[1,2] It has been established that the color of materials containing colloidal metal is the consequence of the absorption and scattering of light by the metal particles. However, the study of metal colloids continues because their optical properties have great potential for emerging technological applications. For example, the wavelength of light absorbed by metal colloids is determined by the environment around the metal particle. Sensors may be developed

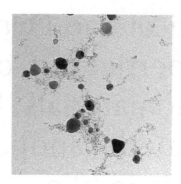

FIGURE 8.1 Transmission electron microscopy (TEM) micrograph of Au nanoparticles ranging from ~5 to ~80 nm in size.

exploiting this property.[3] In addition, Raman scattering from molecules adsorbed onto the surface of metal particles is enhanced by several orders of magnitude, and this enhancement allows for the collection of Raman spectrum from dilute samples.[4-11] Finally, the light-scattering properties of metal nanoparticles could allow them to be used in waveguides or other opto-electronic devices.[12]

In this chapter we will discuss some of the unique features of metal nanomaterials, with emphasis on their optical and dynamic properties. Systems of interest include isolated particles with different shapes as well as aggregates and assemblies or superlattices of such metal nanoparticles. Specifically, the issues of particle size, shape, and particle–particle interaction will be addressed. The focus will be on gold nanomaterials, though the underlying theory governing the optical properties applies to copper, silver, and other metal nanosystems as well. Gold was chosen because it is one of the most studied nanoparticle systems and its dynamic and optical properties are the best characterized.

8.2 ORIGINS OF THE SURFACE PLASMA RESONANCE

The dominant feature in the visible absorption spectrum of metal nanoparticles is the surface plasma resonance (SPR) (Figure 8.2). For gold nanoparticles in water, the SPR occurs at ~520 nm; however, the exact position is dependent upon the embedding medium, shape, and structure of the gold particles. The SPR can best be understood as the resonant interaction of the conduction band electrons near the surface of the nanoparticles with incident light. Theoretically, conduction band electrons in a metal can be treated as a plasma, i.e., a gas composed of charged particles. This plasma oscillates on the surface of metal nanoparticles, and strong absorption of light occurs when the frequency of the incident light matches the resonant frequency of the plasma.

Extensive theoretical work has been done to develop equations to calculate the absorption spectrum of metal nanoparticles.[13-15] Mie theory is the basis for most of these equations, though other theories have been developed.[1,2,13,16,17] Essentially, Mie theory is the solution of Maxwell's equations for the scattering of light off a small

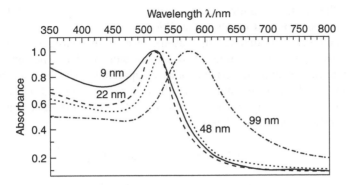

FIGURE 8.2 UV-visible spectrum of isolated spherical gold nanoparticles. The position and width of the SPR band are functions of the particle size. (From Link, S. and El-Sayed, M.A., *J. Phys. Chem. B*, 103, 8410–8426, 1999. With permission.)

charged sphere in a dielectric medium. The absorption cross section (α) for metal nanoparticles is predicted to be

$$\alpha = 18\pi V\varepsilon_m^{3/2}\varepsilon_2/\lambda[(\varepsilon_1 + 2\varepsilon_m)^2 + \varepsilon_2^2] \tag{8.1}$$

where V is the volume of the particles, λ is the wavelength of the incident light, ε_m is the dielectric constant of the surrounding medium, and ε_1 and ε_2 are the real and imaginary parts of the complex permittivity of the metal, respectively. As can be seen from Equation 8.1, maximum absorption occurs when $-2\varepsilon_m = \varepsilon_1$, which is the resonance condition. In practice, the resonance peak position is influenced by the wavelength dependence of not only ε_m and ε_1 but also V and ε_2. The dielectric properties of the surrounding medium along with ε_2 determine the width of the plasma resonance.

8.3 THE SPR AND HOT ELECTRON RELAXATION DYNAMICS IN SPHERICAL Au NANOPARTICLES

The Mie theory can be used to provide a qualitative understanding of the response of the SPR in gold nanoparticles under various conditions. To begin, the size of the particle strongly influences the position and width of the SPR (Figure 8.2). Particles larger than ~25 nm display a red-shifted and broadened SPR,[13,18] whereas particles less than ~10 nm in size display a broadened and blue-shifted SPR.[19,20] Larger-size particles experience inhomogeneous polarization as the wavelength of light approaches the particles' size. This changes ε_1 and ε_2, which leads to the observed line width broadening. Also, as the volume of the particle increases, its contribution to the determination of the position of SPR increases. The red shift in SPR for larger particles is thus an effect of the increase in particle volume. When the particle size decreases, it begins to approach the mean free path of electrons in bulk gold; thus, more collisions between the electrons and surface occur. This has the effect of dampening the electron motion, which causes a change in ε_1 and ε_2.[21] This has an effect on both position and width of the SPR.

Quantum mechanics dictates that as the size of the particles decreases, the conduction band, which is composed of a continuum of states, becomes discrete. The energy of intraband transitions becomes comparable to kT, the ambient thermal energy. This also leads to dampening of the electron motion and a change in the dielectric properties. Using the Drude model,[22] the intraband contribution to the dielectric constant can be calculated via

$$\varepsilon_1^{\text{intraband}}(\omega) = 1 - [\omega_p^2/(\omega^2 + \Gamma^2)] \tag{8.2}$$

$$\varepsilon_2^{\text{intraband}}(\omega) = \omega_p^2\Gamma/[\omega(\omega^2 + \Gamma^2)] \tag{8.3}$$

where ω_p is the plasma frequency and Γ is a damping constant. The damping constant, Γ, also has a size dependence

$$\Gamma = \Gamma_0 + Av_F/R \tag{8.4}$$

where Γ_0 is the frequency of inelastic collisions, A is an electron–surface interaction constant (typically unity), v_F is the Fermi velocity of electrons, and R is the particle radius. The effects of intraband transitions must be taken into consideration when calculating the position of the PR for the transitions occurring within the same spectral region as the SPR. This is not the case, however, for other metal nanoparticles, such as Ag and Al.[13]

Electron–phonon interaction is a fundamental property of solid materials. For metals, this interaction determines not only static properties such as optical properties but also dynamic properties such as hot electron relaxation. Hot electron relaxation in metals occurs on ultrafast timescales due to strong electron–phonon interactions. One basic question is whether the electron relaxation time in nanoparticles is different from that of the corresponding bulk metal. In other words, does particle size affect the electronic relaxation time and, if so, how? The ultrafast relaxation of electrons in bulk or nanoparticulate metals can be monitored using ultrafast laser techniques.[23] For example, the electronic relaxation dynamics in gold nanoparticles can be monitored via femtosecond pump-probe laser spectroscopy. Hot electrons are created by photoexcitation, and the relaxation process of the electrons (cooling) on the femtosecond timescale is dominated by electron–phonon coupling, which results in transfer of energy from excited electrons into the lattice vibrational motion (phonons). The relaxation of hot electrons in a metal nanoparticle can be described by a two-temperature model for energy exchange between electrons and the lattice

$$C_e\left(T_e\right)\frac{\partial T_e}{\partial t} = -g\left(T_e - T_l\right)$$
$$C_l\frac{\partial T_l}{\partial t} = g\left(T_e - T_l\right) \tag{8.5}$$

where T_e and T_l are the temperatures of the electrons and lattice, respectively, $C_e(T_e)$ is the temperature-dependent electronic heat capacity, C_l is the lattice heat capacity, and g is the electron–phonon coupling constant.[24–26] Since the electronic heat capacity depends on the temperature of the electron distribution, the effective rate constant for the relaxation is $g/C_e(T_e)$, and it decreases as the electronic temperature increases. Higher excitation intensities produce higher electronic temperatures, which then yield longer relaxation times. The relaxation time also reflects how strongly the excited electrons couple to the phonons. The electron and phonon density of states, as well as the spectral overlap between electrons and phonons, is expected to decrease with decreasing particle size. At the same time, surface effects become more significant with decreasing size. The change in density of states (DOS) and surface properties may have competing effects on the electronic relaxation dynamics, which may obscure the effect of the change in DOS on the measured electronic relaxation of gold nanoparticles.[23]

Early electronic relaxation measurements by Smith et al.[27] suggested a possible size dependence of the electronic relaxation in the 1- to 40-nm range. However, subsequent measurements by Logunov et al.[28] and Hodak et al.,[29] as well as by Link and El-Sayed,[30] found no size dependence of the electronic relaxation for particle sizes down to 2.5 nm. The relaxation time in Au nanoparticles was reported to be the same as that in bulk gold (~1 psec). The discrepancy between the two measurements could be the result of the power dependence of the relaxation process. The excitation intensity used in the transient absorption measurements of Zhang et al.[31] is likely much higher than that used in the transient bleach experiments of Hodak et al.[79] and El-Sayed et al.[26] Another complication is the solvent surface interaction, which has been shown to have a profound influence on the relaxation time.[31,32]

As particle size becomes smaller than 1 nm, the electronic relaxation time seems to become significantly longer, suggesting that the particles are becoming more molecular in nature; that is, the conduction band can no longer be described as a continuum of states. This has the effect of increasing the lifetime of the excited state. The measured relaxation time for Au_{13} clusters is ~300 psec,[27] whereas Au_{28} clusters capped with glutathione have a biexponential decay with subpicosecond and nanosecond components.[33] The fast component in Au_{28} is attributed to relaxation from a higher-lying excited state populated by the 400-nm excitation pulse to a lower electronic state, whereas the longer nanosecond component is attributed to the radiative lifetime of the Au_{28} clusters. Preliminary data on electronic relaxation in Au_{11} clusters (~0.7-nm diameter) reveal a similar long-lived lifetime component (C. Grant, unpublished research, 2002), supporting the notion that for very small metal clusters, the electronic relaxation is much slower due to weak electron–vibration coupling.

Oscillations in the electronic relaxation have been observed when the probe wavelength matches that of the SPR[20,34,35] (Figure 8.3). The oscillations have been attributed to a coherent excitation of the radial breathing vibrational modes of the particles. After the initial excitation, the electrons transfer their energy into the lattice heating up the particle, causing a rapid expansion of the particle volume. This changes the electron density of the particle, manifesting itself as a modulation in the transient absorption signal. The frequency of the oscillation has been found to be dependent on particle size, with the frequency increasing linearly with decreasing

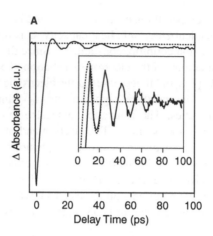

FIGURE 8.3 Transient absorption of ~60-nm gold nanoparticles. The oscillations are due to the coherent excitation of the radial breathing modes of the nanoparticle. (From Hodak, J.H. et al., *J. Phys. Chem. B*, 104, 9954–9965, 2000. With permission.)

particle size. In addition, the degree of damping of the oscillations depends on the polydispersity of the nanoparticle sample. In a monodisperse system, upon excitation all of the particles vibrate at the same frequency, giving rise to oscillations in the transient absorption/bleach spectrum. However, as the size distribution of the sample increases, the different-sized particles vibrate with different frequencies, leading to damping of the observed oscillations. Thus, the transient absorption technique can be used to determine polydispersity of metal nanoparticle samples. Similar oscillatory behavior has been observed in the dynamics of bimetallic core-shell[20,34,35] nanoparticles.

8.4 OPTICAL EMISSION OF METAL NANOPARTICLES AND NANOCLUSTERS

In 1998 Wilcoxon et al.[36] first reported the existence of photoluminescence in small metal nanoparticles. When excited by 230 nm of light, gold nanoparticles with diameters of <5 nm display a broad emission band centered around 440 nm. The quantum yield of the emission is low, but the observation is significant because photoemission is seldom observed in bulk metals.[37,38] Similar to semiconductors, photoemission in metals and metal nanoparticles is due to the recombination of electrons with holes. The emission is influenced by the size, shape, and surrounding medium of the particle.[36] For example, the emission quantum yield of gold nanorods has been found to be several orders of magnitude greater than that of small nanoparticles, even though the size of the rods is an order of magnitude larger than spherical nanoparticles that emit.[39] It has been theorized that SPR amplifies the adsorbed and emitted light from the gold nanoparticles, thus enhancing the emission from gold nanoparticles. Femtosecond time–resolved emission studies of 25-nm gold nanoparticles suggest that the emission occurs on the same timescale as thermal

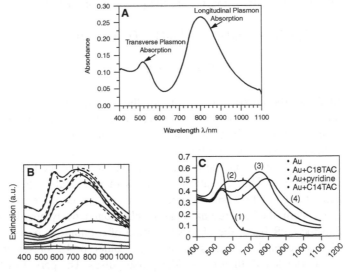

FIGURE 8.4 (A) UV-visible spectra of gold nanorods. (From Link, S., and El-Sayed, M.A., *J. Phys. Chem. B*, 103, 8410–8426, 1999. With permission.) (B) SiO_2/Au core-shell, where the solid line is data and the dashed line is the theoretical absorption. (From Oldenburg, R.D. et al., *Chem. Phys. Lett.* 288: 243–247, 1998. With permission.) (C) Gold nanoparticle aggregates prepared from pyridine and different surfactants. (From Norman, Jr., T.J., *J. Phys. Chem. B* 106: 7005–7012, 2002.)

relaxation and has the same decay kinetics.[40] This suggests that the SPR plays a role in the mechanism for emission.

8.5 THE SPR AND HOT ELECTRON RELAXATION DYNAMICS IN SPHEROIDAL, ROD-SHAPED, AND Au NANOSHELL PARTICLES

Deviations from spherical particle geometry affect the SPR as well. Spheroidal, rod-shaped, and gold nanoshells have two plasma resonances: one for the transverse and one for the longitudinal vibrational mode[14,17,41] (Figure 8.4). Typically in these structures, the transverse plasma resonance (TPR) maintains its position at ~520 nm, but the longitudinal plasma resonance (LPR) is significantly red shifted. Shifts on the order of hundreds on nanometers are not unusual, particularly for gold nanorods, where it has been demonstrated that the ratio between the length of the rod and the rod's diameter determines the position of the LPR.[41] Semiconducting nanoparticles with a surface gold layer (gold nanoshells) show similar behavior, with the position of the LPR depending on the thickness of the shell and diameter of the dielectric core.[42] These shifts are the result of a polarization dependence of the plasma vibrational modes that occurs when particle geometry is no longer spherical. Taking this polarization into account, absorption cross section α becomes

$$\alpha_{1,t} = [2\pi N V \varepsilon_m^{3/2}/3\lambda]\Sigma P_{1,t}^{-2}\varepsilon_2/[(\varepsilon_1 + (1 - P_{1,t})/P_{1,t})^2 + \varepsilon_2^2] \qquad (8.6)$$

where N is the particle density and $P_{l,t}$ are the polarization factors

$$P_l = [(1 - D^2)/D^2]\{2D^{-1}\ln[(1 + D)/ (1 - D)] - 1\} \qquad (8.7)$$

$$P_t = (1 - P_l)/2 \qquad (8.8)$$

where P_l and P_t are the polarization terms for the longitudinal plasma mode and transverse plasma mode, respectively, and D is a function of the particle dimension. The electronic relaxation in gold nanorods is similar to that of spherical nanoparticles.[30,43] Interestingly, the oscillatory component of the relaxation dynamics gives valuable information about the homogenous line width of the LPR, just as the oscillatory component of the relaxation dynamics of spherical particles can give information about the polydispersity of that system. As a result of the polydispersity of the nanorod system, the longitudinal band is inhomogeneously broadened. When probing within the LPR band, the observed period of oscillations becomes shorter at bluer probe wavelengths. This is because the bluer probe pulses interrogate nanorods of a shorter length than the redder probe pulses. The period of the oscillations can be calculated using

$$\tau \approx 2L/c_t \qquad (8.9)$$

where L is the length of the rod and c_t is the transverse speed of sound of bulk gold. Thus, from the oscillation period, the LPR homogenous line width can be extrapolated.

8.6 OPTICAL ABSORPTION AND HOT ELECTRON DYNAMICS IN Au NANOPARTICLE SUPERLATTICES AND AGGREGATES

An ensemble of individual gold nanoparticles near each other (separations on the order of angstroms) in an ordered array is referred to as a superlattice, whereas an ensemble of individual gold nanoparticles near each other with no ordered arrangement is referred to as an aggregate. Superlattices and aggregates can consist of tens to thousands of gold nanoparticles arranged in a variety of different structures. It is this structural arrangement that has a profound influence on the PR of these gold nanoparticle systems. However, analogies can be drawn between the behavior of the PR of superlattices and aggregates and that of discrete gold nanoparticles to achieve a qualitative understanding of the PR in these structures. For example, gold superlattices display a red-shifted plasma resonance (Figure 8.5), which can be viewed as a consequence of interparticle interaction.[44] The collective plasma oscillations of the individual gold nanoparticles in the lattice determine ε_m, and it is this ε_m that is responsible for the red shift in the PR. The PR in gold nanoparticle aggregates displays a similar behavior. Generally, aggregates of gold nanoparticles can be divided into two groups: weakly interacting aggregates (WIAs) and strongly interacting aggregates (SIAs). These two types of structures display different optical characteristics.

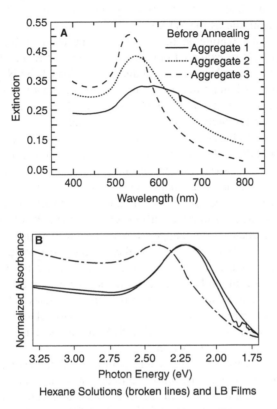

FIGURE 8.5 (A) Absorption spectra of DNA/Au nanoparticle aggregates connected with DNA of 24 (1), 48 (2), and 72 (3) base pairs. (From Mayya, K.S. et al., *Bull. Chem. Soc. Jpn.*, 73, 1757–1761, 2000. With permission.) (B) Au superlattices with the dotted line corresponding to the absorbance in the isolated system nanoparticle in hexane. (From Lazarides, A.A., and Schatz, G.C., *J. Phys. Chem. B*, 104, 460–467, 2000. With permission.) These data show weak or moderate interparticle interaction.

   Typically in WIA structures, the gold nanoparticles are linked to each other by a spacer molecule such as DNA[45,46] or a dithiol[47] (Figure 8.5). Both the density of gold nanoparticles in the WIAs and the number of particles in the WIAs affect the position and width of the PR. A broader and redder PR occurs in WIA structures, with a greater number and higher density of gold nanoparticles. As in superlattices, the collective plasma oscillations of the gold nanoparticles determine ε_m. This causes the observed shift in the PR to redder wavelengths and coupling of the individual particle TPR. The broadening can be viewed as a consequence of dampening of the plasma oscillations, which leads to changes in ε_1 and ε_2. In large gold nanoparticles, inhomogeneous polarization by wavelengths of light on the order of the particle size can cause changes in the dielectric constants, which leads to broadening of the PR. As a consequence of the coupling of the PR of the individual particles in the aggregate, WIA structures can be viewed as a giant gold nanoparticle. Therefore, the broadening seen in larger and denser aggregates can be viewed as the effect of inhomogeneous polarization of the WIAs by light, whose wavelengths are of the same order of magnitude as the WIA size.

The characteristic feature in the absorption spectrum of SIA structures is a distinct TPR and LPR or extended plasmon resonance (EPR) band. An example of an SIA system is that produced by the reduction of HAuCl$_4$ by Na$_2$S.[48] During this reaction, a broad EPR band emerges in the near infrared that subsequently shifts to bluer wavelengths as the reaction progresses. Most SIA systems studied, however, are prepared by first producing gold nanoparticles and then introducing an agent to partially remove the stabilizing agent[49,50] (Figure 8.4). In these cases, the EPR band red shifts from the TPR as the SIA structures form. The factors affecting the position of the EPR band in SIA structures are not fully understood. However, experiments have shown that aggregate size, shape, density, and method of preparation play a key role in determining the nature of this band.[49] A small-angle x-ray-scattering (SAX) experiment suggests that the shift observed in the EPR of the HAuCl$_4$–Na$_2$S system is the consequence of a change from a globular to a stringy SIA structure.[51] A structural feature of many SIA systems that distinguishes them from WIA structures is a physical connection between some of the particles in the aggregate. This would imply that the individual particles in the aggregates cannot be treated as spherical particles, but as spheroidal or rod-shaped particles. Thus, the P_l and P_t terms must be considered when predicting the behavior of the PR in SIA structures.[13,15] From this perspective, the EPR in SIA structures is then a consequence of the strong particle–particle interaction.

Qualitative arguments such as these may not fully explain the origin of the position or broadness of the EPR in SIA structures. Continuing with the analogy of SIA structures as extended nanorods, the position of the EPR would be a function of the length or stringiness of the SIAs, and the line width of the EPR is a reflection of the size or structure dispersion of the SIAs. A hole-burning experiment supports this view of SIA structures.[52] However, physically, SIA and WIA structures are a collection of gold nanoparticles near each other. In this respect, SIA structures, like WIA structures, may be similar to large gold nanoparticles. Therefore, the position and width of the EPR may be a function of the number of particles and density of particles in the aggregate. Recently, an attempt has been made to model the position of the EPR in metal nanoparticle aggregates by treating the surface electrons in the aggregates as if they were a network of RLC circuits, electronic circuits composed of resistors (R), inductors (L), and capacitors (C).[53,54] Though this may be an accurate description of SIA systems, this model currently does not adequately predict observed experimental data.

Periodic oscillations in the transient absorption profiles also occur in the SIA aggregates when probed at the EPR band[52] (Figure 8.6). The observation of oscillations supports the idea that the broad EPR band consists of subbands from different-sized or -shaped aggregates. Further evidence for this idea is the probe wavelength dependence of the oscillations. As the probe wavelength is varied from red to blue, the period of the oscillations becomes longer, which is the same as what occurs in Au nanorods. As in isolated gold nanoparticles, the vibrational oscillations observed for Au nanoparticle aggregates were assigned to their radial breathing modes. However, the vibrational behavior is somewhat different from that of purely isolated spherical particles. The Lamb equation[55]

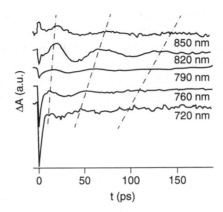

FIGURE 8.6 Transient oscillation of gold nanoparticle aggregates. (Data from Grant, C.D. et al., *J Am. Chem. Soc.* 125: 549–553, 2003.)

$$R = \frac{\eta \tau}{2\pi} c_1 \qquad (8.10)$$

where R is the radius of the aggregate or particle, η is the vibrational eigenvalue, τ is the period of the oscillations, and c_1 is the longitudinal speed of sound in bulk gold, can be used to accurately predict the period of oscillation for isolated spherical metal particles. For the Au nanoparticle aggregates, however, the observed vibrational period from transient absorption is longer than that predicted based on Equation 8.9 using the radius of the aggregates measured from dynamic light scattering. A possible explanation for the discrepancy is that the vibrational motion of the aggregates is softer than that of the isolated hard spherical particles due to a lower speed of sound in the aggregates than in bulk gold.[52] This seems plausible because the aggregates are not solid structures like the elastic nanoparticles. The transient absorption SAX and hole-burning experiments suggest that the EPR band in SIA structures is, in fact, composed of subbands, with each subband originating from an aggregate of different size or structures. Since the vibrational frequency of the aggregates is on the order of 1 cm^{-1}, the transient absorption technique provides a unique method for probing such low-frequency vibrational modes of condensed matter systems, such as metal nanoparticle aggregates that are difficult to determine by other means.

8.7 CONCLUSION

The SPR in metal nanoparticles such as gold is affected by the size, shape, density, structure, and particle–particle interaction. This leads to a tunable absorption band in the visible and near-infrared region. Mie theory can be used to accurately calculate the absorption spectrum of most metal nanoparticle systems and as the basis to qualitatively understand the SPR in the various gold nanoparticle systems. Furthermore, the study of electronic relaxation of metal nanoparticles using ultrafast laser techniques provides valuable information about fundamental properties such as electron–phonon interaction and particle–particle interaction. Particularly, while the relaxation time gives an indication of the bulk or molecular nature of the nanoparticles, as well as the strength of electron–phonon interaction, the oscillations observed in the transient absorption profiles provide a sensitive and unique probe of the size, structure, polydispersity, and interparticle interaction of metal nanomaterials.

REFERENCES

1. J.C. Maxwell-Garnett, Colours in metal glasses and in metallic films, *Philos. Trans. R. Soc. London* 203:385–420, 1904.
2. G. Mie, Contribution to optical properties of turbulent media, specifically colloidal metal dispersions, *Ann. Physik* 25:377–445, 1908.
3. F. Meriaudeau, T. Downey, A. Wig, A. Passian, M. Buncick, and T.L. Ferrell, Fiber optic sensors based on gold island plasmon resonance, *Sens. Actuators B* 54:106–117, 1999.
4. A.M. Ahern and R.L. Garrell, Protein–metal interactions in protein–colloid conjugates probed by surface-enhanced Raman spectroscopy, *Langmuir* 7:254–261, 1991.
5. R.G. Freeman, M.B. Hommer, K.C. Grabar, M.A. Jackson, and M.J. Natan, Ag-clad Au nanoparticles: novel aggregation, optical, and surface-enhanced Raman scattering properties, *J. Phys. Chem.* 100:718–724, 1996.
6. M. Moskovits, Surface-enhanced spectroscopy, *Rev. Mod. Phys.* 57:783–826, 1985.
7. M. Hidalgo, R. Montes, J.J. Laserna, and A. Ruperez, Surface-enhanced resonance Raman spectroscopy of 2-pyridylhydrazone and 1,10-phenanthroline chelate complexes with metal ions on colloidal silver, *Anal. Chim. Acta* 318:229–237, 1996.
8. S.M. Barnett, B. Vlckova, I.S. Butler, and T.S. Kanigan, Surface-enhanced Raman scattering spectroscopic study of 17-alpha-ethinylestradiol on silver colloid and in glass-deposited Ag-17-alpha-ethinylestradiol film, *Anal. Chem.* 66:1762–1765, 1994.
9. P. Matejka, The role of surfactants in the surface enhanced Raman scattering spectroscopy, *Chem. Listy* 86:875–883, 1992.
10. K. Cermakova, O. Sestak, P. Matejka, V. Baumruk, and B. Vlckova, Surface-enhanced Raman scattering (SERS) spectroscopy with borohydride-reduced silver colloids: controlling adsorption of the scattering species by surface potential of silver colloid, *Collect. Czech Chem. Commun.* 58:2682–2694, 1993.
11. S. Schneider, P. Halbig, H. Grau, and U. Nickel, Reproducible preparation of silver sols with uniform particle size for application in surface-enhanced Raman spectroscopy, *Photochem. Photobiol.* 60:605–610, 1994.
12. S.J. Oldenburg, G.D. Hale, J.B. Jackson, and N.J. Halas, Light scattering from dipole and quadrupole nanoshell antennas, *Appl. Phys. Lett.* 75:1063–1065, 1999.
13. U. Kreibig and M. Vollmer, *Optical Properties of Metal Clusters*, Berlin: Springer, 1995.
14. J.A. Creighton and D.G. Eadon, Ultraviolet-visible absorption spectra of the colloidal metallic elements, *J. Chem. Soc. Faraday Trans.* 87:3881–3891, 1991.
15. M. Quinten, The color of finely dispersed nanoparticles, *Appl. Phys. B* 73:317–326, 2001.
16. U. Kreibig and L. Genzel, Optical absorption of small metallic particles, *Surf. Sci.* 156:678–700, 1985.
17. R. Gans, Über die Form ultramikroskopischer Silberteilchen, *Ann. Physik.* 47:270–284, 1915.

18. M. Kerker, *The Scattering of Light and Other Electromagnetic Radiation*, New York: Academic Press, 1969.
19. M.M. Alvarez, J.T. Khoury, T.G. Schaaff, M.N. Shafigullin, I. Vezmar, and R.L. Whetten, Optical absorption spectra of nanocrystal gold molecules, *J. Phys. Chem. B* 101:3706–3712, 1997.
20. J.H. Hodak, A. Henglein, and G.V. Hartland, Photophysics of nanometer sized metal particles: electron–phonon coupling and coherent excitation of breathing vibrational modes, *J. Phys. Chem. B* 104:9954–9965, 2000.
21. A. Taleb, C. Petit, and M.P. Pileni, Optical properties of self-assembled 2D and 3D superlattices of silver nanoparticles, *J. Phys. Chem. B* 102:2214–2220, 1998.
22. N.W. Ashcroft and N.D. Mermin, *Solid State Physics*, Philadelphia: Saunders College, 1976.
23. J.Z. Zhang, Ultrafast studies of electron dynamics in semiconductor and metal colloidal nanoparticles: effects of size and surface, *Account. Chem. Res.* 30:423–429, 1997.
24. R.W. Schoenlein, W.Z. Lin, J.G. Fujimoto, and G.L. Easley, Femtosecond studies of nonequilibrium electronic processes in metals, *Phys. Rev. Lett.* 58:1680–1683, 1987.
25. C.K. Sun, F. Vallee, L.H. Acioli, E.P. Ippen, and J.G. Fujimoto, Femtosecond-tunable measurement of electron thermalization in gold, *Phys. Rev. B* 50:15337–15348, 1994.
26. H.E. El Sayed-Ali, T.B. Norris, M.A. Pessot, and G.A. Mourou, Time-resolved observation of electron–phonon relaxation in copper, *Phys. Rev. Lett.* 58:1212–1215, 1987.
27. B.A. Smith, J.Z. Zhang, U. Giebel, and G. Schmid, Direct probe of size-dependent electronic relaxation in single-sized Au and nearly monodisperse Pt colloidal nanoparticles, *Chem. Phys. Lett.* 270:139–144, 1997.
28. S.L. Logunov, T.S. Ahmadi, M.A. El-Sayed, J.T. Khoury, and R.L. Whetten, Electron dynamics of passivated gold nanocrystals probed by subpicosecond transient absorption spectroscopy, *J. Phys. Chem. B* 101:3713–3719, 1997.
29. J. Hodak, I. Martini, and G.V. Hartland, Ultrafast study of electron–phonon coupling in colloidal gold particles, *Chem. Phys. Lett.* 284:135–141, 1998.
30. S. Link and M.A. El-Sayed, Spectral properties and relaxation dynamics of surface plasmon electronic oscillations in gold and silver nanodots and nanorods, *J. Phys. Chem. B* 103:8410–8426, 1999.
31. J.Z. Zhang, B.A. Smith, A.E. Faulhaber, J.K. Andersen, and T.J. Rosales, Femtosecond Studies of Colloidal Nanoparticles: Dependence of Electronic Energy Relaxation on the Liquid–Solid Interface and Particle Size, paper presented at Ultrafast Processes in Spectroscopy, Trieste, Italy, 1995.
32. A.E. Faulhaber, B.A. Smith, J.K. Andersen, and J.Z. Zhang, Femtosecond electronic relaxation dynamics in metal nano-particles: effects of surface and size confinement, *Mol. Cryst. Liq. Cryst. Sci. Technol. A* 283:25–30, 1996.
33. S. Link, M.A. El-Sayed, T.G. Schaaff, and R.L. Whetten, Transition from nanoparticle to molecular behavior: a femtosecond transient absorption study of a size-selected 28 atom gold cluster, *Chem. Phys. Lett.* 356:240–246, 2002.
34. M. Nisoli, S. DeSilvestri, A. Cavalleri, A.M. Malvezzi, A. Stella, G. Lanzani, P. Cheyssac, and R. Kofman, Coherent acoustic oscillations in metallic nanoparticles generated with femtosecond optical pulses, *Phys. Rev. B* 55:13424–13427, 1997.
35. J.H. Hodak, I. Martini, and G.V. Hartland, Observation of acoustic quantum beats in nanometer sized Au particles, *J. Chem. Phys.* 108:9210–9213, 1998.
36. J.P. Wilcoxon, J.E. Martin, F. Parsapour, B. Wiedenman, and D.F. Kelley, Photoluminescence from nanosize gold clusters, *J. Chem. Phys.* 108:9137–9143, 1998.
37. A. Mooradain, Photoluminescence of metals, *Phys. Rev. Lett.* 22:185–187, 1969.

38. G.T. Boyd, T. Rasing, J.R.R. Leite, and Y.R. Shen, Local-field enhancement on rough surfaces of metals, semimetals, and semiconductors with the use of optical second-harmonic generation, *Phys. Rev. B* 30:519–526, 1984.

39. M.B. Mohamed, V. Volkov, S. Link, and M.A. El-Sayed, The "lightning" gold nanorods: fluorescence enhancement of over a million compared to the gold metal, *Chem. Phys. Lett.* 317:517–523, 2000.

40. Y.N. Hwang, D.H. Jeong, H.J. Shin, D. Kim, S.C. Jeoung, and G. Cho, Femtosecond emission studies on gold nanoparticles, *J. Phys. Chem. B* 106:7581–7584, 2002.

41. S. Link, M.B. Mohamed, and M.A. El-Sayed, Simulation of the optical absorption spectra of gold nanorods as a function of their aspect ratio and the effect of the medium dielectric constant, *J. Phys. Chem. B* 103:3073–3077, 1999.

42. S.J. Oldenburg, R.D. Averitt, S.L. Westcott, and N.J. Halas, Nanoengineering of optical resonances, *Chem. Phys. Lett.* 288:243–247, 1998.

43. G.V. Hartland, M. Hu, O. Wilson, P. Mulvaney, and J.E. Sader, Coherent excitation of vibrational modes in gold nanorods, *J. Phys. Chem. B* 106:743–747, 2002.

44. J.R. Heath, C.M. Knobler, and D.V. Leff, Pressure/temperature phase diagrams and superlattices of organically functionalized metal nanocrystal monolayers: the influence of particle size, size distribution, and surface passivant, *J. Phys. Chem. B* 101: 189–197, 1997.

45. A.A. Lazarides and G.C. Schatz, DNA-linked metal nanosphere materials: structural basis for the optical properties, *J. Phys. Chem. B* 104:460–467, 2000.

46. J.J. Storhoff, A.A. Lazarides, R.C. Mucic, C.A. Mirkin, R.L. Letsinger, and G.C. Schatz, What controls the optical properties of DNA-linked gold nanoparticle assemblies? *J. Am. Chem. Soc.* 122:4640–4650, 2000.

47. K.S. Mayya, V. Patil, and M. Sastry, An optical absorption investigation of cross-linking of gold colloidal particles with a small dithiol molecule, *Bull. Chem. Soc. Jpn.* 73:1757–1761, 2000.

48. T.J. Norman, Jr., C.D. Grant, D. Magana, J.Z. Zhang, J. Liu, D. Cao, F. Bridges, and A. van Buuren, Near infrared optical absorption of gold nanoparticle aggregates, *J. Phys. Chem. B* 106:7005–7012, 2002.

49. A.N. Shipway, M. Lahav, R. Gabai, and I. Willner, Investigations into electrostatically induced aggregation of Au nanoparticles, *Langmuir* 16:8789–8795, 2000.

50. C.G. Blatchford, J.R. Campbell, and J.A. Creighton, Plasma resonance-enhanced Raman scattering by adsorbates on gold colloids: the effects of aggregation, *Surf. Sci.* 120:435–455, 1982.

51. T.J. Norman, Jr., C.D. Grant, D. Magana, R. Anderson, J.Z. Zhang, D. Cao, F. Bridges, J. Liu, and T. van Buuren, Longitudinal plasma resonance shifts in gold nanoparticle aggregates, *SPIE Proc.* 4807:51–58, 2002.

52. C.D. Grant, A.M. Schwartzberg, T.J. Norman, Jr., and J.Z. Zhang, Ultrafast electronic relaxation and coherent vibrational oscillation of strongly coupled gold nanoparticle aggregates, *J. Am. Chem. Soc.* 125:549–553, 2003.

53. V.N. Pustovit and G.A. Niklasson, Observability of resonance optical structure in fractal metallic clusters, *J. Appl. Phys.* 90:1275–1279, 2001.

54. I.H.H. Zabel and D. Stroud, Metal clusters and molecular rocks: electromagnetic properties of conducting fractal aggregates, *Phys. Rev. B* 46:8132–8138, 1992.

55. J.H. Hodak, A. Henglein, and G.V. Hartland, Size dependent properties of Au particles: coherent excitation and dephasing of acoustic vibrational modes, *J. Chem. Phys.* 111:8613–8621, 1999.

9 Synthesis and Characterization of PbSe Nanocrystal Assemblies

M. Bashouti, A. Sashchiuk, L. Amirav, S. Berger,
M. Eisen, M. Krueger, U. Sivan, and E. Lifshitz

CONTENTS

9.1 Introduction 207
9.2 Results and Discussion 209
    9.2.1 Wire-Like Assembly Grown by Molecular Template
        Mechanism 209
    9.2.2 Spherical and Wire-Like Assemblies Grown
        with TOPO/TBP Surfactant 212
9.3 Summary 221
Acknowledgment 221
References.. 221

9.1 INTRODUCTION

Colloidal semiconductor nanocrystals (NCs) can be considered artificial atoms. Numerous colloidal syntheses have been developed in the last two decades, producing nearly monodispersed spherical NCs (with <5% size distribution), capped with organic ligands or inorganic epitaxial shells.[1–5] These artificial atoms exhibit unique physical properties, associated with the quantum size effect and the surface coating.[6–11] The effect of the shape of the individual NCs has received less attention, although various recent works reported the synthesis of rod-like NCs with various aspect (length/width) ratios, using selective facet stabilization by chemical reagents,[12,13] coordinating molecular templates,[14–16] electrochemical means,[17] or the vapor–liquid–solid technique.[18,19]

Currently, there are several proofs demonstrating the possibility of integrating the spherical or rod-like semiconductor NCs into light-emitting diodes,[20] photovoltaic cells,[21–24] electronic circuitry,[25,26] and biological markers.[27,28] These proofs show that the NCs may indeed play a major role in future technologies, such as nanoelectronics, photonics, telecommunication, and biological and environmental sensors.

However, most of these applications will require multi-NC channels with well-defined spatial separation and morphology. Recent developments showed the ability to prepare one-, two-, and three-dimensional assemblies of NCs, using the colloidal NCs as the building blocks.[29-34] These assemblies may show tunable optical and electrical properties, which either preserve the individual nature of the NCs or exhibit new collective effects. Promising results have already been achieved with Ag colloidal NCs, compressed as a monolayer on a Langmuir trough, demonstrating a reversible metal–Mott insulator transition.[35] Likewise, Takagahara[36] predicted a large enhancement of the nonlinear optical properties of semiconductor NC assemblies, accompanying the decrease of the inter-NCs' distance, based on an efficient dipole–dipole interaction. Although the synthesis and physical properties of metal NC assemblies (e.g., Au and Ag)[30,37-39] and magnetic NCs (e.g., Co, PtFe, Fe_2O_3)[40] have been studied extensively, only a few studies of semiconductor NC assemblies have been carried out. Those include two- and three-dimensional superlattices of spherical CdS,[41] CdSe,[42] InP,[43] PbSe,[44] or CdS and PbS NCs linked to ordered organic templates.[33,45-47]

The present work focuses on the synthesis and characterization of one- and three-dimensional assemblies of PbSe NCs, using two different colloidal procedures. The unique properties of the individual and assembled structures of the PbSe semiconductor draw wide interest. A bulk PbSe has a cubic (rock salt) crystal structure and a narrow direct band gap (0.21 eV, at 300 K) at the L point of the Brillouin zone. The high dielectric constant (ε_∞ =24.0) and the small electron and hole effective mass (<0.1 m*)[48] create an exciton with a relatively large effective Bohr radius (a_B =46 nm), eight times larger than that of CdSe. Thus, size quantization effects are strongly pronounced in PbSe NCs. Recent interband optical studies of colloidal PbSe quantum dots (QDs) exhibited well-defined band-edge excitonic transitions tuning between 0.9 and 2.0 eV, a small Stokes shift, and a submicrosecond lifetime.[49-51] The $1S_{e,h}$–$1P_{e,h}$ or $1P_{e,h}$–$1D_{e,h}$ intraband transitions were also explored recently, revealing a short lifetime and obscuring the phonon bottleneck.[51] These results suggest that PbSe NCs can be used for opto-electronic devices in the infrared (IR) spectral regime and in telecommunication applications.

Recently, Murray et al.[52] synthesized individual spherical PbSe NCs, with narrow size distribution and band gap tuning at the near-IR spectral regime (0.9 to 2.1 eV). In parallel, we reported another synthetic colloidal procedure for the preparation of individual spherical PbSe NCs, with average size ranging between 2.5 and 7 nm and passivated surfaces.[53] The synthesis induces a reaction between tributyl-phosphine-selenide (TBP-Se) and lead 2-ethyl-hexanoate, in trioctyl-phosphine-oxide (TOPO) surfactant at a constant elevated temperature and inert atmosphere. Furthermore, we presented a synthetic route for the preparation of PbSe/PbS core-shell NCs. The core-shell structures consist of an inner semiconductor PbSe core, covered by an epitaxial layer of PbS semiconductor, with relatively close crystal matching. These structures exhibit an enhanced quantum yield of the photoluminescence (PL) emission.

The work presented in this chapter demonstrates the use of additional colloidal procedures, leading to the formation of glassy and ordered assemblies with a spherical or wire-like morphology. These include the use of alkyldiamine solvent, controlling an assembly formation by a coordinating molecular template mechanism, and further adjustment of the temperature between 10 and 117°C. Alternatively, a TOPO/tributyl-phosphine (TBP) surfactant was used, permitting the creation of monodispersed PbSe NCs, which assemble into micron-sized glassy spheres or ordered wires upon an increase of the precursors' concentration and with temperature interruptions (between 130 and 180°C) during the growth. In this chapter, the assembly growth mechanism and the electrical properties of the wire-like assemblies are discussed.

9.2 RESULTS AND DISCUSSION

9.2.1 Wire-Like Assembly Grown by Molecular Template Mechanism

Stock solutions of $PbCl_2$ (1 to 5 mmole) or PbO_2 (1 to 5 mmole) and Se (1 to 5 mmole) precursors, dissolved in ethylenediamine (EtDA), propylenediamine (PrDA), and cyclohexanediamine (cHxDA) (all with technical grade) were prepared at room temperature, under standard inert conditions in a glove box. The Se stock solution had a greenish color. A 70-ml mother solution of alkyldiamine was placed in a three-neck flask, under inert conditions (with Argon flow). Ten milliliters of KBH_4/alkyldiamine (1 mmole) were added to the mother solution. Then, 20 ml of the Se stock solution were injected rapidly into the mother solution, which was followed by an immediate change of its color from green-like to deep brown. The reaction was preceded by a slower addition of 20 ml of Pb stock solution to the three-neck flask. The reaction was carried out at different temperatures, varying from 10 to 117°C, and a time duration ranging from 15 min to 4 h. Aliquots from the reaction were removed every 15 min by a syringe and were injected into a toluene solution, which quenched any further growth of the NCs and their assembly. The products were separated from the toluene solution by the addition of an ethanol and centrifugation.

High-resolution scanning electron microscopy (HR-SEM) observations were carried out on a Zeiss Leo 982 instrument operated at 4 kV. Transmission electron microscopy (TEM) studies, combined with energy-dispersive analyses of x-ray emission (EDAX) and selected area electron diffraction (SAED), were carried out on a JEOL 2000FX instrument, operated at 200 kV. An HR-TEM study was done in a JEOL 3010 instrument operated at 300 kV. The TEM specimens were prepared by injecting small liquid droplets of the solution on an amorphous Formvar-coated copper grid (300 mesh) stabilized with evaporated carbon film and then drying at room temperature.

Various morphological assembly PbSe NCs were grown by a reaction between $PbCl_2$ or PbO_2 and dissolved Se in colloidal solution, depending on the solvent, temperature, and duration of the reaction. Representative HR-SEM micrographs of NCs prepared with PbO_2 in EtDA solution at 10°C (~1 h), 40°C (~1 h), and 117°C (~1 h) are shown in Figure 9.1A to C, respectively. This figure reveals that a 2-μm wire-like assembly (with 100% efficiency) is achieved at the lowest temperature and evolution of those wires into rods (with aspect ratio of ~5) and cubes (with ~100-nm edge) at elevated temperatures. It should be indicated that multipod structures, with triple- and quadrapods, were occasionally observed at 10°C, sustaining the reaction over a few hours. A representative multipod structure is shown in Figure

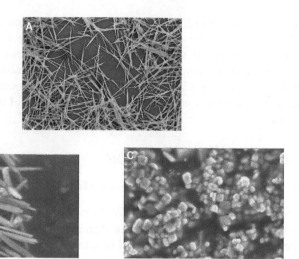

FIGURE 9.1 HR-SEM micrographs of PbSe NCs synthesized with a PbO$_2$ precursor in EtDA solvent (A at 10°C; B at 40°C; C at 117°C) for about an hour.

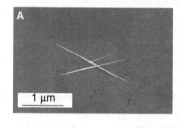

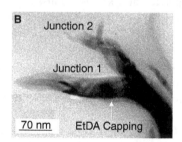

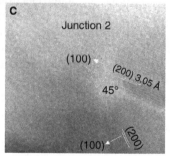

FIGURE 9.2 (A) HR-SEM micrograph of a multipod prepared with a PbCl$_2$ precursor in EtDA solution. (B) HR-TEM micrograph of PbSe multipod, exhibiting two junctions. (C) Magnified HR-TEM image of a junction, presenting the formation of a new arm, merging in the [100] growth direction, at an angle of 45° with respect to the main wire.

9.2A, and the corresponding HR-TEM images of the junctions are shown in Figure 9.2B and C. These images represent the growth of a side arm from a main wire at a 45° angle and a growth direction [100]. Most probably, the side arm is created at a point of a defect in the organic capping or in crystallinity.

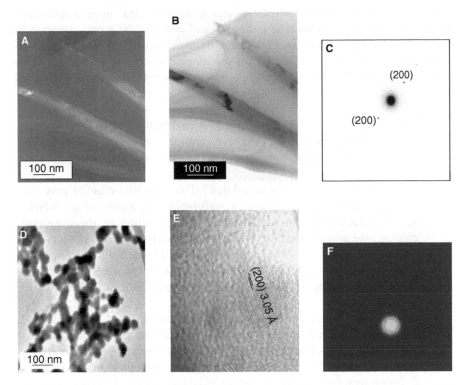

FIGURE 9.3 Dark- and bright-field TEM micrographs of PbSe wire-like assembly (A and B, respectively), prepared in EtDA solvent, and the corresponding SAED picture (C). HR-TEM of PbSe assembly, prepared in cHxDA solvent (D and E), and the corresponding SAED picture (F).

The solvent characteristic has a substantial influence on the morphology of the assembly. Bright- and dark-field images of the wire-like structures, shown in Figure 9.3A and 9.3B, respectively, prepared with $PbCl_2$ precursor in EtDA solvent, show that each wire consists of nanometer-sized PbSe grains (more than 100 nm long) arranged as a chain along a growth axis (z-axis). A thin amorphous layer capping shown in the TEM pictures, most likely corresponds to the diamine organic ligands at the outer surface of the assembly. A selected electron diffraction pattern, taken from the PbSe grains, shown in Figure 9.3C, corresponds to a face-centered-cubic (fcc) phase. Furthermore, the EDAX measure confirms the existence of a one-to-one Pb:Se atomic ratio. Figure 9.3D shows a bright-field image of a PbSe assembly prepared in cHxDA solvent, leading to a glassy structure. Figure 9.3E presents a lattice image of a few PbSe grains observed at a corner of this glassy assembly, and Figure 9.3F represents the corresponding SAED picture. Overall, the images in Figure 9.3 suggest that short diamine chain (e.g., EtDA) molecules promote the formation of ordered stacking of PbSe NCs along the z-axis (Figure 9.3A and B), whereas longer chains produce a glassy assembly (Figure 9.3D and F).

The described investigations conceal that there is a delicate balance, which determines the morphology of the assembly, controlled by the solvent, temperature, and reaction duration. The solvent nature should receive special attention. Control synthesis (not shown here) carried out at low temperatures, using single-amine molecules, led to the formation of individual and spherical NCs only, whereas the experimental results, shown in Figure 9.1 through Figure 9.3, suggest that the alkyldiamine molecules are essential for the formation of PbSe assemblies at low temperatures. The EtDA molecules, in particular, have several important character-istics:

1. These molecules enable formation of a chemically stable coordinating complex with Pb ions, in a *trans* configuration,[54] as shown schematically in Figure 9.4A. Such a complex will act as an intermediate species in the reaction.

2. Li et al.[54] and Yu et al.[16] suggested that EtDA molecules dissolve pure S_8, forming various sulfur imides or sulfur–nitrogen anions (e.g., S_7NH, S_4N^-), including S^{-2}. In regard to the stock solution of the selenium element, we propose that the EtDA molecules dissolve the selenium via hydrogen bonding, as shown schematically in Figure 9.4B (leading to a greenish solution), whereas the reduction of this element to Se^{-2} occurs only after its mixing with a KBH_4-reducing agent (leading to a deep brown solution)

3. A preliminary approach of Se^{-2} ions toward the $Pb(EtDA)_2^{+2}$ complexes is partially hindered in the x–y plane by the EtDA molecules, whereas the move in the z-direction is unlimited, promoting a dominated reaction in this direction. Thus, the EtDA molecules are acting as a molecular directing template for the growth of elongated grains (shown schemati-cally in Figure 9.4C and D), which can further be stacked along the z-axis to an ordered wire-like assembly. In fact, longer alkyldiamine chains do not determine the growth direction in an efficient way, leading to the formation of a glassy structure (Figure 9.3D and F).

4. Eventually, the alkyldiamine molecules are absorbed on the external sur-face of the assembly, either coordinating to unsaturated Pb sites or, by Van der Waals forces, forming a condensed capping, as shown by the HR-TEM in Figure 9.2 and Figure 9.3.

9.2.2 Spherical and Wire-Like Assemblies Grown with TOPO/TBP Surfactant

Individual PbSe NCs were prepared by a chemical reaction between lead-cyclohex-anebutirate (Pb-cHxBu) and selenium precursors (STREM or Aldrich with technical grade). A stock solution was prepared by dissolving those precursors in TBP solution at room temperature, under standard inert conditions in a glove box, using the following Se:Pb-cHxBu:TBP mass ratios: (1) 0.25:0.6:50, (2) 0.5:1.2:50, and (3) 1:2.5:50.

A mother solution of TOPO surfactant (6.0 g, Aldrich 90 or 99% purity) was placed in a three-neck flask, under inert conditions (with Argon flow) at 150°C. Then the stock solution was injected rapidly into the mother solution, followed by an immediate drop of the temperature to 118°C. The reaction proceeded in two different

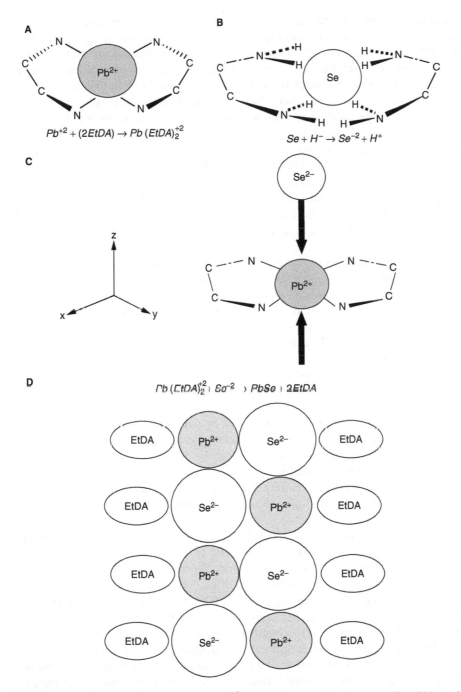

FIGURE 9.4 Schematic drawing of the $Pb^{2+}/(EtDA)_2$ coordinating complex (A) and $Se^{2-}/(EtDA)_2$ solvation (B). The approach of Se^{2-} ions toward $Pb^{2+}/(EtDA)_2$ complexes, promoting a predominant growth along the z-axis (C), and a crystalline grain growth with EtDA surface capping (D).

fashions: (1) retaining the reaction at the last temperature and termination of the reaction after 5 min by rapid cooling to 70°C, and (2) gradual heating of the solution to 150°C and extension of the reaction duration at this temperature up to 150 min. Aliquots from the reaction were removed every few minutes by a syringe and were injected into a 1-ml methanol solution, which quenched any further growth of the NCs. The products were separated from the TOPO–methanol solution by centrifugation. Further purification and monodispersity of the product were achieved by a few redissolving–centrifuging cycles in methanol–butanol solution. The structural and morphological properties of the monodispersed PbSe NCs and their assemblies were characterized by HR-SEM, TEM, HR-TEM, SAED, and EDAX, using the instruments indicated in Section 9.2.1. The absorbance curves of the intermediate aliquots were recorded with a Shimatzu VIR spectrometer.

The electrical properties of selective wire-like assemblies were examined by the integration of a single wire to an electronic circuit by the following steps:

1. p-doped Si substrate, covered with a 200-nm oxide layer, was treated with trimetylacolorosilane to obtain a hydrophobic surface.
2. Markers were drawn on the substrate with optical lithography.
3. PbSe NC wires were deposited on the substrate by spin coating.
4. A single wire and its position with respect to the markers were selected by SEM inspection; poly(methylmethacrylate) (PMMA) resist was deposited by spin coating, followed by an annealing stage.
5. Writing and developing were done by the electron beam lithography technique, with subsequent evaporation of Ti/Au contact leads. This process was resumed by the removal of photoresist.
6. The outcome circuit was examined by the SEM technique (*vide infra*). Some precaution was taken regarding the need to strip the PbSe NCs' wire organic coating on the surface at the electrical contact points and an unintentional modification of the wires' surface upon the removal of the photoresist. Thus, a number of options were examined (e.g., annealing temperature, dissolving solvents) until optimal conditions were reached.

The reaction between the indicated precursors, starting with a stock solution of Se:Pb-cHxBu:TBP with the lowest mass ratio of 0.25:0.6:50, at a constant temperature of 118°C, led to the formation of monodispersed spherical NCs with average sizes of 3.5 to 5.5 nm (size distribution < 10%), depending on the reaction duration. The individual PbSe NCs were capped with TOPO/TBP surfactant. A representative HR-TEM image of a single NC after 5 min of synthesis preparation is shown in Figure 9.5A, revealing a high crystallinity, indexed as rock salt cubic phase [100].

Increasing the precursor concentration and raising the temperature of the reaction (from 118 to 150°C) after the precursors' injection, accompanied by an extension of the reaction duration, enabled formation of NC assemblies with various morphologies. A TEM image of an intermediate aliquot, taken from a mother solution containing Se:Pb-cHxBu:TBP with a mass ratio of 0.5:1.2:50 (at 150°C), drawn after 5 min, is shown in Figure 9.5B. This figure clearly shows the preliminary aggregation

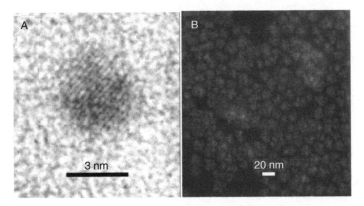

FIGURE 9.5 (A) HR-TEM images of a spherically shaped individual PbSe NC, using a mass ratio of 0.25:0.6:50 of the Se:Pb-cHxBu:TBP precursors at 118°C. (B) TEM image of a preliminary stage in the formation of spherical assembly, exhibiting an initial aggregation of a few NCs, using a mass ratio of 0.5:1:50 of the Se:Pb-cHxBu:TBP precursors at 150°C. The reaction time in (A) and (B) was about 5 min.

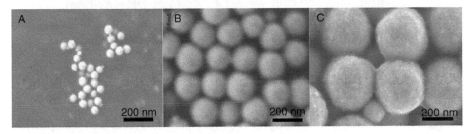

FIGURE 9.6 HR-SEM images of PbSe NC assemblies prepared with a 0.25:0.6:50 mass ratio of the Se:Pb-cHxBu:TBP precursors at 150°C, with a reaction time of 10 min (A), 25 min (B), and 40 min (C).

of a few individual NCs. Representative HR-SEM micrographs of NC assemblies observed after 10, 25, and 40 min at 150°C are shown in Figure 9.6A to C, respectively. In fact, reaction duration between 10 and 60 min led to the formation of spherical assemblies, with a surprisingly uniform diameter between 50 and 450 nm. The growth rate of the spherical assemblies is plotted in Figure 9.7, showing a threshold time for the assembly formation of about 5 min at 150°C. Prolonging the reaction time beyond 60 min at 150°C led to the formation of cubes and rod- and wire-like structures, as shown in Figure 9.8A and B. Furthermore, use of a TOPO surfactant with 90% purity increased the production of anisotropic assemblies to ~15%.

TEM images and the corresponding SAED patterns of spherical and wire-like assemblies prepared with precursor concentrations of Se:Pb:TBP, 1:2.5:50, at 150°C for 60 min, are compared in Figure 9.9. The TEM and SAED pictures of the spherical structures are shown in Figure 9.9A and B, and those of the wire structure are shown

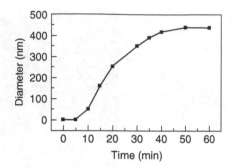

FIGURE 9.7 Plot of the spherical PbSe assembly's diameter vs. the reaction time (minutes).

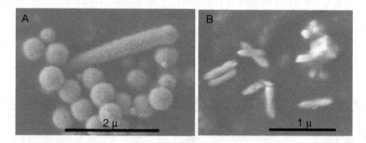

FIGURE 9.8 Representative HR-SEM micrographs of PbSe assemblies shaped as spheres and a wire (A) and rods and cubes (B), prepared with a 0.5:1:50 mass ratio of the Se:Pb-cHxBu:TBP precursors at 150°C, with a reaction time of 90 and 150 min, respectively.

in Figure 9.9C and D, respectively. The SAED patterns reveal that the spherical assembly has a glassy morphology, whereas the wire-like assembly has an ordered structure, with a diffraction pattern in the [100] zone axis, revealing a rock salt cubic structure with a lattice constant of 6.1 Å. A representative bright-field HR-TEM image of a wire assembly is shown in Figure 9.10. The dark spots correspond to NC grains, which are close to the zone axis, whereas the brighter spots relate to those NCs that deviate from the zone axis. Interestingly, by rotating the wire slowly, these darker parts of the wire move continuously along the wire. The superstructure appears to have a helix-like shape. The nanocrystals' nearest neighbors differ by only a fraction of a degree in their orientation, adding up to ~1 to 2° when moving along the wire by tens of nanometers.

Occasionally, bent wires are formed, as presented by the HR-TEM micrograph in Figure 9.11B. Higher-magnification HR-TEM images of a junction and of an edge are shown in Figure 9.11A and C, respectively. Panels B and C show well-resolved lattice fringes (in the [100] zone axis) of a cubic PbSe lattice, supporting the single-crystalline structure of the bent wire. Figure 9.11 suggests that the wire consists of individual cubic NCs with a 10-nm edge.

The change in the optical properties upon the growth of the assemblies was examined by recording the absorption spectra of intermediate aliquots during the reaction. Representative absorbance curves of spherical NC assemblies, with preparation times

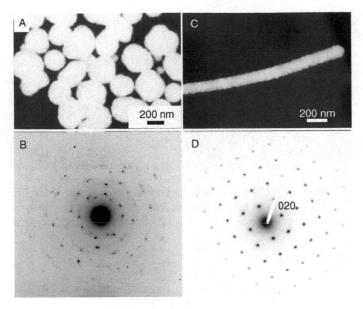

FIGURE 9.9 A TEM micrograph of spherical assemblies (A) and the corresponding SAED pattern (B), revealing a glassy morphology. A TEM micrograph of a wire-like assembly (C) and the corresponding SAED pattern (D), showing a single-crystal diffraction in [100] zone axis, with an ordered rock salt cubic structure and lattice constant of 6.1 Å.

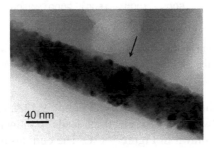

FIGURE 9.10 Bright-field HR-TEM micrograph of wire-like assembly. The arrow points to a darker part composed of individual NCs, close to the zone axis. By rotating the wire slowly, these darker parts move continuously along the wire.

of 30, 35, and 40 min, are shown in Figure 9.12. Each absorbance curve exhibits two pronounced exciton bands between 1000 and 1600 nm (0.775 and 1.240 eV), blue shifted with respect to the bulk energies. This suggests that the assemblies preserve the individual quantum size effect of the individual NCs; however, they are slightly red shifted, one with respect to the other, due to the collective properties among the NCs within an assembly. The relatively broad exciton bands may be a consequence of the NC–NC interactions, as discussed below.

Preliminary conductivity measurements of ordered wire-like PbSe NC assemblies were carried out by depositing a single wire on a hydrophobic surface of an

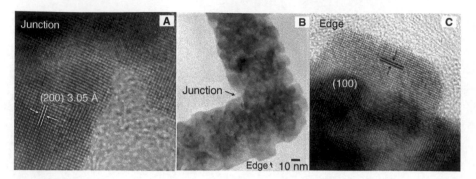

FIGURE 9.11 (B) An HR-TEM micrograph of a bent wire-like assembly. Magnified HR-TEM images of a junction (A) and an edge (C) reveal the high crystallinity of the bent wire, with NC building blocks of about 10 nm, with a rock salt structure and a lattice constant of 6.1 Å.

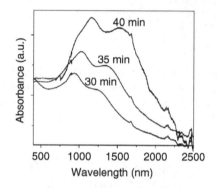

FIGURE 9.12 Representative absorbance curves of three spherical assemblies, prepared with a 0.5:1:50 mass ratio of the Se:Pb-cHxBu:TBP precursors at 150°C, and a reaction time as indicated in the figure.

Si/SiO$_2$ substrate, as shown schematically in Figure 9.13A. An SEM picture of the single wire coated between two Au electrical leads is shown in Figure 9.13B, and the electrical circuit is drawn in Figure 9.13C. The current-to-voltage [I(V)] curves of three different wires, with dimensions of 110 × 700 nm, 60 × 1000 nm, and 150 × 1200 nm (measured at room temperature), are plotted by the squares, circles, and triangles in Figure 9.13D. The calculated specific resistivity of those wires is 0.15 Ωcm. The HR-TEM (Figure 9.11) showed that the wires examined were built from 10-nm individual NCs.

Organization of the individual PbSe NCs into ordered or disordered assemblies is governed by the NC's uniformity in size and shape, surface coating and inter-NCs' distance, growth temperature, the NC's concentration, and reaction solvents.[55] The results discussed in this section suggest that PbSe NCs' glassy assembly, with only short-range order and random NC orientation, provides an isotropic spherical structure, whereas ordered assembly with crystalline coherence over a micron, with preferred NC orientations, produces a highly anisotropic wire-like structure.

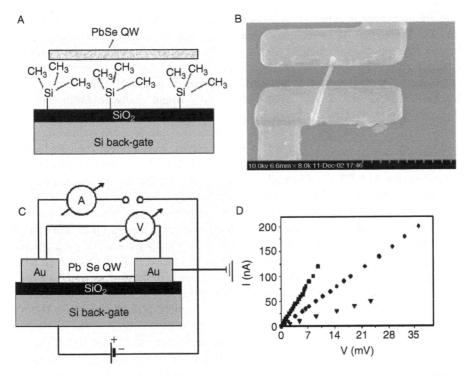

FIGURE 9.13 (A) Schematic drawing of a single wire-like assembly deposited on a hydrophobic surface. (B) HR-SEM picture of a single PbSe wire between Ti/Au leads. (C) Schematic drawing of an electrical circuit, containing a single wire and Si back-gate. (D) I-V curves of three different PbSe wires, with width and length as mentioned in the text, measured at room temperature.

The inter-NCs' interactions play a major role in the assembly stability, which is controlled predominately by the inter-NCs' distance. The distance can vary from intimate contact to about 5 nm (depending on the length of the organic capping). The inter-NCs' distance in the present case, considering the TOPO/TBP capping, is ~1.1 nm, although interdigitative bundling of the surface ligands can reduce the inter-NCs' distance. Previous works suggested that an inter-NCs' distance of <0.5 nm permits an exchange interaction between proximal NCs, allowing a wave function overlap and electronic delocalization, approaching those of the corresponding bulk materials.[56] On the contrary, at an inter-NCs' distance of >0.5 nm, the interactions are dominated by dipolar coupling. Electric dipoles can be either permanent dipoles, resulting from permanent displacement of electric charge,[57] particularly found in highly polarizable NCs, or transition dipoles, occurring when an excited NC is placed in the near field of a ground-state NC.[58]

It is anticipated that the inter-NCs' interactions in PbSe assemblies are controlled mainly by the permanent dipole–dipole interactions, due to retention of a quantum size property of individual NCs, as pronounced in the absorption spectra (Figure 9.12), and due to the high dielectric constant of PbSe ($\varepsilon_\infty = 24.0$), permitting instantaneous polarization and creation of a permanent dipole.[30] The organization of PbSe NCs that are proximate is expected to induce mutual polarization, creating a permanent electric dipole. This dipole can be characterized as a surface-localized charge given by the expression $\mu = 6er\ \varepsilon_{2\infty}/(\varepsilon_{2\infty} + 2_{1\infty})$,[59] where $\varepsilon_{2\infty}$ corresponds to the dielectric constant of a PbSe NC, $\varepsilon_{1\infty}$ to the dielectric constant of the external medium ($\varepsilon_{1\infty\ (TOPO)} = 2.5$), and r to the individual NC's diameter. Thus, $\mu = 500$ Debye is estimated by the indicated relation for a PbSe NC with a 10-nm diameter. Furthermore, a dipole–dipole interaction energy between adjacent NCs can be calculated with the classical formula $E = -\mu/2\ \pi\varepsilon_0 r\ (r^2 - D^2)$, where D is the NC–NC mutual distance and $\varepsilon_0 = 9.9 \times 10^{-12}$ C^2 J^{-1}m^{-1} is the vacuum dielectric permitivity. Then evaluation of the dipole–dipole energy for 10-nm NCs is 28 kJ/mol, larger than a typical molar kinetic energy with RT = 2.4 kJ/mol (T = 150°C, R = Rydberg constant), supporting the strong dipole–dipole interactions and the stable assembly structures.

As indicated above, PbSe NC assemblies are formed when the concentration of the precursors in the mother solution is relatively high, permitting the formation of individual NCs that are relatively proximate. This is further enhanced by a gradual heating to 150°C, which enables the inter-NCs' mutual diffusion and the creation of aggregation.[60] Thus, a fast inter-NC approach leads to the formation of a glassy structure, whereas extended duration allows the NCs to find an equilibrium site and form an ordered superstructure. Because of the mutual polarization, NCs tend to aggregate along the polarization axis, forming anisotropic rod- or wire-like shapes. The existence of chemical impurities in the mother solution (TOPO, 90%), such as phosphonic acid, is known to selectively stabilize certain NC facets, leading to the formation of elongated individual NCs.[12] Those tend to be stacked along the unique axis of the NCs, increasing the formation of wire-like ordered assemblies.

The electrical properties of the NC assembly depend intimately on the mutual orientation of adjacent NCs and on the overall anisotropy of the assembly. In fact, the dipole–dipole interactions act as a wireless substitution for the electrical contacts between neighboring NCs. In such a case, the charge transport among the NCs follows either a resonant tunneling process among the discrete states[61] or activated hopping from a surface of one NC to the other.[62] At $k_B T > E_C$, the I-V characteristics, under the tunneling mechanism, should have a linear behavior, overcoming the Coulomb blockage threshold. More generally, the I-V characteristics should follow the scaling law $I \propto (V/V_T - 1)^\xi$, where V_T is the threshold potential. ξ can have various values, depending on the dimensionality of the assembly.[63] An activated-hopping model was used recently to explain the electrical transport mechanism in carbon nanotubes and semiconductor wires at low temperatures, showing a linear I(V) dependence at relatively weak electric fields and nonlinear behavior under the influence of strong electric fields.[64]

The I-V curves, shown in Figure 9.13D, show a linear behavior that may be associated with either a hopping mechanism at low electric fields in the range of 0.0 to 5.0 10^2 V/cm or a resonant tunneling mechanism at $k_B T > E_C$. Thus, a distinction between the two possible mechanisms can be expressed by a measure of the conductivity vs. temperature. These measurements will be resumed in the near future. In any event, the wire-like assemblies of PbSe measured so far, consisting of nearly monodispersed 10-nm NCs, show a relatively low specific resistivity of 0.15 Ωcm, dominated by the size of the constituent 10-nm NCs. The specific resistivity value is close to the specific resistivity of carbon nanotube, suggesting that the conductivity of the PbSe NCs, ordered in a one-dimensional structure, is enhanced along the long axis of the assembly, implying its usefulness for nanoelectronic devices.

9.3 SUMMARY

Individual PbSe NCs were prepared by a chemical reaction between Pb-CHxBu and TBP:Se precursors in a TOPO mother solution, at 118°C. A relatively low precursor concentration led to the formation of separate NCs, whereas an increase in this concentration, accompanied by an increase in the temperature to 150°C, enabled the formation of glassy spherical assemblies at relatively short time intervals. Longer reaction times permitted the growth of wire-like assemblies. Wire-like assemblies were also formed by the replacement of the TOPO solution with EtDA, acting as a coordinating molecular template, blocking the growth of certain crystallographic facets, and permitting a dominated growth along a selective axis. In both cases, it is anticipated that the assemblies are stabilized by the NC–NC dipole–dipole interactions, with an estimated energy of 28 kJ/mol, substantially higher than a typical molar kinetic energy (RT = 2.5 kJ/mol). Furthermore, the measured specific resistivity of a wire-like assembly was $0.15\,\Omega$cm, determined by the size of the PbSe NC building blocks. The stabilized PbSe NC assemblies may be used in nanoelectronic and opto-electronic devices.

ACKNOWLEDGMENT

This research was support by the Deutsch-Israel Program (DIP) and by the Israel Academy of Science and Humanities.

REFERENCES

1. S.A. Empedocles, D.J. Norris, and M.G. Bawendi, *Phys. Rev. Lett.* 77 (1996) 3873.
2. J. Tittle, W. Gohde, F. Koberling, A. Mews, A. Kornowski, H. Weller, A. Eychmuller, and Th. Basche, *Ber. Bunsenges. Phys. Chem.* 101 (1997) 1626.
3. M. Nirmal, B.O. Dabbousi, M.G. Bawendi, J.J. Macklin, J.K. Trautman, T.D. Harris, and L.E. Brus, *Nature* 383 (1996) 802.
4. D. Norris and M. Bawendi, *Phys. Rev. B* 53 (1996) 16338.
5. D. Bertram, O. Micic, and A. Nozik, *Phys. Rev. B* 57 (1998) R4265.
6. A.P. Alivisatos, *J. Phys. Chem.* 100 (1996) 13266 (and references therein).
7. J.H. Fendler and F. Meldrum, *Adv. Mater.* 7 (1995) 607.
8. Al.L. Efros, M. Rosen, M. Kuno, M. Nirmal, D.J. Norris, and M.G. Bawendi, *Phys. Rev. B* 53 (1996) 4843.
9. M. Nirmal, D.J. Norris, M. Kuno, M.G. Bawendi, Al. L. Efros, and M. Rosen, *Phys. Rev. Lett.* 75 (1995) 3728.
10. S. Schmitt-Rink, D.A.B. Miller, and D.S. Chemla, *Phys. Rev. B* 35 (1987) 8113.
11. (a) L.E. Brus, *J. Phys. Chem.* 80 (1984) 4403. (b) L.E. Brus, *IEEE J. Quantum Electron.* QE-22 (1986) 1909. (c) L.E. Brus, *J. Phys. Chem.* 90 (1986) 2555.

12. (a) L. Manna, E.C. Scher, and A.P. Alivisatos, *J. Am. Chem. Soc.* 122 (2000) 12700. (b) Z.A. Peng and X. Peng, *J. Am. Chem. Soc.* 123 (2001) 1389. (c) X. Peng, L. Manna, W. Yang, J. Wickham, E. Scher, A. Kadavanich, and A.P. Alivisatos, *Nature* 404 (2000) 59.

13. (a) J.M. Petroski, Z.L. Wang, and M.J. El-Sayed, *J. Phys. Chem. B* 102 (1998) 102, 3316. (b) T.S Ahmadi, Z.L. Wang, T.C. Green, A. Henglein, and M.A. El-Sayed, *Science* 272 (1996) 1924.

14. Y. Li, H. Liao, Y. Ding, Y. Fan, Y. Zhang, and Y. Aian, *Inorg. Chem.* 38 (1999) 1382.

15. W. Wang, Y. Geng, Y. Qian, M. Ji, and X. Liu, *Adv. Mater.* 10 (1998) 1479.

16. D. Yu, D. Wang, Z. Meng, J. Lu, and Y. Qian, *J. Mater. Chem.* 12 (2002) 403.

17. S.S. Chang, C.W. Shih, C.D. Chen, and W.C. Lai, *C.R.C. Langmuir* 15 (1999) 701.

18. (a) R.S. Wagner and W.C. Ellis, *Appl. Phys. Lett.* 4 (1964) 89. (b) T.J. Trentler, K.M. Hickman, S.C. Goel, A.M. Viano, P.C. Givvons, and W.E. Buhro, *Science* 270 (1995) 1791. (c) J. Westwater, D.P. Gosain, S. Tomiya, and S. Usui, *J. Vac. Sci. Technol. B* 15 (1997) 554.

19. X. Duan, J. Wang, and C.M. Lieber, *Appl. Phys. Lett.* 76 (2000) 1116.

20. V.L. Colvin, M.C. Schlamp, and A.P. Alivisatos, *Nature* 370 (1994) 354.

21. N.C. Greenham, X. Peng, and A.P. Alivisatos, *Synth. Metals* 84 (1997) 545.

22. G. Hodes, *Isr. J. Chem.* 33 (1993) 95.

23. H. Weller, A. Eyechmuller, R. Vogel, L. Katsikas, A. Hasselbarth, and M. Gicrsig, *Isr. J. Chem.* 33 (1993) 107.

24. W.U. Huynh, J.J. Dittmer, N. Teclemarian, D.J. Milliron, A.P. Alivisatos, K.W.J. Barnham, *Phys. Rev. B* 67 (2003) 115326.

25. D.L. Klein, R. Roth, A. Lim, A.P. Alivisatos, and P.L. McEuen, *Nature* 389 (1997) 679.

26. (a) T. Vossmeyer, S. Jia, E. Delonno, M.R. Diehl, S.H. Kim, X. Peng, A.P. Alivisatos, and J.R. Heath, *J. Appl. Phys.* 84 (1998) 3664. (b) S.H. Kim, G. Markovich, S. Rezvani, S.H. Choi, K.L. Wang, and J.R. Heath, *Appl. Phys. Lett.* 74 (1999) 317.

27. M. Dahan, T. Laurence, F. Pinaud, D.S. Chemla, A.P. Alivisatos, M. Suer, and S. Weiss, *Phys. Rev. B* 6320 (2001) 5309.

28. C.J. Murphy, E.B. Brauns, L. Gearheart, in *Advances in Microcrystalline and Nano-crystalline Semiconductors*, Mater. Res. Soc., Pittsburgh, PA (1997), p. 597.

29. V. Colvin, M. Schlamp, and A.P. Alivisatos, *Nature* 370 (1994) 354.

30. C.P. Collier, T. Vossmeyer, and J.R. Heath, *Ann. Rev. Phys. Chem.* 49 (2000) 371 (and references within).

31. C.B. Murray, C.R. Kagan, and M.G. Bawendi, *Science* 270 (1995) 1335.

32. R. Leon, P.M. Petroff, D. Leonard, and S. Fafard, *Science* 267 (1995) 1966.

33. P.V. Braun, P. Osenar, and S.I. Stupp, *Nature* 380 (1996) 325.

34. (a) C.B. Murray, C.R. Kagan, and M.G. Bawendi, *Annu. Rev. Mater. Sci.* 30 (2000) 545 (and references within). (b) A.L. Rogach, D.V. Talapin, E.V. Shevchenko, A. Kornowski, M. Haase, and H. Weller, *Adv. Funct. Mater.* 12 (2002) 653. (c) M.P. Pilleni, *J. Phys. Chem. B* 105 (2001) 3358.

35. C.P. Colleir, R.J. Saykally, J.J. Shiang, S.E. Henrichs, and J.R. Heath, *Science* 277 (1997) 1978.

36. T. Takagahara, *Surf. Sci.* 367 (1992) 310.

37. (a) T. Vossmeyer, E. Delonno, and J.R. Heath, *Angew. Chem. Int. Ed. Engl.* 36 (1997) 1080. (b) G. Markovich, C.P. Collier, S.E. Henrichs, F. Remacle, R.D. Levine, and J.R. Heath, *Acc. Chem. Res.* 32 (1999) 415.

38. (a) S. Link, L. Wang, and M.A. El-Sayed, *J. Phys. Chem. B* 103 (1999) 3529. (b) S. Lind, C. Bruda, Z.L. Wang, and M.A. El-Sayed, *J. Chem. Phys.* 111 (1999) 1255.

39. (a) A. Taleb, C. Petit, and M.P. Pileni, *Chem. Mater.* 9 (1997) 950. (b) M.P. Pileni, *Langmuir* 13 (1997) 266.
40. (a) A.T. Ngo and M.P. Pileni, *J. Appl. Phys.* 92 (2002) 4649. (b) C.T. Black, C.B. Murray, R.L. Sandstrom, and S. Shouheng, *Science* 290 (2000) 1131. (c) V.R. Puntes, K.M. Krishnan, and A.P. Alivisatos, *Science* 291 (2001) 2115. (d) S. Sun, C.B. Murray, D. Weller, L. Folks, and A. Moser, *Science* 287 (2000) 1989.
41. (a) T. Vossmeyer, G. Reck, L. Katsikas, E.T.K. Haupt, B. Schulz, and H. Weller, *Science* 267 (1995) 1476. (b) T. Vossmeyer, L. Katsikas, M. Giersig, I.G. Popovic, K. Diesner, A. Chemseddine, A. Eychmüller, and H. Weller, *J. Phys. Chem.* 98 (1994) 7665. (c) H. Döllefeld, H. Weller, and A. Eychmüller, *Nano Lett.* 1 (2001) 267.
42. (a) C.R. Kagan, C.B. Murray, M. Nirmal, and M.G. Bawendi, *Phys. Rev. Lett.* 76 (1996) 1517. (b) C.R. Kagan, C.B. Murray, and M.G. Bawendi, *Phys. Rev. B* 54 (1996) 8633.
43. O.I. Micic, K.M. Jones, A. Cahill, and A.J. Nozik, *J. Phys. Chem. B* 102 (1998) 9791.
44. J. Tang, G. Ge, and L.E. Brus, *J. Phys. Chem. B* 106 (2002) 5653.
45. F.C. Meldrum, F. Flath, and W. Knoll, *Thin Solid Films* 348 (1999) 188.
46. A.P. Alivisatos, K.P. Johnsson, X. Peng, T.E. Wilson, C.J. Lowth, M.P. Bruchez, Jr., and P.G. Schultz., *Nature* 382 (1996) 609.
47. M. Sirota, I. Minkin, V. Hensel, M. Lahav, E. Lifshitz, *J. Phys. Chem. B* 105 (2001) 6792.
48. A. Santoni, G. Paolucci, G. Santoro, K.C. Prince, and N.I. Christensen, *J. Phys. Condens. Mater.* 4 (1992) 6759.
49. A.D. Andreev and A.A. Lipovskii, *Phys. Rev. B* 59 (1999) 15402.
50. H. Du, C. Chen, R. Krishnan, T.D. Kraus, J.M. Harbold, F.W. Wise, M.G. Thomas, and J. Silcox, *Nano Lett.* 2 (2002) 1321.
51. B.L. Wehrenberg, C. Wang, and P. Guyot-Sionnest, *J. Phys. Chem. B* 106 (2002) 10634.
52. C.B. Murray, S. Sun, W. Gaschler, H. Doyle, T.A. Betley, and C.R. Kagan, *IBM J. Res. Dev.* 45 (2001) 47.
53. A. Sashchiuk, L. Langot, R. Chaim, and E. Lifshitz, *J. Cry. Growth* 240 (2002) 431.
54. Y. Li, Z. Wang, and Y. Ding, *Inorg. Chem.* 38 (1999) 4737.
55. (a) N. Herron, J.C. Calabrese, W.E. Farneth, and Y. Wang, *Science* 259 (1993) 1426. (b) T. Vossemyer, G. Reck, L. Katsikas, E.T.K. Haupt, B. Schulz, and H. Weller, *Science* 267 (1995) 1476. (c) J. Rockengerger, L. Troger, A. Kornowski, T. Vossmeyer, and A. Eychmuller, *J. Phys. Chem. B* 101 (1997) 2691.
56. (a) M.V. Artemyev, A.I. Bibik, L.I Gurinovich, S.V. Gaponenko, and U. Woggon, *Phys. Rev. B* 60 (1999) 1504. (b) G. Markovich, C.P. Collier, S.E. Henrichs, F. Remacle, R.D. Levine, and J.R. Heath, *Ann. Chem. Res.* 32 (1999) 415. (c) C.P. Collier, R.J. Syakally, J.J. Shinag, S.E. Nerichs, and J.R. Heath, *Science* 277 (1997) 1978.
57. P. Bartlett, R.H. Ottewill, and P.N. Pusey, *J. Chem. Phys.* 93 (1990) 1299.
58. C.R. Kagan, C.B. Murray, M. Nirmal, and M.G. Bawendi, *Phys. Rev. Lett.* 76 (1996) 1517.
59. M. Shim and P. Guyot Sionnest, *J. Chem. Phys.* 111 (1999) 6955.
60. S. Connolly, S. Fullam, B. Korgel, and D. Fitzmaurice, *J. Am. Chem. Soc.* 120 (1998) 2969.
61. J.R. Heath, P.J. Kuekes, G.S. Snider, and R.S. Williams, *Science* 280 (1998) 1716.
62. K.C. Baverly, J.F. Sanpao, and J.R. Heath, *J. Phys. Chem. B* 106 (2002) 2131.
63. A.A. Middleton and N.S. Wingreen, *Phys. Rev. Lett.* 71 (1993) 3198.

64. (a) S.M. Grannan, A.E. Lange, E.E. Haller, and J.W. Beeman, *Phys. Rev. B* 45 (1992) 4516. (b) J. Zhang, W. Cui, M. Juda, D. McCammon, R.L. Kelley, S.H. Moseley, E.K. Stahle, and A.E. Szymkowiak, *Phys. Rev. B* 48 (1993) 2312. (c) P. Sheng, E.K. Sichel, and J.I. Gittleman, *Phys. Rev. Lett.* 40 (1978) 1197. (d) G.T. Kim, S.H. Jhang, J.G. Park, and S. Roth, *Synth. Metals* 117 (2001) 123.

*Biological Methods
of Nanoparticle Organization*

10 Biomolecular Functionalization and Organization of Nanoparticles

Christof M. Niemeyer

CONTENTS

10.1 Introduction 227
10.2 Biomolecule-Directed Nanoparticle Organization.....................................230
    10.2.1 Protein-Based Recognition Systems 231
    10.2.2 DNA-Based Assemblies...233
        10.2.2.1 DNA Directed Oligomerization of Nanoparticles 234
        10.2.2.2 Individual Nanoscaled Particle Assemblies 236
        10.2.2.3 DNA–Protein Conjugates as Model Systems...........238
        10.2.2.4 Construction of DNA–Protein Conjugates
                for Biotemplating...240
10.3 Applications..248
    10.3.1 Protein-Coated Nanoparticles as Biolabels 248
    10.3.2 Nucleic Acid–Coated Particles...251
    10.3.3 Therapeutic Applications..257
    10.3.4 Nanoparticles as Model Systems 257
10.4 Conclusions and Perspectives 258
Acknowledgments...259
References..259

10.1 INTRODUCTION

The essence of chemical science finds its full expression in the words of that epitome of the artist–scientist Leonardo da Vinci: "Where Nature finishes producing its own species, man begins, using natural things and with the help of this nature, to create an infinity of species." Nobel laureate Jean-Marie Lehn has used this quotation in giving his outlook on the future and perspectives of supramolecular chemistry.[1] Supramolecular chemistry concerns both the investigation of nature's principles to produce fascinating complex and functional molecular assemblies and the utilization of such principles to

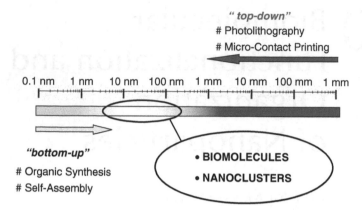

FIGURE 10.1 A gap currently exists in the engineering of small-scale devices. Whereas conventional top-down processes hardly allow the fabrication of structures smaller than about 200 to 100 nm, the limits of regular bottom-up processes are in the range of about 2 to 5 nm. Due to their own dimensions, two different types of compounds appear to be suited for addressing that gap: biomolecular components, such as proteins and nucleic acids, and colloidal nanoparticles composed of metal and semiconductor materials. (From Niemeyer, C.M., *Angew. Chem. Int. Ed.*, 40, 4128–4158, 2001. With permission.)

generate novel devices and materials, potentially useful for sensing, catalysis, transport, and other applications in medicinal or engineering sciences.[1–3]

Examples from nature have strongly motivated the developments of supramolecular chemistry because natural evolution has led to incredibly functional assemblages of proteins, nucleic acids, and other macromolecules to perform complicated tasks that are still daunting for us to try to emulate. As an example, the 20-nm ribosome particle is a most effective supramolecular nanomachine, which spontaneously self-assembles from its more than 50 individual protein and nucleic acid building blocks, thereby impressively demonstrating the power of biologically programmed molecular recognition.[4–6] Starting with the discovery of the DNA double-helix structure, biology has meanwhile grown from a purely descriptive and phenomenological discipline to a molecular science. Recombinant DNA technology brought insights into the basic principles of many biochemical processes, and it has also opened the door to modern biotechnology. Today, we are able to genetically engineer relatively simple bacterial cells, and we are on our way toward tailoring complex organisms. In view of such revolutionary developments in molecular biosciences, it seems particularly challenging to fuse biotechnology with materials science, and in particular with the research on inorganic nanoparticles, composed of metal or semiconductor materials.[7] The combination of these two disciplines will allow us to take advantage of the evolutionary improved biological components for generating new smart materials, and vice versa, to apply today's advanced materials and physicochemical techniques to solve biological problems.

Both biomolecules and nanoparticles meet at the same length scale (Figure 10.1). On the one hand, biomolecular components have typical size dimensions in the range of about 5 to 200 nm. Thus, to exploit and utilize the concepts administered in natural nanometer-scaled systems, the development of a nanochemistry is crucial.[8] On the other hand, commercial requirements to produce increasingly miniaturized microelectronic devices strongly motivate the elaboration of nanoscale systems. The structural dimensions of computer microprocessors are currently in the range of about 200 nm. They are just available by conventional top-down photolithographic processes, but such technologies hardly allow for the large-scale production of parts significantly smaller than 100 nm. Since "there is (still) plenty of room at the bottom," as Nobel physicist Richard Feynman pointed out more than 40 years ago,[9] today's nanotechnology research puts great emphasis on the development of bottom-up strategies that concern the self-assembly of (macro)molecular and colloidal building blocks to create larger, functional devices.[10] The individual components to be employed in self-assembling processes should fulfill some basic requirements. First, distinct intrinsic functionality, such as steric, optical, electronic, or catalytic properties, are desired to allow for the particular applications envisaged. Second, the modules need to be programmable through a specific constitution, configuration, conformation, and dynamic properties to enable specific recognition for self-assembly. Third, the building blocks have to have an appropriate size to bridge the gap between the submicrometer dimensions reachable by classical top-down engineering and the dimensions addressable by classical bottom-up approaches, such as chemical synthesis and supramolecular self-assembly.

Inorganic nanoparticles are particularly attractive building blocks for the generation of larger superstructures, and a number of outstanding monographs and reviews already exist, reviewing the rapidly increasing area of basic and applied nanoparticle research.[11–17] Nanoparticles can be readily prepared from various materials in large quantities by relatively simple methods, and their dimensions can be controlled from one to about several hundred nanometers, with a fairly narrow size distribution. Most often, the particles are composed of metals, metal oxides, and semiconductor materials, such as Ag_2S, CdS, CdSe, and TiO_2. The nanoparticles have highly interesting optical, electronic, and catalytic properties, which are very different from those of the corresponding bulk materials and which often depend strongly on the particles' size in a highly predictable way. Examples include the wavelength of light emission from semiconductor nanocrystals, which can be utilized for biolabeling[18,19] or lasers,[20] as well as the external magnetic field required to switch a magnetized particle, which is highly relevant in magnetotactic bacteria[21] and hard disk drives.[22] Moreover, nanoparticles are obtainable as highly perfect nanocrystals, which can be used as building blocks for the assembly of larger two- and three-dimensional structures.[23]

This chapter is intended to summarize approaches that currently emerge at the intersection of materials science and molecular biotechnology[7,24–26] and which go beyond the questions of pure materials science. This novel and highly interdisciplinary field of research is closely associated with the surface chemistry and physical properties of inorganic nanoparticles, the topics of bioorganic and bioinorganic chemistry, and the various aspects of molecular biology, recombinant DNA technology, recombinant protein expression, and immunology. In particular, this chapter reviews the utilization of nucleic acids and proteins as programmable recognition units to modify and improve the nanoparticle's assembly capabilities and, thus, to

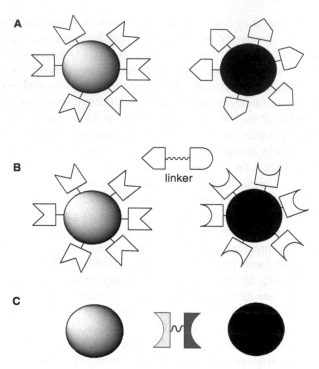

FIGURE 10.2 Solution phase chemical coupling using biomolecular recognition elements. (A) Two sets of nanoparticles are functionalized with individual recognition groups that are complementary to each other. (B) The particle-bound recognition groups are not complementary to each other but can be bridged through a bispecific linker molecule. (C) A bivalent linker that directly recognizes the surfaces of the nanoparticles is used for oligomerization. (From Niemeyer, C.M., *Angew. Chem. Int. Ed.*, 40, 4128–4158, 2001. With permission.)

allow for the organization of nanoparticles, enabling the fabrication of a nanostructured and mesoscopic supramolecular hybrid architecture.

10.2 BIOMOLECULE-DIRECTED NANOPARTICLE ORGANIZATION

Many approaches have been described to yield two- and three-dimensional arrays of metal and semiconductor nanoparticles,[13–15,17] for instance, using random inclusion into gel and glassy matrices, deposition on structured surfaces, or solution phase chemical coupling by means of bivalent cross-linker molecules. An example of the latter is the alkyldithiol-directed aggregation of gold colloids,[27] leading to extended networks of cross-linked nanoparticles. Similar results can be achieved with biomolecules (Figure 10.2). Initially, two sets of nanoparticles to be organized are functionalized with individual recognition groups that either are directly complementary to each other or are complementary to a molecular linker (Figure 10.2A and B, respectively). Driven by the biospecific interaction, the two particle batches assemble to extended nanoparticle networks, which in some cases can grow to the size of macroscopic materials. Rather than just exploring novel, unconventional coupling systems, the major motivation for this approach is based on the unique properties of biomolecular linkers that promise new applications, such as tunable and switchable materials. In particular, the tremendous recognition capabilities of biomolecules, and their potential to be addressed and manipulated through biochemical procedures using designed tools, such as enzymes, should provide the basis for entirely novel routes to rationally construct advanced materials.

10.2.1 Protein-Based Recognition Systems

The coupling approaches depicted in Figure 10.2 have been experimentally realized using protein-based recognition systems. One of the first examples took advantage of the specific interaction between antibodies and low-molecular-weight organic compounds.[28] The latter, often termed hapten groups, can be used to cross-link nanoparticles, previously coated with antibody molecules that specifically recognize that particular hapten. Shenton et al.[28] functionalized gold and silver nanoparticles, with immunoglobulins of classes G and E (IgG and IgE, respectively), which had a specificity directed against either the d-biotin or the dinitrophenyl (DNP) group. To allow for the cross-linking of the IgG-functionalized nanoparticles, bivalent linkers had been synthesized that contained two hapten groups in terminal positions. Either the monospecific or bispecific linkers were used, allowing for the directed assembly of homo-oligomeric or hetereodimeric aggregates, respectively. The precipitate formed was characterized by transmission electron microscopy (TEM), revealing large disordered three-dimensional networks of discrete nanoparticles. In the case of bimetallic Ag-Au networks, however, the degree of integration was not as high as expected for an oligomerization process regulated by highly specific biomolecular recognition. This suggests that further progress will require more sophisticated engineering of the antibody–antigen interface.[28]

Other approaches of using protein-based recognition systems to organize inorganic nanoparticles made use of the unique interaction between the biotin-binding protein streptavidin (STV) and its low-molecular-weight ligand D-biotin.[29,30] The remarkable biomolecular recognition of the water-soluble molecule biotin (vitamin H) through the homotetrameric protein STV is characterized by the extraordinary affinity constant of the STV–biotin interaction of about 10^{14} dm^{3}mol^{-1}, indicating the strongest ligand–receptor interaction currently known.[31] Another great advantage of STV is its extreme chemical and thermal stability. Since biotinylated materials are often commercially available or can be prepared with a variety of mild biotinylation procedures, biotin–STV conjugates form the basis of many diagnostic and analytical tests.[32]

Connolly and Fitzmaurice[30] have used the STV–biotin interaction to organize gold colloids functionalized by chemisorptive coupling of a disulfide-bridged bisbiotin compound. According to the generalized scheme depicted in Figure 10.2B, the subsequent cross-linking was achieved by the addition of STV as a linker. The immediate change of the sol's color from red to blue was indicative of the formation of oligomeric networks. This process was also monitored by dynamic light scattering (DLS), allowing the determination that the average hydrodynamic radius of all particles in solution rapidly increased upon STV-directed assembly. TEM images revealed networks containing on average about 20 interconnected particles, which are separated by about 5 nm, correlating with the diameter of an STV molecule. Since the deposition of the colloid aggregates on the graphite substrate for TEM characterization might affect aggregation or alteration of existing structures, small-angle x-ray scattering (SAXS) was employed to probe the solution structure of the networks. The angular dependence of the intensity of the scattered x-ray radiation quantitatively depends on the size and shape of the assemblies measured. This allows calculation of the pair-distance distribution function (PDDF), which describes the number of pairs of scattering centers as a function of their separation. SAXS of nonaggregated samples revealed dispersed particles indicated by a single maximum in the PDDF. In the case of STV-coupled nanoparticles, however, SAXS indicated the presence of an average trimer of 16-nm-diameter hard spheres (the gold colloids), separated by a minimum distance of 4 nm.[30]

In agreement with earlier studies from our group,[29] the above findings suggest that the STV–biotin system is most versatile for developing novel strategies of assembling nanoparticles in solution or on a substrate. The applicability of the STV–biotin system for generating supramolecular aggregates is enhanced by the availability of various biotin analogs and recombinant STV mutants, revealing a wide range of rate and equilibrium constants and, thus, allowing for fine-tuning of the dynamic and structural properties of colloidal hybrid materials. Conceptual work has noted that extensive use of protein-based assembly might lead to a "factory of the future," in which nanofabrication of spatially defined aggregates takes place via the bottom-up assembly of nanoparticles, directed by multiple highly specific bio-molecular recognition elements.[33] Assuming that such a scenario might be realized some time in the future, it will dramatically benefit from the advances of modern biotechnology, providing us with the technical skills for tailoring novel protein components by means of directed evolution strategies.[34,35] Recombinant DNA tech-nology has already led to the development of high-affinity artificial linker systems, such as antibodies with specificity against digoxigenin and fluorescein haptens or short peptides, such as the 9-mer FLAG sequence (i.e., MDYKDDDD). These coupling systems have proven their applicability in various diagnostic assays, and it is likely that they are also convenient for generating inorganic nanoparticle networks.

The power of today's biotechnological methods was recently applied to *de novo* generate protein linker units that directly recognize distinct surfaces of semiconduc-tor materials, thereby avoiding the necessity of previous chemical functionalization (Figure 10.2C). Whaley et al.[36] used phage display techniques to select 12-mer peptides with binding specificity for distinct semiconductor surfaces. Based on a combinatorial library of random 12-mer peptides, composed of about 10^9 different peptides, phage clones were selected for their specific binding capabilities to five different single-crystal semiconductors, GaAs(100), GaAs(111)A, GaAs(111)B, InP(100), and Si(100). These substrates were chosen to allow for systematic evalu-ation of the peptide–substrate interactions. Specific peptide binding was found that was selective for the crystal composition (for example, binding to GaAs, but not to Si) and crystal face (for example, binding to GaAs(100), but not to GaAs(111)B). The clone G12-3, for example, specifically binds to a GaAs pattern but not to the SiO_2 regions on the same substrate. This clone was even capable of differentiating between two zinc blende crystal structures, GaAs and AlAs, which have almost identical lattice constants of 5.66 and 5.65 Å, respectively. Binding occurred specifically to the GaAs substrate.

The basis for this selectivity, whether it is chemical, structural, or electronic, is still under investigation. The various substrates offer individual chemical reactivities, and thus the phage selectivity is not particularly surprising. Even in the case of different GaAs surfaces, the composition of surface oxide layers is expected to be different, which in turn might also influence the peptide–surface interaction. However, sequence analysis of various clones has revealed the presence of, on average, about 40% polar functional groups and about 50% Lewis base functional groups. The statistical abundance of the latter is only 34%, suggesting that the interactions between peptide Lewis bases and Lewis acid sites on the substrates are important for specific recognition. The peptides evolved may be applied to the generation of bispecific linker molecules with binding capabilities for two different semiconductor or metal surfaces (Figure 10.2C), thereby paving the way to novel bottom-up approaches for fabricating complex nanoparticle heterostructures.[36] In a recent continuation, this approach was expanded on the fabrication of a highly ordered composite material from genetically engineered M13 bacteriophage and ZnS nanocrystals.[37] The bacteriophages were coupled with ZnS solution precursors and spontaneously evolved a self-supporting hybrid film material that was ordered at the nanoscale and at the micrometer scale into approximately 70-μm domains, which were continuous over a centimeter-length scale. Besides opening up novel ways for generating unconventional materials, this work has implications on the understanding of the fundamental processes of biomineralization, in which protein components control the formation of materials by specifically interacting with growing crystallites.

In recent developments, mechanical hybrid elements composed of nanostructured inorganic components and biomolecules were constructed, taking advantage of motor proteins.[38,39] The fabrication of the latter was achieved from a recombinant F_1-ATPase, immobilized at 80-nm-diameter Ni posts and coupled with a nanopropeller of ca. 1000 nm in length. The ATPase molecule, which is capable of producing about 100 pN *nm of rotary torque by hydrolysis of ATP,[40] rotates the propeller with a velocity of about five rounds per second.

10.2.2 DNA-Based Assemblies

With respect to developing nanotechnology devices, researchers have early suggested the use of biological macromolecules as components of nanostructured systems.[41–43] DNA is a particularly promising candidate to serve as a construction material in nanosciences.[44,45] Despite its simplicity, the enormous specificity of the A-T and G-C Watson–Crick hydrogen bonding allows the convenient programming of artificial DNA receptor moieties. The power of DNA as a molecular tool is enhanced by our ability to synthesize virtually any DNA sequence by automated methods and to amplify any DNA sequence from microscopic to macroscopic quantities by PCR. Another very attractive feature of DNA is the great mechanical rigidity of short double helices, so that they behave effectively like a rigid rod spacer between two tethered functional molecular components on both ends. Moreover, DNA displays a relatively high physicochemical stability. Finally, nature provides a variety of highly specific enzymes, such as endonucleases and ligases, which allow for processing of the DNA material with atomic precision and accuracy on the Ångström level. No other (polymeric) material offers these advantages, which are ideal for molecular constructions in the range of about 5 nm up to the micrometer scale.

DNA has already extensively been used for constructions on the nanometer-length scale,[44,45] and recently even nanomechanical devices have been fabricated from this genetic material.[46] This chapter will focus on the use of DNA as a construction material for the fabrication of nanoparticular assemblies.

10.2.2.1   DNA-Directed Oligomerization of Nanoparticles

Mirkin and colleagues[47] have used DNA hybridization to generate repetitive nano-cluster materials (Figure 10.3). Following the generalized scheme in Figure 10.2B, two noncomplementary oligonucleotides were coupled in separate reactions with 13-nm gold particles via thiol adsorption.[47] A DNA duplex molecule containing a double-stranded (ds) region and two cohesive single-stranded (ss) ends, which are complementary to the particle-bound DNA, was used as a linker. The addition of the linker duplex to a mixture of the two oligonucleotide-modified colloids led to the aggregation and slow precipitation of a macroscopic DNA–colloid material. The reversibility of this process was demonstrated by the temperature-dependent changes of the UV-visible (UV/VIS) spectrophotometric properties. Since the colloids contained multiple DNA molecules, the oligomerized aggregates were well ordered and three-dimensionally linked, as judged from the TEM analysis. Images of two-dimensional, single-layer aggregates reveal closely packed assemblies of the colloids with uniform particle separations of about 6 nm, corresponding to the length of the DNA linker duplex.[47]

More recent work of the Mirkin group[47] concerned the generation of binary networks. For example, two types of gold clusters of either 40 or 5 nm diameter were modified with individual 12-mer oligonucleotides and were assembled using a complementary 24-mer oligonucleotide linker.[48] Due to the specificity of Watson–Crick base pairing, only heterodimeric A-B composites with alternating particle sizes are formed (Figure 10.3). In the case of an excess of one particle, satellite-like aggregate structures are generated and characterized by TEM. Using a similar approach, oligonucleotide-functionalized CdSe/ZnS quantum dots and gold nanoparticles have been incorporated into binary nanoparticle networks.[49] TEM analysis revealed that the hybrid metal/semiconductor assemblies exhibited the A-B structure expected, and fluorescence and electronic absorption spectroscopy indicated cooperative optical and electronic phenomena within the network materials.

Further studies of DNA-linked gold nanoparticle assemblies concerned the influence of the DNA spacer length on the optical[50] and electrical[51] as well as the melting properties[52] of these networks. The experiments provided evidence that the linker length kinetically controls the size of the aggregates and that the optical properties of the nanoparticle assemblies are governed by aggregate size.[50] In contrast, it was found that the electrical properties of dried nanoparticle aggregates are not influenced by the linker length. Although SAXS clearly indicated the linker length-dependent distances between particles in solution, the networks collapse upon drying, thereby

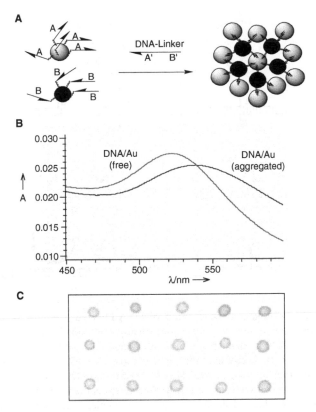

FIGURE 10.3 Assembly of gold colloids using DNA hybridization. (A) Two batches of gold particles are derivatized with noncomplementary oligonucleotides, via either 5′- or 3′-thiol groups. The nanoparticles are oligomerized using a single-stranded nucleic acid linker molecule. Note that exclusively heterodimeric A-B linkages are present within the network, thereby allowing the generation of binary particle networks composed of particles of different sizes[48] and different materials.[49] (B) Network formation leads to a characteristic change in the plasmon absorption. (C) The color change can be used to macroscopically detect DNA hybridization using an assay in which DNA–Au conjugates and a sample with potential target DNA are spotted on a hydrophobic membrane. Red spots indicate the absence of fully matched DNA, whereas blue spots are indicative of the presence of complementary target DNA. (Data in (C) from Storhoff, J.J. and Mirkin, C.A., *Chem. Rev.*, 99, 1849–1862, 1999. With permission. Copyright © 1999 American Chemical Society.)

forming bulk materials composed of nanoparticles covered with an insulating film of DNA. These materials show semiconductor properties not influenced by the linker lengths.[51] The strongly cooperative melting effect of the nanoparticle networks results from two key factors: the presence of multiple DNA linkers between each pair of nanoparticles and a decrease in the melting temperature as DNA strands melt due to a concomitant reduction in local salt concentration. This mechanism, originating from short-range duplex-to-duplex interactions, is independent of DNA base sequences and should be universal for any type of nanostructured probe that is heavily functionalized with oligonucleotides.[52]

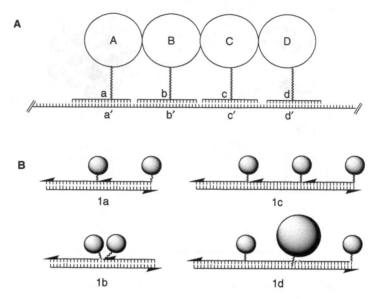

FIGURE 10.4 (A) Schematic representation of DNA-directed assembly of four different nanoscaled building blocks to form a stoichiometrically and spatially defined supramolecular aggregate. (B) Assembly of nanocrystal molecules using DNA hybridization.[57] Conjugates from gold particles (represented as shaded spheres) and 3'- or 5'-thiolated oligonucleotides allowed the fabrication of head-to-head (1a) or head-to-tail (1b) homodimers. A template containing the complementary sequence in triplicate effects the formation of the trimer 1c. Aggregate 1d reveals a limited flexibility, due to the unnicked double helical backbone.[58] Note that, according to the scheme in (A), the DNA-directed assembly has been applied to the organization of proteins,[56] beta-galactosyl carbohydrates,[170] and organometallic compounds.[171] (From Niemeyer, C.M., *Angew. Chem. Int. Ed.*, 40, 4128–4158, 2001. With permission.)

10.2.2.2 Individual Nanoscaled Particle Assemblies

The work of Mirkin and coworkers described above leads to novel hybrid materials with promising electronic and optical properties, potentially useful for applications in materials sciences,[24,53–55] and the results also have a tremendous impact on immediate diagnostic applications, in particular DNA microarray-based nucleic acid analyses (Section 10.3.2). Due to the enormous potential of these novel compounds, it is particularly important to overcome limitations that may result from the lack of stoichiometric control during the assembly process. To control the architecture of nanomaterials, spatially defined arrangements of molecular devices are required. For example, to organize metal and semiconductor nanocrystals into ultrasmall electronic devices, one may consider a linear aggregate of several individual components (Figure 10.4A). Similarly, as previously demonstrated for proteins,[56] gold nanoclusters containing a single nucleic acid moiety[57,58] have been rationally assembled into stoichiometrically defined nanoscale aggregates (Figure 10.4B).

To control the stoichiometry and architecture of nanomaterials, Alivisatos and coworkers[57,58] synthesized well-defined mono-adducts from commercially available

1.4-nm gold clusters containing a single reactive maleimido group and thiolated 18-mer oligonucleotides. Subsequent to purification, these conjugates allowed the rational construction of well-defined nanocrystalline molecules by DNA-directed assembly using a single-stranded template that contained the complementary sequence stretches (Figure 10.4B). Depending on the template, the DNA–nanocluster conjugates were assembled to generate the head-to-head and head-to-tail homodimeric target molecules 1a and 1b in approximately 70% purity, as determined by TEM analysis. Consistent with model calculations, the center-to-center distances observed in the two isomers were about 3 to 10 and 2 to 6 nm in 1a and 1b, respectively. TEM images also revealed that the trimeric molecules (1c) showed the structure expected. More recently, multiple gold nanocrystal aggregates were generated by DNA-directed assembly.[58] The preparations, containing up to three nanoclusters of different sizes organized in multiple fashions, were purified by electrophoresis subsequent to the self-assembly. TEM characterization indicated that the nanocrystal molecules have a high flexibility when the DNA backbone is nicked (e.g., 1c), whereas an unnicked double helix (e.g., 1d) significantly lowers the flexibility. UV/VIS absorbance measurements indicated changes in the spectral properties of the nanoparticles as a consequence of the supramolecular organization.[58]

The initial demonstration of using DNA as a framework for the precise spatial arrangement of molecular components was carried out with covalent conjugates of single-stranded DNA oligomers and the STV protein.[56] Using the STV–DNA conjugates 2 (Figure 10.5) as model systems, a variety of essential basic studies on the DNA-directed assembly of macromolecules were carried out, which have been reviewed earlier.[59] In addition to their model character, however, the covalent DNA–STV conjugates are also convenient as versatile molecular adapters in the nanofabrication of supramolecular assemblies. The covalently attached oligonucleotide moiety supplements the STV's four native biotin-binding sites with a specific recognition domain for a complementary nucleic acid sequence. This bispecificity allows use of the DNA–STV conjugates as adapters to assemble basically any biotinylated compound along a nucleic acid template.[59] As shown in Figure 10.5, for example, the strong biotin–STV interaction and the specific nucleic acid hybridization capabilities of the DNA–STV conjugates 2 were utilized to organize gold nanoclusters.[29] In this work, 1.4-nm gold clusters containing a single amino substituent were derivatized with a biotin group and, subsequently, the biotin moiety was used to organize the nanoclusters into the tetrahedral superstructure, defined by the biotin-binding sites of the STV. Subsequently, the nanocluster-loaded proteins self-assembled in the presence of a complementary single-stranded nucleic acid carrier molecule, thereby generating novel biometallic nanostructures, such as 3 (Figure 10.5). Since the DNA–STV conjugates can be used like a molecular construction kit, this approach even allows the combined assembly of inorganic and biological components to fabricate functional biometallic aggregates, such as 4, containing an immunoglobulin molecule (Figure 10.5). The latter biocomponent can be used for targeting the biometallic nanostructures toward certain tissues or other surfaces, recognized by the antibody. This approach impressively demonstrates the applicability of protein–ligand interaction and DNA hybridization for the nanoconstruction of novel inorganic/bioorganic hybrid systems.[29]

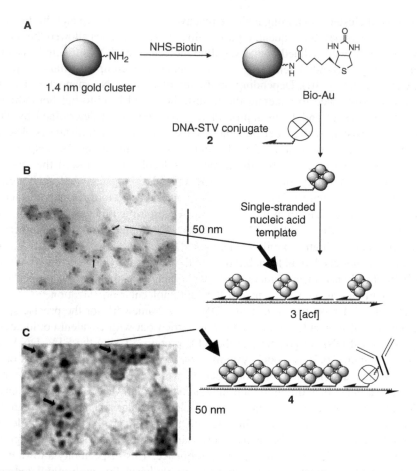

FIGURE 10.5 (A) Schematic representation of DNA-directed assembly of biotinylated gold nanoclusters using covalent DNA–STV conjugates 2 as molecular adapters. The nanocluster-loaded STV conjugates self-assemble in the presence of a single-stranded nucleic acid carrier molecule, containing complementary sequence stretches, to form biometallic aggregates 3. TEM images of the biometallic aggregate 3[acf.] (the letters in brackets indicate the protein components bound to the carrier) (B) and an antibody-containing biometallic construct 4 (C). The latter was fabricated from gold-labeled conjugates 2 and a conjugate of 2 and a biotiny-lated immunoglobulin, previously coupled in separate reactions. Note that the four gold clusters bound to 2 cannot be optically resolved by TEM. (From Niemeyer, C.M., *Angew. Chem. Int. Ed.*, 40, 4128–4158, 2001. With permission.)

10.2.2.3 DNA–Protein Conjugates as Model Systems

Despite these advances, still very little is known on the manipulation and tailoring of such nanoparticle networks, for instance, on ways to influence the structure and topography of the DNA hybrid materials subsequent to their formation by self-assembly. Within the context of basic studies on DNA-linked nanoparticle networks, the oligomeric aggregates 5 (Figure 10.6) generated from bioorganic STV particles

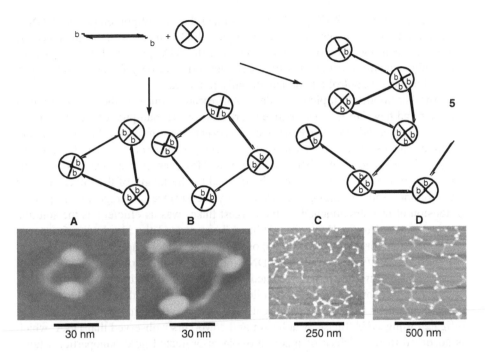

FIGURE 10.6 DNA–protein 5 conjugates as model systems for DNA-interconnected nanoparticles. The synthesis of the networks was achieved by using dsDNA as spaces. The dsDNA fragments contain two binding sites attached to the two 5′-ends of the dsDNA (b = binding site). Either the binding sites are biotinyl groups allowing cross-linking of the biotin-binding protein streptavidin as a model nanoparticle (A–C), or the binding sites are thiol groups allowing the connection of 5-nm gold colloids (D). The assembly of the two components leads to the formation of oligomeric networks (C, D), but also individual particle aggregates are formed that can be isolated by electrophoretic or chromatographic methods (A, B). The images at the bottom were obtained by scanning force microscopy analysis. (From Niemeyer, C.M., *Angew. Chem. Int. Ed.*, 40, 4128–4158, 2001. With permission.).

and bisbiotinylated dsDNA are suitable model systems to gain insight into the properties of complex particle networks. The STV functions as a 5-nm model particle that can realize only a limited number of interconnects to other particles within the network. Either one, two, three, or four biotinylated DNA fragments can be conjugated with the STV by means of the high-affinity STV–biotin interaction. This simplifies the complexity of the supramolecular particle networks and thus allows for the convenient analysis[60,61] and modeling[62] of effects occurring from variations, e.g., in the assembly and immobilization parameters. In addition, the size of the dsDNA linker fragments, which are typically about 30 to 170 nm in length, allows for convenient direct observation by scanning force microscopy (SFM) (Figure 10.6).[59,63–65] These findings are convenient for both disciplines involved: the use of the SFM allows the gaining of insights into biomolecular structure and assembly dynamics, and, in turn, the well-defined and clearly distinguishable DNA–STV nanostructures can be used as soft materials calibration standards for the detailed study of tip convolution effects and the influences of measuring parameters in SFM analyses.[60,61]

In a recent work, Park et al.[66] demonstrated that the oligomerization of DNA-coated gold nanoparticles, depicted in Figure 10.3, can as well be mediated by conjugates composed of oligonucleotides and the STV protein. Interestingly, the thermodynamically most stable oligomeric networks are only formed upon heating of the kinetically controlled adducts initially produced.

Another important topic of fundamental research concerns the investigation of whether and how DNA-functionalized particles can be manipulated by DNA-modifying enzymes. Nicewarner Pena and coworkers[67] demonstrated that the oligonucleotides chemisorbed at the surface of gold nanoparticles serve as templates for the enzymatic extension with DNA polymerases. However, it was observed that the efficiency of the reaction was strongly affected by the length of the linker between the oligomer and the particle as well as by the surface coverage of the primer. Extension of primers attached by the longest linker was as efficient as the solution phase reaction. Yun and coworkers[68] have demonstrated the absolute maintenance of biofunction for biomaterials composed of duplex DNA appended with 1.4-nm Au particles. Enzyme activity and DNA-binding affinities of restriction endonucleases and methyltransferases seemed to be unaltered by the nanoparticle–DNA conjugates. Recently, Kanaras and colleagues[69] have shown that large DNA–nanoparticle networks can be site specifically cleaved by restriction endonucleases and religated using DNA ligase. MacIntosh and coworkers[70] observed that DNA, which is adsorbed to the surface of mixed-monolayer-protected gold nanoparticles functionalized with tetraalkylammonium ligands, no longer can function as a template for transcription by T7 RNA polymerase *in vitro*. Interestingly, similar conjugates efficiently transfect mammalian cell cultures, as determined by beta-galactosidase plasmid transfer.[71] The success of these transfection assemblies depended on several variables, including the ratio of DNA to nanoparticle during the incubation period, the number of charged substituents in the monolayer core, and the hydrophobic packing surrounding these amines.

10.2.2.4 Construction of DNA–Protein Conjugates for Biotemplating

The electrostatic and topographic properties of biological macromolecules and supramolecular complexes comprised thereof can be used for the synthesis and assembly of organic and inorganic components. In this biological templating approach, regular two-dimensional lattices of bacterial cell surface proteins,[72] hollow biomolecular compartments, such as virus particles, and nano- and micrometer-sized nucleic acid components have already been exploited for the generation of nanoparticles and supramolecular architecture.[7]

In the DNA-templated assembly, the electrostatic and topographic properties of the DNA molecule can be used for the templated synthesis of supramolecular aggregates of inorganic and organic building blocks.[45] Pioneering work on the DNA-templated generation of metal and semiconductor nanoparticle arrays was carried out by Coffer and coworkers.[73-76] They used the negatively charged phosphate backbone of the DNA double helix to accumulate Cd^{2+} ions, which were subsequently treated with Na_2S to form CdS nanoparticles. For this, solutions of calf thymus DNA and Cd^{2+} ions were mixed and, subsequently, molar amounts of Na_2S

initiated the formation of CdS nanoparticles. As indicated from TEM analysis, the particles were fairly monodisperse, with an average diameter of about 5.6 nm.[73] Since the actual role of the DNA was vague from this study, later experiments concerned potential influences of the DNA's sequence, employed as a template.[74] Strikingly, it was found that, in particular, the adenine content of the DNA affects the size of the nanoparticles, thereby providing indirect evidence for the template mechanism suggested. Subsequent studies concerned the synthesis of surface-bound mesoscopic nanoparticle aggregates.[75,76] For this purpose, the 3455-base pair circular plasmid DNA pUCLeu4 was used as a template, forming a circle of about 375 nm in diameter. The DNA was mixed with Cd^{2+} ions in solution and, subsequently, the Cd^{2+}-loaded DNA was adsorbed to an amino-modified glass substrate. The formation of the CdS nanostructures was then achieved by H_2S treatment.[75] TEM analysis revealed that CdS particles of about 5 nm in diameter had developed, which were assembled close to the circular DNA backbone. Measurements of its circumference indicated the intactness of some of the DNA molecules; however, various other aggregates with irregular shapes were also observed.

Richter et al.[77] reported on the use of λ-DNA for the templated growth of Pd clusters. This was achieved by activating the DNA with Pd acetate and subsequent reduction using a sodium citrate/lactic acid/dimethylamine borane solution. This led to the formation of DNA-associated one- and two-dimensional arrays of well-separated 3- to 5-nm Pd clusters.[77] Subsequent work has focused on the preparation of highly conductive nanowires using DNA templates[78] as well as on the mechanisms of the initial nucleation of the metal clusters.[79] The DNA-templated fabrication of mesoscale structures from preformed, positively charged 3-nm CdS particles with a thiocholine-modified surface was reported by Torimoto et al.[80] Due to the electrostatic interaction between positively charged nanoparticle surfaces and the phosphate groups of DNA, the particles were found to be assembled to chains, as determined by TEM analysis. The CdS nanoparticles were arranged in a quasi-one-dimensional dense packing, which revealed interparticle distances of about 3.5 nm, correlating with the height of one helical turn of the DNA double strand.[80] The electrostatic assembly of lysine-passivated gold particles on DNA[81] or intercalator-modified gold particles on DNA and polylysine[82] was also reported.

In addition to its use as a nanometer-scaled template for generating nanoparticles, DNA can also be employed to fabricate micrometer-scaled elements, potentially useful in microelectronics. A descriptive example was published by Braun et al.[83] (Figure 10.7), who used a 16-μm λ-DNA molecule containing two cohesive ends to bridge the approximately 12- to 16-μm distance between two gold microelectrodes, prepared by standard photolithography. Subsequent to functionalization of the electrodes with individual capture oligonucleotides, the DNA fragment was allowed to hybridize. Successful interconnection of the electrodes was confirmed by optical microscopy of the fluorescently labeled λ-DNA. Next, the sodium ions bound to the phosphate backbone of the λ-DNA were exchanged with Ag^+ ions, and the latter were chemically reduced by hydroquinone. The small silver aggregates formed along the DNA backbone were then used as catalysts for further reductive deposition of silver, eventually leading to the formation of a silver nanowire. This micrometer-sized element with a typical width of 100 nm had a granular morphology,

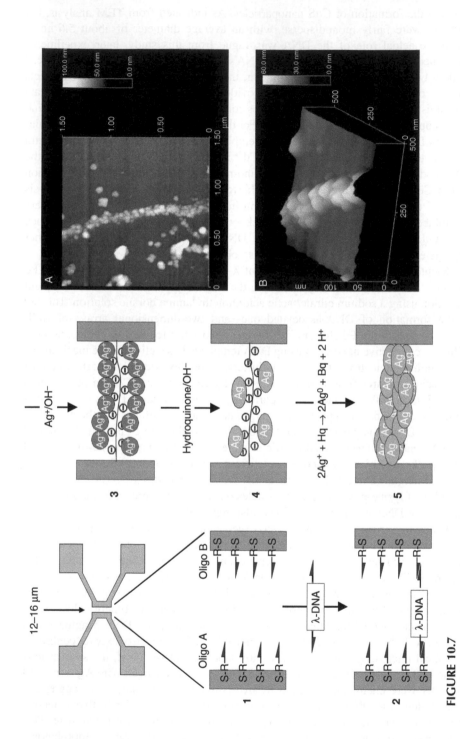

FIGURE 10.7

as determined by SFM (Figure 10.7). Two terminal electrical measurements of the Ag nanowire revealed nonlinear, history-dependent I-V curves, possibly a result of polarization or corrosion of the individual 30- to 50-nm Ag grains that constitute the wires.[83]

The examples presented above give initial impressions on how DNA can be utilized as a template in the synthesis of nanometric and mesoscopic aggregates. However, the studies emphasize the importance of fundamental research on how the interaction between DNA and the various binders, such as metal and organic cations, influences the structure and topology of the nucleic acid components involved. Potential applications include the use of DNA strands[83] and networks[84] to prepare microelectronic devices or the use of DNA circles[64,85,86] to fabricate nanoscaled Aharonov–Bohm rings.[87] Moreover, the above examples clearly indicate the necessity for generating more complex biomolecular architecture, which can be used for the generation of inorganic components by means of biotemplating.

An important step in this direction has recently been carried out by Keren and coworkers.[88] They introduced a novel concept, termed biomolecular lithography, which takes advantage of the sequence-specific binding of conjugates composed of ssDNA and RecA protein. In molecular lithography, the DNA–protein complex is treated with silver ions that are reduced by aldehyde groups, previously generated within the dsDNA target, thereby forming small silver grains to be subsequently utilized for the wet chemical deposition of gold. This procedure leads to the formation of a conductive wire, whereas contains an isolating gap, precisely at the position where RecA was bound (Figure 10.8A). Therefore, in the molecular lithography, the information encoded in the DNA molecules has replaced the masks used in conventional lithography, whereas the RecA protein serves as the resist. This approach should work with high resolution over a broad range of length scales from nanometers to many micrometers. Moreover, molecular lithography enables the generation of branch points (three-way junctions) within linear DNA fragments by employing a dsDNA fragment containing a single-stranded end in the initial RecA polymerization step (Figure 10.8B). The homologous recombination between RecA conjugate and double-stranded DNA can also be utilized for the sequence-specific positioning of molecular objects (Figure 10.8C). For this, the ssDNA used for polymerization of the RecA monomers is modified with molecular entities, which are functional devices by themselves, or else allow for subsequent binding of such. For instance, gold nanoclusters have been selectively positioned by this approach.[88] Due to the broad scope of molecular lithograhy, a wide variety of applications might be foreseen.[89]

Our group has constructed well-defined DNA–protein nanostructures, which might serve as templates for generating inorganic structures by means of metallization

---

FIGURE 10.7 (opposite page) DNA-templated synthesis of a conductive silver wire. A 16-μm DNA molecule was used to bridge the gap between two gold microelectrodes. The sodium ions bound to the DNA's phosphate backbone were exchanged with Ag⁺ ions, and the latter were chemically reduced by hydroquinone to form small silver aggregates. Further reductive deposition of silver led to the formation of a silver nanowire. SFM images revealed the granular structure of the wire. (From Braun, E. et al., *Nature*, 391, 775–778, 1998. With permission.)

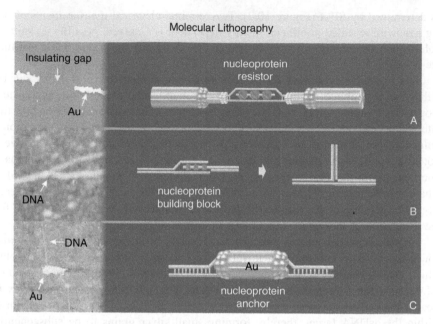

FIGURE 10.8 Molecular lithography. (From Niemeyer, C.M., *Science*, 297, 62–63, 2002. With permission. Copyright © 2002 American Association for the Advancement of Science.)

reactions. To this end, we have established that the oligomeric DNA–STV conjugates 5, described above (see Figure 10.6), can be effectively transformed into well-defined supramolecular DNA–STV nanocircles 6 by thermal treatment (Figure 10.9).[64] Due to their ready availability and well-defined stoichiometry and structure, the nanocircles 6 form the basis of a supramolecular construction kit for generating hapten–DNA conjugates. For example, functionalization of 6 with biotinylated haptens allows one to generate hapten conjugates 7, applicable as reagents in a novel competitive immuno-PCR (cIPCR) assay for the ultrasensitive detection of low-molecular-weight analytes.[90] Model studies suggested that this cIPCR allows for an ~10- to 1000-fold improvement of the detection limit of conventional antibody-based assays, such as

---

FIGURE 10.9 (opposite page) Construction of supramolecular DNA and STV conjugates. (A) Schematic representation of the self-assembly of oligomeric DNA–STV conjugates 5 from 5′,5′-bis-biotinylated DNA and STV.[63] Note that the schematic structure of 5 is simplified since a portion of the STV molecules function as tri- and tetravalent linker molecules between adjacent DNA fragments (see SFM, image (B)). The supramolecular networks of 5 can be disrupted by thermal treatment, leading to the efficient formation of DNA–STV nanocircles 6 (see SFM, image (C)). The nanocircles 6 can be functionalized by the coupling of biotinylated hapten groups, such as fluorescein (Fsc), to yield nanocircle 5. For simplification, complementary DNA strands are drawn as parallel lines. The 3′-ends are indicated by the arrowheads. (Adapted from Niemeyer, C.M. et al., *Angew. Chem. Int. Ed.*, 39, 3055–3059, 2000. With permission.)

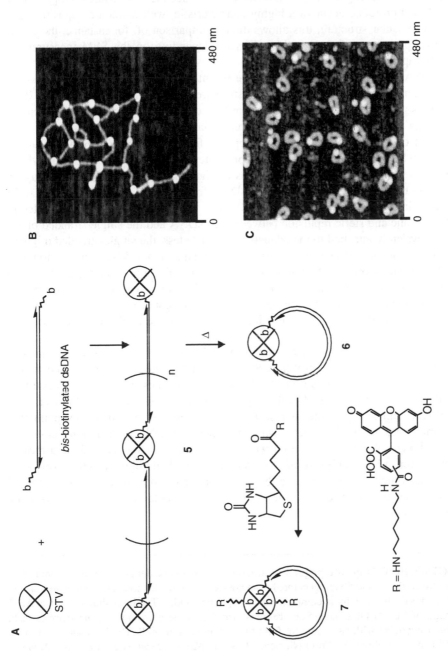

FIGURE 10.9

the competitive enzyme-linked immunosorbent assay (ELISA).[90] With respect to biomolecular nanotechnology, 5 and 6 can also be used as soft materials topography standardization reagents for SFM analyses.[60,61] Since the two different biopolymers, DNA and proteins, occur in a highly characteristic, well-defined composition and supramolecular structure, this allows direct comparison of, for instance, the deformation properties of the two biopolymers, depending on the SFM measurement modes applied.[61]

DNA–STV conjugates 5 and 6 have recently been used as building blocks for generating complex biomolecular nanostructures.[91,92] As indicated in Figure 10.10, networks 5a and 5b were generated from the covalent oligonucleotide–STV conjugates 2a and 2b, respectively, and bis-biotinylated dsDNA. Interestingly, the comparison of analogous oligomers 5 (prepared from native STV) revealed that the covalent STV–oligonucleotide hybrid conjugates 5a,b react with the bis-biotinylated DNA to generate oligomeric aggregates of significant smaller size, containing on average only about 2.5 times fewer dsDNA fragments per aggregate[91] (see AFM images in Figure 10.9 and Figure 10.10). This phenomenon has been attributed to electrostatic and steric repulsion between the dsDNA and the single-stranded oligomer covalently attached to the adducts 2. Nevertheless, the single-stranded oligonucleotide moieties can be used for further functionalization of 5 by hybridization with complementary oligonucleotide-tagged macromolecules. For instance, oligomers 5a were transformed into the corresponding nanocircles 6a, which can hybridize with analogous circles 6b, containing a complementary oligonucleotide sequence, to form the nanocircle dimers 8. The SFM image of the dimeric conjugate 8 indicates the predicted structure.[92]

The above examples indicate that the modular construction using specific nucleic acid hybridization and high-affinity STV–biotin interaction opens up a wide variety of new routes to the utilization of DNA–STV conjugates in life sciences and biomolecular nanotechnology. Similar approaches should even allow the fabrication of highly complex supramolecular structures. It should be noted at this point that Seeman[44] has impressively demonstrated the power of DNA in the rational construction of a complex molecular framework. For instance, he synthesized the truncated octahedron, a DNA polyhedron containing 24 individual oligonucleotide arms at its vertices, which can, in principle, be used as a framework for the selective spatial positioning of 24 different proteins, inorganic nanoclusters, or other functional molecular devices.

FIGURE 10.10 (opposite page) Construction of supramolecular DNA and STV conjugates. (A) Networks 5a and 5b are generated from covalent oligonucleotide–STV conjugates 2a and 2b, respectively, containing complementary oligonucleotides. The inset shows a typical AFM image of 5a. (B) Oligomers 5a,b are transformed into the corresponding nanocircles 6a,b, which hybridize with each other to form the nanocircle dimers 8. The inset shows a typical AFM image of dimer 8. (From Niemeyer, C.M., in *NanoBiotechnology: Concepts, Methods and Applications*, Niemeyer, C.M. and Mirkin, C.A., Eds., Wiley-VCH, Weinheim, Germany, 2004, 227–243. With permission.)

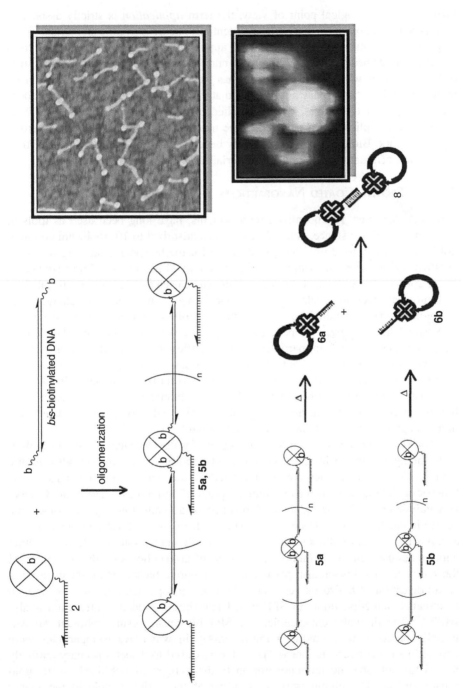

FIGURE 10.10

10.3 APPLICATIONS

From a biotechnological point of view, the term *application* is strictly associated
with commercial products that (at least potentially) address significant markets. With
respect to this stringent definition, real applications of the bioinorganic hybrid
materials described above are rare at the current stage. However, the combination
of biotechnology with materials sciences has great potential for generating advanced
materials and, in turn, today's advanced materials and devices, as well as modern
physicochemical techniques, can be applied to solve biological problems. Thus,
bioanalytical applications are currently the most practical outcome of the interdisciplinary work. Biological sciences greatly benefit from the refinement of analytical
techniques, originally developed for materials and physicochemical sciences.

10.3.1 PROTEIN-COATED NANOPARTICLES AS BIOLABELS

Gold nanoparticles, functionalized with proteins, have long been used as tools in
biosciences.[93] For instance, antibody molecules adsorbed to 10- to 40-nm colloidal
gold are routinely used in histology, allowing for the biospecific labeling of distinguished regions of tissue samples and subsequent TEM analysis. More advanced,
small gold clusters of 0.8 or 1.4 nm diameter can be used for the site-specific labeling
of biological macromolecules.[94-96] These labels have a number of advantages over
colloidal probes, including better resolution, stability, and uniformity. Moreover,
their small size improves the penetration and allows more quantitative labeling of
antigenic sites. The 1.4-nm particles are the smallest size of gold clusters that can
be seen directly in the conventional electron microscope, allowing for a spatial
resolution of about 7 nm when covalently attached to antibody fragments. In addition,
the clusters' visibility can be improved by a wet chemical silver enhancement step
for use in electron or light microscopy for histological purposes or to detect less
than picogram amounts of antigens in immunoblots.[97]

The silver enhancement can also be applied for the electrical sensing of biological binding events, for instance, via the short-circuiting of microelectrodes (Figure
10.11). Gold nanoparticles are immobilized in the gaps of microelectrodes via
biospecific interaction, such as immunosorption of antibodies.[98] The colloids serve
as catalytic cores for the reductive deposition of a conducting layer of silver that
short-circuits the two electrodes. The resulting decrease in ohmic resistance is used
as a positive signal for the sensing of the biospecific interaction to be detected. Other
approaches used electrode-immobilized layers of gold colloids for the adsorption of
the redox enzyme horseradish peroxidase to prepare a biosensor for the electrocatalytic detection of hydrogen peroxide.[99-101] Gold nanoparticles can also be applied
to enhance detection limits in SPR-based real-time biospecific interaction analysis.[102,103] The dramatic enhancement of SPR biosensing using colloidal Au was
initially observed in a sandwich immunoassay in which Au nanoparticles were
coupled to a secondary antibody. The latter was used to detect a primary antibody
bound through specific immunosorption to the antigen immobilized on the gold
sensor surface. The immunosorptive binding of the colloidal gold to the sensor
surface led to a large shift in plasmon angle, a broadened plasmon resonance, and

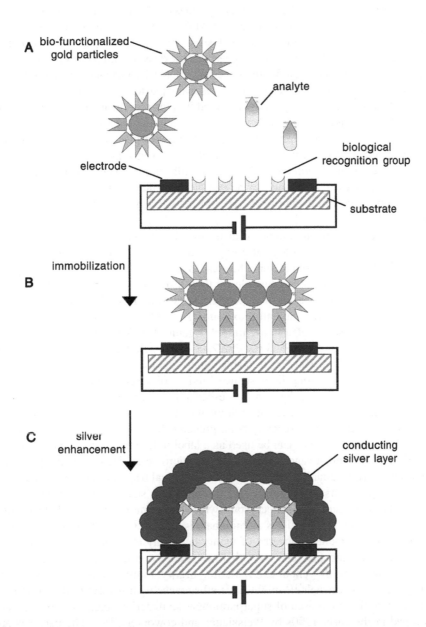

FIGURE 10.11 Electrical sensing of biorecognition events. Receptor groups, such as antibodies or DNA oligomers, immobilized in the gap between two microelectrodes, are used as a capture agent to specifically bind complementary target molecules. In a sandwich-type assay, the captured analyte is tagged with colloidal gold through a second biological recognition unit. Subsequent reductive deposition of silver leads to the formation of a conducting metal layer that shortcuts the two electrodes. (From Niemeyer, C.M., *Angew. Chem. Int. Ed.*, 40, 4128–4158, 2001. With permission.)

an increase in minimum reflectance, thereby allowing for picomolar detection of the antigen.[102] Similarly, an approximately 1000-fold improvement in sensitivity was obtained in nucleic acid hybridization analysis, using a colloidal gold–oligonucleotide conjugate as a probe.[103] These results suggest that the detection limit of SPR now begins to approach that of traditional fluorescence-based DNA hybridization detection.

The use of colloidal silver plasmon-resonant particles as optical reporters in biological assays has recently been described.[104] Plasmon-resonant particles can be readily observed individually with a microscope configured for dark-field microscopy, with standard white-light illumination. The use of plasmon-resonant particles, surface coated with bioactive ligands, was illustrated in a model sandwich immunoassay for goat anti-biotin antibody. Because plasmon-resonant particle labels are nonbleaching and bright enough to be rapidly identified and counted, they might soon be applicable in ultrasensitive assay formats based on single-target molecule detection.[104]

Semiconductor nanoparticles are also powerful fluorescent probes that can be used for the labeling of biological components.[18,19,105–110] The quantum dots have several advantages over conventional fluorescent dyes. Their fluorescence absorption and emission are conveniently tunable by their size and material composition, and the emission peaks have a narrow spectral line width. Typically, emission widths are 20 to 30 nm (full width at half maximum (FWHM)), which is only one third of the emission line width of a conventional organic dye. The high quantum yields often range from 35 to 50% for CdSe/ZnS core-shell nanoparticles. Moreover, quantum dots are about 100-fold as stable against photobleaching as an organic dye, and they often reveal a long fluorescence lifetime of several hundreds of nanoseconds. This allows for time-delayed fluorescence measurements, which can be used to suppress the autofluorescence of biological matrices. Stable, water-soluble oligonucleotide conjugates have recently been prepared from hydroxyl-capped quantum dots.[111] These nanoparticles can be used as a label in fluorescence *in situ* hybridization (FISH). Recent advances made use of micrometer-sized polymer particles, tagged with various quantum dots to achieve an optical bar code for biomolecules.[112]

Besides problems of solubility, physicochemical stability, and quantum efficiency of the semiconductor nanocrystals, which still remain to be solved, the routine application of such nanoparticles as biolabels is still controversial, in particular due to general environmental concerns of using the highly toxic cadmium compounds in biomedical diagnostics. Due to their lower toxicity, colloids composed of other materials, such as the chemically stable, highly luminescent, and less toxic $LaPO_4$:Ce,Tb nanocrystals,[113] might have advantages with respect to this problem.

Nanoparticles composed of superparamagnetic materials, e.g., iron oxide, were developed in the early 1990s by Weissleder and coworkers.[114,115] The particles are typically 5 to 10 nm in diameter, they possess a half-life of about 90 min in blood, and the major potential applications are as intravenous contrast agents for the lymphatic system,[114] as bone marrow contrast agents,[116] as long-half-life perfusion agents for the brain[117] and heart,[118] and as magnetic moieties in organ-targeted superparamagnetic contrast agents for magnetic resonance imaging (MRI).[119] With respect to the latter type of applications, in particular, attachment of biomolecular recognition groups to the surface of the inorganic nanoparticles greatly enhances the specificity of targeting. For instance, dextran-coated superparamagnetic iron oxide particles were derivatized with a peptide sequence from the HIV-tat protein to improve intracellular magnetic labeling of different target cells. The particles were internalized into lymphocytes more than 100-fold more efficiently than nonmodified particles. Labeled cells were highly magnetic, were detectable by nuclear magnetic resonance (NMR) imaging, and could be retained on magnetic separation columns. This method has great potential for *in vivo* tracking of magnetically labeled cells by MRI and for recovering intracellularly labeled cells from organs.[120,121] In another example, conjugation of superparamagnetic monocrystalline iron oxide nanoparticles to the transferrin. This protein allowed for targeting and cellular accumulation of the particles to tumors overexpressing engineered human transferrin receptor and, thus, enabled high-spatial-resolution imaging of gene expression *in vivo* and microscopical mapping of transgene expression in tumor sections by means of MRI.[122]

10.3.2 Nucleic Acid–Coated Particles

Whereas protein-coated gold colloids have long been used in bioanalytical techniques, applications of DNA-functionalized Au particles were just introduced 6 years ago by Mirkin and coworkers.[47,53] The DNA-directed nanoparticle aggregation depicted in Figure 10.3 can be used for simple and cheap sensors in biomedical diagnostics, e.g., for the detection of nucleic acids from pathogenic organisms. Although detailed studies of the optical phenomena associated with the supramolecular colloid assembly are still in progress,[50] several analytical applications has already been reported, allowing for the detection of nucleic acids in homogenous solutions.[53,123–126] In addition, the use of DNA-functionalized nanoparticles in the heterogeneous nucleic acid hybridization, with capture oligonucleotides attached to solid supports, has recently attracted much attention. Taton and colleagues[127] as well as other groups[128 133] have reported on the DNA-directed immobilization of gold nanoparticles to form supramolecular surface architecture. Alivisatos and coworkers[57] have recently adopted the DNA-directed immobilization for the specific sorting of oligonucleotide-functionalized semiconductor quantum dots.[133] Moreover, particles of different morphology, rods and sticks, have also been assembled by means of DNA hybridization.[134,135]

The specific nucleic acid–mediated immobilization of gold nanoparticles can be utilized for the topographic labeling of surface-bound DNA targets. This readily allows for the highly sensitive scanometric detection of nucleic acids in DNA chip analyses.[136] The latter approach is based on the gold particle–promoted reduction of silver ions (Figure 10.12), which allows for an approximately 100-fold increased sensitivity compared to conventional fluorescent DNA detection.[137] In a recent continuation, this approach was adapted to the array-based electronic detection of nucleic acid analytes using a setup similar to that depicted in Figure 10.10.[138]

The size-selective light scattering of 50- and 100-nm-diameter Au nanoparticles has been used for the two-color labeling of oligonucleotide arrays.[139] An alternative for the simultaneous determination of multiple DNA targets, the use of Raman-active dye-labeled gold nanoparticles, has also been explored by Cao's group.[140] The gold nanoparticles facilitate the formation of a silver coating that acts as a surface-enhanced

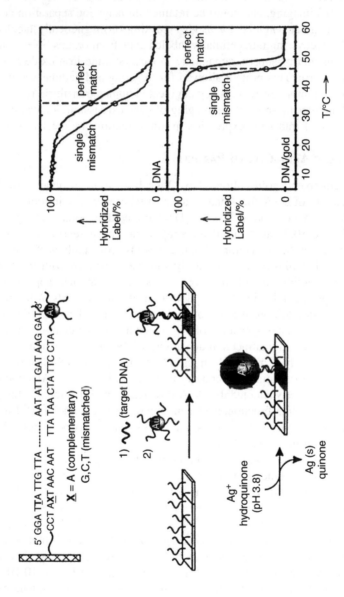

FIGURE 10.12

Raman-scattering promoter for the dye-labeled particles that have been captured by target molecules and an underlying chip in microarray format. This approach provides the high sensitivity and high selectivity attributes of gray-scale scanometric detection but adds multiplexing and ratioing capabilities because a very large number of probes can be designed based on the concept of using a Raman tag as a narrowband spectroscopic fingerprint. Six dissimilar DNA targets with six Raman-labeled nanoparticle probes were distinguished with a 20-femtomolar detection limit.[140] Wang and coworkers[141] have recently demonstrated an electrochemical coding technology for the parallel multicolor detection of multiple DNA targets. Three encoding inorganic nanocrystal tracers (ZnS, CdS, PbS) are used to differentiate the signals of three DNA targets in connection to stripping-voltammetric measurements of the heavy metal dissolution products. These products yield well-defined and resolved stripping peaks at the mercury-coated glassy-carbon electrode, thereby yielding femtomole detection limits.

The use of colloidal gold nanoparticles also allows for signal enhancement in the DNA hybridization detection using quartz crystal microbalance,[142–145] angle-dependent light scattering,[146] and surface plasmon resonance (see above).[103] Willner and coworkers[147] have recently reported on the use of DNA-functionalized 2.6-nm CdS nanoparticles for the photoelectrochemical detection of nucleic acid hybridization. An array of quantum dots was assembled at the surface of an electrode using the specific hybridization of complementary DNA fragments. Upon irradiation, a photocurrent is generated that is proportional to the amount of target DNA.

Examples of the employment of both protein and nucleic acid moieties to functionalize nanoparticles have already been discussed in Section 10.2.2.3. With respect to bioanalytical applications, Figure 10.13 illustrates that DNA-tagged proteins can be readily immobilized at DNA-coated gold nanoparticles.[148] This approach has several advantages over the conventional methods to adsorb proteins to colloidal gold. The biofunctionalized nanoparticles not only retain the undisturbed recognition properties of the proteins immobilized but also reveal an extraordinary stability, which even allows for regeneration of the DNA–Au particles. The bioinorganic hybrid components have been used as reagents in a sandwich immunoassay for the detection of proteins. Using a gold particle–promoted silver development, this technique allowed analysis of femtomolar amounts of antigens. This demonstrates

FIGURE 10.12 (opposite page) Scanometric detection of nucleic acids in DNA chip analyses. Capture oligonucleotides immobilized on glass slides are used for specific binding of target nucleic acids. Oligonucleotide-functionalized gold nanoparticles are employed as probes in solid-phase DNA hybridization detection. Subsequent to a silver enhancement step, the immobilization of the colloidal gold probe is detected by scanning the glass substrate with a conventional flatbed scanner. Despite its simplicity, this technique allows for an approximately 100-fold increased sensitivity compared to fluorescent DNA detection. Moreover, due to the extraordinary sharp melting behavior of the immobilized DNA–nanoparticle networks (see melting curves at the right), this method allows for single-mismatch detection. (From Taton, T.A. et al., *Science*, 289, 1757–1760, 2000. With permission. Copyright © 2000 American Association for the Advancement of Science.)

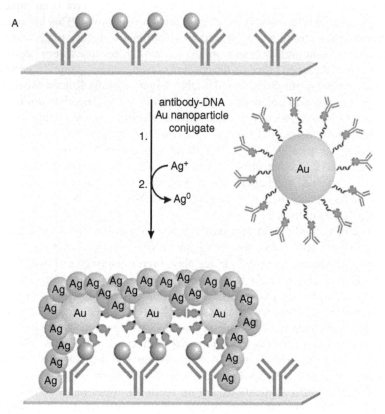

FIGURE 10.13 Array-based scanometric detection of proteins using DNA- and protein-modified nanoparticle probes. (A) Schematic drawing of the utilization of antibody/DNA-functionalized gold nanoparticles as reagents in a sandwich immunoassay. Surface-attached capture antibodies are used to bind the antigen (gray circles) through specific immunosorption. Subsequently, the antigen is labeled with anti-mouse IgG-modified gold nanoparticles. The IgG–DNA particle conjugate was previously assembled from Au nanoparticles modified with ssDNA and IgG–oligonucleotide conjugates. For signal generation, a silver development is carried out. (B) The generation of silver precipitate is monitored spectrophotometrically, or by imaging with a CCD camera or flatbed scanner. (C) Typical dose–response curve obtained from the sandwich assay, depicted in (A). The absorbance at 490 nm of the silver film depends on the amount of antigen present in various samples analyzed. The black sets of signals were obtained using 34-nm gold particles (Au_{34}-5), whereas the gray histograms represent signals obtained with 13-nm gold particles (Au_{13}-5). (From Niemeyer, C.M., in *NanoBiotechnology: Concepts, Methods and Applications*, Niemeyer, C.M. and Mirkin, C.A., Eds., Wiley-VCH, Weinheim, Germany, 2004, 227–243. With permission.)

the possibility of the spatially addressable detection of chip-immobilized antigens.[148] Nam and colleagues[149] have shown that by utilizing oligonucleotide-modified Au nanoparticles encoded with sequences that act as bio-bar codes, one can screen for multiple-target polyvalent proteins simultaneously in one solution. This concept was demonstrated with two types of detection formats, a homogeneous assay and one based

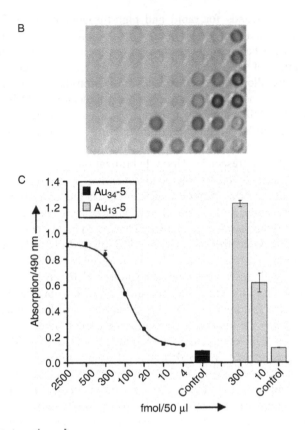

FIGURE 10.13 (continued)

on oligonucleotide microarrays. With such an approach, one can prepare a large number of bar codes from synthetically accessible oligonucleotides. For instance, a 12-mer sequence offers 4^{12} possible bar codes.

Recently, Weissleder's group developed superparamagnetic iron oxide nanoparticles that can be used as magnetic relaxation switches to detect molecular interactions in the reversible self-assembly of disperse magnetic particles into stable nanoassemblies[150] (see also reference 151). Monodisperse magnetic nanoparticles conjugated with complementary oligonucleotide sequences self-assemble into stable magnetic nanoassemblies, resulting in a decrease of the spin–spin relaxation times of neighboring water protons.[150] When these nanoassemblies are treated with a DNA cleaving agent, such as restriction enzymes, the nanoparticles become dispersed, switching the T2 of the solution back to original values.[151] As shown for four different types of molecular interactions (DNA–DNA, protein–protein, protein–small molecule, and enzyme reactions), magnetic relaxation measurements can be used to detect these interactions with high efficiency and sensitivity.[152] Furthermore, the magnetic changes are detectable in turbid media and in whole-cell lysates without protein purification and, thus, magnetic nanosensors might be used in a variety of biological applications, such as in homogeneous assays, as reagents in miniaturized microfluidic systems, as affinity ligands for rapid and high-throughput magnetic readouts of arrays, as probes for magnetic force microscopy, and for *in vivo* imaging.

A magnetic field has recently been used for the remote electronic control of DNA hybridization of gold nanoparticle–DNA molecular beacon conjugates.[153] Molecular beacons allow one to utilize hybridization-induced changes in DNA conformation for the macroscopic detection of intermolecular DNA hybridization.[154] DNA beacons are ssDNA molecules that, due to their nucleotide sequence, form an intramolecular hairpin-loop structure. Both ends are chemically modified with a fluorophor and a quencher dye, the latter of which, due to spatial proximity, effectively quenches the fluorescence. Upon hybridization with nucleic acid targets, containing a sequence stretch complementary to the loop region of the beacon, the quencher is spatially removed from the fluorophor, leading to a strong enhancement in fluorescence. Recently, a new class of molecular beacons has been introduced, in which the organic quencher dye was replaced by a gold metal cluster.[126] This hybrid construct was synthesized from a 3′-amino-5′-thiol-modified ssDNA oligomer by subsequent coupling with an amino-reactive fluorophore and with commercially available 1.4-nm gold clusters containing a single maleimido group in their ligand shells. The quenching efficiency, i.e., the difference in fluorescence between the open duplex and the intramolecularly closed beacon, was determined to be about 99.5%, indicating that the fluorescent signal of the beacon increases about 200-fold upon hybridization with complementary target. This type of gold–oligomer–dye hybrid beacon was applied in the detection of single mismatches in DNA, and competitive hybridization assays revealed that the ability to detect single-base mutations is about 8-fold greater than with conventional molecular beacons, whereas the sensitivity of detection is enhanced up to 100-fold.[126] Similar gold nanoparticle–DNA molecular beacon conjugates have been exploited for the remote electronic control of DNA hybridization.[153] Inductive coupling of a radio frequency magnetic field (RFMF) with a frequency of 1 GHz to the 1.4-nm metal nanocrystal, which functions as an antenna in the DNA constructs, causes the local temperature of the bound DNA to increase, thereby inducing denaturation while leaving surrounding molecules relatively unaffected. Although inductive heating has already been applied to macroscopic samples, as well as in the treatment of cancer cells with magnetic field-induced excitation of biocompatible superparamagnetic nanoparticles,[155] the use of gold nanocrystal–DNA conjugates should allow an extension of this concept. For instance, complex operations, such as gene regulation, biomolecular assembly, and enzymatic activity, of distinct portions of nucleic acids or proteins might be controlled, whereas the rest of the molecule and neighboring species would remain unaffected. Moreover, because the addressing is not optical, this technology would even be applicable in highly scattering media.[153]

The magnetic switching of biomolecular functionality has been demonstrated by Willner and colleagues[156–158] in a couple of examples using micrometer-sized superparamagnetic iron oxide particles. As an example, magnetic field-stimulated ON and OFF biochemiluminescence was accomplished by electrocatalyzed reduction of naphthoquinone-functionalized magnetic particles in the presence of a biocatalytic peroxidase/luminol system.[157] The dual biosensing by magnetocontrolled bioelectrocatalysic analysis of glucose and lactate by the enzymes glucose ocidase and lactate dehydrogenase was conducted by using ferrocene-functionalized electrodes and enzyme-modified iron oxide magnetic particles.[156] Wang et al.[159] have demonstrated the magnetic triggering of the elecrical DNA detection, realized through a magnetic collection of assemblies, formed by specific hybridization of oligonucleotide–Au nanoparticles with microbeads containing complementary oligonucleotides. The assemblies are subjected to silver development using hydrochinon, and the resulting conglomerate is positioned on an electrode by an external magnetic field. The chronopotentiometric hybridization signals allowed the detection of small amounts of target DNA. It can be envisioned that smaller nanometer-sized functionalized magnetic particles[160] will allow for a variety of applications in nanobiotechnology and biomedical diagnostics.

10.3.3 THERAPEUTIC APPLICATIONS

It is well established that gold particles can be used as carriers for the delivery of double-stranded DNA in the so-called gene gun technology.[161] In this method, plasmid DNA of a typical length of several thousand base pairs is adsorbed to Au or tungsten colloids, which have a typical size of about 500 nm to several micrometers. The DNA- or RNA-coated particles are loaded into a gun-like device in which a low-pressure helium pulse delivers the particles into virtually any target cell or tissue. As an advantage of this technique, one does not have to remove cells from tissue in order to transform cells. The particles penetrate the cells and release the DNA, which is diffused into the nucleus and, for instance, incorporated into the chromosomes of the organism. Since its development in the mid-1980s, the gene gun technology has led to numerous examples demonstrating the tremendous potential of this method for biolistic transfection of organisms, as well as for DNA immunization. The latter technique, also known as DNA vaccination or genetic immunization, is a vaccine technology that can be used to stimulate protective immunity against many infectious pathogens, malignancies, and autoimmune disorders in animal models. Plasmid DNA encoding a polypeptide antigen is introduced into host cells, for instance, via gene gun particle bombardment, where it serves as a template for the high-efficiency translation of its antigen. This leads to an immune response, which is mediated by the cellular or humoral immune system and is specific for the plasmid-encoded antigen.[162]

Paramagnetic nanoparticles are currently being investigated for biomedical applications concerning the treatment of cancer.[155,163,164] Biocompatible dextran- or silan-coated superparamagnetic magnetite nanoparticles can be incorporated into malignant cells by endocytosis. The intracellular magnetic fluids are excited with alternating magnetic fields, leading to an increase in the local temperature and thereby inducing hyperthermia. It has been demonstrated that magnetic fluid hyperthermia (MFH) affects mammary carcinoma cells *in vitro* and *in vivo*.

10.3.4 NANOPARTICLES AS MODEL SYSTEMS

Two general aspects of metal and semiconductor nanoparticles make them suitable model systems for the study of basic biological phenomena. On one hand, their size matches with the macromolecular components employed in living systems. Proteins, nucleic acid fragments and their supramolecular complexes, such as the nucleosome, the various complexes involved in DNA replication and transformation, and the ribosome have typical dimensions in the range of 2 to 200 nm. On the other hand, nanoparticles of inorganic materials are the basic building blocks in biomineralization, a fundamental biological process in which nature chemically generates form by means of genetic instructions. It is thus not surprising that the current advances in the study of metal and semiconductor nanoparticles have led to their consideration as model systems for both the study of interactions between proteins and nucleic acids and the understanding of the basic principles of biomineralization. These types of applications have been recently reviewed elsewhere.[7]

10.4 CONCLUSIONS AND PERSPECTIVES

This chapter summarizes a young and rapidly increasing research field located at the crossroads of materials research, nanosciences, and molecular biotechnology. In particular, advances are described concerning the functionalization of inorganic nanoparticles with evolutionary optimized biological components. Since the nanoparticles and biomolecules typically meet at the same nanometer-length scale, this interdisciplinary approach will contribute to the establishment of a novel field, descriptively termed biomolecular nanotechnology, or nanobiotechnology.[45]

Although the chemical coupling of biomolecules and inorganic materials can be achieved with a variety of methods, there is still a great demand for mild and selective coupling techniques that allow for the preparation of thermodynamically stable, kinetically inert, and stoichiometrically well-defined bioconjugate hybrid nanoparticles. Nature's examples suggest that enzymatic procedures combined with modern methods of bioorganic syntheses[165] will open novel routes for the generation of well-defined inorganic/biomolecular components, useful as building blocks for the bottom-up assembly of sophisticated nanostructured architecture. The first steps to the latter have already been achieved, using protein- or nucleic acid-based recognition elements (Section 10.2 and Chapter 1). Since in both systems a large number of complementary binding pairs with a wide variety of free energies of association are available, it remains to be seen whether protein-based conjugation might offer advantages over nucleic acid-based assembly.[33] A priori, one should expect that nucleic acids are superior since the physicochemical properties of a single 20-mer oligonucleotide is representative of 4^{20} ($=10^{12}$) different recognition elements. The binding strength can be conveniently adjusted by varying the lengths and gas chromatography (GC) contents of native DNA or by using nonnatural DNA analogs with altered backbone properties or base-pairing specificity. In addition, directed evolution, already applied to tailor protein linkers,[36] should be applicable to generate novel nucleic acid-based linkers. It seems safe to predict that the tremendous advances in the development of artificial nucleic acid receptors, aptamers selected to specifically recognize any target structure,[166] will soon be applied to the de novo generation of nucleic acid receptors, capable of recognizing inorganic surfaces. Consequently, combinations of the two fundamental biological systems — nucleic acids and proteins — are very promising to synergistically cooperate and allow for novel functions and applications.[59]

Another crossover of biotechnology and materials science concerns the utilization of the chemical and topographical properties of biological components for templating the spatial assembly of organic and inorganic components to form nanometer- and micrometer-sized elements. There is still a tremendous need for systematic fundamental research on the interaction between biomolecules and the abiotic components, and future perspectives have to address these and will also concern the tailoring of the biocomponent's shape and physicochemical properties by genetic engineering and bioconjugate chemistry. Although commercial applications are rare at the current state, the interdisciplinary work has great potential for generating advanced materials, which might lead to novel devices for sensing, signal transduction, and catalysis, as well as for new biocompatible materials and interfaces, currently being developed for biomedical sciences and tissue engineering.[167] In addition to long-term perspectives, today's nanoparticles are already used in bioanalytical applications or as model systems to solve biological problems (Section 10.3 and Chapter 1).

The future development of the joint venture of biotechnology and materials sciences will profit from the rapid current advances of chemistry as a central science. In addition, the current genome and proteome research will also be beneficial for this field, because it provides data that will allow us to produce even more suitable biocomponents. This might open the door to nanosensors, catalytic and light-harvesting devices, ultrafast molecular switches and transistors, supramolecular mediators between electrical and living systems, and other bio- and opto-electronic parts. Clearly, we are still far from the fabled autonomous nanomachines that heal wounds and perform surgery in a living organism. However, at the beginning of this new century, one should remember the dramatic development that total organic synthesis has undergone from the early 1900s to the race for molecular summits in the late 1990s.[168,169] A similar rate of progress in supramolecular sciences at the interface of biotechnology and materials research promises plenty of excitement from future developments.

ACKNOWLEDGMENTS

I thank Deutsche Forschungsgemeinschaft and Fonds der Chemischen Industrie for financially supporting our work.

REFERENCES

1. J-M Lehn. *Supramolecular Chemistry: Concepts and Perspectives*. Weinheim, Germany: VCH Verlagsgesellschaft, 1995.
2. F Vögtle. *Supramolecular Chemistry*. Chichester, England: Wiley, 1991.
3. JL Atwood, JED Davies, DD MacNicol, F Vögtle. *Comprehensive Supramolecular Chemistry*. Oxford: Elsevier, 1996.
4. P Nissen, J Hansen, N Ban, PB Moore, TA Steitz. The structural basis of ribosome activity in peptide bond synthesis. *Science* 289: 920–930, 2000.
5. N Ban, P Nissen, J Hansen, PB Moore, TA Steitz. The complete atomic structure of the large ribosomal subunit at 2.4 A resolution. *Science* 289: 905–920, 2000.
6. DMJ Lilley. The ribosome functions as a ribozyme. *ChemBioChem* 2: 31–35, 2001.

7. CM Niemeyer. Nanoparticles, proteins, and nucleic acids: biotechnology meets materials science. *Angew. Chem. Int. Ed.* 40: 4128–4158, 2001.
8. D Philp, JF Stoddart. Selbstorganisation in natürlichen und in nichtnatürlichen Systemen. *Angew. Chem. Int. Ed.* 35: 1154–1196, 1996.
9. RP Feynman. In *Miniaturization*, HD Gilbert, Ed. New York: Reinhold Publishing Corp., 1961, pp. 282–296.
10. GM Whitesides, JP Mathias, CT Seto. Molecular self-assembly and nanochemistry: a chemical strategy for the synthesis of nanostructures. *Science* 254: 1312–1319, 1991.
11. G Schmid. *Clusters and Colloids.* Weinheim: VCH, 1994.
12. U Kreibig. *Optical Properties of Metal Clusters.* New York: Springer, 1995.
13. H Weller. Selbstorganisierte Strukturen aus Nanoteilchen. *Angew. Chemie* 108: 1159–1161, 1996.
14. M Antonietti, C Göltner. Überstrukturen funktioneller Kolloide: eine Chemie im Nanometerbereich. *Angew. Chemie* 109: 944–964, 1997.
15. AP Alivisatos. Semiconductor clusters, nanocrystals, and quantum dots. *Science* 271: 933–937, 1996.
16. DL Feldheim, CD Keating. Self-assembly of single-electron transistors and related devices. *Chem. Soc. Rev.* 27: 1–12, 1998.
17. AN Shipway, E Katz, I Willner. Nanoparticle arrays on surfaces for electronic, optical, and sensor applications. *ChemPhysChem* 1: 19-52, 2000.
18. M Bruchez, Jr, M Moronne, P Gin, S Weiss, AP Alivisatos. Semiconductor nanocrystals as fluorescent biological labels. *Science* 281: 2013–2015, 1998.
19. WCW Chan, SM Nie. Quantum dot bioconjugates for ultrasensitive nonisotopic detection. *Science* 281: 2016–2018, 1998.
20. D Bimberg, NN Ledentsov, M Grundmann, F Heinrichsdorff, IM Ustinov, PS Kopiev, MV Maximov, Zh I Alferov, JA Lott. Applications of self-organized quantum dots to edge emitting and vertical cavity lasers. *Physica E* 3(1–3)129–136, 1998.
21. RE Dunin-Borkowski, MR McCartney, RB Frankel, DA Bazylinski, M Posfai, PR Buseck. Magnetic microstructure of magnetotactic bacteria by electron holography. *Science* 282: 1868–1870, 1998.
22. S Sun, CB Murray, D Weller, L Folks, A Moser. Monodisperse FePt nanoparticles and ferromagnetic FePt nanocrystal superlattices. *Science* 287: 1989–1992, 2000.
23. AP Alivisatos. Naturally aligned nanocrystals. *Science* 289: 736–737, 2000.
24. RF Service. DNA ventures into the world of designer materials. *Science* 277: 1036–1037, 1997.
25. CM Niemeyer. DNA as a material for nanotechnology. *Angew. Chem. Int. Ed. Engl.* 36: 585–587, 1997.
26. CA Mirkin, TA Taton. Semiconductors meet biology. *Nature* 405: 626–627, 2000.
27. M Brust, D Bethell, DJ Schiffrin, C Kiely. Novel gold-dithiol nano networks with non-metallic electronic properties. *Adv. Mater.* 7: 795–797, 1995.
28. W Shenton, SA Davies, S Mann. Directed self-assembly of nanoparticles into macroscopic materials using antibody-antigen recognition. *Adv. Mater.* 11: 449–452, 1999.
29. CM Niemeyer, W Bürger, J Peplies. Covalent DNA–streptavidin conjugates as building blocks for the fabrication of novel biometallic nanostructures. *Angew. Chem. Int. Ed. Engl.* 37: 2265–2268, 1998.
30. S Connolly, D Fitzmaurice. Programmed assembly of gold nanocrystals in aqueous solution. *Adv. Mater.* 11: 1202–1205, 1999.
31. PC Weber, DH Ohlendorf, JJ Wendoloski, FR Salemme. Structural origins of high-affinity biotin binding to streptavidin. *Science* 243: 85–88, 1989.

32. M Wilchek, EA Bayer. Avidin–biotin technology. *Methods Enzymol.* 184: 51–67, 1990.
33. S Mann, W Shenton, M Li, S Connolly, D Fitzmaurice. Biologically programmed nanoparticle assembly. *Adv. Mater.* 12: 147–150, 2000.
34. MT Reetz. Kombinatorische und evolutionsgesteuerte Methoden zur Bildung enantioselektiver Katalysatoren. *Angew. Chemie* 113: 292–320, 2001.
35. KA Powell, et al. Gerichtete Evolution und Biokatalyse. *Angew. Chemie* 113: 4068–4080, 2001.
36. SR Whaley, DS English, EL Hu, PF Barbara, AM Belcher. Selection of peptides with semiconductor binding specificity for directed nanocrystal assembly. *Nature* 405: 665–666, 2000.
37. SW Lee, C Mao, CE Flynn, AM Belcher. Ordering of quantum dots using genetically engineered viruses. *Science* 296: 892–895, 2002.
38. RK Soong, GD Bachand, HP Neves, AG Olkhovets, HG Craighead, CD Montemagno. Powering an inorganic nanodevice with a biomolecular motor. *Science* 290: 1555–1558, 2000.
39. GD Bachand, RK Soong, HP Neves, AG Olkhovets, HG Craighead, CD Montemagno. Precision attachment of individual F1-ATPase biomolecular motors on nanofabricated substrates. *Nano Lett.* 1: 42–45, 2001.
40. JK Lanyi, A Pohorille. Proton pumps: mechanism of action and applications. *Trends Biotechnol.* 19: 140–144, 2001.
41. KE Drexler. Molecular engineering: an approach to the development of general capabilities for molecular manipulation. *Proc. Natl. Acad. Sci. U.S.A.* 78: 5275–5278, 1981.
42. NC Seeman. Nucleic acid junctions and lattices. *J. Theor. Biol.* 99: 237–247, 1982.
43. KE Drexler. *Nanosystems: Molecular Machinery, Manufacturing, and Computation.* New York: Wiley, 1992.
44. NC Seeman. DNA engineering and its application to nanotechnology. *Trends Biotechnol.* 17: 437–443, 1999.
45. CM Niemeyer. Self-assembled nanostructures based on DNA: towards the development of nanobiotechnology. *Curr. Opin. Chem. Biol.* 4: 609–618, 2000.
46. CM Niemeyer, M Adler. Nanomechanical devices based on DNA. *Angew. Chem. Int. Ed.* 41: 3779–3783, 2002.
47. CA Mirkin, RL Letsinger, RC Mucic, JJ Storhoff. A DNA-based method for rationally assembling nanoparticles into macroscopic materials. *Nature* 382: 607–609, 1996.
48. RC Mucic, JJ Storhoff, CA Mirkin, RL Letsinger. DNA-directed synthesis of binary nanoparticle network materials. *J. Am. Chem. Soc.* 120: 12674–12675, 1998.
49. GP Mitchell, CA Mirkin, RL Letsinger. Programmed assembly of DNA functionalized quantum dots. *J. Am. Chem. Soc.* 121: 8122–8123, 1999.
50. JJ Storhoff, AA Lazarides, RC Mucic, CA Mirkin, RL Letsinger, GC Schatz. What controls the optical properties of DNA-linked gold nanoparticle assemblies? *J. Am. Chem. Soc.* 122: 4640–4650, 2000.
51. S-J Park, AA Lazarides, CA Mirkin, PW Brazis, CR Kannewurf, RL Letsinger. The electrical properties of gold nanoparticle assemblies linked by DNA. *Angew. Chem. Int. Ed.* 39: 3845–3848, 2000.
52. R Jin, G Wu, Z Li, CA Mirkin, GC Schatz. What controls the melting properties of DNA-linked gold nanoparticle assemblies? *J. Am. Chem. Soc.* 125: 1643–1654, 2003.
53. JJ Storhoff, CA Mirkin. Programmed materials synthesis with DNA. *Chem. Rev.* 99: 1849–1862, 1999.

54. KJ Watson, S-J Park, J-H Im, S-BT Nguyen, CA Mirkin. DNA-block copolymer conjugates. *J. Am. Chem. Soc.* 123: 5592–5593, 2001.

55. RB Fong, Z Ding, CJ Long, AS Hoffman, PS Stayton. Thermoprecipitation of streptavidin via oligonucleotide-mediated self-assembly with poly(N-isopropylacrylamide). *Bioconjug. Chem.* 10: 720–725, 1999.

56. CM Niemeyer, T Sano, CL Smith, CR Cantor. Oligonucleotide-directed self-assembly of proteins: semisynthetic DNA-streptavidin hybrid molecules as connectors for the generation of macroscopic arrays and the construction of supramolecular bioconjugates. *Nucleic Acids Res.* 22: 5530–5539, 1994.

57. AP Alivisatos, KP Johnsson, X Peng, TE Wilson, CJ Loweth, MP Bruchez, Jr, PG Schultz. Organization of "nanocrystal molecules" using DNA. *Nature* 382: 609–611, 1996.

58. CJ Loweth, WB Caldwell, X Peng, AP Alivisatos, PG Schultz. DNA als Gerüst zur Bildung von Aggregaten aus Gold-Nanokristallen. *Angew. Chem. Int. Ed.* 38: 1808–1812, 1999.

59. CM Niemeyer. Bioorganic applications of semisynthetic DNA–protein conjugates. *Chem. Eur. J.* 7: 3188–3195, 2001.

60. S Gao, LF Chi, S Lenhert, B Anczykowsky, CM Niemeyer, M Adler, H Fuchs. High quality mapping of DNA–protein complex by dynamic scanning force microscopy. *ChemPhysChem* 2: 384–388, 2001.

61. B Pignataro, LF Chi, S Gao, B Anczykowsky, CM Niemeyer, M Adler, H Fuchs. Dynamic scanning force microscopy study of self-assembled DNA–protein oligomers. *Appl. Phys. A* 74: 447–452, 2002.

62. J Richter, M Adler, CM Niemeyer. Monte-Carlo-simulation of the assembly of bis-biotinylated DNA and streptavidin. *ChemPhysChem* 79–83, 2003.

63. CM Niemeyer, M Adler, B Pignataro, S Lenhert, S Gao, LF Chi, H Fuchs, D Blohm. Self-assembly of DNA–streptavidin nanostructures and their use as reagents in immuno-PCR. *Nucleic Acids Res.* 27: 4553–4561, 1999.

64. CM Niemeyer, M Adler, S Gao, LF Chi. Supramolecular nanocircles consisting of streptavidin and DNA. *Angew. Chem. Int. Ed.* 39: 3055–3059, 2000.

65. CM Niemeyer, M Adler, S Lenhert, S Gao, H Fuchs, LF Chi. Nucleic-acid super-coiling as a means for construction of switchable DNA–nanoparticle networks. *ChemBioChem* 2: 260–265, 2001.

66. S-J Park, AA Lazarides, CA Mirkin, RL Letsinger. Directed assembly of periodic materials from protein and oligonucleotide-modified nanoparticle building blocks. *Angew. Chem. Int. Ed.* 40: 2909–2912, 2001.

67. SR Nicewarner Pena, S Raina, GP Goodrich, NV Fedoroff, CD Keating. Hybridization and enzymatic extension of Au nanoparticle-bound oligonucleotides. *J. Am. Chem. Soc.* 124: 7314–7323, 2002.

68. CS Yun, GA Khitrov, DE Vergona, NO Reich, GF Strouse. Enzymatic manipulation of DNA–nanomaterial constructs. *J. Am. Chem. Soc.* 124: 7644–7645, 2002.

69. AG Kanaras, Z Wang, AD Bates, R Cosstick, M Brust. Towards multistep nanostructure synthesis: programmed enzymatic self-assembly of DNA/gold systems. *Angew. Chem. Int. Ed. Engl.* 42: 191–194, 2003.

70. CM MacIntosh, EA Esposito, AK Boal, JM Simard, CT Martin, VM Rotello. Inhibition of DNA transcription using cationic mixed monolayer protected gold clusters. *J. Am. Chem. Soc.* 123: 7626–7629, 2001.

71. KK Sandhu, CM McIntosh, JM Simard, SW Smith, VM Rotello. Gold nanoparticle-mediated transfection of mammalian cells. *Bioconjug. Chem.* 13: 3–6, 2002.

72. UB Sleytr, P Messner, D Pum, M Sara. Kristalline Zelloberflächen-Schichten prokaryotischer Organismen (S-Schichten): von der supramolekularen Zellstruktur zur Biomimetik und Nanotechnologie. *Angew. Chem. Int. Ed.* 38: 1034–1054, 1999.
73. JL Coffer, SR Bigham, RF Pinizzotto, H Yang. Formation of CdS cluster using calf thymus DNA. *Nanotechnology* 3: 69–76, 1992.
74. SR Bigham, JL Coffer. The influence of adenine content on the properties of Q-CdS clusters stabilized by polynucleotides. *Colloids Surfaces A* 95: 211–219, 1995.
75. JL Coffer, SR Bigham, X Li, RF Pinizzotto, YG Rho, RM Pirtle, IL Pirtle. Dictation of the shape of mesoscale semiconductor nanoparticle assemblies by plasmid DNA. *Appl. Phys. Lett.* 69: 3851–3853, 1996.
76. JL Coffer. Approaches for generating mesoscale patterns of semiconductor nanoclusters. *J. Cluster Sci.* 8: 159–179, 1997.
77. J Richter, R Seidel, R Kirsch, M Mertig, W Pompe, J Plaschke, HK Schackert. Nanoscale palladium metallization of DNA. *Adv. Mater.* 12: 507–510, 2000.
78. J Richter, M Mertig, W Pompe, I Monch, HK Schackert. Construction of highly conductive nanowires on a DNA template. *Appl. Phys. Lett.* 78: 536–538, 2001.
79. LC Ciacchi, W Pompe, A De Vita. Initial nucleation of platinum clusters after reduction of K_2PtCl_4 in aqueous solution: a first principles study. *J. Am. Chem. Soc.* 123: 7371–7380, 2001.
80. T Torimoto, M Yamashita, S Kuwabata, T Sakata, H Mori, H Yoneyama. Fabrication of CdS nanoparticle chains along DNA double strands. *J. Phys. Chem. B* 103: 8799–8803, 1999.
81. A Kumar, M Pattarkine, M Bhadbhade, AB Mandale, KN Ganesh, SS Datar, CV Dharmadhikari, M Sastry. Linear superclusters of colloidal gold particles by electrostatic assembly on DNA templates. *Adv. Mater.* 13: 341–344, 2001.
82. F Patolsky, Y Weizmann, O Lioubashevski, I Willner. Au-nanoparticle nanowires based on DNA and polylysine templates. *Angew. Chem. Int. Ed.* 41: 2323–2327, 2002.
83. E Braun, Y Eichen, U Sivan, G Ben-Yoseph. DNA-templated assembly and electrode attachment of a conducting silver wire. *Nature* 391: 775–778, 1998.
84. E Di Mauro, CP Hollenberg. DNA technology in chip construction. *Adv. Mater.* 5: 384–386, 1993.
85. W Han, SM Lindsay, RE Dlakic, RE Harrington. AFM of nanocircles. *Nature* 386: 563, 1997.
86. J Shi, DE Bergstrom. Neuartige DNA-Ringe mit starren tetraedrischen Spacern. *Angew. Chemie* 109: 70–72, 1997.
87. Y Aharonov, D Bohm. Aharonov–Bohm circles. *Phys. Rev.* 115: 485, 1959.
88. K Keren, M Krueger, R Gilad, G Ben-Yoseph, U Sivan, E Braun. Sequence-specific molecular lithography on single DNA molecules. *Science* 297: 72–75, 2002.
89. CM Niemeyer. Nanotechnology: tools for the biomolecular engineer. *Science* 297: 62–63, 2002.
90. CM Niemeyer, R Wacker, M Adler. Hapten-functionalized DNA–streptavidin nanocircles as supramolecular reagents in a novel competitive immuno-PCR assay. *Angew. Chem. Int. Ed.* 40: 3169–3172, 2001.
91. CM Niemeyer, M Adler, S Gao, LF Chi. Nanostructured DNA–protein aggregates consisting of covalent oligonucleotide–streptavidin conjugates. *Bioconjug. Chem.* 12: 364–371, 2001.
92. CM Niemeyer, M Adler, S Gao, LF Chi. Supramolecular DNA–streptavidin nanocircles with a covalently attached oligonucleotide moiety. *J. Biomol. Struct. Dyn.* 20(2): 223–230, 2002.

93. J Kreuter. In *Microcapsules and Nanoparticles in Medicine and Pharmacy*, M Donbrow, Ed. Boca Raton, FL: CRC Press, 1992, pp. 125–148.
94. DE Safer, J Hainfeld, JS Wall, JE Reardon. Biospecific labeling with undecagold: visualization of the biotin-binding site on avidin. *Science* 218: 290–291, 1982.
95. DE Safer, L Bolinger, JS Leigh. Undecagold clusters for site-specific labeling of biological macromolecules: simplified preparation and model applications. *J. Inorg. Biochem.* 26: 77–91, 1986.
96. JF Hainfeld, FR Furuya. A 1.4-nm gold cluster covalently attached to antibodies improves immunolabeling. *J. Histochem. Cytochem.* 40: 177–184, 1992.
97. JF Hainfield. Uranium-loaded apoferritin with antibodies attached: molecular design for uranium neutron-capture therapy. *Proc. Natl. Acad. Sci. U.S.A.* 89: 11064, 1992.
98. OD Velev, EW Kaler. *In situ* assembly of colloidal particles into miniaturized biosensors. *Langmuir* 15: 3693–3698, 1999.
99. J Zhao, RW Henkens, J Stonehuerner, JP O'Daly, AL Crumbliss. *J. Electroanal. Chem.* 327: 109–119, 1992.
100. Y Xiao, H-X Ju, H-Y Chen. A reagentless hydrogen peroxide sensor based on incorporation of MRP in poly(thionine) film on a monomer-modified electrode. *Anal. Chim. Acta* 391: 73–82, 1999.
101. F Patolsky, T Gabriel, I Willner. Controlled electrocatalysis by microperoxidase-11 and Au-nanoparticle superstructures on conductive supports. *J. Electroanal. Chem.* 479: 69–73, 1999.
102. LA Lyon, MD Musick, MJ Natan. Colloidal Au-enhanced surface plasmon resonance immunosensing. *Anal. Chem.* 70: 5177–5183, 1998.
103. L He, MD Musick, SR Nicewarner, FG Salinas, SJ Benkovic, MJ Natan, CD Keating. Colloidal Au-enhanced surface plasmon resonance for ultrasensitive detection of DNA hybridization. *J. Am. Chem. Soc.* 122: 9071–9077, 2000.
104. S Schultz, DR Smith, JJ Mock, DA Schultz. Single-target molecule detection with nonbleaching multicolor optical immunolabels. *Proc. Natl. Acad. Sci. U.S.A.* 97: 996–1001, 2000.
105. H Mattoussi, JM Mauro, ER Goldmann, GP Anderson, VC Sundar, Mikulec F.V., MG Bawendi. Self-assembly of CdSe–ZnS quantum dot bioconjugates using an engineered recombinant protein. *J. Am. Chem. Soc.* 122: 12142–12150, 2000.
106. AP Alivisatos. Colloidal quantum dots. From scaling laws to biological applications. *Pure Appl. Chem.* 72: 3–9, 2000.
107. ER Goldman, GP Anderson, PT Tran, H Mattoussi, PT Charles, JM Mauro. Conjugation of luminescent quantum dots with antibodies using an engineered adaptor protein to provide new reagents for fluoroimmunoassays. *Anal. Chem.* 74: 841–847, 2002.
108. ER Goldman, ED Balighian, H Mattoussi, MK Kuno, JM Mauro, PT Tran, GP Anderson. Avidin: a natural bridge for quantum dot–antibody conjugates. *J. Am. Chem. Soc.* 124: 6378–6382, 2002.
109. JK Jaiswal, H Mattoussi, JM Mauro, SM Simon. Long-term multiple color imaging of live cells using quantum dot bioconjugates. *Nat. Biotechnol.* 21: 47–51, 2003.
110. X Wu, H Liu, J Liu, KN Haley, JA Treadway, JP Larson, N Ge, F Peale, MP Bruchez. Immunofluorescent labeling of cancer marker Her2 and other cellular targets with semiconductor quantum dots. *Nat. Biotechnol.* 21: 41–46, 2003.
111. S Pathak, SK Choi, N Arnheim, ME Thompson. Hydroxylated quantum dots as luminescent probes for *in situ* hybridization. *J. Am. Chem. Soc.* 123: 4103–4104, 2001.
112. M Han, X Gao, JZ Su, SM Nie. Quantum-dot-tagged microbeads for multiplexed optical coding of biomolecules. *Nat. Biotechnol.* 19: 631–635, 2001.
113. K Riwotzki, H Meyssamy, H Schnablegger, A Kornowski, M Haase. Synthese von Kolloiden und redispergierbaren Pulvern stark lumineszierender LaPO$_4$:Ce,Tb-Nanokristalle. *Angew. Chem. Int. Ed.* 40: 573–576, 2001.

114. R Weissleder, G Elizondo, J Wittenberg, AS Lee, L Josephson, TJ Brady. Ultrasmall superparamagnetic iron oxide: an intravenous contrast agent for assessing lymph nodes with MR imaging. *Radiology* 175: 494–498, 1990.

115. R Weissleder, G Elizondo, J Wittenberg, CA Rabito, HH Bengele, L Josephson. Ultrasmall superparamagnetic iron oxide: characterization of a new class of contrast agents for MR imaging. *Radiology* 175: 489–493, 1990.

116. E Seneterre, R Weissleder, D Jaramillo, P Reimer, AS Lee, TJ Brady, J Wittenberg. Bone marrow: ultrasmall superparamagnetic iron oxide for MR imaging. *Radiology* 179: 529–533, 1991.

117. EA Neuwelt, R Weissleder, G Nilaver, RA Kroll, S Roman-Goldstein, J Szumowski, MA Pagel, RS Jones, LG Remsen, CI McCormick et al. Delivery of virus-sized iron oxide particles to rodent CNS neurons. *Neurosurgery* 34: 777–784, 1994.

118. R Weissleder, AS Lee, BA Khaw, T Shen, TJ Brady. Antimyosin-labeled monocrystalline iron oxide allows detection of myocardial infarct: MR antibody imaging. *Radiology* 182: 381–385, 1992.

119. JB Mandeville, J Moore, DA Chesler, L Garrido, R Weissleder, RM Weisskoff. Dynamic liver imaging with iron oxide agents: effects of size and biodistribution on contrast. *Magn. Reson. Med.* 37: 885–890, 1997.

120. L Josephson, CH Tung, A Moore, R Weissleder. High-efficiency intracellular magnetic labeling with novel superparamagnetic–Tat peptide conjugates. *Bioconjug. Chem.* 10: 186–191, 1999.

121. M Zhao, MF Kircher, L Josephson, R Weissleder. Differential conjugation of tat peptide to superparamagnetic nanoparticles and its effect on cellular uptake. *Bioconjug. Chem.* 13: 840–844, 2002.

122. R Weissleder, A Moore, U Mahmood, R Bhorade, H Benveniste, EA Chiocca, JP Basilion. *In vivo* magnetic resonance imaging of transgene expression. *Nat. Med.* 6: 351–355, 2000.

123. R Elghanian, JJ Storhoff, RC Mucic, RL Letsinger, CA Mirkin. Selective colorimetric detection of polynucleotides based on the distance-dependent optical properties of gold nanoparticles. *Science* 277: 1078–1081, 1997.

124. JJ Storhoff, R Elghanian, RC Mucic, CA Mirkin, RL Letsinger. One-pot colorimetric differentiation of polynucleotides with single base imperfections using gold nanoparticle probes. *J. Am. Chem. Soc.* 120: 1959–1964, 1998.

125. RA Reynolds, CA Mirkin, RL Letsinger. Homogeneous, nanoparticle-based quantitative colorimetric detection of oligonucleotides. *J. Am. Chem. Soc.* 122: 3795–3796, 2000.

126. B Dubertret, M Calame, AJ Libchaber. Single-mismatch detection using gold-quenched fluorescent oligonucleotides. *Nat. Biotechnol.* 19: 365–370, 2001.

127. TA Taton, RC Mucic, CA Mirkin, RL Letsinger. The DNA-mediated formation of supramolecular mono- and multilayered nanoparticle structures. *J. Am. Chem. Soc.* 122: 6305–6306, 2000.

128. J Reichert, A Csaki, JM Köhler, W Fritzsche. Chip-based optical detection of DNA hybridization by means of nanobead labeling. *Anal. Chem.* 72: 6025–6029, 2000.

129. R Möller, A Csaki, JM Köhler, W Fritzsche. DNA probes on chip surfaces studied by scanning force microscopy using specific binding of colloidal gold. *Nucleic Acids Res.* 28: E91, 2000.

130. CM Niemeyer, B Ceyhan, S Gao, LF Chi, S Peschel, U Simon. Site-selective immobilization of gold nanoparticles functionalized with DNA oligomers. *Colloid Polym. Sci.* 279: 68–72, 2001.

131. A Csaki, R Moller, W Straube, JM Kohler, W Fritzsche. DNA monolayer on gold substrates characterized by nanoparticle labeling and scanning force microscopy. *Nucleic Acids Res.* 29: E81–81, 2001.

132. S Han, J Lin, F Zhou, RL Vellanoweth. Oligonucleotide-capped gold nanoparticles for improved atomic force microscopic imaging and enhanced selectivity in polynucleotide detection. *Biochem. Biophys. Res. Commun.* 279: 265–269, 2000.
133. D Gerion, WJ Parak, SC Williamson, D Zanchet, CM Micheel, AP Alivisatos. Sorting fluorescent nanocrystals with DNA. *J. Am. Chem. Soc.* 124: 7070–7074, 2002.
134. JKN Mbindyo, BD Reiss, BR Martin, CD Keating, MJ Natan, TE Mallouk. DNA-directed assembly of gold nanowires on complementary surfaces. *Adv. Mater.* 13: 249–254, 2001.
135. E Dujardin, LB Hsin, CRC Wang, S Mann. DNA-driven self-assembly of gold nanorods. *Chem. Commun.* 14: 1264–1265, 2001.
136. TA Taton, CA Mirkin, RL Letsinger. Scanometric DNA array detection with nanoparticle probes. *Science* 289: 1757–1760, 2000.
137. CM Niemeyer, D Blohm. DNA-microarrays. *Angew. Chem. Int. Ed.* 38: 2865–2869, 1999.
138. SJ Park, TA Taton, CA Mirkin. Array-based electrical detection of DNA with nanoparticle probes. *Science* 295: 1503–1506, 2002.
139. TA Taton, G Lu, CA Mirkin. Two-color labeling of oligonucleotide arrays via size-selective scattering of nanoparticle probes. *J. Am. Chem. Soc.* 123: 5164–5165, 2001.
140. YC Cao, R Jin, CA Mirkin. Nanoparticles with Raman spectroscopic fingerprints for DNA and RNA detection. *Science* 297: 1536–1540, 2002.
141. J Wang, G Liu, A Merkoci. Electrochemical coding technology for simultaneous detection of multiple DNA targets. *J. Am. Chem. Soc.* 125: 3214–3215, 2003.
142. F Patolsky, KT Ranjit, A Lichtenstein, I Willner. Dendritic amplification of DNA analysis by oligonucleotide-functionalized Au-nanoparticles. *Chem. Commun.* 12: 1025–1026, 2000.
143. L Lin, H Zhao, J Li, J Tang, M Duan, L Jiang. Study on colloidal Au-enhanced DNA sensing by quartz crystal microbalance. *Biochem. Biophys. Res. Commun.* 274: 817–820, 2000.
144. SB Han, JQ Lin, M Satjapipat, AJ Baca, FM Zhou. A three-dimensional heterogeneous DNA sensing surface formed by attaching oligodeoxynucleotide-capped gold nanoparticles onto a gold-coated quartz crystal. *Chem. Commun.* 7: 609–610, 2001.
145. XD Su, SFY Li, SJ O'Shea. Au nanoparticle- and silver-enhancement reaction-amplified microgravimetric biosensor. *Chem. Commun.* 8: 755–756, 2001.
146. GR Souza, JH Miller. Oligonucleotide detection using angle-dependent light scattering and fractal dimension analysis of gold-DNA aggregates. *J. Am. Chem. Soc.* 123: 6734–6735, 2001.
147. I Willner, F Patolsky, J Wasserman. Photoelectrochemistry with controlled DNA-cross-linked CdS nanoparticle arrays. *Angew. Chem. Int. Ed.* 40: 1861–1864, 2001.
148. CM Niemeyer, B Ceyhan. DNA-directed functionalization of colloidal gold with proteins. *Angew. Chem. Int. Ed.* 40: 3685–3688, 2001.
149. JM Nam, SJ Park, CA Mirkin. Bio-barcodes based on oligonucleotide-modified nanoparticles. *J. Am. Chem. Soc.* 124: 3820–3821, 2002.
150. L Josephson, JM Perez, R Weissleder. Magnetic nanosensors for the detection of oligonucleotide sequences. *Angew. Chem. Int. Ed.* 40: 3204–3206, 2001.
151. JM Perez, T O'Loughin, FJ Simeone, R Weissleder, L Josephson. DNA-based magnetic nanoparticle assembly acts as a magnetic relaxation nanoswitch allowing screening of DNA-cleaving agents. *J. Am. Chem. Soc.* 124: 2856–2857, 2002.
152. JM Perez, L Josephson, T O'Loughlin, D Hogemann, R Weissleder. Magnetic relaxation switches capable of sensing molecular interactions. *Nat. Biotechnol.* 20: 816–820, 2002.

153. K Hamad-Schifferli, JJ Schwartz, AT Santos, S Zhang, JM Jacobson. Remote electronic control of DNA hybridization through inductive coupling to an attached metal nanocrystal antenna. *Nature* 415: 152–155, 2002.
154. S Tyagi, FR Kramer. Molecular beacons: probes that fluoresce upon hybridization. *Nat. Biotechnol.* 14: 303–308, 1996.
155. A Jordan, R Scholz, P Wust, H Fähling. Cancer treatment with AC magnetic field induced excitation of biocompatible superparamagnetic nanoparticles. *J. Magn. Magn. Mater.* 201: 413–419, 1999.
156. E Katz, L Sheeney-Haj-Ichia, AF Bückmann, I Willner. Dual biosensing by magneto-controlled bioelectrocatalysis. *Angew. Chem. Int. Ed.* 41: 1343–1346, 2002.
157. L Sheeney-Haj-Ichia, E Katz, J Wasserman, I Willner. Magneto-switchable electrogenerated biochemiluminescence. *Chem. Commun.* 2: 158–159, 2002.
158. E Katz, I Willner. Magneto-stimulated hydrodynamic control of electrocatalytic and bioelectrocatalytic processes. *J. Am. Chem. Soc.* 124: 10290–10291, 2002.
159. J Wang, D Xu, R Polsky. Magnetically-induced solid-state electrochemical detection of DNA hybridization. *J. Am. Chem. Soc.* 124: 4208–4209, 2002.
160. T Hyeon, SS Lee, J Park, Y Chung, HB Na. Synthesis of highly crystalline and monodisperse maghemite nanocrystallites without a size-selection process. *J. Am. Chem. Soc.* 123: 12798–12801, 2001.
161. MT Lin, L Pulkkinen, J Uitto, K Yoon. The gene gun: current applications in cutaneous gene therapy. *Int. J. Dermatol.* 39: 161–170, 2000.
162. S Gurunathan, DM Klinman, RA Seder. DNA vaccines: immunology, application, and optimization. *Annu. Rev. Immunol.* 18: 927–974, 2000.
163. A Jordan, P Wust, R Scholz, B Tesche, H Fähling, T Mitrovics, T Vogl, J Cervos-Navarro, R Felix. Cellular uptake of magnetic fluid particles and their effects in AC magnetic fields on human adenocarcinoma cells *in vitro*. *Int. J. Hyperthermia* 12: 705–722, 1996.
164. A Jordan, P Wust, R Scholz, T Mitrovics, H Fähling, J Gellermann, T Vogl, J Cervos-Navarro, R Felix. Effects of magnetic fluid hyperthermia (MFH) on C_3H mammary carcinoma *in vivo*. *Int. J. Hyperthermia* 13: 587–605, 1997.
165. D Kadereit, J Kuhlmann, H Waldmann. Linking the fields: the interplay of organic synthesis, biophysical chemistry, and cell biology in the chemical biology of protein lipidation. *ChemBioChem* 1: 145–169, 2000.
166. M Famulok, G Mayer, M Blind. Nucleic acid aptamers: from selection *in vitro* to applications *in vivo*. *Acc. Chem. Res.* 33: 591–599, 2000.
167. Tissue engineering. *Nat. Biotechnol.* 18 (Suppl. IT): 56–58, 2000.
168. RF Service. Race for molecular summits. *Science* 285: 184–185, 1999.
169. KC Nicolaou, D Vourloumis, N Winssinger, PS Baran. Der Stand der Totalsynthese zu Beginn des 21. Jahrhunderts. *Angew. Chem. Int. Ed.* 39: 44–122, 2000.
170. K Matsuura, M Hibino, Y Yamada, K Kobayashi. Construction of glyco-clusters by self-organization of site-specifically glycosylated oligonucleotides and their cooperative amplification of lectin-recognition. *J. Am. Chem. Soc.* 123: 357–358, 2001.
171. SM Waybright, CP Singleton, K Wachter, CJ Murphy, UHF Bunz. Oligonucleotide-directed assembly of materials: defined oligomers. *J. Am. Chem. Soc.* 123: 1828–1833, 2001.
172. CM Niemeyer. In *NanoBiotechnology: Concepts, Methods and Applications*, CM Niemeyer, CA Mirkin, Eds. Weinheim, Germany: Wiley-VCH, 2004.

11 Nature-Inspired Templated Nanoparticle Superstructures

Shu-Hong Yu and Helmut Cölfen

CONTENTS

11.1 Introduction 270
11.2 Small-Molecule Additive-Mediated Shape Control 271
    11.2.1 Inorganic Species 271
    11.2.2 Organic Species...273
    11.2.3 Solvent Effects 275
11.3 Surfactant-Mediated Templating and Crystallization 280
    11.3.1 Micelles 280
        11.3.1.1 Mineralization of Inorganic Minerals and
                Mesoscale Self-Assembly 281
        11.3.1.2 Shape Control of Low-Dimensional
                Semiconductor Nanocrystals......................................282
        11.3.1.3 Shape Control of Metal Nanocrystals.......................283
    11.3.2 Microemulsions 284
    11.3.3 Monolayers 285
    11.3.4 Vesicles 288
11.4 Supramolecular Self-Assembly from Artificial Organic Templates..........288
    11.4.1 Supramolecular Polymers 288
    11.4.2 Transcription from Organic Matrix Template 289
11.5 Polymer-Controlled Crystallization and Morphosynthesis 291
    11.5.1 Biopolymers 292
    11.5.2 Dendrimers 298
    11.5.3 Polyamide..298
    11.5.4 Double Hydrophilic Block Copolymers 299
        11.5.4.1 Selective Adsorption and Crystal Growth.................299
        11.5.4.2 Morphosynthesis and Hierarchical Structures
                of Inorganic Crystals 301
        11.5.4.3 Morphosynthesis of Metal Carbonate Minerals 310
        11.5.4.4 Fine-Tuning of the Nanostructure of Inorganic
                Nanocrystals 315

        11.5.5   Polymeric Templates 316
11.6   Oriented Attachment Growth and Spontaneous Organization 320
        11.6.1   Oriented Attachment Growth 320
        11.6.2   Spontaneous Organization 323
11.7   Conclusions and Outlook 324
Acknowledgments 325
References 325

11.1 INTRODUCTION

Exploration of novel bioinspired strategies for self-assembling or surface-assembling molecules or colloids to generate materials with controlled morphologies with structural specialty, complexity, and related unique properties is among the hottest current research subjects.[1–8] Biomineralization and bioinspired morphosynthesis have been rapidly developing into one of the central objectives of biomimetic chemistry.[9,10] It has been recent intensive research activity for materials scientists and chemists to learn how to create unusual superstructures resembling biominerals with their specialty and complexity existing in nature.[11–14] An intensive recent review gives the latest advances in biomineralization of unicellular organisms with complex mineral structures from the viewpoint of molecular biology.[15] The biospecific interaction and coupling of biomolecules with inorganic nanosized building blocks have shown the potential assembly capabilities for generating new organized materials with strong application impact.[16]

Synthesis of inorganic crystals or hybrid inorganic–organic materials with specific size, shape, orientation, organization, complex form, and hierarchy has been a focus of recent research[12–28] because of the importance and potential to design new materials and devices in various fields such as catalysis, medicine, electronics, ceramics, pigments, and cosmetics. For example, shape control and structural specialty (size, shape, phase, dimensionality, assembly, etc.) have significant relevance for the properties of semiconductor nanocrystals,[23] metal nanocrystals,[24–27] and other inorganic materials,[20–22] which may add additional variables for tailoring the properties of nanomaterials.[29] Much effort has been made for fabricating one-dimensional nanowires or nanorods[30–36] by using hard templates such as carbon nanotubes[30,31] and porous aluminum templates[32–34] or by laser-assisted catalytic growth (LCG),[35,36] as well as controlled solution growth at room temperature or elevated temperature.[23,24,37–40]

As an alternative strategy, bioinspired morphosynthesis has been emerging as an important enviromentally friendly route to generate inorganic materials with controlled morphologies by using self-assembled organic superstructures, inorganic or organic additives, and templates with complex functionalization patterns.[4–22]

In this chapter, we will give a general overview of the current state of the art in the area of biomimetic mineralization and morphosynthesis of various inorganic crystals with structural specialty and complexity by emerging new nature-inspired approaches developed in recent years. General strategies for morphogenesis and biomimetic mineralization of inorganic crystals may be classified as follows: (1) additive-mediated shape control, (2) surfactant- or polymer-mediated templating and crystallization, (3) supramolecular self-assembly from designed artificial organic templates, and (4) oriented attachment growth and spontaneous organization. The latest advances in morphosynthesis of various technically important inorganic crystals via simple solution processes are reviewed with a focus on how to generate the inorganic crystals with unusual structural specialty and complexity and sizes spanning from the nanoscale to the macroscopic scale, which are expected to have potential applications in the multidisciplinary fields of advanced materials.

11.2 SMALL-MOLECULE ADDITIVE-MEDIATED SHAPE CONTROL

The shape of inorganic crystals can be affected by various factors, i.e., inorganic anions, organic additives, or solvents. The anisotropic growth of the particles can be explained by the specific adsorption of ions or organic additives to particular faces, therefore inhibiting the growth of these faces.

This strategy of crystal morphogenesis has been known for a long time and has even found industrial application, mainly based on empirical observations. A thermodynamic treatment of the crystal morphology changes based on an energy-minimized total surface area was already described by Wulff[41] as early as 1901. This is the basis for the thermodynamic consideration of morphology changes upon additive adsorption based on growth velocity changes of specific crystal surfaces, which are proportional to their surface energy.[41] Although it is nowadays well known that this purely thermodynamic treatment cannot always predict the crystal morphology as crystallization, and the resulting morphology often also relies on kinetic effects as well as on defect structures like screw dislocations and kinks, it formed the foundation for the consideration of additive-mediated crystal morphology changes.

In the following sections, we will give specific typical examples to illustrate the striking influence of small additives on the shape and structures of inorganic crystals.

11.2.1 INORGANIC SPECIES

It is well known that the adsorption of inorganic ions on specific crystal faces has been found to exert a strong influence on the shape of inorganic crystals. Earlier studies show that simple inorganic ions such as Mg^{42} and Sr,[42] Li cations,[43,44] and sulfate anions[42] have obvious effects on $CaCO_3$ crystallization.

Sugimoto et al.[45] designed a gel-sol method for general synthesis of monodisperse particles. The idea is to use an extremely condensed precursor gel as a matrix of the subsequent particles and also to provide the reservoir of the metal (and hydroxide) ions, which could prevent the coagulation of the particles by fixing them in the gel matrix for the subsequent growth, even at a high concentration of electrolyte, and keep them growing without renucleation.

A systematic shape control of hematite particles can be achieved by choosing different anions in the gel-sol process,[46] which can exert strong inhibition on the growth of the specific faces. The presence of sulfate and phosphate anions can retard the growth in the direction normal to the c-axis, leading to the formation of ellipsoidal or peanut-type particles (Figure 11.1A). In contrast, the presence of chloride ions leads to the formation of pseudocubic-shaped hematite particles by retarding the

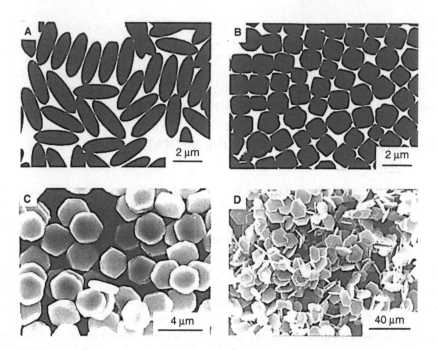

FIGURE 11.1 SEM images showing α-Fe_2O_3 particles with different shapes in the presence of different anions. (A) Ellipsoid in the presence of SO_4^{2-}. (B) Pseudocube in the presence of Cl^-. (C) Thick platelet. (D) Thin platelet in the absence of anions. (From Sugimoto, T. and Wang, Y., *J. Colloid Interface Sci.*, 207, 137–149, 1998. With permission. Copyright © 1998 Elsevier.)

growth of the (012) faces (Figure 11.1B). In the absence of these anions, the aging of α-FeOOH under alkaline conditions leads to the formation of thick or thin platelets (Figure 11.1C and D). The influence of sulfide anions on the shape of hematite is of particular interest.

Monodisperse hematite peanuts can be produced after aging the amorphous β-FeOOH gel produced from $Fe(OH)_3$.[47] The morphology evolution of the peanuts relied on the concentration of SO_4^{2-} anions and aging time, accompanying the tim-dependent two-step phase transformation.[47] The selective adsorption of such simple anions onto the specific habit planes of the hematite particles leads to the shape change. Similarly, the shape of $Al_3(SO_4)_2(OH)_5\cdot2H_2O$[48] and TiO_2 particles[49,50] can also be changed by anion adsorption.

The biological calcite in skeletons often contains Mg with varied contents from 10 to 40 mol-%,[51,52] which is much higher than the 10 mol-% of thermodynamically stable magnesium calcite, so that under such conditions, aragonite formation should be expected. It is interesting to note that laboratory experiments on calcite crystallization from supersaturated solutions containing a 4:1 Mg^{2+}/Ca^{2+} molar ratio in the absence or presence of polyasparate, polyacrylate, or macromolecules extracted from the high Mg-calcite skeleton of a coralline alga showed in each case that amorphous calcium carbonate (ACC) phase transformation resulted in calcite crystals with Mg levels of approximately 20 mol-%.[53] Several reports on the remarkable influence of

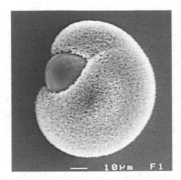

FIGURE 11.2 Scanning electron micrograph of a mixed particle with a core of polycrystalline magnesian calcite and an outer shell of polycrystalline aragonite grown in the presence of a 4:1 Mg/Ca solution. (From Raz, S. et al., *Adv. Mater.*, 12, 38–42, 2000. With permission. Copyright © 2000 Wiley-VCH.)

Mg^{2+} ions on the morphology and polymorphs of $CaCO_3$ minerals have also appeared.[54–57] Spherulites were formed in a limited window in the presence of both Mg^{2+} and SO_4^{2-} ions as impurities,[54] whereas the presence of Sr^{2+} ions leads to the formation of dumbbells.[55] A recent report shows that the presence of Mg^{2+} ions results in aragonite rather than calcite at high Mg^{2+} concentrations.[56] The examination of the effects of Mg^{2+} as well as citric or malic acid shows that both Mg^{2+} and its combination with organic additives can cooperatively affect the calcite nucleation and growth.

Spherulite and dumbbell morphologies were also observed for high magnesian calcites with magnesium contents much greater than the thermodynamically maximum possible amount for calcite of 10 mol-% Mg.[53] For this extreme situation, most unusual core-shell composite particles consisting of two different $CaCO_3$ polymorphs in direct contact to each other were observed, which formed via an amorphous $CaCO_3$ precursor (Figure 11.2).

In addition, surface-functionalized Au nanocolloids were used as nuclei for the crystallization of $CaCO_3$ by Küther et al.[58,59] and micrometer-sized (10-μm) spherules with well-defined surface structures were obtained. Mercaptobenzoic acid-capped gold particles can also be used to direct the crystallization of $CaCO_3$ for the nucleation of vaterite and aragonite particles.[60] The inclusion of the gold particles makes the described materials an interesting metal–organic–inorganic hybrid material.

11.2.2 ORGANIC SPECIES

It is well known that small molecular organic species can also exert a significant influence on the growth of either organic crystals[61] or inorganic crystals.[62–66] An earlier report by Kitano and Hood.[62] shows that small organic additives can influence the polymorphic calcium carbonate.

A series of carboxylates, phosphonates, and sulfonates were tested as possible inhibitors for crystal growth of barium sulfate (barite), and diphosphonates were found to have specific effects on the morphology of the barite crystals.[64,66] Molecular modeling results showed that a good geometrical and stereochemical match of the inhibitor to the growing face is required for effective growth modification, in addition to the electrostatic effect. Coveney et al.[67] designed a series of small, flexible molecules for the purpose of recognizing and binding to all of the important growing faces of barite in order to promote isotropic growth on the basis of molecular modeling, resulting in spherical crystalline barite. A family of macrocyclic aminomethylphosphonates was confirmed as good inhibitors since the phosphonate group has shown extreme affinity for sulfate sites in the barium sulfate lattice.[64,68] Computer simulation of energy minimization and molecular dynamics calculations of the binding of the macrocycles to all eight growth-determining faces of barium sulfate suggested that the macrocycle 1,7-dioxo-4,10-diaza-12-crown-4-N,N'-dimethylenephosphonate was the best candidate as a universal face-binding agent for isotropic growth.[67]

The influence of small organic additives on the crystallization of calcium carbonate was also examined. Nygren et al.[69] showed that the diphosphonates (1-hydroxyethylidene-1,1-diphosphonate (HEDP)) could enter the surface of calcite or growth steps by replacing two CO_3^{2-} ions if the distance between two carbonate ions matched the distance of the bisphosphonate groups well.

The effect of various organic additives on the morphologies of other inorganic minerals has also been investigated from the viewpoint of colloid chemistry. The shape control of the alpha iron oxide (hematite) has been previously demonstrated by Sugimoto et al.[70,71] A variety of equiaxial hematite morphologies can be synthesized.[70] It was shown that adsorption of the organics is a necessary condition for shape modification in the hematite system. Bell and Adair[72] have shown that several organic acids and bases have an effect on alpha-alumina's particle shape. Sugimoto et al.[73] also investigated the effects of different chelates such as ethylenediamine-N,N,N',N'-tetraacetic acid (EDTA), nitrilotriacetic acid (NTA), aspartic acid (AA), trimethylenediamine (TMD), N,N-dimethylethylenediamine (DMED), diethylenetriamine (DETA), triethylenetetramine (TETA), and tris(2-aminoethyl)amine (TAEA) on the synthesis of various monodisperse sulfide particles. The result shows that it is generally impossible to synthesize uniform particles in highly concentrated solutions of chelates on the order of 0.1 mol l^{-1} or more because of serious coagulation.[74] However, gelatin has been found to be a very powerful anticoagulant as a protective colloid in the synthesis of monodisperse sulfide particles. The separation of the nucleation and growth stages, which is crucial for producing the monodisperse particles, could be achieved. Interestingly, monodisperse hollow CuS particles can be obtained due to the fact that the internal sulfide ions of the CuS particles covered with an adsorption layer of gelatin are preferentially oxidized by O_2 and dissolved into the solution phase with Cu^{2+} ions upon adding ammonia.[70]

The structure details of single crystals can be fine-tuned by growth or dissolution in the presence of organic additives. A striking example was reported for the generation of chiral calcite morphologies in the presence of aspartic acid (Asp) enantiomers.[75] Specific binding of an Asp enantiomer to those surface step edges that offer the best geometric and chemical fit was found to change the step-edge free energies, giving rise to direction-specific binding energies for each enantiomer. This results in the propagation of chiral crystal modifications from the atomic length scale to the macroscopic one, as shown in Figure 11.3.

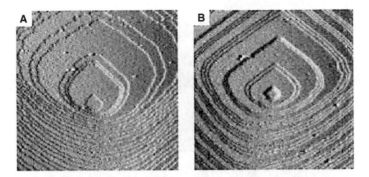

FIGURE 11.3 Images showing the effect of amino acid enantiomers on calcite morphology. Growth hillocks following addition of supersaturated solutions with 0.01 M L-aspartic acid (A) and 0.01 M D-aspartic acid (B). (From Orme, C.A., et al., *Nature*, 411, 775–779, 2001. With permission. Copyright © 1999 Nature Publishing Group.)

The above small organic molecule-mediated crystallization and orientation growth could be extended to other mineral systems. Surprisingly, Tian et al.[76] showed a first example for growing unusual oriented and extended helical semiconductor ZnO nanostructures by a synthetic ceramic method, which is very similar to the growth morphology of nacreous calcium carbonate. The helical ZnO nanostructures were grown at room temperature on the glass substrate containing oriented ZnO nanorod arrays,[77] which were placed in a solution containing $Zn(NO_3)_2$, hexomethylenetetramine, and sodium citrate and reacted at 95°C for 1 day. Figure 11.4 shows the similarity of scanning electron microscopy (SEM) images of the helical ZnO nanocolumns (Figure 11.4A and B) and the growth tip of a young abalone shell containing oriented columns of aragonite nanoplates (Figure 11.4C), although the structural dimensions are different. As a result of secondary growth on the helical nanorods, aligned and well-defined nanoplates are formed, as in nacre. The side-width growth of the ZnO nanoplates leads to hexagonal ZnO plates that begin to overlap with one another.

The organic species were reported to act as simple physical compartments, to control nucleation, or to terminate crystal growth by surface poisoning in biomineralization of nacre.[78,79] The similar biomimetic structures of helical ZnO rods as the nacreous $CaCO_3$ indicated that helical growth might play some role in the formation of organized nacreous calcium carbonate. However, the organic species citrate ion still plays critical roles in the formation of such structures, because the citrate ion has a tendency to inhibit the growth of the (002) surfaces, possibly through selective surface adsorption.[76] Such highly resembled growth-patterned superstructures imply the possibilities to control the morphology and orientation of inorganic crystals by the simplest synthetic way.

11.2.3 SOLVENT EFFECTS

It has been of great interest in synthetic chemistry and materials science to seek alternative soft solution reactions[80–83] to replace the traditional high-temperature solid-state reaction for the formation of technologically important crystals in order

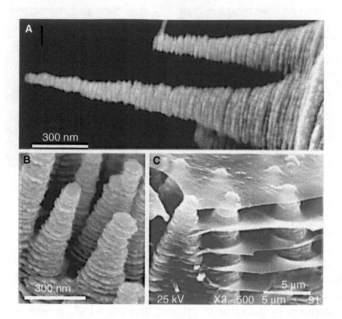

FIGURE 11.4 Comparison of ZnO helical structures with nacre. (A, B) High-magnification image of oriented ZnO helical columns. (C) Nacreous calcium carbonate columns and layers near the growth tip of a young abalone. (From Tian, Z.R. et al., *J. Am. Chem. Soc.*, 124, 12954–12955, 2002. With permission. Copyright © 2002 American Chemical Society.)

to remove the diffusion control barrier. Hydrothermal/solvothermal processes have been emerging as an important synthetic method for the synthesis of various inorganic crystals with well-defined shapes, sizes, and structures. Especially the solvothermal process, in which the reaction was done in different organic solvents, has shown remarkable advantages and versatility in controlled synthesis of inorganic nanomaterials.[83]

Recent advances show that the physical properties of solvents, such as the chelating ability and the interaction strength of the solvents with inorganic species, play a key role in the shape evolution of one-dimensional semiconductor nanorods. Previously, micrometer-sized ZnO rods were produced in aqueous solution containing hexamethylenetetramine.[84] A remarkable solvent-dependent morphological change of CdS nanocrystals formed in different solvent was observed before.[85–87] The simple polyamines, which contain either polydentate or monodentate, can be elegantly used to act as *shape controller* for the synthesis of a variety of semiconductor nanocrystals.[83] The possible chelating mode is illustrated in Scheme 11.1. The close interaction between anchor atoms in ligands (Lewis base) and metal ions on the surface (Lewis acid) is another prerequisite for the formation of nanorods, because this weak interaction will result in the effect that ligand molecules will not impose effective influence on the nucleation and growth of nanoparticles.[85] Altering this kind of interaction strength between ligands and metal ions could control the shape and size of various chalcogenide nanocrystals, as demonstrated previously in the synthesis of CdS nanocrystals.[85,88]

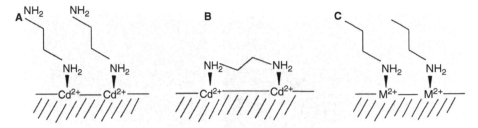

SCHEME 11.1 Illustration of the possible coordination mode on the surface. (A) Monodentate mode of ethylenediamine (en) molecules. (B) Polydentate mode of en molecules. (C) Monodentate mode of n-butylamine. (From Yang, J. et al., *Angew. Chem. Ed. Int.*, 41, 4697–4700, 2002. With permission. Copyright © 2002 Wiley-VCH.)

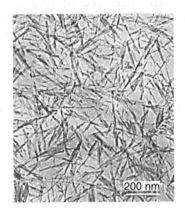

FIGURE 11.5 Wurtzite ZnS nanorods synthesized in n-butylamine at 250°C for 12 h.

The solvothermal method is both flexible and reproducible for controlling the phase, shape, and size of ZnS nanocrystals. By the similar solvothermal reaction at 120°C using ethanol instead of ethylenediamine as the solvent, sphalerite ZnS (S) nanoparticles with a size of ~3 nm can be easily synthesized.[89] Furthermore, wurtzite ZnS, ZnSe nanorods can also be synthesized by using n-butylamine, a monodentate amine, as the solvent. The nanorods form and tend to form bundle structures or arrays within a limited experimental window, as shown in Figure 11.5.

Wurtzite single-crystal ZnS nanosheets with rectangle lateral dimensions in a range of 0.3 to 2 μm and preferential growth direction along the a- and c-axes can be templated from removing the ethylenediamine (en) by thermal decompostion from a lamellar precursor ZnS·(en)$_{0.5}$ produced by solvothermal reaction of zinc salts such as Zn(CH$_3$COO)$_2$·2H$_2$O or ZnCl$_2$ and with thiourea in ethylenediamine (en) at 120 to 180°C (Figure 11.6). The oxidization of the lamellar molecular sheets in air produce wurtzite ZnO flake-like dendrites,[89] as indicated by the selected area electron diffraction (SAED) pattern in Figure 11.6A. The high-resolution transmission electron microscopy (HRTEM) image shows well-resolved (100) lattice planes with lattice spacing of 3.3 Å, indicating the preferential orientation along the a-axis (Figure 11.6B).

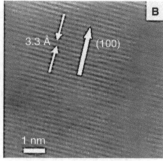

FIGURE 11.6 (A) TEM images and SAED patterns for wurtzite ZnS single-crystal nanosheets obtained by thermal decomposition of lamellar $ZnS \cdot (en)_{0.5}$ precursor at 500°C for 0.5 h in vacuum; inserted SEAD pattern recorded along <010> zone. (B) Well-resolved HRTEM image of the wurtzite ZnS single-crystal nanosheet. (From Yu, S.H. and Yoshimura, M., *Adv. Mater.*, 14, 296–300, 2002. With permission. Copyright © 2002 Wiley-VCH.)

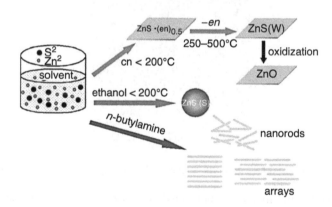

SCHEME 11.2 Illustration of controlled synthesis of ZnS nanocrystals, with different dimensionalities and phases via a solvothermal approach: ZnS dots, nanosheets, nanorods, and bundles. (Partially taken from Yu, S.H. and Yoshimura, M., *Adv. Mater.*, 14, 296–300, 2002. With permission. Copyright © 2002 Wiley-VCH.)

The general synthetic strategies for ZnS nanocrystals, with different phases and shapes, and ZnO dendrites are summarized in Scheme 11.2, showing that both shape and phase of the ZnS nanocrystals can be well controlled by choosing the solvents. A rational design of the solvothermal reaction allows for the possibility to control both the dimensionality (dots, rods, and sheets) of ZnS nanocrystals and the phase.

The formation of CdS nanorods in *n*-butylamine indicates that one anchor atom in a ligand is necessary and already adequate for the formation of nanorods, even though more anchor atoms may be present in a ligand. A family of II-VI group semiconductor nanorods has been synthesized by use of *n*-butylamine as the *shape controller*.[88]

Besides the nanorods/nanowires and nanosheets, other novel nanostructures can be fabricated. Amazingly, trigonal tellurium nanobelts, nanotubes, and nanohelices

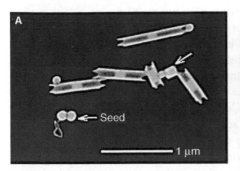
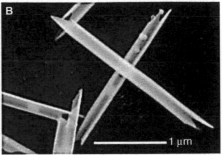

FIGURE 11.7 SEM images of the Te nanotubes synthesized by refluxing a solution of orthotelluric acid in ethylene glycol for (A) 4 and 6 min, respectively. The white arrow indicates the presence of seeds. (From Mayers, B. and Xia, Y.N., *Adv. Mater.*, 14, 279–282, 2002. With permission. Copyright © 2002 Wiley-VCH.)

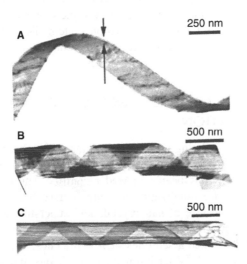

FIGURE 11.8 TEM images of a typical tellurium nanobelt (A), a helical nanobelt (B), and a typical helical nanobelt within a nanobelt-roll nanotube (C). (From Mo, M.S. et al., *Adv. Mater.*, 14, 1658–1662, 2002. With permission. Copyright 2002 © Wiley-VCH.)

can be synthesized by solution approaches.[90,91] The tellurium nanotubes were synthesized by reflexing a solution of orthotelluric acid in ethylene glycol at ~197°C (Figure 11.7). The early stage shows that tubular structures obviously grew from cylindrical seeds, as shown in Figure 11.7A.[90] More complex Te nanostructures such as nanobelts, nanotubes, and nanohelices can be created by a very simple controlled hydrothermal reaction that involves a reaction of sodium tellurite (Na_2TeO_3) in aqueous ammonia solution at 180°C.[91] The nanobelts have a thickness of about 8 nm, with a width of 30 to 500 nm and lengths up to several hundreds of micrometers, as shown in Figure 11.8A. The nanobelts obviously tend to twist and form helices, as shown in Figure 11.8B. When the nanobelts are further twisted and rolled, the nanotubes form. In addition, an interesting nanostructure, the so-called coaxed nanobelt-within-nanotube structure, is observed, as shown in Figure 11.8C. The template-roll-growth and template-twist-joint-growth mechanisms were proposed for the explanation of the formation of such special nanostructures.[91] However, the detailed mechanism still needs to be further investigated.

Recently, the influence of alcohols like ethanol, isopropanol, or diethylene glycol was also examined for crystallization of $CaCO_3$ crystals, and the results showed that a $CaCO_3$ polymorph selection could be achieved by preventing the transformation of complex-shaped vaterite aggregates to calcite.[92]

The above solvent-mediated shape control mechanisms have proven to be very versatile in the synthesis of a variety of semiconductor nanocrystals with unusual shape and complex structures. In this mechanism, ligands control the shape of the nanocrystals through the interaction between ligands and metal ions on the surface of the nuclei. The further understanding of the detailed mechanism will allow solution approaches to be very promising for the synthesis of low-dimensional nanocrystals with more complexity. Recent advances have undoubtedly confirmed the special role of solvents played in the control of the shape and phase (polymorph) of the inorganic crystals in the solution reaction system. The further understanding of solvent effects on the crystallization process and detailed shape control mechanism should be explored more broadly and intensively.

11.3 SURFACTANT-MEDIATED TEMPLATING AND CRYSTALLIZATION

Use of surfactant for templating mesoporous materials has been rapidly developed since the pioneer work by Beck et al.[93] in 1992. The surfactant-mediated templating of mesoporous materials has been reviewed recently.[94] The emerging surfactant-mediated templating and crystallization techniques have been widely exploited in controlling the shape and size of nanocrystals and superstructures.[12,95] A common feature of this kind of templating is that ordered inorganic structures have been replicated from self-organized assemblies such as micelles, microemulsions, foams, monolayers, and vesicles.[1,96] Water-in-oil microemulsions and vesicles formed by amphiphilic surfactants were two of the first systems used to mimic biomineralization and show promise in templating inorganic materials.[96–98] In addition, lipids also play an important role in biomineralization and other biological processes, which has recently been intensively reviewed by Collier and Messersmith.[99] In these systems, the mineralization reaction occurs on the surface of the vesicles and the water droplets. The size, shape, and curvature of the vesicles and droplets are determined by the surface tension difference between the immiscible water and oil, which controls the form of the mineralized materials.[4]

11.3.1 MICELLES

Reverse micelles provide good well-defined nanoreactors for the formation of aniso-tropic nanoparticles. The reverse micelle reaction medium usually consists of either an anionic surfactant such as AOT (sodium bis(2-ethylhexyl)sulfosuccinate) or nonionic surfactants. Reverse micelles and microemulsion techniques have been widely used for the preparation of various inorganic crystals, such as copper nanorods,[26] higher-order structures such as $BaCrO_4$ chains and filaments,[22] $BaSO_4$ filaments and cones,[100,101] as well as $CaSO_4$,[102] $BaCO_3$,[103] $CaCO_3$ nanowires,[104] $BaWO_4$ nanorods/nanowires,[105,106] CdS, CdSe nanotubes/nanowires,[107] CdS nanorods,[108,109] CdS nanotrangles,[110] Cu nanorods,[111] and Ag nanodisks.[112]

11.3.1.1 Mineralization of Inorganic Minerals and Mesoscale Self-Assembly

The strong binding interactions between surfactants and inorganic nuclei can effectively inhibit the crystal growth and lead to the spontaneous structure reconstruction and self-organization of the primary nanoparticles. Micrometer-long twisted bundles of $BaSO_4$ and $BaCrO_4$ nanofilaments in water-in-oil microemulsions can be prepared from the anionic surfactant, AOT.[22] The reaction occurs at room temperature in unstirred isooctane containing a mixture of $Ba(AOT)_2$ reverse micelles and NaAOT microemulsions with encapsulated sulfate (or chromate) anions. The reverse micelles are about 2 nm in diameter and consist of a spherical cluster of about 10 Ba^{2+} ions strongly associated with the sulfonic acid head groups of the surfactant, along with hydration water.[22] In contrast, the microemulsions are larger (4.5 nm across) because they contain bulk water (aqueous Na_2SO_4 or Na_2CrO_4) at a water-to-surfactant molar ratio, $w = 10$. When mixed together, the two reaction fields interact so that the constituents are slowly exchanged and $BaSO_4$ or $BaCrO_4$ nanoparticles nucleate and grow within the delineated space. With time, other filaments are formed parallel to the original thread to produce a small bundle of coaligned inorganic nanofilaments held together by surfactant bilayers. The locking in of new filaments by surfactant interdigitation generates a bending force in the nonattached segment of the longer primary thread. This results in the coiling of the bundle into a characteristic spiral-shaped structure several hundred nanometers in size that becomes self-terminating at one end because further addition of the primary nanoparticles is prevented by spatial closure.

The construction of higher-order structures from inorganic nanoparticle building blocks was successfully demonstrated by achieving sufficient informational content in the preformed inorganic surfaces to control long-range ordering through interactive self-assembly.[22] The prismatic nanocrystals have flat surfaces with low curvature so that the hydrophobic driving force for assembly can be strengthened through the intermolecular interaction, resulting in the formation of a bilayer between adjacent particles by the interdigitation of surfactant chains attached to nanoparticle surfaces. When the $[Ba^{2+}]:[SO_4^{2-}]$ (or $[Ba^{2+}]:[CrO_4^{2-}]$) molar ratio is equal to 1.0, remarkable linear chains of individual $BaSO_4$ or $BaCrO_4$ nanoparticles are formed, as shown in Figure 11.9. Similar arrays of monodisperse hydrophobic Prussian Blue (ferric hexacyanoferrate(II)) nanoparticles were prepared by the photoreduction of $[Fe(C_2O_4)_3]^{3-}$ in the presence of $[Fe(CN)_6]^{3-}$ ions confined within the nanoscale water droplets of reverse micelles.[113]

A very efficient way to systematically change the crystal morphology of $CaCO_3$ was described via water-induced mesoscale transformation of surfactant-stabilized

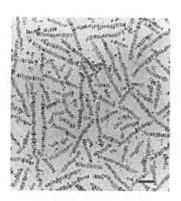

FIGURE 11.9 TEM image showing ordered chains of prismatic $BaSO_4$ nanoparticles prepared in AOT microemulsions at $[Ba^{2+}]:[SO_4^{2-}]$ molar ratio = 1 and w = 10. Scale bar = 50 nm. (From Li, M. et al., *Nature*, 402, 393–395, 1999. With permission. Copyright © 1999 Nature Publishing Group.)

ACC in reverse microemulsions to generate self-assembling vaterite superstructures.[114] The structures could be adjusted between monodisperse spheroidal aggregates at low water amounts via spindle-shaped aggregates and coaligned high-aspect-ratio nanofilaments to large vaterite nanoparticles at high water contents. This shows that the water-triggered $CaCO_3$ crystallization and the electrostatic mineral surfactant interactions lead to coupled mesoscopic processes resulting in the *in situ* higher-order assembly of these complex morphologies.

11.3.1.2 Shape Control of Low-Dimensional Semiconductor Nanocrystals

Reverse microemulsions/micelles could also be applied to synthesize $CaCO_3$ and $BaCO_3$ nanowires with extremely high aspect ratios of 1000 or more by directed aggregation of nanoparticles in a mesoscopic transformation.[104] Although the precise role of the applied Triton X-100 surfactant could not be elucidated, the nanowire formation has been shown to proceed via the steps of individual nanoparticles — fractal aggregates — short wires to long wires, which clearly indicates the importance of mesoscale transformation in the structural development over several length scales.

Semiconductor nanotubes and nanowires have been recently reported by employing nonionic surfactants such as t-octyl-$(OCH_2CH_2)_xOH$, x = 9, 10 (Triton-X) and anionic surfactant AOT.[107] Nanowires of sulfides and selenides of Cu, Zn, and Cd with high aspect ratios can be prepared by using Triton X-100. The results show that it is possible to obtain both nanotubes and nanowires of CdSe and CdS by this surfactant-assisted synthesis. The nanotubes are generally long, with lengths up to 5 μm, as shown in Figure 11.10. The outer diameter of the nanotubes is in the 15- to 20-nm range, and the diameter of the central tubule is in the 10- to 15-nm range. The wall thickness is therefore about 5 nm. The formation mechanism of the nanotubes in the presence of surfactant is still not clear. In addition, unusual CdS nanotrangles have been prepared in a $Cu(AOT)_2$/isooctane/H_2O reverse micelle system.[110]

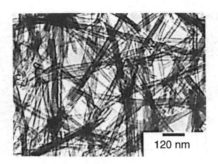

FIGURE 11.10 TEM image of CdSe nanotubes obtained by using Triton X-100 as the surfactant. (From Rao, C.N.N. et al., *Appl. Phys. Lett.*, 78, 1853–1855, 2001. With permission. Copyright © 2002 American Institute of Physics.)

11.3.1.3 Shape Control of Metal Nanocrystals

A reverse micelle system has also been applied for the preparation of metal nanoparticles with unusual shape. Pileni et al.[111,112] used a $Cu(AOT)_2$/isooctane/H_2O reverse micelle system to synthesize Cu nanorods and Ag nanodisks with a different precursor. Previously, gold nanorods have been synthesized electrochemically in the presence of a concentrated CTAB surfactant and cosurfactant solution mixture that is known to form rod-shaped micelles.[115] A gold sheet acts as the sacrificial electrode (anode), and a platinum electrode acts as the cathode at which the gold ions are reduced. The shape distribution of the rods depends on the current and the ratio of the concentration of the surfactant to that of the cosurfactant used. Recently, an elegant wet chemical approach, called seed-mediated micellar template, for controlling the shape and aspect ratio of noble metal nanorods/nanowires was reported by Murphy et al.[116–118] In this strategy, small citrate-capped Ag or Au nanospheres with a diameter of 3 to 5 nm were initially prepared by the reduction of Au or Ag salts with the strong reduction agent sodium borohydride and were used as the seeds in the secondary growth of anisotropic nanostructures. These seeds were added to a solution containing a rod-like micellar template formed by cetyltrimethylammonium bromide (CTAB) and more metal salt in the presence of the weaker reduction agent ascorbic acid. The uniform nanorods tend to self-assemble upon gradual solvent evaporation to form a chain-like structure,[118] which resembles a two-dimensional smectic liquid crystal, as shown in Figure 11.11A. The aspect ratio of the nanorods can be controlled by varying the ratio of seed to metal salt, and the long rods can be isolated from spherical particles by centrifugation. High aspect ratios of Ag and Au nanowires can be controlled by fine-tuning of the solution conditions (Figure 11.11B).[116,117] The detailed nanorod growth mechanism is still not clear. In contrast, Chen et al.[119] showed that using a higher concentration of NaOH in a similar system can produce triangular Ag nanodisks, which evolved from the truncated triangular silver nanoplates obtained through seed-mediated growth of silver particles in the presence of concentrated CTAB. The single-crystal nanodisks with basal plane (111) have a thickness on the order of 20 to 30 nm and a diameter about 60 nm. Similar

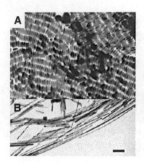
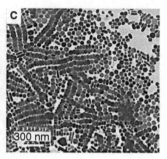

FIGURE 11.11 TEM images of shape-separated silver nanorods (A) and nanowires (B). (From Jana, N.R. et al., *Chem. Commun.*, 617–618, 2001. With permission. Copyright © 2001 The Royal Society of Chemistry.) (C) TEM image shows the formation of necklace-like structures due to the nanodisks partially stacked together. Scale bar = 100 nm. (From Chen, S. et al., *J. Phys. Chem. B, 106,* 10777–10781, 2002. With permission. Copyright © 2002 American Chemical Society.)

to the case of the nanorods, the nanodisks tend to self-assemble into large-scale necklace-like nanostructures, as shown in Figure 11.11C. The self-assembled mono-layer of CTAB on the basal plane is suggested to account for both the anisotropic growth and the self-assembly. The self-assembled chain-like superstructures are very similar to the ordered chains of surfactant-coated prismatic $BaSO_4$ nanoparticles obtained in the AOT microemulsion system,[22] suggesting that the surfactant-medi-ated mesoscale self-assembly process could be effective in different systems.

A catanionic reverse micelle system consisting of water, decane, and an equimo-lar mixture of two surfactants, i.e., undecylic acid and decylamine, was prepared, where the cationic surfactant was produced by the protonation of the amine by undecylic acid. High-aspect-ratio, single-crystal $BaWO_4$ nanowires with diameters as small as 3.5 nm and lengths up to more than 50 μm were synthesized in such a catanionic reverse micelle system.[106] A reverse micelle system formed by the mixture of AOT and a zwitterionic surfactant lecithin was adopted for the formation of CdS nanorods with an average diameter of 4.1 nm and length up to 150 nm.[109]

It should be pointed out that the poor crystallinity of the nanostructures obtained by this approach as well as the application of a large surfactant excess could restrict the applicability of this approach.

11.3.2 MICROEMULSIONS

Microemulsions have been widely applied for morphosynthesis of inorganic mate-rials. Bicontinuous microemulsions,[97] emulsion foams,[98,120] and emulsion droplets[121] have been explored for replication of these organized reaction mini-environments.

The morphosynthesis of calcium carbonate microsponges was demonstrated by using a microemulsion of an octane/SDS/$CaHCO_3$ system (Figure 11.12).[121] The slow CO_2 evaporation from a microemulsion, formed from a CO_2-saturated aqueous solution of calcium bicarbonate in a volatile oil (octane) stabilized by sodium dodecyl sulfate (SDS), leads to the formation of textured spheres of vaterite, which is a metastable polymorph of calcium carbonate.[121] The organic component (SDS) plays

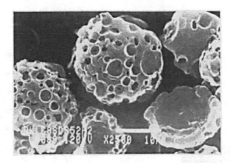

FIGURE 11.12 Vaterite spheroids formed from a microemulsion of octane:SDS:CaHCO$_3$ (71:4:25 wt%). Note the complex surface patterning and uniform size of the spheres. Scale bar = 10 μm. (From Walsh, D. et al., *Adv. Mater.*, 11, 324–328, 1999. Copyright © 1999 Wiley-VCH.)

a passive role in the mineralization of the vaterite spheres, principally acting to stabilize the water droplets and, therefore, the spherical morphology. Hollow silica microspheres,[122] organoclay microspheres with a sponge-like or hollow interior,[123] were also templated in water-in-oil emulsions.

In addition, hydrophobic nanoparticles of cobalt hexacyanoferrate, cobalt pentacyanonitrosylferrate, and chromium hexacyanochromate were synthesized by coprecipitation reactions involving mixtures of water-in-oil microemulsions containing the appropriate reactants.[124]

11.3.3 MONOLAYERS

Monolayers have been employed as a two-dimensional matrix to mimic the biomineralization process for growing inorganic thin films or crystals. The main concept behind this approach is the view pioneered by Lowenstam[125] that protein layers such as the β-sheets in nacre have an epitaxial arrangement of functional groups to specific crystal faces. The principles and specific examples on monolayer-directed crystallization of various inorganic minerals have been reviewed by Heywood.[126] Mann and Heywood et al.[127–131] systematically investigated the influence of surfactant head groups with different functionalities on the crystallization of undercompressed monolayers, which are formed from saturated long-alkyl-chain carboxylates, sulfates, amines, and alcohols. The controlled nucleation of CaCO$_3$[127–129] and BaSO$_4$ [130–132] under the control of monolayers was demonstrated. The different polymorphs of CaCO$_3$ can be induced from a supersaturated solution under the control of compressed monolayers of a saturated long-alkyl-chain carboxylate, and the selection of the polymorphs depends on the Ca^{2+} concentration in the subphase.[127,128] Differently oriented vaterite crystals were obtained by switching the polarity of monolayer to positive (i.e., octadecylamine monolayers).[128,129]

The monolayer-directed crystallization has also been employed for preparation of well-defined nanocrystals. Hanabusa et al.[147] showed that either the thin films or the nanocrystals can be induced by arachidic acid (AA) monolayers. Epitaxial growth of semiconductor nanocrystals under Langmuir monolayers by epitaxial matching of the crystals' faces and the head groups of the surfactants led to the formation of

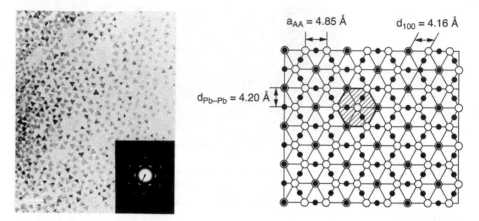

FIGURE 11.13 TEM image of a PbS particulate film. Scale bar = 200 nm. The film was obtained by the infusion of H_2S to an AA monolayer, floating on an aqueous 5.0×10^{-4} *M* $Pb(NO_3)_2$ solution in a circular trough, for 45 min. The PbS film was deposited on an amorphous carbon-coated copper grid. The right part shows schematically the match of the proposed overlap between Pb^{2+} ions and AA head groups; o is the AA head group and ● is the Pb^{2+} ions. The dotted line area is a unit cell. (From Fendler, J.H. and Meldrum, F.C., *Adv. Mater.*, 7, 607–632, 1995. With permission. Copyright © 1995 Wiley-VCH.)

well-shaped nanocrystals exhibiting specific faces. The potential of this approach and various examples have been reviewed by Fendler et al.[133]

The PbS thin films contained uniform equilateral triangular PbS crystals, which were obtained by exposing the solution to an AA monolayer in a sealed system (Figure 11.13, left). The size of the crystals can be varied by controlling the reaction time.[133] The perfect orientation growth from the {111} basal planes can be well explained by matching the AA monolayer and the cubic PbS structures, as illustrated in Figure 11.13 (right). Synchrotron x-ray studies of AA monolayers in the solid state showed that they contain fully extended molecules with a planar zigzag conformation. The AA molecules are oriented approximately normal to the liquid surface in a hexagonal closely packed array with lattice parameter $a = 4.85$ Å. The epitaxial growth of PbS from the {111} face resulted from the geometric complementarity between the monolayer and the {111} face. The Pb-Pb and S-S interionic distances of 4.2 Å in the PbS {111} plane geometrically matched the d{111} spacing of 4.16 Å for the AA monolayer, as shown in Figure 11.13 (right).[133] The size and preferential orientation of PbS nanocrystals could be controlled by doping the AA monolayers with octadecylamine (ODA).[138] Similarly, crystallization of CdS under AA monolayers generated rod-like CdS nanorods with lengths of 50 to 300 nm and widths of 5 to 15 nm.[139] The results demonstrated that the monolayers are good crystallization templates for inducing the orientation growth of nanocrystals.

An interesting result that the monolayer of amphiphilic tricarboxyphenylporphyrin iron(III) μ-oxo dimers can produce highly patterned excavations resulting in chiral calcite crystals (Figure 11.14) was demonstrated by Lahiri et al.[136] The porphyrin presents a complex semirigid surface array of carboxylate group intermediates between protein matrices and simple molecules. The monolayers of the μ-oxo

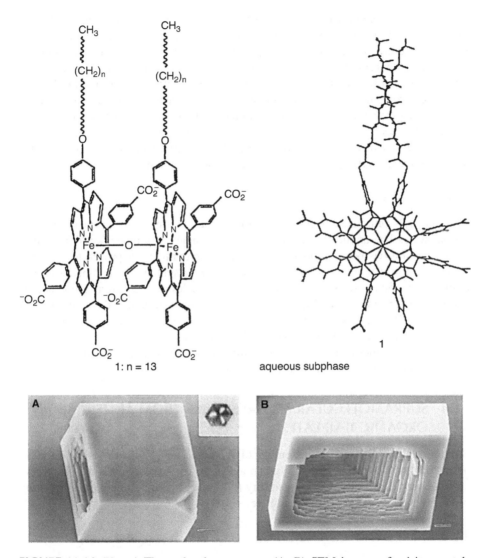

FIGURE 11.14 (Upper) The molecular structure. (A, B) SEM images of calcite crystals obtained from nucleation under **1**. (A) Image of a single calcite crystal showing the truncated corner and cavity on a distal {10.4} plane, obtained by the dipping method. Inset: An *in situ* optical micrograph of a calcite crystal nucleated under **1**, viewed from above. (B) View of the rectangular projections and a terraced gallery, respectively, on adjoining {10.4} planes distal to the truncated corner. Scale bar = 5 μm. (From Lahiri, J. et al., *J. Am. Chem. Soc.*, 119, 5449–5450, 1997. With permission. Copyright © 1997 American Chemical Society.)

iron(III) porphyrin dimer **1**[141] were formed at the air–water interface using a solution of **1** in 3:1 chloroform/methanol (1 mg/ml) to spread the film. Strikingly, about 20% of the three symmetry-related distal {10.4} calcite faces contained impressive rectangular cavities, and some had two or three cavities on adjoining distal {10.4} faces, within which one could see clearly the layered and terraced galleries, as depicted in Figure 11.14. Significantly, the crystals with a layered excavation on one {10.4} face had rectangular projections on the other, suggesting different views of the same internal structure. The results clearly show that these excavations produce an intrinsically chiral morphology, even though calcite has a nonenantiomorphic crystal lattice.[136] The observed enantiomorphs of the crystal with chirality could only have originated from the porphyrin dimer template 1, which must have staggered but asymmetric porphyrin planes according to molecular modeling. The molecules were found to be incorporated inside the crystal by adsorption of porphyrin from accessible surfaces, which confirmed that the highly textured laminations could be due to the anisotropic adsorption of the template 1 on specific planes of calcite crystals.

11.3.4 Vesicles

The materials templated from vesicle mineralization are spherical on the macroscopic scale. The application of phospholipid vesicles as a model system for biomineralization has been reported.[142] The earliest biomimetic approach to vesicular mineralization was reported by Ozin et al.,[96,139] who synthesized a lamellar aluminophosphate phase in a solution containing vesicles formed from tetraethylene glycol, amphiphilic alkylamines, and water. The templated material is composed of millimeter-dimension solid spheroids with impressive morphology and patterned surfaces. These micrometer-scaled surface patterns contain bowl, honeycomb, quilted shapes, mesh, and not well-defined structural features,[139] which closely resemble the natural biomineralized structures.

11.4 SUPRAMOLECULAR SELF-ASSEMBLY FROM ARTIFICIAL ORGANIC TEMPLATES

Supramolecular directed self-assembly of inorganic and inorganic–organic hybrid nanostructures has been emerging as an active area of recent research. The recent advance shows the remarkable feasibility to mimic a natural mineralization system by a designed artificial organic template.

11.4.1 Supramolecular Polymers

A supramolecular functional polymer can be directly employed as a mineralization template for the synthesis of novel inorganic nanoarchitectures.[144] For example, CdS helices were successfully templated from supramolecular nanoribbons.[145] This novel inorganic nanostructure has a coiled morphology with a pitch of 40 to 60 nm, which can be rationalized in terms of the period of the twisted organic ribbons. A triblock architecture termed *dendron rod coil* (DRC) can form hydrogen bonds in a head-to-head fashion through dendron segments and self-assemble into nanoribbons.[141] The mineralization of the helical structures has been done in both 2-ethylhexyl methacrylate (EHMA) and ethyl methacrylate (EMA). Figure 11.15 shows a typical CdS helix with a zigzag pattern and a pitch of 40 to 50 nm, which was isolated from a 1-wt% gel of the DRC in EHMA, to which a solution of cadmium nitrate in THF was added prior to exposure to hydrogen sulfide gas.[141] The polycrystalline zinc

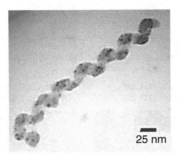

25 nm

FIGURE 11.15 TEM micrograph of a typical CdS helix with a pitch of 40 to 50 nm precipitated in gels of the DRC in EHMA. (From Sone, E.D. et al., *Angew. Chem. Int. Ed.*, 41, 1705–1709, 2002. With permission. Copyright © 2002 Wiley-VCH.)

blende CdS consists of small domains with grain sizes of about 4 to 8 nm. The results suggest that it is possible to achieve good control over the morphology of the templated product by using extremely uniform, stable, nonaggregated supramolecular objects as templates.

A peptide-amphiphile (PA) molecule was designed for mineralization of PA nanofibers and hydroxyapatite (HAp) nanofibers.[142] This amphiphile can assemble in water into cylindrical micelles because of the amphiphile's overall conical shape, resulting in the alkyl tails packed in the center of the micelle, whereas the peptide segments are exposed to the aqueous environment.[142] The PA molecules have been found to self-assemble at acidic pH but disassemble at neutral and basic pHs. After self-assembling into nanofibers, the nanofibers were cross-linked by the oxidation of the cysteine thiol groups through treatment with $0.01~M~I_2$. The cross-linked PA fibers contained intermolecular disulfide bonds, and intact fiber structures were kept. These cross-linked fibers with negatively charged surfaces are able to direct mineralization of HAp to form a composite material in which the crystallographic *c*-axes of hydroxyapatite are coaligned with the long axes of the fibers. The alignment of HAp nanofibers resembles the lowest level of hierarchical organization of bone.[143]

11.4.2 Transcription from Organic Matrix Template

The various examples of novel inorganic structures obtained by direct transcription from organic templates have been intensively reviewed recently.[14] A supramolecular self-assembly process using organogelators as templates, which resembles the organic matrix in biomineralization used by organic systems, to transcribe inorganic nanostructures has been intensively studied. Organogelators are low-molecular-weight compounds that can gelate organic fluids at low concentrations[144–149] based on hydrogen bonding, π–π stacking, or charge transfer interactions. Shinkai et al.[150–157,159] employed such a system first to mineralize inorganic structures by transcribing the chiral gelator, fibers, and other morphologies to form unique superstructures. The formation of a three-dimensional network based on fibrous aggregates in organic fluids could be responsible for the gelling phenomenon.

A helically structured silica has been successfully templated by a sol-gel transcription in chiral organogel systems.[150–156] The results show that certain cholesterol

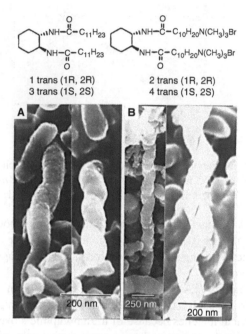

1 trans (1R, 2R)
3 trans (1S, 2S)

2 trans (1R, 2R)
4 trans (1S, 2S)

FIGURE 11.16 SEM pictures of the silica helices obtained by sol-gel transcription in left-handed 1+2 (1:1 wt%) (A) and right-handed 3+4 (1:1 wt%) (B) organogels. (From Jung, J.H. et al., *J. Am. Chem. Soc.*, 122, 5008–5009, 2000. With permission. Copyright © 2000 American Chemical Society.)

derivatives can gelate even tetraethoxysilane (TEOS), which can be used to produce silica by sol-gel polymerization.[150] Sol-gel polymerization of TEOS solutions produces silica with a novel hollow fiber structure due to the template effect of the organogel fibers.[150–156] Both right- and left-handed helical silica structures can be created by transcription of right- and left-handed structures in chiral diaminocyclo-hexane-based organogel systems, as shown in Figure 11.16.[155]

This approach can also be extended for producing other metal oxide nanofibers or nanotubes with chiral structures.[157–159] TiO_2 fibers were fabricated by using an amphiphilic compound *trans*-(1R,2R)-1,2-cyclohexanedi(11-aminocarbonylunde-cylpyridinium) hexafluorophosphate as a template.[157] The amphiphilic compound containing cationic charge moieties is expected to electrostatically interact with anionic titania species under basic conditions in the sol-gel polymerization process of $Ti[OCH(CH_3)_2]_4$. Transition metal (Ti, Ta, V) oxide fibers with chiral, helical, and nanotubular structures can be prepared by the sol-gel polymerization of metal alkoxides using *trans*-(1R,2R)- or *trans*-(1S,2S)-1,2-di(undecylcarbonylamino)cyclohex-ane as structure-directing agents.[158]

A recent report shows that well-defined TiO_2 helical ribbons and nanotubes can be prepared through sol-gel polymerization of titanium tetraisopropoxide [$Ti(O_iPr)_4$] in gels of a neutral dibenzo-30-crown-10-appended cholesterol gelator.[159] The transcription effect can be directly verified after removal of 1 by calcination (Figure 11.17A). The TiO_2 structures obtained from the gel of 1 in 1-butanol clearly show

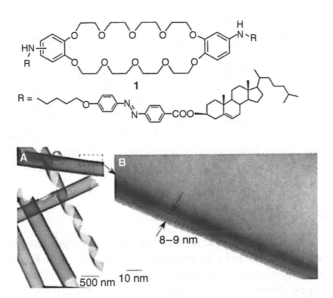

FIGURE 11.17 The chemical structure and the TEM images of the TiO$_2$ helical ribbon and the double-layered nanotube (after calcination) obtained through transcription of the 1-butanol gel of 1: low magnification (A) and high magnification (B). (From Jung, J.H. et al., *Chem. Mater.*, 14, 1445–1447, 2002. With permission. Copyright © 2002 American Chemical Society.)

an interesting inner hollow structure with a diameter identical to that of the original organogel template. Two TiO$_2$ layers of the TiO$_2$ tubules with 8- to 9-nm spaces were proved by a high-magnification TEM image (Figure 11.17B), indicating that Ti(O$_i$Pr)$_4$ molecules (or oligomeric TiO$_2$ particles) were adsorbed onto both sides of the 8- to 9-nm-thick organogel layer, which existed as a bilayer between these two developing TiO$_2$ layers.

These chiral helical tubes could have various applications in electronics, optics, and photocatalysts.

11.5 POLYMER-CONTROLLED CRYSTALLIZATION AND MORPHOSYNTHESIS

Polymers, as a class of surfactants, stabilizers, functional additives, or a soft template, have been found to play key roles in the bioinspired approaches for mimicking the biomineralization process. These surfactants are not always surface active in the classical sense (i.e., hydrophilic–hydrophobic, being attracted by the air–water or oil–water interface) but can become selectively attracted by mineral or metallic surfaces, only in aqueous solution.[95,160] Despite their broad application for templating purposes — especially with sol-gel-derived materials — we will not consider hydrophilic–hydrophobic block copolymers here, because they can be used in analogy to the low-molar-mass surfactants (see Section 11.3), with the difference of their lower critical micelle concentration (CMC) and longer chain length. For a review of the use of block copolymer templates, see references 161 and 162.

The principle of controlled crystallization and morphosynthesis in the presence of a polymer is totally different from the transcriptive template effect provided by the predesigned artificial polymer matrix or templates (see Section 11.4). The structure setup and evolution certainly do not rely on transcription or a straightforward template effect, but they do rely on a synergistic effect of the mutual interactions between functionalities of the polymers and inorganic species, and the subsequent reconstruction in a programmed or coded way.[2,8,95] In this section, we will discuss the recent advance in the area of controlled crystallization and morphosynthesis of various inorganic minerals by use of biopolymers and nonclassical block copolymers — so-called double hydrophilic block copolymers, low-molecular-weight polyelectrolytes, and rigid polymeric templates.

11.5.1 BIOPOLYMERS

Biopolymers are a natural soluble additive choice for morphogenesis of complex superstructures, which exist in living organisms due to their role in biomineralization, although *in vitro* experiments so far have commonly failed to reproduce the complex biomineral shapes. There have been intensive studies on understanding how nature created many fascinating complex structures in living organisms through biomineralization processes. However, the highly synergistic character of these processes still leaves many unknown variables in its description. In the following part, commonly widely existing minerals such as $CaCO_3$, biosilica, and hydroxyapatite have been chosen as examples to illustrate how biopolymers can control the shape and mediate the mineralization of these minerals.

Biopolymers, which are not arranged in an oriented fashion, can influence the $CaCO_3$ morphology in an unpredictable manner such as cationic, anionic, and nonionic dextran[163] or show a systematic influence on the $CaCO_3$ morphology such as soluble collagen.[164] With increasing collagen concentration, a transition of well-faceted rhombohedral calcite crystals via multiple-layer crystals to spherulitic calcite aggregates has been observed, which was explained by a model of preferential collagen adsorption.[165] A more subtle control of the $CaCO_3$ morphology and modification was found for soluble mollusk shell proteins extracted from nacre forming aragonite/calcite if extracted from the prism in coincidence with earlier reports.[165] The morphologies were found to be determined by the soluble proteins rather than the insoluble matrix on which they were nucleated, suggesting the role of the latter as a specialized nucleating sheet, whereas the soluble proteins actively control the crystal growth process determining the crystal morphology. Similar findings, but with the extension to ACC, were reported for proteins extracted from an ascidian skeleton where amorphous and crystalline $CaCO_3$ coexist in defined domains that are separated by an organic sheath.[167] The stabilization of spherical ACC, where the inner cores contain much more calcium, was also possible with extracted soluble macromolecules from coralline algae and, in this case, most remarkably also in the dried state.[53]

As one kind of amazing biosilica-producing organisms, diatoms exhibit complicated and highly organized superstructures existing in living organisms. Amorphous silica in diatom cell walls is intimately relevant for organic species, which have been hypothesized to act as regulating molecules in biosilicification. Recently, Kröger et

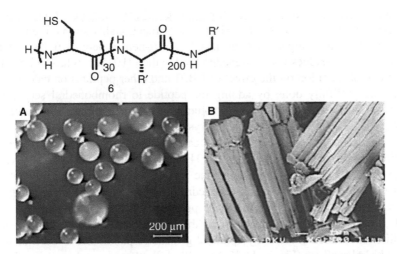

FIGURE 11.18 Different-ordered silica shapes obtained using block copolypeptide poly($_L$-cysteine$_{30}$-b-$_L$-lysine$_{200}$). (A) Optical micrograph of silica spheres obtained from synthesis under nitrogen. Scale bar = 200 μm. (B) Scanning electron micrograph of packed silica columns obtained from synthesis under air. Scale bar = 1 μm. (From Cha, J.N. et al., *Nature*, 403, 289–292, 2000. With permission. Copyright © 2000 Nature Publishing Group.)

al.[167] reported that the complex morphology of the nanopatterned silica diatom cell walls is related to species-specific sets of polycationic peptides, so-called silaffins, which have been isolated from diatom cell walls. Silaffins contain covalently modified lysine–lysine, which bear polyamines consisting of *N*-methylpropylamine units and ε-*N*,*N*-dimethyllysine groups.[167] Silica nanospheres and their networks can form *in vitro* within seconds when silaffins are added to a solution of silicic acid. The morphologies of precipitated silica can be controlled by changing the chain lengths of the polyamines as well as by a synergistic action of long-chain polyamines and silaffins.[168] The results imply that similar mixtures are used *in vivo* by the diatoms to create their intricate mineralized cell walls. More recently, the same group has developed a method for the gentle extraction of silaffins in their native state, which contain additional modifications, besides polyamine moieties, in order to avoid the use of harsh anhydrous hydrofluoride (HF) treatment for dissolving biosilica during extraction.[169] The phosphorylation of the serine residue is essential for biological activity, and a plastic siliaffin–silica phase was detected through time-resolved analysis of silica morphogenesis *in vitro*, which could represent building materials for diatom biosilica.[169] Cha et al.[170] have recently shown another example for biosilification. They isolated the proteins, so-called silicateins that form the insoluble, organic filaments found inside spicules, from a marine sponge.[171] Similar to the silaffins, silicateins can also catalyze the formation of silica at ambient conditions, but from TEOS rather than silicic acid. The oxidation of the cysteine sulfydryl groups, which is known to affect the assembly of the block copolypeptide, allows production of hard silica spheres and well-defined columns of amorphous silica (Figure 11.18). These results emphasize the importance of the self-assembled block copolypeptide architecture in the formation of complex silica shapes.

Laursen and DeOliveira have shown an excellent example of using protein secondary structures to control the orientation of functionality so that the functional groups can bind to a targeted crystal face.[172] An α-helical peptide (CBP1) with an array of aspartyl residues was designed to bind to the $\{1\bar{1}0\}$ prism faces of calcite.[172] The careful observation on the effect of CBP1 and other peptides on calcite crystal growth was skillfully done by adding the peptide to rhombohedral seed crystals growing from a saturated $Ca(HCO_3)_2$ solution at Ca^{2+}/peptide ratios of about 20/1 to 30/1. When CBP1 was added to seed crystals, as shown in Figure 11.19A, at 3°C, and growth was allowed to continue, the calcite crystals elongated along the [001] direction (c-axis) with rhombohedral $\{104\}$ caps (Figure 11.19B). After washing the crystals with water and replacing the mother solution with freshly saturated $Ca(HCO_3)_2$ solution, a regular rhombohedron formed by a sort of repair process occurred with subsequent growth on the putative prism surfaces (Figure 11.19C). CBP1 is only about 40% helical when the temperature is 25°C, leading to the formation of studded crystals by epitaxial growth perpendicular to each of the six rhomobhedral surfaces (Figure 11.19D and E). After these crystals were washed and regrown in fresh $Ca(HCO_3)_2$ solution, repair of the nonrhombohedral surfaces was again observed. In each study, a new rhombohedron was formed and, thus, six regular rhombohedra overgrew the original seed (Figure 11.19F).

A combinatorial approach to select peptides from a 15-mer peptide library as additives for $CaCO_3$ crystallization yielded active patterns containing acidic Asp and Glu, as well as hydroxyl-functionalized Ser and Thr, which can be phosphorylated by casein kinase C/II, indicating the importance of acidic groups in coincidence with biomineralization proteins like dentin phosphoryn or sialoprotein.[173] Such combinatorial approaches have a high potential to identify active amino acid patterns as well as to supply specific additives to tailor $CaCO_3$ crystallization, which in the above case yielded vaterite hollow spheres.

FIGURE 11.19 (opposite) Left: The footprint of two α-helical peptide (CBP1) molecules binding to the $\{1\bar{1}0\}$ prism faces of calcite. The filled circles are Ca^{2+} ions, and open circles are CO_3^{2-} ions. Large circles are ions in the plane of the surface, and small circles are 1.28 Å behind this plane. The hexagons indicate that peptide carboxylate ions occupy CO_3^{2-} sites on the corrugated surface. Right: SEM micrographs showing the effect of CBP1 on the growth of calcite crystals. (A) Calcite seed crystals showing typical rhombohedral morphology. (B) Elongated calcite crystals formed from seed crystals at 3°C in saturated $Ca(HCO_3)_2$ containing about 0.2 mM CBP1, showing expression of new surfaces corresponding to the prism faces and rhombohedral $\{104\}$ caps. Growth is along the c-axis, which extends through the vertices of the end caps. (C) Repair and reexpression of rhombohedral surfaces when crystals from (B) are allowed to grow in saturated $Ca(HCO_3)_2$ after removal of CBP1 solution. (D, E) Respective earlier and later stages of growth of calcite crystals from rhombohedral seed crystals at 25°C in saturated $Ca(HCO_3)_2$ containing about 0.2 mM CBP1, showing epitaxial growth from the rhombohedral surfaces. (E) inset: Fluorescence micrograph of epitaxial studded calcite crystals grown under similar conditions in the presence of fluorescein-labeled CBP1; the dark spot in the center corresponds to a (104) face, indicating that the peptide is binding to the newly expressed surfaces and not $\{104\}$. (F) Repair and reexpression of rhombohedral surfaces when crystals from (E) are allowed to grow in saturated $Ca(HCO_3)_2$ after removal of CBP1. (From DeOliveira, D.B. and Laursen, R.A., *J. Am. Chem. Soc.*, 119, 10627–10631, 1997. With permission. Copyright © 1997 American Chemical Society.)

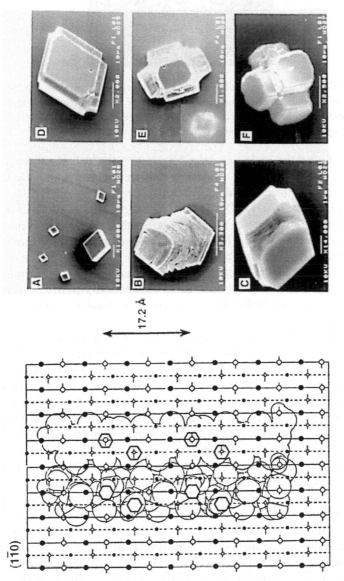

FIGURE 11.19

FIGURE 11.20 Helical $CaCO_3$ structures produced by the PILP process. Occasionally, the helices are partially hollow. The hollow helix is fractured by micromanipulation. Scale bar = 10 μm. (From Gower, L.A. and Tirrell, D.A., *J. Cryst. Growth*, 191, 153–160, 1998. With permission. Copyright © 1998 Elsevier Sciences.)

The use of other low-molecular-mass polyelectrolytes, such as polyacrylic acid or polyaspartic acid crystal growth modifiers or structure directing agents, can lead to the formation of unusual hierarchical superstructures such as helical $CaCO_3$,[174] hollow octacalcium phosphate $(Ca_8H_2(PO_4)_6 \cdot H_2O)$,[175] $BaSO_4$,[176] and $BaCrO_4$ fiber bundles, or superstructures with complex repetitive patterns,[177] although it must be pointed out that the experimental window for the formation of these superstructures is narrow.

Unusual complex structures of calcite helices or hollow helices can be obtained in the presence of a chiral poly(aspartate), which can produce helical protrusions or occasionally hollow helices, as observed by Gower and Tirrell[174] (Figure 11.20), showing that it is not essential that the polypeptide be chiral to produce these helical structures.

Gower et al.[178] proposed a so-called polymer-induced liquid precursor (PILP) process to illustrate the possible mechanism for the formation of the complex morphologies. In this mechanism, the addition of small polymer amounts (μg ml^{-1} range) to the crystallizing solution leads to a liquid–liquid phase separation to precursor droplets, which can adapt complex shapes before they crystallize. As the resulting crystalline products retain the shape of the precursor, nonequilibrium morphologies can be generated. Such liquid precursor droplets were independently observed with anionic dextran sulfate as an additive.[179] The role of the polymer appears to be the sequestering and concentration of the calcium ions while simultaneously delaying crystal nucleation and growth to form a metastable solution. The spherulitic vaterite aggregates with helical extensions with spiral pits were suggested to grow inward from the deposited $CaCO_3$ films caused by the polymer, which could act as the membranous substrates for the further growth of $CaCO_3$ crystals. The observed interesting phenomenon is quite similar to that which occurs in some biogenic minerals, such as the otoliths of fish. Here, the spherulitic growth is induced by nucleation sites distributed along a surface, and the aragonite or vaterite crystals grow with their *c*-axes approximately perpendicular to the surface.[180]

The amazing helical shape of the vaterite crystals was suggested to be the result of elasticity effects of a gelatinous precursor membrane or of a disclination effect

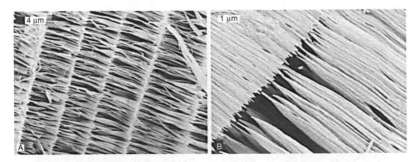

FIGURE 11.21 Complex forms of $BaSO_4$ bundles and superstructures produced in the presence of 0.11 mM sodium polyacrylate (Mn = 5100), at room temperature, [$BaSO_4$] = 2 mM, pH = 5.3, 4 days. (A) The detailed superstructures with repetitive patterns. (B) A zoomed SEM image of the well-aligned bundles. (From Yu, S.H. et al., *Nano Lett.*, 3, 379–382, 2003. With permission. Copyright © 2003 American Chemical Society.)

in the partially crystalline membrane. However, what seems to be clear is that the PILP process enables production of crystalline nonequilibrium morphologies resembling solidified melts or liquid crystals as they form from a fluid-like precursor state. As a result of the shape flexibility of the liquid precursors, the PILP process could be successfully used to produce fairly monodisperse $CaCO_3$-coated oil core-shell particles.[181] $CaCO_3$ mineralization of collagen via a PILP process resulted in fibrous aggregates growing from isotropic gelatinous globules, and the fibers became single crystalline with time, so that it was suggested that PILP processes may play a role in biomineralization.[182]

Recently, the crystallization of octacalcium phosphate ($Ca_8H_2(PO_4)_6$·5 H_2O) (OCP) from aqueous solutions containing polyaspartate (M_w = 11,000 g mol^{-1}) or polyacrylate (M_w ~ 2100 g mol^{-1}) at 60°C and pH 5 resulted in the direct assembly of a complex inorganic-polymer spherical-shell architecture in the absence of an external template.[175] The hollow microstructures consist of a thin porous membrane of oriented OCP crystals that are highly interconnected, which could be similar to that observed by Gower and Odom.[178] The hierarchical growth processes involve the radial growth of dense, multilayered spheroids, secondary overgrowth of a porous thin-shell precursor, and anisotropic dissolution of the spheroidal cores, producing this remarkable hollow structure.[175]

In addition to the above-reported helical or hollow spherical morphology, elegant nanofiber bundles and their superstructures of inorganic crystals could also be generated. Polyacrylic acid sodium salt (M_w = 5100) can serve as a very simple structure-directing agent for the room-temperature, large-scale synthesis of highly ordered cone-like crystals[176] or very long, extended nanofibers made of $BaCrO_4$ or $BaSO_4$ with hierarchical and repetitive growth patterns[177] (Figure 11.21), where temperature and concentration variation allows control of the finer details of the architecture, namely, length, axial ratio, opening angle, and mutual packing.[177] The formation of interesting hierarchical and repetitive superstructures by simple polyelectrolyte additives is worth further exploration for other mineral systems.

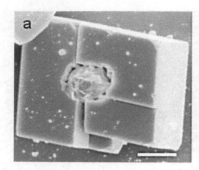
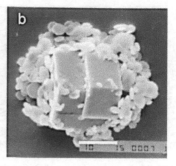

FIGURE 11.22 Left: Scanning electron microscopy of the $CaCO_3$ crystallization assay with octadecylamine-doped poly(propylene imine) dendrimers as additive after 14 days. Right: $CaCO_3$ crystallization assay with a CTAB-doped dendrimer after 4 days. (From Donners, J.J.J.M. et al., *Chem. Commun.*, 1937–1938, 2000. With permission. Copyright © 2000 Royal Society of Chemistry.)

11.5.2 DENDRIMERS

Dendrimers were recently discovered as active additives for the controlled $CaCO_3$ precipitation. Anionic starburst dendrimers were found to stabilize spherical vaterite particles up to a week, with controllable particle size in the range of 2.3 to 5.5 nm, via the dendrimer generation number.[183] Besides the macromolecules extracted from biominerals, poly(propylene imine) dendrimers interacting with octadecylamine are the only known polymeric species, which has been reported to stabilize kinetically formed ACC for periods exceeding 14 days in water.[184] The spherical ACC particles partly transformed into rhombohedral calcite over the course of 4 days, and it is remarkable that about 80% of the rhombohedra enveloped an ACC sphere, which serves as a material depot (Figure 11.22A). ACC formation was explained by the failure of the rigid dendrimers with low charge density to match a specific crystal face, followed by partial ACC dissolution and recrystallization to calcite around the spherical ACC particles (Figure11.22A).

The surfactant associated with the dendrimer was found to be of significant influence on the crystal morphology, where CTAB led to calcite rhombohedra covered with disk-shaped vaterite crystals, which subsequently dissolved (Figure 11.22B), and SDS led to polydisperse spherical aggregates mainly of organic material, with a minor amount of calcite.[185] As the applied dendrimer allows the interaction with a large amount of surfactants, this system appears extremely versatile with respect to chemical functionality and shape of the dendrimer aggregates and, thus, $CaCO_3$ morphology.

11.5.3 POLYAMIDE

A series of novel carboxlate-containing polyamides were synthesized from 2,6-diaminobenzoic acid and dimethylmalonyl dichloride, isophthalyl dichloride, or fumaryl dichloride.[186,187] Those polymers have seven and eight carboxylate units in a polymer chain, and their Ca(II) complexes possess NH···O hydrogen bonds, which prevent the metal–carboxlate bond from hydrolysis due to the shift of the pKa of the conjugated carboxylic acid.[188] The polymer was identified to be incorporated into $CaCO_3$ crystals crystallized in the presence of this kind of polymer by use of ^{13}C cross-polarization/magic-angle-spinning CP/MAS solid-state nuclear magnetic resonance (NMR). $CaCO_3$ crystals were grown on these polyamides or sodium salt of the polyamides.[186] The polyamide $[\{NHC_6H_3(COO^-)NHCO\text{-}m\text{-}C_6H_4CO\}]_n$ has carboxylate parallel-oriented moieties with adjacent carboxylate distances almost twice the distance of calcium ions, showing that this parallel fashion is favorable for the interaction with calcium ions. However, the polyamide $[\{NHC_6H_3(COO)NHCOCH=CHCO\}]_n$ is difficult to interact with $CaCO_3$ crystals due to the *trans*-geometry of the fumaryl spacer.[186]

11.5.4 DOUBLE HYDROPHILIC BLOCK COPOLYMERS

11.5.4.1 Selective Adsorption and Crystal Growth

The crystal shape is usually, in fact, the outside embodiment of the unit cell structure. The diverse crystal morphologies of the same compound are due to the difference of the crystal faces in surface energy and its external growth environment.[41] Generally speaking, the growth rate of a crystal face is usually related to its surface energy if the same growth mechanism acts on each face. The fast-growing faces usually have high surface energies, and finally they will disappear in the final morphology. It is well known that the adsorption of ions or low-molar-mass additives onto the specific crystal surfaces can influence the crystal morphologies[28] (see also Section 11.2.1), as shown in Figure 11.23A (top). However, if nanoparticles shall be used for higher-order assembly, it is necessary to at least temporarily stabilize the primary nanoparticle building blocks for further structure development to avoid the uncontrolled aggregation. This is only possible to a limited extent by electrostatic stabilization or low-molar-mass additives. The pure electrostatic stabilization is often not sufficient for the application of crystalline nanoparticles as building blocks for superstructures, as the Debye length is usually in the range of the attractive Van der Waals forces.[189] On the other hand, the stabilization by adsorption of a polyelectrolyte is very effective for electrosteric stabilization, as shown for iron oxyhydroxide, which was immediately stabilized by cellulose-sulfate and κ-carrageenan to avoid the precipitation of an amorphous solid.[190] However, the limit for selective adsorption can be exceeded, as the concentration of adsorbing groups is high in a polyelectrolyte of sufficient chain length, so that the nonselective adsorption occurs at all faces, as shown in Figure 11.23B (top), leading to nondirected crystal growth.[191] Therefore, the ideal choice is to combine the requirement for stabilization and the selective adsorption functions together within one molecule, a block copolymer with a polyelectrolyte block that is short enough for selective adsorption but long enough for sufficient interaction with the crystal surface, and a stabilizing block for steric stabilization, as shown in Figure 11.23C (top).

Block copolymers usually consist of a hydrophilic and a hydrophobic block so that they can be used as polymeric surfactants, which show a similar behavior to low-molecular-weight surfactants. Double hydrophilic block copolymers (DHBCs)

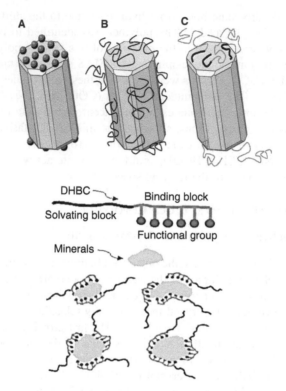

FIGURE 11.23 Top: (A) Specific adsorption of ions of low-molar-mass additives onto a crystal. (B) Unspecific adsorption of polyelectrolytes. (C) Specfic adsorption of block copolymers with a short polyelectrolyte block. Bottom: Illustration scheme of the concept of DHBCs.

are a new class of polymers, which consist of two water-soluble blocks of different chemical natures. These polymers are typically rather small, having block lengths between 10^3 and 10^4 g mol^{-1}. The solvating block shows good solubility in water. The binding block contains variable chemical patterns, which show strong affinity to minerals and have a strong interaction with inorganic crystals, as illustrated in Figure 11.23 (bottom). In this way, not only is the required lowering of the interface energy by adsorption obtained, but it is also possible to create a high specificity to block distinct crystalline surfaces by adjustment of the functional polymer pattern.[95]

Recently, it has been shown that DHBCs[192–194] can exert a strong influence on the external morphology or crystalline structure of inorganic particles, such as calcium carbonate,[195–198] calcium phosphate,[199] barium sulfate,[200–202] barium chromate,[203,204] zinc oxide,[205,206] and cadmium tungstate.[207] The results show that not only is the shape control of primary nanocrystals achieved, but also complicated superstructures can be produced through the mesoscale self-assembly of the programmed primary units.

Different kinds of DHBCs with different functional patterns were designed and used as crystal modifiers.[191,192] The carboxylic acid groups of a commercial block copolymer poly(ethylene glycol)-*block*-poly(methacrylic acid) (PEG-*b*-PMAA)

(PEG = 3000 g mol^{-1}, 68 monomer units; PMAA = 700 g mol^{-1}, 6 monomer units; Th. Goldschmidt AG, Essen, Germany) were partially phosphonated (with a phosphonation degree of 1 to 21%) to give a copolymer with carboxyl and phosphonated groups, PEG-b-PMAA-PO$_3$H$_2$, according to reference 195. A block copolymer containing a poly(EDTA)-like carboxy-functionalized block, poly(ethylene glycol)-$block$-poly(ethylene imine)-poly(acetic acid), PEG-b-PEI-(CH$_2$CO$_2$H)$_n$ (PEG-b-PEIPA) (PEG = 5000 g mol^{-1}; PEIPA = 1800 g mol^{-1}) was synthesized as described elsewhere.[182]

The copolymers, which are based on PEG-b-PEI with various functional acidic groups, –COOH, –PO$_3$H$_2$, –SO$_3$H, and –SH, were synthesized by further functionalization of the PEI block by polymer analogous reactions in a modular approach.[182] This approach is especially useful as it allows the varying of the chemical functionalization pattern on identical polymer blocks. In addition, the partially phosphorylated poly(hydroxyethyl ethylene) block copolymers with PEG (PEG-b-PHEE– PO$_4$H$_2$(30%)) were synthesized.[208]

11.5.4.2 Morphosynthesis and Hierarchical Structures of Inorganic Crystals

Systematic morphosynthesis of different inorganic minerals with controlled morphology and novel superstructures by using DHBCs with varying patterns of functional groups will be described in this section. In addition, the influence of physicochemical parameters such as the crystallization sites, temperature, concentration of reactants, and copolymers, as well as the introduction of foreign cationic colloidal structures on the morphology, crystallization, and superstructure, will be discussed, demonstrating that the ability of the copolymer to interact with the inorganic crystals and the rather flexible morphosynthesis of inorganic crystals can be fine-tuned.

11.5.4.2.1 Calcium Phosphates

A biomimetic approach to the precipitation of calcium phosphates has been reported by using double hydrophilic block copolymers poly(ethylene oxide)-b-alkylated poly(methacrylic acid) (PEO-b-PMAA-C$_{12}$)[209] as well as PEO-b-PMAA as additives.[199] The precipitation of calcium phosphates at pH 6.3 without adding additives or pure PEO produces similar-looking plate-like crystals, underscoring that the pure PEO part is insufficient to exert influence on crystallization. In contrast, the presence of unmodified PEO-b-PMAA will lead to the formation of spherical colloids.[199] The partial alkylation of unmodified PEO-b-PMAA leads to loose aggregation of the polymers in water solution, and the aggregates are used as mineralization sites. By controlling the pH of the Ca^{2+}-loaded polymer solution and, thus, the crystal modification (brushite and hydroxyapatite), delicate mesoskeletons of interconnected calcium phosphate nanofibers with star-like, neuron-like, and more complex nested forms can be obtained, as shown in Figure 11.24.[199] The composition of hydroxyapatite and calcium hydrogen phosphate dihydrate (brushite) depends on the initial pH value in solution. These structures result from a feedback loop between selective polymer adsorption on all faces parallel to the hydroxyapatite c-axis and the deformation of the polymer aggregate by the polymer adsorption onto the growing filaments.

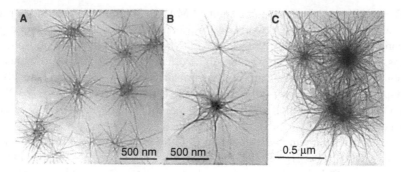

FIGURE 11.24 TEM images of calcium phosphate block copolymer nested colloids. (A) Star-like form at early stage at pH 3.5. (B) Later stage showing a complex central core and very long and thin (2 to 3 nm) filaments. (C) Neuron-like tangles produced at pH 5. (From Antonietti, M. et al., *Chem. Eur. J.*, 4, 2493–2500, 1998. With permission. Copyright © 1998 Wiley-VCH.)

11.5.4.2.2 BaCrO₄ and BaSO₄

In the absence of polymeric additives, the precipitation process of $BaCrO_4$ occurs very quickly and the obtained products are found to be well-crystallized hashemite crystals. The dendritic X-shaped crystals with an average length of about 10 μm were obtained in the absence of additives. However, nanofiber bundles were obtained in the presence of the partially phosphonated derivative PEG-*b*-PMAA-PO₃H₂ (21% phosphonation degree), as shown in Figure 11.25A. The particles obtained in the presence of PEG-*b*-PEIPA were found to be egg-shaped, with an overall size of 100 to 180 nm. Egg-shaped particles with a size of 1 μm were also obtained with PEG-*b*-PMAA as a modifier (Figure 11.23B). The inhibition effect of PEG-*b*-PMAA for hashemite crystallization was found to be comparable to that of PEG-*b*-PEIPA, which coincides with the results reported for barite crystallization.[176] As expected, a more pronounced partial hydrophobic modification of the functional polymer block results in mutual aggregation of the DHBCs and, therefore, an altered modification pattern, resulting in altered mineralized superstructures. When the PEI backbone of PEG-*b*-PEI-(CH₂-CH₂-PO₃H₂) was additionally modified with lauroyl-COC₁₁H₂₃ moieties (PEG-*b*-PEI(CH₂CH₂-PO₃H₂)-COC₁₁H₂₃), the hybrid superstructures were composed of smaller aggregated spheres clustered together, and the rod-like structure obtained with the nonhydrophobically modified polymers was lost (Figure 11.25C). Uniform, well-defined lancet-like particles with a length of 20 μm and a width of 5 μm were obtained in the presence of PEG-PEI(CSHNCH₃)ₙ-COC₁₁H₂₃ (Figure 11.25D). Apparently, in this case, the hydrophobic modification does not lead to clustered aggregated spheres, as found for the phosphonated polymer analog, suggesting a site-selective block copolymer adsorption. The aspect ratio of *a/b* is as large as 5 to 5.4, which is much higher than that previously observed in the case of barite,[210] an observation that supports the site-selective block copolymer adsorption. When another phosphorus-containing copolymer, the partially phosphorylated poly(hydroxyethyl ethylene) block copolymer (PEG-*b*-PHEE-OPO₃H₂ (30%)),[208]

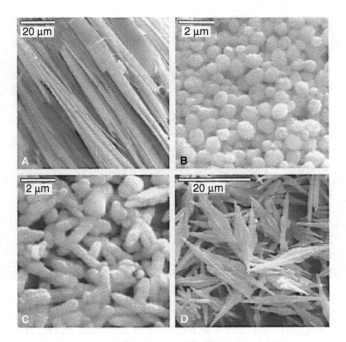

FIGURE 11.25 The controlled morphologies of $BaCrO_4$ by DHBCs. The polymer concentration was kept at 1 g l^{-1} ($[BaCrO_4]$ = 2 mM), pH 5. (A) PEG-b-PMAA-PO_3H_2 (21% phosphonation degree). (B) PEG-b-PMAA. (C) PEG-b-PEI-$(CH_2$-CH_2-$PO_3H_2)_n$. (D) PEG-b-PEI(CSHN-$CH_3)_n$-$C_{11}H_{23}$. (From Yu, S.H. et al., *Chem. Eur. J.*, 8, 2937–2345, 2002. With permission. Copyright © 2002 Wiley-VCH.)

was used, the morphology changed to smaller, trapezoidal aggregates with no clear expression of faces. It is delineated that this polymer is not purely hydrophilic anymore but shows amphiphilic character and surface activity.[208]

All DHBCs based on a bare PEI binding site give comparatively little control over the oriented growth of the crystals under the applied acidic conditions at pH 5, underscoring that the growing crystals rather prefer to interact with the negatively charged $-PO_3H_2$ and $-COOH$ groups instead of the positively charged PEI block.

More complex morphologies of hashemite can be formed in the presence of PEG-b-PMAA-PO_3H_2. This polymer is a strong growth inhibitor and was found to effectively produce fibers for the isomorphic $BaSO_4$.[176]

Figure 11.26A and B show SEM images of fibrous superstructures with sharp edges composed of densely packed, highly ordered parallel nanofibers of $BaCrO_4$. The TEM micrograph with higher resolution in Figure 11.26C clearly shows the self-organized nature of the superstructure. Whereas the majority of the fibers appear to be aligned in a parallel fashion, gaps between the single fibers can form but are also closed again. An electronic diffraction pattern taken from such an oriented planar bundle, as shown in Figure 11.26D, confirmed that the whole structure scatters as a well-crystallized single crystal, where scattering is along the [001] direction and the fibers are elongated along [210], suggesting a two-dimensional order of the nanofibers.

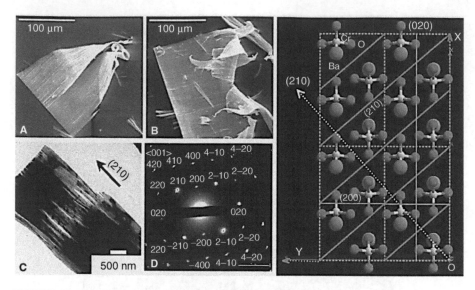

FIGURE 11.26 Left: SEM and TEM images of highly ordered BaCrO₄ nanofiber bundles obtained in the presence of PEG-b-PMAA-PO₃H₂ (21% phosphonation degree) (1 g l⁻¹) ([BaCrO₄] = 2 mM), pH 5. (A) Two fiber bundles with a cone-like shape and length of about 150 μm. (B) Very thin fiber bundles. (C) TEM image of the thin part of the fiber bundles — inserted electronic diffraction pattern taken along the <001> zone — showing that the fiber bundles are well-crystallized single crystals and elongated along [210]. Right: Plot of BaCrO₄ crystal structures viewed perpendicular to the [210] axis (indicated by arrow). (From Yu, S.H. et al., *Chem. Eur. J.*, 8, 2937–2345, 2002. With permission. Copyright © 2002 Wiley-VCH.)

The atomic surface structure modeling data for the surface cleavage of the hashemite crystal shows that the faces (1$\bar{2}$2), (1$\bar{2}$1), (1$\bar{2}$0), and ($\bar{1}$20), which are parallel to the [210] axis, contain slightly elevated barium ions, indicating that the negatively charged –PO₃H₂, –COOH groups of PEG-b-PMAA-PO₃H₂ can preferentially adsorb on these faces by electrostatic attraction and block these faces from further growth. In contrast, the surface cleavage of the (210) face shows no attackable barium ions on the surface (Figure 11.26, right), suggesting that the negatively charged functional polymer group does not favorably adsorb on this face, leading to a fast growth rate. This is favorable for the orientated growth along the [210] direction.

All structures always grow from a single starting point, implying that the fibers grow against the glass wall or other substrates, such as TEM grids, which obviously provide the necessary heterogeneous nucleation sites. The growth front of the fiber bundles is always very smooth, suggesting the homogeneous joint growth of all single nanofilaments, with the ability to cure occurring defects in line with the earlier findings for BaSO₄.[176] The opening angle of the cones is always rather similar, which seems to depend on temperature, degree of phosphonation of the polymer, and polymer concentration.[203] The control experiments show that the higher the temperature, the more linear the structures become. The superstructure develops more clearly, and to a much larger size, when the mineralization temperature is lowered

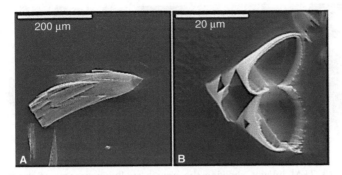

FIGURE 11.27 (A) Typical cone-like superstructure containing densely packed $BaCrO_4$ nanofiber bundles in the presence of PEG-*b*-PMAA-PO$_3$H$_2$ (21% phosphonation degree) (1 g l^{-1}) ([$BaCrO_4$] = 2 m*M*, pH 5), 2 days, 4°C. (B) Multi-funnel-like superstructures with remarkable self-similarity in the presence of PEG-*b*-PMAA-PO$_3$H$_2$ (1% phosphonation degree) (1 g l^{-1}) ([$BaCrO_4$] = 2 m*M*, pH 5). (From Yu, S.H. et al., *Chem. Eur. J.*, 8, 2937–2345, 2002. With permission. Copyright © 2002 Wiley-VCH.)

to 4°C, as shown in Figure 11.27A. A lower phosphonation degree (~1%) of a PEG-*b*-PMAA is already powerful enough to induce the formation of the fiber bundles and the superstructures, as shown in Figure 11.27B.

Interestingly, secondary cones can nucleate from either the rim or defects onto the cone, thus resulting in the tree-like structures (Figure 11.27A). The fact that a cone stops growing once a second cone has nucleated at one spot on the rim shows that the growth presumably is slowing down with time, favoring the growth of the secondary cone. The cone-like superstructures tend to grow further into a self-similar multicone tree structure, which has been observed for barite mineralized in the presence of polyacrylates, but only under very limited experimental conditions.[176] This is explained as the consequence of the influence of electrostatic dipole and multipole fields of the superstructure on approaching ions and colloidal building blocks, resulting in branching instead of further growth.[191,211-213] In this particular case, the dipole moment of a fiber with charged ends increases linearly with its length so that a certain critical length exists where it is energetically more favorable to nucleate a new fiber rather than continue the growth of the existing one. The detailed structures of the cone-like bundles has been found to be in the form of cone-in-a-cone *matrioshka* structure. The ability of the edges to attract each other and to fuse leads to the simultaneous occurrence of thin, very filigree fused cones (Figure 11.27B).

The initially formed particles are amorphous with sizes up to 20 nm, which can aggregate to larger clusters. Evidently, this state of matter is the typical starting point for all types of highly inhibited reactions. The very low solubility product of barium chromate (K_{sp} = 1.17 × 10^{-10}) shows that the superstructures do not really grow from a supersaturated ion solution but by aggregation/transformation of the primary clusters formed. This is in perfect agreement with earlier findings that in an ion solution with concentrations far above the saturation level, amorphous clusters are formed first, which then produce the crystalline nuclei at a later stage.[214] The amorphous

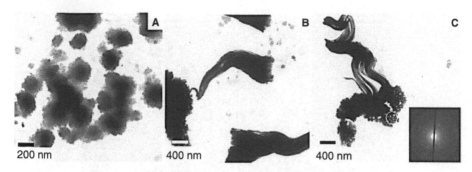

FIGURE 11.28 TEM images showing the growth evolution process of the fiber bundles, which formed on TEM grids in the presence of PEG-*b*-PMAA-PO$_3$H$_2$ (21% phosphonation degree), 1 g l^{-1}, [BaCrO$_4$] = 2 mM, in a PP tube after reaction for 4 h (A) and 24 h (B). (C) TEM images and inserted electronic diffraction patterns showing the evidence of growth of the fiber bundles from the aggregation and connection of amorphous particles, which formed on TEM grids in a PP tube after reaction for 4 days: a typical fiber bundle showing that the bundle grew starting from a start point and with many amorphous particles on the end. PEG-*b*-PMAA-PO$_3$H$_2$ (21% phosphonation degree), 1 g l^{-1}, [BaCrO$_4$] = 1 mM. (From Yu, S.H. et al., *Chem. Eur. J.*, 8, 2937–2345, 2002. With permission. Copyright © 2002 Wiley-VCH.)

aggregates formed after 4 h are depicted in Figure 11.28A. After 1 day, the first conical-like fiber bundles were found on the TEM grids, as shown in Figure 11.28B. Indeed, it is seen that the bundles start their life from a single nucleation site and grow/multiply from the beginning in the way seen at larger scales by SEM. Figure 11.28C shows another typical event with a starting point and developing and growing fiber bundles, and many amorphous particles at the end (growing point).

From the structural viewpoint, the hashemite/barite crystals have a mirror plane perpendicular to the *c*-axis, implying that a homogeneous nucleation will always result in crystals with identical charges of the opposite faces and no dipole crystals can be formed. However, a dipole crystal may be favored for a heterogeneous nucleation, as one end of the crystal is determined by the heterogeneous surface and the other by the solution/dispersion. The crystal has a dipole moment $\mu = Q \cdot l$ (Q = charge, l = length of the crystal), which increases while the crystal is growing. This implies self-limiting growth so that a new heterogeneous nucleation event on the rim should become favorable after the dipole moment has reached a critical value due to energy minimization.[203]

Based on the above experimental observation and understanding, a self-limiting growth mechanism was proposed to explain the remarkable similarity of the super-structures,[203] as shown in Scheme 11.3:

1. At the beginning, amorphous nanoparticles are formed that are stabilized by the DHBCs (stage 1).
2. Heterogeneous nucleation of fibers occurs on glass substrates and the fibers grow under control of the functional polymer, presumably by multipole field-directed aggregation of amorphous nanoparticles, as proposed

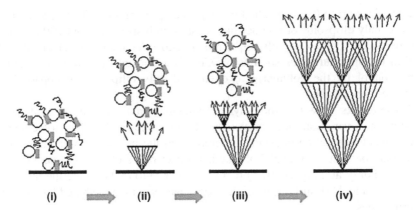

SCHEME 11.3 Proposed mechanism for the formation of a complex superstructure with self-similarity. (From Yu, S.H. et al., *Chem. Eur. J.*, 8, 2937–2345, 2002. With permission. Copyright © 2002 Wiley-VCH.)

    in reference 176 or by oriented attachment growth, which will be discussed later.

3. The growth is continuously slowed down until secondary nucleation or overgrowth becomes more probable than the continuation of the primary growth. The secondary cone will grow as the first one has done.

4. The secondary heterogeneous nucleation event taking place on the rim can repeatedly occur, depending on the mass capacity of amorphous nanoparticles in the system.

The whole suggested mechanism relies on the absence of flow, which would disturb the directed aggregation according to local dipole/multipole fields or oriented attachment. No fiber bundles or cones are obtained if the solution is stirred continuously at room temperature after mixing the reactants. Instead, only irregular and nearly spherical particles are obtained. This is again in agreement with a recently published mechanism in which the fiber formation is due to the directed self-assembly of the primary particles,[176,177] which is suppressed by stirring. The glass wall or other surface provides the necessary heterogeneous nucleation site for the birth of the fibers. In contrast, only spherical particles can be obtained when the same reaction is done in polypropylene (PP) or plastic bottles.[177]

The concentration of the reactants, ratio of cations and anions, pH, copolymer concentration, and temperature have a strong influence on the precipitation of barium chromate. When the concentration of both ions was equimolar ($[Ba^{2+}]:[CrO_4^{2-}]$ = 1:1) and was as low as 0.025 M (in presence of 1 g l^{-1} of copolymer), no precipitation occurred, even after aging for 3 weeks at room temperature, although the reference experiment without polymer showed crystallization. This means that the polymers are able to stabilize the precursor particles at this concentration level without the ability to undergo further structural transitions. When the concentration increased to 0.05 M, fiber formation occurred after 1 to 4 days. Further increase of the ion concentrations up to 0.2 M resulted in very quick precipitation. TEM shows that in this case, only ellipsoids or boulder-like crystals with sizes of about 200 to 600 nm were produced. If the $[Ba^{2+}]:[CrO_4^{2-}]$ molar ratio is changed to 1:5 or 2.5:1, the solution becomes turbid as soon as Na_2CrO_4 is added into the system. Again, no strong control by the polymer is exerted, and only ellipsoidal nanoparticles or aggregates are obtained.

These results show that the polymer essentially interacts with neutral crystals that show no overall surface enrichment of either one or the other ionic species. This mode of interaction is clearly different from the previous observation from a microemulsion system, in which the higher aspect ratios of $BaCrO_4$ nanofilaments were found only at a higher $[Ba^{2+}]:[CrO_4^{2-}]$ molar ratio (5:1), where the interaction with the anionic AOT surfactants of this particular system is mediated by the excess of barium ions.[22]

When the concentration of the polymer was lowered to 0.2 g l^{-1}, the solution quickly became turbid, and the precipitate consisted of nanoparticles of mostly spherical shape. This result suggests that the interaction capacity of the phosphonated copolymer at lower concentrations is not high enough to stabilize the fibers with their high internal surface but still shows strong interaction with all of the crystal faces. When the concentration of the polymer increased to 1 g l^{-1}, only fiber bundles were obtained, as discussed before. Upon increase of the polymer concentration to 2 g l^{-1}, the solution stayed transparent and yellow, and a very loose flocculate was obtained after 2 or 3 days, which was found to be composed of aggregated, but not oriented, and therefore more dispersed fiber bundles.

This ties in well with the idea of a strong interaction of the copolymer with the crystals, where the increased concentration leads to a higher binding and, consequently, a higher polymer content in the precipitate and makes the hybrid morphology more dispersed. These results indicate that only a limited concentration window exists for the formation of three-dimensionally highly orientated fiber bundles. Apparently, the force that is responsible for the three-dimensional packing can be overcompensated by a high polymer surface coverage and the resulting steric and possible electrostatic repulsion.

When the pH value was varied from 5.5 to 3.5 and 5.5 to 2 respectively, no precipitate was found, even after aging for 2 weeks. The suitable pH range for the formation of $BaCrO_4$ fibers was between 5 and 7. In contrast, when the pH was changed to pH 10, the precipitate was found to be composed of nearly spherical particles in the micrometer range that formed a superstructure or aggregates of nanoparticles. This suggests that a selective adsorption of the block copolymer can take place in only a limited pH range, whereas beyond the pH limits, the polymer adsorbs in a nonselective fashion.

Increasing the temperature from 4°C (cone-like fiber bundles with a large opening angle) or 25°C (more linear and extended fiber bundles) to 50°C results in faster nucleation and less specificity of the growth process. Small twinned or peanut-shaped nanoparticles were found at 50°C. Further increase of the temperature to 100°C shows a continuation of this trend. Again, only peanut-shaped aggregates were found.

This observation holds for both glass and polypropylene bottles, meaning that higher temperatures indeed allow homogeneous nucleation to prevail over the heterogeneous nucleation at lower temperatures, which leads to fibers. Here, a different species, the peanuts, where a nucleus can grow in two opposite directions, was found. Heterogeneous nucleation presumably has not stopped but has lost importance because of the acceleration of the homogeneous nucleation rate and the dominance of this stage for the crystallization process. Peanuts are formed at a much higher overall rate than fibers under this condition.

It should be pointed out that the growth of fiber hybrid superstructures is promoted by the addition and recrystallization of primary colloidal particles rather than the buildup by single ions, and this is in agreement with a number of previous literature reports on biomineralization and biomimetic mineralization, as reviewed in reference 8. The reason that the accelerated homogeneous nucleation, by applying increased temperature, resulted in peanut or dumbbell-like particles, which stopped growth, could be related to the characteristics of the primary particles. These particles carry the character of an electrostatic multipole field, which is perpetuated and amplified throughout the growth process organized by the polymer. The detailed mechanism, however, needs further elucidation, as the alternative explanation of an oriented attachment growth mechanism (see Section 11.6.1) could explain the experimental results equally well. The further modification of the glassware (negative and positive charge) or the addition of highly polar colloidal nuclei could result in better control of the heterogeneous nucleation.

Interestingly, single uniform $BaCrO_4$ nanofibers can also be synthesized by a combination of a polymer-controlled crystallization process and controlled nucleation by colloidal species locally producing a high supersaturation of both DHBCs and Ba^{2+}.[204] The addition of a minor amount of cationic colloidal structures, such as PSS/PAH (poly(styrene sulfonate, sodium salt)/polyallylamine hydrochloride) polyelectrolyte capsules, in the same reaction system can promote the independent growth of a number of fibers, thus leading to the dynamic discrimination of side nucleation and the resulting altered superstructures (Figure 11.29).

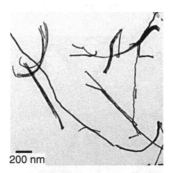

200 nm

FIGURE 11.29 TEM image of the separated and very long $BaCrO_4$ nanofibers with high aspect ratio obtained in the presence of a small amount of PSS/PAH capsules (20 µl), PEG-*b*-PMAA-PO$_3$H$_2$, 1 g l^{-1}, [$BaCrO_4$] = 2 m*M*, pH 5, 25°C. (From Yu, S.H. et al., *Adv. Mater.*, 15, 133–136, 2003. With permission. Copyright © 2003 Wiley-VCH.)

A recently reported striking example showed that flower-like $BaSO_4$ particles with a forbidden 10-fold symmetry can be produced by a multistep growth process in the presence of a sulfonated PEG-*b*-PEI.[202] For the formation of these particles, an elongated primary nanoparticle formed by selective polymer adsorption was proposed that has 10 side faces.[202] This particle serves as a seed for the heterogeneous nucleation of 10 single-crystalline petals, which are overgrown in a later stage, as shown in Figure 11.30. Thus, a structure forms that is single crystalline (the petals) but which does not violate the laws of crystal symmetry due to the primary particle with 10 side faces.

11.5.4.3 Morphosynthesis of Metal Carbonate Minerals

Calcium carbonate ($CaCO_3$) has been widely used as model system for studying the biomimetic process due its abundance in nature as well as its important industrial applications in paints, plastics, rubber, and paper.[215] Biomimetic synthesis of $CaCO_3$ crystals in the presence of organic templates or additives has been intensively investigated in recent years.[216,217]

11.5.4.3.1 Sphere-to-Rod-to-Dumbbell-to-Sphere Transformation Mechanism

Rhombohedral-shaped calcite crystals consisting of six (104) faces are easily produced under equilibrium conditions in the absence of additives, which is believed to be the macroscopic expression of the unit cell. Biomimetic crystallization of large $CaCO_3$ spherules with controlled surface structures and sizes has recently been realized by a slow gas–liquid diffusion reaction of the CO_2 released from ammonia carbonate in a closed desiccator in the presence of DHBCs.[198] Contrary to previous work, in which calcite composites were prepared by a rapid supersaturation/precipitation technique (the double-jet injection),[195–201] the gas diffusion reaction technique is much slower and closer to the speed of calcium carbonate morphogenesis in biomineralization, which can range from a few hours for eggshells to many days.[218]

The morphology evolution of $CaCO_3$ and $BaCO_3$ in the presence of block copolymers is shown in Figure 11.31. When PEG-*b*-PMAA was used as a crystal

FIGURE 11.30 SEM micrographs of $BaSO_4$ precipitated in the presence of 1 g l^{-1} PEO-*b*-PEI-SO$_3$H at pH 5. (From Cölfen, H. et al., *Cryst. Growth Des.*, 2, 191–196, 2002. With permission. Copyright © 2003 American Chemical Society.)

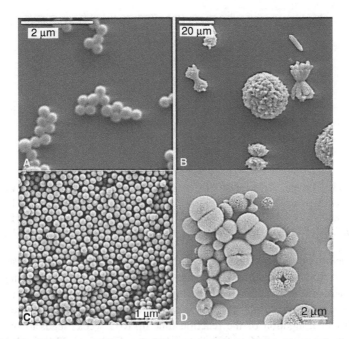

FIGURE 11.31 SEM images of the $CaCO_3$ and $BaCO_3$ particles obtained after different reaction times. (A) $CaCO_3$ nanospheres with uniform sizes (310 nm) formed on a glass substrate grown for 3 h in the presence of PEG-*b*-PMAA (1 g l^{-1}), $[CaCl_2]$ = 10 m*M*. (B) The morphology evolution process was nicely captured by stopping the reaction after 3 days, in the presence of PEG-*b*-PEIPA (1 g l^{-1}), showing that big spherules are developed starting from small nanoparticles and grow further to form elongated rods, peanuts, and dumbbells, and are finally assembled into complete spherules. (C, D) The time-dependent morphology evolution of the $BaCO_3$ crystals obtained in the presence of PEG-*b*-PMAA, $[BaCl_2]$ = 10 m*M*, 1 g l^{-1}, on a glass slip. (C) Primary monodisperse nanospheres with a size of 350 nm, 3 h. (B, D) One day. (B) The coexistence of the different stages of rods, dumbbells, quadrupolar particles, and nearly fully grown spheres.

growth modifier, uniform metastable vaterite nanospheres with a diameter of about 310 nm were obtained in the very early stages after 3 h of reaction (Figure 11.31A). Subsequent aggregation of these nanospheres and the transformation to stable calcite beginning on the surface may lead to the calcite rhombohedra on the particle surface. Further overgrowth by the process of Oswald ripening can lead to the observed radial outgrowth to big spherules, with sizes up to 93 μm (Figure 11.32A), at the expense of the smaller particles. The surface structure is very similar to that reported by Tremel et al.,[219,220] where Au nanocolloids were used as nuclei for crystallization of calcite spherules with sizes of about 10 to 15 μm. The crystals were reported to organize radially and grow radiating out from the center, with the direction of growth along the crystallographic [001] direction.[219,220] This is consistent with the wide-angle x-ray diffraction scattering (WAXS) results, which show no or only little exposition of the (006) face. The radial outgrowth of the inner part of the crystals is also consistent with that reported by Jennings et al.,[54] where such crystals were grown only in the presence of both Mg^{2+} and SO_4^{2-} ions in the solution.

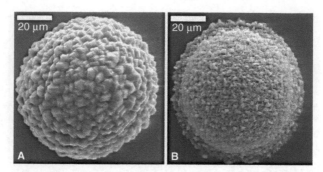

FIGURE 11.32 Typical SEM image of typical $CaCO_3$ spherules grown for 2 weeks by gas diffusion reaction at room temperature in the presence of PEG-*b*-PMAA, 1 g l⁻¹, [CaCl₂] = 10 m*M* (A) and PEG-*b*-PEIPA, 1 g l⁻¹, [CaCl₂] = 10 m*M* (B). (From Yu, S.H. et al., *Adv. Funct. Mater.*, 12, 541–545, 2002. With permission. Copyright © 2002 Wiley-VCH.)

With the time prolonged up to 3 days, the precipitate is already composed of bigger particles with both spherical and twin superstructures, with sizes in the range of 30 to 60 μm. Again, the intermediates of the big spherules are overgrown. The surface is found to be very rough, with 5- to 10-μm-sized calcite units establishing the superstructure. Further prolongation of the reaction time up to 2 weeks leads to big spherules with sizes in the range of 66 to 93 μm, as shown in Figure 11.32A. Such big $CaCO_3$ calcite spherules with a size of about several tens of micrometers and complex surface structures of calcite rhombohedra and resulting cavities cannot be easily generated by conventional solution growth methods. The surface of the spherules is composed of densely packed polygonal calcite subunits, which closely resemble the fracture surface of the prismatic calcite layers of the shell of the mollusk *Atrina serrata*.[11] The inner part of the spherules displays clearly layered radial growth structure,[198] whereas on the outside, the same crystals adopt a more equilibrated rhombohedral appearance.

Similarly, the slow precipitation process finally leads to uniform big spherules in the presence of poly(ethylene glycol)-*block*-poly(ethylene imine)-poly(acetic acid) (PEG-*b*-PEIPA), which after 3 days still contain a variety of intermediates with shapes such as rods, small spheres, peanuts, dumbbells, and overgrown bigger spherules (Figure 11.32B). The surface structure of the big spherules obtained in the presence of PEG-*b*-PEIPA is obviously different from that obtained in the presence of PEG-*b*-PMAA. The surface is very rough and is composed of smaller truncated and randomly oriented calcite *cornerstones*, with an average diameter of about 3 to 5 μm, and many cavities among them, which is similar to the formation of snowball-like cobalt oxide superstructures.[221] This implies a multistep growth mechanism where, at first, the rod-to-sphere transition occurs, with subsequent overgrowth of the spheres once they are formed. When the DHBCs are partially modified with hydrophobic moieties, interestingly, the morphology of the obtained nanoparticle aggregates remains unaffected with respect to the nonmodified DHBCs, suggesting the predominant role of the ionic functional groups for $CaCO_3$ structure development.[194]

Similar to the case of $CaCO_3$, highly monodispersed $BaCO_3$ nanospheres with a size of 350 nm were obtained after the start of the gas diffusion reaction for 3 h (Figure 11.31C). With increasing reaction time, these structures disappear and a

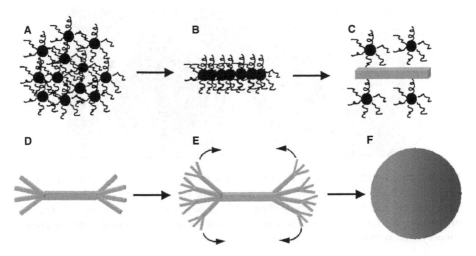

SCHEME 11.4 The proposed sphere-to-rod-to-dumbell-to-sphere transformation mechanism. (A) Homogeneous nanospheres. (B) Aggregation of nanospheres. (C) Rod formed from spherical precursor structures. (D, E) Dumbbell intermediates. (F) Final spherical structure.

coexisting family of rods, dumbbells, quadrupole-shaped crystals, and fully grown spheres appears, as shown in Figure 11.31D. Although the crystal growth is already dendritic without additives, the sphere–rod–dumbbell–sphere transition (Figure 11.31C and D) is observed only in the presence of the block copolymer.

The morphology evolution process of $CaCO_3$ and $BaCO_3$ particles under control of DHBCs suggested that the final stage of big spherules results from a complex growth mechanism, first found for the growth of fluoroapatite in gelatin gels by Kniep et al.[211] and Busch et al.[212] (see also Section 11.5.5). At the first stage, rod-like particles are formed that can grow at their ends, resulting in dumbbell-like particles. These dumbbells further grow into spherical particles consistent with earlier findings on the same polymer, which resulted in the formation of hollow spheres,[195] and consistent with a particle aggregation along electric field lines.[211,213] The observation of particles in different stages of their development indicates that nucleation and subsequent aggregation are slow. This is typical for all slow-evaporation techniques, opposite of the double-jet procedure. The sphere-to-rod-to-dumbbell-to-sphere morphology transition upon growth is illustrated in Scheme 11.4.

Although the sphere-to-rod-to-dumbbell-to-sphere transition seems to be a general growth phenomenon and has been observed for several systems ($CaCO_3$, $BaCO_3$, $MnCO_3$, $CdCO_3$),[222] the exact growth mechanism is still unknown, although some explanation has been given in the literature based on the role of intrinsic electric fields that direct the growth of dipole crystals.[212,213]

11.5.4.3.2 Formation Mechanism of the Calcite Hollow Spheres

For the slow precipitation of the $CaCO_3$ presence of PEG-*b*-PMAA by the gas diffusion method, another growth mechanism is suggested besides that found for $BaCO_3$ (Scheme 11.4) that yields faster $CaCO_3$ growth in the presence of the same block copolymer.[196] From the spherical shape of the early stage of $CaCO_3$ nanoparticles, it can be deduced

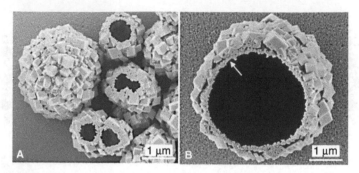

FIGURE 11.33 SEM images of $CaCO_3$ particles grown on a glass slip in the early stage: PEG-*b*-PMAA, $[CaCl_2]$ = 10 m*M*, 1 g l^{-1}, 5 h. (A) The $CaCO_3$ particles with either spherical or hollow structures. (B) The calcite rhombohedral subunits with a grain size of about 320 nm grew on the surface of the hollow structure, and the inner part consisted of tiny primary nanocrystals, as indicated by the arrow.

that these precursor particles are composed of nanocrystalline vaterite. With prolonged time, the small spheres are directed to self-assemble into approximately 3-μm spherical structures, and a phase transition of vaterite to calcite sets in. After 5 h, hollow spheres are obtained that consist of rhombohedral-shaped calcite crystals with well-expressed (104) faces with a unit size of about 300 to 400 nm, as shown in Figure 11.33. Figure 11.33A obviously shows that all those structures are hollow at this stage. The inner part of the hollow structures consists of smaller nanoparticles. The size of the particles on the inner wall of the cavity is remarkably similar to that of the 310-nm spherical precursor particles. Further prolonging the reaction time leads to the formation of big spherules.

The following growth mechanism is shown in Scheme 11.5. At first, vaterite nanospheres form, which subsequently aggregate to approximately 3-μm spherical particles. Later, transformation to calcite beginning on the surface leads to the calcite rhombohedra on the particle surface. These rhomboedra then grow at the cost of the dissolving vaterite particles by Oswald ripening, so that the center of the sphere acts as a sacrificed material depot for the outer calcite particles, resulting in a hollow sphere. Further overgrowth then can lead to the observed radial outgrowth to big spherules, with sizes up to 90 μm, at the expense of the smaller particles (Figure 11.32).

It is remarkable that hollow calcite spheres, but with less expressed surface rhomboeders, have been reported[223] only when the same block copolymer was used together with sodium dodecyl sulfate (SDS) as the second additive. The above-suggested aggregation-based mechanism is different from the growth scenarios found for the same system but for faster crystallization conditions that started at a more alkaline pH,[196,213] so that the crystallization kinetics as well as the starting conditions demonstrate a serious influence on the crystal habitus, even in the presence of the same structure-directing block copolymer. This is not unexpected, as the pH is an important variable for $CaCO_3$ polymorph selection, where the higher pH values favor calcite over vaterite.[224] The more acidic starting conditions in this study may favor initial vaterite growth, whereas the steadily increasing pH will lead to calcite at a later experimental stage.

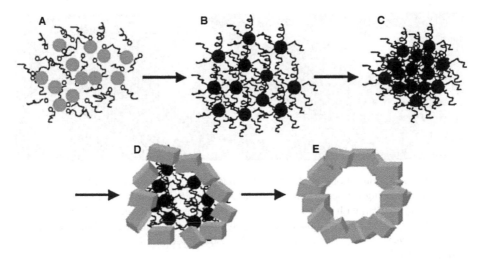

SCHEME 11.5 The proposed formation mechanism of the calcite hollow spheres. (A) The polymer-stabilized amorphous nanoparticles. (B) Formation of spherical vaterite precursors. (C) Aggregation of the vaterite nanoparticles. (D) Vaterite–calcite transformation starting on the outer sphere of the particles. (E) Formation of calcite hollow spheres under consumption of the vaterite precursors.

11.5.4.3.3   Other Complex Structures by Modified Crystallization

If enough time is provided for the controlled nanoparticle aggregation in crystallization via CO_2 evaporation, complex morphologies can be obtained from temporary block copolymer-stabilized nanoparticles that start the aggregation at gas bubbles at the air–solution interface and can be tuned in their morphology by a simple variation of the solution surface tension.[208] This outlines the possibility to use temporarily stabilized nanoparticles as building units for aggregation processes, which are controlled by external templates (Figure 11.34).

When the crystallization conditions of $CaCO_3$ in the presence of a DHBC are modified by applying water–alcohol solvent mixtures with varying solvent compositions — and thus solvent quality changes for the block copolymer and $CaCO_3$ — further handles are available to tune the higher-order assembly of $CaCO_3$ nanoparticle aggregates, including well-defined elongated or spherical morphologies.[225] In addition, hollow calcite aggregates and hollow vaterite disks can be produced if the crystallization is done in the presence of a DHBC and other surfactants, due to the formation of complex micelles as templates in combination with free block copolymers as inhibitors.[223]

11.5.4.4   Fine-Tuning of the Nanostructure of Inorganic Nanocrystals

In addition to the above-reported superstructure formation under control of block copolymers, DHBCs can be used to fine-tune the nanostructure details of other inorganic nanocrystals.

Very thin one- and two-dimensional $CdWO_4$ nanocrystals with controlled aspect ratios were conveniently fabricated at ambient temperature or by hydrothermal

FIGURE 11.34 Control of complex CaCO₃ nanoparticle superstructures with block copolymers with increasing surface activity (from A to C). (A) PEG$_{45}$-PGL$_{27}$ 100% phosphorylated. (B, C) PEG$_{84}$-PHEE$_{13}$ with 40 and 10% phosphorylation degrees, respectively. PEG, poly(ethylene glycol); PGL, poly(glycidol); PHEE, poly(hydroxyethyl ethylene). (From Rudloff, J. et al., *Macromol. Chem. Phys.*, 203, 627–635, 2002. With permission. Copyright © 2002 Wiley-VCH.)

ripening under the control of DHBCs.[207] The TEM image in Figure 11.35A shows very thin, uniform CdWO₄ nanorods/nanobelts with lengths in the range of 1 to 2 μm and a uniform width of 70 nm along their entire length (aspect ratio of about 30). The thickness of the nanobelts is about 6 to 7 nm. The slow and controlled reactant addition by the double-jet technique under stirring maintains formation of intermediate amorphous nanoparticles at the jets,[195] so that nanoparticles are the precursors for the further particle growth rather than ionic species. Successive hydrothermal ripening after the double-jet reaction leads to a rearrangement of the rods into two-dimensional lens-shaped, raft-like superstructures with a resulting lower aspect ratio, as shown in Figure 11.35B. In contrast, very thin and uniform nanofibers with a diameter of 2.5 nm, a length of 100 to 210 nm, and an aspect ratio of 40 to 85, as shown in Figure 11.35C, can be readily obtained when the DHBC PEG-*b*-PMAA is added to the solvent reservoir before the double-jet crystallization process, and the mixture is then hydrothermally ripened at 80°C. When the partly phosphonated DHBC PEG-*b*-PMAA-PO₃H₂ (21%) (1 g l⁻¹) is added at an elevated temperature of 130°C, even without using the double jets, but at higher concentrations and coupled supersaturation, very thin platelet-like particles with a width of 17 to 28 nm, a length of 55 to 110 nm, and an aspect ratio of 2 to 4 are obtained by a direct hydrothermal process, as shown in Figure 11.35D. The nanoparticles display an interesting shape-dependent evolution of the luminescent properties, which may be of interest for applications, since tungstate materials have important luminescence behavior, structural properties, and potential applications.[226]

Polymer-controlled crystallization in water at ambient temperatures provides an alternative and promising tool for morphogenesis of inorganic nanocrystals and superstructures, which could be extended to various systems.

11.5.5 POLYMERIC TEMPLATES

The use of relatively rigid polymeric templates is another principal approach for the generation of complex CaCO₃ morphologies. These can be monolayers of

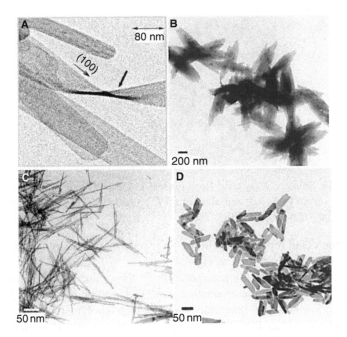

FIGURE 11.35 TEM images of the $CdWO_4$ obtained under different conditions. (A, B) No additives. (A) pH = 5.3, double jet, $[Cd^{2+}]:[WO_4^{2-}] = 8.3 \times 10^{-3}$ M (final solution), at room temperature. (B) pH = 5.3, double jet, $[Cd^{2+}]:[WO_4^{2-}] = 8.3 \times 10^{-3}$ M (final solution); then hydrothermal crystallization, 80°C, 6 h. (C) pH = 5.3, in the presence of PEG-b-PMAA (1 g l^{-1}), 20 ml, double jet, $[Cd^{2+}]:[WO_4^{2-}] = 8.3 \times 10^{-3}$ M (final solution); then hydrothermal crystallization, 80°C, 6 h. (D) Direct hydrothermal treatment of 20 ml of solution containing equal molar $[Cd^{2+}]:[WO_4^{2-}] = 8.3 \times 10^{-2}$ M, pH = 5.3, at 130°C, 6 h, in the presence of 1 g l^{-1} PEG-b-PMAA-PO$_3$H$_2$ (21%). (From Yu, S.H. et al., *Angew. Chem. Int. Ed.*, 41, 2356–2360, 2002. With permission. Copyright © 2002 Wiley-VCH.)

hydrogen-bonded molecular ribbons, which result in the precipitation of calcite and vaterite with altered morphology,[227] or shape-persistent rigid polymers for the modification of calcite morphologies, which allows the investigation of the relation between the structure of the polymeric template and the developing crystal by molecular modeling approaches.[228]

In addition, a β-chitin scaffold was applied in a double-diffusion experiment where the location of the three $CaCO_3$ polymorphs with different morphology depended on the local supersaturation and kind of polymorph.[229] In addition, molecular imprinting approaches that were originally developed for organic molecules were applied for the control of crystal morphology.[230] By this approach, spherical clusters of calcite rhomboeders could be synthesized, but fine morphological details of the initial template crystal could not be reproduced. Similar considerations apply if a constrained volume, like the pores in a polycarbonate membrane, is used as the crystallization environment.[231] Although the found ACC precursor allows for curvature, with time, the typical calcite rhomboeders are formed so that the cylindrical membrane pores determine only the size and rough shape of the formed crystal aggregates, not allowing for the reproduction of fine details. It appears that the

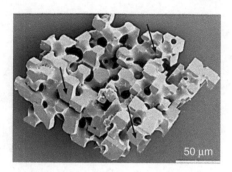

FIGURE 11.36 Single CaCO$_3$ crystal precipitated in the polymeric replica of a sea urchin skeletal plate. (From Park, R.J. and Meldrum, F.C., *Adv. Mater.*, 14, 1167–1169, 2002. With permission. Copyright © 2002 Wiley-VCH.)

quality of fine detail reproduction in the polymeric matrix is strongly dependent on the reagent concentration for CaCO$_3$.

An interesting work by Bianconi et al.[232–234] shows that inorganic–organic composite materials with specific orientation, size, and morphology can be produced by a mineralization reaction of inorganic crystals within a polymer matrix. CdS crystals were crystallized within PEO thin films by immersing CdCl$_2$ containing PEO thin films in a solution of THF containing S(SiMe$_3$)$_2$ for 10 days.[232] Strong binding of metal ions by the polymer matrix has been shown to be crucial in controlling mineralization within polymer films.[233] Recently, novel oriented high-pressure phases of PbS and PbS$_2$ could be crystallized within poly(ethylene oxide) matrices through a similar approach.[234] This biomimetic synthetic approach appears to be general for the production of various oriented inorganic crystals within a PEO matrix under optimal reaction conditions and even favors the formation and stabilization of high-pressure or completely new crystalline phases.[234]

Meldrum et al.[235] reported that they could obtain a perfect single calcite crystal reproduction of a sea urchin skeletal plate via its polymeric replica when the reagent concentrations were very low (Figure 11.36), in contrast to imperfect replications at higher salt concentrations. It is most remarkable that the single-crystal replica could be obtained with fine details and be curved as well as planar surfaces; this was suggested to have formed from an isolated nucleation side with subsequent slow propagation throughout the porous polymer template, requiring a delicate control of the crystallization conditions. This method is potentially very versatile because it could be used to produce any desired CaCO$_3$ shape via a polymer matrix.

Gels can be used as a less rigid organic matrix compared to the above polymer matrices for the morphogenesis of CaCO$_3$. Gelatin gels are of special interest, as these polymers are a degradation product of collagen, which is an important part of the organic matrix of bone and skin. The bimimetic morphogenesis of inorganic crystals in gelatin matrices has attracted much recent interest.

Addition of polypeptide salts like polyaspartate or polyglutamate, mimicking acidic biomineralization proteins, to a gelatin matrix results in polymorph and morphology control of CaCO$_3$, also with added Mg^{2+} in dependence of the polypeptide concentration and uniaxial deformation of the gelatin film.[236] The CaCO$_3$ polymorph

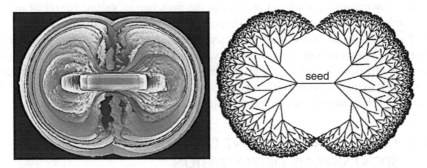

FIGURE 11.37 Left: Successive development stages of the morphogenesis of a fluorapatite–gelatine composite visualized as a superposition. Each stage is surrounded by a black line. (A) Hexagonal seed. (B) Dumbbell. (C) Closed sphere. The size of the sphere corresponds to 100 μm. Right: Image of a simulation using a two-dimensional model for the crystal growth of the fluorapatite–gelatine composites. (From Busch, S. et al., *Chem. Mater.*, 13, 3260–3271, 2001. With permission. Copyright © 2001 American Chemical Society.)

selectivity is related to the local supersaturation inside the gel[237] in coincidence with results in a β-chitin scaffold.[229] However, even without polypeptide additives, gelatin gels were found active in $CaCO_3$ morphogenesis[238] in analogy to an earlier-reported fluoroapatite example. In a poly(acrylamide) hydrogel, a different growth morphology was obtained, whereas addition of polyaspartate led to the same results in both gel systems, indicating that the role of the polymer network is mainly to provide a spatial microenvironment for crystallization, whereas the crystallization process can be controlled by functional additives or functional groups in the gel.

Busch and Kniep et al.[211,212,239,240] reported that the growth of anisotropic fluorapatite spheres by double diffusion in denatured collagen matrices begins with elongated hexagonal-prismatic seeds with an aspect ratio of 5 and a length up to 30 μm. The dumbbells form through progressive stages of directed self-assembled upgrowths of the needle-like prismatic seeds at both ends. These dumbbells further grow by successive and self-similar upgrowths to produce equatorially notched spheres. Typical development stages are visualized in a superposition (Figure 11.37, left) to give an impression of the observed growth. A fractal model was proposed for the formation of anisotropic fluorapatite spheres.[211] The simulated surface of the spherocrystal consists of very small needle-like units, as shown in two-dimensional simulation with the umbrella tree model (Figure 11.37, right), which is consistent with the SEM observation. The intrinsic electric field line model was proposed to explain the formation of the dumbbells.[212] Even though there exists evidence that in nature bones as well as dentine indeed show a piezoelectric and pyroelectric effect due to their polar components,[241–243] and the piezoelectric effect becomes stronger with increasing amounts of the organic component,[244] it should be pointed out that there is no direct experimental evidence for a strong permanent intrinsic electric field of the fractal aggregates, as the claim of pyroelectric behavior was not experimentally established until now.[212] The formation of dumbbells was also found in various systems such as barium sulfate[200] and calcium carbonate[196] by immediate precipitation in the presence of DHBCs or $CaCO_3$ by double-diffusion experiments in gels,[238] hematite by a sol-gel process in the presence of sulfate ions,[45] or micro-wave-induced formation of peanut-shaped calcite crystals in the presence of cit-rate.[245] Recent work shows that the fractal structures of dumbbells and spherical superstructures were indeed obtained in various metal carbonate systems through a slow gas diffusion reaction in the presence of double hydrophilic block copoly-mers.[222] Despite the increasing number of examples of observed dumbbells, the detailed mechanism still needs to be further investigated.

11.6 ORIENTED ATTACHMENT GROWTH AND SPONTANEOUS ORGANIZATION

Recently, natural aggregation-based crystal growth mechanisms and spontaneous self-organization of unstabilized nanoparticles in a surfactant/ligand-free solution system have been gradually recognized as important strategies in designing inorganic crystals. In this section, the recent advances in this special emerging field will be overviewed.

11.6.1 ORIENTED ATTACHMENT GROWTH

Traditional solution synthetic methods have been widely used for the controlled synthesis of various colloidal nanoparticles.[17,18] One basic crystal growth mechanism in solution systems is the well-known Oswald ripening process.[260] In this process, the formation of tiny crystalline nuclei in a supersaturated medium occurs first and then is followed by crystal growth, in which the larger particles grow at the cost of the small ones due to the energy difference between large particles and smaller particles of a higher solubility based on the Gibbs–Thompson law.[247] Another growth mechanism involving mostly linearly oriented particle aggregation,[248–251] which was termed *oriented attachment* by Penn and Banfield et al.,[252–255] has emerged recently with an increasing number of examples, such as α-Fe_2O_3,[248,249] Au,[250] hydroxyapatite ($Ca_{10}(PO_4)_6(OH)_2$),[256] TiO_2,[251,252] FeOOH and CoOOH,[253,255] and ZnS.[257] In these systems, the bigger particles are grown from small primary nanoparticles through an oriented attachment mechanism, in which the adjacent nanoparticles are self-assembled by sharing a common crystallographic orientation and docking these particles at a planar interface.[253] The driving force for this spontaneous oriented attachment is that the elimination of the pairs of high-energy surfaces will lead to a substantial reduction in the surface free energy from the thermodynamic view-point.[254,258]

Averback et al.[259,260] identified a similar spontaneous self-assembly process, which they called contact epitaxy, during their study of the deposition of Ag nano-particles onto Cu substrate. The initial randomly oriented Ag nanocrystals can align epitaxially with the substrate, which was explained as the rotation of the nanopar-ticles within the aggregates driven by short-range interactions between adjacent surfaces.[259]

Increasing evidence in several systems has been gradually observed for either directed or undirected particle aggregation. This kind of growth mode could lead to the formation of faceted particles, or anisotropic growth if it occurs near equilibrium

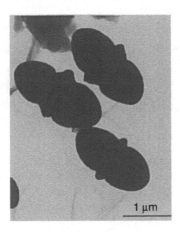

FIGURE 11.38 TEM image showing α-Fe$_2$O$_3$ double ellipsoids produced by aging a solution (0.018 M FeCl$_3$, 0.05 M HCl) at 100°C for 1 week. (From Bailey, J.K. et al., *J. Colloid Interface Sci.*, 157, 1–13, 1993. With permission. Copyright © 1993 Elsevier Sciences.)

and there is sufficient difference in the surface energies of different crystallographic faces.[255] It is possible to form perfect highly anisotropic crystals that clearly show the growth via oriented attachment, as impressively evidenced for TiO$_2$.[252]

An early interesting finding by Bailey et al.[248] shows that a rather complex morphology of α-Fe$_2$O$_3$ double ellipsoids (Figure 11.38) can be easily obtained through aging a solution containing β-FeOOH rods at 100°C for 1 week. The particles were obtained by a hydrolysis reaction of FeCl$_3$ in HCl solution. These complex double-ellipsoid structures were clearly developed from the heterogeneous nucleation of hematite on β-FeOOH rods formed in the early stage. The α-Fe$_2$O$_3$ (hematite) particles not only grow outward but also use these rods as templates, and a collar forms growing along the rods, leading to the formation of such a double-ellipsoid shape.[248] This example shows that the size and aggregation of the metastable phase can influence the final morphology.

Penn and Banfield confirmed that both anatase[252] and iron oxide nanoparticles[253,254] with sizes of a few nanometers can coalesce under hydrothermal conditions by the oriented attachment process, as shown in Figure 11.39. The crystal lattice planes are almost perfectly aligned, showing that dislocations at the contact areas between the adjacent particles that lead to defects in the finally formed bulk crystals can be avoided. A recent observation shows that the oriented attachment can also control the coarsening of mercaptoethanol-capped ZnS nanoparticles, even though the organic ligands will control the aggregation state of the nanoparticles during the hydrothermal treatment.[257] A two-stage growth kinetics was proposed for ZnS crystal growth: the primary particle quickly increases in volume in the first stage, controlled by oriented attachment; next, the growth stops until surface-bound ligands are removed, and then Oswald ripening dominates and the oriented attachment becomes less important with increasing particle size.[257]

A very recent report by Weller et al.[261] provided some strong evidence that perfect ZnO nanorods can be conveniently self-assembled from small ZnO quasi-spherical nanoparticles based on the oriented attachment mechanism by the evaporation and

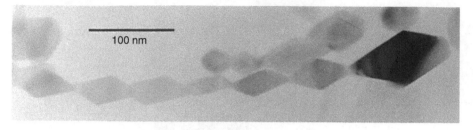

FIGURE 11.39 TEM micrograph of a single crystal of anatase that was hydrothermally coarsened in 0.001 M HCl, showing that the primary particles align, dock, and fuse to form oriented chain structures. (From Penn, R.L. and Banfield, J.F., *Geochim. Cosmochim. Acta*, 63, 1549–1557, 1999. With permission. Copyright © 1998 Elsevier Sciences.)

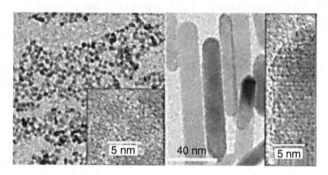

FIGURE 11.40 TEM images of ZnO. (A) Starting sol. (B) After 1 day of reflux of the concentrated sol. The insets show high-resolution TEM images of individual nanoparticles. (From Pacholski, C. et al., *Angew. Chem. Int. Ed.*, 41, 1188–1191, 2002. With permission. Copyright © 2002 Wiley-VCH.)

reflux of a solution containing 3- to 5-nm nanoparticles. In previous studies, the self-assembly of nanoparticles capped by ligands was mainly driven by the interactions of the organic ligands rather than by the interaction of the particle cores, as reported by Weller et al.[261] A ZnO sol with an average particle size of approximately 3 nm, as shown in Figure 11.40A, was easily prepared by dropwise adding a 0.03-M solution of KOH (65 ml) in methanol into a solution of zinc acetate dihydrate (0.01 M) in methanol (125 ml) under vigorous stirring at about 60°C. Refluxing of the concentrated solution leads to the formation of rod-like nanoparticles. Prolonging the heating time mainly leads to an increase of the elongation of the particle along the c-axis. After refluxing for 1 day, single-crystalline nanorods with average lengths of 100 nm and widths of approximately 15 nm are formed, as shown in Figure 11.40B. This is a perfect model case for the formation of well-defined one-dimensional nanostructures based on the oriented attachment mechanism without using any ligands or surfactants.[261]

Aggregation-based growth mechanisms provide a route for the incorporation of defects, such as edge and screw dislocations, in stress-free and initially defect-free nanocrystalline materials,[253] which could improve the reactivity of the nanocrystalline

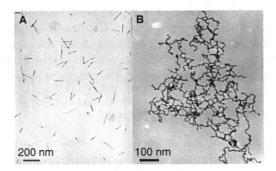

FIGURE 11.41 (A) TEM image of thioglycolic acid–stabilized nanowires self-organized from CdTe nanoparticles. (B) Nanoparticle chain-like structures showing the intermediate state of nanoparticle–nanowire transition. (Courtesy of Prof. Nicholas A. Kotov of Oklahoma State University.)

subunits. Although so far there are only a few reports on the oriented attachment of nanoparticles, this new growth mechanism could offer an additional tool to design advanced materials with anisotropic material property and could be used for the synthesis of more complex crystalline three-dimensional structures in which the branching sites could be added as individual nanoparticles.[262] This could even lead to the formation of highly perfect crystals. In addition, the oriented attachment approach could describe the nanofiber formation, as reported in Section 11.5.4.2.

11.6.2 Spontaneous Organization

Previous observations of self-organization or directed aggregation of nanoparticles into ordered structures rely mainly on a special binding interaction of the bridging ligands capped on the surface of nanoclusters. For example, the self-alignment of acetate and thiolate capping agent-protected CdS nanoclusters with the empirical formula $Cd_4S_1RS_{2.9}(CH_3COO)_{1.32}$ into chain structures or two-dimensional thin films was observed by Chemseddine et al.[262] However, recent advances show that even unstabilized nanoparticles can spontaneously self-organize into perfect nanostructures.

A strikingly elegant result reported by Kotov et al.[263] undoubtedly confirmed that such spontaneous organization of small nanoparticles into much elongated nanorods/nanowires (Figure 11.41A) can be achieved at room temperature in the case of CdTe when the protective shell of organic stabilizer thioglycolic acid on the surface of the initial CdTe nanoparticles is partially removed by methanol addition and redispersed in water at pH 9.[263]

Figure 11.41B clearly depicts the presence of pearl-necklace agglomerates as intermediates of the nanoparticle–nanowire transition, suggesting that the growth mechanism should be related to a special interaction/attraction between the nanoparticles.[263] A dipole–dipole interaction was believed to be the driving force for such self-organization of the nanoparticles.[263] The photoluminescence spectrum shows the high quality of the CdTe nanowires. This approach provides a new pathway to high-quality luminescent CdTe nanowires at room temperature, even though the yield is still low. In addition, room temperature crystallization and transformation of amorphous nanoparticles obtained from a fast double-jet reaction into highly anisotropic belt-like $CdWO_4$ nanorods have also been achieved.[207] These results imply that the spontaneous self-organization with subsequent fusion into crystallographically aligned particles indeed could act in various inorganic systems.

11.7 CONCLUSIONS AND OUTLOOK

In this chapter, the state of the art in the area of biomineralization/biomimetic mineralization and morphosynthesis of various inorganic crystals by recently emerging nature-inspired approaches has been overviewed, including small-molecule additives or solvent-mediated shape control, surfactant-mediated templating and crystallization, supramolecular self-assembly from designed artificial organic templates, polymer-controlled crystallization, oriented attachment growth, and spontaneous organization. The latest advances in morphosynthesis and biomineralization/biomimetic mineralization of various technically important inorganic crystals were summarized, with a focus on how to generate inorganic crystals with unusual structural specialty and complexity.

The basic principles and potentials of these new approaches were illustrated with specific examples, which seem very promising for controlling morphology, microstructure, complexity, and length scales of various inorganic crystals. It is especially clearly demonstrated that the morphosynthesis of various inorganic crystals can be realized by manipulating the special interaction between the soft–hard interfaces of polymers/surfactants (or organic species) and the inorganic crystals. Interestingly, the emerging oriented attachment observed under near-natural crystallization conditions and spontaneous self-organization of unstabilized nanoparticles could shed new light on the understanding of the detailed interaction mechanism between inorganic nanoparticles and their subsequent higher-order self-assembly mechanism. The current progress indubitably emphasizes that probably all inorganic crystals will be amenable to morphosynthetic control by use of flexible molecule templates and suitable self-assembly mechanisms.

These emerging nature-inspired solution routes may open alternative doorways toward low-dimensional nanocrystals and more complex superstructures. However, there is still a lack of understanding of the nucleation, crystallization, self-assembly, and growth mechanism of complex superstructures in these solution systems. Further exploration in this area should provide new possibilities for rationally designing various kinds of inorganic materials with ideal hierarchy and controllable length scale. These unique hierarchical materials of structural specialty and complexity, with a size range spanning from nanometers to micrometers, are expected to find potential applications in various fields. It is undoubtedly necessary to investigate the relationship between the structural specialty/complexity (shape, size, phase, dimensionality, hierarchy, etc.) of the materials and their properties, which could shed light on their novel applications and fundamental properties as building blocks in various fields of materials science and related fields.

ACKNOWLEDGMENTS

S.-H. Yu thanks the special funding support by the Century Program of the Chinese Academy of Sciences and the Natural Science Foundation of China. We acknowledge financial support by the Max Planck Society and the DFG (SFB 448). S.H. Yu thanks the Alexander von Humboldt Foundation and the Max Planck Society for supporting his research stay in Germany. H.C. thanks the Dr. Hermann Schnell foundation for financial support. Prof. Dr. M. Sedlak and Dr. Jan Rudloff are acknowledged for the synthesis of the phosphonated and phosphorylated double hydrophilic block copolymers, which are involved in this research. Prof. Nicholas A. Kotov of Oklahoma State University is acknowledged for leaving Figure 11.41 at our deposition. We are very thankful to the director of the Max Planck Institute of Colloids and Interfaces, Prof. Dr. Markus Antonietti, for his sound advice and continuing interest in this project, as well as his critical reading of this chapter.

REFERENCES

1. S Mann, GA Ozin. Synthesis of inorganic materials with complex form. *Nature* 382:313–318, 1996.
2. S Mann. The chemistry of form. *Angew Chem Int Ed* 39:3393–3406, 2000.
3. E Dujardin, S Mann. Bio-inspired materials chemistry. *Adv Mater* 14:775–788, 2002.
4. LA Estroff, AD Hamilton. At the interface of organic and inorganic chemistry: bioinspired synthesis of composite materials. *Chem Mater* 13:3227–3235, 2001.
5. GA Ozin. Morphogenesis of biomineral and morphosynthesis of biomimetic forms. *Acc Chem Res* 30:17–27, 1997.
6. GA Ozin. Panoscopic materials: synthesis over "all" length scales *Chem Commun* 419–432, 2000.
7. T Kato, A Sugawara, N Hosoda. Calcium carbonate-organic hybrid materials. *Adv Mater* 14:869–877, 2002.
8. H Cölfen, S Mann. Emergence of higher-order organization by mesoscale self-assembly and transformation of hybrid nanostructures. *Angew Chem Int Ed* 42: in press, 2003.
9. S Mann. *Biomimetic Materials Chemistry.* New York: VCH Publishers, 1996.
10. SA Davis, M Breulmann, KH Rhodes, B Zhang, S Mann. Template-directed assembly using nanoparticle building blocks: a nanotectonic approach to organized materials. *Chem Mater* 13:3218–3226, 2001.
11. SW Weiner, L Addadi. Design strategies in mineralized biological materials. *J Mater Chem* 7:689–702, 1997.
12. DM Dabbs, IA Aksay. Self-assembled ceramic products by complex fluid templation. *Annu Rev Phys Chem* 51:601–622, 2000.
13. S Mann. Biomineralization: the form(id)able part of bioinorganic chemistry. *J Chem Soc Dalton Trans* 3953–3961, 1997.
14. KJC van Bommel, A Friggeri, S Shinkai. Organic templates for the generation of inorganic materials. *Angew Chem Int Ed* 42:980–999, 2003.
15. E Bäuerlein. Biomineralization of unicellular organisms: an unusual membrane biochemistry for the production of inorganic nano- and microstructures. *Angew Chem Int Ed* 42:614–641, 2003.

16. CM Niemeyer. Nanoparticles, proteins, and nucleic acids: biotechnology meets materials science. *Angew Chem Int Ed* 40:4128–4158, 2001.
17. E Matijevi. Preparation and properties of uniform size colloids. *Chem Mater* 5:412–426, 1993.
18. E Matijevi. Controlled colloid formation. *Curr Opin Colloid Interface Sci* 1:176–183, 1996.
19. M Antonietti, C Göltner. Superstructures of functional colloids: chemistry on the nanometer scale. *Angew Chem Int Ed* 36:910–928, 1997.
20. DD Archibald, S Mann. Template mineralization of self-assembled anisotropic lipid microstructures. *Nature* 364:430–433, 1993.
21. H Yang, N Coombs, GA Ozin. Morphogenesis of shapes and surface patterns in mesoporous silica. *Nature* 386:692–695, 1997.
22. M Li, H Schnablegger, S Mann. Coupled synthesis and self-assembly of nanoparticles to give structures with controlled organization. *Nature* 402:393–395, 1999.
23. XG Peng, L Manna, WD Yang, J Wickham, E Scher, A Kadavanich, AP Alivisatos. Shape control of CdSe nanocrystals. *Nature* 404:59–61, 2000.
24. TS Ahmadi, ZL Wang, TC Green, A Henglein, MA EI-Sayed. Shape-controlled synthesis of colloidal platinum nanoparticles. *Science* 272:1924–1926, 1996.
25. CP Gibson, K Putzer. Synthesis and characterization of anisometric cobalt nanoclusters. *Science* 267:1338–1340, 1995.
26. MP Pileni, BW Ninham, T Gulik-Krzywicki, J Tanori, I Lisiecki, A Filankembo. Direct relationship between shape and size of template and synthesis of copper metal particles. *Adv Mater* 11:1358–1362, 1999.
27. SJ Park, S Kim, S Lee, ZG Khim, K Char, T Hyeon. Synthesis and magnetic studies of uniform iron nanorods and nanospheres. *J Am Chem Soc* 122:8581–8582, 2000.
28. JH Adair, E Suvaci. Morphological control of particles. *Curr Opin Colloid Interface Sci* 5:160–167, 2001.
29. SH Yu, J Yang, YT Qian. Low dimensional nanocrystals. In *Encyclopedia of Nanoscience and Nanotechnology*, HS Nalwa, Ed. American Scientific Publishers, 2004.
30. CM Lieber. One-dimensional nanostructures: chemistry, physics and applications. *Solid State Commun* 107:607–616, 1998.
31. JT Hu, TW Odom, CM Lieber. Chemistry and physics in one dimension: synthesis and properties of nanowires and nanotubes. *Acc Chem Res* 32:435–445, 1999.
32. JD Klein, RD Herrick, D Palmer, MJ Sailor, CJ Brumlik, CR Martin. Electrochemical fabrication of cadmium chalcogenide microdiode arrays. *Chem Mater* 5:902–904, 1993.
33. CR Martin. Nanomaterials: a membrane-based synthetic approach. *Science* 266:1961–1966, 1994.
34. D Routkevitch, T Bigioni, M Moskovits, JM Xu. Electrochemical fabrication of CdS nanowire arrays in porous anodic aluminum oxide templates. *J Phys Chem* 100:14037–14047, 1996.
35. XF Duan, CM Lieber. General synthesis of compound semiconductor nanowires. *Adv Mater* 12:298–302, 2000.
36. XF Duan, CM Lieber. Laser-assisted catalytic growth of single crystal GaN nanowires. *J Am Chem Soc* 122:188–189, 2000.
37. Y Zhou, SH Yu, CY Wang, XG Li, YR Zhu, ZY Chen. A novel ultraviolet irradiation photoreduction technique for the preparation of single-crystal Ag nanorods and Ag dendrites. *Adv Mater* 11:850–852, 1999.
38. A Stein, SW Keller, TE Mallouk. Turning down the heat-design and mechanism in solid-state synthesis. *Science* 259:1558–1564, 1993.

39. TJ Trentler, KM Hickman, SC Goel, AM Viano, PC Gibbons, WE Buhro. Solution–liquid–solid growth of crystalline III-V semiconductors: an analogy to vapor–liquid–solid growth. *Science* 270:1791–1794, 1995.

40. SH Yu, M Yoshimura. Shape and phase control of ZnS nanocrystals: template fabrication of wurtzite ZnS single-crystal nanosheets and ZnO flake-like dendrites from a lamellar molecular precursor $ZnS-(NH_2CH_2CH_2NH_2)_{0.5}$. *Adv Mater* 14:296–300, 2002.

41. G. Wulff. Zur Frage der Geschwindigkeit des Wachstums und der Auflösung der Krystallflächen. *Z Krystallogr* 34:449–530, 1901.

42. MM Reddy, GH Nancollas. Crystallization of calcium-carbonate. 4. Effect of magnesium, strontium and sulfate-ion. *J Cryst Growth* 35:33–38, 1976.

43. JO Titiloye, SC Parker, DJ Osguthorpe, S Mann. Predicting the influence of growth additives on the morphology of ionic-crystals. *J Chem Soc Chem Commun* 1494–1496, 1991.

44. S Rajam, S Mann. Selective stabilization of the (001) face of calcite in the presence of lithium. *J Chem Soc Chem Commun* 1789–1791, 1990.

45. T Sugimoto, MM Kahn, A Muramatsu. Preparation of monodisperse peanut-type alpha-Fe_2O_3 particles from condensed ferric hydroxide gel. *Colloids Surf A* 70:167–169, 1993.

46. T Sugimoto, Y Wang. Mechanism of the shape and structure control of monodispersed alpha-Fe_2O_3 particles by sulfate ions. *J Colloid Interface Sci* 207:137–149, 1998.

47. T Sugimoto, MM Khan, A Muramatsu, H Itoh. Formation mechanism of monodisperse peanut-type alpha-Fe_2O_3 particles from condensed ferric hydroxide gel. *Colloids Surfaces A Physicochem Eng Aspects* 79:233–247, 1998.

48. T Sugimoto, H Itoh, H Maiyake. Formation of monodisperse microcrystals of basic aluminum sulfate by the gel-sol method. *J Colloid Interface Sci* 188:101–114, 1997.

49. T Sugimoto, K Okada, H Itoh. Synthesis of uniform spindle-type titania particles by the gel-sol method. *J Colloid Interface Sci* 193:140–143, 1997.

50. T Sugimoto, K Okada, H Itoh. Formation mechanism of uniform spindle-type titania particles in the gel-sol process. *J Disper Sci Technol* 19:143–161, 1998.

51. JH Schroeder, EJ Dwornik, JJ Papike. Primary protodolomite in echinoid skeletons. *Geol Soc Am Bull* 80:1613, 1969.

52. JR Goldsmith, DL Graf, OI Joehnsu. The occurrence of magnesian calcites in nature. *Geochim Cosmochim Acta* 7:212, 1955.

53. S Raz, S Weiner, L Addadi. Formation of high magnesian calcites via an amorphous precursor phase: possible biological implications. *Adv Mater* 12:38–42, 2000.

54. SL Tracy, CJP François, HM Jennings. The growth of calcite spherulites from solution. I. Experimental design techniques. *J Cryst Growth* 193:374–381, 1998.

55. SL Tracy, DA Williams, HM Jennings. The growth of calcite spherulites from solution. II. Kinetics of formation. *J Cryst Growth* 193:382–388, 1998.

56. FC Meldrum, ST Hyde. Morphological influence of magnesium and organic additives on the precipitation of calcite. *J Cryst Growth* 231:544–558, 2001.

57. N Wada, K Yamashita, T Umegaki. Effects of carboxylic acids on calcite formation in the presence of Mg^{2+} ions. *J Colloid Interface Sci* 212:357–364, 1999.

58. J Küther, R Seshadri, W Tremel. Crystallization of calcite spherules around designer nuclei. *Angew Chem Int Ed* 37:3044–3047, 1998.

59. J Küther, R Seshadri, G Nelles, W Assenmacher, HJ Butt, W Mader, W Tremel. Mercaptophenol-protected gold colloids as nuclei for the crystallization of inorganic minerals: templated crystallization on curved surfaces. *Chem Mater* 11:1317–1325, 1999.

60. I Lee, SW Han, HJ Choi, K Kim. Nanoparticle-directed crystallization of calcium carbonate. *Adv Mater* 13:1617–1620, 2001.

61. I Weissbuch, L Addadi, M Lahav, L Leiserowitz. Molecular recognition at crystal interface. *Science* 253:637–645, 1991.

62. Y Kitano, DW Hood. Influence of organic material on polymorphic crystallization of calcium carbonate. *Geochim Cosmochim Acta* 29:29–41, 1965.

63. S Mann, JM Didymus, NP Sanderson, BR Heywood, EJA Samper. Morphological influence of functionalized and non-functionalized α,φ-dicarboxylates on calcite crystallization. *J Chem Soc Faraday Trans* 86:1873–1880, 1990.

64. RJ Davey, SN Black, LA Bromley, D Cottier, B Dobbs, JE Rout. Molecular design based on recognition at inorganic surfaces. *Nature* 353:549–550, 1991.

65. MP Kelley, B Janssens, B Kahr, WM Vetter. Recognition of dyes by K₂SO₄ crystals: choosing organic guests for simple salts. *J Am Chem Soc* 116:5519–5520, 1994.

66. M Rifani, YY Yin, DS Elliott, MJ Jay, S Jang, MP Kelley, L Bastin, BJ Kahr. Solid state dye lasers from stereospecific host–guest interactions. *J Am Chem Soc* 117:7572–7573, 1995.

67. PV Coveney, R Davey, JLW Griffin, Y He, JD Hamlin, S Stackhouse, A Whiting. A new design strategy for molecular recognition in heterogeneous systems: a universal crystal-face growth inhibitor for barium sulfate. *J Am Chem Soc* 122:11557–11558, 2000.

68. AL Rohl, DH Gay, RJ Davey, CRA Catlow. Interactions at the organic/inorganic interface: molecular modeling of the interaction between diphosphonates and the surfaces of barite crystals. *J Am Chem Soc* 118:642–648, 1996.

69. MA Nygren, DH Gay, CRA Catlow, MP Wilson, AL Rohl. Incorporation of growth-inhibiting diphosphonates into steps on the calcite cleavage plane surface. *J Chem Soc Faraday Trans* 94:3685–3693, 1998.

70. T Sugimoto, H Itoh, T Mochida. Shape control of monodisperse hematite particles by organic additives in the gel-sol system. *J Colloid Interface Sci* 205:42–52, 1998.

71. T Sugimoto, Y Wang, A Muramatsu. Systematic control of size, shape and internal structure of monodisperse alpha-Fe₂O₃ particles. *Colloids Surfaces A Physicochem Eng Aspects* 134:265–279, 1998.

72. NS Bell, JH Adair. Adsorbate effects on glycothermally produced alpha-alumina particle morphology. *J Cryst Growth* 203:213–226, 1999.

73. T Sugimoto, S Chen, A Muramatsu. Synthesis of uniform particles of CdS, ZnS, PbS and CuS from concentrated solutions of the metal chelates. *Colloid Surf A Physicochem Eng Aspects* 135:207–226, 1998.

74. T Sugimoto. Preparation of metal sulfides from chelates. In *Fine Particles: Synthesis, Characterization, and Mechanisms of Growth*, T Sugimoto, Ed. New York: Marcel Dekker, 2000, pp. 199–216.

75. CA Orme, A Noy, A Wierzbicki, MT McBride, M Grantham, HH Teng, PM Dove, JJ DeYoreo. Formation of chiral morphologies through selective binding of amino acids to calcite surface steps. *Nature* 411:775–779, 2001.

76. ZR Tian, JA Voigt, J Liu, B Mckenzie, MJ Mcdermott. Biomimetic arrays of oriented helical ZnO nanorods and columns. *J Am Chem Soc* 124:12954–12955, 2002.

77. BC Bunker, PC Rieke, BJ Tarasevich, AA Campbell, GE Fryxell, GL Graff, L Song, J Liu, JW Virden. Ceramic thin-film formation on functionalized interfaces through biomimetic processing. *Science* 264:48–55, 1994.

78. G Falini, S Albeck, S Weiner, L Addadi. Control of aragonite or calcite polymorphism by mollusk shell macromolecules. *Science* 271:67– 69, 1996.

79. AM Belcher, XH Wu, PK Christensen, PK Hansma, GD Stucky, DE Morse. Control of crystal phase switching and orientation by soluble mollusc-shell proteins. *Nature* 381:56–58, 1996.

80. M Yoshimura, W Suchanek, K Byrappa. Soft solution processing: a strategy for one-step processing of advanced inorganic materials. *MRS Bull* 9:17–25, 2000.

81. A Stein, SW Keller, TE Mallouk. Turning down the heat — design and mechanism in solid-state synthesis. *Science* 259:1558–1564, 1993.

82. WE Buhro, KM Hickman, TJ Trentler. Turning down the heat on semiconductor growth: solution–chemical syntheses and the solution–liquid–solid mechanism *Adv Mater* 8:685–688, 1996.

83. SH Yu. Hydrothermal/solvothermal processing of advanced ceramic materials. *J Ceram Soc Jpn* 109:S65–S75, 2001.

84. MA Verges, A Mifsud, CJ Serna. Formation of rod-like zinc-oxide microcrystals in homogeneous solutions. *J Chem Soc Faraday Trans* 86:959–963, 1990.

85. SII Yu, J Yang, ZH Han, Y Zhou, RY Yang, Y Xie, YT Qian, YH Zhang. Controllable synthesis of nanocrystalline CdS with different morphologies and particle sizes by a novel solvothermal process. *J Mater Chem* 9:1283–1287, 1999.

86. SH Yu, YS Wu, J Yang, ZH Han, Y Xie, YT Qian, XM Liu. A novel solventothermal synthetic route to nanocrystalline CdE (E = S, Se, Te) and morphological control. *Chem Mater* 10:2309–2312,1998.

87. J Yang, JH Zeng, SH Yu, L Yang, GE Zhou, YT Qian. Formation process of CdS nanorods via solvothermal route. *Chem Mater* 12:3259–3263, 2000.

88. J Yang, C Xue, SH Yu, JH Zeng, YT Qian. General synthesis of semiconductor chalcogenide nanorods by using the monodentate ligand *n*-butylamine as a shape controller. *Angew Chem Ed Int* 41:4697–4700, 2002.

89. SH Yu, M Yoshimura. Shape and phase control of ZnS nanocrystals: template fabrication of wurtzite ZnS single-crystal nanosheets and ZnO flake-like dendrites from a lamellar molecular precursor ZnS·$(NH_2CH_2CH_2NH_2)_{0.5}$. *Adv Mater* 14:296–300, 2002.

90. B Mayers, YN Xia. Formation of tellurium nanotubes through concentration depletion at the surfaces of seeds. *Adv Mater* 14:279–282, 2002.

91. MS Mo, JH Zeng, XM Liu, WC Yu, SY Zhang, YT Qian. Controlled hydrothermal synthesis of thin single-crystal tellurium nanobelts and nanotubes. *Adv Mater* 14:1658–1662, 2002.

92. F Manoli, E Dalas. Spontaneous precipitation of calcium carbonate in the presence of ethanol, isopropanol and diethylene glycol. *J Cryst Growth* 218:359–364, 2000.

93. CT Kresge, ME Leonowicz, WJ Roth, JC Vartuli, JS Beck. Ordered mesoporous molecular sieves synthesized by a liquid-crystal template mechanism. *Nature* 359:710–712, 1992.

94. S Polarz, M Antonietti. Porous materials via nanocasting procedures: innovative materials and learning about soft-matter organization. *Chem Commun* 2593–2604, 2002.

95. M Antonietti. Surfactants for novel templating applications. *Curr Opin Colloid Interface Sci* 6:244–248, 2001.

96. S Oliver, A Kuperman, N Coombs, A Lough, GA Ozin. Lamellar aluminophophates with surface patterns that mimic diatom and ratiolarian microskeletons. *Nature* 378:47–50, 1995.

97. D Walsh, JD Hopwood, S Mann. Crystal tectonics-construction of reticulated calcium-phosphate frameworks in biocontinuous reverse microemulsions. *Science* 264:1576–1578, 1994.

98. D Walsh, S Mann. Fabrication of hollow porous shells of calcium carbonate from self-organization media. *Nature* 377:320–323, 1995.

99. JH Collier, PB Messersmith. Phospholipid strategies in biomineralization and biomaterials research. *Annu Rev Mater Res* 31:237–263, 2001.

100. JD Hopwood, S Mann. Synthesis of barium sulfate nanoparticles and nanofilaments in reverse micelles and microemulsions. *Chem Mater* 9:1819–1828, 1997.

101. M Li, S Mann. Emergence of morphological complexity in $BaSO_4$ fibers synthesized in AOT microemulsions. *Langmuir* 16:7088–7094, 2000.

102. GD Rees, R Evans-Gowing, SJ Hammond, BH Robinson. Formation and morphology of calcium sulfate nanoparticles and nanowires in water-in-oil microemulsions. *Langmuir* 15:1993–2002, 1999.

103. LM Qi, J Ma, H Cheng, Z Zhao. Reverse micelle based formation of $BaCO_3$ nanowires. *J Phys Chem B* 101:3460–3463, 1997.

104. D Kuang, AW Xu, YP Fang, HD Ou, HQ Liu. Preparation of inorganic salts ($CaCO_3$, $BaCO_3$, $CaSO_4$) nanowires in the Triton X-100/cyclohexane/water reverse micelles. *J Cryst Growth* 244:379–383, 2002.

105. S Kwan, F Kim, J Akana, P Yang. Synthesis and assembly of $BaWO_4$ nanorods. *Chem Commun* 447–448, 2001.

106. HT Shi, LM Qi, JM Ma, HM Cheng. Synthesis of single crystal $BaWO_4$ nanowires in catanionic reverse micelles. *Chem Commun* 1704–1705, 2002.

107. CNN Rao, A Govindaraj, FL Deepak, NA Gunari. Surfactant-assisted synthesis of semiconductor nanotubes and nanowires. *Appl Phys Lett* 78:1853–1855, 2001.

108. CC Chen, CY Chao, ZH Lang. Simple solution-phase synthesis of soluble CdS and CdSe nanorods. *Chem Mater* 12:1516–1618, 2000.

109. BA Simmons, S Li, VT John, GL McPherson, A Bose, W Zhou, J He. Morphology of US nanocrystals synthesized in a mixed surfactant system. *Nano Lett* 2:263–268, 2002.

110. N Pinna, K Weiss, J Urban, MP Pileni. Triangular CdS nanocrystals: structural and optical studies. *Adv Mater* 13:261–264, 2001.

111. J Tanori, MP Pileni. Control of the shape of copper metallic particles by using a colloidal system as template. *Langmuir* 13:639–646, 1997.

112. M Maillard, S Giorgio, MP Pileni. Silver nanodisks. *Adv Mater* 14:1084–1086, 2002.

113. S Vaucher, M Li, S Mann. Synthesis of Prussian Blue nanoparticles and nanocrystal superlattices in reverse microemulsions. *Angew Chem Int Ed* 39:1793–1796, 2000.

114. M Li, S Mann. Emergent nanostructures: water-induced mesoscale transformation of surfactant-stabilized amorphous calcium carbonate nanoparticles in reverse microemulsions. *Adv Funct Mater* 12:773–779, 2003.

115. YY Yu, SS Chang, CL Lee, CRC Wang. Gold nanorods: electrochemical synthesis and optical properties. *J Phys Chem B* 101:6661–6664, 1997.

116. NR Jana, CJ Murphy. Controlling the aspect ratio of inorganic nanorods and nanowires. *Adv Mater* 14:80–82, 2002.

117. NR Jana, L Gearheart, CJ Murphy. Wet chemical synthesis of high aspect ratio cylindrical gold nanorods. *J Phys Chem B* 105:4065–4067, 2001.

118. NR Jana, L Gearheart, CJ Murphy. Wet chemical synthesis of silver nanorods and nanowires of controllable aspect ratio. *Chem Commun* 617–618, 2001.

119. S Chen, Z Fan, DL Carroll. Silver nanodisks: synthesis, characterization, and self-assembly. *J Phys Chem B* 106:10777–10781, 2002.

120. A Imhof, DJ Pine. Ordered macroporous materials by emulsion templating. *Nature* 389:948–951, 1997.

121. D Walsh, B Lebeau, S Mann. Morphosynthesis of calcium carbonate (vaterite) micro-sponges. *Adv Mater* 11:324–328, 1999.
122. CE Fowler, D Khushalani, S Mann. Facile synthesis of hollow silica microspheres. *J Mater Chem* 11:1968–1971, 2001.
123. E Muthusamy, D Walsh, S Mann. Morphosynthesis of organoclay microspheres with sponge-like or hollow interiors. *Adv Mater* 14:969–972, 2002.
124. S Vaucher, J Fielden, M Li, E Dujardin, S Mann. Molecule-based magnetic nanopar-ticles: synthesis of cobalt hexacyanoferrate, cobalt pentacyanonitrosylferrate, and chromium hexacyanochromate coordination polymers in water-in-oil microemul-sions. *Nano Lett* 2:225–229, 2002.
125. HA Lowenstam. Minerals formed by organisms. *Science* 211:1126–1131, 1981.
126. BR Heywood. Template-directed nucleation and growth of inorganic materials. In *Biomimetic Materials Chemistry*, S Mann, Ed. New York: VCH Publishers, 1996, pp. 143–173.
127. S Mann, BR Heywood, S Rajam, JD Birchall. Controlled crystallization of $CaCO_3$ under stearic acid monolayers. *Nature* 334:692–695, 1988.
128. S Rajam, BR Heywood, JBA Walker, S Mann, RJ Davey, JD Birchall. Oriented crystallization of $CaCO_3$ under compressed monolayers. 1. Morphological-studies of mature crystals. *J Chem Soc Faraday Trans* 87:727–734, 1991.
129. BR Heywood, S Rajam, S Mann. Oriented crystallization of $CaCO_3$ under compressed monolayers. 2. Morphology, structure and growth of immature crystals. *J Chem Soc Faraday Trans* 81:735–743, 1991.
130. BR Heywood, S Rajam, S Mann. Organic template-directed inorganic crystallization-oriented nucleation of $BaSO_4$ under compressed Langmuir monolayers. *J Am Chem Soc* 114:4681–4686, 1992.
131. BR Heywood, S Mann. Template-directed inorganic crystallization-oriented nucle-ation of barium-sulfate under Langmuir monolayers of an aliphatic long-chain phos-phate. *Langmuir* 8:1492–1498, 1992.
132. BR Heywood, S Mann. Crystal recognition at inorganic organic interfaces: nucleation and growth of oriented $BaSO_4$ under compressed Langmuir monolayers. *Adv Mater* 4:278–282, 1992.
133. JH Fendler, FC Meldrum. Colloids chemistry and nanostructured materials. *Adv Mater* 7:607–632, 1995.
134. J Yang, JH Fendler. Morphology control of PbS nanocrystallites, epitaxitally under mixed monolayers. *J Phys Chem* 99:5505–5511, 1995.
135. J Yang, JH Fendler. Epitaxial-growth of size-quantized cadmium sulfide crystals under arachidic acid monolayers. *J Phys Chem* 99:5500–5504, 1995.
136. J Lahiri, G Xu, DM Dabbs, N Yao, IA Aksay, JT Groves. Porphyrin amphiphiles as templates for the nucleation of calcium carbonate. *J Am Chem Soc* 119:5449–5450, 1997.
137. J Lahiri, GF Fate, SB Ungashe, JT Groves. Multi-heme self-assembly in phospholipid vesicles. *J Am Chem Soc* 118:2347–2358, 1996.
138. S Mann, JP Hannington, RJP Williams. Phospholipid-vesicles as a model system for biomineralization. *Nature* 324:565–567, 1986.
139. GA Ozin, S Oliver. Skeletons in the beaker: synthetic hierarchical inorganic materials. *Adv Mater* 7:943–947, 1995.
140. JD Hartgerink, ER Zubarev, SI Stupp. Supramolecular one-dimensional objects. *Curr Opin Colloid Interface Sci* 5:355–361, 2001.
141. ED Sone, ER Zubarev, SI Stupp. Semiconductor nanohelices templated by supramo-lecular ribbons. *Angew Chem Int Ed* 41:1705–1709, 2002.

142. JD Hartgerink, E Beniash, SI Stupp. Self-assembly and mineralization of peptide-amphiphile nanofibers. *Science* 294:1684–1688, 2001.

143. W Traub, S Weiner. 3-dimensional ordered distribution of crystals in turkey collagen-fibers. *Proc Natl Acad Sci USA* 86:9822–9826, 1989.

144. P Terech, RG Weiss. Low molecular mass gelators of organic liquids and the properties of their gels. *Chem Rev* 97:3133–3160, 1997.

145. S Shinkai, K Murata. Cholesterol-based functional tectons as versatile building-blocks for liquid crystals, organic gels and monolayers. *J Mater Chem* 8:485–495, 1998.

146. O Gronwald, S Shinkai. Sugar-integrated gelators of organic solvents. *Chem Eur J* 7:4329–4334, 2001.

147. K Hanabusa, K Hiratsuka, M Kimura, H Shirai. Easy preparation and useful character of organogel electrolytes based on low molecular weight gelator. *Chem Mater* 11:649–655, 1999.

148. K Hanabusa, M Yamada, M Kimura, H Shirai. Prominent gelation and chiral aggregation of alkylamides derived from trans-1,2-diaminocyclohexane. *Angew Chem Int Ed* 35:1949–1951, 1996.

149. K Murata, M Aoki, T Suzuki, T Harada, H Kawabata, T Komori, F Ohseto, K Ueda, S Shinkai. Thermal and light control of the sol-gel phase-transition in cholesterol-based organic gels: novel helical aggregation modes as detected by circular-dichroism and electron-microscopic observation. *J Am Chem Soc* 116: 6664–6676,1994.

150. Y Ono, K Nakashima, M Sano, Y Kanekiyo, K Inoue, J Hojo, S Shinkai. Organic gels are useful as a template for the preparation of hollow fiber silica. *Chem Commun* 9: 1477–1478, 1998.

151. JH Jung, Y Ono, S Shinkai. Novel preparation method for multi-layered, tubular silica using an azacrown-appended cholesterol as template and metal-deposition into the interlayer space. *J Chem Soc Perkin Trans* 2:1289–1291, 1999.

152. Y Ono, K Nakashima, M Sano, J Hojo, S Shinkai. Template effect of cholesterol-based organogels on sol-gel polymerization creates novel silica with a helical structure. *Chem Lett* 1119–1120, 1999.

153. JH Jung, Y Ono, S Shinkai. Sol-gel polycondensation of tetraethoxysilane in a cholesterol-based organogel system results in chiral spiral silica. *Angew Chem Int Ed* 39:1862–1865, 2000.

154. JH Jung, Y Ono, S Shinkai. Sol-gel polycondensation in a cyclohexane-based organogel system in helical silica: creation of both right- and left-handed silica structures by helical organogel fibers. *Chem Eur J* 6:4552–4557, 2000.

155. JH Jung, Y Ono, K Hanabusa, S Shinkai. Creation of both right-handed and left-handed silica structures by sol-gel transcription of organogel fibers comprised of chiral diaminocyclohexane derivatives. *J Am Chem Soc* 122:5008–5009, 2000.

156. Y Ono, K Nakashima, M Sano, J Hojo, S Shinkai. Organogels are useful as a template for the preparation of novel helical silica fibers. *J Mater Chem* 11:2412–2419, 2001.

157. S Kobayashi, K Hanabusa, N Hamasaki, M Kimura, H Shirai, S Shinkai. Preparation of TiO$_2$ hollow-fibers using supramolecular assemblies. *Chem Mater* 12:1523–1525, 2000.

158. S Kobayashi, N Hamasaki, M Suzuki, M Kimura, H Shirai, K Hanabusa. Preparation of helical transition-metal oxide tubes using organogelators as structure-directing agents. *J Am Chem Soc* 124:6550– 6551, 2002.

159. JH Jung, H Kobayashi, KJC Bommel, S Shinkai, T Shimizu. Creation of novel helical ribbon and double-layered nanotube TiO$_2$ structures using an organogel template. *Chem Mater* 14:1445–1447, 2002.

160. H Cölfen. Double-hydrophilic block copolymers: synthesis and application as novel surfactants and crystal growth modifiers. *Macromol Rapid Commun* 22:219–252, 2001.

161. S Förster, T Plantenberg. From self-organizing polymers to nanohybrid and biomaterials. *Angew Chem Int Ed* 41:689–714, 2002.

162. JW Kriesel, MS Sander, TD Tilley. Block copolymer assisted synthesis of mesoporous, multicomponent oxides by nonhydrolytic, thermolytic decomposition of molecular precursors in nonpolar media. *Chem Mater* 13:3554–3563, 2001.

163. VV Hardikar, E Matijevic. Influence of ionic and nonionic dextrans on the formation of calcium hydroxide and calcium carbonate particles. *Colloids Surf A* 186:23–31, 2001.

164. FH Shen, QL Feng, CM Wang. The modulation of collagen on crystal morphology of calcium carbonate. *J Cryst Growth* 242:239–244, 2002.

165. QL Feng, G Pu, Y Pei, FZ Cui, HD Li, TN Kim. Polymorph and morphology of calcium carbonate crystals induced by proteins extracted from mollusk shell. *J Cryst Growth* 216:459–465, 2000.

166. J Aizenberg, G Lambert, S Weiner, L Addadi. Factors involved in the formation of amorphous and crystalline calcium carbonate: a study of an ascidian skeleton. *J Am Chem Soc* 124:32–39, 2002.

167. N Kröger, R Deutzmann, M Sumper. Polycationic peptides from diatom biosilica that direct silica nanosphere formation. *Science* 286:1129–1132, 1999.

168. N Kröger, R Deutzmann, C Bergsdorf, M Sumper. Species-specific polyamines from diatoms control silica morphology. *Proc Natl Acad Sci USA* 97:14133–14138, 2000.

169. N Kröger, S Lorenz, E Brunner, M Sumper. Self-assembly of highly phosphorylated silaffins and their function in biosilica morphogenesis. *Science* 298:584–586, 2002.

170. JN Cha, GD Stucky, DE Morse. Biomimetic synthesis of ordered silica structures mediated by block copolypeptides. *Nature* 403:289–292, 2000.

171. JN Cha, K Shimizu, Y Zhou, SC Christiansen, BF Chmelka, GD Stucky, DE Morse. Silicatein filaments and subunits from a marine sponge direct the polymerization of silica and silicones *in vitro*. *Proc Natl Acad Sci USA* 96:361–365, 1999.

172. DB DeOliveira, RA Laursen. Control of calcite crystal morphology by a peptide designed to bind to a specific surface. *J Am Chem Soc* 119:10627–10631, 1997.

173. CM Li, GD Botsaris, DL Kaplan. Selective *in vitro* effect of peptides on calcium carbonate crystallization. *Cryst Growth Design* 2:387–393, 2002.

174. LA Gower, DA Tirrell. Calcium carbonate films and helices grown in solutions of poly(aspartate). *J Cryst Growth* 191:153–160, 1998.

175. A Bigi, E Boanini, D Walsh, S Mann. Morphosynthesis of octacalcium phosphate hollow microspheres by polyelectrolyte-mediated crystallization. *Angew Chem Int Ed* 41:2163–2166, 2002.

176. LM Qi, H Cölfen, M Antonietti, M Lei, JD Hopwood, AJ Ashley, S Mann. Formation of $BaSO_4$ fibers with morphological complexity in aqueous polymer solutions. *Chem Eur J* 7:3526–3532, 2001.

177. SH Yu, M Antonietti, H Cölfen, J Hartmann. Growth and self-assembly of $BaCrO_4$ and $BaSO_4$ nanofibers toward hierarchical and repetitive superstructures by polymer-controlled mineralization reactions. *Nano Lett* 3:379–382, 2003.

178. LB Gower, DJ Odom. Deposition of calcium carbonate films by a polymer-induced liquid-precursor (PILP) process. *J Cryst Growth* 210:719–734, 2000.

179. VV Hardikar, E Matijevic. Influence of ionic and nonionic dextrans on the formation of calcium hydroxide and calcium carbonate particles. *Colloids Surf A* 186:23–31, 2001.

180. L Addadi, S Weiner. Control and design principles in biological mineralization. *Angew Chem Int Ed* 31:153–169, 1992.

181. VM Patel, P Sheth, A Kurz, M Ossenbeck, DO Shah, LB Gower. Synthesis of calcium carbonate coated emulsion droplets for drug detoxification. *ACS Symposium Series: Concentrated Colloidal Dispersions: Theory, Experiments, and Applications 2002*, 223rd National Meeting, B Markovic, P Somansundaran, Eds. American Chemical Society, in press.

182. MJ Olszta, DJ Odom, EP Douglas, LB Gower. A new paradigm for biomineral formation: mineralization via an amorphous liquid-phase precursor. *Conn Tissue Res* 44 (Suppl. 1):326–334, 2003.

183. K Naka, Y Tanaka, Y Chujo. Effect of anionic starburst dendrimers on the crystallization of $CaCO_3$ in aqueous solution: size control of spherical vaterite particles. *Langmuir* 18:3655–3658, 2002.

184. JJJM Donners, BR Heywood, EW Meijer, RJM Nolte, C Roman, APLHJ Schenning, NAJM Sommerdijk. Amorphous calcium carbonate stabilised by poly(propylene imine) dendrimers. *Chem Commun* 1937–1938, 2000.

185. JJJM Donners, BR Heywood, EW Meijer, RJM Nolte, NAJM Sommerdijk. Control over calcium carbonate phase formation by dendrimer/surfactant templates. *Chem Eur J* 8:2561–2567, 2002.

186. N Ueyama, T Hosoi, Y Yamada, M Doi, T Okamura, A Nakamura. Calcium complexes of carboxylate-containing polyamide with sterically disposed NH···O hydrogen bond: detection of the polyamide in calcium carbonate by ^{13}C cross-polarization/magic angle spinning spectra. *Macromolecules* 31:7119–7126, 1998.

187. N Ueyama, H Kozuki, M Doi, Y Yamada, K Takahashi, A Onoda, T Okamura, H Yamamoto. Secure binding of alternately amidated poly(acrylate) to crystalline calcium carbonate by NH···O hydrogen bond. *Macromolecules 34:*2607–2614, 2001.

188. N Ueyama, J Takeda, Y Yamada, A Onoda, T Okamura, A Nakamura. Dinuclear calcium complexes with intramolecularly NH···O hydrogen-bonded dicarboxylate ligands. *Inorg Chem* 38:475–478, 1998.

189. DH Napper. *Polymeric Stabilization of Colloidal Dispersions*. London: Academic Press, 1983, pp 1–30.

190. F Jones, H Cölfen, M Antonietti. Iron oxyhydroxide colloids stabilized with polysaccharides. *Colloid Polym Sci* 278:491–501, 2000.

191. H Cölfen. Biomimetic Mineralization Using Hydrophilic Copolymers: Synthesis of Hybrid Colloids with Complex Form and Pathways towards Their Analysis in Solution, Habilitation thesis, Potsdam, Germany, 2000.

192. H Cölfen. Double-hydrophilic block copolymers: synthesis and application as novel surfactants and crystal growth modifiers. *Macromol Rapid Commun* 22:219–252, 2001.

193. M Sedlak, M Antonietti, H Cölfen. Synthesis of a new class of double-hydrophilic block copolymers with calcium binding capacity as builders and for biomimetic structure control of minerals. *Macromol Chem Phys* 199:247–254, 1998.

194. M Sedlak, H Cölfen. Synthesis of double-hydrophilic block copolymers with hydrophobic moieties for the controlled crystallization of minerals. *Macromol Chem Phys* 202:587–597, 2001.

195. H Cölfen, M Antonietti. Crystal design of calcium carbonate microparticles using double-hydrophilic block copolymers. *Langmuir* 14:582–589, 1998.

196. H Cölfen, LM Qi. A systematic examination of the morphogenesis of calcium carbonate in the presence of a double-hydrophilic block copolymer. *Chem Eur J* 7:106–116, 2001.

197. JM Marentette, J Norwig, E Stockelmann, WH Meyer, G Wegner. Crystallization of CaCO$_3$ in the presence of PEO-block-PMAA copolymers. *Adv Mater* 9:647–651, 1997.
198. SH Yu, H Cölfen, J Hartmann, M Antonietti. Biomimetic crystallization of calcium carbonate spherules with controlled surface structures and sizes by double-hydrophilic block copolymers (DHBC). *Adv Funct Mater* 12:541–545, 2002.
199. M Antonietti, M Breulmann, C Göltner, H Cölfen, KK Wong, D Walsh, S Mann. Inorganic/organic mesostructures with complex architectures: precipitation of calcium phosphate in the presence of double-hydrophilic block copolymers. *Chem Eur J* 4:2493–2500, 1998.
200. LM Qi, H Cölfen, M Antonietti. Crystal design of barium sulfate using double-hydrophilic block copolymers. *Angew Chem Int Ed* 39:604–607, 2000.
201. LM Qi, H Cölfen, M Antonietti. Control of barite morphology by double-hydrophilic block copolymers. *Chem Mater* 12:2392–2403, 2000.
202. H Cölfen, LM Qi, Y Mastai, L Borger. Formation of unusual 10-petal BaSO$_4$ structures in the presence of a polymeric additive. *Cryst Growth Des* 2:191–196, 2002.
203. SH Yu, H Cölfen, M Antonietti. Control of the morphologenesis of barium chromate by using double-hydrophilic block copolymers (DHBCs) as crystal growth modifiers. *Chem Eur J* 8:2937–2345, 2002.
204. SH Yu, H Cölfen, M Antonietti. The combination of colloid-controlled hetergeneous nucleation and polymer-controlled crystallization: facile synthesis of separated, uniform high-aspect-ratio single-crystalline BaCrO$_4$ nanofibers. *Adv Mater* 15:133–136, 2003.
205. M Öner, J Norwig, WH Meyer, G Wegner. Control of ZnO crystallization by a PEO-b-PMAA diblock copolymer. *Chem Mater* 10:460–463, 1998.
206. A Taubert, D Palms, O Weiss, MT Piccini, DN Batchelder. Polymer-assisted control of particle morphology and particle size of zinc oxide precipitated from aqueous solution. *Chem Mater* 14:2594–2601, 2002.
207. SH Yu, M Antonietti, H Cölfen, M Giersig. Synthesis of very thin 1D and 2D CdWO$_4$ nanoparticles with improved fluorescence behavior by polymer control crystallization. *Angew Chem Int Ed* 41:2356–2360, 2002.
208. J Rudloff, M Antonietti, H Cölfen, J Pretula, K Kaluzynski, S Penczek. Double hydrophilic block copolymers with monophosphate ester moieties as crystal growth modifiers of CaCO$_3$. *Macromol Chem Phys* 203:627–635, 2002.
209. M Antonietti, M Weissenberger. Amphiphilic derivatives of poly(acrylic acid) as stabilizer in emulsion polymerization. *Macromol Chem Rapid Commun* 18:295–302, 1997.
210. LA Bromley, D Cottier, RJ Davey, B Dobbs, S Smith, BR Heywood. Interactions at the organic/inorganic interface: molecular design of crystallization inhibitors for barite. *Langmuir* 9:3594–3599, 1993.
211. R Kniep, S Busch. Biomimetic growth and self-assembly of fluorapatite aggregates by diffusion into denatured collagen matrices. *Angew Chem Int Ed* 35:2624–2626, 1996.
212. S Busch, H Dolhaine, A DuChesne, S Heinz, O Hochrein, F Laeri, O Podebrad, U Vietze, T Weiland, R Kniep. Biomimetic morphogenesis of fluorapatite–gelatin composites: fractal growth, the question of intrinsic electric fields, core/shell assemblies, hollow spheres and reorganization of denatured collagen, *Eur J Inorg Chem* 10:1643–1653, 1999.
213. H Cölfen, LM Qi. The mechanism of the morphogenesis of CaCO$_3$ in the presence of poly(ethylene glycol-b-poly(methacrylic acid). *Progr Colloid Polym Sci* 117:200–203, 2001.

214. J Rieger, E Hädicke, IU Rau, D Boeckh. A rational approach to the mechanisms of incrustation inhibition by polymeric additives. *Tens Surf Deterg* 34:430–435, 1997.

215. E Dalas, P Klepetsanis, PG Koutsoukos. The overgrowth of calcium carbonate on poly(vinyl chloride-co-vinyl acetate-co-maleic acid). *Langmuir* 15:8322–8327, 1999.

216. G Falini, S Albeck, S Weiner, L Addadi. Control of aragonite or calcite polymorphism by mollusk shell macromolecules. *Science* 271:67–69, 1996.

217. DB DeOliveira, RA Laursen. Control of calcite crystal morphology by a peptide designed to bind to a specific surface. *J Am Chem Soc* 119:10627–10631, 1997.

218. S Mann. *Biomineralization, Principles and Concepts in Bioinorganic Materials Chemistry.* New York: Oxford University Press, 2001.

219. J Küther, R Seshadri, W Tremel. Crystallization of calcite spherules around designer nuclei. *Angew Chem Int Ed* 37:3044–3047, 1998.

220. J Küther, R Seshadri, G Nelles, W Assenmacher, HJ Butt, W Mader, W Tremel. Mercaptophenol-protected gold colloids as nuclei for the crystallization of inorganic minerals: templated crystallization on curved surfaces. *Chem Mater* 5: 1317–1325, 1999.

221. F Jones, H Cölfen, M Antonietti. Interaction of kappa-carrageenan with nickel, cobalt, and iron hydroxides. *Biomacromolecules* 1:556–563, 2000.

222. SH Yu, H Cölfen, M Antonietti. Polymer controlled morphosyhthesis and mineralization of metal carbonate superstructures. *J Phys Chem B*, submitted.

223. LM Qi, J Li, JM Ma. Biomimetic morphogenesis of calcium carbonate in mixed solutions of surfactants and double-hydrophilic block copolymers. *Adv Mater* 14:300–303, 2002.

224. CY Tai, FB Chen. Polymorphism of $CaCO_3$ precipitated in a constant-composition environment. *AlChE J* 44:1790–1798, 1998.

225. LM Qi, J Li, JM Ma. Morphological control of $CaCO_3$ particles by a double-hydrophilic block copolymer in mixed alcohol–water solvents. *Chem J Chin Univ* 23:1595–1597, 2002.

226. N Saito, N Sonoyama, T Sakata. Analysis of the excitation and emission spectra of tungstates and molybdate. *Bull Chem Soc Jpn* 69:2191–2194, 1996.

227. S Champ, JA Dickinson, PS Fallon, BR Heywood, M Mascal. Hydrogen-bonded molecular ribbons as templates for the synthesis of modified mineral phases. *Angew Chem Int Ed* 39:2716–2719, 2000.

228. JJJM Donners, RJM Nolte, NAJM Sommerdijk. A shape persistent polymeric crystallization template for $CaCO_3$. *J Am Chem Soc* 124:9700–9701, 2002.

229. G Falini, S Fermani, A Ripamonti. Crystallization of calcium carbonate salts into beta-chitin scaffold. *J Inorg Biochem* 91:475–480, 2002.

230. SM D'Souza, C Alexander, MJ Whitcombe, AM Waller, EN Vulfson. Control of crystal morphology via molecular imprinting. *Polym Int* 50:429–432, 2001.

231. E Loste, FC Meldrum. Control of calcium carbonate morphology by transformation of an amorphous precursor in a constrained volume. *Chem Commun* 901–902, 2001.

232. PA Bianconi, J Lin, A Strzelecki. Crystallization of an inorganic phase controlled by a polymer matrix. *Nature* 349:315–317, 1991.

233. J Lin, E Cates, PA Bianconi. A synthetic analog of the biomineralization process: controlled crystallization of an inorganic phase by a polymer matrix. *J Am Chem Soc* 116:4738–4745, 1994.

234. MW Pitcher, E Cates, L Raboin, PA Bianconi. A synthetic analogue of the biomineralization process: formation of novel lead sulfide phases. *Chem Mater* 12:1738–1742, 2000.

235. RJ Park, FC Meldrum. Synthesis of single crystals of calcite with complex morphologies. *Adv Mater* 14:1167–1169, 2002.

236. G Falini. Crystallization of calcium carbonates in biologically inspired collagenous matrices. *Int J Inorg Mater* 2:455–461, 2000.

237. G Falini, S Fermani, M Gazzano, A Ripamonti. Polymorphism and architectural crystal assembly of calcium carbonate in biologically inspired polymeric matrices. *J Chem Soc Dalton Trans* 3983–3987, 2000.

238. O Grassmann, G Müller, P Löbmann. Organic–inorganic hybrid structure of calcite crystalline assemblies grown in a gelatin hydrogel matrix: relevance to biomineralization. *Chem Mater* 14:4530–4535, 2002.

239. S Busch, U Schwarz, R Kniep. Morphogenesis and structure of human teeth in relation to biomimetically grown fluorapatite–gelatine composites. *Chem Mater* 13:3260–3271, 2001.

240. S Busch, U Schwarz, R Kniep. Chemical and structural investigations of biomimetically grown fluorapatite–gelatin composite aggregates. *Adv Funct Mater* 13:189–198, 2003.

241. C Andrew, L Bassett. Biologic significance of piezoelectricity. *Calcif. Tissue Res* 1:252–272, 1968.

242. E Fukada, I Yasuda. On the piezoelectric effect of bone. *J Phys Soc Jpn* 12:1158–1162, 1957.

243. GVB Cochran, RJ Pawluk, CAL Bassett. Stress generated electric potentials in mandible and teeth. *Arch Oral Biol* 12: 917–920, 1967.

244. R Ogolnik, A Kleinfinder, A Fremaux, D Geiger. Piezoelectrical activity of dental tissues. *J Biol Buccale* 4:117–122, 1976.

245. R Rodríguez-Clemente, J Gómez-Morales. Microwave precipitation of $CaCO_3$ from homogeneous solutions. *J Cryst Growth* 169:339–346, 1996.

246. T Sugimoto. Preparation of monodispersed colloidal particles. *Adv Colloid Interface Sci* 28:65–108, 1987.

247. JW Mullin. *Crystallization*, 3rd ed. Oxford: Butterworth-Heinemann, 1997.

248. JK Bailey, CJ Brinker, ML Mecartney. Growth mechanisms of iron oxide particles of differing morphologies from the forced hydrolysis of ferric chloride solutions. *J Colloid Interface Sci* 157:1–13, 1993.

249. M Ocana, MP Morales, CJ Serna. The growth mechanism of α-Fe_2O_3 ellipsoidal particles in solution. *J Colloid Interface Sci* 171:85–91, 1995.

250. V Privman, DV Goia, J Park, E Matijevic. Mechanism of formation of monodispersed colloids by aggregation of nanosize precursors. *J Colloid Interface Sci* 213:36–45, 1999.

251. A Chemseddine, T Moritz. Nanostructuring titania: control over nanocrystal structure, size, shape, and organization. *Eur J Inorg Chem* 235–245, 1999.

252. RL Penn, JF Banfield. Morphology development and crystal growth in nanocrystalline aggregates under hydrothermal conditions: insights from titania. *Geochim Cosmochim Acta* 63:1549–1557, 1999.

253. RL Penn, JF Banfield. Imperfect oriented attachment: dislocation generation in defect-free nanocrystals. *Science* 281:969–971, 1998.

254. F Banfield, SA Welch, H Zhang, TT Ebert, RL Penn. Aggregation-based crystal growth and microstructure development in natural iron oxyhydroxide biomineralization products. *Science* 289:751–754, 2000.

255. RL Penn, G Oskam, TJ Strathmann, PC Searson, AT Stone, DR Veblen. Epitaxial assembly in aged colloids. *J Phys Chem B* 105:2177–2182, 2001.

256. K Onuma, A Ito. Cluster growth model for hydroxyapatite. *Chem Mater* 10:3346–3351, 1998.

257. F Huang, HZ Zhang, and JF Banfield. Two-stage crystal-growth kinetics observed during hydrothermal coarsening of nanocrystalline ZnS. *Nano Lett* 3:373–378, 2003.
258. AP Alivisatos. Perspectives: biomineralization: naturally aligned nanocrystals *Science* 289:736–737, 2000.
259. HL Zhu, RS Averback. Sintering processes of two nanoparticles: a study by molecular dynamics simulations. *Philos Mag Lett* 73:27–33, 1996.
260. M Yeadon, M Ghaly, JC Yang, RS Averback, JM Gibson. "Contact epitaxy" observed in supported nanoparticles. *Appl Phys Lett* 73:3208–3210, 1998.
261. C Pacholski, A Kornowski, H Weller. Self-assembly of ZnO: from nanodots to nanorods. *Angew Chem Int Ed* 41:1188–1191, 2002.
262. A Chemseddine, H Jungblut, S Boulmaaz. Investigation of the nanocluster self-assembly process by scanning tunneling microscopy and optical spectroscopy. *J Phys Chem* 100:12546–12551, 1996.
263. ZY Tang, NA Kotov, M Giersig. Spontaneous organization of single CdTe nanoparticles into luminescent nanowires. *Science* 297:237–240, 2002.

Assembly of Magnetic Particles

12 Magnetic Nanocrystals and Their Superstructures

Elena V. Shevchenko, Dmitri V. Talapin,
Andrey L. Rogach, and Horst Weller

CONTENTS

12.1   Introduction 342
12.2   Colloidal Synthesis of Magnetic Nanocrystals 342
      12.2.1   Size Tuning..343
              12.2.1.1   General Considerations 343
              12.2.1.2   Effect of Heating Time..346
              12.2.1.3   Effect of the Reaction Temperature 346
              12.2.1.4   Effect of the Nature and Concentration
                         of the Capping Ligands...348
              12.2.1.5   Growth via Additional Injections of Precursors 349
              12.2.1.6   Postpreparative Size-Selective Fractionation............350
      12.2.2   FePt and $CoPt_3$ Magnetic Alloy Nanocrystals:
             Varying the Molar Ratio between Components 350
      12.2.3   Shape Control..351
      12.2.4   Postpreparative Annealing...352
              12.2.4.1   Improvement of the Particle Crystallinity.................352
              12.2.4.2   Transition from fcc to $L1_0$ Phase in FePt
                         Nanocrystals 355
      12.2.5   Surface Ligand Exchange and Phase Transfer 355
12.3   Superstructures of Magnetic Nanocrystals...356
      12.3.1   Two- and Three-Dimensional Ordered Superlattices 356
      12.3.2   Colloidal Crystals of Nanocrystals 361
      12.3.3   Template Synthesis of Inverse Opals..361
12.4   Conclusions 364
Acknowledgments...364
References...364

12.1 INTRODUCTION

At the nanometer scale, many chemical and physical properties of solids become dramatically dependent on size. For ferromagnetic materials, this means that below some critical size, each particle can contain only a single magnetic domain. Changes in the magnetization can no longer occur through the motion of domain walls and instead require the coherent rotation of spins, resulting in larger coercivities. For a particle size smaller than the single domain size, the spins are increasingly affected by the thermal fluctuations and originally ferromagnetic material becomes super-paramagnetic.[1]

Nanometer-sized magnetic materials attract special attention because of their potential applications in ultra-high-density storage devices. Advanced ultra-high-density recording media require uniform particles with room temperature coercivity H_c of ~3000 to 5000 Oe. It is expected that H_c values close to this value can be achieved in intermetallic nanocrystals such as FePt or CoPt.[2] To date, the monodisperse nanocrystals of CoPt and FePt magnetic alloys are the best candidates for the permanent magnetic applications because of their large uniaxial anisotropy[3–5] and high chemical stability.[6,7]

This contribution provides an overview of the recent progress in preparation of magnetic nanocrystals utilizing colloidal chemistry approaches. The nanocrystals synthesized by this route consist of a magnetic core capped with a shell of surface ligands, which prevents agglomeration of nanocrystals in a colloidal solution and determines the interparticle spacing in their ordered two- and three-dimensional superstructures. The syntheses of monodisperse highly crystalline nanoparticles with tunable size, shape, and variable surface capping are described. The assembly of monodisperse magnetic particles into superstructures is further discussed as an important step toward fabrication of functional devices made from individual nano-crystals as building blocks.

12.2 COLLOIDAL SYNTHESIS OF MAGNETIC NANOCRYSTALS

An ideal synthesis of nanocrystals has to provide the achievement of desired particle sizes over the largest possible range, narrow size distributions, high crystallinity, control of shape, and desired surface properties. For potential application of nano-crystals, a relatively low cost of the final product becomes an additional requirement. Numerous methods such as vacuum deposition techniques,[8] chemical vapor deposition,[9] normal incidence pulsed laser deposition,[10,11] colloidal syntheses utilizing thermal[2,3,5,7,12–17] or sonochemical[18,19] decomposition of organometallic precursors, high-temperature reduction of metal salts,[20–22] and chemical reactions inside reverse micelles[23] have been applied for preparation of magnetic metal nanoparticles. In a further discussion, we will concentrate mainly on different modifications of the colloidal organometallic synthesis proposed by Murray et al.[2,3] In this approach, magnetic nanocrystals consisting of either a single metal or an alloy are formed in solution from organometallic precursors in the presence of organic molecules (so-called stabilizing agents, or capping ligands) binding to the nanocrystal surface and regulating their growth. The stabilizing agents prevent uncontrollable growth and fusion of nanocrystals in a colloidal solution and provide their solubility in different solvents.[24] The proper choice of the capping ligands is of great importance, as the attractive magnetic interactions between nanocrystals tend to decrease an overall stability of magnetic colloids. A relatively high reaction temperature ($\sim$150 to 300°C) is essential for the high crystallinity of the particles formed. Several successful syntheses based on the colloidal organometallic approach were developed for Fe,[16] Ni,[17] Co,[2,12–15,22] FePt,[3] CoPt$_3$,[7] g-Fe$_2$O$_3$,[25] and Fe$_3$O$_4$[26] magnetic nanocrystals. Table 12.1 provides an overview of the reported precursors, stabilizing agents, growth conditions, and resulting sizes and shapes.

A typical colloidal organometallic synthesis is carried out as shown in Figure 12.1 on an example of CoPt$_3$ nanocrystals.[7] It is performed under an inert gas atmosphere using a standard Schlenk line technique. Injection of the cobalt precursor into a hot solution containing platinum acetylacetonate and a stabilizing agent (adamantane carboxylic acid (ACA)) in a mixture of hexadecylamine (HDA) and diphenyl ether induces fast simultaneous decomposition of platinum and cobalt precursors, yielding CoPt$_3$ nanocrystals. Subsequent annealing of the nanoparticles at $\sim$280°C is necessary to improve their crystallinity. After the reaction mixture is cooled to room temperature, the nanocrystals are precipitated by addition of pro-panol-2 and isolated from the crude solution by centrifugation. The isolated nano-crystals are easily redispersable in a variety of nonpolar organic solvents, such as chloroform, hexane, or toluene.

Figure 12.2 shows size evolution of the x-ray diffraction (XRD) pattern of CoPt$_3$ nanocrystals. The reflexes indicating the chemically disordered face-centered-cubic (fcc) CoPt$_3$ phase become narrower upon increasing the particle size from $\sim$3 to 17 nm.

12.2.1 Size Tuning

12.2.1.1 General Considerations

In the case of crystalline nanoparticles with a size in the range of 3 to 10 nm and consisting of $\sim$10^2 to 10^4 atoms, nearly continuous size tunability is possible, as an addition or removal of a unit cell requires only a small variation of the nanocrystal free energy. The method used to control the size depends on peculiarities of the concrete synthesis. For instance, MnFe$_2$O$_4$ nanoparticles with a rather narrow size distribution can be synthesized inside water-in-toluene reverse micelles playing in the role of nanoreactors. The size of micelles depends on the water-to-toluene volume ratio and determines the size of spinel ferrite nanoparticles.[27] Below we discuss different methods of size tuning for magnetic nanocrystals prepared by organome-tallic synthesis.

Any synthesis of colloidal particles employing soluble precursors involves two consecutive stages: nucleation and growth.[28] The narrow size distribution of the resulting nanocrystals can be achieved only if the nucleation stage is temporally separated from the further growth.[29] The nuclei grow by consuming dissolved molec-ular species (monomers) from the surrounding solution. When all molecular precur-sors are consumed, the reaction mixture has two scenarios for further evolution. If the addition of a monomer to a nanocrystal is a reversible process (i.e., a nanocrystal can both consume and release the monomer), the nanocrystals grow via the so-called

TABLE 12.1
Examples of Magnetic Nanocrystals Synthesized by Colloidal Organometallic Routes

Material	Stabilizing Agents	Size Range, nm	Shape	Reaction Temperature, °C	References
Fe	TOPO	2	Spheres	340	16
	TOPO–TOP–DDAB	2×11 to 2×27	Rods	320	16
	OA, TBP	6	Spheres	250	2
e-Co	OA-alkylphosphine	3–13	Spheres	200–250	2, 22
hcp-Co	OA, lauric acid, TOP	3–17	Spheres	180	15
	OA–TOPO, OA–primary amine, primary amine–TOPO	2×4 to 4×90	Discs	182	13, 14
g-Fe$_2$O$_3$ (maghemite)	OA	4–13	Spheres	100, 300	25
Fe$_3$O$_4$ (magnetite)	OA–oleyl amine	3–20	Spheres	265	26
FePt	OA–oleyl amine	3–10	Spheres	100, 300	3
CoPt$_3$	ACA, primary amine–ACA, ACA–TDPA	1.5–18	Spheres, Wires, Cubes	145–225	7

Note: OA = oleic acid; TOPO = trioctylphosphine oxide; TOP = trioctylphosphine; TBP = tributylphosphine; DDAB = dodecyldimethylammonium bromide; ACA = adamantane carboxylic acid; TDPA = tetradecylphosphonic acid.

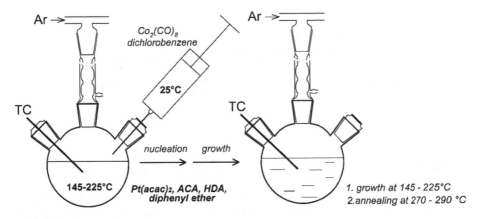

FIGURE 12.1 Organometallic synthesis of CoPt₃ nanocrystals.

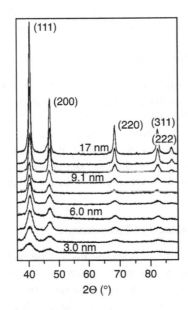

FIGURE 12.2 Size evolution of the XRD pattern of CoPt₃ nanocrystals. Average sizes were calculated by the Debye–Scherrer equation.

Ostwald ripening, when smaller particles evolve monomers and dissolve, whereas larger particles consume monomers and grow.[30–32] If the transformation of molecular precursors to nanoparticles is irreversible, no changes of the particle size via Ostwald ripening is possible and the reaction comes to saturation.

If the first scenario is realized, the Ostwald ripening provides a possibility of a slow and controllable increase of the mean particle size in the course of heating.[32] Moreover, Ostwald ripening of nanometer-sized particles results in much narrower size distributions than those observed for micrometer-sized particles.[33] If the Ostwald ripening in a given system is impossible or very slow, the tuning of the nanocrystal

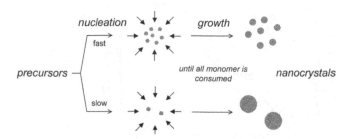

SCHEME 12.1 Synthesis of nanoparticles in the absence of Ostwald ripening.

size can be achieved only by control over nucleation and growth rates, as illustrated in Scheme 12.1. Fast nucleation provides a high concentration of nuclei, finally yielding smaller nanocrystals, whereas slow nucleation provides a low concentration of seeds consuming the same amount of monomer, resulting in larger particles.[34]

Below we consider the experimental results reported for different magnetic nanocrystals from the point of the influence of various parameters on the particle size.

12.2.1.2   Effect of Heating Time

The Ostwald ripening growth regime is not typical in the case of metal nanocrystals.[24] The influence of the reaction time on the size of magnetic nanocrystals is discussed in several works.[7,14,22,35] For practically important FePt and CoPt$_3$ magnetic alloy nanocrystals, no pronounced influence of the reaction time on size was found (Figure 12.3). In contrast, prolonged heating resulted in progressive decomposition of the stabilizing agent molecules rather than in nanocrystal growth. On the other hand, excessive nanocrystal growth in the course of heating was reported for e-Co nanocrystals.[22]

12.2.1.3   Effect of the Reaction Temperature

The reaction temperature affects both the nucleation and the growth rates in any colloidal synthesis of nanocrystals. In the absence of Ostwald ripening, the variation of the reaction temperature was used for tuning the size of Co[15,22] and CoPt$_3$[7] nanocrystals. An increase of the nanocrystal size, with an increase in the reaction temperature from 170 to 200°C, was reported for Co nanocrystals prepared via reduction of the CoCl$_2$ superhydride.[22] The opposite behavior was reported for Co nanocrystals prepared via thermal decomposition of Co$_2$(CO)$_8$.[12] Decomposition of cobalt carbonyl at 180°C resulted in ~9-nm Co nanocrystals, whereas the reaction at 100°C yielded ~25 nm particles.[12]

A detailed study on the dependence of nanocrystal size on the reaction temperature was performed for CoPt$_3$ magnetic alloy nanocrystals[34] (Figure 12.4 and Figure 12.5). An increase of temperature from 145 to 220°C led to a decrease of average final particle size from ~10 to 3 nm. This can be explained by taking into account the temperature dependencies of the nanocrystal nucleation and growth rates. The activation energy for the nucleation process is usually much higher than that for the

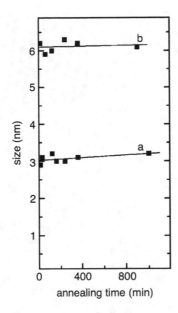

FIGURE 12.3 Dependence of FePt (a) and $CoPt_3$ (b) nanocrystal size on duration of heating during the synthesis. FePt and $CoPt_3$ nanocrystals were prepared via organometallic route in OA–oleyl amine[3] and ACA–HDA[34] coordinating mixtures, respectively. The size of nanocrystals was calculated from the width of the XRD reflexes according to the Debye–Scherrer equation.

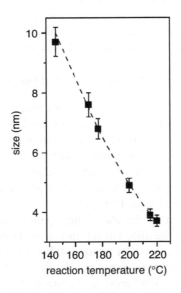

FIGURE 12.4 Dependence of $CoPt_3$ nanocrystal size on the reaction temperature.

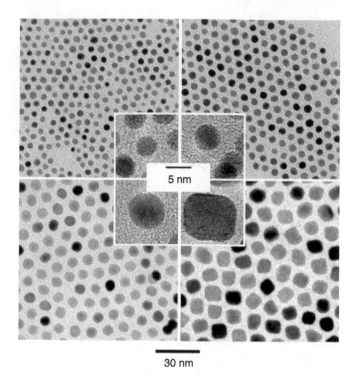

30 nm

FIGURE 12.5 TEM and HRTEM images showing effect of the reaction temperature on the mean size and size distribution of CoPt₃ nanocrystals: 3.7-, 4.9-, 6.3-, and 9.3-nm large particles were prepared at 220, 200, 170 and 145°C, respectively.

particle growth,[28] which makes the nucleation rate more sensitive to changes in temperature than the growth rate. The reaction temperature can thus be used to adjust the balance between nucleation and growth rates. At higher temperatures, more nuclei are formed and the final particle size is smaller (Scheme 12.1).

12.2.1.4 Effect of the Nature and Concentration of the Capping Ligands

Bulky capping ligands are usually favorable for preparation of smaller particles, whereas less bulkiness leads to larger nanocrystals. For instance, the synthesis of Co nanocrystals in the presence of bulky alkylphosphine ligand $P(C_8H_{17})_3$ resulted in ~3- to 7-nm particles, whereas a similar synthesis in the presence of $P(C_4H_9)_3$ yielded ~6- to 11-nm nanocrystals.[20,22] Similar behavior was reported for Ni nanocrystals stabilized with oleic acid (OA) and tributylphosphine (TBP) (~12 to 13 nm) or trioctylphosphine (TOP) (~8 to 10 nm).[2]

The effect of variation of the stabilizing agent concentration on the size of magnetic nanocrystals can be different for various systems. An increase in concentration of ligands resulted in smaller nanocrystals for Co,[2,35] Ni,[2] and cobalt ferrite prepared via oxidation of a uniform iron-cobalt alloy,[36] whereas the opposite behavior was observed for CoPt₃ nanocrystals (Figure 12.6)[34] and maghemite particles prepared via oxidation of iron seeds.[25]

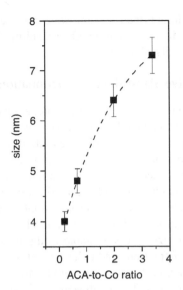

FIGURE 12.6 Influence of the amount of stabilizing agent (ACA) on CoPt$_3$ nanocrystal size.

For the growth of nanocrystals in a kinetic regime, the decrease of their size with the increase of the stabilizer concentration can take place if a strong passivation of the nanocrystal surface results in slowing down of the growth rate, whereas the nucleation rate remains less influenced (Scheme 12.1). The opposite situation can also be realized, as studies on CoPt$_3$ nanocrystals show.[34] Stabilizing agents such as ACA and amines, used to passivate the surface of growing nanocrystals, can also form stable complexes with individual metal atoms of a molecular precursor. Thus, platinum precursor Pt acetylacetonate decomposed in the HDA–diphenyl ether–1,2-hexadecandiol mixture at ~130°C, whereas addition of ACA increased the decomposition temperature up to ~220°C. A similar behavior was observed for the cobalt precursor: addition of ACA drastically enhanced the stability of the cobalt carbonyl solution against thermodecomposition. Formation of these complexes precedes the nucleation step[37] and decreases the monomer reactivity; therefore, an increase of the amount of ligand is expected to suppress the nucleation rate. In accordance with Scheme 12.1, slowing the nucleation rate results in a larger final nanocrystal size, as was also observed experimentally (Figure 12.6).

12.2.1.5 Growth via Additional Injections of Precursors

For some reaction schemes, it is rather difficult to increase the size of nanocrystals to the desirable value by varying the experimental conditions mentioned above. In this case, additional amounts of precursors can be introduced into a reaction mixture containing presynthesized nanocrystals, which then act as nucleation seeds for the deposition of additional monomer. Control of the final nanocrystal size can be achieved by varying the amount of precursors. This universal approach was successfully applied to increase the size of both magnetic and semiconductor nanocrystals.[3,7,38,39] Thus, the syntheses of FePt nanocrystals with a size reaching 10 nm and of -Fe$_2$O$_3$ nanocrystals larger than 11 nm were reported.[3,25] A problem of this approach may be an additional nucleation effect resulting in the formation of nanocrystals with a broad size distribution.[7]

12.2.1.6 Postpreparative Size-Selective Fractionation

Another general approach for size tuning is a postpreparative separation of a colloidal solution of polydisperse nanocrystals into a series of fractions with narrower size distribution. This technique is widely applied to colloids of semiconductor nanoparticles.[40,41] One of the commonly used methods of size fractionation is a so-called size-selective precipitation based on stepwise addition of a nonsolvent into a solution of nanocrystals in order to induce a gentle destabilization of the colloid. Since larger nanocrystals exhibit greater attractive Van der Waals forces,[24] the addition of a nonsolvent results in aggregation of the largest particles in a given sample, whereas the particles with smaller sizes are still stable in the form of colloidal dispersion. Sequential precipitation of fractions and collection of them by centrifugation allow the isolation of particles with different sizes with a narrowed size distribution. If the nanoparticles are sensitive to oxidation, all steps of size-selective precipitation have to be carried out under an inert atmosphere.

For magnetic nanocrystals, the size-selective precipitation technique can also be performed by applying an external magnetic field, as larger particles usually exhibit stronger attractive magnetic forces and precipitate in the magnetic field before smaller nanocrystals.[2] Thus, magnetic separation was performed to separate hcp-Co nanodisks from -Co nanospheres.[14]

12.2.2 FePt and CoPt$_3$ Magnetic Alloy Nanocrystals: Varying the Molar Ratio between Components

The synthesis of monodisperse FePt nanocrystals is based on simultaneous thermal decomposition of iron carbonyl and reduction of platinum salt in the presence of OA–oleyl amine stabilizing agents.[3] The control of the molar ratio of iron and platinum precursors in the reaction mixture can be used to tune the composition of resulting Fe$_x$Pt$_{1-x}$ alloy nanocrystals. Thus, a 3-to-2 molar ratio of iron carbonyl to platinum salt resulted in the formation of Fe$_{48}$Pt$_{52}$ nanocrystals, a 2-to-1 ratio provided Fe$_{52}$Pt$_{48}$, and a 4-to-1 molar ratio led to Fe$_{70}$Pt$_{30}$.[3] In contrast, CoPt$_3$ nanocrystals prepared from Pt(acac)$_2$ and Co$_2$(CO)$_8$ in the presence of ACA and HDA showed nearly constant composition upon variation of the molar ratio between cobalt and platinum precursors.[7,34]

On the other hand, a strong influence of the molar ratio between Co and Pt precursors on the final CoPt$_3$ nanocrystal size was observed.[34] An increase of the concentration of the Co precursor resulted in a decrease of the particle size (Figure 12.7A), whereas an increase of concentration of the Pt precursor led to an increase of size of CoPt$_3$ nanocrystals (Figure 12.7B). A possible explanation suggests that concentrations of the platinum and cobalt precursors affect the nucleation rate in different ways. We assume that the growth of CoPt$_3$ nanocrystals starts from the formation of some polynuclear Co clusters such as Co$_4$(CO)$_{12}$ and Co$_6$(CO)$_{16}$,[14]

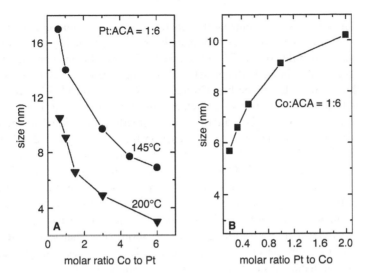

FIGURE 12.7 (A) Influence of the molar ratio between Co and Pt precursors (with the Pt concentration kept constant) on the CoPt₃ nanocrystal size at different reaction temperatures. (B) Influence of the molar ratio between Pt and Co precursors (with the Co concentration kept constant) on the CoPt₃ nanocrystal size for the reaction temperature 200°C. The amount of stabilizer (ACA) was the same in all experiments.

serving as nuclei for further growth. When the concentration of the cobalt precursor increases, the number of these Co seeds also increases, finally leading to formation of smaller CoPt₃ nanocrystals, whereas concentration of the platinum precursor does not affect the amount of Co seeds.

12.2.3 SHAPE CONTROL

Earlier studies on magnetic nanocrystals were focused on the preparation of particles with nearly spherical shapes. However, the shape anisotropy of magnetic nanocrystals can result in advanced magnetic properties.[1] Preparation of colloidal nanocrystals of different shapes has attracted much attention during recent years. Approaches to control the shape of nanoparticles were proposed for several semiconductor (CdSe,[42,43] PbS,[44] ZnO[45]) and metal (Pt,[46] Au[47]) materials. Some shape-controlled syntheses were also proposed for magnetic nanocrystals.[7,12–14,16,35] Thus, ~2-nm spherical Fe nanoparticles prepared by thermal decomposition of Fe(CO)₅ were transformed into nanorods with a diameter of ~2 nm and different aspect ratios, depending on the concentration of dodecyldimethylammonium bromide (DDAB) in the reaction mixture.[16] Shape transformation of spherical iron nanoparticles into rods was explained by their mutual oriented attachment followed by fusion. The iron nanorods formed possessed a body-centered cubic structure of -Fe and exhibited enhanced magnetic properties compared to the spherical nanocrystals. Thus, the blocking temperature (the temperature of transition from the superparamagnetic to ferromagnetic state) of iron nanorods was found to be ~10 times higher than that for spherical nanoparticles.[16]

A possibility of shape-controlled preparation was also reported for Co nano-crystals.[12–14,35] The formation of hcp Co nanodisks was explained by selective adsorption of trioctylphosphine oxide (TOPO) on the (101) face of Co nanoparticles, which inhibited its growth relative to the growth of other faces, leading to a disk-like shape. The length and diameter of the hcp Co nanodisks were controlled by the variation of reaction time and precursor-to-surfactant ratio. For instance, at a fixed OA concentration, the diameter of Co nanodisks was directly proportional to the concentration of TOPO.[13] The increase of the reaction time finally yielded spherical Co nanocrystals. This shape transformation with the increase of the reaction time was explained by the dual role of TOPO acting as a selective absorber and altering the relative growth rates of different faces of the nanocrystals, on the one hand, and promoting atom exchange with the monomer during the growth, on the other hand.[14] The use of linear amines improved the reaction yields of Co nanodisks, whereas trialkylamines completely prohibited their formation. Note that a side view of the nanodisks in transmission electron microscopy (TEM) images strongly resembles nanorods, so the nanodisks can be confused with real nanorods.[14] To verify the disk shape, TEM tilting experiments were necessary.

CoPt$_3$ nanowires were synthesized in ACA–HDA–diphenyl ether mixture by performing the reaction at a relatively low temperature (~100°C).[7] The CoPt$_3$ nano-wire-reach fraction (Figure 12.8A) was separated from spherical nanoparticles by high-speed centrifugation of colloidal solution. The length of wires varied from 6 nm up to 150 to 200 nm. The wires were typically composed of two heads at the ends, which were approximately the same size as the spherical particles presenting in the same sample, and a body with some bulges (Figure 12.8B). The concentration of wires and their length decreased drastically upon prolonged heating of the reaction mixture. Under continuous refluxing (~3 h), the amount of nanowires became negligible (Figure 12.8C).The mechanism of formation and decomposition of CoPt$_3$ nanowires is still not clear and requires further investigation.

The TEM and high-resolution TEM (HRTEM) images of CoPt$_3$ nanocrystals with sizes smaller than ~7 nm commonly indicate their nearly spherical shape (Figure 12.5). However, further increase of the nanocrystal size results in a rather abrupt transition from quasi-spherical to strongly faceted (as well as cubic) CoPt$_3$ nano-crystals (Figure 12.9). The cubic particles can form if the growth rate in the [111] direction is much higher than that in the [100] direction.[48] It was also observed that annealing at ~275°C resulted in smoothing the edges of cubic nanocrystals formed at 145°C (cf. Figure 12.5D and Figure 12.9). The ratio between growth rates in different directions can be varied by specific adsorption of organic surfactant at particular crystallographic facets, inhibiting the growth in a particular direction.[14] The phenomena responsible for the abrupt transition from spherical to faceted particles during growth in a colloidal solution require further studies.

12.2.4 POSTPREPARATIVE ANNEALING

12.2.4.1 Improvement of the Particle Crystallinity

Magnetic properties of nanocrystals are strongly dependent not only on particle size but also on the crystal structure and the presence of faults.[20] Thus, internal structural

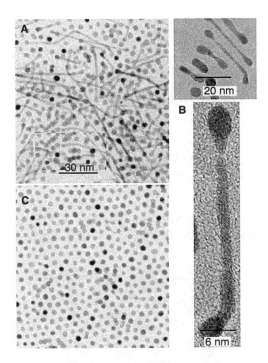

FIGURE 12.8 TEM images of $CoPt_3$ nanocrystals grown in the ACA–HDA–diphenyl ether mixture: nanowire-rich $CoPt_3$ fraction (A), enlarged view of nanowires (B), and spherical $CoPt_3$ nanocrystals formed from the nanowire-rich sample after its refluxing for 3 h (C). (From Shevchenko, E.V. et al., *J. Am. Chem. Soc.*, 124, 11480–11485, 2002. With permission. Copyright © 2002 American Chemical Society.)

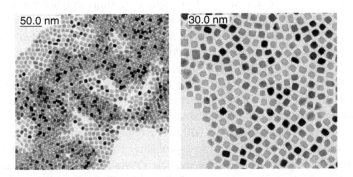

FIGURE 12.9 TEM overview images of faceted $CoPt_3$ nanocrystals prepared at 145°C.

defects (stacking faults, twinned planes) considerably reduce coercivity, the parameter determining the applicability of a material in the magnetic recording and storage.[2]

Information about nanoparticle crystallinity can be obtained from comparison of the average particle sizes estimated by powder XRD and TEM methods. The width of the XRD reflexes provides information about the x-ray coherence length,

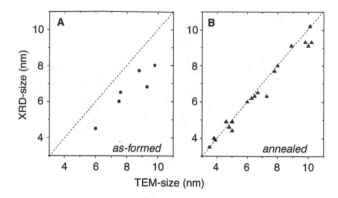

FIGURE 12.10 Comparison of the XRD and TEM sizes of CoPt₃ nanocrystals before (A) and after (B) annealing at 275 to 285°C.

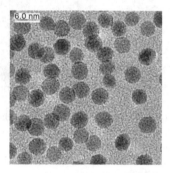

FIGURE 12.11 HRTEM image of annealed CoPt₃ nanocrystals.

which is close to the average size of the single crystalline domain inside the nanocrystal, whereas TEM images show the total size of a nanoparticle. Nanocrystal sizes estimated by these methods will be further referred to as XRD size and TEM size, correspondingly. For CoPt₃ nanocrystals prepared at 140 to 200°C, a large difference between XRD and TEM sizes was observed (Figure 12.10A), with an XRD size (x-ray coherence length) being considerably smaller. Annealing of these nanocrystals at the boiling point of the crude solution (~275 to 285°C) for ~30 min resulted in an increase of the XRD size, whereas no considerable change of the TEM size was observed (Figure 12.10B). The observed decrease in the difference between XRD and TEM sizes is most probably a result of annealing of defects accompanied by improvement of the particle crystallinity. The annealing process requires relatively high temperatures (200 to 300°C), necessary to trigger atom diffusion inside the nanocrystals.[49] After annealing, the XRD size becomes nearly equal to the TEM size (Figure 12.10B), indicating that most nanocrystals are single crystallites with a nearly perfect lattice. HRTEM images additionally confirmed the excellent crystallinity of annealed CoPt₃ nanocrystals (Figure 12.11). Almost all nanocrystals possessed the lattice fringes without stacking faults and other defects.

12.2.4.2 Transition from fcc to L1$_0$ Phase in FePt Nanocrystals

As prepared, FePt nanocrystals possess the chemically disordered face-centered-cubic (fcc) structure with relatively low magnetocrystalline anisotropy. The annealing of fcc-FePt nanocrystals allowed their conversion into the face-centered tetragonal L1$_0$ phase with high magnetic anisotropy.[3] The phase transition temperature for FePt nanoparticles was significantly lower than that reported for bulk FePt alloy. Thus, depending on Fe$_x$Pt$_{1-x}$ stoichiometry and nanocrystal size, the transition temperature was in the range of 500 to 700°C, whereas for the bulk material the reported value was 1300°C.[50] However, considerable coalescence of nanocrystals took place under this thermal treatment. In order to reduce the phase transition temperature, it was proposed that FePt nanocrystals be doped with silver.[51] Incorporation of Ag promoted the transition from the fcc to the tetragonal L1$_0$ phase, reducing the transition temperature by ~100 to 150°C compared to the pure FePt nanocrystals. It was suggested that Ag atoms leave the FePt lattice at temperatures less than 400°C, and the vacancies formed increase the mobility of the Fe and Pt atoms, accelerating the phase transformation.

12.2.5 SURFACE LIGAND EXCHANGE AND PHASE TRANSFER

The nanocrystal surface can be modified by postpreparative exchange of the capping ligands.[3,24] This procedure provides an additional degree of freedom in further handling the nanocrystals because the chemistry of the surface ligands determines a number of important nanocrystal properties: solubility in polar or nonpolar solvents, stability against oxidation, interparticle spacing in self-assembled superstructures, etc. Moreover, specific surface groups of the stabilizers can be used for linking magnetic nanocrystals to different species, such as other nanoparticles or biomolecules. For instance, Fe$_2$O$_3$–Fe$_3$O$_4$ magnetic nanocrystals functionalized with amine groups were coupled with oligonucleotides and used as a magnetic nanosensor for the detection of specific oligonucleotides.[52] It was even shown that species adsorbed at the surface of magnetic nanoparticles, e.g., electron donors such as pyridine, can change the magnetization and decrease the surface contribution into magnetic behavior of nanocrystals.[53] Poly(ethylenimine)-functionalized FePt nanocrystals were packed into uniform films with controllable dimension and thickness using the layer-by-layer deposition technique.[54]

Some potential applications of nanocrystals, such as biolabeling, require magnetic particles to be water soluble and stable in water, with no loss of properties over long periods. Through the surface ligand exchange of the initial stabilizers — ACA–tetradecylphosphonic acid (TDPA) in the case of CoPt$_3$[7] and OA–oleyl amine in the case of FePt[3] nanocrystals — with long-chain 11-mercaptoundecanoic acid, CoPt$_3$ and FePt nanocrystals can be successfully transferred from nonpolar solvents to water. Correspondent aqueous colloids of magnetic nanocrystals revealed essential stability. Figure 12.12 presents TEM and HRTEM images of water-soluble FePt and CoPt$_3$ nanocrystals possessing the same size and size distribution that they had before the surface exchange.

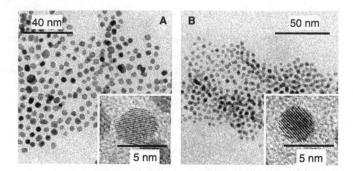

FIGURE 12.12 TEM and HRTEM (insets) images of FePt (A) and CoPt$_3$ (B) nanocrystals after their transfer from toluene to water using 11-mercaptoundecanoic acid.

12.3 SUPERSTRUCTURES OF MAGNETIC NANOCRYSTALS

Monodisperse magnetic nanocrystals can be used as building blocks for fabrication of two- and three-dimensional superstuctures. The particles organize themselves in the ordered structures because of attractive interactions through the Van der Waals or magnetic forces.[24,55,56] However, too-strong interparticle interactions can disturb self-assembly of nanocrystals. Thus, slow evaporation of a highly boiling solvent resulted in assembly of 9-nm e-Co nanocrystals into a hexagonal two-dimensional superlattice, whereas no ordered assemblies were observed for larger Co nanocrystals due to the strong dipole–dipole interactions (the superparamagnetic transition for ~12-nm e-Co nanocrystals takes place at room temperature).[15]

Potentially, self-assembly of colloidal nanocrystals can reach the precision of lithographically produced arrays within the nanometer scale. Magnetostatic particle interactions in the closely packed arrays can result in a broad transition from super-paramagnetic to ferromagnetic behavior.[20] Assemblies of metal and semiconductor nanoparticles can also be considered as a model system for investigating collective transport and optical phenomena in ordered structures.[57–59]

12.3.1 TWO- AND THREE-DIMENSIONAL ORDERED SUPERLATTICES

Self-assembly of monodisperse nanocrystals under different conditions of destabilization of a colloidal solution can result in the formation of either closely packed glassy films or highly ordered superlattices. Closely packed glassy films possessing a short-range order only are usually formed if the evaporation rate of a solvent is relatively high.[24] Superlattices with a low concentration of lattice defects and a long-range order of nanocrystals are formed if the particles are mobile enough and have enough time to find their lowest energy sites in the superstructure. Thus, slow evaporation of a highly boiling carrier solvent at increased temperatures (50 to 60°C) is favorable for formation of ordered superstructures. In addition to two- and three-dimensional superlattices known for other systems, such as monodisperse semiconductor nanocrystals and latex spheres, some specific self-organized structures inherent only to magnetic nanocrystals were reported. Thus, spherical Co nanocrystals

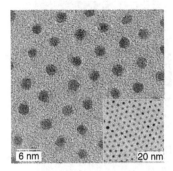

FIGURE 12.13 Hexagonal closely packed two-dimensional array of FePt nanocrystals on a TEM grid. (From Shevchenko, E. et al., *Adv. Mater.*, 14, 287–290, 2002. With permission. Copyright © 2002 Wiley-VCH.)

can organize themselves into pearl-like aggregates and even form bracelet-like rings in order to minimize the magnetostatic energy.[13,60]

The wetting properties of the substrate can influence the superlattice morphology.[2] If the colloidal solution of nanoparticles wets the substrate, a two-dimensional superlattice grows, preferably forming a monolayer. As the surface coverage increases, the nanocrystals adsorb to the ledges and kinks of the growing structure, forming terraces that extend laterally across the substrate. If the colloidal solution does not wet the substrate, a three-dimensional superlattice preferably grows, with facets reflecting the packing symmetry inherent for the nanocrystals.

FePt nanocrystals with narrow size distributions prepared by the method of Sun et al.[3] tend to organize themselves into ordered arrays whose quality strongly depends on the deposition conditions.[56,61] Figure 12.13 shows an image of a hexagonal two-dimensional array of 2.5-nm large FePt nanoparticles on a TEM grid. The regular interparticle spacing of ~40 Å is determined by the capping ligands (OA–oleyl amine). The ligands maintain the interparticle distance within self-assembled superstructures and determine the spacing between the nearest neighbors. The possibility of exchange of OA and oleyl amine with short-chain ligands in order to vary interparticle distances within a superlattice was demonstrated.[3]

Owing to their narrow particle size distribution and the uniform spherical shape, the $CoPt_3$ nanocrystals have a strong tendency to self-organize into two- and three-dimensional superstructures.[7,56] Spontaneous self-assembly was observed when colloidal solutions of $CoPt_3$ nanocrystals were spread onto a substrate with subsequent slow evaporation of the solvent. Depending on the particle size and conditions of the solvent evaporation, several kinds of self-organized superstructures were observed. Figure 12.14A shows a TEM overview image of a closely packed monolayer of $CoPt_3$ nanocrystals. If the surface coverage by nanocrystals was higher than one monolayer, nanocrystals of the second layer occupied positions between the nanocrystals in the first layer (Figure 12.14B and C). This structure can be formed due to the relatively large interparticle spacing of ~2.5 nm maintained by the bulky ACA capping ligands. The third layer of nanocrystals occupied the positions typical for the cubic close packing ccp phase, as it is seen through the difference in phase

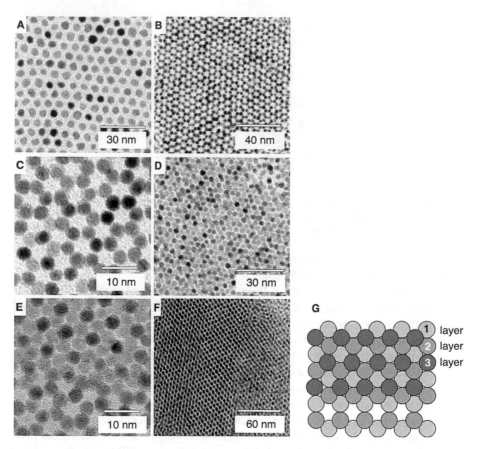

FIGURE 12.14 TEM images of a monolayer of 4.8-nm $CoPt_3$ nanoparticles (A); two layers of 3.6- and 4.0-nm $CoPt_3$ nanoparticles (B, C); three layers of 3.6-nm $CoPt_3$ nanoparticles (D); three layers of 4.0-nm $CoPt_3$ nanoparticles (E); and multiple layers of 4.5-nm $CoPt_3$ nanoparticles (F). (G) Graphical illustration of a three-layer arrangement of $CoPt_3$ nanocrystals. (From Shevchenko, E.V. et al., *J. Am. Chem. Soc.*, 124, 11480–11485, 2002. With permission. Copyright © 2002 American Chemical Society.)

contrast between two underlying layers and a darker third layer (Figure 12.14D and E). Figure 12.14F presents a TEM image of a multilayer three-dimensional superlattice where the $CoPt_3$ nanocrystals are arranged in a nearly defect-free three-dimensional structure exhibiting long-range order. A graphical illustration (Figure 12.14G) clarifies the three-layer arrangement of $CoPt_3$ nanocrystals.

Two-dimensional superlattices built from monodisperse nanoparticles of two different sizes were recently reported.[62,63] Au or Ag and Au nanoparticles organized themselves in regular AB_2 or AB alloy superstructures, depending on the relative amount of each kind of nanocrystals and the ratio of particle diameters. By mixing of two monodisperse colloids of magnetic $CoPt_3$ nanocrystals (4.5 and 2.6 nm diameter), followed by slow evaporation of the solvent, a three-dimensional alloy

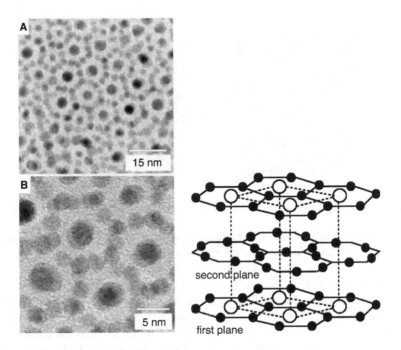

FIGURE 12.15 TEM images illustrating three-dimensional superlattices of $CoPt_3$ nanoparticles formed in a mixture of two different sizes (4.5 and 2.6 nm). Scheme: Illustrative drawing of the AB_5 structure. (From Shevchenko, E.V. et al., *J. Am. Chem. Soc.*, 124, 11480–11485, 2002. With permission. Copyright © 2002 American Chemical Society.)

superlattice was obtained (Figure 12.15A and B). The images correspond to the structure of the same type (AB_5) as a $CaCu_5$ intermetallic compound[64] (Figure 12.15, scheme). The same arrangement was also reported for a superlattice formed by latex globules of two different sizes.[65,66] In the first plane, each 4.5-nm $CoPt_3$ nanocrystal is surrounded by a hexagon formed from 2.6-nm nanocrystals. The second plane consists only of hexagons of small particles that are rotationally shifted by 30° relative to the first plane, and the third plane repeats the first one.

Two- or three-dimensional superlattices consisting of both magnetically hard and soft phases are promising for advanced permanent magnetic applications. Self-assembly of FePt and Fe_3O_4 nanocrystals was used to build exchange-coupled nanocomposite magnets. The magnetic properties of resulting nanocomposites were dependent on the initial mass ratios of FePt and Fe_3O_4 nanocrystals. Depending on the size of the nanoparticles, the formation of different structures was reported. In the case of 4-nm large FePt and 4-nm large Fe_3O_4 nanocrystals, a hexagonal lattice of randomly distributed metal and oxide nanocrystals was obtained, whereas self-assembly of 4-nm large FePt and 8-nm large Fe_3O_4 nanocrystals led to a locally ordered structure in which each Fe_3O_4 nanoparticle was surrounded by six to eight FePt nanocrystals. A large difference in particle sizes resulted in a phase segregation of metal alloy and oxide nanocrystals.

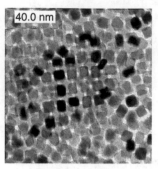

FIGURE 12.16 TEM images of self-assembled structures of cubic CoPt₃ nanocrystals.

The shape of individual nanocrystals can strongly affect the morphology of self-assembled structures. In the case of spherical particles, the crystalline lattices of individual nanocrystals are randomly oriented within a superstructure, as a rule. However, faceted and, in particular, cubic nanocrystals can form lattice-matched superstructures, where each nanocrystal has the same orientation as the neighbors (Figure 12.16). Such superstructures are of interest for creating materials with a high magnetic anisotropy constant because of the possibility to align the magnetization axes of individual nanocrystals.[67] The magnetic anisotropy of CoPt₃ nanocrystals[5] makes them superior candidates for realizing this new kind of artificial solids. Another example of shape-driven self-assembly of magnetic nanocrystals is long ribbons of the hcp-Co nanodisks stacked face-to-face and lying on the edge.[14]

Self-assembly of colloidal nanocrystals can also be affected by applying an external magnetic film. When it was applied in the plane of the substrate during deposition of colloidal nanocrystals, spindle-shaped superlattices grew with their long axes aligned along the external magnetic field.[2] Within these spindle-shaped deposits, individual magnetic nanocrystals were arranged into nearly perfect arrays showing a long-range order. The direction of the external magnetic field influenced the morphology of two-dimensional self-assembled structures of magnetic nanocrystals, as was reported for Co nanoparticles.[68] If the magnetic film was applied perpendicular to the substrate, uniform two-dimensional ordered hexagonal superlattices were obtained, whereas the magnetic field applied parallel to the substrate led to the formation of stripes.

The annealing of superlattices can change the packing symmetry of the self-assembled magnetic nanocrystals. Under annealing of two-dimensional arrays of -Co nanoparticles under vacuum or inert atmosphere at ~300°C, the packing symmetry of three-dimensional ordered regions was converted from originally hexagonal to cubic.[22] The internal transition of the nanocrystal structure from cubic -Co to hcp-Co was responsible for this change. Further annealing of Co nanoparticles led to a dramatic change of the crystalline core, fusing and sintering of nanoparticles, and destruction of the superlattice. The temperature of nanocrystal coalescence may depend on the substrate. Thus, no obvious aggregation of FePt nanocrystals in a monolayer was observed even at 700°C on a SiO₂ substrate, whereas on an amorphous carbon surface the particles coalesced at ~530°C.[3,69]

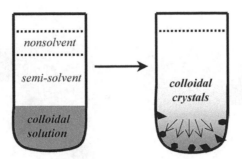

SCHEME 12.2 Outline illustrating the concept of the oversaturation technique used for growing colloidal crystals consisting of nanoparticles.

12.3.2 COLLOIDAL CRYSTALS OF NANOCRYSTALS

Crystallization of monodisperse nanocrystals into macroscopic solids (colloidal crystals) is the next step in preparation of ordered superstructures. In these colloidal crystals, monodisperse nanocrystals can be considered as artificial atoms in the next level of hierarchy. Monodisperse $CoPt_3$ nanocrystals have a strong tendency to form three-dimensional superlattices, as evidenced from Figure 12.14. Applying the recently introduced technique of controlled oversaturation,[70] micrometer-sized colloidal crystals consisting of monodisperse FePt [61] and $CoPt_3$ [7] nanoparticles were grown. Crystallization of nanoparticles was induced by slow diffusion of a nonsolvent (methanol) into a solution of FePt or $CoPt_3$ nanocrystals through a buffer layer of propanol-2 (Scheme 12.2). Slow destabilization of a colloidal dispersion of nanoparticles in toluene resulted in nucleation of colloidal crystals, preferentially on the walls of a glass tube. Thicker black and thinner brownish-colored colloidal crystals visible to the naked eye were formed in approximately 2 weeks.

Figure 12.17A and B shows optical micrographs of FePt colloidal crystals adhering to the glass. The crystals grew preferably in the form of faceted triangular or hexagonal platelets, 10 to 30 μm on a side. The crystals with pyramidal-like shapes were also observed. Scanning electron microscopy (SEM) provides a closer look at the morphology of colloidal crystals. In the case of FePt, clearly distinguishable were faceted triangular platelets, incomplete hexagonal platelets, and tetrahedronal colloidal crystals with imperfect sides showing terraces and cleaved ledges (Figure 12.17C to E). Figure 12.18 presents SEM images of well-faceted hexagonal, triangular, and pyramid-like colloidal crystals built from $CoPt_3$ nanocrystals.

Arrangement of nanocrystals at the surface of colloidal crystals can be studied using TEM and HRSEM techniques (Figure 12.19). TEM images were taken from small fragments of the FePt colloidal crystals obtained by mechanical grinding and treatment in an ultrasonic bath. Hexagonal arrangement of nanocrystals, i.e., building blocks of the colloidal crystals, is clearly seen at the edges of the crystalline pieces and in the HRSEM image.

12.3.3 TEMPLATE SYNTHESIS OF INVERSE OPALS

From the moment of the introduction of the photonic band gap concept by Yablonovitch[71] and John,[72] materials possessing a three-dimensional periodicity of

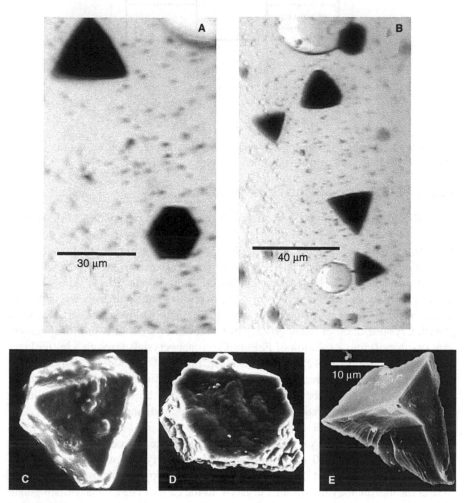

FIGURE 12.17 Optical micrographs (A, B) and SEM images (C–E) of colloidal crystals of FePt nanoparticles. (From Shevchenko, E. et al., *Adv. Mater.*, 14, 287–290, 2002. With permission. Copyright © 2002 Wiley-VCH.)

the dielectric constant have attracted much attention. The periodically varying index of refraction in photonic crystals causes a redistribution of the density of photonic states due to the Bragg diffraction, which is associated with stop bands for light propagation. Chemical self-assembly methods have been utilized for making three-dimensional photonic crystals (so-called artificial opals) from colloidal silica or polymer micrometer-sized spheres. Using artificial opals as templates, so-called inverse opals can be prepared by impregnation of voids with different materials, followed by subsequent removal of the spheres by annealing or dissolution, which can possess a complete photonic band gap.[73]

FIGURE 12.18 SEM images of colloidal crystals of $CoPt_3$ nanoparticles. (From Shevchenko, E.V. et al., *J. Am. Chem. Soc.*, 124, 11480–11485, 2002. With permission. Copyright © 2002 American Chemical Society.)

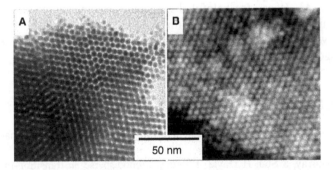

FIGURE 12.19 (A) TEM image of a fragment of an FePt colloidal crystal. (B) HRSEM image of the surface of an FePt colloidal crystal. (From Rogach, A.L. et al., *Adv. Funct. Mater.*, 12, 653–664, 2002. With permission. Copyright © 2002 Wiley-VCH.)

Magnetic nanocrystals can be used for fabrication of metal-dielectric inverse opals addressable through an external magnetic field. Figure 12.20 shows an SEM image of an inverse FePt colloidal crystal obtained by complete impregnation of a three-dimensional template of monodisperse latex microspheres with FePt nanocrystals, followed by dissolution of polymer spheres in toluene.

FIGURE 12.20 SEM image of an inverse colloidal crystal made from PtFe nanocrystals. (Courtesy of Dr. N. Gaponik.)

12.4 CONCLUSIONS

Colloidal chemistry provides powerful tools to produce highly monodisperse magnetic nanocrystals with variable size, shape, composition, and surface properties. Tuning these parameters, in turn, allows variation of magnetic properties of nanoparticles in a broad range. The assembly of monodisperse magnetic nanocrystals in superstructures is already a step toward fabrication of functional devices consisting of individual particles as building blocks, whose appearance can be envisaged for the near future.

ACKNOWLEDGMENTS

We thank Dr. M. Haase for helpful discussions. Special appreciation goes to A. Kornowski and S. Bartholdi-Nawrath for assistance with the TEM measurements. This work was supported by the EU research project BIOAND and the Deutsche Forschungsgemeinschaft through SFB 508.

REFERENCES

1. DL Leslie-Pelecky, RD Rieke. Magnetic properties of nanostructured materials. *Chem Mater* 8: 1770–1783, 1996.
2. CB Murray, S Sun, H Doyle, T Betley. Monodisperse $3d$ transition-metal (Co, Ni, Fe) nanoparticles and their assembly into nanoparticle superlattices. *MRS Bull* 26(2): 985–991, 2001.
3. Sun, CB Murray, D Weller, L Folks, A Moser. Monodisperse FePt nanoparticles and ferromagnetic FePt nanocrystal superlattices. *Science* 287: 1989–1992, 2000.
4. D Weller, H Brändle, C Chappert. Relationship between Kerr effect and perpendicular magnetic anisotropy in $Co_{1-x}Pt_x$ and $Co_{1-x}Pd_x$ alloys. *J Magn Magn Mater* 121: 452–460, 1993.
5. F Wiekhorst, E Shevchenko, H Weller, J Kötzler. Anisotropic superparamagnetism of monodispersive cobalt–platinum nanocrystals. *Phys Rev B* 67(22): 224416/1-224416/11, 2003.

6. TA Tyson, SD Conradson, RFC Farrow, BA Jones. Observation of internal interfaces in Pt_xCo_{1-x} (x ~ 0.7) alloy films: a likely cause of perpendicular magnetic anisotropy. *Phys Rev B* 54: R3702–R3705, 1996.

7. EV Shevchenko, DV Talapin, AL Rogach, A Kornowski, M Haase, H Weller. Colloidal synthesis and self-assembly of $CoPt_3$ nanocrystals. *J Am Chem Soc* 124: 11480–11485, 2002.

8. RA Ristau, K Barmak, LH Lewis, KR Coffey, JK Howard. On the relationship of high coercivity and $L1_0$ ordered phase in CoPt and PtFe thin films. *J Appl Phys* 86: 4527–4534, 1999.

9. SW-K Choi, J Puddephatt. Cobalt–palladium and cobalt–platinum bilayer films formed by chemical vapor deposition. *Chem Mater* 9: 1191–1195, 1997.

10. IJ Jeon, D Kim, JS Song, JH Her, DR Lee, KB Lee. Normal-incidence pulsed-laser deposition: better method for fabrication of multilayer structures. *Appl Phys A Mater Sci Proc* 70: 235–238, 2000.

11. IJ Jeon, DW Kang, DE Kim, DH Kim, SB Choe, SC Shin. Control of the magnetic anisotropy of a Co/Pt nanomultilayer with embedded particles. *Adv Mater* 14: 1116–1120, 2002.

12. VF Puntes, K Krishnan, AP Alivisatos. Synthesis of colloidal cobalt nanoparticles with controlled size and shapes. *Top Catalysis* 19: 145–148, 2002.

13. VF Puntes, K Krishnan, AP Alivisatos. Colloidal nanocrystal shape and size control: the case of cobalt. *Science* 291: 2115–2117, 2001.

14. VF Puntes, D Zanchet, CK Erdonmez, AP Alivisatos. Synthesis of hcp-Co nanodisks. *J Am Chem Soc* 124: 12874–12880, 2002.

15. VF Puntes, K Krishnan, AP Alivisatos. Synthesis, self-assembly, and magnetic behavior of a two-dimensional superlattice of single-crystal Co nanoparticles. *Appl Phys Lett* 78: 2187–2189, 2001.

16. SJ Park, S Kim, S Lee, ZG Khim, K Char, T Hyeon. Synthesis and magnetic studies of uniform iron nanorods and nanospheres. *J Am Chem Soc* 122: 8581–8582, 2000.

17. TO Ely, C Amiens, B Chaudret, E Snoeck, M Verelst, M Respaud, JM Broto. Synthesis of nickel nanoparticles. Influence of aggregation induced by modification of poly(vinylpyrrolidone) chain length on their magnetic properties. *Chem Mater* 11: 526–529, 1999.

18. K Suslick, M Fang, T Hyeon. Sonochemical synthesis of iron colloids. *J Am Chem Soc* 118: 11960–11961, 1996.

19. DN Srivastava, N Perkas, A Gedanken, I Felner. Sonochemical synthesis of mesoporous iron oxide and accounts of its magnetic and catalytic properties. *J Phys Chem B* 106: 1878–1883, 2002.

20. S Sun, CB Murray. Synthesis of monodisperse cobalt nanocrystals and their assembly into magnetic superlattices. *J Appl Phys* 85: 4325–4330, 1999.

21. C Petit, A Taleb, MP Pileni. Cobalt nanosized particles organized in a 2D superlattice: synthesis, characterization, and magnetic properties. *J Phys Chem B* 103: 1805–1810, 1999.

22. S Sun, CB Murray, H Doyle. Controlled assembly of monodisperse e-cobalt-based nanocrystals. *Mater Res Soc Symp Proc* 577: 385–398, 1999.

23. C Liu, B Zou, AJ Rondinone, ZJ Zhang. Reverse micelle synthesis and characterization of superparamagnetic $MnFe_2O_4$ spinel ferrite nanocrystallites. *J Phys Chem B* 104: 1141–1145, 2000.

24. CB Murray, CR Kagan, MG Bawendi. Synthesis and characterization of monodisperse nanocrystals and closed-packed nanocrystal assemblies. *Annu Rev Mater Sci* 30: 545–581, 2000.

25. T Hyeon, SS Lee, J Park, Y Chung, HB Na. Synthesis of highly crystalline and monodisperse maghemite nanocrystallites without a size-selection process. *J Am Chem Soc* 123: 12798–12801, 2001.

26. S Sun, H Zeng. Size-controlled synthesis of magnetite nanoparticles. *J Am Chem Soc* 124: 8204–8205, 2002.

27. C Liu, B Zou, AJ Rondinone, ZJ Zhang. Reverse micelle synthesis and characterization of superparamagnetic $MnFe_2O_4$ spinel ferrite nanocrystallites. *J Phys Chem B* 104: 1141–1145, 2000.

28. T. Sugimoto. *Monodisperse Particles.* Amsterdam: Elsevier, 2001, 793 pp.

29. VK La Mer, RH Dinegar. Theory, production and mechanism of formation of monodispersed hydrosols. *J Am Chem Soc* 72: 4847–4854, 1950.

30. IM Lifshitz, VV Slyozov. The kinetics of precipitation from supersaturated solid solutions. *J Phys Chem Solids* 19: 35–49, 1961.

31. C Wagner. Theorie der Alterung von Niederschlägen durch Umlösen. *Z Elektrochem* 65: 581–591, 1961.

32. AS Kabalnov, ED Shchukin. Ostwald ripening theory: applications to fluorocarbon emulsion stability. *Adv Colloid Interfac Sci* 38: 69–97, 1992.

33. DV Talapin, AL Rogach, M Haase, H Weller. Evolution of an ensemble of nanoparticles in a colloidal solution: theoretical study. *J Phys Chem B* 105: 12278–12285, 2001.

34. EV Shevchenko, DV Talapin, H Schnablegger, A Kornowski, M Haase, H Weller. Study of nucleation and growth in the hot organometallic synthesis of magnetic alloy nanocrystals: the role of nucleation rate in size control of $CoPt_3$ nanocrystals. *J Am Chem Soc* 125(30): 9090–9101, 2003.

35. JI Park, NJ Kang, Y Jun, SJ Oh, HC Ri, J Cheon. Superlattice and magnetism directed by the size and shape of nanocrystals. *Chem Phys Chem* 3: 543–547, 2002.

36. T Hyeon, Y Chung, J Park, SS Lee, YW Kim, BH Park. Synthesis of highly crystalline and monodisperse cobalt ferrite nanocrystals. *J Phys Chem B* 106: 6831–6833, 2002.

37. WW Yu, X Peng. Formation of high-quality CdS and other II-VI semiconductor nanocrystals in noncoordinating solvents: tunable reactivity of monomers. *Angew Chem Int Ed* 41: 2368–2371, 2002.

38. X Peng, J Wickham, AP Alivisatos. Kinetics of II-VI and III-V colloidal semiconductor nanocrystal growth: "focusing" of size distributions. *J Am Chem Soc* 120: 5343–5344, 1998.

39. DV Talapin, AL Rogach, A Kornowski, M Haase, H Weller. Highly luminescent monodisperse CdSe and CdSe/ZnS nanocrystals synthesized in a hexadecylamine–trioctylphosphine oxide–trioctylphospine mixture. *Nano Lett* 1: 207–211, 2001.

40. A Chemseddine, H Weller. Highly monodisperse quantum sized CdS particles by size selective precipitaion. *Ber Bunsen-Ges Chem* 97: 636–637, 1993.

41. CB Murray, DJ Norris, MG Bawendi. Synthesis and characterization of nearly monodisperse CdE (E = sulfur, selenium, tellurium) semiconductor nanocrystallites. *J Am Chem Soc* 115: 8706–8715, 1993.

42. X Peng, L Manna, W Yang, J Wickham, E Scher, A Kadavanich, AP Alivisatos. Shape control of CdSe nanocrystals. *Nature* 404: 59–61, 2000.

43. L Manna, EC Scher, AP Alivisatos. Synthesis of soluble and processable rod-, arrow-, teardrop-, and tetrapod-shaped CdSe nanocrystals. *J Am Chem Soc 122:* 12700–12706, 2000.

44. SM Lee, YW Jun, SN Cho, J Cheon. Single-crystalline star-shaped nanocrystals and their evolution: programming the geometry of nano-building blocks. *J Am Chem Soc* 124: 11244–11245, 2002.

45. C Pacholski, A Kornowski, H Weller. Self-assembly of ZnO: from nanodots to nanorods. *Angew Chem Int Ed* 41: 1188–1191, 2002.
46. TS Ahmadi, ZL Wang, A Henglein, MA El-Sayed. "Cubic" colloidal platinum nanoparticles. *Chem Mater* 8: 1161–1163, 1996.
47. NR Jana, L Gearheart, CJ Murphy. Wet chemical synthesis of high aspect ratio cylindrical gold nanorods. *J Phys Chem B* 105: 4065–4067, 2001.
48. ZL Wang. Structural analysis of self-assembling nanocrystals superlattices. *Adv Mater* 10: 13–30, 1998.
49. MH Yang, CP Flynn. Growth of alkali halides from molecular beams: global growth characteristics. *Phys Rev Lett* 62: 2476–2479, 1989.
50. M Hansen. *Constitution of Binary Alloys*. New York: McGraw-Hill, 1958, pp. 492–494.
51. S Kang, JW Harrell, DE Nikles. Reduction of the fcc to $L1_0$ ordering temperature for self-assembled FePt nanoparticles containing Ag. *Nano Lett* 2: 1033–1036, 2002.
52. L Josephson, JM Perez, R Weissleder. Magnetic nanosensors for the detection of oligonucleotide sequences. *Angew Chem Int Ed* 40: 3204–3206, 2001.
53. DA van Leeuwen, JM van Ruitenbeek, LJ de Jongh. Quenching of magnetic moments by ligand–metal interactions in nanosized magnetic metal clusters. *Phys Rev Lett* 73: 1432–1435, 1994.
54. S Sun, S Anders, HF Hamann, JU Thiele, JEE Baglin, T Thomson, EE Fullerton, CB Murray, BD Terris. Polymer mediated self-assembly of magnetic nanoparticles. *J Am Chem Soc* 124: 2884–2885, 2002.
55. CP Collier, T Vossmeyer, JR Heath. Nanocrystal superlattices. *Ann Rev Phys Chem* 49: 371–404, 1998.
56. AL Rogach, DV Talapin, EV Shevchenko, A Kornowski, M Haase, H Weller. Organisation of matter on different size scales: monodisperse nanocrystals and their superstructures. *Adv Funct Mater* 12: 653–664, 2002.
57. SH Kim, G Medeiros-Ribeiro, DAA Ohlberg, RS Williams, JR Heath. Individual and collective electronic properties of Ag nanocrystals. *J Phys Chem B 103:* 10341–10347, 1999.
58. CT Black, CB Murray, RL Sandstrom, S Sun. Spin-dependent tunneling in self-assembled cobalt-nanocrystal superlattices. *Science* 290: 1131–1134, 2000.
59. G Markovich, CP Collier, JR Heath. Reversible metal-insulator transition in ordered metal nanocrystal monolayers observed by impedance spectroscopy. *Phys Rev Lett* 80: 3807–3810, 1998.
60. SL Tripp, SV Pusztay, AE Ribbe, A Wei. Self-assembly of cobalt nanoparticle rings. *J Am Chem Soc* 124: 7914–7915, 2002.
61. E Shevchenko, D Talapin, A Kornowski, F Wiekhorst, J Kötzler, M Haase, A Rogach, H Weller. Colloidal crystals of monodisperse FePt nanoparticles grown by a three-layer technique of controlled oversaturation. *Adv Mater* 14: 287–290, 2002.
62. CJ Kiely, J Fink, M Brust, D Bethell, DJ Schiffrin. Spontaneous ordering of bimodal ensembles of nanoscopic gold clusters. *Nature* 396: 444–446, 1998.
63. CJ Kiely, J Fink, JG Zheng, M Brust, D Bethell, DJ Schiffrin. Ordered colloidal nanoalloys. *Adv Mater* 12: 640–643, 2000.
64. WB Pearson. *Crystal Chemistry and Physics of Metals and Alloys*. London: Wiley-Interscience, 1972, 643 pp.
65. S Hachisu, S Youshimura. Optical demonstration of crystalline superstructures of latex globules. *Nature* 283: 188–189, 1980.
66. S Youshimura, S Hachisu. Order formation in binary-mixture of hard sphere colloids. *Prog Colloid Polym Sci* 68: 59–70, 1983.

67. H Zeng, J Li, JP Liu, ZL Wang, S Sun. Exchange-coupled nanocomposite magnets by nanoparticle self-assembly. *Nature* 420: 395–398, 2002.
68. M Giersig, M Hilgendorff. On the road from single, nanosized magnetic clusters to multi-dimensional nanostructures. *Colloids Surfaces A Physicochem Eng Aspects* 202: 207–213, 2002.
69. ZR Dai, S Sun, ZL Wang. Phase transformation, coalescence, and twinning of monodisperse FePt nanocrystals. *Nano Lett* 1: 443–447, 2001.
70. DV Talapin, EV Shevchenko, A Kornowski, N Gaponik, M Haase, AL Rogach, H Weller. A new approach to crystallization of CdSe nanoparticles into ordered three-dimensional superlattices. *Adv Mater* 13: 1868–1871, 2001.
71. E Yablonovitch. Inhibited spontaneous emission in solid-state physics and electronics. *Phys Rev Lett* 58: 2059–2062, 1987.
72. S John. Strong localization of photons in certain disordered dielectric superlattices. *Phys Rev Lett* 58: 2486–2489, 1987.
73. A Blanco, E Chomski, S Grabtchak, M Ibisate, S John, SW Leonard, C Lopez, F Meseguer, H Miguez, JP Mondia, GA Ozin, O Toader, HM van Driel. Large-scale synthesis of a silicon photonic crystal with a complete three-dimensional bandgap near 1.5 micrometres. *Nature* 405: 437–440, 2000.

13 Synthesis, Self-Assembly, and Phase Transformation of FePt Magnetic Nanoparticles

Shishou Kang, Xiangcheng Sun, J.W. Harrell, and David E. Nikles

CONTENTS

13.1 FePt for Granular Thin-Film Magnetic Media..369
13.2 FePtCu Ternary Alloy Nanoparticles...371
13.3 FePtAg and FePtAu Ternary Alloy Nanoparticles 375
13.4 Magnetic Properties of FePtAg and FePtAu Nanoparticle Arrays...........378
13.5 Significance of This Work 381
Acknowledgments...382
References..382

13.1 FePt FOR GRANULAR THIN-FILM MAGNETIC MEDIA

The information storage industry has a 50-year track record of increasing the data storage capacity of its magnetic disk drive products at an exponential pace[1] that matches or exceeds Moore's law.[2,3] The state-of-the-art data storage density is 100 gigabits per square inch,[4] and the industry is now pursuing basic research aimed at densities beyond 1 terabit per square inch. These high data storage densities require new recording media with high uniaxial magnetic anisotropy, high moment, and small grain sizes.[5] The $L1_0$ phase of FePt has $K_u = 6.6 - 10 \times 10^7$ erg/cm^3 and $M_s = 1140$ emu/cm^3,[6] making it an excellent candidate, provided that films containing very small grains can be prepared.

A team from IBM first reported the synthesis of nearly monodisperse 3- to 4-nm-diameter FePt nanoparticles by the simultaneous polyol reduction of platinum acetylacetonate and the thermal decomposition of iron pentacarbonyl.[7] As prepared, the FePt particles had a face-centered-cubic (FCC) structure and were superparamagnetic. They had a coating of organic surfactants and could be dispersed in hexane. When the dispersion was cast onto a solid substrate and the hexane evaporated

369

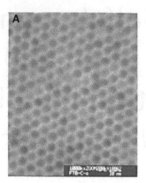 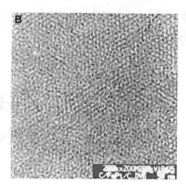

FIGURE 13.1 TEM images of self-assembled films containing FePt nanoparticles before (A) and after (B) heat treatment at 600°C.

slowly, the particles self-assembled into films containing closely packed arrays. After heat treatment at temperatures above 500°C in an inert atmosphere, the particles transformed to the $L1_0$ (tetragonal) phase, having high magnetocrystalline anisotropy, giving films with high coercivity. The coercivity depended on the heating temperature, and for complete phase transformation, the films had to be heated to nearly 580°C. One is able to record on the films with a linear density of 127 kfcpi and demonstrated that the thermal stability factor ($KV/k_BT > 48$) was suitable for a thin-film magnetic recording medium.

We have repeated the synthesis of FePt nanoparticles and prepared self-assembled films consisting of closely packed arrays of particles (Figure 13.1A).[8] After heat treatment, the particles transformed to the $L1_0$ phase with the degree of transformation dependent on the heat treatment temperature. Detailed magnetic characterizations, particularly magnetic susceptibility curves, revealed a distribution of magnetic anisotropies, which may arise from a distribution of degrees of phase transformation. Indeed, the distribution was narrower for films heat treated at higher temperatures, as the phase transformation was more complete. However, even for the films heat treated at the highest temperatures, the distribution of magnetic anisotropies was still rather broad. At the higher heat treatment temperatures, there was also considerable particle coalescence, giving rise to a distribution of particle volumes.

Self-assembled FePt nanoparticles are attractive candidates for granular thin-film recording media. Before this application can be realized, some basic materials science issues must be resolved. The films must be heat treated to transform the particles to the $L1_0$ phase. This causes degradation of the organic surfactants, leading to loss in particle positional order, as seen in Figure 13.1B. Furthermore, as the heat treatment temperature exceeds 550°C, there is extensive particle coalescence.[9] Particle coalescence leads to increased switching volumes, which defeats the purpose of making small particles. The solution to this problem could be to heat treat at lower temperatures, thereby decreasing the amount of particle coalescence. However, heat treatment at lower temperatures results in incomplete phase transformation, giving films with a distribution of order parameters and, in turn, a distribution of magnetic anisotropies, which is detrimental to high-density magnetic recording. It would be highly desirable if the FCC to tetragonal phase transformation could be carried out at lower temperatures, at least below temperatures at which the particles coalesce and preferably below temperatures at which the surfactants decompose.

It has been shown that adding copper into sputter-deposited FePt films greatly reduces the $L1_0$ ordering temperature.[10,11] Maeda and coworkers[10] showed that for films containing $[FePt]_{85}Cu_{15}$, the coercivity was 5000 Oe after annealing at 300°C, whereas H_C for films containing FePt was only a few hundred Oe after annealing at 300°C. Takahashi and coworkers[11] found that adding 4 atomic percent Cu into FePt sputtered films decreased the temperature required for FCC to $L1_0$ ordering from 500 to 400°C. They suggested that the addition of Cu lowered the melting point for the alloy, increasing the atomic diffusivity, thereby enhancing the kinetics of ordering. These reports suggest the possibility of lowering the temperature needed to bring about the FCC-to-$L1_0$ phase transformation for FePt nanoparticles by adding copper, which motivated us to prepare FePtCu ternary alloy nanoparticles. Adding Cu to sputtered FePt films simply requires using the appropriate Cu target; however, adding Cu to FePt nanoparticles required finding a chemical procedure that gave small FePtCu particles with a nearly monodisperse size distribution.

13.2 FePtCu TERNARY ALLOY NANOPARTICLES

FePtCu ternary alloy nanoparticles were successfully prepared by the simultaneous polyol reduction of platinum acetylacetonate and copper bis(2,2,6,6-tetramethyl-3,5-heptadionate) and the thermal decomposition of iron pentacarbonyl.[12] The chemistry was quite similar to that used by Sun et al.[7] to prepare FePt nanoparticles, except the solvent was diphenyl ether instead of dioctyl ether. Attempts to use copper acetylacetonate and dioctyl ether severely limited the amount of copper that could be incorporated into the particles. Copper acetylacetonate had limited solubility in dioctyl ether, thereby limiting the amount of copper that could be put into the reaction. Using the more soluble copper bis(2,2,6,6-tetramethyl-3,5-heptadionate) and using diphenyl ether as a solvent allowed the copper content of the particles to be systematically varied over a wide range of compositions. For this work, we prepared $Fe_{55}Pt_{35}Cu_{10}$, $Fe_{45}Pt_{44}Cu_{11}$, $Fe_{48}Pt_{40}Cu_{12}$, and $Fe_{35}Pt_{50}Cu_{15}$ nanoparticles.

As prepared, the particles were single crystals with a disordered FCC structure. The average diameter was 3.5 nm with a very narrow distribution of particle diameters. The particles were dispersed in a mixture of hexane and octane using a mixture of oleic acid and oleyl amine. When the dispersion was deposited onto a solid substrate, either a silicon wafer or a carbon-coated copper TEM grid, and the solvent was allowed to evaporate slowly, the particles assembled into a closely packed array (Figure 13.2).

The films were heated in a tube furnace in an atmosphere containing 4% H_2 in Ar, and x-ray diffraction was used to determine the amount of phase transformation. Results for the composition $Fe_{45}Pt_{44}Cu_{11}$ (Figure 13.3) were typical of the other compositions. Only after heat treatment above 550°C did the particles begin to transform to the $L1_0$ phase. The copper remained in the particles; otherwise, the (111) peak for FCC Cu would have appeared near $2\theta = 43°$. Only after heating at 700°C did the $L1_0$ phase nearly completely form, as seen by the sharp (001) and

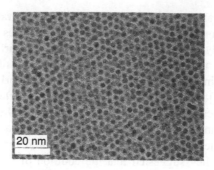

FIGURE 13.2 TEM image of self-assembled films containing FePtCu nanoparticles.

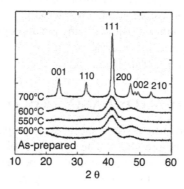

FIGURE 13.3 Effect of heat treatment temperature on the x-ray diffraction curves for $Fe_{45}Pt_{44}Cu_{11}$ nanoparticles.

(110) diffraction peaks and the splitting of the FCC (200) peak into the $L1_0$ (200) and (002) peaks. The ternary Fe-Pt-Cu phase diagram shows a range of compositions for FePtCu in which a single $L1_0$ phase is formed and other compositions in which the $L1_0$ phase and the FCC phase coexist.[13] In both phases, Cu is substituted for iron. Careful examination of the x-ray diffraction curves for the films heat treated at 700°C (Figure 13.4) revealed an extra peak near $2\theta = 48.5°$, between the $L1_0$ (200) and the (002) peaks, which may be due to the presence of a small amount of the FCC phase. This extra peak was not seen for $Fe_{48}Pt_{40}Cu_{12}$.

The hysteresis curve in Figure 13.5 was measured for the $Fe_{45}Pt_{44}Cu_{11}$ particles heat treated at 700°C and is typical of the curves for the other compositions also heat treated at 700°C. The shape of the curve suggests a mixture of hard and soft magnetic phases, and it could be decomposed into hard and soft loops. The coercivities of the hard and soft components are listed in Table 13.1, along with the coercivities of the full curves. The hard component coercivities of $Fe_{45}Pt_{44}Cu_{11}$, $Fe_{48}Pt_{40}Cu_{12}$, and $Fe_{35}Pt_{50}Cu_{15}$ were all comparable to that of $Fe_{48}Pt_{52}$. In all cases, the shape of the loops suggests that the maximum field is not sufficient to saturate the loops and that the actual coercivities are higher. These results are consistent with the x-ray diffraction (XRD) curves that suggest phase separation of the FePtCu

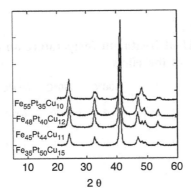

FIGURE 13.4 X-ray diffraction curves for films containing the different FePtCu compositions after heat treatment at 700°C.

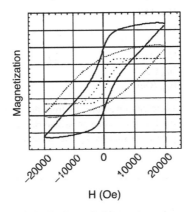

FIGURE 13.5 Magnetic hysteresis curve for $Fe_{45}Pt_{44}Cu_{11}$ heat treated at 700°C (solid curve). The curve was decomposed into a hard magnetic component (dotted curve) and a soft magnetic component (dashed curve).

particles into $L1_0$ and disordered FCC phases. The hysteresis loop for the $Fe_{48}Pt_{40}Cu_{12}$ particles had the smallest soft magnetic fraction (<20%), which was consistent with the fact that the XRD showed no evidence of a cubic phase. The soft magnetic phases may also consist of iron oxides, which, if disordered, would not exhibit well-defined XRD peaks. The effect of heat treatment temperature on coercivity is plotted in Figure 13.6, where the hard component coercivity was used at the highest annealing temperature for the films containing FePtCu particles. The coercivity for the $Fe_{48}Pt_{52}$ nanoparticles began to increase after heat treatment above 450°C, whereas the coercivity for the FePtCu nanoparticles increased only after heating to 550°C. Only after heat treatment at 650°, where the x-ray diffraction curves indicated that the particles had largely transformed to the $L1_0$ phase, did the coercivity for the hard component for the FePtCu nanoparticles match that for the FePt nanoparticles.

TABLE 13.1
Effect of Heat Treatment Temperature on the
Coercivity of the Films

		500°C	550°C	600°C	700°C
$Fe_{48}Pt_{52}$	H_C (Oe)		3970	6500	11,600[a]
$Fe_{55}Pt_{35}Cu_{10}$	H_C (Oe)	32	518	1149	3530
	H_{CS} (Oe)				1430
	H_{CH} (Oe)				6800
$Fe_{45}Pt_{44}Cu_{11}$	H_C (Oe)	7	105	347	4150
	H_{CS} (Oe)				1330
	H_{CH} (Oe)				10,400
$Fe_{48}Pt_{40}Cu_{12}$	H_C (Oe)	11	80	109	8920
	H_{CS} (Oe)				660
	H_{CH} (Oe)				12,065
$Fe_{35}Pt_{50}Cu_{15}$	H_C (Oe)	11	20	30	3120
	H_{CS} (Oe)				930
	H_{CH} (Oe)				11,500

Note: H_C is the coercivity taken directly from the hysteresis curves, whereas the values of H_{CS} (soft component) and H_{CH} (hard component) were obtained by decomposing the hysteresis curves.

[a] Minor loop.

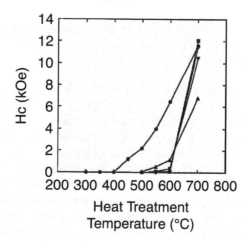

FIGURE 13.6 Effect of heat treatment temperature on the coercivity for films containing $Fe_{48}Pt_{52}$ (●), $Fe_{55}Pt_{35}Cu_{10}$ (▲), $Fe_{45}Pt_{44}Cu_{11}$ (▼), $Fe_{48}Pt_{40}Cu_{12}$ (■), or $Fe_{35}Pt_{50}Cu_{15}$ (◆) nanoparticles.

The addition of Cu to FePt nanoparticles did not lower the temperature required for the FCC-to-L1$_0$ phase transformation. This was unexpected because the transformation occurs at temperatures in the range of 300 to 400°C in FePtCu sputtered films. Instead, the FePtCu nanoparticles had to be heated to temperatures as high as or even higher than the temperature required to transform the FePt nanoparticles. The reason for the different effect of additive Cu in chemically synthesized nanoparticles and sputtered films is not well understood. One factor may be grain size. It has been shown that additive Cu promotes grain growth in sputtered FePt films, and chemical ordering has been shown to increase with grain size.[14] There has also been a recent report that for FePt sputtered films, relief of film stresses occurs in the course of the phase transformation process,[15] whereas there is no film stress for the self-assembled FePtCu nanoparticles.

Our FePtCu results led us to try to incorporate other metals into FePt particles. A recent report compared the effect of additive Cu, Ag, and Au on the chemical ordering of sputtered FePt films.[16] In agreement with previous reports,[10,11] Cu was found to promote the L1$_0$ ordering for FePt. During annealing, the Cu did not appear to phase segregate and large grain growth occurred. On the other hand, Ag and Au additives tended to phase segregate from the FePt, thus limiting the FePt grain size. Adding silver to CoPt/SiO$_2$ granular films was found to promote the CoPt L1$_0$ ordering process, thereby reducing the ordering temperature by 100°C, compared to CoPt.[17] Ag reduced the activation energy for the ordering process. This result encouraged us to prepare FePt nanoparticles containing Ag.

13.3 FePtAg AND FePtAu TERNARY ALLOY NANOPARTICLES

FePtAg nanoparticles were made by the coreduction of platinum acetylacetonate and silver acetate and the thermal decomposition of iron pentacarbonyl in diphenyl ether in the presence of oleic acid and oleyl amine.[18] Earlier attempts to make FePtAg using silver acetylacetonate and dioctyl ether as the solvent were unsuccessful. Silver acetylacetonate was not adequately soluble in dioctyl ether. The procedure using silver acetate and diphenyl ether gave 3.5-nm diameter particles with a disordered FCC structure having an expanded lattice (a = 229 pm), as expected for FCC FePt having large Ag atoms incorporated into the particles. By varying the relative amount of Fe, Pt, and Ag, we obtained compositions [FePt]$_{100-x}$Ag$_x$, where x = 0, 4, 15, 18, or 30. As with FePt and FePtCu nanoparticles, the FePtAg nanoparticles could be dispersed in hydrocarbon solvents and cast into self-assembled films (Figure 13.7A).

The films were heat treated under 2% H$_2$ in Ar for 30 min at temperatures ranging from 300 to 500°C. The x-ray diffraction pattern for the film containing [FePt]$_{88}$Ag$_{12}$ heat treated at 350°C (Figure 13.8) showed a hint of the (001) and (110) L1$_0$ peaks. After heat treatment at 400°C, the peaks due to the L1$_0$ phase were clearly present, and there was also a shoulder on the low 2θ side of the (111) peak. After heat treatment at higher temperatures, this shoulder became a separate peak consistent with the Ag (111) diffraction peak. For the sample heat treated at 500°C, the Ag (200) diffraction peak could also be seen. This showed that the Ag atoms had phase separated from the FePt particles. Presumably, the diffusion of the Ag atoms from

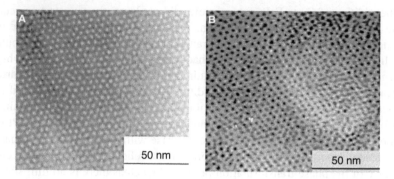

FIGURE 13.7 TEM images of films containing [FePt]$_{88}$Ag$_{12}$ nanoparticles before (A) and after (B) heat treatment at 400°C for 30 min.

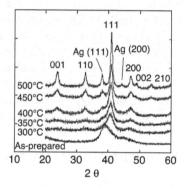

FIGURE 13.8 Effect of heat treatment temperature on the x-ray diffraction curves for films containing [FePt]$_{88}$Ag$_{12}$ nanoparticles.

the nanoparticles leaves lattice vacancies, which allow the Fe and Pt atoms to move to their lattice positions at lower temperatures.

The amount of Ag in the FePt nanoparticles could be easily varied by varying the amount of silver acetate added to the particle synthesis. Figure 13.9 shows x-ray diffraction curves for FePt nanoparticles containing different amounts of Ag after heat treatment at 500°C. The curves for [FePt]$_{88}$Ag$_{12}$ and [FePt]$_{85}$Ag$_{15}$ showed the greatest degree of transformation to the L1$_0$ phase, indicating there is an optimum Ag content of 12 to 15 atomic percent.

Our success with adding silver to FePt particles led us to examine the possibility of adding gold to the particles. Using the same procedure as for the preparation of FePtAg nanoparticles, we prepared [FePt]$_{92}$Au$_8$, [FePt]$_{88}$Au$_{12}$, [FePt]$_{85}$Au$_{15}$, and [FePt]$_{76}$Au$_{24}$ nanoparticles by substituting gold acetate for silver acetate.[19] As with the FePt, FePtCu, and FePtAg nanoparticles, the FePtAu particles were superparamagnetic, had an FCC structure, had an average diameter of 3.5 nm, and could be cast to form highly ordered, closely packed films (Figure 13.10). The particles were annealed in a tube furnace for 30 min under 5% H$_2$ in Ar. The x-ray diffraction curves for [FePt]$_{88}$Au$_{12}$ in Figure 13.11 show the effect of heat treatment temperature

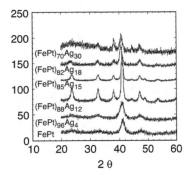

FIGURE 13.9 X-ray diffraction curves for films containing the different FePtAg compositions after heat treatment at 500°C.

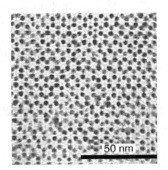

FIGURE 13.10 TEM image of a film containing [FePt]$_{85}$Au$_{15}$ nanoparticles, as prepared.

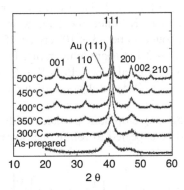

FIGURE 13.11 Effect of heat treatment temperature on the x-ray diffraction curves for films containing [FePt]$_{88}$Au$_{12}$ nanoparticles.

on the FCC-to-L1$_0$ phase transformation. The L1$_0$ (001) and (110) diffraction peaks were apparent after heat treatment at 350°C and increased in intensity when the films were heated to higher temperatures. After heating at 500°C, the Au (111)

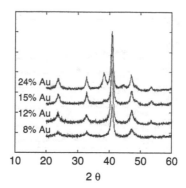

FIGURE 13.12 X-ray diffraction curves for films containing different FePtAu compositions after heat treatment at 450°C.

diffraction peak appeared at $2\theta = 38.2°$, which indicated that the gold had phase separated from the FePt particles. X-ray diffraction curves are shown in Figure 13.12 for FePt nanoparticles containing different amounts of Au after heat treatment at 450°C. Generally, the higher the gold content, the greater the degree of transformation to the $L1_0$ phase, indicating that the optimum gold content has not been determined.

13.4 MAGNETIC PROPERTIES OF FePtAg AND FePtAu NANOPARTICLE ARRAYS

The coercivity, H_C, is a good measure of the degree of chemical ordering in the nanoparticle arrays. In weakly interacting single-domain particles, H_C will depend on the magnetic anisotropy energy, K, and the volume, V, of the particle. For $KV <$ 25 $k_B T$, the particles will be superparamagnetic and $H_C = 0$. In general, K will have contributions due to bulk, surface, and shape effects. For the FCC phase, surface effects will probably dominate K, since the bulk K is quite low. For FCC Co nanoparticles, K has been found to be ~3 × 10^6 erg/cc. If this value were assumed for FCC FePt, then $KV/k_B T$ ~ 2.4 at room temperature for 4-nm diameter particles. The bulk anisotropy of fully ordered $L1_0$ FePt is K ~ 6 – 10 × 10^7 erg/cc,[6] which would give $KV/k_B T$ ~ 48 – 81. Ordered 4-nm particles would thus be expected to exhibit room temperature coercivity.

For weakly interacting single-domain particles with randomly oriented easy axes, the coercivity would be expected to follow the Sharrock equation, given by[20]

$$H_c = H_0 \left\{ 1 - \left[\frac{k_B T}{KV} \ln(f_0 t) \right]^{2/3} \right\}$$

(13.1)

where f_0 is an attempt frequency on the order of 10^9 to 10^11 Hz and t is the effective measuring time. H_0 is the low temperature or short time coercivity given by H_0 ~ K/M_s, where M_s is the saturation magnetization. For highly ordered FePt nanoparticles,

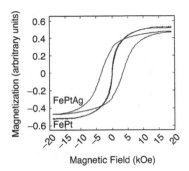

FIGURE 13.13 Magnetic hysteresis curves for films containing $Fe_{48}Pt_{52}$ or $[FePt]_{88}Ag_{12}$ nanoparticles after heat treatment at 400°C.

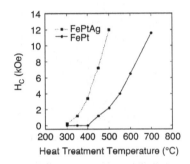

FIGURE 13.14 Effect of heat treatment temperature on the coercivity for films containing $Fe_{48}Pt_{52}$ or $[FePt]_{88}Ag_{12}$ nanoparticles.

Equation 13.1 predicts high coercivities; however, H_C will depend on temperature and particle size.

The degree of chemical ordering, and hence H_C, will depend on a number of factors, including annealing temperature and time and annealing conditions. For three-dimensional arrays of ~4-nm FePt nanoparticles, Sun and coworkers[7] have shown that H_C begins to increase rapidly as a function of annealing temperature at ~500°C and reaches a maximum value at ~600°C of >10 kOe, depending on the annealing time and conditions.

Coercivity measurements on annealed FePtAg and FePtAu arrays support our findings from XRD measurements that the addition of Ag and Au significantly reduces the required annealing temperature for obtaining chemical order. Figure 13.13 compares the hysteresis loops for films containing $[FePt]_{88}Ag_{12}$ and $Fe_{48}Pt_{52}$ nanoparticles annealed at 400°C for 30 min. The $Fe_{48}Pt_{52}$ film was still superparamagnetic ($H_C = 0$), whereas H_C for the $[FePt]_{88}Ag_{12}$ film was 3400 Oe. A plot of film coercivity as a function of heat treatment temperature (Figure 13.14) shows a clear benefit of adding Ag into the FePt nanoparticles. The $[FePt]_{88}Ag_{12}$ films show an increase in coercivity at heat treatment temperatures some 100 to 150°C lower than those for the $Fe_{48}Pt_{52}$ films. The effect of additive Ag depends on the concentration.

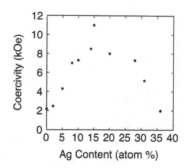

FIGURE 13.15 Effect of silver content on the coercivity for FePtAg films heated at 500°C for 30 min.

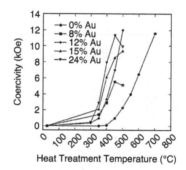

FIGURE 13.16 Effect of heat treatment temperature on the coercivity for films containing FePt with different amounts of Au.

Figure 13.15 shows H_C as a function of Ag concentration for films heat treated at 500°C for 30 min. The coercivity enhancement occurs over a wide range of concentrations and is largest in the range of 12 to 15%.

Figure 13.16 shows the coercivity as a function of annealing temperature for different Au concentrations. The coercivity increases as a function of Au concentration up to 24% Au. The effect of Au decreases with increasing Au concentration above 24% (not shown). It can be seen that the effect of 15 and 24% Au is even greater than Ag up to an annealing temperature of 450°C. The coercivity enhancement for 24% Au occurs at about 150°C below that for FePt with no additive.

As previously discussed, one of the concerns in annealing FePt nanoparticle arrays in order to achieve chemical ordering is that particle aggregation occurs at high annealing temperatures.[9] This destroys the size uniformity and positional order of the self-assembled arrays, which are important for applications such as ultra-high-density magnetic recording. Particle aggregation leads to exchange interactions between grains in multigrain particles, which are reflected in magnetic measurements. A commonly used method for characterizing magnetic interactions is to measure δM curves, given by[21]

$$\delta M = 2M_r - 1 + M_d \qquad (13.2)$$

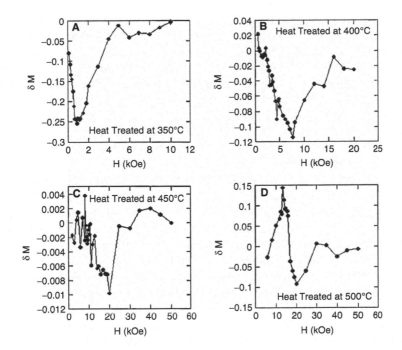

FIGURE 13.17 δM curves for films containing $[FePt]_{88}Ag_{12}$ nanoparticles heat treated at 350°C (A), 400°C (B), 450°C (C), and 500°C (D).

where M_r is the so-called isothermal remanence curve, obtained starting from an alternating current or thermally demagnetized state, and M_d is the direct current remanence curve, obtained starting from a reverse saturation state. Exchange interactions give rise to positive values of δM, whereas magnetostatic interactions between isolated particles give negative values of δM. Measurements on $[FePt]_{88}Ag_{12}$ arrays show negative δM values up to an annealing temperature of 400°C but positive values at 500°C (Figure 13.17). These results are consistent with plan view TEM images showing weak aggregation after annealing at 400°C. More work is needed, however, to determine the extent of aggregation in three-dimensional arrays on silicon substrates, which were used for the magnetic measurements.

13.5 SIGNIFICANCE OF THIS WORK

The FePt nanoparticle chemistry developed by Sun et al.[7] is remarkably robust. We have made minor changes in their procedure to prepare FePtCu,[12] FePtAg,[18] FePtAu,[19] and FePtCo[22] nanoparticles. We have also made FePd and CoPt nanoparticles either by replacing platinum acetylacetonate with palladium acetylacetonate or by replacing iron pentacarbonyl with cobalt tricarbonyl nitrosyl.[23] This chemistry provides great versatility in tailoring the particle properties.

If $L1_0$ FePt nanoparticles are to be used in granular magnetic recording media, a number of materials problems must be solved. Methods of self-assembly must be found in which highly ordered arrays of particles have their magnetic easy axes (the crystallographic c-axis) aligned either in the plane of the films or perpendicular to the plane of the film. It is quite easy to assemble the superparamagnetic FePt particles; however, maintaining that high degree of particle positional ordering after heat treatment at temperatures above 500°C has been a problem. The organic surfactants decompose and allow the particles to move from their lattice positions, thereby degrading the degree of positional ordering. To get a sense of the problem, compare the TEM images in Figure 13.1. The image in Figure 13.1A shows the closely packed, self-assembled FePt particles before heat treatment. The TEM image in Figure 13.1B shows the film after heat treatment at 600°C. Notice the considerable loss in positional ordering and particle agglomeration.

Adding either Ag or Au to FePt nanoparticles decreases the temperature required for the phase transformation from FCC to the $L1_0$ phase by some 100 to 150°C, whereas adding copper gives no benefit. For both Ag and Au, the mechanism appears to be the same. During heating, the Ag or Au atoms leave the FePt particles, giving rise to lattice vacancies that provide pathways for the Fe and Pt atoms to move to their $L1_0$ lattice positions. This gives films with high coercivity at temperatures at which the loss in the degree of particle positional order was much less. The two TEM images in Figure 13.7 show that by heat treating the FePtAg film at 400°C, a high degree of particle positional ordering is maintained while achieving a ferromagnetic film with $H_C = 3400$ Oe. Magnetic characterizations, particularly δM plots, suggest little or no particle agglomeration when the films are heat treated at lower temperatures. This ability to transform the particles at lower temperatures is an important advance in the use of FePt nanoparticles in magnetic recording media.

ACKNOWLEDGMENTS

This research was sponsored by the NSF Materials Research Science and Engineering Center awards DMR-9809423 and DMR-0213985.

REFERENCES

1. C.D. Mee, E.D. Daniel, M.H. Clark. *Magnetic Recording: The First 100 Years*. Piscataway, NJ: IEEE Press, 1999.
2. G.E. Moore. Cramming more components into integrated circuits. *Electronics* 38: 1965.
3. P.P. Gelsinger, P.A. Gagini, G.H. Parker, A.Y.C. Yu. Microprocessors circa 2000. *IEEE Spectrum*, October 1989, pp. 43–47.
4. Z. Zhang, Y.C. Feng, T. Clinton, G. Badran, N.-H. Yeh, G. Tarnopolsky, E. Girt, M. Munteanu, S. Harkness, H. Richter, T. Nolan, R. Tanjan, S. Hwang, G. Rauch, M. Ghaly, D. Larsan, E. Singleton, V. Vas'ko, J. Ho, F. Stageberg, V. Kong, K. Duxstad, S. Slade. Magnetic recording demonstration over 100 Gb/in². *IEEE Trans. Magn.* 38: 1861–1866, 2002.
5. R. Wood. The feasibility of magnetic recording at 1 terabit per square inch. *IEEE Trans. Magn.* 36: 36–42, 2000.
6. D. Weller, A. Moser, L. Folks, M.E. Best, W. Lee, M.F. Toney, M. Schwickert, J.-U. Thiele, M.F. Doerner. High K_u materials approach to 100 Gbits/in². *IEEE Trans. Magn.* 36: 10–15, 2000.

7. S. Sun, C.B. Murray, D. Weller, L. Folks, A. Moser. Monodisperse FePt nanoparticles and ferromagnetic FePt nanocrystal superlattices. *Science* 287: 1989–1992, 2000.

8. J.W. Harrell, S. Wang, D.E. Nikles, M. Chen. Thermal effects in self-assembled FePt nanoparticles with partial chemical ordering. *Appl. Phys. Lett.* 79: 4393–4395, 2001.

9. Z.R. Dai, S. Sun, Z.L. Wang. Phase transformation, coalescence and twinning of monodisperse FePt nanocrystals. *Nano Lett.* 1: 443–447, 2001.

10. T. Maeda, T. Kai, A. Kikitsu, T. Nagase, J. Akiyama. Reduction of ordering temperature of an FePt-ordered alloy by addition of Cu. *Appl. Phys. Lett.* 80: 2147–2149, 2002.

11. Y.K. Takahashi, M. Ohnuma, K. Hono. Effect of Cu on the structure and magnetic proerties of FePt sputtered film. *J. Magn. Magn. Mater.* 246: 259–265, 2002.

12. X. Sun, S. Kang, J.W. Harrell, D.E. Nikles. Synthesis, self-assembly and magnetic properties of FePtCu nanoparticle arrays. *J. Appl. Phys.* 93: 7237–7239, 2003.

13. M. Shahmiri, S. Murphy, D.J. Vaughan. *Mineralogical Mag.* 49: 547–554, 1985.

14. Y.K. Takahashi, T. Ohkubo, M. Ohnuma, K.Y. Hono. Size effect on the ordering of FePt granular films. *J. Appl. Phys.* 93: 7166–7168, 2003.

15. K.W. Wierman, C.L. Platt, K.J. Howard, F.E. Spada. Evolution of stress with $L1_0$ ordering in FePt and FeCuPt thin films *J. Appl. Phys.* 93: 7160–7162, 2003.

16. C.L. Platt, K.W. Wierman, E.B. Svedberg, R. van de Veerdonk, A.G. Roy, D.E. Laughlin. $L1_0$ ordering and microstructure of FePt thin films with Cu, Ag and Au additive. *J. Appl. Phys.* 92: 6104–6109, 2002.

17. C. Chen, O. Kitakami, S. Okamoto, Y. Shimada. Ordering and orientation of $CoPt/SiO_2$ granular films with additive Ag. *Appl. Phys. Lett.* 76: 3218–3220, 2000.

18. S. Kang, D.E. Nikles, J.W. Harrell. Reduction of ordering temperature of self-assembled FePt nanoparticles by the addition of Ag. *Nano Lett.* 2: 1033–1036, 2002.

19. S. Kang, Z. Jia, D.E. Nikles, J.W. Harrell. Synthesis, self-assembly and magnetic properties of $[FePt]_{1-x}Au_x$ nanoparticles. *IEEE Trans. Magn.* 39: 2753–2757, 2003.

20. M.P. Sharrock, J.T. McKinney. Kinetic effects in coercivity measurements. *IEEE Trans. Magn.* 17: 3020–3022, 1981.

21. P.E. Kelley, K. O'Grady, P.L. Mayo, R.W. Chantrell. Switching mechanisms in cobalt–phosphorus films. *IEEE Trans. Magn.* 25: 3881–3883, 1989.

22. M. Chen, D.E. Nikles. Synthesis, self-assembly and magnetic properties of $Fe_xCo_yPt_{100-x-y}$ nanoparticles. *Nano Lett.* 2: 211–214, 2002.

23. M. Chen, D.E. Nikles. Synthesis of spherical FePd and CoPt nanoparticles. *J. Appl. Phys.* 91: 8477–8479, 2002.

14 Assemblies of Magnetic Particles: Properties and Applications

Michael Hilgendorff and Michael Giersig

CONTENTS

14.1 Introduction 385
14.2 Theoretical Aspects 387
    14.2.1 Syntheses and Assembly 387
    14.2.2 Magnetism 389
14.3 Methods...391
    14.3.1 Ferrofluid Syntheses...391
    14.3.2 Assembly 394
14.4 Applications...398
    14.4.1 Data Storage Devices 398
    14.4.2 Carbon Nanotube Growth 400
    14.4.3 Photonic Crystals...401
14.5 Concluding Remarks..402
Acknowledgment 403
References..403

14.1 INTRODUCTION

Nanostructured materials have attracted intense research interest over recent years, as they provide the critical building blocks for nanoscience and nanotechnology and exhibit new and enhanced properties compared to bulk materials. In the field of nanoscale building blocks, physical parameters such as ε_0, μ_0, and ρ defined as constants in descriptions of bulk materials properties, become size dependent. In other words, changing the size of nanoscale materials produces new materials.

For the design of novel nanostructured devices, one needs a technique for preparing an industrial-scale highly symmetric periodic particle array (PPA) within the range of several microns. Different routes have been investigated in the past to obtain small particles that can be assembled in large areas of high symmetry. These approaches typically apply scale-down techniques, e.g., ball milling of bulk materials, and bottom-up techniques, e.g., cluster or nanoparticle growth from precursors

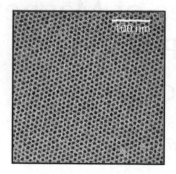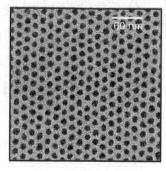

FIGURE 14.1 TEM images at different magnifications of monodisperse 11-nm Co particles stabilized by oleic acid, obtained after size-selective precipitation, and self-assembled on a carbon-coated Cu grid by spraying. (From Giersig, M. and Hilgendorff, M., in *Nanoscale Materials*, Liz-Marzán, L.M. and Kamat, P., Eds., Kluwer Academic Publishers, Boston, 2002, pp. 335–371. With permission. Copyright © 2002 Kluwer Academic Publishers.)

in gaseous or liquid phases. The latter, i.e., the wet chemical synthesis of inorganic colloidal particle fluids, is most elegant from a chemists' point of view, and the primary motivation of this chapter is created. For an overview, see references 1 to 7.

Inorganic colloidal fluids are in the first kinetically stable dispersions of nanoscale clusters or fine particles — which can be crystalline or amorphous — in a solvent.[8] If the dispersed material is known to exhibit ferromagnetic behavior in the bulk material, colloidal suspensions thereof are generally called ferrofluids or magnetic fluids.[9,10]

One can develop PPA with ferrofluids having a standard particle size distribution of <10% by simply drying a drop of solution on suitable substrates. This self-assembly technique has been improved through use of different coating techniques (e.g., spin coating, dip coating, or spraying) in combination with applied external forces. The origin of these external forces is either mechanical, used for the preparation of Langmuir–Blodgett (LB) films,[11,12] or electrostatic, which is used for layer-by-layer (LbL) assembly-applying polyelectrolytes.[13–16] The use of external electric and magnetic fields to improve the self-assembly of charged and magnetic nanoparticles, respectively, has also been reported.[17–19] Figure 14.1 shows typical TEM images of self-assembled monodisperse Co particles prepared by spraying a 10-μl drop of ferrofluid (~10^{-2} M) onto a carbon-coated copper grid. Hexagonal ordering and high symmetry of the particle layer are obvious from both images taken at different magnifications.

To facilitate the preparation of nanocrystalline particles appropriate for technological applications, easy methods must exist for the preparation of nonaggregated, monodisperse colloids as well as highly ordered layers on different substrates. Naturally, it is preferable to have inexpensive materials and efficient syntheses as well as state-of-the-art industrial-scale preparation techniques. One must strive to lower the costs associated with scaling up new laboratory techniques for industrial requirements.

Unfortunately, most requirements are limited by the possibilities offered by chemical preparation techniques in the field of colloid chemistry. For example, different applications require suspensions of colloidal particles in different solvents (e.g., aqueous solutions for biocompatability or nonpolar solvents for development of well-ordered self-assembled monolayers), just as various materials may require preparation methods tailored to their chemical behaviors.

Actual experimental and theoretical aspects of ferrofluid syntheses and assembling methods thereof have been published previously and will be presented in a summary. The same holds for descriptions of magnetic properties of assembled layers.[20-26] This chapter will conclude with the presentation of recent advances in magnetic particle assemblies, prepared from ferrofluids in view of different applications.

14.2 THEORETICAL ASPECTS

14.2.1 SYNTHESES AND ASSEMBLY

Discussion of assemblies of colloidal magnetic particles requires the discussion of some major items, most notably the theoretical description of nanoparticle interactions in a colloidal suspension. Equations describing the interacting behavior of colloidal nanoparticles were in principle developed by integrating equations describing atomic, ionic, or molecular interactions over the sum of all atoms in a nanoparticle.

The first item of interest is the origin of colloid stability, which involves the control of repulsive (F_R) and attractive (F_A) forces acting on colloidal particles as a function of the temperature (Brownian motion). Repulsive electrostatic (F_{RES}) and steric (F_{RS}) forces in aqueous media, and mainly steric repulsion in hydrophobic media, must be in equilibrium with attractive Van der Waals (F_{AW}) and magnetic dipole–dipole (F_{AM}) interactions. The total interaction force F_T can be written as

$$F_T = F_A + F_R = (F_{AW} + F_{AM}) + (F_{RES} + F_{RS}) \qquad (14.1)$$

The total energy V_T plotted against the separation distance D (distance between the center of neighboring particles) is commonly comparable to the well-known Lennard–Jones potential, which is valid in the absence of applied external fields.

Actually, the most popular preparations concerning self-assembled two- and three-dimensional layers of magnetic particles deal with the synthesis of ferrofluids in apolar solvents. In apolar media of low dielectric constant, electrostatic repulsion can be neglected (as well as magnetic dipole–dipole interaction, assuming superparamagnetic particles at room temperature and the absence of an external magnetic field), but it is considered that attractive Van der Waals forces between the particles are omnipresent. The steric repulsion of adsorbed layers on the particle surface is therefore necessary to overcome these attractions.[27]

Secondly, in discussing the theory of drying, we must establish a theory to explain the interactions that take place between colloidal solutions and different substrates as well as the forces between all interfaces in the system under observation.

Forces influencing the colloid stability are Van der Waals attraction, attractive magnetic dipole–dipole interaction, electrostatic repulsion, and steric repulsion. The

London Van der Waals attraction is one of the most important forces of colloidal chemistry, and it was developed about 70 years ago. It describes the induction of dipoles in neighboring atoms by temporary formed electric dipoles within an atomic nucleus, resulting in an attractive energy V_{AW}, inversely proportional to the sixth power of the separation distance D (B = London constant):

$$V_{AW} = -B/D^6 \tag{14.2}$$

Equation 14.2 is also valid for colloidal particles consisting of thousands of atoms, because integration of Equation 14.2 over the sum of the attractive energies of all atoms includes both neighboring nanoparticles, each of which induces an electric dipole in the opposite particle.

The description of the attractive magnetic dipole–dipole interaction between magnetic particles in a ferrofluid follows, in principle, the description of the electric dipole–dipole Van der Waals attraction. One has to reflect that colloid chemistry generally produces particles whose magnetic properties are well described by laws describing paramagnetic behavior, with the difference that magnetic susceptibilities are orders of magnitudes enlarged. This property is called superparamagnetism. As a result, the magnetic dipole–dipole interaction of superparamagnetic particles dispersed in a solvent is also inversely proportional to the sixth power of the separation distance D.[28,29]

Both attractive potentials have been further developed for different geometries, resulting in expressions that describe colloidal particles covered with a protecting shell.[30,31]

The simplest model describing electrostatic repulsion is the electric double-layer model at the interfaces of two phases; it was developed by Helmholtz in 1879. Since then, several refinements have been published to describe experimental results, leading to a three-layer model. Most recently, a four-layer model was developed by Charmas.[32] However, it was later mentioned that Equation 14.3, valid for the three-layer model, is the most appropriate equation for nanosized particles[33] (D is the distance between layers, R is the particles radius, including any shell, φ is the electric potential, ε_0 is the vacuum permittivity, and ε_r is the permittivity of the liquid medium):

$$V_{RES} = 2\pi\varepsilon_0\varepsilon_r R\varphi_0^2 \exp(-\kappa D) \tag{14.3}$$

$$\kappa = (2N_A e^2 c z^2/\varepsilon_0\varepsilon_r kT)^{1/2} \tag{14.4}$$

$$\varphi = \varphi_0 \exp(-\kappa D) \tag{14.5}$$

The steric repulsion of stabilizers adsorbed on the surface of colloidal particles is necessary to overcome attractive forces. Buske et al.[27] showed that chemisorption (beyond physisorption) of the surfactants is required to prevent instabilities that may occur from adsorption/desorption equilibrium reactions. This requirement renders the possibility of ligand exchange.

Discussion of the assembly of particles further requires a description of the theory of drying. Ferrofluids can be dried on substrates using different methods, such as spin or dip coating. The simplest method involves the drying of a drop of solution on a substrate. This method is commonly used to prepare samples for TEM investigations, which is the most favorable technique for the investigation of colloids. Contrary to the investigation of bulk materials, TEM investigations of colloids are extremely fast because it is only necessary to dry a drop of solution on a grid, which can then be viewed and analyzed immediately.

Following descriptions in common textbooks of physical chemistry, a drop of solution with a concentration c (or the volume fraction ϕ of colloidal particles, which may be coated with a layer of surfactants), dynamic viscosity η, vapor pressure p_d, and surface tension σ_{lg} between the drop and atmosphere will have a contact angle θ on a substrate. This angle is related to the interface tensions σ_{sl} acting between the solution and the substrate and σ_{sg} acting between the substrate and the atmosphere. The relationship of these forces with the contact angle θ can be described by Equation 14.6:

$$(\sigma_{sg} - \sigma_{sl})/\sigma_{lg} = \cos \theta \qquad (14.6)$$

The drying of colloidal solutions in an environment with a given temperature T and a pressure p is naturally a time-dependent process, not reflected by the theoretical calculations described above. Forces that induce ordering phenomena in colloids dried on substrates need a minimum time t_0 to fulfill final conditions. Thus, the choice of solvents evaporating in a time $t < t_0$ will give different results compared to solvents evaporating in a time $t > t_0$. For a more detailed discussion of the theory of drying, see Scherer.[34]

14.2.2 MAGNETISM

Many possible applications of ferrofluids necessitate two materials prerequisites: high remanent magnetization and high anisotropy at room temperature for the dispersed particles. A large product of remanent magnetization and anisotropy, the *energy product*, is, for instance, a prerequisite of permanent magnets. High remanent magnetization is a characteristic property of soft ferromagnetic materials. Moreover, high anisotropy is a characteristic feature of ferromagnetic 3d elements alloyed with noble metal elements of the platinum group (hard magnetic materials).[35-37] Furthermore, dipole–dipole interactions increase with increasing concentration because of collective interaction and reach a maximum for highly symmetric assemblies.

Magnetic properties of colloidal particles assembled on a substrate are important mainly in case of data storage applications. Recent results in applications for carbon nanotube growth catalyzed by nanoparticles assembled from ferrofluids did not reveal the necessity for use of *ferromagnetic* materials.[38-42] The latest results show that the assembled particles' surface properties are of main importance in catalyzing carbon nanotube growth. Photonic crystals are usually prepared by assembly of micron-sized materials into highly symmetric two- and three-dimensional arrays. Thus, magnetic particles containing or loaded with latex spheres have been used for preparation of ordered crystals made from magnetized latex spheres while using the magnetic properties to vary size and symmetry. Those crystals can then further be used as a host for depositing materials of high refractive index, forming inverse opals after removal of the host material. Some publications concerning the magnetic properties of magnetized latex particles are available, but they are of less importance in the field of photonic crystal research.

In the following, the main principle — magnetic properties — important for data storage device developments, i.e., magnetization, anisotropy, and magnetic relaxation, will be qualitatively summarized.

Standard magnetic measurements are M vs. H hysteresis loops as well as temperature-dependent M (or χ) FC/ZFC curves. Hysteresis loops allow the determination of H_C, M_R, and M_S, which all depend on the applied temperature. One observation in magnetic colloid science is that magnetization often remains unsaturated, even at high fields. Furthermore, hysteresis loops sometimes remain unclosed within the range of the applied magnetic field. Values of M_R and H_C obtained from those loops (minor or inner loops) are then decreased compared to values obtained from closed loops because of incomplete alignment of magnetic domains within the sample. The determination of the susceptibility

$$\chi \equiv dM/dB, \ B \rightarrow 0 \qquad (14.7)$$

and χ vs. T easily allows one to distinguish between ferro-, para-, and diamagnetic properties of materials. Susceptibilities of diamagnetic materials are negative and temperature independent. Susceptibilities of paramagnetic materials are positive and decrease with increasing temperature according to the Curie law. Both susceptibilities are independent of the applied magnetic field strength and very small ($|\chi| \ll 1$). In ferromagnetic materials, the susceptibility is defined differently, since dM/dB is infinite for a single-domain ferromagnet; in other words, there exists a remanent magnetization without an applied magnetic field B. It is generally accepted to use M/B as a measure of the susceptibility, which is positive. The ferromagnetic susceptibility decreases with increasing temperature, depends on the applied magnetic field strength, and is orders of magnitudes enlarged compared to susceptibilities of dia- and paramagnetic materials ($|\chi| \gg 1$).

Magnetic nanoparticles typically behave superparamagnetically above and ferromagnetically below a blocking temperature T_B. Temperature-dependent ZFC curves are usually determined in order to define the blocking temperature T_B at which the ZFC magnetization shows a maximum. T_B decreases with increasing applied magnetic field and vice versa with interparticle distance, and it is commonly determined to calculate the effective anisotropy constant K_{eff} from the Néel equation:

$$K_{eff}V = \ln(\tau/\tau_0)k_B T_B \qquad (14.8)$$

where τ is the relaxation time of magnetization over an energy barrier $\Delta E = K_{eff}V$ (assuming uniaxial anisotropy), τ_0 is an attempt time, typically set between 10^{-12} and 10^{-9} sec, and V is the magnetic volume. In direct current (dc) susceptibility experiments over a time range of t = 100 sec, $\ln(t/\tau_0)$ has values in the range 25 of 32. Hence, the

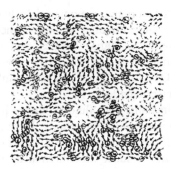

FIGURE 14.2 Micromagnetic simulation of transitions showing that percolation occurs for a modest concentration of grains of lower anisotropy (10%). (From [49]. With permission. Copyright © 1998 Henley Publishing.)

magnetization of a material will relax within the measurement time t if $\tau \le$ t. More detailed information is available in Leslie-Pelecky and Rieke[25] and Binns.[43]

It is a common observation that the compounds used for a ferrofluid preparation determine not only the shape but also the stability of the final particles. Small changes in preparation conditions can give completely different crystalline phases. Moreover, it is well known in the field of colloid chemistry that crystal structures different from bulk materials are available.[18,19,44–48] Different crystal structures possess unique properties, i.e., anisotropies, which must be considered before the application of such particles for memory devices. Consider a monolayer of magnetic crystals. It has been simulated (Figure 14.2) that if 10% of this array were to consist of crystals of a different structure (a lower anisotropy will cause the percolation of distorted spins by exchange coupling), then this fact alone would be the main reason for media noise and an unknown loss of stored data.[49] Nevertheless, the preparation of exchange-coupled two-component films may have some advantages in developing permanent magnets with large energy products.[50–52]

The following section provides an overview of successfully applied methods for ferrofluid syntheses and assembled layer preparation thereof.

14.3   METHODS

14.3.1   Ferrofluid Syntheses

Wet chemical syntheses based on a wet chemical bottom-up approach of monometallic magnetic particles such as Fe,[20,53–56] Co,[19,20,27,46–48,53,57–65] or Ni[20,53,66–69] have been successfully synthesized by electrochemical reduction[61,69]; chemical reduction by Li,[68] hydroborides,[53,62–64] or polyols[70] of metal salts; and thermal decomposition (initiated conventionally[19,27,46,56–60,65–67] or by ultrasound[55]) of zero-valent metal–organic compounds in organic solvents in the presence of bulky stabilizers, such as (1) fatty acids (in combination with organic amines or phosphines), (2) polymers, or (3) tensides (often called surfactants). In the case of number 1, the fatty acids act as stabilizing compounds to overcome oxidation and dipole–dipole interactions, whereas the amines or phosphines control the particle growth. Tensides can be used as stabilizers for inorganic nanomaterials dispersed in organic solvents or as compounds forming stable micelles in heterogeneous oil-in-water systems, e.g., sodium dodecyl sulfate (SDS). Other tensides, e.g., alkyl ammonium bromides (didodecyldimethylammonium bromide [DDAB]) cetyltrimethylammonium bromide (CTAB) or Aerosol OT (AOT), form a stable emulsion of water droplets in hydrophobic solvents (reverse micelles).

The development of bimetallic ferrofluid syntheses has been based on existing synthetic procedures of monometallic ferrofluids. In general, they have been produced by thermal decomposition of bimetallic zero-valent metal–organic compounds[71-76] or by *in situ* or successive thermal decomposition of zero-valent monometallic metal–organic compounds,[40] as well as by *in situ* or successive reduction of metal salts,[39,77-93] or combinations thereof.[94-109] In summary, bimetallic particles form alloys, core-shell particles, or cluster-in-cluster particles,[70] all dependent largely, but not exclusively, on reaction kinetics. Applied methods and properties of these fluids have been summarized in Hilgendorff.[24]

The most common preparation of magnetite (Fe_3O_4) ferrofluids was developed by Massart about 20 years ago and described in detail in 1987.[110] The synthesis was based on the coprecipitation of Fe(II) and Fe(III) salts in aqueous solutions stabilized by repulsive electrostatic forces. TEM investigations typically show aggregated particles, consisting usually of a mixture of ferrimagnetic magnetite, maghemite (γ-Fe_2O_3), and paramagnetic hematite (αFe_2O_3). The ratio of the different iron oxide phases is directly compared to the expense associated with performing the synthesis under oxygen-free conditions, because γFe_2O_3 and αFe_2O_3 are oxidation products of Fe_3O_4.

Since these reports, several optimizations have been discovered to produce ferrofluids useful for new applications. One of them was concerned with the possibility of changing the magnetic properties of the inverse Fe_3O_4 spinell by replacing Fe(II) with Co(II), Ni(II), Mn(II), or Zn(II) ions.[111-113]

Additionally, many reports dealt with solving the classic problem of nanoparticle aggregation in aqueous solutions. To overcome such aggregation, new preparations have involved micellar solutions[21,114,115] as well as the transfer of iron oxide particles from aqueous to nonpolar solvents by hydrophobizing the surface by adsorbing bulky stabilizers such as fatty acids.[116-118]

Different methods have been applied to decrease particle size distribution. One method that has been successfully applied involves synthesis of iron oxides in reverse micelles. A second method deals with a size-selective precipitation after transfer of iron oxide particles to nonpolar organic solvents. Most recently, Sun and Zeng[119] developed a very successful method to achieve monodisperse Fe_3O_4 nanoparticles. The method is based on the reduction of iron(III)-acetylacetonate by 1,2-hexadecanediol and further thermal decomposition at high temperatures (solvent, diphenylether; bp 265°C) in the presence of oleic acid and oleylamine as stabilizers. As a result, small Fe_3O_4 nanoparticles (diameter, ~4 nm) having a particle size distribution of ~5% are formed and can be further grown by the seed-mediated growth method.[120-122]

Table 14.1 provides an overview of successful syntheses and surface properties of various magnetic colloids, determining the applicability of different assembling methods.

TABLE 14.1

Overview of Successful Syntheses and Surface Properties of Various Magnetic Colloids

Synthesis	Materials	Properties	References
D *Decomposition:* Thermal decomposition of zero-valent metal–organic compounds in organic solvents stabilized by the surfactants Duomeen, Solsperse, Manoxol polymer (generally poly(vinylpyridine) (PVP)), or bulky organic molecules such as fatty acids. Thermal decomposition of iron(II)pentanedionate	Metals and their alloys (and oxides, prepared by postoxidation) Magnetite	Uncharged, hydrophobic	19, 27, 40, 46, 56–60, 65–67, 71–76, 119
E *Electrolysis:* Electrochemical reduction of metal ions in tetrahydrofuran stabilized by tetraoctylammonium ions	Metals and their alloys	Charged, hydrophobic	61, 69, 77
F *Reduction in apoferritin:* Sodium borhydride reduction or hydroxide-initiated hydrolysis and condensation of metal salts in the cavities of demineralized ferritin (apoferritin)	Metals, their alloys, and oxides	Charged, hydrophilic	94
R *Reverse micelles:* Sodium borhydride reduction or hydroxide-initiated hydrolysis and condensation of metal salts in water-in-oil reverse micelles formed by different surfactants (AOT, CTAB, DDAB)	Metals, their alloys, and oxides	Charged, hydrophobic	21, 38, 39, 54, 62, 64, 86–93
M *Micelles:* Sodium borhydride coreduction or hydroxide-initiated hydrolysis and condensation of metal salts in oil-in-water micelles formed by the surfactant SDS	FeCu alloy Magnetic latex spheres formed by emulsion polymerization	Charged, hydrophilic	90
O *Organohydroborate reduction:* Reduction of metal salts by tetraoctylammonium triethylborhydride (TOAEt$_3$BH) or superhydride in tetrahydrofuran	Metals and their alloys	Charged, hydrophobic	47, 48, 53, 63, 78
P *Polyol reduction:* Reduction of metal salts stabilized by polymers (e.g., PVP) or bulky stabilizers (e.g., fatty acids)	Metals and their alloys	Various	70, 81–85
C *Combination:* Decomposition/polyol reduction, decomposition/transmetallation, or organoborhydride reduction/transmetallation	Metal alloys	Uncharged, hydrophobic	94–109
H *Hydrolysis and condensation* of iron(II) and iron(III) salts in aqueous solution stabilized by charge	Iron oxides	Charged, hydrophilic	110–118

General observations for ferrofluids are increasing particle sizes with decreasing concentrations and decreasing chain length of the stabilizers, and increasing size distributions with increasing particle size. As one challenge of syntheses is to achieve a narrow size distribution, the reverse micelle technique and the so-called hot-injection technique have been successfully applied, especially for the preparation of larger particles (diameters, >5 to 15 nm). Using the latter method, standard size deviations of <10% have been experimentally demonstrated (standard size deviations of about 5% have been predicted theoretically by choosing a proper combination of solvent and stabilizers[123]). Moreover, the hot-injection technique has been employed to facilitate further growth of primary prepared particles (seed particles) by reactive precipitation of precursors on the surface of seed particles.

14.3.2 ASSEMBLY

Concerned with the control of particle evolution during syntheses via kinetic rather than thermodynamic control of particle size and particle size distribution is the phenomenon of self-assembly, which was first investigated using submicron spheres of polystyrene (PS)-latex or SiO_2 dispersed in water. This work concluded that self-assembly is conditional on the narrow size distribution of the particles (standard deviation, ~5%), i.e., a narrow distribution of the sum of repulsive and attractive forces between particles in a colloidal solution. Thus, this conclusion is not limited to well-separated single particles. It holds also for aggregates of particles often found in aqueous systems, assuming that equal forces interact between aggregates of similar size.

Various methods have been applied for assembling magnetic colloids in recent years. The different methods and their requirements are summarized in Table 14.2 (in view of Table 14.1).

The simplest particle layer preparation method, which allows application of external forces, is drying a drop of solution on a flat substrate solid or liquid. This procedure results in a particle (R ~ 5 nm) monolayer if the inorganic material concentration of the fluid is on the order of about 10^{-4} M. The monolayer typically consists of particle *sheets* of high symmetry interrupted by *empty* areas containing excess stabilizer (Figure 14.3). The different sizes of ordered arrays compared to Figure 14.1 reflect the fact that different deposition techniques follow different laws. The particle sheets can self-assemble into hexagonal ordered arrays, as shown in Figure 14.1, or cubic ordered arrays, depending on the stabilizer's properties. The particle area density on a substrate can be increased by increasing the ferrofluid concentration. Increasing the concentration results in multilayers forming three-dimensional crystals of various symmetry, e.g., hexagonal close-packed (hcp), face centered cubic (fcc), body-centered cubic (bcc), or tetragonal.

Applying an external magnetic field during drying (magnetophoretic deposition technique (MDT)) results in an increased particle sheet size (for monolayers) having increased symmetry.[18,19,24] Ordering phenomena in ferrofluids placed in external magnetic fields have been observed and theoretically calculated in the literature. More recently, the formation of ordered stripes and hexagonal sheets has been experimentally observed while drying a thin liquid film of a ferrofluid on a substrate in an applied external magnetic field.[19,24] Assuming a lattice gas model and the

TABLE 14.2
Overview of Successful Assembling Methods, Their Requirements, and Their Applicability for Colloids Prepared via Methods Summarized in Table 14.1

Assembling Method	Requirements	Applicable for	References
SA *Self-assembly*: Self-organization of colloidal particles into symmetric arrays during their deposition on flat substrates, i.e., flat solid or liquid surfaces, or their precipitation into large particulate crystals from oversaturated liquids	Narrow distribution of interacting forces between the colloidal particles	D, F, M (latex), R, and C	Most of the cited references
IA *External field improved assembly*:			
LB: Mechanical force improved assembly on aqueous surfaces	Hydrophobic properties	D, E, R, O, P, and C	11, 12
MDT: Magnetic force improved assembly on flat surfaces	Ferro- or superparamagnetic properties	D, R, M, C, and H[a]	18, 19, 24, 138, 139
EDT: Electric force improved assembly on flat surfaces	Charged colloids	R, M, and H[a,b]	13–16, 129, 138
LbL *Layer-by-layer assembly*: Alternative deposition of polymers and colloids on solid surfaces of various geometries	Charged colloids and aqueous solvents by use of polyelectrolytes	All[c]	

[a] Particles available via the methods E, F, O, and P are typically too small to be influenced by external electric and magnetic fields during deposition at room temperature.
[b] The EDT approach has not yet been demonstrated for magnetic colloids, to the authors' knowledge.
[c] Applicability is dependent on the solubility of polymers in different solvents.

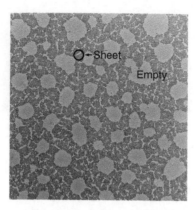

FIGURE 14.3 Typical TEM image of a sheet-like self-assembled film obtained by drying a drop of ferrofluid on a carbon-coated Cu grid.

FIGURE 14.4 TEM image constructed from 11 micrographs. The mesh is 45×45 μm. The insert demonstrates that the structure is the same on at least four meshes.

Carnahan–Starling model, theoretical calculations by Lacoste and Lubensky[124] have shown, for a given film thickness and a given applied external magnetic field normal to the plane of thin films, that the ordering parameters are a function of magnetization and concentration of particles having a size distribution close to zero. These calculations showed that the shapes of ordered structures, which are obtainable by drying a thin layer of ferrofluid under the influence of an external magnetic field applied in an identical plane to the liquid layer, can be completely different. Stripes, as well as hexagonal sheets, are available and can be made to coexist simply by changing the strength of the applied magnetic field or the concentration of the particles.[124]

Figure 14.4 is shown for two reasons. First, it demonstrates the coexistence of stripes (triple layers in this case) and sheets (monolayers, not really visible at this magnification) and confirms theoretical calculations. Second, it has been constructed from 11 single images to demonstrate large area ordering within several square micrometers, which is one major problem of TEM investigations of nanoscale materials. (One mesh of a carbon-coated copper grid, typically used in TEM investigations of colloids, 45×45 μm in size, is shown.) The insert in Figure 14.4 reveals

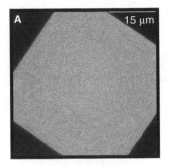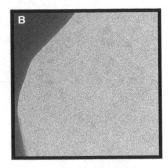

FIGURE 14.5 (A) One mesh of a completely covered Cu grid. The structure results from islands of a double layer, as can be seen in (B). (B) Higher-magnified image showing the degree of ordering.

that the ordering phenomenon is not limited to one mesh. Detailed investigations of MDT showed that stripes as well as sheets of high symmetry are interrupted by isotropic regions of low (or no) symmetry and empty regions.[19] Moreover, the maximum size of noninterrupted well-ordered arrays available by applying MDT is limited to about 1 μm^2.[19]

One major requirement for technological applications of ferrofluids is their assembly into ordered structures of several square millimeters in size. Self-assembling of colloids on aqueous surfaces allows for the preparation of continuous layers of several square millimeters in size, typically consisting of well-ordered *cores* interrupted by grain boundaries between cores of different orientation. There is no principal difference between nanosized inorganic and micron-sized organic colloids. The core size can be improved by adding tensides (or simply soap) to the aqueous phase, which will change its surface tension or the interaction of the particulate depletion layer. The latter refers to theoretical descriptions developed by Debye, Hückel, and Onsager (see common textbooks of physical chemistry).

Figure 14.5 presents a typical result of particle assembly on a water surface. The visible structure results from isolated second-layer islands. A perfect monolayer would not be visible at this magnification, again demonstrating the problem of TEM investigations to show both symmetry and large area substrate coverage in the case of nanoscale materials. It is obvious from Figure 14.5 that the symmetry of the assembled layer is decreased compared to the layer shown in Figure 14.1. Nevertheless, assembling on water surfaces allows preparation of layers of, say, narrow symmetry distribution over several square millimeters. Furthermore, this technique allows the application of external magnetic fields during monolayer formation — static as well as rotating magnetic fields. The latter may have some advantages in preparing assembled layers of circular geometry.

Further optimizations have been available by applying the LB technique, dealing with mechanical force to increase the core size of self-assembled particle layers on water surfaces.[11,12] The preparation of large-scale three-dimensional colloidal crystals grown from oversaturated fluids has also been reported.[104,105]

Different from other methods is the LbL technique. It is in principle applicable for ferrofluids prepared by all of the methods presented in Table 14.1, with one restriction.

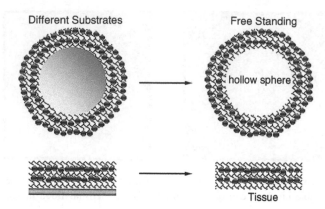

FIGURE 14.6 Schematic illustration of available LbL films. Coatings on substrates of different geometries and freestanding films are shown.

Dealing with polyelectrolytes (the most common technique) requires typically aqueous solutions and charged colloids and is, therefore, generally limited to ferrofluids containing iron oxide. The advantages of the LbL technique are the availability of assembled layers (mainly layers of aggregates in case of aqueous iron oxide colloids) on curved substrates, i.e., latex spheres, and freestanding films. The LbL technique is based on polymer-mediated self-assembly by alternatively adsorbing polymer and nanoparticles on solid substrates.[13–16,125–128] As a result, nanoparticle layers of different thicknesses alternatively divided by layers of polymers have been published. A schematic illustration of available LbL films is presented in Figure 14.6.

Sun et al.[129] have published the first LbL preparation of magnetic FePt films from uncharged colloids and poly(ethylenimine). The authors explained that the particle-protecting stabilizer (oleic acid) is replaced by the polymer. With respect to the chemisorbing nature of oleic acid, i.e., a covalent-like bonding, we suggest a different explanation of the authors' observation: the polymer forms a second stabilizing shell around the oleic acid-stabilized particles, thus changing their surface properties. Further investigations will be necessary to clarify this.

14.4 APPLICATIONS

14.4.1 DATA STORAGE DEVICES

The interest in syntheses of ferrofluids and the preparation of highly symmetric layers thereof has been renewed since the expectation that such layers may be suitable as a new generation of magnetic storage devices.[130] The requirement of *high anisotropy* of each *bit* in such devices resulted in syntheses of bimetallic, rather than monometallic, ferrofluids, because of higher anisotropy, and in posttreatments of assembled layers, i.e., annealing under different conditions. The latter have been necessary because the kinetically controlled colloidal crystal growth typically produces amorphous or crystalline phases of less anisotropy compared to thermodynamically stable crystal modifications. Moreover, exchange-decoupled single nanocrystals are

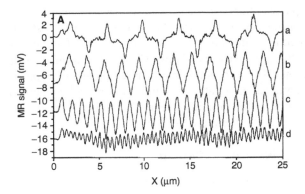

FIGURE 14.7 Magnetoresistive (MR) read-back signals from written bit transitions in a 120-nm-thick assembly of 4-nm-diameter $Fe_{48}Pt_{52}$ nanocrystals. The individual line scans reveal magnetization reversal transitions at linear densities of 500 fc/mm (a), 1040 fc/mm (b), 2140 fc/mm (c), and 5000 fc/mm (d). (From Sun, S. et al., *Science*, 287, 1989–1992, 2000. With permission. Copyright © 2003 American Association for the Advancement of Science.)

required to be present in final devices. Thus, investigations concerning transformation of morphology and related magnetic properties have been done intensely in the past.[20,51,52,63,95,129–134]

One major result is the observation that an upper temperature limit does exist, above which the assembled particles within a layer tend to aggregate and start to grow. This limit is dependent on the material itself, its crystalline state, the composition of stabilizers, and the parameters of posttreatment. Organic stabilizers typically decompose below a temperature of 300°C, depending on their mass. Posttreatments in air result in CO_2 formation of organic stuff and, thus, in complete removal of the protecting shell. Furthermore, air annealing of oxidation-sensitive materials produces particles covered with an oxide shell (or complete oxidation). If the oxide shell behaves antiferromagnetically, such as CoO, it will induce an *intraparticle* exchange-coupled enhanced anisotropy.[135] Annealing under inert gas conditions (Ar, N_2) results in the formation of a carbon layer around the particles or their aggregates. Annealing in the H_2 atmosphere will promote the formation of CH_x and is therefore also suitable to completely remove the stabilizers.

Phase transformations in colloids or their precursors usually occur above the decomposition temperatures of organics. For FePt particle layers, the most investigated system, the postannealing temperature to achieve the preferred crystalline phase without the occurrence of aggregation lies around 530°C. Figure 14.7 presents magnetoresistive read-back signals written from a 120-nm-thick multilayer film, prepared by self-assembly of monodisperse FePt ferrofluid, after posttreatment (560°C, 30 min, Ar).[95] The highest recording density obtained (curve d) was 5000 flux changes per mm (fc/mm).

One major problem in developing high-density data storage media is the preparation of monolayer films. A second is interparticle exchange coupling. The latter could be prevented by formation of a separation layer around the individual particles, e.g., a carbon layer made from the organic stabilizing shell by postannealing in suitable atmospheres. NanoMagnetics Ltd. (www.nanomagnetics.com) reports on data storage densities of 12 Gbit/in.^{2} by assembling and postannealing colloidal particles prepared in apoferritin, i.e., self-assembling apoferritin, containing precursors that will form nanoparticles of desired composition after posttreatment under suitable conditions. This procedure results in exchange-decoupled layers of single ferromagnetic particles of suitable anisotropy separated by a carbon shell resulting from decomposition of the protein micelle. Unfortunately, results are only scarcely available in the scientific literature.[94,136]

14.4.2 CARBON NANOTUBE GROWTH

The availability of monodisperse nanoparticles of *controlled* different sizes has made nanoparticle layers assembled from ferrofluids of interest for investigating the wall number vs. catalyst dependency of carbon nanotube (CNT) growth. Moreover, the easy possibility of varying the surface properties of magnetic nanoparticles allows investigation of these influences on growth kinetics of carbon nanotubes and, thus, the availability of aligned carbon nanotube devices. First, published results will be shown below.

Carbon nanotubes have been successfully grown by chemical vapor deposition (CVD) on Co[38] and bimetallic CoMo colloids[39] prepared by the reverse micelle technique, from iron oxide prepared in apoferritin,[42] and from Fe[41] and FeMo colloids[40] prepared by decomposition of respective metal carbonyls. The results clearly reveal a complex dependency of CNT growth on many parameters. These are nanoparticle size, posttreatment of assembled colloidal layers prior to CNT growth, and the latter's parameters. The main observations are:

1. Successful CNT preparation requires posttreatment of assembled catalysts.
2. An upper limit of the nanoparticle's diameter exists to catalyze the single-wall carbon nanotube's (SWNT) growth (typically between 4 and 12 nm).
3. CNT diameters did not necessarily equal the catalyst diameter (decreased diameters have been observed).
4. The degree of alignment of CNT devices is dependent on the particle's catalytic behavior.

All observations are directly influenced by posttreatments of assembled nanoparticle layers and CNT growth parameters.

Ago et al.[38] have published the preparation of well-aligned multiple-wall carbon nanotube (MWNT) devices catalyzed by assembled Co colloids prepared in reverse micelles. Figure 14.8 shows the result of different posttreatments of the catalyst on the degree of alignment of CNTs. CNTs in Figure 14.8A have been grown from H$_2$-treated nanoparticles, whereas the well-aligned device in Figure 14.8B could be obtained from H$_2$S-treated particles.

These observations reveal the strong catalytic influence of the surface properties of nanoparticles on CNT growth behavior. Moreover, investigations of the same group on bimetallic colloidal catalysts showed that the degree of sulfidation controls the number of CNT walls.[39]

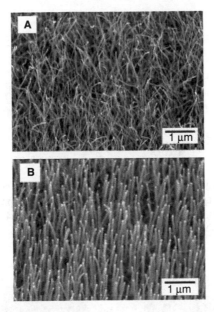

FIGURE 14.8 SEM micrographs of MWNTs grown from Co nanoparticles activated in H_2 gas prior to the pyrolysis (A) and MWNTs grown from H_2S-pretreated Co (B). (From Ago, H. et al., *Appl. Phys. Lett.*, 77, 79–81, 2000. With permission. Copyright © 2000 American Physical Society.)

Devices of aligned MWNT grown from metal islands, CVD deposited through a nanosphere mask, have recently been published to show photonic crystal band gap behavior.[137] Some approaches in photonic crystal research applying magnetic colloids will be presented below.

14.4.3 PHOTONIC CRYSTALS

A few articles concerning the preparation of photonic crystals from magnetized latex are available.[138,139] Two different routes were used to prepare magnetic organic–inorganic hybrid spheres. One dealt with the synthesis of magnetite incorporated in polymeric microspheres produced by emulsion polymerization.[139–141] The second applied the LbL approach to prepare latex particles covered with magnetite.[15,16]

Xu et al.[139] have published the preparation of iron oxide–containing latex spheres by emulsion polymerization. Figure 14.9 shows a typical TEM image of successful synthesis. These microspheres then must be used for the preparation of three-dimensional crystals. The photonic band gap of these crystals could be varied by varying the distance between the microspheres within a colloidal crystal by applying MDT. Increasing the applied magnetic field resulted in decreasing particle distances and blue shift of the Bragg diffraction wavelength. Unidirectional shifts have been observed with increasing NaCl concentration in aqueous solutions and decreasing dielectric constants of nonaqueous solutions.

Bizdoaca et al.[16] have used the LbL approach to prepare magnetite-covered latex particles. These particles then have to be further coated with SiO_2-covered Au

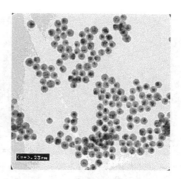

FIGURE 14.9 TEM image of monodisperse polystyrene particles containing superparamagnetic nanoparticles. The aggregates of iron oxide nanoparticles appear as black dots in the larger polystyrene spheres. (From Xu et al. *Adv Mater* 13:1681–1684, 2001. With permission.) Copyright 2001, WILEY-VCH.

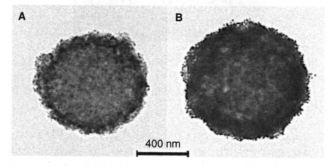

FIGURE 14.10 TEM images of a 640 nm PS microsphere covered (a) with two layers of magnetic Fe_3O_4 nanoparticles ($d = 750 \pm 20$ nm) and (b) after additional coverage with tree layers of Au nanoparticles ($d = 860 \pm 20$ nm). (From Bizdoaca et al. *J Vac Sci Technol A* 21:1515–1518, 2003. With permission.) Copyright 2003, American Vacuum Society.

particles.[138] The use of SiO_2-covered Au particles became necessary because of the requirement of a negatively charged surface of Au colloids (because prepared Au particles via the applied method are positively charged). This approach means a coating of microspheres with a high refractive material prior to the colloid crystal formation is necessary. Figure 14.10 demonstrates successful preparation of latex–magnetite–gold core-shelled nanostructured microspheres. Properties of photonic band gap behavior of colloidal crystals prepared from these particles are under investigation.

14.5 CONCLUDING REMARKS

We have presented some principles of syntheses of colloidal magnetic particles, their properties, and recent advances in preparing assembled layers thereof on different substrates. The results, of course, are not necessarily limited to magnetic particles. The only restriction for nonmagnetic materials is the applicability of MDT.

Furthermore, we have shown that colloid chemistry is an interesting tool for the development of new materials, especially concerning the development of devices reduced in size. As reduced size means increased reactivity, colloid chemistry cannot provide solutions in all fields of device development. The future will show the limitation of colloid chemistry and where it will be placed within the field of materials research.

ACKNOWLEDGMENT

The authors gratefully acknowledge the financial support provided by the German Science Foundation (DFG): II C10 SPP 1072. We also acknowledge Dr. M. Farle, Universität Duisburg-Essen, Germany, for helpful discussions about magnetism and T. Kempa, Boston College, U.S., for revising the manuscript.

REFERENCES

1. CJ Brinker, GW Scherer. *Sol-Gel Science*. San Diego: Academic Press, 1990.
2. H-D Dörfler. *Grenzflächen- und Kolloidchemie*. Weinheim: VCH, 1994 (in German).
3. G Schmid, Ed. *Clusters and Colloids: From Theory to Applications*. Weinheim: VCH, 1994.
4. J-H Fendler, Ed. *Nanoparticles and Nanostructured Films*. Weinheim: Wiley-VCH, 1998.
5. HS Nalwa, Ed. *Handbook of Surfaces and Interfaces of Materials*. San Diego: Academic Press, 2001.
6. P Moriarty. Nanostructured materials. *Rep Prog Phys* 64:297–381, 2001.
7. LM Liz-Marzan, D Norris, Guest-Eds. New aspects of nanocrystal research. *Mater Res Soc Bull* 26, 2001.
8. W Ostwald. *Die Welt der vernachlässigten Dimensionen*. Dresden: Steinkopf, 1915 (in German).
9. RE Rosensweig. Magnetic fluids. *Sci Am* 247:124–132, 1982.
10. RE Rosensweig. *Ferrohydrodynamics*. New York: Dover Publishing, 1998.
11. SA Iakovenko, AS Trifonov, M Giersig, A Mamedov, DK Nagesha, VV Hanin, EC Soldatov, NA Kotov. One- and two-dimensional arrays of magnetic nanoparticles by the Langmuir–Blodgett technique. *Adv Mater* 11:388–392, 1999.
12. T Fried, G Shemer, G Markovich. Ordered two-dimensional arrays of ferrite nanoparticles. *Adv Mater* 13:1158–1161, 2001.
13. MA Correa-Duarte, M Giersig, NA Kotov, LM Liz-Marzán. Control of packing order of self-assembled monolayers of magnetite nanoparticles with and without SiO_2 coating by microwave irradiation. *Langmuir* 14:6430–6435, 1998.
14. FG Aliev, MA Correa-Duarte, A Mamedov, JW Ostrander, M Giersig, LM Liz-Marzán, NA Kotov. Layer-by-layer assembly of core-shell magnetite nanoparticles: effect of silica coating on interparticle interactions and magnetic properties. *Adv Mater* 11:1006–1010, 1999.
15. F Caruso, M Spasova, A Susha, M Giersig, RA Caruso. Magnetic nanocomposite particles and hollow spheres constructed by a sequential layering approach. *Chem Mater* 13:109–116, 2001.

16. EL Bizdoaca, M Spasova, M Farle, M Hilgendorff, F Caruso. Magnetically directed self-assembly of submicron spheres with a Fe_3O_4 nanoparticle shell. *J Magn Magn Mater* 240:44–46, 2002.
17. M Giersig, P Mulvaney. Ordered two-dimensional gold colloid lattices by electrophoretic deposition. *J Phys Chem* 97:6334–6336, 1993.
18. M Giersig, M Hilgendorff. The preparation of ordered colloidal magnetic particles by magnetophoretic deposition. *J Phys D Appl Phys* 32:L111–L113, 1999.
19. M Hilgendorff, B. Tesche, M Giersig. Creation of 3-d crystals from cobalt nanoparticles in external magnetic fields. *Aust J Chem* 54:497–501, 2001.
20. CB Murray, S Sun, H Doyle, T Betley. Monodisperse 3d transition-metal (Co, Ni, Fe) nanoparticles. *Mater Res Soc Bull* 26:985–991 2001.
21. M-P Pileni. Magnetic fluids: fabrication, magnetic properties, and organization of nanocrystals. *Adv Funct Mater* 11:323–336, 2001.
22. M Giersig, M Hilgendorff. Assemblies of magnetic particles. In *Nanoscale Materials*, LM Liz-Marzán, P Kamat, Eds. Boston: Kluwer Academic Publishers, 2002, pp. 335–371.
23. M Hilgendorff, M Giersig. Synthesis of colloidal magnetic nanoparticles: properties and applications. In *Low-Dimensional Systems: Theory, Preparation, and Some Applications*, NATO Science Ser. II, Vol. 91, LM Liz-Marzán, M Giersig, Eds. Dordrecht: Kluwer Academic Publishers, 2003, pp. 151–162.
24. M Hilgendorff. Bimetallic ferrofluids. In *Encyclopedia of Nanoscience and Nanotechnology*, Vol. 1, HS Nalwa, Ed. Stevenson Ranch: American Scientific Publishers, 2004, pp. 213–233.
25. DL Leslie-Pelecky, RD Rieke. Magnetic properties of nanostructured materials. *Chem Mater* 8:1770–1783, 1996.
26. JI Martin, J Nogues, K Liu, JL Vincent, IK Schuller. Ordered magnetic nanostructures: fabrication and properties. *J Magn Magn Mater* 256:449–501, 2003.
27. N Buske, H Sonntag, T Götze. Magnetic fluids: their preparation, stabilization and applications in colloid science. *Colloids Surfaces* 12:195–202, 1984.
28. DYC Chan, D Henderson, J Barojas, AM Homola. The stability of a colloidal suspension of coated magnetic particles in an aqueous solution. *J Res Dev* 29:11–17, 1985.
29. AP Philipse, MPB van Bruggen, C Pathmamanoharan. Magnetic silica dispersions: preparation and stability of surface-modified silica particles with a magnetic core. *Langmuir* 10:92–99, 1994.
30. JN Israelachvili. *Intermolecular and Surface Forces*. San Diego: Academic Press, 1992.
31. R Tadmor. The London–Van der Waals interaction energy between objects of various geometries. *J Phys Condens Matter* 13:L195–L202, 2001.
32. R Charmas. Four-layer complexation model for ion adsorption at energetically heterogeneous metal oxide/electrolyte interfaces. *Langmuir* 15:5635–5648 1999.
33. LM Liz-Marzán, MA Correa-Duarte, I Pastoriza-Santos, P Mulvaney, T Ung, M Giersig, NA Kotov. Core-shell nanoparticles and assemblies thereof. In *Handbook of Surfaces and Interfaces of Materials*, Vol. 3, HS Nalwa, Ed. San Diego: Academic Press, 2001, pp. 189–237.
34. GW Scherer. Theory of drying. *J Am Chem Soc* 73:3–14, 1990.
35. RD Shull, LH Bennet. Nanocomposite magnetic materials. *Nanostruct Mater* 1:83–88, 1992.
36. GR Harp, SSP Parkin, WL O'Brian, BP Tonner. Induced Rh magnetic moments in Fe-Rh and Co-Rh alloys using x-ray magnetic circular dicroism. *Phys Rev B* 51:12037–12040, 1995.

37. G Moraïtis, H Dreyssé, MA Khan. Band theory of induced magnetic moments in CoM (M = Rh, Ru) alloys, *Phys Rev B* 54:7140–7142, 1996.
38. H Ago, T Komatsu, S Ohshima, Y Kuriki, M. Yumura. Dispersion of metal nanoparticles for aligned carbon nanotube arrays. *Appl Phys Lett* 77:79–81, 2000.
39. H Ago, S Ohshima, K Uchida, M Yumura. Gas-phase synthesis of single wall carbon nanotubes from colloidal solution of metal nanoparticles. *J Phys Chem B* 105:10453–10456, 2001.
40. Y Li, J Liu, Y Wang, ZL Wang. Preparation of monodispersed Fe-Mo nanoparticles as the catalyst for CVD synthesis of carbon nanotubes. *Chem Mater* 13:1008–1014, 2001.
41. CL Cheung, H Kurtz, H Park, CM Lieber. Diameter controlled synthesis of carbon nanotubes. *J Phys Chem B* 106:2429–2433, 2002.
42. Y Li, W Kim, Y Zhang, M Rolandi, D Wang, H Dai. Growth of single walled carbon nanotubes from discrete catalytic nanoparticles of various sizes. *J Phys Chem B* 105:11424–11431, 2001.
43. C Binns. Magnetism at surfaces and interfaces. In *Handbook of Surfaces and Interfaces of Materials*, Vol. 2, HS Nalwa, Ed. San Diego: Academic Press, 2001, pp. 357–392.
44. L Katsikas, A Eichmüller, M Giersig, H Weller. Discrete exitonic transitions in quantum-sized CdS particles. *Chem Phys Lett* 172:201–204, 1990.
45. IG Dance, RG Garbutt, TD Bailey. Aggregated structures of the compounds $Cd(SC_6H_4X_4)_2$ in DMF solution. *Inorg Chem* 29:603–608, 1990.
46. M Respaud, JM Broto, H Rakoto, AR Fert, L Thomas, B Barbara, M Verelst, E Snoeck, P Lecante, A Mosset, J Osuna, T Ould Ely, C Amiens, B Chaudret. Surface effects on the magnetic properties of ultrafine cobalt particles. *Phys Rev B* 57:2925–2935, 1998.
47. DP Dinega, MG Bawendi. Eine aus der Lösung zugängliche neue Kristallstruktur von Cobalt. *Angew Chem* 111:906–1909, 1999; *Angew Chem Int Ed* 38:1788–1791, 1999.
48. VF Puntes, KM Krishnan, P Alivisatos. Synthesis, self-assembly, and magnetic behavior of a two-dimensional superlattice of single-crystal ε-Co nanoparticles. *Appl Phys Lett* 78:2187–2189, 2001.
49. H Laidler, K O'Grady. Crystallographic effects in Co alloy media. www.datatech-online.com, 1:93–97, 1998.
50. R Skomski, JMD Coey. Giant energy product in nanostructured two-phase magnets. *Phys Rev B* 48:15812–15816, 1993.
51. H Zeng, S Sun, TS Vedantam, JP Liu, Z-R Dai, Z-L Wang. Exchange-coupled FePt nanoparticle assembly. *Appl Phys Lett* 80:2583–2585, 2002.
52. H Zeng, J Li, JP Liu, ZL Wang, S Sun. Exchange-coupled magnets by nanoparticle self-assembly. *Nature* 420:395–398, 2002.
53. H Bönnemann, W Brijoux, R Brinkmann, E Dinjus, T Joussen, B Korall. Erzeugung von kolloidalen Übergangsmetallen in organischer Phase und ihre Anwendung in der Katalyse. *Angew Chem* 103:1344–1346, 1991; *Angew Chem Int Ed* 30:1312–1314, 1991.
54. J Rivas, MA López-Quintela, MG Bonome, RJ Duro, JM Greneche. Production and characterization of FeB amorphous particles. *J Magn Magn Mater* 122:1–5, 1993.
55. KS Suslick, M Fang, T Hyeon. Sonochemical synthesis of iron colloids. *J Am Chem Soc* 118:11960–11961, 1996.
56. TW Smith, D Wychick. Colloidal iron dispersions prepared via the polymer-catalyzed decomposition of iron pentacarbonyl. *J Phys Chem* 84:1621–1629, 1980.
57. PH Hess, PH Parker Jr. Polymers for stabilization of colloidal cobalt particles. *J Appl Polym Sci* 10:1915–1927, 1966.

58. E Papirer, P Horny, H Balard, R Anthore, C Petipas, A Martinet. The preparation of a ferrofluid by decomposition of dicobalt octacarbonyl. Parts I and II. *J Colloid Interface Sci* 94:207–228, 1983.

59. J Osuna, D de Caro, C Amiens, B Chaudret, E Snoeck, M Respaud, J-M Broto, A Fert. Synthesis, characterization, and magnetic properties of cobalt nanoparticles from an organometallic precursor. *J Phys Chem* 35:14571–14574, 1996.

60. C Pathmamanoharan, AP Philipse. Preparation and properties of monodisperse magnetic cobalt colloids grafted with polyisubutene. *J Colloid Inter Sci* 205:340–353, 1998.

61. JA Becker, R Schäfer, R Festag, W Ruland, JH Wendorff, J Pebler, SA Quaiser, W Helbig, MT Reetz. Electrochemical growth of superparamagnetic cobalt clusters. *J Chem Phys* 103:2520–2527, 1995.

62. GN Glavee, KJ Klabunde, CM Sørensen, GC Hadjipanayis. Sodium borohydride reduction of cobalt ions in nonaqueous media. Formation of ultrafine particles (nanoscale) of cobalt metal. *Inorg Chem* 32:474–477, 1993.

63. S Sun, CB Murray. Synthesis of monodisperse cobalt nanocrystals and their assembly into magnetic superlattices. *J Appl Phys* 85:4325–4330, 1999.

64. C Petit, A Taleb, MP Pileni. Cobalt nanosized particles organized in a 2D superlattice: synthesis, characterisation, and magnetic properties. *J Phys Chem B* 103:1805–1810, 1999.

65. JP Stevenson, M Rutnakornpituk, M Vadala, AR Esker, SW Charles, S Wells, JP Dailey, JS Riffle. Magnetic cobalt dispersions in poly(dimethylsiloxane) fluids. *J Magn Magn Mater* 225:47–58, 2001.

66. ASR Hoon, M Kilner, GJ Russell, BK Tanner. Preparation and properties of nickel ferrofluids. *J Magn Magn Mater* 39:107–110, 1983.

67. N Cordente, M Respaud, F Senocq, M-J Casanove, C Amiens, B Chaudret. Synthesis and magnetic properties of nickel nanorods. *Nano Lett* 1:565–568, 2001.

68. DL Leslie-Pelecky, S-H Kim, M Bonder, XQ Zhang, RD Rieke. Structural properties of chemically synthesized nanostructured Ni and Ni:Ni$_3$C nanocomposites. *Chem Mater* 10:164–171, 1998.

69. MT Reetz, M Winter, R Breinbauer, T Thurn-Albrecht, W Vogel. Size-selective electrochemical preparation of surfactant stabilized Pd-, Ni-, and Pt/Pd colloids. *Chem Eur J* 7:1084–1094, 2001.

70. N Toshima, T Yonezawa. Bimetallic nanoparticles: novel materials for chemical and physical applications. *New J Chem* 22:1179–1201, 1998.

71. DB Lambrick, N Mason, NJ Harris, GJ Russell, SR Hoon, M Kilner. An iron–cobalt "alloy" magnetic fluid. *IEEE Trans Magn* MAG-21:1891–1893, 1985.

72. DB Lambrick, N Mason, SR Hoon, M Kilner. Preparation and properties of Ni-Fe magnetic fluids. *J Magn Magn Mater* 65:257–260, 1987.

73. DB Lambrick, N Mason, SR Hoon, M Kilner, JN Chapman. The preparation of Co/Fe alloy fine particles from HFeCo$_3$(CO)$_{12}$. *IEEE Trans Magn* 24:1644–1646, 1988.

74. T Ould Ely, C Pan, C Amiens, B Chaudret, F Dassenoy, M-J Casanove, A Mosset, M Respaud, J-M Broto. Nanoscale bimetallic Co$_x$Pt$_{1-x}$ particles dispersed in poly(vinylpyrrolidone): synthesis from organometallic precursors and characterization. *J Phys Chem B* 104:695–702, 2000.

75. MC Fromen, A Serres, D Zitoun, M Respaud, C Amiens, B Chaudret, P Lecante, MJ Casanove. Structural and magnetic study of bimetallic Co$_x$Rh$_{1-x}$ particles. *J Magn Magn Mater* 242–245:610–612, 2001.

76. D Zitoun, M Respaud, MC Fromen, MJ Casanove, P Lecante, C Amiens, B Chaudret. Magnetic enhancement in nanoscale CoRh particles. *Phys Rev Lett* 89:037203, 2002.

77. MT Reetz, W Helbig, SA Quaiser. Electrochemical preparation of nanostructured bimetallic clusters. *Chem Mater* 7:2227–2228, 1995.
78. PM Paulus, H Bönnemann, AM van der Kraan, F Luis, J Sinzig, LJ de Jongh. Magnetic properties of nanosized transition metal colloids: the influence of noble metal coating. *Eur Phys J D* 9:501–504, 1999.
79. H Bönnemann, W Brijoux, R Brinkmann, M Wagener. World patent. To Studiengesellschaft Kohle mbH. WO 99/41758, 1999.
80. H Bönnemann. Nanostructured metal colloids: chemistry and potential applications. In *Handbook of Surfaces and Interfaces of Materials*, Vol. 3., HS Nalwa, Ed. San Diego: Academic Press, 2001, pp. 41–64.
81. W Yu, Y Wang, H Liu, W Zheng. Preparation and characterization of polymer-protected Pt/Co bimetallic colloids and their catalytic properties in the selective hydrogenation of cinnamaldehyde. *J Mol Catal A* 112:105–113, 1996.
82. N Nunomura, T Teranishi, M Miyake, A Oki, S Yamada, N Toshima, H Hori. Magnetization of nano-fine particles of Pd/Ni alloys. *J Magn Magn Mater* 177–181:947–948, 1998.
83. N Nunomura, H Hori, T Teranishi, M Miyake, S Yamada. Magnetic properties of nanoparticles in Pd/Ni alloys. *Phys Lett A* 249:524–530, 1998.
84. T Teranishi, M Miyake. Novel synthesis of monodispersed Pd/Ni nanoparticles. *Chem Mater* 11:3414–3416, 1999.
85. P Lu, T Teranishi, K Asakura, M Miyake, N Toshima. Polymer-protected Ni/Pd bimetallic nano-clusters: preparation, characterization, and catalysis for hydrogenation of nitrobenzene. *J Phys Chem B* 103:9673–9682, 1999.
86. J Rivas, RD Sánchez, A Fondado, C Izco, AJ García-Bastida, J García-Otero, J Mira, D Baldomir, A Gonzáles, I Lado, MA López-Quintela, SB Oseroff. Structural and magnetic characterization of Co particles coated with Ag. *J Appl Phys* 76:6564–6566, 1994.
87. AJ García-Bastida, RD Sánchez, J García-Otero, J Rivas, A González-Penedo, J Solla, MA López-Quintela. Influence of the annealing temperature in different atmospheres on the magnetic behavior of Co/Ag nanoparticles fabricated by microemulsions. *Mater Sci Forum* 269–272:919–924, 1998.
88. RD Sánchez, MA López-Quintela, J Rivas, A González-Penedo, AJ García-Bastida, CA Ramos, RD Zysler, S Ribeiro-Guevara. Magnetization and electron paramagnetic resonance of Co clusters embedded in Ag nanoparticles. *J Phys Condens Matter* 11:5643–5654, 1999.
89. N Duxin, N Brun, P Bonville, C Colliex, MP Pileni. Nanosized Fe-Cu-B alloys and composites synthesized in diphasic systems. *J Phys Chem B* 101:8907–8913, 1997.
90. N Duxin, N Brun, C Colliex, MP Pileni. Synthesis and magnetic properties of elongated Fe-Cu alloys. *Langmuir* 14:1984–1989, 1998.
91. EE Carpenter, CT Seip, CJ O'Connor. Magnetism of nanophase metal and metal alloy particles formed in ordered phases. *J Appl Phys* 85:5184–5186, 1999.
92. CJ O'Conner, V Kolesnichenko, EE Carpenter, C Sangregorio, W Zhou, A Kumbhar, J Sims, F Agnoli. Fabrication and properties of magnetic particles with nanometer dimensions. *Synth Met* 122:547–557, 2001.
93. B Ravel, EE Carpenter, VG Harris. Oxidation of iron in iron/gold core/shell nanoparticles. *J Appl Phys* 91:8195–8197, 2002.
94. B Warne, OI Kasyutich, EL Mayes, JAL Wiggins, KKW Wong. Self assembled nanoparticulate Co:Pt for data storage applications. *IEEE Trans Magn* 36:3009–3011, 2000.
95. S Sun, CD Murray, D Weller, L Folks, A Moser. Monodisperse FePt nanoparticles and ferromagnetic FePt nanocrystal superlattices. *Science* 287:1989–1992, 2000.

96. S Yamamuro, D Farrell, KD Humfeld, SA Majetich. Structure of self-assembled Fe and FePt nanoparticle arrays. *Mater Res Soc Symp Proc* 636:D.10.8.1, 2001.

97. D Farrell, S Yamamuro, SA Majetich. Magnetic properties of self-assembled Fe nanoparticle arrays. *Mater Res Soc Symp Proc* 674:U.4.4.1, 2001.

98. KD Humfeld, AK Giri, EL Venturini, SA Majetich. Time dependent properties of iron nanoparticles. *IEEE Trans Magn* 37:2194–2196, 2001.

99. RV Chamberlin, KD Humfeld, D Farrell, S Yamamuro, Y Ijiri, SA Majetich. Magnetic relaxation of iron particles. *J Appl Phys* 91:6961–6963, 2002.

100. S Yamamuro, D Farrell, SA Majetich. Direct imaging of self-assembled magnetic nanoparticle arrays: phase stability and magnetic effects on morphology. *Phys Rev B* 65:224431, 2002.

101. M Chen, DE Nikles. Synthesis, self-assembly, and magnetic properties of $Fe_xCo_yPt_{100-x-y}$ nanoparticles. *Nano Lett* 2:211–214, 2002.

102. M Chen, DE Nikles. Synthesis of spherical FePd and CoPt nanoparticles. *J Appl Phys* 91:8477–8479, 2002.

103. K Ono, Y Kakefuda, R Okuda, Y Ishii, S Kamimura, A Kitamura, M Oshima. Organometallic synthesis of magnetic properties of ferromagnetic Sm-Co nanoclusters. *J Appl Phys* 91:8480–8482, 2002.

104. EV Shevchenko, DV Talapin, A Kornowski, F Wiekhorst, J Kötzler, M Haase, AL Rogach, H Weller. Colloidal crystals of monodisperse FePt nanoparticles grown by a three-layer technique of controlled oversaturation. *Adv Mater* 14:287–290, 2002.

105. EV Shevchenko, DV Talapin, AL Rogach, A Kornowski, H Weller. Colloidal synthesis and self assembly of $CoPt_3$ nanocrystals. *J Am Chem Soc* 124:11480–11485, 2002.

106. J-I Park, J Cheon. Synthesis of "solid solution" and "core-shell" type cobalt–platinum magnetic nanoparticles via transmetallation reactions. *J Am Chem Soc* 123:5743–5746, 2001.

107. NS Sobal, M Hilgendorff, H Möhwald, M Giersig, M Spasova, T Radetic, M Farle. Synthesis and structure of colloidal bimetallic nanocrystals: the non-alloying system Ag/Co. *Nano Lett* 2:621–624, 2002.

108. M Spasova, T Radetic, N Sobal, M Hilgendorff, U Wiedwald, M Farle, M Giersig, U Dahmen. Structure and magnetism of Co and CoAg nanocrystals. *Mater Res Soc Symp Proc* 721:195–200, 2002.

109. U Wiedwald, M Ulmeanu, M Farle, NS Sobal, M Giersig, A Fraile-Rodriguez, D Arvanitis. Ordered arrays of magnetic nanoscale particles: composition dependence of the orbital magnetic moment. *MAX-lab Activity Rep* 2001:372–373, 2002.

110. R Massart, V Cabuil. Synthèse en milieu alcalin de magnétite colloïdale: contrôle du rendement et de la taille des particules. *J Chim Phys* 84:967–973, 1987 (in French).

111. RV Upadhyay, KJ Davies, S Wells, SW Charles. Preparation and characterization of ultra-fine $MnFe_2O_4$ and $Mn_xFe_{1-x}O_4$ spinel systems. II. Magnetic fluids. *J Magn Magn Mater* 139:249–254, 1995.

112. KJ Davies, S Wells, RV Upadhyay, SW Charles, K O'Grady, M El Hilo, T Meaz, S Mørup. The observation of multi-axial anisotropy in ultrafine cobalt ferrite particles used in magnetic fluids. *J Magn Magn Mater* 149:14–18, 1995.

113. PC Fannin, SW Charles, JL Dormann. Field dependence of the dynamic properties of colloidale suspensions of $Mn_{0.66}Zn_{0.34}Fe_2O_4$ and $Ni_{0.5}Zn_{0.5}Fe_2O_4$ particles. *J Magn Magn Mater* 201:98–101, 1999.

114. L Liz, MA López Quintela, J Mira, J Rivas. Preparation of colloidal Fe_3O_4 ultrafine particles in microemulsions. *J Mater Sci* 29:3797–3801, 1994.

115. JA López Pérez, MA López Quintela, J Mira, J Rivas, SW Charles. Advances in the preparation of magnetic nanoparticles by the microemulsion method. *J Phys Chem B* 101:8045–8047, 1997.

116. KJ Davies, S Wells, SW Charles. The effect of temperature and oleate adsorption on the growth of maghemite particles. *J Magn Magn Mater* 122:24–28, 1993.

117. I Mlescu, L Gabor, F Claici, N Tefu. Study of some magnetic properties of ferrofluids filtered in magnetic field gradient. *J Magn Magn Mater* 222:8–12, 2000.

118. T Fried, G Shemer, G Markovich. Ordered two-dimensional arrays of ferrite nanoparticles. *Adv Mater* 13:1158–1161, 2001.

119. S Sun, H Zeng. Size-controlled synthesis of magnetite nanoparticles. *J Am Chem Soc* 124:8204–8205, 2002.

120. KR Brown, MJ Natan. Hydroxylamine seeding of colloidal Au nanoparticles in solution and on surfaces. *Langmuir* 14:726–728, 1998.

121. NR Jana, L Gearheart, CJ Murphy. Evidence for seed-mediated nucleation in the chemical reduction of gold salts to gold nanoparticles. *Chem Mater* 13:2313–2322, 2001.

122. H Yu, PC Gibbons, KF Kelton, WE Buhro. Heterogeneous seeded growth: a potentially general synthesis of monodisperse metallic nanoparticles. *J Am Chem Soc* 123:9198–9199, 2001.

123. DV Talapin, AL Rogach, M Haase, H. Weller. Evolution of an ensemble of nanoparticles in a colloidal solution: theoretical study. *J Phys Chem B* 105:12278–12285, 2001.

124. D Lacoste, TC Lubensky. Phase transition in a ferrofluid at magnetic-field induced microphase separation. *Phys Rev E* 64:41506, 2001.

125. JH Fendler. Self-assembled nanostructured materials. *Chem Mater* 8:1616–1624, 1996.

126. T Cassagneau, TE Mallouk, JH Fendler. Layer-by-layer assembly of thin film Zener diodes from conducting polymers and CdSe nanoparticles. *J Am Chem Soc* 120:7848–7859, 1998.

127. J Schmitt, P Mächtle, D Eck, H Möhwald, CA Helm. Preparation and optical properties of colloidal gold monolayers. *Langmuir* 15:3256–3266, 1999.

128. BH Sohn, BH Seo. Fabrication of the multilayerd nanostructure of alternating polymers and gold nanoparticles with thin films of self-assembling diblock copolymers. *Chem Mater* 13:1752–1757, 2001.

129. S Sun, S Anders, HF Hamann, J-U Thiele, JEE Baglin, T Thomson, EE Fullerton, CB Murray, BD Terris. Polymer mediated self-assembly of magnetic nanoparticles. *J Am Chem Soc* 124:2884–2885, 2002.

130. D Weller, A Moser. Thermal effect limits in ultra-high density magnetic recording. *IEEE Trans Magn* 35:4423–4439, 1999.

131. ZR Dai, S Sun, ZL Wang. Phase transformation, coalescence, and twinning of monodisperse FePt nanocrystals. *Nano Lett* 1:443–447, 2001.

132. JW Harrell, S Wang, DE Nikles, M Chen. Thermal effects in self assembled FePt nanoparticles with partial chemical ordering. *Appl Phys Lett* 79:4393–4395, 2001.

133. K Sato, B Bian, Y Hirotsu. Fabrication of oriented L1$_0$-FePt and FePd nanoparticles with large coercivity. *J Appl Phys* 91:8516–8518, 2002.

134. CT Black, CB Murray, RL Sandstrom, S Sun. Spin dependent tunneling in self-assembled cobalt nanocrystal superlattices. *Science* 290:1131–1134, 2000.

135. WH Meiklejohn, CP Bean. New magnetic anisotropy. *Phys Rev* 105:904–913, 1957.

136. E Mayes, K Wong. Novel system for producing magnetic thin films. www.datatech-online.com, Edition 3, 1999.

137. K Kempa, B Kimball, J Rybczynsky, ZP Huang, PF Wu, D Steeves, M Sennett, M Giersig, DVGLN Rao, DL Carnahan, DZ Wang, JY Lao, WZ Li, ZF Ren. Photonic crystals based on periodic arrays of aligned carbon nanotubes. *Nano Lett* 3:13–18, 2003.

138. EL Bizdoaca, M Spasova, M Farle, M Hilgendorff, LM Liz-Marzan, F Caruso. Self-assembly and magnetism in novel core-shell microspheres. *J Vac Sci Technol A* 21:1515–1518, 2003.

139. X Xu, G Friedman, KD Humfeld, SA Majetich, SA Asher. Superparamagnetic photonic crystals. *Adv Mater* 13:1681–1684, 2001.

140. K. Wormuth. Superparamagnetic latex via emulsion polymerization. *J Colloid Interface Sci* 241:366–377, 2001.

141. Y Deng, L Wang, W Yang, S Fu, A Elaïssari. Preparation of magnetic polymeric particles via inverse microemulsion polymerization process. *J Magn Magn Mater* 257:69–78, 2002.

Layered Nanoparticle Assemblies

15 Template Synthesis and Assembly of Metal Nanowires for Electronic Applications

Sarah K. St. Angelo and Thomas E. Mallouk

CONTENTS

15.1 Introduction 413
    15.1.1 Nanowire Synthesis...414
    15.1.2 Nanowire Assembly 414
15.2 Template Growth of Nanowires...415
    15.2.1 Electrodeposition of Nanowires...415
    15.2.2 Chemical Modification of Nanowire Surfaces 417
15.3 Electrical Measurements 418
15.4 Electronic Properties of Nanowires Containing Functional
    Components..421
15.5 Nanowires at Interfaces: Properties and Assembly Strategies 426
    15.5.1 Controlling Particle Surface Interactions 426
    15.5.2 Nanowire Assembly in Lithographically Defined Wells..........427
    15.5.3 Diffusion of Nanowires on Surfaces.......................................428
    15.5.4 Biomolecular Recognition in Nanowire Assembly 430
15.6 Conclusions 431
Acknowledgments...432
References...432

15.1 INTRODUCTION

In their efforts to move the seemingly remote goal of molecular-scale electronics ever closer to demonstrable reality, chemists, physicists, materials scientists, and engineers are looking to their diverse skills for the tools to build functional circuits from the bottom up. The pressure to make faster, smaller, more densely arranged computing elements is pushing traditional lithographic technologies to their theoretical and practical limits. While the life span of the top-down approach is being extended, the potential advantages of assembling molecular and nanoscale devices are coming into focus. A wealth of organic, inorganic, and biological materials make up the palette of nanoscale building blocks available to researchers. Many of these materials are relatively inexpensive compared to the exponential cost increase for construction of new lithographic fabrication facilities. Although the complete picture of what a nano/molecular electronics-based computer will be or how it will function is not yet clear, many researchers are converging on possible solutions.

15.1.1 NANOWIRE SYNTHESIS

There are now many approaches to producing nanowires and other nanoelectronic components. Soft chemical reduction of metal ions in micellar templates has been used to make Au[1,2] and Ag[1] nanowires. Ag nanowires may also be synthesized in bulk quantities from silver salts by chemical reduction in the presence of ethylene glycol and poly(vinyl pyrrolidone).[3,4] Anisotropic Au particles may be photochemically reduced from the metal salt, using surfactant micelles as soft templates.[5] High-aspect-ratio crystalline silver nanowires have been grown by electrochemical reduction in a reverse hexagonal liquid crystalline phase.[6] Se nanowires are made using similar soft chemical techniques but without micellar or other templates.[7] Although wires produced using these and comparable soft chemical techniques are capable of providing large numbers of nanowires, there is often polydispersity in length and diameter and a contamination of the sample with smaller spherical or polyhedral metal particles. Consequently, solution syntheses of nanowires will probably be most useful when they are coupled to advanced separation methods that allow particular lengths and diameters to be selected from mixtures.

Vapor phase techniques provide an interesting alternative to solution methods and provide access to different compositions and physical forms. Many kinds of high-aspect-ratio nanowires have been synthesized by the vapor–liquid–solid (VLS) method. Generally, Au nanoparticles are used as a catalyst to grow wires of Si, Ge,[8–10] GaN,[11] GaAs and other III-V semiconductors,[12,13] and Au_2Si-SiO_2.[14] There are also many examples of using chemical vapor deposition (CVD) to produce nanowires. Some recent cases include the growth of MgO,[15] ZnO,[16] SiO_2, Si_3N_2,[17] and Si/SiGe.[18] Nanowires produced by CVD are often grown directly on a silicon surface[15,16,18] but may also be grown in a porous template.[17] Nanowires of up to millimeter lengths can be deposited on the step edges of highly ordered pyrolitic graphite.[19,20] The step edges are decorated with metal oxides that are subsequently reduced. The wires then lie along the step edges and may be transferred to a secondary polymeric substrate. Carbon nanotubes are also being intensively studied as wires and devices for nanoscale electronics.[21–24]

15.1.2 NANOWIRE ASSEMBLY

The ordering of anisotropic nanowires has not received nearly as much attention as the assembly of isotropic particles.[25] With the increasing interest in the use of anisotropic particles in nanoelectronics, more studies in this area are beginning to appear. Carbon nanotubes have been assembled on surfaces using electric field-assisted deposition.[26] Nanotubes may also be aligned and assembled into useful structures by flowing them through microfluidic channels.[27] The nanotubes tend to align along the length of the channel, forming parallel arrays. Crossing nanotubes may be assembled by depositing one set of nanotubes, rotating the flow channels by 90°, and depositing a second set. Continuous chains of nanowires of other materials may be deposited from solution into microchannels, leaving wires on a substrate.[28] Capillary forces have been used to assemble carbon nanotubes on a substrate[29] in a manner similar to that often used in growing colloidal crystals.[30] The nanotubes tend to deposit on the substrate parallel to the air–water–substrate triple line, and this results in a directionally, though not positionally, ordered array. Langmuir–Blodgett techniques have also been used to induce ordering in nanowire films.[31,32] The surface pressure on the particles can induce positional as well as directional order in a relatively dense particle film. The ordered particle arrays can then be transferred to solid substrates.

15.2 TEMPLATE GROWTH OF NANOWIRES

The nanowires used in our experiments, which will be described in detail below, were all produced by replication of porous membrane templates using techniques adapted from the original work of Al-Mawlawi et al.[34] and Martin et al.[35] Typically, aluminum oxide (purchased from Whatman, Inc., or synthesized in-house) or polycarbonate (purchased from Osmonics) is used as the membrane material. The inner pore diameter, which varies from about 10 to 300 nm, defines the diameter of the nanowire, whereas the charge passed during electrodeposition defines the length.

15.2.1 ELECTRODEPOSITION OF NANOWIRES

In general, the template growth of nanowires proceeds according to the scheme shown in Figure 15.1A. The first step is to provide a conductive connection to the template. This is done by thermally evaporating several hundred nanometers of Ag onto one side of the membrane. Pinholes in the evaporated metal layer are closed by electrochemical reduction of more silver in the pores. Additionally, in the aluminum oxide templates, a region of branched pores is present on one face of the membrane. The electroplated silver fills in this region, allowing nanowires to be plated in the unbranched pores. Several materials, including Au, Pt, Pd, Ni, Sn, Co, Cu, Pb, Bi, Se, poly(pyrrole), Ag_2Se, CdSe, and CdTe, may be electroplated onto the Ag plug. Stripes of two or more different metals are created by rinsing out the previous plating solution and adding a new one. An example of striped nanowires that are still confined by the template is shown in Figure 15.1B. The dark stripes are Au, the bright stripes are Ag, and unfilled membrane is visible at the top of the optical micrograph.

When the desired metal or semiconductor segments have been electroplated in the appropriate lengths, the nanowires are released from the membrane by first dissolving the thermally evaporated/electroplated Ag with 50% HNO_3 (aq). If nanowires made from a metal soluble in HNO_3 are desired, a thin spacer plug of Au can be added between the sacrificial Ag and the rest of the nanowire. The spacer protects the rest of the nanowire from etching with a short exposure to HNO_3. The aluminum

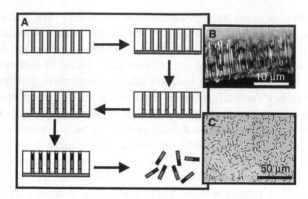

FIGURE 15.1 Electrochemical growth of nanowires in a porous membrane. (A) Schematic of the nanowire growth process. Ag is thermally evaporated onto one side of an alumina membrane, and more Ag is electrochemically deposited to seal the pores. A metal of interest is then deposited onto the Ag. Another metal or material may then be deposited onto the previous one, forming a stripe. The nanowires are released by dissolving the Ag and Al$_2$O$_3$. (B) Optical micrograph showing Ag-Au-Ag-Au-Ag nanowires still encased in the membrane on top of the evaporated Ag layer. (C) Optical micrograph showing Au nanowires released from the membrane. (Adapted from Martin, B.R. et al., *Adv Mater,* 11, 1021–1025, 1999. With permission.)

oxide templates are usually dissolved in 5 *M* NaOH (aq). The base is removed by first sedimenting the nanowires via centrifugation and removing the supernatant. The wires are resuspended in water with brief immersion in an ultrasonic bath. This process is repeated five times to remove base and the dissolved template from the suspension. Au nanowires that had been removed from the template and Ag contact in this way were dispersed on a microscope slide and are shown in Figure 15.1C. After rinsing, the nanowires may be transferred to another solvent or otherwise prepared for the addition of self-assembled monolayers (SAMs) or other surface coatings.

When polycarbonate templates are used, about 5-nm Cr and 100-nm Ag or Au are serially evaporated onto one face of the membrane. We generally do not electrodeposit sacrificial Ag when using polycarbonate because the pores are smaller (10 to 30 nm diameter, as opposed to hundreds of nanometers) and not as likely to remain open after thermal evaporation, and there is no branched pore region to avoid. The metallic nanowire body is electrochemically deposited, as described above; however, a different procedure is used to free the nanowires from the polycarbonate template. The thermally evaporated metals and any metal electrochemically deposited on the surface of the membrane may be removed by wiping repeatedly with a methanol-dampened paper tissue. This effectively removes any backing metal as well as any overplated metal on the top side of the membrane. The polycarbonate membrane is then dissolved in CH$_2$Cl$_2$. The nanowires are rinsed several times by sedimentation, supernatant removal, and resuspension in fresh solvent in order to remove as much dissolved polymer as possible. The nanowires may be transferred to a polar solvent by rinsing the nanowires with a solvent of intermediate polarity, e.g., passing the wires through isopropanol before suspending in water.

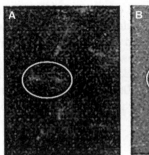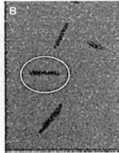

FIGURE 15.2 (A) Fluoresce micrograph of Au-Pt-Au nanowires in which the Au segments are derivatized with Rhodamine B isothiocyanate. The Pt segments are not fluorescent and are not visible in this view. (B) Bright-field micrograph of the same nanowires shown in (A). The full lengths of the wires are visible in this view. (Adapted from Martin, B.R. et al., *Adv Mater,* 11, 1021–1025, 1999. With permission.)

15.2.2 CHEMICAL MODIFICATION OF NANOWIRE SURFACES

One example of the chemical versatility afforded by multimetal nanowires is orthogonal self-assembly on the surface of Au-Pt-Au particles.[35] In work originally described by Whitesides and Martin et al.,[35] isonitriles form SAMs on Pt and Au; however, thiols replace isonitriles on Au at sufficiently high concentration. By using this orthogonal self-assembly approach, one can define different surface chemistries on the Au and Pt stripes of individual nanowires and fluorescently tag the Au sections. This was accomplished by treating Au-Pt-Au nanowires with 1-butaneisocyanide (BIC), then with 2-mercaptoethylamine (MEA), and finally with Rhodamine B isothiocyanate (details are given in reference 3). MEA displaces the BIC from the Au sections of the nanowire, and the Rhodamine B isothiocyanate forms a thiourea bond with MEA. When Au-Pt-Au wires thus treated were dispersed on a microscope slide for fluorescence imaging, the Au tips were fluorescent, as seen in Figure 15.2A. The corresponding image to the right (Figure 15.2B) is a bright-field micrograph in which the entirety of the wires is visible. Control experiments with single-metal Au and Pt nanowires derivatized as above behaved as expected: Au wires fluoresced along the entire wire, and Pt wires did not fluoresce at all.

This orthogonal assembly on striped nanowires allows for the use of two (or potentially more) SAMs on a single particle. Different electrostatic properties or degrees of hydrophilicity on individual particles could be used to position them on substrates. Electroactive molecules may be placed along a particular portion of the nanowires, whereas the other sections may be used as contacts or aid in multiparticle assembly.

We have used slightly different techniques to control the surface chemistry of template-grown carbon nanowires and nanotubules.[37] Alumina templates were used as before; however, the body of the nanowires was first defined by filling the pores with 3,3',4,4'-tetra(phenylethynyl)biphenyl (TPEB) as a precursor to poly(naphthalene). TPEB was spread on the membrane and then heated to 1000°C in Ar. Carbon nanowires may also be synthesized from bis-*o*-divinylarene (BODA) monomers.

The alumina is dissolved in 25% NaOH (aq) in a Teflon-lined reactor for 12 h at 200°C. The carbon surface was made hydrophilic by treatment with sodium nitrite, sulfanilic acid (sodium salt hydrate), and concentrated HCl in H_2O for 12 h. This procedure decorates the carbon surface with sulfonate groups via a diazo linkage, and elemental analysis shows the presence of N and S. Hydrophobic carbon nano-wires were produced by heating at 1000°C for 2 h. Elemental analysis of the hydrophobic carbon indicated a loss of hydroxyl and carbonyl groups, which implies the formation of additional C–C bonds.

15.3 ELECTRICAL MEASUREMENTS

In order to evaluate electrical properties of individual nanowires, we use a two-terminal measurement technique based on electric field-assisted assembly.[38] As shown in Figure 15.3A, interdigitated Ti/Au electrode structures are photolithograph-ically defined on a layer of SiO_2. An Si_3N_4 layer was used to isolate the electrodes from the metallic nanowires. The Si_3N_4 layer thickness was varied from 500 to 100 nm directly above the electrode fingers. A thinner layer over the interdigitated area increased the local field strength and promoted deposition of the wires in these regions.

Electrofluidic assembly was conducted by dispersing a dilute suspension of nanowires in isopropanol onto the electrode structures. The electrodes were biased with alternating voltages ranging from 5 to 70 V_{rms} (electric field strength = 1.0×10^4 to 1.4×10^5 V/cm). For example, the optical micrograph in Figure 15.3C shows 5-μm-long (350-nm-diameter) Au nanowires aligned onto the electrode structure in Figure 15.3A. By applying a voltage of 30 V_{rms} at 1 kHz to one of the electrodes (the other is grounded), the nanowires aligned approximately perpendicular to the electrode fingers and preferentially on the thinner Si_3N_4 layer. We observed nanowire alignment at voltages greater than 25 V_{rms}, with the time necessary for alignment depending directly on the applied voltage (i.e., 9 sec for 25 V_{rms} and 5 sec for 70 V_{rms}).

This assembly occurs because of the high polarizability of metal nanowires in the dielectric medium. The polarized nanowires are drawn to regions of increasing field strength because of the dielectrophoretic force acting on them. As the nanowires approach the electrodes, the dielectrophoretic force increases, and the wires are aligned. When a given nanowire aligns to the interdigitated electrodes, the local field strength is decreased. The reduction in field strength in the region of an aligned nanowire decreases the likelihood that another nanowire will be aligned near (within 2 μm) the first. Additional nanowires align along the electrodes in a random fashion until they are depleted from the suspension.

In order to promote regular and predictable placement of nanowires on the electrodes, 4×4 μm isolated top electrodes were added to the electrode structure (Figure 15.3B). The top electrodes were capacitively coupled to the buried elec-trodes; therefore, the local electric field strength was further increased. As shown in Figure 15.3D, nanowires are aligned to the top electrodes by applying 20 V_{rms} at 1 kHz to the buried electrodes. At potentials above 25 V_{rms}, additional nanowires align to the buried electrodes, as in the previous configuration. The alternating nanowire alignment observed in the presence of top electrodes may be accounted for by noting

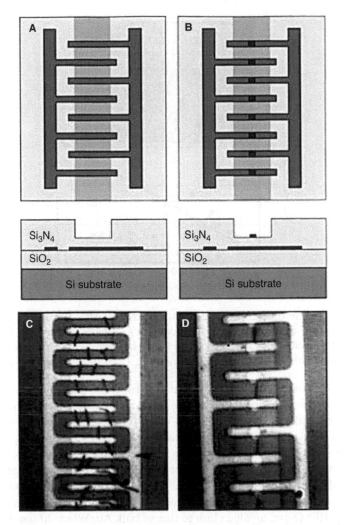

FIGURE 15.3 Electrode structures used for electrophoretic assembly of nanowires. (A) Top and cross-sectional view of the electrodes used for assembly and electrical measurements. (B) The same electrode structure, but with addition of 4 × 4 μm top electrodes. (C) Optical micrograph of the electrode structure shown in (A). Au nanowires (200 nm diameter, 5 μm long) are aligned to the electrode fingers. The wires were aligned by applying 30 V at 1 kHz. (D) Optical micrograph of the electrode structure shown in (B). The top electrodes improve the precision of nanowire alignment. The wires were aligned by applying 20 V at 1 kHz. (Adapted from Smith, P.A. et al., *Appl Phys Lett,* 77, 1399–1401, 2000. With permission.)

that the nanowire eliminates the electric field associated with the bridged electrodes. Additionally, the electric field generated by the adjacent electrodes is reduced two-fold. Again, the weaker field decreases the tendency for nanowires to align in these areas.

By adding large area pads to the alignment structure, one can measure the electrical properties of individual wires. In initial experiments, to ensure adequate

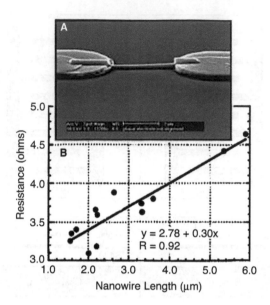

FIGURE 15.4 (A) SEM of an aligned 300-nm-diameter Au wire. Top metal was evaporated onto the ends of the nanowire and the electrodes. (B) Resistance vs. nanowire length for Au wires at room temperature. A resistivity of 2.9×10^{-6} Ω-cm was calculated from the slope of the line. (From Smith, P.A. et al., *Appl Phys Lett,* 77, 1399–1401, 2000. With permission.)

contact, an additional 10 nm of Ti/500 nm of Au was thermally evaporated over the wire tips. An Au wire aligned for electrical measurement is shown in Figure 15.4A. The room temperature current–voltage (I-V) characteristics of 350-nm-diameter Au nanowires were determined as a function of probe pad spacing. The results of a series of these experiments are shown in Figure 15.4B as a plot of resistance vs. nanowire length. The measured resistance is the sum of contact resistance and the resistance of the nanowire itself. By using the slope of this line and the cross-sectional area of the nanowire, the resistivity of 350-nm Au wires was determined to be $\sim 2.9 \times 10^{-6}$ Ω-cm, a value close to that of bulk Au. When 70-nm-diameter Au rods were measured, the resistivity was found to be $\sim 4.5 \times 10^{-6}$ Ω-cm. The same method has been used to measure nanowires of various compositions, as is described below.

Additional control over the electrophoretic assembly of nanowires can be achieved by varying the structure of the electrodes used for alignment.[39] By increasing the local field strength associated with particular electrodes, nanowires deposit in a predictable sequence. Cr/Au bus bars were defined on a glass surface, and a Si_3N_4 layer was deposited to cover the bus bars. Cr/Au electrodes were then patterned over the bus bars. The capacitive coupling between the bus bars and electrodes is defined by the overlap between them. The field strength between a given pair of electrodes, and therefore the probability of nanowire alignment, may be varied either by changing the area of overlap between the electrodes and the bus bars or by changing the distance between electrode tips. Stronger fields were generated with larger overlap areas and with shorter interelectrode distances. Since there is a threshold

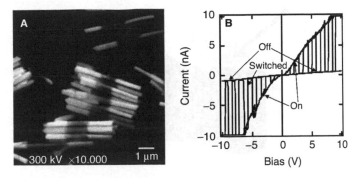

FIGURE 15.5 (A) SEM of Au-CdSe-Au nanowires. (B) I-V curves taken from an Au-CdSe-Au wire while the wire was illuminated, while in the dark, and while the light was switched on and off at 5-sec intervals. (Adapted from Peña, D.J. et al., *J Phys Chem B*, 106, 7458–7462, 2002. With permission.)

of field strength required to align nanowires to electrodes, the electrodes that have an inherently stronger field at a given potential will cause nanowires to align first. Nanowires may be aligned one at a time by simply increasing the voltage applied to the bus bars. Alternately, different types of nanowires may be selectively aligned by simply changing the suspension composition at the desired point in the assembly.

15.4 ELECTRONIC PROPERTIES OF NANOWIRES CONTAINING FUNCTIONAL COMPONENTS

Metal–CdSe–metal nanowires can be synthesized by methods similar to those described above.[40] Au and Ni were grown at the ends of the CdSe-containing nanowires, and both 350- and 70-nm wires were produced by using alumina and polycarbonate membranes, respectively. CdSe was electrodeposited from a home-made plating solution composed of 0.3 M CdSO$_4$, 0.25 M H$_2$SO$_4$, and 0.7 mM SeO$_2$. CdSe was grown in alumina using a cyclic voltammetric technique by sweeping the potential between –350 or –400 mV and –800 mV at 750 mV/sec (vs. saturated calomel or Ag/AgCl). Nanowires in polycarbonate were grown using a similar method. A scanning electron micrograph of 350-nm-diameter Au-CdSe-Au nanowires is shown in Figure 15.5A.

When CdSe-containing nanowires were aligned to evaluate their electrical properties, they showed a pronounced photoconductivity. For example, in Figure 15.5B, the I-V curves of a 350-nm Au-CdSe-Au wire are shown for three conditions: in the dark, illuminated with white light via a microscope objective, and switching between dark and light at 5-sec intervals. The curve showing the lower current values was obtained in the dark, and the curve with higher current values was obtained under illumination. Both curves are nonlinear and have a conductance gap at 0 V. This conductance gap is expected from a symmetrically constructed nanowire containing a high work function metal, Au, and an n-type semiconductor, CdSe. At the extremes of the applied bias, there is a 15-fold change in conductivity between light and dark. The effects of switching the light on and off while varying the potential

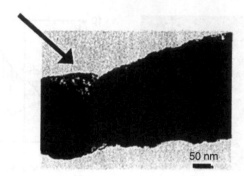

FIGURE 15.6 HRTEM of an Au-Ag-(TiO$_2$-PSS)$_{10}$-TiO$_2$-Au nanowire. The junction can be seen due to the focused electron beam inducing metal melting around the junction. (Adapted from Kovtyukhova, N.I. et al., *J Phys Chem B,* 105, 8762–8769, 2001. With permission.)

are also shown in Figure 15.5B. When the light was switched on, the current increased rapidly to the level observed during continuous illumination. Likewise, when the light was switched off, the current rapidly fell to the level measured in the dark. Similar photoeffects on conductivity were observed for 350-nm-diameter CdSe wires that were capped with Ni as well as for 70-nm-diameter wires of both compositions.

Oxide semiconductors and other electroactive materials may be incorporated into the nanowires. TiO$_2$ (4 to 10 nm) or ZnO (3 to 5 nm) particles and polymer/polyelectrolyte films can be built up in or on Au nanowires in a layer-by-layer fashion.[41,42] The oxide nanoparticles were prepared by literature methods.[43,44] Alternate layers may consist of polystyrenesulfonate (PSS; 1 wt%) or polyaniline (PAN). PSS is deposited from H$_2$O, and PAN is deposited from dimethyl formamide (DMF; 0.015 wt%). Layered assembly in nanowires was accomplished by first growing sacrificial Ag, followed by an Au segment in the membrane. The Au surface was primed for multilayer adsorption by treating the nanowire tip with 5% MEA (in ethanol). The multilayer was grown by alternately immersing the membrane in the nanoparticle solution and the polymer/polyelectrolyte solution. Each immersion step was 15 to 30 min, and after the immersion, the membrane was rinsed thoroughly with either 0.01 *M* HCl for nanoparticles, H$_2$O for polymer/polyelectrolyte, or DMF for PAN. After the desired number of multilayers was grown, the second half of the nanowire was electrodeposited on top of the multilayer structure. The backing Ag layer and the Al$_2$O$_3$ membrane were removed as usual, and this treatment did not damage the multilayer. As seen in the high-resolution transmission electron micrograph (HR-TEM) in Figure 15.6, particulate and polymeric materials are visible at the metal–metal junction. This particular wire, Au/Ag/(TiO$_2$/PSS)$_{10}$Au, was melted by the electron beam during several seconds of focusing. This melting makes the junction more apparent and more clearly displays the film materials.

Monolayer molecular junctions may also be grown within metal nanowires. In one case, a 16-mercaptohexadecanoic acid SAM was sandwiched between 70-nm-diameter Au segments.[45] Here, polycarbonate membranes were used to template the 70-nm-diameter wires. First, a segment of Au was electrodeposited in the membrane.

A SAM of 16-mercaptohexadecanoic acid was applied to the open Au tip by immersing the membrane in a 0.5 mM ethanolic solution for 24 h (Ar atmosphere). The second half of the Au nanowire was grown by first seeding the top of the SAM with Sn^{2+}.[46] The ions were adsorbed from 1:1 $CH_3OH:H_2O$ that was acidified with 0.007 M trifluoroacetic acid. Ag nanoparticles were generated at the Sn^{2+} sites upon addition of ammoniacal $AgNO_3$. A continuous metallic layer was formed by the electroless reduction of a 0.04 M Au^+ solution with formaldehyde. The remainder of the nanowire was grown either by continued electroless deposition of Au or by electrochemical deposition of Au (or another metal). Because these nanowires were grown in polycarbonate, CH_2Cl_2 was used to free the wires from the membrane and rinse them, and it was the solvent in which the wires were electrophoretically aligned for I-V characterization.

Current–voltage characteristics for a 70-nm-diameter Au-SAM-Au nanowire are shown in the upper portion of Figure 15.7A. This semilogarithmic plot shows the

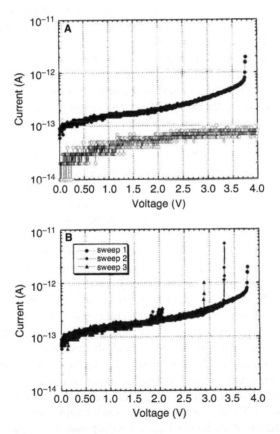

FIGURE 15.7 (A) I-V curve for the fifth positive cycle of an Au-SAM-Au nanowire (after four cycles to 3.5 V). The lower curve is the background current (no nanowire spanning electrodes). (B) I-V curves showing degradation over several cycles to the breakdown voltage. (From Mbindyo, J.K.N. et al., *J Am Chem Soc*, 124, 4020–4026, 2002. With permission.)

fifth positive potential cycle. The first four cycles were swept to a voltage less than 3.5 V and are not shown since they simply overlay the data shown. The lower plot is a control experiment in which the leakage current between the electrode fingers (here, a 3-μm gap) was measured. No nanowire was present, and the resulting current is due to leakage through the dielectric material under which the electrodes are buried. When comparing the two plots, it is seen that in the low-bias region, the current of the nanowire is two to three times the leakage current. As the bias is increased, the nanowire current increases more quickly than the background. Between approximately 2.5 and 3.5 V, there is a nearly exponential dependence of current on voltage. At 3.5 V, there is a sharp current increase, and at 3.6 V, the measurement current compliance of 10 pA is reached. Figure 15.7B shows the results of repeatedly sweeping the same SAM junction nanowire while limiting the current compliance to 10 pA. The current did not change appreciably at low bias; however, the potential that induced breakdown decreased by about 0.3 V with each cycle. In a similar experiment with another nanowire, the current compliance was chosen to be 100 pA. Under multiple bias sweeps, the breakdown voltage was measured to be <1 V. This indicates a more rapid breakdown at the SAM junction at higher current densities. Although there is some variation in the value of breakdown voltage, most nonshort junctions break down between 3.5 and 3.8 V. Occasionally, breakdown voltages near 7 V are observed. This relatively large bias may be indicative of SAM bilayers or multilayers having formed at the junction in some devices.

The same techniques can be used to grow in-wire molecular layers in order to study the behavior of molecules possessing unusual electronic properties.[47] One molecule that exhibits nonlinear differential resistance (NDR) is 4-{[2-nitro-4-(phenylethynyl)phenyl]-ethynyl}benzenethiol. This NDR molecule was incorporated into the SAM of 16-mercaptohexadecanoic acid as described above. A schematic representation of the NDR molecule and part of the SAM is shown in Figure 15.8. The molecules are of similar height, and the metal seeding, followed by electroless plating of the remainder of the Au wire (as described above), is used to make contact with the top of the mixed SAM layer.

When a positive bias was applied to these wires, negative differential resistance was observed, as shown in Figure 15.9A. Upon applying a negative bias scan from 0 to −2.0 V, diode behavior was observed (Figure 15.9B). These effects are most likely due to the NDR molecules present in the mixed monolayer because much lower current is observed with the 16-mercaptohexadecanoic acid SAM under the same conditions. Approximately 10% of nanowires prepared in this way showed NDR I-V curves. Among this 10%, similar peak shapes and positions were observed, and values were similar to those seen in other studies.[48] Many of the nanowires tested in this manner had ohmic I-V characteristics, which are likely due to imperfections in the SAM layer, possibly incurred during the growth of the top metal. Still others showed I-V characteristics of the spacer molecules only, suggesting that in some cases the NDR molecule was not incorporated into the junction. This is possibly due to microphase separation of the mixed SAM on the length scale of the nanowire diameter.

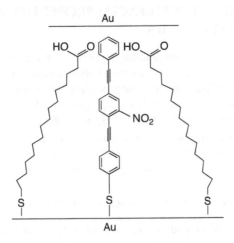

FIGURE 15.8 Schematic showing the NDR molecule (4-{[2-nitro-4-(phenylethynyl)-phe-nyl]ethynyl}benzenethiol) assembled in a matrix of 16-mercaptohexadecanoic acid. (Adapted from Kratochvilova, I. et al., *J Mater Chem,* 12, 2927–2930, 2002. With permission.)

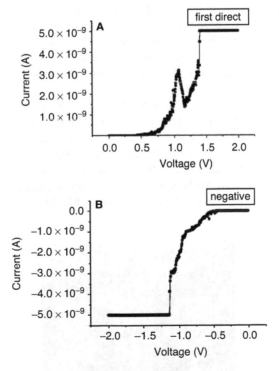

FIGURE 15.9 (A) I-V characteristics for an Ag-NDR-Au nanowire from 0 to 2.0 V. (B) I-V curve of the device in (A) with potential sweep from 0 to –2.0 V. (Adapted from Kratochvilova, I. et al., *J Mater Chem,* 12, 2927–2930, 2002. With permission.)

15.5 NANOWIRES AT INTERFACES: PROPERTIES AND ASSEMBLY STRATEGIES

Balancing the challenges of constructing working nanoscale devices are the tasks of assembling the components into useful configurations and then accessing them with conventional or self-assembled address lines. Whereas assembly of spherical particles continues to be a field rich in theoretical and experimental studies, the assembly of anisotropic particles has not been so thoroughly explored. In recent years, several groups have undertaken studies of this kind. By necessity, we have also investigated the properties of our metallic nanowires on a variety of surfaces and interfaces that are potential substrates for circuit assembly.

15.5.1 CONTROLLING PARTICLE–SURFACE INTERACTIONS

The nanowires described here are more properly referred to as a suspension than as a colloid. Au wires that are 350 nm in diameter sink from solution in a matter of minutes, and 30-nm-diameter wires sink in a matter of days. This solution instability encourages us to use a surface- or interface-bound approach to studying nanowire motion and assembly.[49] Generally, we have used Au nanowires and Au or SiO_2 surfaces to study nanowire interactions because of the variety of surface chemistries available. By choosing appropriate surface chemistries, we can manipulate the behavior of the nanowires. For example, a suspension of 2-mercaptoethane sulfonic acid (MESA)-derived Au wires in water deposited onto a MESA-derivatized Au surface remains mobile for hours. Brownian motion causes the wires to diffuse along the surface as long as the water does not evaporate. The particles remain mobile due to electrostatic repulsion between the particles and the substrate. In contrast, MESA-coated wires placed on a positively charged MESA surface are immobile. Wires may be selectively immobilized on a surface that has two different surface chemistries. As shown in Figure 15.10, particles adhere to a pattern defined on the substrate. On this surface, a gold pattern was defined on the native oxide of an Si wafer (the Au of the pattern is not visible because the image was recorded in dark field). The Au was treated with MEA to render it attractive to MESA-coated wires. The SiO_2

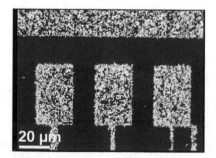

FIGURE 15.10 Dark-field optical micrograph showing MESA-coated Au nanowires bound to MEA-derivatized features on the substrate. (Adapted from Martin, B.R. et al., *Adv Funct Mater,* 12, 759–765, 2002. With permission.)

of the surrounding surface was also hydrophilic, but not electrostatically attractive, and MESA-coated nanowires were mobile on that surface. The mobile wires were rinsed away by passing the substrate through an air–water interface.

15.5.2 Nanowire Assembly in Lithographically Defined Wells

Nanowires may be selectively placed on surfaces by patterning wells in a hydrophilic surface. We have defined features commensurate with nanowire dimensions by using standard photolithographic and electron beam writing techniques. For example, about 600 nm of divinylsiloxane-bis(benzocyclobutene) (BCB) was spin coated on a Si wafer covered with 5 nm of Cr/100 nm of Au. Photoresist was then spun on and patterned into an array of wells. The well pattern was transferred to the BCB underlayer by etching to the Au in O_2/CF_4 plasma. The remaining photoresist was removed with acetone, and the BCB surface was rendered hydrophilic with O_2 plasma treatment.

The exposed Au at the bottom of the wells allowed us to differentiate the chemistry at the bottom; however, in this example, Au was derivatized with MESA, making it hydrophilic. When an aqueous suspension of hydrophilic MESA-coated nanowires was placed on this surface, they were mobile on all surfaces and electrostatically repulsive at the well bottoms. The 300-nm-diameter particles readily sank to the surface and remained mobile on the surface. When the nanowires encountered a well, they sank to the lower surface, were trapped there, and were mobile within the confines of the well. At relatively low surface coverages, the nanowires filled the wells one at a time. Because the wires in the wells are mobile, they moved in order to accommodate additional wires. In this manner, an approximately energy-minimized conformation was achieved in the wells, as shown in the upper panel of Figure 15.11A.

The final degree of well filling may be controlled by the initial particle concentration. For example, by reducing the nanowire concentration, an average of one

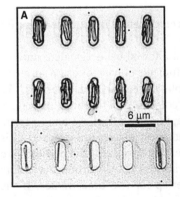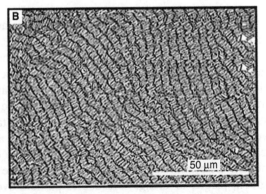

FIGURE 15.11 (A) Bright-field optical micrographs of Au nanowires trapped in lithographically defined topographic features. (Adapted from Martin, B.R. et al., *Adv Funct Mater,* 12, 759–765, 2002. With permission.) (B) Bright-field optical micrograph of ordered two-dimensional ribbons formed at high density of MESA-coated Au nanowires on a surface.

wire may assemble in each well (Figure 15.11A, lower panel). Long-range order may be induced by crowding. In the same manner that nanowires are packed into the confines of a well, increasing the particle density on a surface causes the nanowires to align themselves in parallel (Figure 15.11B). Long ribbons consisting of hundreds of nanowires can form an ordered phase resembling the smectic phase observed in liquid crystals. Similar ordering of nanowires has been observed by confining the particles to the air–water interface on a Langmuir–Blodgett trough.[31]

15.5.3 DIFFUSION OF NANOWIRES ON SURFACES

The dynamics of spherical particles have been much more thoroughly studied than those of anisotropic particles.[25] If nanowires are to be used as building blocks for nanoelectronics, we must understand their basic physical properties. There is a significant literature of theoretical work involving the dynamics and collective behavior of anisotropic particles[50–54] that further drives experimental investigations. By examining the dynamic behavior of nanowires, one can hope to achieve better control over the types of structures and the efficiency with which they are assembled.

When a suspension of nanowires is placed on a surface, the nanowires are often observed to move under the influence of Brownian motion. Under certain conditions, the particles twitch and travel along the surface for hours or days. Diffusion coefficients may be obtained by measuring ensemble or trajectory diffusion.[55] Ensemble measurements are usually made when individual molecules are not observable, so a sample of many particles is measured, usually by spectroscopic or electrochemical techniques. This is generally the case for molecular diffusion. Trajectory diffusion is directly observable in the Brownian motion of particles accessible to optical microscopy, such as our nanowires. The diffusion coefficient, D, is defined as

$$D = \frac{\left\langle x^2 \right\rangle}{2t}$$

where $<x^2>$ is the average of the square of displacement and t is the time lapse for each displacement.

We recorded a series of time-lapse images that allowed us to visualize moving particles[49] and measure experimental diffusion coefficients of nanowires on surfaces. First, we digitally recorded the optical micrographs at regular time intervals. We artificially recolored the micrographs one of three colors: cyan, magenta, or yellow. Starting with image t_1 as cyan, image t_2 as magenta, image t_3 as yellow, and so forth, we recolored the entire series. When three time-adjacent images are overlaid, it is readily apparent whether the particles are mobile on the surface. If the particles are mobile, the cyan, magenta, and yellow colors are all visible. Any secondary colors (green, purple, or orange) are due to particle overlap for two time frames. If the particles are immobile, the three primary colors for three time frames all overlap and appear black. In order to measure diffusion coefficients for mobile particles, we gathered groups of three temporally adjacent images. Following several particles, we measured the center-to-center displacements for each time step. In this manner,

TABLE 15.1
Surface Diffusion Coefficients
for Several Nanowire–Surface Combinations

Diameter × Length Rod SAM Surface SAM	Interaction	Diffusion Coefficient, D (μm²/sec)
300 nm × 6 μm MESA MESA surface	Anionic/anionic	0.19 ± 0.01
300 nm × 2 μm MESA MESA surface	Anionic/anionic	0.41 ± 0.13
90 nm × 6 μm MESA MESA surface	Anionic/anionic	0.34 ± 0.06
300 nm × 6 μm MESA $C_{16}H_{33}SH$ surface	Anionic/neutral	0.16 ± 0.01
300 nm × 2 μm MESA $C_{16}H_{33}SH$ surface	Anionic/neutral	0.44 ± 0.09
90 nm × 6 μm MESA $C_{16}H_{33}SH$ surface	Anionic/neutral	0.00
300 nm × 6 μm MESA Acid-rinsed MEA surface	Anionic/cationic	0.00
300 nm × 6 μm MESA Water-rinsed MEA surface	Anionic/neutral	0.14 ± 0.06
300 nm × 6 μm MESA Base-rinsed MEA surface	Anionic/neutral	0.16 ± 0.01

Source: St. Angelo, S.K. et al., *Adv Mater,* 15, 400–402, 2003.

we accumulated the diffusion coefficients for several nanowire and surface combinations.[55]

A summary of diffusion coefficients obtained for different-size nanowires and different substrate chemistries is given in Table 15.1.[56] MESA-coated Au wires on a MESA–Au surface showed a greater dependence on length than diameter for the diffusion coefficient. The Stokes–Einstein theory for prolate ellipsoids indicates that scaling of the diffusion coefficient should vary as $l^{-1/3}d^{-2/3}$. We observed a stronger length dependence than that predicted by the Stokes–Einstein theory, which may be due to a hydrodynamic coupling of the nanowire to the surface, in effect causing friction and decreasing the rate of diffusion. This frictional component should have a stronger length dependence (rather than diameter) and likely accounts for the deviation.

Electroneutral hydrophobic substrates were prepared by treating clean Au films with hexadecanethiol ($C_{16}H_{33}SH$). MESA-coated wires diffusing on these surfaces showed similar results as the MESA–MESA experiments. There was, however, one unusual feature of these results: the 90-nm-diameter wires showed no detectable diffusion after making contact with the surface. For larger-diameter particles, the long-range electrostatic force should dominate; however, for smaller diameters, short-range Van der Waals forces become more important. The size regimes investigated here must be at a balance between electrostatic repulsive and Van der Waals attractive forces.

Au substrates were also derivatized with MEA, which should have a pH-dependent surface charge density. After derivatization, the substrates were each rinsed with one of three aqueous solutions, above, below, and approximately at the pK_a of the amine head groups of the SAM. A dilute suspension of MESA-coated nanowires (6 μm × 300 nm) was applied to each surface, and the diffusion coefficients are given in Table 15.1. Not surprisingly, the surface on which most of the amine groups were protonated showed no nanowire diffusion. The electrostatic attraction between the particles and surface yielded a contact map of their interaction. For the other two surfaces, above and at the amine pK_a, diffusion coefficients similar in magnitude to the electroneutral surface experiment (for the same nanowire dimensions) were obtained. The similarity in these values is due to the nanowires and surface being weakly interactive, but the contribution of friction is greater for the less repulsive interactions. By varying the degree of protonation of the surface, we were able to tune the degree of wire–surface interaction. This suggests that it may be possible to allow particles to diffuse to a particular region and then stop them by increasing the pH of the supporting solution.

15.5.4 BIOMOLECULAR RECOGNITION IN NANOWIRE ASSEMBLY

Specific lock-and-key biomolecular interactions represent an interesting strategy for binding nanowires to each other and to particular surfaces. DNA hybridization provides access to the specificity and subtle control afforded by the sequence-dependent binding. It is now well known that DNA can act as a "smart glue" for assembly of inorganic nanostructures.[57–59]

We have studied the assembly of metal nanowires on surfaces using DNA hybridization.[60] Fluorescence spectroscopy (using tetramethyl rhodamine-tagged DNA) was used to quantify the amount of DNA bound to the nanowires. DNA coverages were calculated by measuring the fluorescence of the solution because fluorescence quenching and other optical effects at the nanowire surface would likely interfere with quantification. Nanowires were derivatized by adding 30 μl of 5 μM thiol-modified single-stranded DNA to 50 μl of nanowires in buffer solution. After 4 h reaction time, DNA coverage on the nanowires was 5.7×10^{12} molecules/cm^2 for a 36-mer. By doubling the initial concentration of thiol-modified DNA, the surface concentration increased to $\sim 1.0 \times 10^{13}$ molecules/cm^2 for a 36-mer with a fluorescent tag. This value is approximately 10% of the coverage calculated for a closely packed monolayer of single-stranded DNA on the nanowires. Higher surface coverages may be a hindrance to hybridization due to steric effects.[61] The surface coverage of single-stranded DNA may be fit to the Langmuir adsorption equation:

$$\Gamma_{DNA} = \frac{\Gamma^0_{DNA} K_{ads} \left[DNA \right]}{1 + K_{ads} \left[DNA \right]}$$

where Γ_{DNA} is coverage, Γ^0_{DNA} is saturation coverage, and K_{ads} is the equilibrium constant for adsorption. An approximate K_{ads} value of $1.0 \pm 0.4 \times 10^5\ M^{-1}$ for adsorption of the thiolated single-stranded DNA was obtained by finding the slope of $1/\Gamma_{DNA}$ vs. $1/[DNA]$. Other equilibrium constants for adsorption of other types of DNA were found in the same manner. In order to increase the hybridization efficiency, mercaptohexanol was added as a dilutent molecule for the single-stranded DNA. Mercaptohexanol acts to displace nonspecifically bound DNA and to stand up molecules that may otherwise lie flat on the nanowire surface.[61] Both of these factors improve solution hybridization efficiency, and we observed an increase from $21 \pm 3\%$ to $63 \pm 11\%$ by adding the mercaptohexanol. This hybridization was found to be reversible, as determined by heating the hybridized DNA above its melting temperature and measuring the single-stranded complement by absorbance spectroscopy.

In order to use DNA hybridization to attach nanowires to substrates, the wires and the surface were derivatized with complementary single-stranded DNA. The DNA on the wires and surface was allowed to hybridize in a rotator bath for 14 h at 25°C. After hybridization, the surfaces were gently rinsed with deionized water or buffer. Upon rinsing, nonspecifically bound nanowire aggregates were removed, and the specifically bound nanowires remained. Nonspecific binding is a persistent problem in the biological assembly of particles; however, control experiments using noncomplementary DNA showed about fourfold lower particle coverage than the complementary case ($(2.5 \pm 0.5) \times 10^5$ and $(9 \pm 2) \times 10^5$ particles/cm^2, respectively). This result indicates preferential surface binding based on DNA hybridization.

In order to improve the efficiency of nanowire binding to surfaces, 100-mer single-stranded DNA sequences consisting of 10 repeating units were used. By using longer DNA with repeating units, the entire length of the DNA need not be hybridized to bind particles to the surface. Rather, one or more repeating units of the 100-mer DNA may hybridize and affix the nanowire to the surface. This circumvents the perfect alignment and complete hybridization necessary when using the 36-mer nonrepeating DNA. We observed a modest (1.5-fold) increase in surface coverage with the 100-mer repeating DNA relative to the 36-mer. Again, the complementary surface–nanowire case showed a much higher surface coverage than the noncomplementary case ($(1.3 \pm 0.4) \times 10^6$ vs. $(3 \pm 2) \times 10^5$ particles/cm^2).

15.6 CONCLUSIONS

The technical challenges that attend the development of nanoelectronics are formidable, yet early demonstrations of switching components, memory, and logic functions have been made at the proof-of-concept stage.[13,21,22,24,26,27,62-67] Current research in nanoscience has impressive depth and breadth — many new materials are available and a number of different assembly techniques are being explored. Our approach has been to advance several aspects of the overall problem in parallel. We have demonstrated nanowire synthesis, with control over the length and diameter of the wires. The multicomponent nature of the nanowires obtainable by template synthesis includes a rich surface chemistry. We have investigated the inclusion of functional components into the wires themselves in the form of photoactive materials, molecular layers, and polymer–particle composites. We have demonstrated means to assemble our nanowires onto electrode structures and have evaluated their electronic properties. We have also derivatized them with a variety of SAMs and DNA and have measured their basic interactions with surfaces and two-dimensional diffusion coefficients. Although we have made significant progress in this area, there are still considerable challenges to be overcome. The basic science underpinning device fabrication and circuit assembly must be further developed in order for practical nanowire-based computing to be achieved.

ACKNOWLEDGMENTS

This work has been supported by DARPA and the Office of Naval Research under contract N00014-01-10659. We thank Prof. James Tour for providing a sample of 4-{[2-nitro-4-(phenylethynyl)phenyl]-ethynyl}benzenethiol. We also thank our collaborators and coworkers Benjamin R. Martin, Jeremiah K.N. Mbindyo, Nina I. Kovtyukhova, Peter Smith, Baharak Razavi, James Mattzela, Irena Kratochvilova, Achim Amma, David J. Peña, Christine D. Keating, Theresa S. Mayer, and Thomas N. Jackson, who contributed to various aspects of this work.

REFERENCES

1. CJ Murphy, NR Jana. Controlling the aspect ratio of inorganic nanorods and nanowires. *Adv Mater* 14:80–82, 2002.
2. NR Jana, L Gearheart, CJ Murphy. Wet chemical synthesis of high aspect ratio cylindrical gold nanorods. *J Phys Chem B* 105:4065–4067, 2001.
3. Y Sun, Y Xia. Large-scale synthesis of uniform silver nanowires through a soft, self-seeding, polyol process. *Adv Mater* 14:833–837, 2002.
4. Y Sun, Y Yin, BT Mayers, T Herricks, Y Xia. Uniform silver nanowires synthesis by reducing $AgNO_3$ with ethylene glycol in the presence of seeds and poly(vinyl pyrrolidone). *Chem Mater* 14:4736–4745, 2002.
5. F Kim, JH Song, P Yang. Photochemical synthesis of gold nanorods. *J Am Chem Soc* 124:14316–14317, 2002.
6. L Huang, H Wang, Z Wang, A Mitra, KN Bozhilov, Y Yan. Nanowire arrays electrodeposited from liquid crystalline phases. *Adv Mater* 14:61–64, 2002.
7. B Gates, B Mayers, B Cattle, Y Xia. Synthesis and characterization of uniform nanowires of trigonal selenium. *Adv Funct Mater* 12:219–227, 2002.
8. AM Morales, CM Lieber. A laser ablation method for the synthesis of crystalline semiconductor nanowires. *Science* 279:208–211, 1998.
9. Y Wu, P Yang. Direct observation of vapor–liquid–solid nanowire growth. *J Am Chem Soc* 123:3165–3166, 2001.
10. Y Wu, P Yang. Germanium nanowire growth via simple vapor transport. *Chem Mater* 12:605–607, 2000.
11. X Duan, CM Lieber. Laser-assisted catalytic growth of single crystal GaN nanowires. *J Am Chem Soc* 122:188–189, 2000.
12. X Duan, CM Lieber. General synthesis of compound semiconductor nanowires. *Adv Mater* 12:298–302, 2000.
13. MS Gudlksen, LJ Lauhon, J Wang, DC Smith, CM Lieber. Growth of nanowire superlattice structures for nanoscale photonics and electronics. *Nature* 415:617–620, 2002.
14. JS Wu, S Dhara, CT Wu, KH Chen, YF Chen, LC Chen. Growth and optical properties of self-organized Au_2Si nanospheres pea-podded in a silicon oxide nanowire. *Adv Mater* 14:1847–1850, 2002.

15. Y Yin, G Zhang, Y Xia. Synthesis and characterization of MgO nanowires through a vapor-phase precursor method. *Adv Funct Mater* 12:293–298, 2002.
16. J-J Wu, S-C Liu. Low-temperature growth of well-aligned ZnO nanorods by chemical vapor deposition. *Adv Mater* 14:215–218, 2002.
17. L Zambov, A Zambova, M Cabassi, TS Mayer. Template-directed CVD of dielectric nanotubes. *Chem Vap Deposition* 9:26–33, 2003.
18. Y Wu, R Fan, P Yang. Block-by-block growth of single-crystalline Si/SiGe superlattice nanowires. *Nano Lett* 2:83–86, 2002.
19. MP Zach, K Inazu, KH Ng, JC Hemminger, RM Penner. Synthesis of molybdenum nanowires with millimeter-scale lengths using electrochemical step edge decoration. *Chem Mater* 14:3206–3216, 2002.
20. EC Walter, BJ Murray, F Favier, G Kaltenpoth, M Grunze, RM Penner. Noble and coinage metal nanowires by electrochemical step edge decoration. *J Phys Chem B* 106:11407–11411, 2002.
21. T Rueckes, K Kim, E Joselevich, GY Tseng, C-L Cheung, CM Lieber. Carbon nanotube-based nonvolatile random access memory for molecular computing. *Science* 289:94–97, 2000.
22. M Ouyang, JL Huang, CM Lieber. Fundamental electronic properties and applications of single-walled carbon nanotubes. *Acc Chem Res* 35:1018–1025, 2002.
23. P Avouris. Molecular electronics with carbon nanotubes. *Acc Chem Res* 35:1026–1034, 2002.
24. A Bachtold, P Hadley, T Nakanishi, C Dekker. Logic circuits with carbon nanotube transistors. *Science* 294:1317–1320, 2001.
25. TF Tardos, Ed. *Solid/Liquid Dispersions*. Bracknell, England: Academic Press, 1987.
26. MR Diehl, SN Yaliraki, RA Beckman, M Barahona, JR Heath. Self-assembled, deterministic carbon nanotubes wiring networks. *Angew Chem Int Ed* 41:353–356, 2002.
27. Y Huang, X Duan, Q Wei, CM Lieber. Directed assembly of one-dimensional nanostructures into functional networks. *Science* 297:630–633, 2001.
28. B Messer, JH Song, P Yang. Microchannel networks for nanowire patterning. *J Am Chem Soc* 122.10232–10233, 2000.
29. H Shimoda, SJ Oh, HZ Geng, RJ Walker, XB Zhang, LE McNeil, O Zhou. Self-assembly of carbon nanotubes. *Adv Mater* 14:899–901, 2002.
30. P Jiang, JF Bertone, KS Hwang, VL Colvin. Single-crystal colloidal multilayers of controlled thickness. *Chem Mater* 11:2132–2140, 1999.
31. F Kim, S Kwan, J Akana, P Yang. Langmuir–Blodgett nanorod assembly. *J Am Chem Soc* 123:4360–4361, 2001.
32. D Appell. Wired for success. *Nature* 419:553–555, 2002.
33. CR Martin. Membrane-based synthesis of nanomaterials. *Chem Mater* 8:1739–1746, 1996.
34. D Al-Mawlawi, CZ Liu, M Moskovits. Nanowires formed in anodic oxide nanotemplates. *J Mater Res* 9:1014–1018, 1994.
35. BR Martin, DJ Dermody, BD Reiss, M Fang, LA Lyon, MJ Natan, TE Mallouk. Orthogonal self-assembly on colloidal gold-platinum nanorods. *Adv Mater* 11:1021–1025, 1999.
36. JJ Hickman, PE Laibinis, DI Auerbach, C Zou, TJ Gardner, GM Whitesides, MS Wrighton. Toward orthogonal self-assembly of redox active molecules on Pt and Au: selective reaction of disulfide with Au and isocyanide with Pt. *Langmuir* 8:357–359, 1992.
37. A Amma, B Razavi, SK St. Angelo, TS Mayer, TE Mallouk. Synthesis, chemical modification, and surface assembly of carbon nanowires. *Adv Funct Mater* 13, 365–370, 2003.

38. PA Smith, CD Nordquist, TN Jackson, TS Mayer, BR Martin, J Mbindyo, TE Mallouk. Electric-field assisted assembly and alignment of metallic nanowires. *Appl Phys Lett* 77:1399–1401, 2000.

39. B Razavi, BR Martin, TE Mallouk, TN Jackson. Electric field-assisted nanoparticle assembly. *Intl Semicon Dev Res Symp Proceedings* Dec 5–7, 2001: 318–321.

40. DJ Peña, JKN Mbindyo, AJ Carado, TE Mallouk, CD Keating, B Razavi, TS Mayer. Template growth of photoconductive metal–CdSe–metal nanowires. *J Phys Chem B* 106:7458–7462, 2002.

41. NI Kovtyukhova, BR Martin, JKN Mbindyo, PA Smith, B Razavi, TS Mayer, TE Mallouk. Layer-by-layer assembly of rectifying junctions in and on metal nanowires. *J Phys Chem B* 105:8762–8769, 2001.

42. NI Kovtyukhova, BR Martin, JKN Mbindyo, TE Mallouk, M Cabassi, TS Mayer. Layer-by-layer self-assembly strategy for template synthesis of nanoscale devices. *Mater Sci Eng C* 19:255–262, 2002.

43. N Kovtyukhova, PJ Ollivier, S Chizhik, A Dubravin, E Buzaneva, A Gorchinskiy, A Marchenko, N Smirnova. Self-assembly of ultrathin composite TiO₂/polymer films. *Thin Solid Films* 337:166–170, 1999.

44. L Spanhel, MA Anderson. Semiconductor clusters in the sol-gel process: quantized aggregation, gelation, and crystal-growth in concentrated ZnO colloids. *J Am Chem Soc* 113:2826–2833, 1991.

45. JKN Mbindyo, TE Mallouk, JB Mattzela, I Kratochvilova, B Razavi, TN Jackson, TS Mayer. Template synthesis of metal nanowires containing monolayer molecular junctions. *J Am Chem Soc* 124:4020–4026, 2002.

46. VP Menon, CR Martin. Fabrication and evaluation of nanoelectrode ensembles. *Anal Chem* 67:1920–1928, 1995.

47. I Kratochvilova, M Kocirik, A Zambova, J Mbindyo, TE Mallouk, TS Mayer. Room temperature negative differential resistance in molecular nanowires. *J Mater Chem* 12:2927–2930, 2002.

48. J Chen, W Wang, MA Reed, AM Rawlett, DW Price, JM Tour. Room-temperature negative differential resistance in nanoscale molecular junctions. *Appl Phys Lett* 77:1224–1226, 2000.

49. BR Martin, SK St. Angelo, TE Mallouk. Interactions between suspended nanowires and patterned surfaces. *Adv Funct Mater* 12:759–765, 2002.

50. D Spisak. 2-dimensional diffusion of particles with dipolar-like interaction. *Physica A* 209:42–50, 1994.

51. J Parkinson, KE Kadler, A Brass. Self-assembly of rodlike particles in 2 dimensions: a simple model for collagen fibrillogenesis. *Phys Rev E* 50:2963–2966, 1994.

52. F de J Guevara-Rodriguez, M Medina-Noyola. Long-time tracer diffusion of non-spherical Brownian particles. *Phys Rev E* 61:6368–6374, 2000.

53. JM Lahtinen, T Hjelt, T Ala-Nissila, Z Chvoj. Diffusion of hard disks and rodlike molecules on surfaces. *Phys Rev E* 64:021204-1–021204-7, 2001.

54. A Vincze, L Demko, M Vörös, M Zrínyi, MN Esmail, Z Hórvölgyi. Two-dimensional aggregation of rod-like particles: a model investigation. *J Phys Chem B* 106:2404–2414, 2002.

55. M Takeo. *Disperse Systems.* Weinhein, Germany: Wiley-VCH, 1999, pp. 43–60.

56. SK St. Angelo, CC Waraksa, TE Mallouk. Diffusion of gold nanorods on chemically functionalized surfaces. *Adv Mater* 15:400–402, 2003.

57. CJ Loweth, WB Caldwell, XG Peng, AP Alivisatos, PG Schultz. DNA-based assembly of gold nanocrystals. *Angew Chem Int Ed* 38:1808–1812, 1999.

58. CA Mirkin. Programming the assembly of two- and three-dimensional architectures with DNA and nanoscale inorganic building blocks. *Inorg Chem* 39:2258–2272, 2000.

59. RC Mucic, JJ Storhoff, CA Mirkin, RL Letsinger. DNA-directed synthesis of binary nanoparticle network materials. *J Am Chem Soc* 120:12674–12675, 1998.

60. JKN Mbindyo, BD Reiss, BR Martin, CD Keating, MJ Natan, TE Mallouk. DNA-directed assembly of gold nanowires on complementary surfaces. *Adv Mater* 13:249–254, 2001.

61. TM Hern, MJ Tarlov. Characterization of DNA probes immobilized on gold surfaces. *J Am Chem Soc* 119:8916–8920, 1997.

62. Z Zhong, F Qian, D Wang, CM Lieber. Synthesis of p-type gallium nitride nanowires for electronic and photonic nanodevices. *Nano Lett*, 3:343–346, 2003.

63. Y Cui, Z Zhong, D Wang, WU Wang, CM Lieber. High performance silicon nanowire field effect transistors. *Nano Lett* 3:149–152, 2003.

64. J-O Lee, G Lientschnig, F Wiertz, M Struijk, RA Janssen, R Egberink, DN Reinhoudt, P Hadley, C Dekker. Absence of strong gate effects in electrical measurements on phenylene-based conjugated molecules. *Nano Lett* 3:113–117, 2003.

65. A Bachtold, P Hadley, T Nakanishi, C Dekker. Logic circuits based on carbon nanotubes. *Physica E* 16:42–46, 2003.

66. AR Pease, JO Jeppesen, JF Stoddart, Y Luo, CP Collier, JR Heath. Switching devices based on interlocked molecules. *Acc Chem Res* 34:433–444, 2001.

67. CP Collier, JO Jeppesen, Y Luo, J Perkins, EW Wong, JR Heath, JF Stoddart. Molecular-based electronically switchable tunnel junction devices. *J Am Chem Soc* 123:12632–12641, 2001.

16 Heteronanostructures of CdS and Pt Nanoparticles in Polyelectrolytes: Factors Governing the Self-Assembly and Light-Induced Charge Transfer and Transport Processes

Sara Ghannoum, Jad Jaber, Mariezabel Markarian, Yan Xin, and Lara I. Halaoui

CONTENTS

16.1 Nanoparticles as Building Blocks for Higher-Order Architectures:
Toward a Future Nanotechnology..438
16.2 Layer-by-Layer Assembly of Nanoparticles in Polyelectrolytes
by Electrostatic Interactions..439
    16.2.1 Linear LbL Assembly of Polyacrylate-Modified
Q-CdS and Nano-Pt 439
    16.2.2 Matrix Effect on the Optical Properties 440
16.3 Kinetics of Adsorption and Factors Governing the Self-Assembly:
Case Study of Nano-Pt/PDDA..442
    16.3.1 Kinetics of Adsorption 443
        16.3.1.1 Adsorption via Two Kinetics Regimes 443
        16.3.1.2 Adsorption Mechanism 444
    16.3.2 The Interrupted-Deposition Protocol:
Dependence on Rinsing–Drying 445
    16.3.3 Effect of Salt Content 446
        16.3.3.1 Effect on the Nanoparticles' Surface Density..........446
        16.3.3.2 Effect on the Assembly Dynamics...........................449

16.4   Photogenerated Charge Transfer and Transport at LbL Assemblies:
       Case Study Q-CdS/PDDA 449
       16.4.1   The Q-State: Anodic and Cathodic Photocurrents 451
       16.4.2   Model of Photocurrent Generation at Ensembles
                of Q-Particles...452
       16.4.3   Affecting Light Conversion Efficiency
                by Multilayered Assembly 454
16.5   Conclusions and Final Remarks..455
References..456

16.1   NANOPARTICLES AS BUILDING BLOCKS
FOR HIGHER-ORDER ARCHITECTURES:
TOWARD A FUTURE NANOTECHNOLOGY

"The world of neglected dimensions" that Ostwald wrote about in his book first published in 1917[1] has come into the spotlight: its significance has been fully recognized in the nanoresearch boom we have witnessed in the past two decades. The world of mesoscopia is currently expanding. Research into nanoscopic materials has spanned across the physical and biological sciences, driven by a fundamental interest in the transition that occurs between molecular and macroscopic behavior and by a breadth of potential applications foreseen to see the light based on the unique properties brought forth by spatial confinement and spatial organization of this class of matter, thereby building an infrastructure for a future era of nano-technology in areas as diverse as renewable energy, nanorobotics, and molecular computing.

A cornerstone to these developments consists of devising controlled means for the assembly of molecular zero- and one-dimensional materials into higher-order two- and three-dimensional nanoarchitectures and understanding the fundamental characteristics of these systems. Henceforth, materials scientists are building complex structures from molecular and nanometer-size building blocks and asking the vital question of how the properties of such systems evolve from those of the constituent nanoelements as the degree of complexity increases. By this is perhaps an attempt to mimic the path nature has taken in building from atomic and molecular entities the complex systems that constitute life.

Interest in metal and semiconductor nanoparticles, in particular, stems from their potential usefulness in areas of catalysis, photocatalysis, electronics, (Raman) surface enhancement, and renewable energy. Assembling semiconductor and metal nanoparticles into two- and three-dimensional structures, and understanding the factors that govern the assembly process and the optical, electrical, and photoelectrochemical behavior of these complex systems, constitutes the basic drive of our work. Herein, we focus on recent advances in the electrostatic self-assembly of nanoparticles in polyelectrolytes, the assembly dynamics, the influential parameters in this buildup process, and the use of these arrays for solar energy conversion.

16.2 LAYER-BY-LAYER ASSEMBLY OF NANOPARTICLES IN POLYELECTROLYTES BY ELECTROSTATIC INTERACTIONS

The electrostatic layer-by-layer (LbL) assembly of oppositely charged materials is founded in the pioneering work by Iler[2] in the 1960s regarding the assembly of colloidal particles and later by Decher et al.[3–8] on the assembly of polyelectrolyte multilayers. Such multilayers are built by dipping a charged surface in a solution of oppositely charged polyelectrolyte for a specific length of time, followed by rinsing and drying. The surface is thus primed for the next buildup step, in which it is dipped in an oppositely charged polyion solution for another length of time. The result is a bilayer of oppositely charged polyions. This procedure is repeated for layer-by-layer growth and has reportedly led, under correctly chosen deposition parameters, to a buildup of films up to 1000 layers.[9]

Requiring only beakers and solutions, and allowing the thickness of the deposited layers to be fine-tuned with angstrom resolution by varying the dipping conditions,[8,10] this electrostatic self-assembly was termed *molecular beaker epitaxy* by Keller et al.[11] Driven by low cost, universality, ease of fabrication, and the film's high stability, this protocol has been successfully employed to incorporate a wide range of materials in polyelectrolytes, including DNA,[12] proteins,[13,14] viruses,[15] dendrimers,[16] nanoplatelets,[17] and metal,[18–21] magnetic,[22] semiconductor,[23–26] and silica nanoparticles.[24,27,28]

16.2.1 LINEAR LbL ASSEMBLY OF POLYACRYLATE-MODIFIED Q-CdS AND NANO-Pt

A modification we discuss in this chapter consists of capping semiconductor (viz., CdS) and metal (viz., Pt) nanoparticles with polyacrylate (PAC), an anionic polyelectrolyte, prior to their self-assembly in a cationic polyelectrolyte, poly(diallyldimethylammonium chloride), or PDDA. Polyacrylate (Mwt, 2100)-capped zinc blende CdS nanocrystallites of $<d> = 3.6 \pm 0.5$ nm were synthesized using a PAC:Cd^{2+}:S^{2-} ratio of 96:1.3:1.[26] Polyacrylate-capped Pt face-centered-cubic (fcc) nanocrystallites ($<d> = 2.5 \pm 0.6$ nm) were synthesized by reduction of $PtCl_6^{2-}$ with citrate under reflux in the presence of polyacrylate with a stoichiometric PAC:Pt ratio of 31:1.[29] In addition to trapping the particles in the nanometer dimension and preventing their agglomeration, the anionic polyelectrolyte capping agent imparts a significant negative charge to the nanoparticle's surface and allows for polyelectrolyte–polyelectrolyte electrostatic and hydrophobic interactions.

Polyacrylate-capped CdS and Pt nanoparticles were assembled layer by layer in PDDA (Mwt, ~200,000 to 350,000) by initially dipping surfaces (viz., silicon, quartz, indium tin oxide (ITO) glass, or Au) for 30 min in 10 mM PDDA/0.4 M NaCl (aq) solutions, followed by 2 × 1 min rinsing in water and drying in air. The PDDA-modified surfaces were then dipped for 1 h in Q-CdS or nano-Pt solutions, followed by 1 min of rinsing and 10 to 15 min of drying.[26,29] In order to simultaneously assemble Pt and CdS in one layer, PDDA-modified surfaces were dipped consecutively in nano-Pt and Q-CdS solutions for varying periods of time, separated by rinsing and drying after each dipping step. (PDDA/Q-CdS)$_n$, (PDDA/Pt)$_n$, (PDDA/Q-CdS/PDDA/Pt)$_n$, and (PDDA/Pt,Q-CdS)$_n$ heteronanostructures have been built using

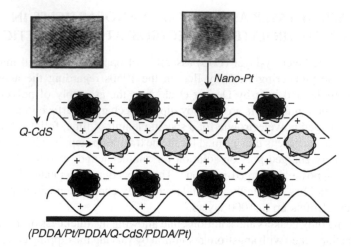

(PDDA/Pt/PDDA/Q-CdS/PDDA/Pt)

FIGURE 16.1 Sketch of a Q-CdS/PDDA/nano-Pt heteronanostructure. The insets show high-resolution TEM (HRTEM) images of a Pt nanocrystal (3.1 and 4.1 nm along the short and long axes, respectively) and a Q-CdS nanocrystal (4.0 and 4.8 nm along the short and long axes, respectively).

this protocol, where *n* designates the number of layer pairs. Figure 16.1 shows an example of these nanoassemblies.

The proper choice of assembly parameters results in a reproducible surface charge buildup per deposition cycle, thus leading to a linear consecutive incorporation of nanoparticles per layer. Figure 16.2A shows a plot of the absorbance of (Q-CdS/PDDA)$_n$ films (at 350 nm) as a function of the number of bilayers, and Figure 16.2B shows the photoluminescence spectra of 2 to 10 Q-CdS/PDDA bilayers. The linear increase in the absorbance and photoluminescence (PL) intensity indicates the reproducible incorporation of Q-CdS nanoparticles per polyelectrolyte layer.[26] Similarly, a linear incorporation of PAC-capped Pt nanoparticles in PDDA resulted from this deposition protocol, as depicted in Figure 16.2C.[29] The linear incorporation of Q-CdS and nano-Pt in PDDA took place despite the submonolayer surface coverage revealed by transmission electron microscopy (TEM) imaging. Such behavior is attributed to the defect-spanning capability of polyelectrolytes, which can bridge over unoccupied sites on the surface.

16.2.2 MATRIX EFFECT ON THE OPTICAL PROPERTIES

Because of the large percentage of atoms at the surface of nanoparticles, the effect of the environment may become significant at this dimension. In the case of Q-CdS assembled in PDDA, the presence of the polyelectrolyte matrix was found to exert no effect on the extent of quantum confinement[30-34] in the nanoparticles but to result in a red shift in the emission attributed to surface states.[26] For instance, Q-CdS (<d> = 3.6 ± 0.5 nm) solutions and (Q-CdS/PDDA)$_n$ films exhibited an identical absorption edge at 460 nm (2.7 eV), corresponding to a 0.2-eV increase in the energy gap relative to the bulk solid as a result of quantum confinement effects (CdS has a band gap of 2.5 eV and a 6.0-nm exciton diameter). This indicates that the extent of

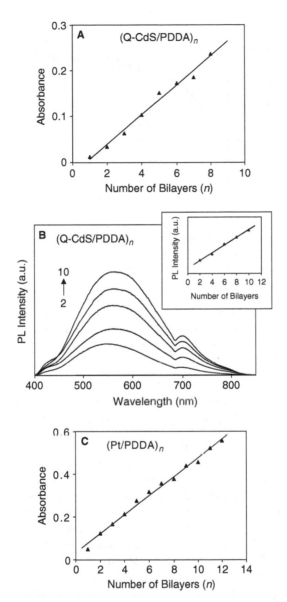

FIGURE 16.2 (A) Absorbance at 350 nm of (Q-CdS/PDDA)$_n$ films vs. the number of bilayers *n*. (B) Photoluminescence spectra of 2, 4, 6, 8, and 10 Q-CdS/PDDA bilayers, with the inset showing the PL peak intensity at ~560 nm vs. *n*. (C) Absorbance at 350 nm of (nano-Pt/PDDA)$_n$ films vs. *n*. (Adapted from Halaoui, L.I., *Langmuir*, 17, 7130–7136, 2001; Ghannoum, S. et al., *Langmuir*, 19, 4804–4811, 2003. With permission.)

quantum confinement in the quantum dots is not affected by electronic interactions with the embedding matrix. On the other hand, photoluminescence measurements of (Q-CdS/PDDA)$_n$ films revealed a 12- to 14-nm red shift in the wavelength attributed to surface states–mediated emission with increasing n[35,36] relative to Q-CdS

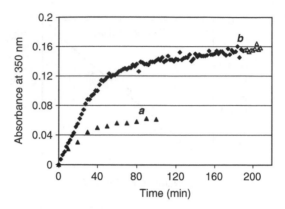

FIGURE 16.3 Absorbance at 350 nm of nano-Pt/PDDA assembled by dipping a PDDA–quartz surface in as-prepared Pt solution for Δt, followed by 1 min of rinsing and 15 min of drying. (a) $\Delta t = 10$ min. (b) $\Delta t = 2$ min. The data points marked (Δ) in (b) were collected after replacing the nano-Pt solution with a new one.

solutions. This red shift was attributed to surface charge interaction between the polyacrylate capping molecules and PDDA, leading to a shift in the positioning of the energy levels responsible for this emission.[26] Similar matrix effects were also observed in the case of larger-size Q-CdS nanocrystallites of <d> = 4.1 ± 0.9 nm assembled in PDDA.[26]

16.3 KINETICS OF ADSORPTION AND FACTORS GOVERNING THE SELF-ASSEMBLY: CASE STUDY OF NANO-Pt/PDDA

Gaining control over the spatial distribution and surface density of nanoparticles self-assembled in polyelectrolytes, including the ratio of multiple nanoparticles simultaneously incorporated in different architectures for a variety of catalytic, energy conversion, optical, and sensing applications, necessitates an understanding of the self-assembly dynamics and a mapping of the parameters governing the multilayers' buildup, viz., the effect of the nanoparticles' surface modifiers, which determine the types of interactions with the embedding polyelectrolyte (hydrophobic, H-bonding, electrostatic forces between opposite charges, etc.), the ionic strength and pH of the nanoparticles' solution, the nanoparticles' concentration, their size, perhaps their shape, the rinsing–drying protocol, and the surface conformation of the priming polyelectrolyte layer.

The dynamics of polymer and polyelectrolyte adsorption on surfaces,[37–39] and the parameters governing polyelectrolyte layer-by-layer assembly, have been extensively studied.[6,8,10,40–44] These studies have established the effects of ionic strength, deposition time, pH, and rinsing–drying procedures on the consecutive multilayered assembly, including their effect on the thickness and structure of the assembled films. On the other hand, the dynamics of nanoparticles' electrostatic self-assembly in polyelectrolytes and the determining factors of this assembly process remain relatively unexplored, apart from a few studies regarding, for instance, the effect of the ionic strength on the density of assembled silica nanoparticles in polyelectrolytes,[24,28] the effect of the polymer solution pH on the degree of incorporation of arylamine–Au nanoparticles in polystyrenesulfonate (PSS) and carboxylic acid–functionalized Au nanoparticles in poly(allylamine),[20] and the effect of grafting organic groups to magnetic nanoparticles on their surface distribution.[22]

In what follows, a case study regarding the kinetics of adsorption of polyacrylate-capped Pt nanoparticles on PDDA, and the effects of the dipping time, ionic strength, and deposition protocol, is presented. In general, the observed behavior is explained by invoking a similar mechanism to polyelectrolyte multilayer buildup driven by electrostatic interactions.

16.3.1 KINETICS OF ADSORPTION

16.3.1.1 Adsorption via Two Kinetics Regimes

The adsorption of PAC-capped Pt nanoparticles on PDDA, followed by *ex situ* UV-visible spectroscopy, was found to take place via two adsorption regimes character-ized with different kinetics in effect at separate timescales.[29] This is depicted in Figure 16.3, which shows plots of Pt/PDDA film absorbance (at 350 nm) as a function of time. Figure 16.3a corresponds to data collected every $\Delta t = 10$ min of dipping in as-prepared Pt solution, followed by 1 min of rinsing in water and 15 min of drying; Figure 16.3b corresponds to data collected every $\Delta t = 2$ min of dipping in the as-prepared Pt solution, followed by the same rinsing–drying protocol.

The assembly dynamics of Pt nanoparticles on PDDA followed a general trend similar to that of the dynamics of adsorption of polymers and polyelectrolytes on surfaces, reported to take place via two kinetics regimes: (1) an initial diffusion-controlled process at low surface coverage and (2) a slower process at high surface coverage, where chain rearrangement on the surface to accommodate the incoming polymer molecules becomes the rate-determining step.[37–39] This has been experi-mentally observed, for instance, by Motschmann et al.[38] for the adsorption of poly-styrene–poly(ethylene oxide) (PS-PE) diblock copolymer and by Hoogeveen et al.[39] for the adsorption of the strong polyelectrolytes PVP+ (quaternized polyvinylpyri-dine) and AMA+ (quaternized dimethylaminoethyl methacrylate) and of the weak polyelectrolyte AMA on TiO_2 and SiO_2 surfaces.

Two kinetics regimes, an initial fast one and a significantly slower adsorption at longer times, are similarly depicted in Figure 16.3 for the adsorption of Pt nanoparticles on PDDA. Frequent interruptions in *ex situ* adsorption studies, how-ever, can affect nanoparticles' transport to the surface and, hence, the adsorption rate in an otherwise diffusion-limited process. When the assembly was interrupted every 10 min, the absorbance (at $\lambda = 350$ nm) of Pt/PDDA films exhibited a linear dependence on $t^{1/2}$ during the first 20 min of deposition ($R^2 = 0.9999$), characteristic of a diffusion-limited adsorption process at low surface coverage.[38,39,45] On the other hand, when the assembly was interrupted every 2 min, the deposition rate was independent of the nanoparticles' concentration and, hence, of the diffusion process. Instead, in this case, the absorbance of Pt/PDDA films increased linearly with time during the initial 40 min, indicative of a zero-order (or pseudo-zero order) kinetics.

In addition to replenishing the surface Pt concentration by interrupting the dipping, some PDDA surface rearrangement possibly takes place concurrently upon rinsing–drying, as discussed in Section 16.3.2, thus creating a more favorable adsorption surface. During the second adsorption stage, the absorbance depended exponentially on time, similar to the behavior reported by Motschmann et al. for the diblock polymer PS-PE.[38]

A strong dependence of the amount of nano-Pt on the dipping time was evident during the first adsorption stage. Within about 1 h, a pseudosaturation in the nanoparticles' incorporation was reached in both studies. When the assembly was interrupted at $\Delta t = 10$ min, 85% of the maximum loading was completed within 40 min. Similarly, when the assembly was interrupted every 2 min, the deposition rate dropped significantly (by a factor of 13) after 1 h in comparison with the first adsorption stage (initial 40 min). Accordingly, it is found that controlling the nanoparticles' density by varying the deposition time is best achieved and optimized during the first kinetics regime. The leveling off in nanoparticles' incorporation at times longer than 1 h is attributed to the film structure and surface charge presenting a high potential barrier to adsorption at high surface coverage, rather than to depleting the Pt solution. Evidence is provided by replacing the Pt solution with a fresh one at $t = 190$ min, which resulted in no measurable effect on the amount of nanoparticles incorporated beyond that point (Figure 16.3b).[29]

16.3.1.2 Adsorption Mechanism

The assembly of polyelectrolyte-modified nanoparticles on an oppositely charged polyion is thought to take place via a mechanism similar to polyelectrolyte adsorption, as described by Hesselink in 1977.[46,47] At low surface coverage, the PAC-capped Pt nanoparticles present at the PDDA–solution interface are believed to immediately anchor to the cationic adsorption sites via the anionic capping agent, driven by electrostatic and some hydrophobic forces and then to relax to a surface conformation of lowest free energy. The entropy loss accompanying nanoparticle-on-polyelectrolyte adsorption is compensated for by the entropy gained upon freeing the counterions and by some Coulombic energy-of-attraction term. The abundance of cationic adsorption sites results in a fast attachment rate relative to the transport rate. As the surface coverage increases, the cationic anchoring sites become scarce, and the increased Pt nanoparticles/surface repulsion caused by negative surface charge buildup results in a potential barrier that slows down the attachment rate. During this second adsorption regime, at high surface coverage, the rate is believed to be limited by the higher-activation energy process of the polyelectrolyte–nanoparticle rearrangement on the surface to accommodate for the incoming nanoparticles. The maximum amount adsorbed on the polyelectrolyte surface is believed to be kinetically determined, which accounts for the reported irreversibility of polyelectrolyte adsorption.[10,48] It should be noted that when nanoparticles are adsorbed from solutions containing free anionic polyelectrolytes, and depending on the salt content and the polyelectrolyte molecular weight and concentration, the anionic polyelectrolyte may compete with the nanoparticles' adsorption, thus contributing to the less-than-complete surface coverage.

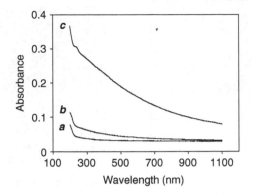

FIGURE 16.4 UV-visible absorption spectra of nano-Pt/PDDA assembled by dipping a PDDA-quartz surface for 4 h in nano-Pt/0.2 M NaCl solution: in a single step (b) and at Δt = 10-min dipping intervals, followed by a 1-min rinse in water and 15 min of drying in air (c). (a) Background spectrum of a PDDA–quartz surface.

16.3.2 THE INTERRUPTED-DEPOSITION PROTOCOL: DEPENDENCE ON RINSING–DRYING

In addition to varying the dipping time, the density of nanoparticles self-assembled on a polyelectrolyte surface can also be controlled by resorting to an interrupted deposition procedure, which consists of inserting multiple rinsing–drying steps in the assembly protocol while maintaining the total dipping time invariant.[29,49] The effect of multiple interruptions on the nanoparticles' density is illustrated in Figure 16.4, which shows UV-visible spectra of Pt/PDDA films assembled by dipping a PDDA-modified quartz surface in nano-Pt/0.2 M NaCl solution for 4 h, either in a single dipping step or interrupted every 10 min by 1 min:15 min rinsing–drying, revealing a 10-times-higher nanoparticles' density in the latter case.* It is also possible to control the nanoparticles' density by varying the dipping time interval. For instance, assembling Pt nanoparticles on PDDA by dipping for 1 h in as-prepared Pt solution at Δt = 2 min (followed by rinsing–drying) resulted in increasing the nanoparticles' density by a factor of ~2 compared to structures assembled at Δt = 10 min (Figure 16.3). Raposo et al.[50] reported a similar effect of the rinsing–drying protocol on the adsorption of poly(o-methoxyaniline) (POMA) on surfaces, where four to five times more POMA was adsorbed upon interrupting the deposition by rinsing–drying every 5 sec compared to 30 sec. This effect was attributed by the authors to freeing adsorption sites previously occupied by water upon drying and perhaps to a change in the polymer conformation, allowing further deposition.

Freeing adsorption sites previously occupied by ions or water, and increasing the nanoparticles' concentration at the PDDA–solution interface by virtue of the frequent interruptions, can account for the increase in the adsorption rate with decreasing Δt during the initial stage of *interrupted deposition* (Figure 16.3a and b). The accompanying increase in nano-Pt incorporation level at saturation, however,

---

* Reaching loading saturation per layer upon dipping in solutions of nano-Pt in 0.1 or 0.2 M NaCl solutions took significantly longer than 1 h, as will be discussed in Section 16.3.3.2.

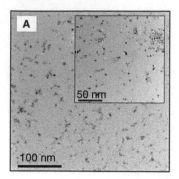 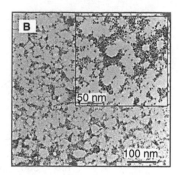

FIGURE 16.5 TEM images of a nano-Pt/PDDA assembled by floating a PDDA/SiO-Cu grid for 1 h on a nano-Pt/0.2 M NaCl solution: in a single step (A) and at Δt = 10-min intervals, followed by a 1-min rinse in water and 15 min of drying in air (B). (Adapted from Markarian, M. et al., in *Self-Assembled Nanostructured Materials*, Lu, Y. et al., Eds., Proceedings of the Materials Research Society, San Francisco, 2003, PV 775, pp. 377–382. With permission.)

is attributed to the cationic polyelectrolyte chains undergoing surface conformational rearrangement induced by rinsing–drying, therefore allowing for stronger electrostatic interactions by providing more favorable adsorption sites. A structural change induced by drying has been reported by Lvov and Decher[6] for multilayers of poly(vinylsulfate)/ poly(allylamine), which exhibited an enhanced film periodicity upon drying, attributed by the authors to a certain preferential orientation of the polymer chains in the dried layers and perhaps to drying inducing parallel plane packing.

Controlling the nanoparticles' density by varying the frequency of the rinsing–drying steps was found to be effective only during the first deposition regime, whereas during the second stage the adsorption rate was independent of the length of dipping time interval. The necessary polyelectrolyte–nanoparticle rearrangement and the potential barrier opposing adsorption at this stage are thus not affected by the rinsing–drying protocol.

TEM imaging showed a fractal distribution of the nanoparticles on PDDA at submonolayer coverage and revealed the increase in the incorporation level upon interrupted deposition, thus corroborating the UV-visible absorption study. Figure 16.5 shows TEM micrographs of Pt nanoparticles assembled on PDDA by floating a PDDA-modified SiO/Cu grid (face down) on the surface of a nano-Pt/0.2 M NaCl solution for 1 h in a single step, or interrupted every 10 min by 1 min:15 min rinsing–drying. The chain-like surface distribution also appears enhanced in this latter case, which can be viewed in support of the proposed model of conformational rearrangement of the priming polyelectrolyte upon rinsing–drying. Such rearrangement can promote a higher density of PDDA trains where the nanoparticles can assemble with minimum potential energy.

16.3.3 Effect of Salt Content

16.3.3.1 Effect on the Nanoparticles' Surface Density

The density of nanoparticles assembled on a polyelectrolyte layer can also be controlled by varying the nanoparticles' solution ionic strength. This effect is illustrated

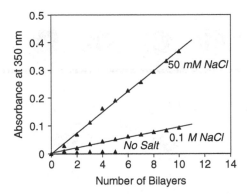

FIGURE 16.6 Absorbance at 350 nm of $(Pt/PDDA)_n$ films vs. the number of bilayers deposited from nano-Pt solutions with 0, 0.05, and 0.1 M NaCl concentrations.

in Figure 16.6, which shows plots of the absorbance of $(Pt/PDDA)_n$ films as a function of the number of bilayers assembled by consecutively dipping films in PDDA and precipitated nano-Pt solutions of different ionic strengths (0 M, 50 mM, 0.1 M), using the same LbL protocol described in Section 16.2.1. The nanoparticles' density increased significantly upon going from a no-salt condition to 50 mM NaCl but then decreased upon further increasing the ionic strength. Increasing the salt content to 0.01 M, for example, resulted in decreasing the nanoparticles' density per layer to 23% of the density of films assembled from nano-Pt/50 mM NaCl solutions. At 0.8 to 1.0 M NaCl concentration, the density of nanoparticles became a mere 9% of that of films assembled from Pt/50 mM NaCl solution.[29]

The effect of ionic strength on polyelectrolyte multilayered assembly, including its role in fine-tuning the multilayers' thickness, has been well established.[8,10,40–44] This effect is due to the known polyelectrolyte conformational rearrangement from a rod-like structure at low ionic strength to a coiled structure at high ionic strength, which results in thicker films in the latter case. At very high salt concentration, a decrease in polyelectrolyte adsorption on an oppositely charged surface takes place due to increased screening of surface charges. On the other hand, there have been only a few studies regarding the effect of ionic strength on the density of nanoparticles assembled on polyelectrolytes.[24,28] Lvov et al.[24] reported an increase in the density of SiO_2 (25 to 78 nm) nanoparticles on PDDA upon increasing the nanoparticles' solution ionic strength from 0.1 to 0.3 M, a behavior attributed to enhanced screening of interparticle repulsion. Recently, Sennerfors et al.[28] reported an increase in the adsorption of SiO_2 nanoparticles (12 nm) on cationic polyelectrolytes upon increasing the ionic strength from 1 to 50 mM, also attributed to screening of interparticle repulsion that allows closer packing (the stratified multilayers were, however, reportedly less stable at the high ionic strength, which was attributed to nanoparticle–polymer desorption at high coverage).

The observed dependence of the density of PAC-Pt nanoparticles (assembled on PDDA) on the Pt solution ionic strength is explained by invoking a similar model of electrostatic interactions and conformational changes in effect for polyelectrolyte polymers,[8,10,47] by taking into account the presence of the soft polymeric capping

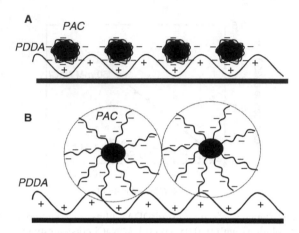

FIGURE 16.7 A cartoon depicting the assembly of polyelectrolyte-modified nanoparticles on the surface of an oppositely charged polyelectrolyte: in the absence of salt (A) and in the presence of a favorable salt concentration (B).

agent. It is assumed that a number of PAC units are anchored to the Pt surface during the synthesis through segments of their chains. In the absence of counterions, segmental repulsion between the polyacrylate units on the Pt surface causes some lengths of the chains to form tails dangling in solution (Figure 16.7A). The small surface charge density (large effective surface area) results in a small attraction term to the PDDA surface, and the large excluded volume and absence of salt result in considerable interparticle repulsion. This and the entropy loss accompanying the nanoparticles-on-PDDA assembly lead to a thermodynamically unfavorable adsorption situation. As the salt concentration increases, intersegmental repulsion is screened, and polyacrylate is pictured, assuming a coiled structure on the nanoparticles' surface (Figure 16.7B). The accompanying increase in surface charge density enhances the attraction to the cationic surface. Furthermore, the decrease in effective volume and the presence of salt result in decreasing interparticle repulsion. The favorable Coulombic energy term and the entropy gained upon liberating the surface counterions (compensating for the entropy loss upon nanoparticles' assembly) result in a favorable adsorption situation, thus accounting for the increase in the nanoparticles' density upon increasing the ionic strength. At increasingly high salt concentrations, however, an unfavorable screening regime is reached in which the surface counterions significantly screen the nanoparticle–cationic polymer attraction, thus hindering the assembly process. This unfavorable screening mode is reached at lower levels of ionic strength, in this case relative to polyelectrolyte assembly.[10,43]

Notably, the linear increase in nanoparticle density per layer was maintained despite the decrease in incorporation levels (Figure 16.6), further indicating that a complete surface coverage is not necessary for a reproducible consecutive incorporation of nanoparticles per deposition cycle. This is attributed to the capability of polyelectrolytes to bridge over unoccupied sites.[8] Similarly, Ostrander et al.[22] reported a linear increase in the incorporation of magnetic yttrium iron garnet nanoparticles assembled in polyelectrolytes, even as the assembly took place via domain formation on the surface.

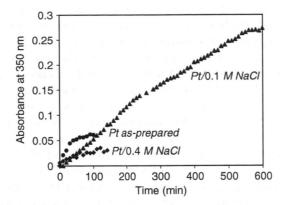

FIGURE 16.8 Absorbance at 350 nm of nano-Pt/PDDA as a function of time taken at $\Delta t =$ 10 min dipping (followed by 1 min of rinsing and 15 min of drying) in nano-Pt/0.1 M NaCl, nano-Pt/0.4 M NaCl, or as-prepared Pt solution.

16.3.3.2 Effect on the Assembly Dynamics

The assembly dynamics and saturation level per layer are also functions of the solution ionic strength and its composition. Figure 16.8 shows plots of the absorbance of Pt/PDDA films as a function of time, collected every 10 min of dipping in precipitated nano-Pt/0.4 M NaCl, nano-Pt/0.1 M NaCl, or as-prepared nano-Pt solutions, followed by a 1-min rinse and 15 min of drying. Although the adsorption rate was initially higher for assembly from as-prepared Pt solutions in comparison to nano-Pt/0.1 M NaCl, in keeping with the higher nanoparticle density in the LbL assembled films (Figure 16.6; cf. Figure 16.2C), saturation took place at lower levels in the former case, which is possibly due to free polyacrylate adsorbing under these conditions. Whereas a PDDA layer was saturated after 1 h of dipping in as-prepared Pt solution, the assembly of Pt nanoparticles from nano-Pt/0.1 M NaCl continued up to about 10 h of interrupted dipping, at which time an ~5-times higher amount of nano-Pt was assembled on PDDA. The long time elapsed before reaching saturation in this case provides considerable dipping time control of the nanoparticles' density per layer, which is significant for the design of nanoassemblies. The low adsorption rate and saturation level resulting from dipping in Pt/0.4 M NaCl are attributed to significant screening of the PDDA-Pt electrostatic attraction, which is consistent with the lower density of Pt per layer in multilayered films assembled LbL at this ionic strength.

The initial high adsorption rate and the higher density of nanoparticles assembled per layer after 1 h of dipping in as-prepared Pt solutions is possibly due to the different solution composition (a higher Pt concentration and a higher ratio of cationic/anionic free species in as-prepared solutions). The effects of the Pt concentration and of the ionic valence and composition on the assembly process are under current investigation.

16.4 PHOTOGENERATED CHARGE TRANSFER AND TRANSPORT AT LBL ASSEMBLIES: CASE STUDY Q-CdS/PDDA

Solar energy conversion at a semiconductor surface in photovoltaic devices and in regenerative or fuel-forming photoelectrochemical cells remains one of the most promising solutions for converting sunlight into electrical and chemical energy. Compared to the natural photosynthetic process, which stores light energy as chemical fuel with a 3% efficiency, semiconductor-based photovoltaic devices can convert light with more than 20% efficiency, and liquid-junction photoelectrochemical cells can produce electrical energy with more than 16% efficiency and chemical energy in excess of 10% efficiency.[51–53] Nevertheless, increasing the power conversion efficiency further while driving down the cost and increasing the lifetime of these devices remains necessary before they become economically viable as alternative energy sources.

Incorporating semiconductor quantum dots[30–34,54–58] — nanoparticles in confined space exhibiting eigenspectrum quantization and a widening of the energy gap with decreased dimensions — in solar cell architectures is believed to hold significant promise for increasing the light conversion efficiency of these devices. This potential stems from a slower theoretical carrier relaxation dynamics at Q-particles as a result of energy level splitting (known as the phonon bottleneck),[59] the impressive increase in surface-to-volume ratio at the nanoscale, and the increase in absorption coefficient with decreased dimensions. The pioneering example of the dye-sensitized nanoparticulate TiO_2 cell from the Swiss Federal Institute group led by O'Regan and Grätzel in the early 1990s,[60] with its 10% efficiency and low cost, undoubtedly spurred a great interest in nanostructured electrodes. Although this cell does not make use of any Q-size effects, the three-orders-of-magnitude increase in surface area has contributed significantly to its high efficiency. Awaiting advancement in this area, a "holy grail" yet to be reached is designing a nanophotocatalyst capable of efficiently splitting water to obtain hydrogen fuel,[53] and another is to exceed the thermodynamic limit of light conversion efficiency at bulk semiconductor solids[61] through the incorporation of quantum semiconductor particles in lightweight and compact photovoltaic and photoelectrochemical device architectures.

Assembling arrays of quantum particles bottom up, gaining control over the order and architecture, and understanding the photoelectrochemical and photovoltaic behavior of these arrays are vital to achieve this potentiality and to gain a fundamental knowledge about this dimension. Thus far, photoelectrochemical studies at quantum dot photoelectrodes remain relatively scarce, and an understanding of interfacial carrier transfer, transport, and relaxation at these electrodes is yet to be established. Studies (at CdS, CdSe, and PbS) involving monolayers of QD semiconductors, nanostructured films electrochemically or chemically deposited on surfaces, and Q-films prepared by the Langmuir–Blodgett (LB) technique have appeared in the literature.[62–73] These studies revealed negative shifts in the photocurrent onset due to Q-size effects, with a couple of reports indicating the feasibility of generating anodic or cathodic photocurrents, depending on the surface[65,66] or on the redox species and electrode potential,[63,74,75] in a departure from the behavior of the bulk solid. Although the Langmuir–Blodgett assembly technique affords a multilayer buildup of separated nanoparticles, favorable for light harvesting, LB films are metastable, which limits their usefulness. The LbL assembly of nanoparticles in polyelectrolytes offers a solution for assembling highly stable three-dimensional arrays as photoelectrodes or photocatalytic structures. Such assemblies can provide a large surface area for harvesting light in a stable matrix that maintains the nanoparticle individuality by minimizing agglomeration. There have been several reports on building photoelectrodes by the LbL technique, including multilayered films of TiO_2 and TiO_2/CdS nanoparticles in PDDA,[23] TiO_2/PbS nanoparticles in PSS,[76] Q-CdS in PDDA,[74–75] and fullerenes in PSS.[77–79] Although these studies have demonstrated the feasibility of charge separation, hopping, and power generation at the nanocomposite polyelectrolyte assemblies, the light conversion efficiencies remain low to date. Better organization and order in these architectures are needed to favor charge separation and charge transport and to minimize recombination losses.

The assemblies, however, also provide an opportunity to study the ensemble behavior of quantum-confined semiconductor particles and how it departs from the behavior of the bulk solid. In what follows, we present some of our results regarding the photoelectrochemical behavior of Q-CdS assembled in PDDA on electrodes.[74,75]

16.4.1 THE Q-STATE: ANODIC AND CATHODIC PHOTOCURRENTS

The photoelectrochemical behavior of multilayers of Q-CdS/PDDA assembled on ITO-coated glass electrodes was investigated in aqueous media. Photoelectrochemical experiments were conducted in a three-electrode photoelectrochemical cell while irradiating the (Q-CdS/PDDA)$_n$/ITO working electrodes with a 300-W Hg lamp and using a Pt auxiliary electrode and an Ag/AgCl reference electrode. Figure 16.9 shows photocurrent–voltage plots at (Q-CdS/PDDA)$_4$ films in aerated phosphate buffer electrolyte and in deoxygenated phosphate buffer in the presence of methyl viologen (pH 7), revealing the generation of anodic and cathodic photocurrents at the Q-films and the feasibility of charge transport through the polyelectrolyte matrix. The generation of anodic and cathodic photocurrents at nanoparticulate films has been previously observed at monolayers of Q-PbS on Au[63] and at CdS and CdSe nanostructured films.[65,66]

In the absence of other redox species (Figure 16.9A), the photogenerated charge transfer reactions to solution species that are thermodynamically feasible at the CdS surface are proton reduction, oxygen reduction, and water oxidation.[80] However, hydrogen and oxygen evolution are known to be kinetically sluggish at CdS in the absence of hydrogen and oxygen evolution catalysts.[81–83] On the other hand, photogenerated holes can attack the CdS lattice by reacting with surface S^{2-}, leading to the photodegradation of CdS in the presence of oxygen.[81,82] This oxidation process and oxygen reduction* are therefore the major reactions at the Q-CdS surface here, assuming similar kinetics as the bulk solid. The mechanism by which these redox reactions lead to the generation of cathodic or anodic photocurrents at the Q-CdS/PDDA films is discussed in the following section.

Photocurrent generation can be affected by using a charge transfer mediator, as studied at these films and by Luo and Guldi and coworkers[77–79] at fullerene/polyelectrolyte LbL films. For example, adding methyl viologen, a strong electron scavenger (Figure 16.9B), resulted in a significant enhancement in anodic photocurrents at Q-CdS/PDDA films in deaerated solutions at potentials positive of the MV^+/MV^{2+} standard redox potential ($E° = -0.67$ V vs. Ag/AgCl).[75] In this case, MV^{2+} is first

---

* Evidence of oxygen reduction was provided by a significant decrease in the cathodic photocurrent upon deoxygenating the solutions by purging with nitrogen.[75]

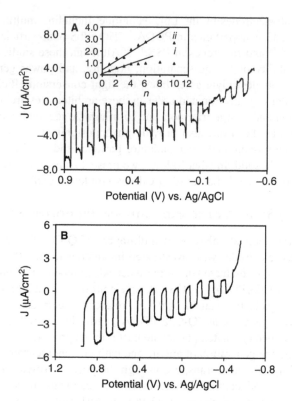

FIGURE 16.9 Photocurrent–voltage plots of four bilayers of Q-CdS/PDDA on ITO in phosphate buffer (pH 7) (A) and in the presence of methyl-viologen (pH 7) (B). The inset of (A) shows plots of anodic (i) and cathodic (ii) photocurrent densities as a function of the number of bilayers.

reduced to MV^+ at the Q-CdS surface by scavenging photogenerated electrons, and then MV^+ is oxidized to MV^{2+} at the ITO collector electrode at potentials positive of $E°(MV^{2+}/MV^+)$. At more negative potentials, dark cathodic currents result from the reduction of MV^{2+} directly at the underlying electrode. The photocurrent can also be increased further by increasing the pH as a result of an increase in the reducing power of the Q-CdS electron,[75] in keeping with the shift in the band-edge positions of single-crystal CdS, colloidal CdS, and other bulk semiconductor materials.[84,85]

16.4.2 MODEL OF PHOTOCURRENT GENERATION AT ENSEMBLES OF Q-PARTICLES

In doped bulk semiconductors, charge separation upon light absorption takes place mainly by charge migration in the space charge region[86] as a result of the field created (or band bending). Under depletion conditions, charge transfer of minority carriers to solution redox species leads to the generation of anodic photocurrents at n-type and cathodic photocurrents at p-type semiconductors. The picture is quite

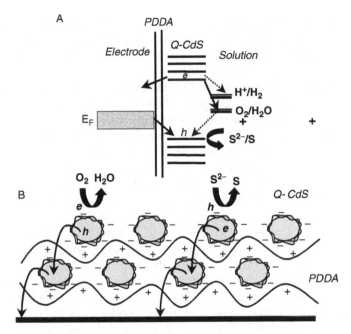

FIGURE 16.10 (A) Processes of photocurrent generation of Q-CdS/PDDA films on an electrode surface leading to the generation of anodic and cathodic photocurrents, in the absence of added redox species. (B) Cartoon depicting charge hopping through the matrix.

different for quantum dots. The quantization of the eigenspectrum and the nanometer dimensions (here, $<d> = 3.6 \pm 0.5$ nm) make band bending and the formation of a space charge region not applicable at this scale. In quantum-confined matter, the photogenerated carrier wave functions extend to the nanoparticle surface and into the medium, rendering both carriers instantly available for charge transfer to solution species or transport through the medium.[75]

The processes leading to bidirectional photocurrent generation at Q-CdS embedded in a polyelectrolyte matrix (aqueous medium, no added redox species) are presented in Figure 16.10,* with oxygen reduction and surface oxidation by holes being the predominant charge transfer reactions. Photocurrent generation is believed to be determined by competing reduction by photogenerated electrons at Q-CdS and electron transport via charge hopping to the underlying electrode on the one hand, and competing oxidation at the Q-particles' surface and hole transport through the polyelectrolyte medium to the underlying electrode on the other hand. In the absence of considerable dark (or photo-) electrochemistry at the underlying electrode, the photocurrent direction and magnitude will be dictated by the net flow of photogenerated e and h transported to the electrode surface, processes that are potential

---

* Charge transfer reactions can be mediated by surface states, which have been omitted from this picture for simplicity. The presence of surface states has been confirmed by photoluminescence measurements of Q-CdS solutions and films.[26] The photocurrent transients observed in the photocurrent–voltage plots in aerated solutions are attributed to surface states–mediated charge transfer, as discussed elsewhere.[74,75]

dependent (Figure 16.9A). Otherwise, the underlying electrode serves as a collector electrode for the photoreduced or photooxidized species produced by reactions at the Q-particles, resulting in the measurement of anodic or cathodic photocurrents, as is the case presented in Figure 16.9B.

The photocurrent generated at the $(Q\text{-}CdS/PDDA)_n/ITO$ electrodes increases with increasing the number of bilayers up to a thickness. Anodic photocurrents increase linearly up to four bilayers and then level off, whereas cathodic photocurrents increase linearly up to eight bilayers and then decrease (inset of Figure 16.9A).[75] As we discussed elsewhere,[75] the leveling off in photocurrent density is not caused by a saturation in the incorporation of Q-CdS or a decreased absorbance by the inner layer, because the absorbance and photoluminescence intensity of $(Q\text{-}CdS/PDDA)_n$ continues to increase linearly with n beyond these film thicknesses. The loss of light conversion is a result of an increase in trapping or recombination of photogenerated charges as films get thicker (17.8 nm for $n = 4$, 35.6 nm for $n = 8$, as measured by ellipsometry)[26]; thus, carriers are lost before being transported to the electrode. A larger density of electron traps relative to the holes can account for the different dependence of anodic and cathodic photocurrents on n. This behavior is typical of semiconductor films because of competing effects of increased light harvesting with increased thickness and an increase in recombination losses as the distance from the electrode increases. In an earlier study, Kotov et al.[23] similarly reported anodic photocurrents at $(TiO_2/PDDA)_n$ films reaching a maximum after 11 bilayers.

Transport in the nanocomposite film is believed to take place via a charge-hopping mechanism between the embedded quantum dots through the intercalating polyelectrolyte, whose thickness is estimated to be larger than 1 nm (in the direction normal to the surface), on average, from ellipsometry measurements after taking the submonolayer Q-CdS surface coverage into account.[26] Laurent and Schlenoff[87] also proposed a charge-hopping mechanism between the viologen sites to explain the feasibility of transport through the poly(styrenesulfonate) matrix when investigating the electrochemistry of poly(butanylviologen)/PSS multilayers. The feasibility of electron transport through films of lamellar solids ($\alpha\text{-}Zr(HPO_4)_2 \cdot H_2O$) and poly(ally-lamine hydrochloride), several nanometers thick, with and without Au colloids embedded in the structure, was also reported by Feldheim et al.[18] Such feasibility is also indicated in several photoactive architectures discussed in the following section.

16.4.3 AFFECTING LIGHT CONVERSION EFFICIENCY BY MULTILAYERED ASSEMBLY

The light conversion efficiency at photovoltaic and liquid-junction photoelectro-chemical cells based on quantum-confined particles remains low to date, despite theoretical expectations of increased efficiencies at quantum dot solar cells. In addition to the photoelectrochemical studies cited earlier at nanoparticulate films, several researchers have reported building photovoltaic devices from Q-dots and Q-rods blended with conducting polymers.[88–92] The best conversion efficiency was recently reported by Huynh et al.[90] at CdSe quantum rods embedded in poly(3-hexylthiophene) as a hole-conducting phase. The structure exhibited a power conversion efficiency of 6.9% at monochromatic 515-nm light and 1.7% at AM 1.5 sunlight.[90]

The major roadblocks that are limiting the light conversion efficiency at these devices are considerable recombination losses at interfaces and low transport properties in the nanocomposite films caused by a significant disorder that results in carrier scattering. The LbL assembly protocol offers the possibility of ordering quantum particles or other chromophores and donor–acceptor functionalities in different layers with specifically designed interlayer separation with angstrom level tunability to enhance photogenerated charge separation and minimize recombination losses. By this, it may afford a way to mimic the spectacularly designed architecture of the natural photosynthetic process with its high level of control over intermolecular spacing, orientation, and local environments, resulting in a series of sequential energy and electron transfer reactions leading to the generation of fuel.

There has been reported success in using the LbL assembly protocol for building multilayered light conversion devices involving different functionalities.[77–79,93–95] Kaschak et al.[95] used the sequential adsorption of polyions to achieve a five-component energy and electron transfer cascade that mimicked some aspects of the natural photosynthesis. In this assembled architecture, absorption took place at coumarin and fluorescein-derivatized poly(allylamine hydrochloride) (PAH) separated from a porphyrin donor, palladium (II)tetrakis(4-N,N,N-trimethylanilinium)porphyrin (PdTAPP^{4+}), by the α-ZrP layer. Another α-ZrP or HTiNbO$_5$ layer separated the porphyrin donor from the poly(viologen) acceptor. The cascade of photon absorption followed by charge separation resulted in quantum yields of 0.47 or 0.61 (for α-ZrP or HTiNbO$_5$ layers).

In other work, Luo et al.[77] assembled fulleropyrrolidinium ion or positively charged ruthenium(II)-polypyridyl-fullerene donor–acceptor dyads in PSS on ITO. The films exhibited photoactivity upon light absorption, indicative of the feasibility of charge separation and transfer to the underlying electrode.[77] Photocurrents increased with increasing the number of bilayers from 1 to 10, but the maximum reported incident photon-to-current conversion efficiency (IPCE) at 10 bilayers did not exceed 1.1% with monochromatic light. A self-organization of a fulleropyrrolidinium–androstane–ferrocene dyad, assembled LbL in polyelectrolytes, into linear nanowire structures was reported by Guldi et al.[78] to result in photocurrent enhancement, compared to films not exhibiting such organization. The enhancement was attributed by the authors to multiple contacts between the fullerene–ferrocene nanowires. Guldi et al.[79] also reported light conversion at a monolayer of tetraphenylporphyrin–fullerene donor–acceptor in PSS, with IPCEs of 0.36 and 0.6% in anaerobic and aerobic conditions, respectively. The increase in photocurrent under aerobic conditions was attributed to mediated electron transfer by oxygen.

Controlling with good precision the order of nanoparticles and other functionalities in polyelectrolyte assemblies, eliminating layer interpenetration, and gaining control over interlayer and interparticle separation are still needed in order to achieve favorable specific environments and gradients within the architectures that result in efficient charge separation and enhanced transport through tunneling junctions between the functional layers, thus minimizing carrier recombination losses and increasing the efficiency of light conversion at such ensembles. Furthermore, the optimal number of layers (or the thickness) has to be experimentally determined in these architectures because of opposing effects of increasing the light absorption with an increasing number of layers and increasing the rate of bulk recombination in thicker films.

16.5 CONCLUSIONS AND FINAL REMARKS

The self-assembly of nanoparticles capped with an anionic polyelectrolyte in a cationic polyelectrolyte by virtue of electrostatic and hydrophobic interactions was presented as a means for the controlled incorporation of semiconductor and metal nanoparticles in assemblies on surfaces. The nanoparticle density in the direction parallel to the surface was found to be a function of several dipping parameters, viz., the dipping time, ionic strength, solution composition, and rinsing–drying protocol; the nanoparticle density normal to the surface can be controlled by varying the number of bilayers and interlayer separation. The feasibility of photogenerated charge separation, charge transport, and hence photocurrent generation was demonstrated at films of quantum dots in polyelectrolytes, albeit with low efficiency. The LbL assembly technique has the potential of allowing the sequential ordered assembly of nanoparticles and other functionalities for different device applications, including solar cells. To this end, minimizing layer interpenetration and a better control over the order and surface density of nanoparticles and other donor–acceptor functionalities are needed to enhance charge separation, favor transport, and minimize carrier recombination losses at these structures. The effects of nanoparticle surface derivatization, solvent properties, and the priming polyelectrolyte surface conformation on the nanoparticle density and distribution on the surface are under current investigation to achieve the needed level of control necessary for these devices. The effect of interlayer thickness, nanoparticle order, and nanoparticle packing density, and the effect of interfacial energetics and kinetics on the light conversion efficiency at quantum dots incorporated in a polyelectrolyte matrix, are also being currently investigated.

REFERENCES

1. CWW Ostwald. *An Introduction to Theoretical and Applied Colloid Chemistry. The World of Neglected Dimensions*, 2nd ed. New York: John Wiley & Sons, 1922.
2. RK Iler. Multilayers of colloidal particles. *J Colloid Interface Sci* 21:569–594, 1966.
3. G Decher, JD Hong, J Schmitt. Buildup of ultrathin multilayer films by a self-assembly process. III. Consecutively alternating adsorption of anionic and cationic polyelectrolytes on charged surfaces. *Thin Solid Films* 210/211:831–835, 1992.
4. G Decher, J Schmitt. Fine-tuning of the film thickness of ultrathin multilayer films composed of consecutively alternating layers of anionic and cationic polyelectrolytes. *Progr Colloid Polym Sci* 89:160–164, 1992.
5. Y Lvov, G Decher, H Mohwald. Assembly, structural characterization, and thermal behavior of layer-by-layer deposited ultrathin films of poly(vinyl sulfate) and poly(allylamine). *Langmuir* 9:481–486, 1993.
6. YM Lvov, G Decher. Assembly of multilayer ordered films by alternating adsorption of oppositely charged macromolecules. *Crystallogr Rep* 39:628–647, 1994.
7. G Decher. Fuzzy nanoassemblies: toward layered polymeric multicomposites. *Science* 277:1232–1237, 1997.

8. G Decher. Layered nanoarchitectures via directed assembly of anionic and cationic molecules. In *Comprehensive Supramolecular Chemistry*, Vol. 9, JF Sauvage, HW Hosseini, Eds. Oxford: Pergamon Press, 1996, pp. 507–528.

9. KM Lenahan, YX Wang, YJ Liu, RO Claus, JR Heflin, D Marciu, C Figura. Novel polymer dyes for nonlinear optical applications using ionic self-assembled monolayer technology. *Adv Mater* 10:853–855, 1998.

10. X Arys, AM Jonas, A Laschewsky, R Legras. Supramolecular polyelectrolyte assemblies. In *Supramolecular Polymers*, A Ciferri, Ed. New York: Marcel Dekker, 2000, pp. 505–563.

11. SW Keller, H-N Kim, TE Mallouk. Layer-by-layer assembly of intercalation compounds and heterostructures on surfaces: toward molecular beaker epitaxy. *J Am Chem Soc* 116:8817–8818, 1994.

12. Y Lvov, G Decher, G Sukhorukov. Assembly of thin films by means of successive deposition of alternate layers of DNA and poly(allylamine). *Macromolecules* 26:5396–5399, 1993.

13. JD Hong, K Lowack, J Schmidtt, G Decher. Layer-by-layer deposited multilayer assemblies of polyelectrolytes and proteins: from ultrathin films to protein arrays. *Prog Colloid Polym Sci* 93:98–102, 1993.

14. YM Lvov, Z Lu, JB Schenkman, X Zu, JF Rusling. Direct electrochemistry of myoglobin and cytochrome P450$_{cam}$ in alternate layer-by-layer films with DNA and other polyions. *J Am Chem Soc* 120:4073–4080, 1998.

15. Y Lvov, H Haas, G Decher, H Möhwald, A Milkhailov, B Mtchedlishvily, E Morgunova, B Vainshtein. Successive deposition of alternate layers of polyelectrolytes and a charged virus. *Langmuir* 10:4232–4236, 1994.

16. S Watanabe, SL Regan. Dendrimers as building blocks for multilayer construction. *J Am Chem Soc* 116:8855–8856, 1994.

17. DW Kim, A Blumstein, J Kumar, LA Samuelson, B Kang, C Sung. Ordered multilayer nanocomposites prepared by electrostatic layer-by-layer assembly between aluminosilicate nanoplatelets and substituted ionic polyacetylenes. *Chem Mater* 14:3925–3929, 2002.

18. DL Feldheim, KC Grabar, MJ Natan, TE Mallouk. Electron transfer in self-assembled inorganic polyelectrolyte/metal nanoparticle heterostructures. *J Am Chem Soc* 118:7640–7641, 1996.

19. T Cassagneau, JH Fendler. Preparation and layer-by-layer self-assembly of silver nanoparticles capped by graphite oxide nanosheets. *J Phys Chem B* 103:1789–1793, 1999.

20. JF Hicks, Y Seok-Shon, RW Murray. Layer-by-layer growth of polymer/nanoparticle films containing monolayer-protected gold clusters. *Langmuir* 18:2288–2294, 2002.

21. N Malikova, I Pastoriza-Santos, M Schierhorn, NA Kotov, LM Liz-Marzán. Layer-by-layer assembled mixed spherical and planar gold nanoparticles: control of interparticle interactions. *Langmuir* 18:3694–3697, 2002.

22. JW Ostrander, AA Mamedov, NA Kotov. Two modes of linear layer-by-layer growth of nanoparticle-polyelectrolyte multilayers and different interactions in the layer-by-layer deposition. *J Am Chem Soc* 123:1101–1110, 2001.

23. A Kotov, I Dékány, JH Fendler. Layer-by-layer self-assembly of polyelectrolyte-semiconductor nanoparticle composite films. *J Phys Chem* 99:13065–13069, 1995.

24. Y Lvov, K Ariga, M Onda, I Ichinose, T Kunitake. Alternate assembly of ordered multilayers of SiO$_2$ and other nanoparticles and polyions. *Langmuir* 13:6195–6203, 1997.

25. Y Liu, A Wang, R Claus. Molecular self-assembly of TiO$_2$/polymer nanocomposite films. *J Phys Chem B* 101:1385–1388, 1997.
26. LI Halaoui. Layer-by-layer assembly of polyacrylate-capped CdS nanoparticles in poly(diallyldimethylammonium chloride) on solid surfaces. *Langmuir* 17:7130–7136, 2001.
27. F Caruso, H Möhwald. Preparation and characterization of ordered nanoparticle and polymer composite multilayers on colloids. *Langmuir* 15:8276–8281, 1999.
28. T Sennerfors, G Bogdanovic, F Tiberg. Formation, chemical composition, and structure of polyelectrolyte-nanoparticle multilayer films. *Langmuir* 18:6410–6415, 2002.
29. S Ghannoum, Y Xin, J Jaber, LI Halaoui. Self-assembly of polyacrylate-capped Pt nanoparticles on a polyelectrolyte surface: kinetics of adsorption, and effect of ionic strength and deposition protocol. *Langmuir* 19:4804–4811, 2003.
30. LE Brus. Electron–electron and electron-hole interactions in small semiconductor crystallites: the size dependence of the lowest excited electronic state. *J Chem Phys* 80:4403–4409, 1984.
31. LE Brus. Electronic wave functions in semiconductor clusters: experiment and theory. *J Phys Chem* 90:2555–2560, 1986.
32. Y Kayanuma. Quantum-size effects of interacting electrons and holes in semiconductor microcrystals with spherical shape. *Phys Rev B* 38:9797–9805, 1988.
33. Rama Krishna, RA Friesner. Quantum confinement effects in semiconductor clusters. *J Chem Phys* 95:8309–8322, 1991.
34. PE Lippens, M Lanoo. Calculation of the band gap for small CdS and ZnS crystallites. *Phys Rev B* 39:10935–10942, 1989.
35. N Herron, Y Wang, H Eckert. Synthesis and characterization of surface-capped, size-quantized cadmium sulfide clusters. Chemical control of cluster size. *J Am Chem Soc* 112:1322–1326, 1990.
36. N Chestnoy, TD Harris, R Hull, LE Brus. Luminescence and photophysics of cadmium sulfide semiconductor clusters: the nature of the emitting electronic state. *J Phys Chem* 90:3393–3399, 1986.
37. K Miyano, K Asano, M Shimomura. Adsorption kinetics of water-soluble polymers onto a spread monolayer. *Langmuir* 7:444–445, 1991.
38. H Motschmann, M Stamm, Ch Toprakcioglu. Adsorption kinetics of block copolymers from a good solvent: a two-stage process. *Macromolecules* 24:3681–3688, 1991.
39. NG Hoogeveen, MA Cohen Stuart, GJ Fleer. Polyelectrolyte adsorption on oxides. I. Kinetics and adsorbed amounts. *J Colloid Interface Sci* 182:133–145, 1996.
40. ST Dubas, JB Schlenoff. Factors controlling the growth of polyelectrolyte multilayers. *Macromolecules* 32:8153–8160, 1999.
41. M Lösche, J Schmidt, G Decher, WG Bouwman, K Kjaer. Detailed structure of molecularly thin polyelectrolyte multilayer films on solid substrates as revealed by neutron reflectometry. *Macromolecules* 31:8893–8906, 1998.
42. W Chen, TJ McCarthy. Layer-by-layer deposition: a tool for polymer surface modification. *Macromolecules* 30:78–86, 1997.
43. D Korneev, Y Lvov, G Decher, J Schmidt, S Yaradaikin. Neutron reflectivity analysis of self-assembled film super lattices with alternate layers of deuterated and hydrogenated polystyrenesulfonate and polyallylamine. *Physica B* 213/214:954–956, 1995.
44. GJ Fleer, MA Cohen Stuart, JMHM Scheutjens, T Cosgrove, B Vincent. Electrostatic effects: charged surfaces and polyelectrolyte adsorption. In *Polymers at Interfaces*. London: Chapman & Hall, 1993, pp. 343–375.
45. BW Greene. The effect of added salt on the adsorbability of a synthetic polyelectrolyte. *J Colloid Interface Sci* 37:144–153, 1971.

46. FTh Hesselink. On the theory of polyelectrolyte adsorption. The effect on adsorption behavior of the electrostatic contribution to the adsorption free energy. *J Colloid Interface Sci* 60:448–466, 1977.

47. FTh Hesselink. Adsorption of polyelectrolytes from dilute solution. In *Adsorption from Solution at the Solid/Liquid Interface*, G Parfit, C Rochester, Eds. London: Academic Press, 1983, pp. 377–412.

48. JB Schlenoff, H Ly, M Li. Charge and mass balance in polyelectrolyte multilayers. *J Am Chem Soc* 120:7626–7634, 1998.

49. M Markarian, J Jaber, Y Xin, S Ghannoum, LI Halaoui. Electrostatic self-assembly of metal and semiconductor nanoparticles in polyelectrolytes: assembly dynamics, and electrochemical behavior. In *Self-Assembled Nanostructured Materials*, Y Lu, CJ Brinker, M Antonietti, C Bai, Eds. Proceedings of the Materials Research Society, San Francisco, 2003, PV 775, pp. 377–382.

50. M Raposo, RS Pontes, LHC Mattoso, ON Oliveira, Jr. Kinetics of adsorption of poly(*o*-methoxyaniline) self-assembled films. *Macromolecules* 30:6095–6101, 1997.

51. NS Lewis. Artificial photosynthesis. *American Scientist* 83:534–541, 1995.

52. A Heller. Conversion of sunlight into electrical power and photoassisted electrolysis of water in photoelectrochemical cells. *Acc Chem Res* 14:154–162, 1981.

53. AJ Bard, MA Fox. Artificial photosynthesis: solar splitting of water to hydrogen and oxygen. *Acc Chem Res* 28:141–145, 1995.

54. GD Stucky, JE MacDougall. Quantum confinement and host/guest chemistry: probing a new dimension. *Science* 247:669–678, 1990.

55. LE Brus, ML Steigerwald. Semiconductor crystallites: a class of large molecules. *Acc Chem Res* 23:183–188, 1990.

56. Y Wang. Nonlinear optical properties of nanometer-sized semiconductor clusters. *Acc Chem Res* 24:133–139, 1991.

57. H Weller. Quantized semiconductor particles: a novel state of matter for materials science. *Adv Mater* 5:88–95, 1993.

58. RW Siegel. Exploring mesoscopia: the bold new world of nanostructure. *Phys Today* 46:64–68, 1993.

59. U Bockelmann, G Bastard. Phonon scattering and energy relaxation in two-, one-, and zero-dimensional electron gases. *Phys Rev B* 42:8947–8951, 1990.

60. B O'Regan, M Grätzel. A low-cost, high-efficiency solar cell based on dye-sensitized colloidal TiO_2 films. *Nature* 353:737–740, 1991.

61. W Shockley, HJ Queisser. Detailed balance limit of efficiency of p–n junction solar cells. *J Appl Phys* 32:510–519, 1961.

62. S Ogawa, F-RF Fan, AJ Bard. Scanning tunneling microscopy, tunneling spectroscopy, and photoelectrochemistry of a film of Q-CdS particles incorporated in a self-assembled monolayer on a gold surface. *J Phys Chem* 99:11182–11189, 1995.

63. S Ogawa, K Hu, F-RF Fan, AJ Bard. Photoelectrochemistry of films of quantum size lead sulfide particles incorporated in self-assembled monolayers on gold. *J Phys Chem B* 101:5707–5711, 1997.

64. T Nakanishi, B Ohtani, K Uosaki. Fabrication and characterization of CdS-nanoparticle mono- and multilayers on a self-assembled monolayer of alkanedithiols on gold. *J Phys Chem B* 102:1571–1577, 1998.

65. L Kronik, N Ashkenasy, M Leibovitch, E Fefer, Y Shapira, S Gorer, G Hodes. Surface states and photovoltaic effects in CdSe quantum dot films. *J Electrochem Soc* 145:1748–1755, 1998.

66. G Hodes, A Albu-Yaron, Optical, structural and photoelectrochemical properties of quantum box semiconductor films. In *Photoelectrochemistry and Electrosynthesis on*

Semiconducting Materials, DS Ginley, K Honda, A Nozik, A Fujishima, N Armstrong, T Sakata, T Kawal, Eds. The Electrochemical Society Proceedings Series, Pennigton, NJ, 1988, PV 88-14, pp. 298–303.

67. T Torimoto, S Nagakubo, M Nishizawa, H Yoneyama. Photoelectrochemical properties of size-quantized CdS thin films prepared by an electrochemical method. *Langmuir* 14:7077–7081, 1998.

68. M Miyake, T Torimoto, T Sakata, H Mori, H Yoneyama. Photoelectrochemical characterization of nearly monodisperse CdS nanoparticles-immobilized gold electrodes. *Langmuir* 15:1503–1507, 1999.

69. S Drouard, SG Hickey, D Riley. CdS nanoparticle-modified electrodes for photoelectrochemical studies. *J Chem Commun* 1:67–68, 1999.

70. S Hickey, DJ Riley. Photoelectrochemical studies of CdS nanoparticle-modified electrodes. *J Phys Chem B* 103:4599–4602, 1999.

71. HS Mansur, F Grieser, MS Marychurch, S Biggs, RS Urquhart, DN Furlong. Photoelectrochemical properties of "Q-state" CdS particles in arachidic acid Langmuir–Blodgett films. *J Chem Soc Faraday Trans* 91:665–672, 1995.

72. HS Mansur, F Grieser, RS Urquhart, DN Furlong. Photoelectrochemical behavior of Q-state $CdS_xSe_{(1-x)}$ particles in arachidic acid Langmuir–Blodgett films. *J Chem Soc Faraday Trans* 91:3399–3404, 1995.

73. T Torimoto, N Tsumura, M Miyake, M Nishizawa, T Sakata, H Mori, H Yoneyama. Preparation and photoelectrochemical properties of two-dimensionally organized CdS nanoparticle thin films. *Langmuir* 15:1853–1858, 1999.

74. LI Halaoui. Characterization and photoelectrochemistry of layer-by-layer assembled films of CdS quantum dots in polyelectrolyte matrix. In *II-VI Compound Semiconductor Photovoltaic Materials*, R Noufi, RW Birkmire, D Lincot, HW Schock, Eds. Proceedings of the Materials Research Society, San Francisco, 2001, PV 668, pp. H8.1–H8.6.

75. LI Halaoui. Photoelectrochemistry in aqueous media at polyacrylate-capped Q-CdS assembled in a polyelectrolyte matrix on electrode surfaces. *J Electrochem Soc* 150:E455–E460, 2003.

76. Y Sun, E Hao, X Zhang, B Yang, J Shen, L Chi, H Fuchs. Buildup of composite films containing TiO_2/PbS nanoparticles and polyelectrolytes based on electrostatic interaction. *Langmuir* 13:5168–5174, 1997.

77. C Luo, DM Guldi, M Maggini, E Menna, S Mondini, NA Kotov, M Prato. Stepwise assembled photoactive films containing donor-linked fullerenes. *Angew Chem Int Ed* 39:3905–3909, 2000.

78. DM Guldi, C Luo, D Koktysh, NA Kotov, TD Ros, S Bosi, M Prato. Photoactive nanowires in fullerence–ferrocene dyad polyelectrolyte multilayers. *Nano Lett* 2:775–780, 2002.

79. DM Guldi, F Pellarini, M Prato, C Granito, L Troisi. Layer-by-layer construction of nanostructured porphyrin–fullerene electrodes. *Nano Lett* 2:965–968, 2002.

80. M Grätzel. Colloidal semiconductors. In *Photocatalysis: Fundamentals and Applications*, N Serpone, E Pelizzetti, Eds. New York: Wiley, 1989. pp. 123–157.

81. A Henglein. Photodegradation and fluorescence of colloidal cadmium sulfides. *Ber Bunsenges Phys Chem* 86:301–305, 1982.

82. A Henglein. Photochemistry of colloidal cadmium sulfide. 2. Effects of adsorbed methyl viologen and of colloidal platinum. *J Phys Chem* 86:2291–2293, 1982.

83. NM Dimitrijevi, S Li, M Grätzel. Visible light-induced oxygen evolution in aqueous cadmium sulfide suspensions. *J Am Chem Soc* 106:6565–6569, 1984.

84. MF Finlayson, BL Wheeler, N Kakuta, K-H Park, AJ Bard, A Campion, MA Fox, SE Webber, JM White. Determination of flat-band position of cadmium sulfide crystals, films, and powders by photocurrent and impedance techniques, photoredox reaction mediated by intragap states. *J Phys Chem* 89:5676–5681, 1985.

85. JR White, AJ Bard. Electrochemical investigation of photocatalysis at cadmium sulfide suspensions in the presence of methylviologen. *J Phys Chem* 89:1947–1954, 1985.

86. B Sapoval, C Hermann. In *Physics of Semiconductors*. New York: Springer-Verlag, 1995, p. 234.

87. D Laurent, JB Schlenoff. Multilayer assemblies of redox polyelectrolytes. *Langmuir* 13:1552–1557, 1997.

88. NC Greenham, X Peng, AP Alivisatos. Charge separation and transport in conjugated-polymer/semiconductor-nanocrystal composites studies by photoluminescence quenching and photoconductivity. *Phys Rev B* 54:17628–17637, 1996.

89. WU Huynh, X Peng, AP Alivisatos. CdSe nanocrystal rods/poly(3-hexylthiophene) composite photovoltaic devices. *Adv Maters* 11:923–926, 1999.

90. WU Huynh, JJ Dittmer, AP Alivisatos. Hybrid nanorod-polymer solar cells. *Science* 295:2425–2427, 2002.

91. WU Huynh, JJ Dittmer, N Teclemariam, DJ Milliron, AP Alivisatos, KWJ Barnham. Charge transport in hybrid nanorod-polymer nanocomposite photovoltaic cells. *Phys Rev B* 67:115326-1-12, 2003.

92. WU Huynh, JJ Dittmer, WC Libby, GL Whittig, AP Alivisatos. Controlling the morphology of nanocrystal-polymer composites for solar cells. *Adv Funct Mater* 13:73–79, 2003.

93. DM Kashak, TE Mallouk. Inter- and intralayer energy transfer in zirconium phosphate-poly(allylamine hydrochloride) multilayers: an efficient photon antenna and a spectroscopic ruler for self-sssembled thin films. *J Am Chem Soc* 118:4222–4223, 1996.

94. SW Keller, SA Johnson, ES Brigham, EH Yonemoto, TE Mallouk. Photoinduced charge separation in multilaycr thin films grown by sequential adsorption of polyelectrolytes. *J Am Chem Soc* 117:12879–12880, 1995.

95. DM Kaschak, JT Lean, CC Waraksa, GB Saupe, H Usami, TE Mallouk. Photoinduced energy and electron transfer reaction in lamellar polyanion/polycation thin films: toward an inorganic "leaf." *J Am Chem Soc* 121:3435–3445, 1999.

17 Layer-by-Layer Assembly Approach to Templated Synthesis of Functional Nanostructures

Nina I. Kovtyukhova

CONTENTS

17.1 Introduction 463
17.2 Layer-by-Layer Self-Assembly of Nanoparticle/Polymer Nanotubes 465
17.3 Layer-by-Layer Synthesis of SiO$_2$ Nanotubes from Molecular
       Precursors 469
17.4 Synthesis of Nanotube-Encapsulated Nanowire Devices.........................473
17.5 Layer-by-Layer Assembly on Top of Nanowire Templates 476
17.6 Electrical Properties 478
17.7 Combined In-Membrane and On-Wire Assembly of Nanowire
       Devices 482
17.8 Conclusions 483
Acknowledgments...483
References..483

17.1 INTRODUCTION

In recent years, high-aspect-ratio inorganic nanoparticles, which possess a unique combination of geometric, electronic, and chemical properties, have been extensively explored toward their potential applications as components of field emitter displays, magnetic media, sensors, resonators, solar cells, and electronic circuits. This work has been stimulated by the availability of new techniques for making nanotubes, wires, scrolls, and ribbons from various materials and by interest in the new physical properties of these unique forms of matter.

The chemical assembly of electronic nanoscale circuits is considered a viable alternative to the conventional lithographic fabrication techniques, which have rapidly been maturing due to both fundamental physical and economic factors. A promising approach to a future generation of ultradense electronic circuits relies upon assembling nanowires or nanotubes as interconnects and device elements into cross-point arrays of addressable functional junctions, which can be further integrated into larger micron-size circuits.[1-6] To date, the possibility of fabricating key computing devices from single-crossed semiconductor nanowires[3] and carbon nanotubes[4,5] has been impressively demonstrated.

Noble metal nanowires are also of interest for circuit assembly.[6-11] The related approach involves the self-assembly of composite nanowires with molecular-scale components embedded within, or grown as coatings on, metals.[2,6] To benefit from the gain in speed and power for the devices operating at the bottom end of the nanometer size regime, advanced interconnect structures with low-signal run-time delays are required. To this end, using high conductive metal nanowire interconnects is more beneficial than semiconductor nanowires and carbon nanotubes. Further reducing of the switching delays is expected from introducing low dielectric constant materials for insulating the nanowire interconnects.[12]

A wide variety of elemental (e.g., metal, carbon, silicon germanium) and compound (e.g., metal chalcogenides, oxides, III-V compounds) nanowires and nanotubes is now available with metallic, semiconducting, and insulating electronic properties. In general, these high-aspect-ratio nanoparticles can be prepared, either from the vapor or solution phase, by one-dimensional growth from a homo- or heterogeneous seed nanocluster or by replicating a template. Additionally, the template-directed methods allow relatively simple preparation of composite concentric structures from components with different electronic properties.[10,11,13-20] Among a variety of templates and template molding techniques, wet chemical replicating membranes with cylindrical pores can be singled out because of a number of advantages useful for electronic applications:

- The availability of robust uniform nanotubes and nanowires in large numbers ($\sim 10^8$ to 10^9 per membrane) and at low cost[10,11,21-26]
- Relatively easy access to mono- and multicomponent tubule-like and solid structures[9-11,21-26]
- The possibility of combining electrochemical and electroless techniques, which provides a precise control over all geometric parameters of the components with different electronic properties[7-11,14,24]
- The availability of concentric composite nanowire structures with open tips accessible to further functionalization and contact fabrication[10,11,14,24]

Nanotubes of different compositions can be grown as thin films on the walls of cylindrical pores of membranes using preparative techniques developed for planar ultrathin films. Conventional sol-gel methods, which involve immersing a membrane in a precursor sol followed by gelation inside the pores, have been used to prepare freestanding TiO_2, ZnO, WO_3, MnO_2, Co_3O_4, and SiO_2 [21,25,26] nanotubes. In this method, internal nanotube diameters can be adjusted by varying concentration and viscosity of the initial sol as well as the immersion time. However, precise control over thickness and morphology of the tubes, especially for those tubes that are only a few nanometers thick, has not been demonstrated yet.

More reliable control over quality of planar ultrathin films has been realized in wet layer-by-layer deposition methods, in which (1) preformed colloidal particles or polymers[27–31,37] or (2) molecular precursors[32–35] successively adsorb a layer at a time onto the growing surface. This is achieved when the alternate (1) adsorption or (2) adsorption and reaction steps are separated by the washing procedure in order to desorb weakly bound particles or molecules, forming additional layers. Film thickness is monitored by simply varying the number of deposition cycles.

The wet layer-by-layer assembly from molecular precursors is a liquid phase analog to the vapor phase atomic layer epitaxy.[36] Chemically, it is a surface analog to a bulk sol-gel synthesis and, therefore, is called the surface sol-gel (SSG) technique.[33] Ideally, the SSG method can limit each adsorption cycle to the extent of a single monolayer. However, in practice, thicker layers have been reported for planar oxide SSG films.[33–35] Nevertheless, SSG allows the finest control over film thickness because a subnanometer-thick layer is deposited in each two-step adsorption–hydrolysis cycle.

Very recently, we demonstrated applicability of both layer-by-layer techniques to template membranes with cylindrical pores. We prepared uniform and smooth freestanding semiconductor/polymer and silicon oxide nanotubes and nanotube-encapsulated metal wires.[10,11,24] This chapter summarizes our accomplishments in the synthesis and characterization of these nanostructures and considers perspectives of their application as nanoelectronic devices.

17.2 LAYER-BY-LAYER SELF-ASSEMBLY OF NANOPARTICLE/POLYMER NANOTUBES

Layer-by-layer synthesis of multilayer nanoparticle/polymer tubules on the membrane pore walls is shown schematically in Figure 17.1, route 1. A template membrane is immersed in a nanoparticle colloidal solution for 5 to 15 min, placed in a suction filtering unit, and washed with a solvent used for the colloid preparation. The membrane is then dried in an Ar stream and immersed in a polymer (e.g., polystyrenesulfonate (PSS)) solution for 5 to 15 min. The same washing procedure is done using deionized water, and the membrane is dried in an Ar stream. Vacuum filtering of the washings allows fast removal of excess reagents from the pores, thus preventing their blocking; drying the membrane in the Ar stream provides for convective transport of the nanoparticle sol and polymer solutions into the pores. Multilayer TiO_2/PSS and ZnO/PSS films were grown on the pore walls by repeating these steps.[10,11] Finally, these nanoparticle/polymer-filled anodic aluminum oxide (AAO) or polycarbonate (PC) membranes are dissolved in aqueous NaOH or methylene chloride, respectively, to give the freestanding nanotubes.

TEM images of multilayer TiO_2/PSS nanotubes prepared on the AAO membrane pore walls are presented in Figure 17.2A to C. The tubes prepared have an outside diameter of about 300 nm (the diameter of the pores) and a length ranging from 5 to 15 μm. The tube ends are always open and their length is less than the thickness of the template membrane (~60 μm). This implies that the tubes are broken during the membrane dissolution. A higher-resolution TEM image (Figure 17.2B and C)

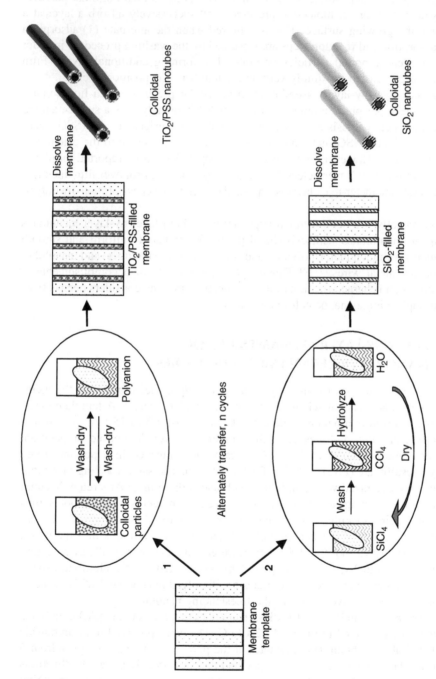

FIGURE 17.1 Scheme for the layer-by-layer synthesis of semiconductor/polymer (route 1) and SiO$_2$ (route 2) nanotubes on the pore walls of a membrane template.

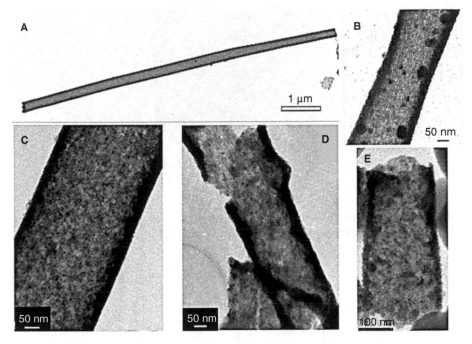

FIGURE 17.2 TEM images of $(TiO_2/PSS)_5TiO_2$ (A, C) and $(TiO_2/PSS)_9TiO_2$ (B) nanotubes prepared in ~300-nm-diameter pores of alumina membrane, and $(TiO_2/PSS)_5TiO_2$ (D) and $(ZnO/PSS)_3ZnO$ (E) nanotubes prepared in 200-nm-diameter (D) and 100-nm-diameter (E) pores of the PC membrane. In (B) the $(TiO_2/PSS)_9TiO_2$ nanotube was activated for electroless gold plating by successive adsorption of $SnCl_3^-$ and $Ag(NH_3)_2^+$ ions. (Adapted from Kovtyukhova, N.I. et al., *J. Phys. Chem. B*, 105, 8762, 2001; Kovtyukhova, N.I. et al., *Mater. Sci. Eng. C*, 19, 255, 2002. With permission.)

shows tube walls of uniform thickness composed of densely packed TiO_2 particles of ~5 nm in diameter (the particles' diameter in the stock TiO_2 solution). The thickness of the walls increases with the number of adsorption cycles and can be approximately estimated as 40 nm for a 10-layer TiO_2/PSS tube (the wall thickness was measured for metal-filled nanotubes, such as those shown in Figure 17.8D and E). This value is consistent with ellipsometric thickness (37 nm) of 10-layer TiO_2/PSS films self-assembled on planar substrates. The outside surface of the tubules is quite smooth, and its geometry completely follows the geometry of the pore walls. One can see round-ended branches on the tube walls, which reflect structure of the alumina membrane pores in their branched parts. These facts suggest strong adsorption of the first TiO_2/PSS layer, which completely covers the pore surface and gives rise to growing the uniformly thick and densely packed tube walls. An atomic force microscopy (AFM) image in Figure 17.3 shows that TiO_2 colloidal particles readily formed a well-packed monolayer on a plane Al/Al_2O_3 substrate.[37]

A Fourier transform infrared (FTIR) spectrum (Figure 17.4, spectrum 1) of the AAO membrane subjected to 10 TiO_2/PSS adsorption cycles reveals new bands at 2935, 2841, 2347, 1637, 1497, 1450, 1412, and 1127 cm^{-1}. These peaks are observed

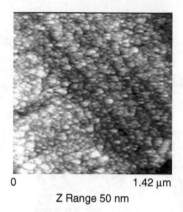

0   1.42 µm
Z Range 50 nm

FIGURE 17.3 AFM image of the first TiO_2 nanoparticle layer adsorbed on aluminum foil. (Adapted from Kovtyukhova, N.I. et al., *Thin Solid Films*, 337, 166, 1999. With permission.)

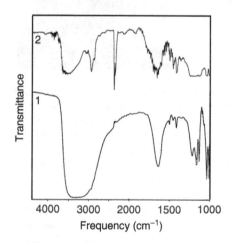

FIGURE 17.4 FTIR transmission spectra of (1) AAO membrane treated with 10 TiO_2/PSS deposition cycles and (2) freestanding PSS film.

in the spectrum of PSS (Figure 17.4, spectrum 2), which confirms PSS presence in the membrane. Apparently, adsorption of PSS polyanions on the positively charged (at pH < 6) TiO_2-terminated surface is driven by electrostatic interactions.

The TiO_2/PSS nanotube growth inside an alumina membrane was monitored by the mass uptake measurements (Figure 17.5, trace 1). After the first cycle, the membrane weight increases linearly with the number of deposited TiO_2/PSS bilayers, and 0.9 mg of the material is added in each cycle. This is consistent with the regular growth of the TiO_2/PSS multilayers on planar substrates (Figure 17.5, trace 2). Taking the average pore diameter at 300 nm and the pore density at ~10^9 cm^{-2}, one can estimate the surface area of an alumina membrane at ~1800 cm^2. An ideally packed monolayer of spherical (diameter, 5 nm) anatase particles adsorbed on this area should have a mass of ~2 mg, which is double the mass uptake per cycle observed

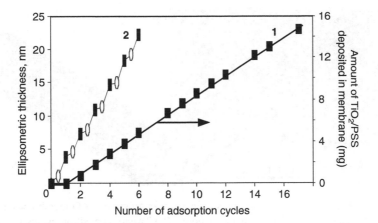

FIGURE 17.5 Plots of (1) amount of TiO_2/PSS deposited into AAO membrane and (2) ellipsometric thickness of a TiO_2/PSS film vs. the number of deposition cycles. Squares correspond to TiO_2-terminated surfaces. (1) Pore diameter 280 ± 20 nm. (2) PSS/TiO_2 film was assembled on a planar Au substrate primed with MEA.

experimentally. Increasing the TiO_2 concentration in the stock sol from 3 to 20 wt% does not result in higher mass uptake values. Close thickness values of the 10-layer TiO_2/PSS tube walls and planar films suggests that the lower-than-expected nano-particle content in the tube walls is not determined by the adsorption kinetics in the pores and is, rather, a characteristic of the TiO_2/PSS multilayers prepared as described above. Indeed, the average ellipsometric thickness of a TiO_2 monolayer in the planar films is 3.2 nm (Figure 17.5, trace 2), which is less than the average size of TiO_2 particles (~5 nm) in the stock sol.

Figure 17.2D and E shows multilayer TiO_2/PSS and ZnO/PSS tubes grown inside 200-nm and 100-nm pores of PC membranes.[11] These tubes also have densely packed walls with a thickness close to that of the tubes prepared inside alumina membranes. The wall thickness values are in reasonable agreement with ellipsometric thicknesses of planar films of the same composition. The tube walls are quite smooth and uniformly thick, and their geometry follows the geometry of the pore walls. As in the case of alumina membranes, we suggest strong adsorption of the first TiO_2/PSS and ZnO/PSS layers on the polycarbonate surface. However, unlike those grown in alumina, the in-PC-grown tubes are heavily damaged and their pieces are never longer than 1 to 1.5 μm. The breaking of the tubules is observed in all cases and may be due to the swelling of PC membranes in solutions used to prepare the tubes and dissolve the membrane.

17.3 LAYER-BY-LAYER SYNTHESIS OF SiO_2 NANOTUBES FROM MOLECULAR PRECURSORS

Silicon oxide nanotubes were grown inside AAO membranes, as shown schematically in Figure 17.1, route 2.[24] In the first step, $SiCl_4$ molecules are adsorbed on the hydrated surface of the alumina membrane. Subsequent washing with CCl_4 removes

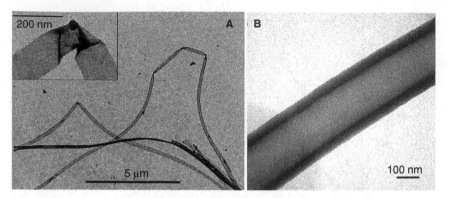

FIGURE 17.6 (A, B) TEM images of SiO_2 nanotubes grown in AAO membranes. (A) Five deposition cycles in 90 ± 20 nm diameter pores. (B) Twenty deposition cycles in 280 ± 20 nm diameter pores. (Adapted from Kovtyukhova, N.I. et al., *Adv. Mater.*, 15, 780, 2003. With permission.)

the unbound adsorbate molecules from the pores. In the second step, the adsorbed $SiCl_4$ is hydrolyzed to give SiO_2. Freestanding nanotubes were obtained by etching the alumina membranes in hot 50% H_2SO_4. Energy-dispersive x-ray (EDX) analysis of the product showed Si and O, with no detectable ($\leq0.5\%$) Cl or Al. Hence, the conversion of $SiCl_4$ to silica is complete, and no alumosilicate phase is present. This allows us to describe chemical composition of the oxide as $(SiO_2)x(SiOH)y$.

TEM images in Figure 17.6A and B show robust SiO_2 nanotubes with smooth and uniform walls. Their shapes clearly replicate the pore structure of AAO, including branches. Tubes 20 to 30 microns long can be found in optical micrographs (not shown), implying the growth of continuous tubules along the pore length. Figure 17.6A illustrates the remarkably high flexibility of these long silica nanotubes. A 100-nm-diameter tube grown in five deposition cycles does not break, even when bent at right angles. The external diameter of the tubes is determined by the pore diameter, and their internal diameter is adjustable by varying the number of deposition cycles or the concentration of $SiCl_4$. Nanotubes with external diameters ranging from 40 to 300 nm, and with wall thicknesses from 2 to 30 nm, were prepared. The dependence of the nanotube wall thickness (which was estimated from TEM images of metal-filled nanotubes (Figure 17.6A and B)) on the number of the deposition cycles is shown in Figure 17.7A, trace 1. The graph is not linear: the amount of SiO_2 deposited per cycle increases gradually as the tubes are grown. For example, under the conditions shown in Figure 17.7A, trace 1, the thickness change per cycle increases from 8.7 Å in the 4th to 10th cycles to 15.3Å in the 10th to 20th cycles. Both values exceed the film thickness increase (~3 Å) expected if only one Si-O-H monolayer is added per cycle. The shape of this film thickness graph closely resembles that of the mass uptake of SiO_2 (Figure 17.7A, trace 2). This correlation allows one to follow the process of nanotube growth by simply weighing the dry membrane after each SSG cycle.

The film growth at different $SiCl_4$ concentrations is shown in Figure 17.7B. At any $SiCl_4$ concentration, a steep rise is observed in the first deposition cycle. After this, the

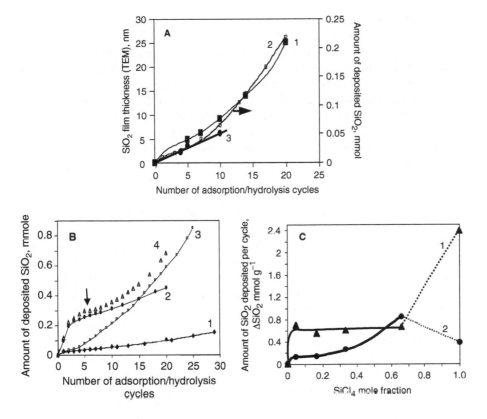

FIGURE 17.7 (A) Plots of SiO_2 (1, 3) wall thickness and (2) amount of SiO_2 deposited from a 67 mol% solution of $SiCl_4$ vs. the number of deposition cycles: pore diameter (1, 2) 280 ± 20 nm and (3) 90 ± 20 nm. (B) Amount of SiO_2 deposited vs. the number of deposition cycles for different concentrations of $SiCl_4$ in CCl_4: (1) 5 mol%, (2) 100 mol%, (3) 67 mol%, and (4) 100% $SiCl_4$ were used in the first 5 cycles and 67% $SiCl_4$ was used in the next 15 cycles; arrow shows the point at which the concentration was changed. Pore diameter, 280 ± 20 nm. (C) Amount of SiO_2 deposited per cycle on (1) alumina and (2) silica-covered membrane surface vs. $SiCl_4$ mole fraction. Pore diameter, 280 ± 20 nm; bare membrane weight, 0.033 ± 2% g. (Adapted from Kovtyukhova, N.I. et al., *Adv. Mater.*, 15, 780, 2003. With permission.)

film growth slows down significantly, but the amount of SiO_2 deposited per cycle gradually increases. Kleinfeld and Ferguson[38] attributed upward curvature in layer-by-layer growth of clay films to island nucleation and growth; however, this model is not consistent with the smooth morphology of SSG silica films seen in the TEM images.

Another possibility is that the upward curvature arises from occlusion of water in the growing film. Water is always present as a surface layer on hydrophilic surfaces under ambient conditions. The extent to which this surface-bound water will result in multilayer deposition depends on the amount of water present, the extent to which H_2O and CCl_4 molecules compete for interaction with $SiCl_4$, and the permeability of the deposited $[SiOCl_x(OH)_y]_n$ film to water, $SiCl_4$, and CCl_4. The coexistence of these three factors may cause the complex character of graphs shown in Figure

17.7B and C. The fact that multiple layers of SiO_2 are grown in each cycle indicates that at high concentrations, a multilayer of $SiCl_4$ molecules is present in the adsorption layer. In the subsequent washing step with CCl_4, all unreacted (unbound) $SiCl_4$ is removed. The following SSG step (immersion in water) completes hydrolysis of any remaining Si-Cl bonds and restores the water surface layer.

We suggest that the amount of SiO_2 deposited in each SSG cycle is determined mainly by the amount of $SiCl_4$ adsorbed in the first step rather than by the hydrolysis step. In this case, a plot of an SiO_2 amount deposited per cycle (ΔSiO_2) vs. $SiCl_4$ mole fraction can be considered a qualitative analog of an $SiCl_4$ adsorption isotherm. Figure 17.7C, traces 1 and 2, shows two such plots taken (1) for the first deposition cycle and (2) as an average for the range of 5th to 10th cycles, in which SiO_2 deposition is linear for all concentrations. The first isotherm (Figure 17.7C, trace 1) characterizes $SiCl_4$ adsorption on the alumina surface, whereas the second one (Figure 17.7C, trace 2) is related to silica surface adsorption because, according to the TEM data, the pore walls are completely covered with SSG film after five cycles. The low concentration region of both plots is concave to the concentration axis, which is characteristic of a strong interaction between the surface and adsorbate[39] and is consistent with the chemical reaction of $SiCl_4$ with H_2O-covered alumina and silica surfaces. It is evident from the low concentration region of the plots that the $SiCl_4$ interaction with the alumina–H_2O surface is stronger. Further flow of the adsorption isotherms for alumina and silica surfaces is different. On the silica–H_2O surface, the amount of SiO_2 deposited, as expected, increases gradually with the $SiCl_4$ mole fraction. However, on the alumina surface, the plot has a long plateau parallel to the concentration axis. We suggest that in the $SiCl_4$ mole fraction range of 0.05 to 0.67, fast formation of the $[SiOCl_x(OH)_y]_n$ layer at the alumina–H_2O–($SiCl_4$+CCl_4) interface occurs, and this layer is dense enough to block further penetration of $SiCl_4$ through the film. An increase in adsorption on the alumina surface is observed when the $SiCl_4$ mole fraction equals 1. Apparently, in the absence of CCl_4, the hydrolysis reaction can continue until all water in the surface layer is consumed and replaced by the $[SiOCl_x(OH)_y]_n$ layer. Hence, the amount of water available for hydrolysis of $SiCl_4$ determines its adsorption value. The fact that film growth is faster on the alumina pore walls (Figure 17.7C, trace 1) than on several layers of the SSG film (Figure 17.7C, trace 2) is consistent with the idea that the more polar alumina surface retains more water. Interestingly, in the plot of film grown on the silica–H_2O surface, a higher adsorption value is found at the $SiCl_4$ mole fraction of 0.67 than at 1 (Figure 17.7C, trace 2). Thus, the silica nanotube growth on the alumina pore walls can be described by the following reaction sequence:

Heat treatment (100°C, 3 to 13 h) of SiO_2-containing membranes dried at ambient temperature in Ar results in a weight loss of ~4%, which can be ascribed to the removal of unbound water from the SiO_2 film. This indicates that the SiO_2 film is relatively porous, and the pore fraction is estimated at ~0.09, taking SiO_2 density at 2.17 g cm^{-3}. This porosity is associated with relatively large pores and does not include micropores (if any) with radii approaching the thickness of an adsorbed water layer, because water in those pores can only be removed at higher temperatures. The porous structure of the SiO_2 film is consistent with the idea of occluded water assisting in multilayer film growth and is likely to be responsible for the gradual increase in the amount of SiO_2 deposited per cycle (Figure 17.7A and B) rather than narrowing the pores during the SSG procedure. The tube growth inside the 90-nm-wide pores is not found to be faster than that inside the 280-nm-wide pores (Figure 17.7A, traces 1 and 3).

The same percent weight loss is observed for initial $SiCl_4$ concentrations of 5, 16, 33, and 100 mol%, whereas for 67 mol% $SiCl_4$ solution, the weight loss is 5.6% after the 10th and 10% after the 20th deposition cycle. Accordingly, estimated porosity is 0.126 for the SiO_2 film deposited in 10 cycles, and it increases to 0.225 during the next 10 cycles. The porosity of the SiO_2 film measured under different growth conditions is consistent with the data shown in Figure 17.3. In a 100% $SiCl_4$ solution, the layer growth is actually slower than it is at 67 mol% $SiCl_4$ (Figure 17.7B, traces 2 and 3). The 100 mol% solution gives lower film porosity (0.09 after five deposition cycles), but upward curvature in film growth is found (Figure 17.7B, trace 4) if the concentration is lowered to 67 mol%, which provides higher porosity (0.111 after 15 cycles). These facts suggest that at relatively low CCl_4 concentration, thicker and more poorly ordered $[SiOCl_x(OH)_y]_n$ layers, which may occlude both H_2O and CCl_4 molecules, are formed at the adsorption step of SSG cycles. Thus, by an appropriate selection of synthesis conditions, one can control not only the thickness of nanotube walls but also their porosity.

17.4 SYNTHESIS OF NANOTUBE-ENCAPSULATED NANOWIRE DEVICES

While still in the membrane, the nanotubes can be electrochemically or chemically filled with metal and then released by etching the membrane to give freestanding wires coated with nanotube shells.[10,11,24]

Electrochemical metal plating inside the nanoparticle/polymer or SiO_2 tubes deposited on the walls of the AAO membrane was performed using a conventional electrochemical plating technique.[22] An alumina membrane with a thin Ag film evaporated on its branched side was used as a cathode in an electrochemical cell equipped with a Pt–wire anode and a saturated calomel electrode (SCE) reference electrode. First, 15- to 20-μm-long Ag wires were grown inside the membrane to fill the branched part of the pores. Then the tubes of desired composition and thickness were grown on the exposed pore walls, and Au nanowires were electroplated inside the tubes in the potentiostatic mode at –0.9V. Initially, the plating current is low and reaches its constant value in ~2 to 3 min, indicating that some

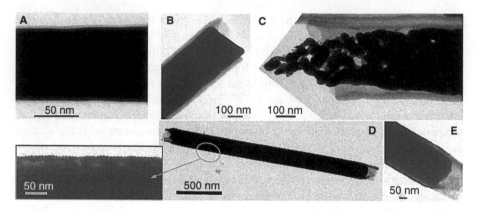

FIGURE 17.8 TEM images of Au wires grown inside the nanotubes. (A) SiO_2 nanotube prepared in five deposition cycles in 90 ± 20 nm diameter pores. (B, C) SiO_2 nanotube prepared in 20 deposition cycles in 280 ± 20 nm diameter pores. (D, E) $(TiO_2/PSS)_9TiO_2$ nanotubes prepared in 280 ± 20 nm diameter pores. (Adapted from Kovtyukhova, N.I. et al., *J. Phys. Chem. B*, 105, 8762, 2001; Kovtyukhova, N.I. et al., *Mater. Sci. Eng. C*, 19, 255, 2002; Kovtyukhova, N.I. et al., *Adv. Mater.*, 15, 780, 2003. With permission.)

amount of material used for the tubes' preparation remains stuck on tips of the Ag wires. Apparently, Au growth starts due to cracks or pores in such a deposit because the bottom part of the Au wires is typically branched (Figure 17.8C). This can be avoided by first plating a sacrificial metal segment.

We have observed that chemical composition of the pore walls influences the quality of the wires, which requires proper adjustment of the plating conditions.[10,11,24] TEM images of gold nanowires prepared in bare alumina membranes show that the growing end has an apparent hollow in the middle (cup shaped).[10] This fact can be understood in terms of adsorption of metal ions on the wall and preferential wetting of the polar alumina surface by the reduced metal. It was recently shown that Au(III) complexes are taken up by hydrous aluminum oxide by formation of Au-O-Al bonds. These adsorbed Au(III) ions were spontaneously reduced to the metallic state without added reducing agents.[40] If the same reaction takes place in alumina membranes, it will promote metal growth on the wall, leading to a cup-shaped tip.

The tip shape can be altered by passivation of the pore walls by alcoxy- or chloroalkylsilanes, which form strong Al-O-Si bonds through the interaction with Al-OH groups on the pore walls. For example, a membrane derivatized with octa-decyltrichlorosilane (OTS) is hydrophobic, and its transmission IR spectrum confirms the OTS adsorption on the pore walls and faces of the membrane. TEM images of Au nanowires grown inside this membrane show a tendency for the previously cup-shaped end to become pointed.[10]

The top ends of the wires prepared inside the SiO_2-coated pores are typically flat or convex (Figure 17.8B),[24] unlike cup-shaped ends of the nanowires grown in unmodified alumina membranes. The interaction of gold with the less polar silica pores is apparently weaker. Thus, the derivatization of the membrane pore walls to some extent allows one to smooth the surface of the wire ends, which facilitates their further functionalization.

Also, the Coulombic efficiency for metal plating depends on the pore walls' treatment. Under the same plating conditions, Au and Ni wires grow about 1.2 times longer in silica-modified pores than they do in unmodified alumina.

When electroplating inside the TiO_2/PSS tubules was conducted galvanostatically under conditions applying to bare AAO membranes, no solid rods of the desired several-micron length were obtained.[10] Instead, breaks inside the metal cores were observed along the tubule length, which finally resulted in short (0.3- to 1-μm) gold rods covered with TiO_2/PSS film. We would explain this by stronger adsorption of Au ions (and hence formation of metal crystallization centers) on the inner surface of TiO_2/PSS tubules vs. the Al_2O_3 surface of the membrane pores or higher electrical conductivity of tubule walls. Both factors can promote gold rod growth starting on such centers randomly distributed along the walls, which finally blocks Au ions' access to the bottom electrode and results in formation of short rod pieces inside the tubes. To overcome this problem, lower current density is required. This results in slower metal growth with preferential crystallization on the growing wire tip and avoids problems associated with charge mobility within the semiconductor-modified membrane walls.

The chemical Au plating inside the TiO_2/PSS tubules[10] was performed by slight modification of the method developed by Menon and Martin[41] for electroless Au plating in pores of polycarbonate membranes. Catalytic activation of the tubules surface was achieved by, first, adsorption of $SnCl_3^-$ ions on the positively charged TiO_2-terminated inner tubule surface. Successful adsorption was confirmed by a membrane color change from white to yellow. Then the membrane was immersed in aqueous silver ammoniate, washed with water, and dried in an Ar stream. It is known that treatment of Sn^{2+}-sensitized surfaces with $Ag(NH_3)_2^+$ complexes leads to reduction of silver cations to the metallic state and formation of Ag aggregates, on which gold crystallization starts.[42] Indeed, evenly distributed metallic nanoparticles of 5 to 50 nm in diameter are clearly seen by TEM on the surface of the TiO_2/PSS tubules (Figure 17.2C). The membrane was then placed in an Au plating bath for 4 days. To enable the gold precipitation inside the tubules rather than in the bulk solution, the bath temperature was kept in the range of 2 to 6°C. The length of the rods prepared this way ranges from 2 to 12 μm.

A typical TEM image of Au-filled $(TiO_2$/PSS$)_9TiO_2$ tubes (Figure 17.8D and E) shows solid metal rods with walls bearing a thin nanoparticle film. The film looks featureless and rather smooth. Its roughness, which is determined by the roughness of the pore walls, can be estimated as commensurable with the average size of starting TiO_2 colloidal particles, ~5 nm.

TEM images of SiO_2-coated Au nanowires (Figure 17.8A and B) show uniformly thick and smooth silica shells around the wires. The tube walls remain defect-free, and no metal penetration of the walls is seen. For both TiO_2/PSS and SiO_2 tube coatings, the nanowires released from the membrane have both ends open, because the tube shells tend to break within 50 nm of ends of the wires (Figure 17.8B and E). This leaves rod ends accessible for further derivatization by wet chemical or CVD techniques and allows one to easily make electrical contact by evaporating metal onto the wire ends.

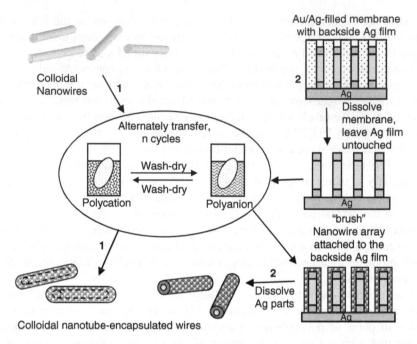

FIGURE 17.9 Scheme for the layer-by-layer synthesis of semiconductor/polymer films on top of nanowires released from a membrane (route 1) or arranged in brush (route 2). (Adapted from Kovtyukhova, N.I. and Mallouk, T.E., *Chem. Eur. J.*, 8, 4355, 2002. With permission.).

17.5 LAYER-BY-LAYER ASSEMBLY ON TOP OF NANOWIRE TEMPLATES

A scheme of the multilayer nanoparticle/polymer films assembly on the surface of template-grown nanowires is shown in Figure 17.9, route 1. An essential advantage of this strategy compared to that described above is its applicability to a wide range of inorganic, organic, and hybrid films, which can be destroyed under conditions applied to dissolve a membrane. This method also allows preparing of nanowires with the ends accessible for further functionalization. This can be achieved by sacrificial metal (e.g., Sn, Ni, Ag) layer deposition on Au rod tips while inside the membrane and its dissolution after the film has been grown on the rod surface, provided the film is stable toward chemicals used for dissolving the silver layer.

Layer-by-layer deposition of (TiO_2, ZnO)/PSS films on the surface of metal nanowires was performed in two ways.[10,11,14]

The first method consisted of gold nanowires being grown electrochemically inside a membrane and transferred into aqueous solution by dissolving evaporated silver and Ag plugs and alumina membranes. Then water was replaced by ethanol, and the gold surface was derivatized with mercaptoethylamine (MEA). Ethanolic ZnO and aqueous PSS were alternately added to the solution of concentrated (by centrifuging) suspension of the wires. Each 5-min adsorption cycle was followed by a centrifuging–washing cycle with three portions of the appropriate solvent, which

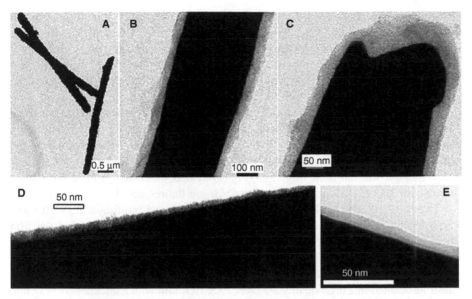

FIGURE 17.10 (A–C) TEM images of Au nanowires that were released from a membrane and subjected to 20 ZnO/PSS adsorption cycles. One can see the nanoparticle film completely covering the walls (B) and ends (C) of the rods. (D, E) TEM images of Au(MEA)(PSS/TiO$_2$)$_{10}$ (D) and Au(MEA)(PSS/PAH)$_{10}$ (E) nanowires prepared by the brush technique. (Adapted from Kovtyukhova, N.I. et al., *J. Phys. Chem. B*, 105, 8762, 2001; Yu, J.S. et al., *Chem. Commun.*, 2445, 2000. With permission.)

then was replaced by the solvent used in the next adsorption cycle. In this way, Au(NH$_2$)/(ZnO/PSS)$_{19}$ZnO nanowires were prepared.[10]

A TEM image of these nanowires shows the nanoparticle film completely covering the walls and ends of the rods (Figure 17.10A to C). The film thickness is estimated at ~40 nm, which is less than expected for the 3-nm-per-ZnO/PSS adsorption cycle observed for the film deposited on a planar Au substrate. Although the film is relatively uniform in thickness, its surface is rather rough and clumpy (see Figure 17.10A), which has never been observed for the planar Au substrate. Perhaps these clumps are formed from ZnO particles and PSS chains in solution (due to technical difficulties in complete removal of reagents during the washing–centrifuging stage), which then precipitate onto the film-covered rod surface.

An alternate way to deposit multilayer semiconductor/polymer or polymer films on the nanowire walls, thereby improving the washing procedure and avoiding clump formation, involves using a nanowire array attached to a silver base.[10,11,14]

Gold nanowires with 0.5 to 0.8 μm of silver on their top were grown electrochemically inside an alumina membrane. The Ag-covered back face of the metal-filled membrane was stuck to a glass slide, and the membrane was dissolved, leaving an Ag/Au/Ag nanowire array attached to the silver base. This Ag/Au/Ag brush was subjected to the typical procedure of layer-by-layer deposition of multilayer films. Briefly, the wires' surface was primed with MEA. On this positively charged Au(NH$_2$) surface, multilayer PSS/TiO$_2$[10,11] or PSS/polyallylamine hydrochloride

(PAH)[14] films were grown by alternately immersing the brush in solutions of the components according to the scheme presented in Figure 17.9, route 2. Importantly, during the film deposition procedure, the brush was always kept wet to prevent the silver base from deformation and breaking caused by drying. Finally, the Ag parts of the brush were dissolved in 1 M HNO_3, leaving $Au(NH_2)/(PSS/TiO_2)_{10}$ or $Au(NH_2)/(PSS/PAH)_{10}$ nanowires in colloidal suspension.

Dissolving the sacrificial Ag layer on the Au wire ends leaves them accessible for chemical functionalization. This was proved by reacting the freestanding $Au(NH_2)/(PSS/PAH)_{10}$ nanowires with MEA to prepare the NH_2-terminated tips' surface, which was then treated with fluorescein thiocyanate to tag this probe molecule through the thiourea link. Imaging so-treated nanowires with fluorescence microscopy shows that only the nanowire tips are fluorescent.[14]

Figure 17.10 depicts TEM images of Au wires bearing multilayer PSS/TiO_2 and PSS/PAH films obtained after the Ag base dissolution. One can see that the film prepared has a relatively rough surface. In the case of $Au(NH_2)/(PSS/TiO_2)_{10}$ wires, the film thickness ranges from 5 to 30 nm, and seldom-uncovered areas of the wire walls are also observed. It should be noted that layer-by-layer assembly on a planar Au substrate resulted in a high-quality 10-layer PSS/TiO_2 film with a thickness of ~37 nm. An experiment on treatment of this film with nitric acid (used for dissolving the Ag parts of the brush) showed that the film was not significantly destroyed. The poor quality of the $(PSS/TiO_2)_{10}$ film on the wire walls probably results from the fact that some Au/Ag wires in the brush are bent and can contact each other, thereby preventing nanoparticles and polymer chains from equal access to the rod surface. However, we believe that this technique can be applicable to relatively short wire brushes.

17.6 ELECTRICAL PROPERTIES

For electrical measurements, nanotube-coated metal nanowires were aligned between two gold pads on a lithographically patterned Si/SiO_2 substrate using an electrofluidic method described elsewhere.[43] Electrical contacts were fabricated by evaporating 200-nm-thick Au pads on top of the wire ends and 50-nm-thick Au or Ag lines on top of the tube coating. Figure 17.11 (bottom, left) shows an optical micrograph of this test structure and an SiO_2-tube encapsulated Au nanowire aligned for conducting electrical measurements. In some experiments, electrical contact with the tube shell was made by electrofluidically positioning the nanowires over the Au or Ag lines.

The I-V characteristics of SSG SiO_2 films of different thicknesses (measured in the $Au@SiO_2/Au$ configuration) are shown in Figure 17.11 (top). The curves show typical insulating behavior with breakdown voltages that increase linearly with film thickness (Figure 17.11, bottom). The hard breakdown field is estimated at 4.8 MV cm^{-1}, which is only slightly less than the breakdown fields (10 to 15 MV cm^{-1}) of the SiO_2 dielectric used in complementary metal-oxide semiconductor (CMOS) integrated circuit technology; this is a surprising result given the fact that a wet chemical deposition method was used.

Because the SiO_2 tubes are porous, their dielectric constant is expected to be lower than that of dense silica ($\varepsilon = 3.8$). Porous silica is a promising dielectric material for nanoelectronics applications because of its relatively low dielectric

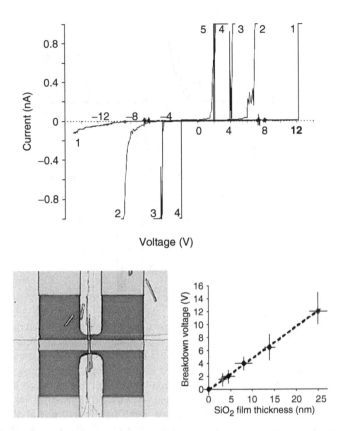

FIGURE 17.11 Top: I-V characteristics of SiO_2-coated gold nanowires with different thicknesses of SiO_2 (nm): (1) 25, (2) 14, (3) 8, (4) 4.5, and (5) 3.5. Bottom: Optical micrograph of a test structure for measuring the electrical properties of SiO_2-coated gold nanowires (left). A plot of the breakdown voltage vs. thickness (right). (Adapted from Kovtyukhova, N.I. et al., *Adv. Mater.*, 15, 780, 2003. With permission.)

constant.[44] Theoretical calculations predict a dielectric constant of 1.8 to 2.8 for a volume pore fraction of 0.4 and an almost linear decrease in ε with increasing porosity. In our experiments, it was difficult to measure the dielectric constant of the SiO_2 films precisely because of the uncertainty in the contact area in the Au@SiO_2/Au configuration. Nevertheless, it is interesting to note that the synthesis offers some control over porosity and that this will probably be reflected in the dielectric constant of the films.

A typical I-V curve for the Au(MEA)@(ZnO/PSS)$_{19}$ZnO/Ag structure, prepared as a concentric device in which the ZnO/PSS film covers the Au nanowire walls and is in contact with an evaporated silver electrode, reveals the expected current-rectifying behavior with turn-on potentials at ~−1.25 and 2.02 V for forward (Ag −) and reverse (Ag +) biases, respectively (Figure 17.12A). The difference between numerical values of the turn-on potentials is in reasonable agreement with the difference between the work functions of Ag (~4.6 eV) and Au (~5.3 eV). The operating voltage is also close to the ZnO band gap (3.28 eV, as estimated from the

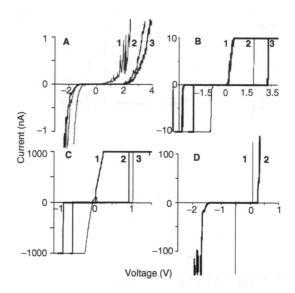

FIGURE 17.12 I-V characteristics of nanowire and planar devices measured in air at ambient temperature. (Adapted from Kovtyukhova, N.I. et al., *Mater. Sci. Eng. C*, 19, 255, 2002. With permission.) (A) Concentric $Au(NH_2)/(ZnO/PSS)_{19}ZnO/Ag$ device sweeps 1 and 2 start from 6 and –3 V, respectively; curve 3 was measured in two sweeps starting from 0 V each. (B) Concentric $Au/(TiO_2/PSS)_9TiO_2/Ag$ device sweeps 1 and 2 start from 4 and –3 V, respectively; curve 3 was measured in two sweeps starting from 0 V each. (C) Planar $Au(NH_2)/(TiO_2/PSS)_9$ TiO_2/Ag device, device area of ~9 μm^2, sweeps 1 and 3 start from 2 and –1 V, respectively; curve 2 was measured in two sweeps starting from 0 V each. (D) Planar (1) Au/PSS/Ag and (2) Au/TiO2/Ag devices, device area of ~9 μm^2, sweeps 1 and 2 start from 1 V.

UV-visible spectrum of starting ZnO sol), and both branches of this curve can be successfully linearized using the Schottky equation ($\ln(I) \propto V$). These observations indicate that rectification is determined by charge injection at the metal/ ZnO/PSS–film interface rather than by a tunneling mechanism, although the I-V curve shows quite satisfactory linearity in Fowler–Nordheim {$\ln(I/V^2) \propto 1/V$} coordinates as well.[10]

An I-V characteristic of a concentric $Au@(TiO_2/PSS)_9TiO_2/Ag$ device is shown in Figure 17.12B. The forward (Ag –) and reverse (Ag +) bias turn-on potentials are ~–2.45 and 3.25 V (0.8 V), respectively. For comparison purposes, an I-V characteristic of the same structure prepared on a planar Au substrate is also presented (Figure 17.12C). (This planar device was prepared by the layer-by-layer self-assembly of a 10-layer PSS/TiO_2 film on a Au substrate primed with MEA, followed by evaporating a top Ag electrode). One can see that both curves are rather similar in their shape, showing a drastic current increase at turn-on potentials. This likeness suggests similar composition and structure of TiO_2/PSS films self-assembled on the planar substrate and inside the alumina membrane pores.

Although for the nanowire device the difference between the turn-on potentials is close to the difference between the work functions of Ag and Au, the charge transport in these TiO$_2$/PSS thin-film devices cannot be described by either the Schottky or Fowler–Nordheim equation and is a subject for further investigations.

Despite the I-V curve shape being similar for both the nanowire and planar Ag/(TiO$_2$/PSS)$_{10}$TiO$_2$/Au devices, numerical values of the turn-on potentials are different (–0.55 V and +0.92V, Δ0.37 V, for the planar device). One of possible explanations for this may be found in the different chemical composition of interface layers, resulting from the device preparation techniques. For the nanowire device, the Au electrode was electroplated into the TiO$_2$/PSS nanotube that, in the ideal case, implies homogeneous Au/TiO$_2$ contact. For the planar device, the TiO$_2$ particle layer is separated from the Au substrate surface by the priming SAM of MEA and first PSS layer. However, the layer-by-layer synthetic technique does not allow one to control the lateral structure of multilayer films. Due to phase boundaries in the organic and inorganic layers (at least in the case of three-dimensional particles), the formation of inhomogeneous contacts, in which both the semiconductor and polymer touch the metal surface, is possible. Also, interface Au species for electroplated and evaporated electrodes may be different. It was found that electrochemical metal deposition and metal evaporation or sputtering resulted in diodes with different barrier heights.[45] Without knowing the details of the interfacial composition, it is difficult to arrive at a complete understanding of the electrical properties.

In some cases, the I-V curve shape suggests breakdown characteristic of insulators, but such breakdowns are quite reproducible in four to six successive voltage sweeps, and this is not usually observed for breakdown in insulators. The comparison between the I-V curves of the Au/(TiO$_2$/PSS)$_9$TiO$_2$/Ag structures (Figure 17.12B and C) and bare Au/TiO$_2$/Ag and Au/PSS/Ag (Figure 17.12D) suggests that the PSS plays a role in the abrupt transition from a nonconducting to a conducting state. The I-V curve of the bare Au/PSS/Ag device shows breakdown at relatively low potentials (~+0.26 and 0.16 V) for both forward and reverse bias modes. It is interesting to note that no breakdowns were observed when polyaniline (PAN) or PAH was used instead of PSS in planar Au(TiO$_2$/polymer)Ag devices.

Among all the polymers investigated by us (PAN, PAH, polyacrylic acid, poly(diallyldimethylammoniumchloride), and polyethylenimine) PSS provides the most regular growth of multilayer TiO$_2$/polymer films, but its electrical properties do not allow the consideration of PSS as the best candidate for diode preparation.

The I-V characteristics of all nanowire and planar Au(MEA)/(TiO$_2$/PSS)$_9$TiO$_2$/Ag devices are quite reproducible, provided a 5-min period is allowed between the sweeps. Shorter periods (1 to 2 min) lead to significantly higher currents and lower turn-on potentials for both forward and reversed bias directions in the second and following sweeps.

This hysteresis is shown in Figure 17.12B for the nanowire device. When the I-V curves are obtained in single sweeps starting from either negative or positive potentials, enhanced asymmetry is seen compared to curves measured in two sweeps starting from 0 V each. The same tendency is found for planar Au(MEA)/(TiO$_2$/PSS)$_9$ TiO$_2$/Ag devices (Figure 17.12C) and bare Au/TiO$_2$/Ag and Au/PSS/Ag structures (Figure 17.12D). One possible explanation for this hysteresis is the localization of injected charge carriers at interface and surface states. Saturation and emptying of surface traps is a slow process that can determine the current flow on this timescale.

This idea is supported by the observation of current immediately after switching the voltage and the subsequent decrease in current by several orders of magnitude. The nature of the traps is not understood at present; they cannot be unambiguously assigned to either TiO_2 particles or PSS chains, because both bare $Au/TiO_2/Ag$ and $Au/PSS/Ag$ structures show similar behavior.

It is interesting to note that the hysteresis for an $Au(MEA)/(ZnO/PSS)_{19}ZnO/Ag$ structure prepared as a concentric nanowire device, although it exists, is much less pronounced than that of the TiO_2/PSS nanowire and planar devices. One can assume that surface states of the semiconductor particles affect the electrical properties of semiconductor/polymer films.

Although the electrical properties of this device are not understood, we note that such hysteresis can be useful in memory devices if persistent high and low conductivity states can be attained at the same potentials.

Although the device characteristics shown here are poor by the standards of the semiconductor industry, it is important to note that only primitive layer sequences (simple polymer/colloid repeats) were used.

17.7 COMBINED IN-MEMBRANE AND ON-WIRE ASSEMBLY OF NANOWIRE DEVICES

To allow better rectification ratios, lower turn-on voltages, and more complex device functions, more structured film shells, in which asymmetry is built in (e.g., with electron- and hole-conducting segments), should be assembled around a metal nanowire. Gaining precise control over the structure, roughness, and thickness of these complex multicomponent shells should be addressed to prepare high-quality interfaces with reproducible electrical properties.

Among the above-described templated layer-by-layer assembly methods, the in-membrane synthesis on the pore walls gives the best-quality, evenly thick tube shells, with the roughness of the outside walls determined by quite smooth alumina pore walls. Metal plating inside the tubes provides immediate metal contact with the appropriate n- or p-conducting particles or molecules and avoids depositing additional priming SAM on the metal surface. However, synthesis in AAO membranes is restricted to components stable under the conditions of the membranes' etching. To extend the range of the components used, one can combine the in-membrane synthesis of tube-coated nanowires (see scheme in Figure 17.1, route 1) with the in-solution deposition of additional layers on the tube walls after releasing the wires from a membrane (Figure 17.9, route 1).

Using this approach, multicomponent $Au@(W_{12}O_{41}/TiO_2)_3W_{12}O_{41}/(PAN/SWNT)_3$ PAN/Au nanowire devices (~300 nm in diameter single-walled carbon nanotubules [SWNT]) have recently been prepared and characterized.[46] The I-V curves (not shown) are asymmetric with turn-on potentials at 1.1 and −2.0 V. The current rectification ratio of ~100 is observed at 2.7 V. The I-V characteristics of both nanowire $Au@(W_{12}O_{41}/TiO_2)_3W_{12}O_{41}/Au$ and planar $Au/(PAN/SWNT)_{19}$ PAN/Au devices are symmetric and the latter one is almost linear in the current range from −1 to 1 nA. These facts suggest the formation of the TiO_2/PAN n-p junction mainly responsible for the current rectification. Our current research addresses understanding the charge transport mechanism and improving electrical characteristics of these diodes.

17.8 CONCLUSIONS

Layer-by-layer assembly of nanoparticles or molecular precursors can be realized in the cylindrical pores of the AAO membrane to prepare robust inorganic oxide and semiconductor/polymer nanotubes. The thickness and porosity of the tubes can be precisely controlled by varying the number of adsorption cycles and the composition of precursor solutions. Because the layer-by-layer thin-film deposition technique is well developed for other classes of materials, including metal oxides, chalcogenides, and phosphates, this method offers a unique route to constructing a wide variety of concentric multicomponent nanostructures with the desired layer thickness and, hence, tunable electrical and optical properties.

It is possible to make nanotube-encapsulated metal nanowires by electrochemical or electroless filling of the tubes with metals. In this way, simple devices that act as rectifiers, as well as insulated metal interconnects, have been realized.

Combining the in-membrane and in-solution layer-by-layer assembly techniques allows preparation of metal nanowires inside structured multicomponent thin-film shells with built-in p–n junctions. The electrical properties of all of the on-wire devices prepared are similar to those of planar thin-film devices of the same composition with evaporated top metal contacts. This means that both organic and nanoparticle components can be introduced into nanowire devices without qualitative changes in their electrical, and most probably chemical, properties.

Thus, template layer-by-layer synthesis is an experimentally simple and manufacturable route to multicomponent nanowires with well-controlled physical and chemical parameters. The availability of concentric rectifying films opens up the possibility of constructing logic gates and cross-point devices by means of electrofluidic or chemically driven nanowire assembly.

ACKNOWLEDGMENTS

I am grateful to T.E. Mallouk for his deep interest in and guidance for this research. I thank my colleagues B.R. Martin, J.-S. Yu, J.E. Kim, S. Lee, J.K.N. Mbindyo, P.A. Smith, M. Cabassi, and T.S. Mayer, who contributed to various aspects of this work. This work was supported by the DARPA "Moletronics" program and the Office of Naval Research.

REFERENCES

1. G Tseng, J Ellenbogen. *Science* 294:1293, 2001.
2. SC Goldstein, M Budiu. In *Proc. 28th Ann. Int. Symp. Computer Architecture 2001*, Göteborg, Sweden. ACM Press, New York, 2001, p. 178.
3. Y Huang, X Duan, Y Cui, L Lauhon, K Kim, CM Lieber. *Science* 294:1313, 2001.

4. T Rueckes, K Kim, E Joselevich, G Tseng, C-L Cheung, CM Lieber. *Science* 289:94, 2000.
5. A Bachtold, P Hadley, T Nakanishi, C Dekker. *Science* 294:1317, 2001.
6. NI Kovtyukhova, TE Mallouk. *Chem Eur J* 8:4355, 2002.
7. JKN Mbindyo, TE Mallouk, JB Mattzela, I Kratochvilova, B Razavi, TN Jackson, TS Mayer. *J Am Chem Soc* 124:4020, 2002.
8. I Kratochvilova, M Kocirik, A Zambova, J Mbindyo, TE Mallouk, TS Mayer. *J Mater Chem* 12:2927, 2002.
9. DJ Peña, JKN Mbindyo, AJ Carado, TE Mallouk, CD Keating, B Razavi, TS Mayer. *J Phys Chem B* 106:7458, 2002.
10. NI Kovtyukhova, BR Martin, JKN Mbindyo, PA Smith, B Razavi, TS Mayer, TE Mallouk. *J Phys Chem B* 105:8762, 2001.
11. NI Kovtyukhova, BR Martin, JKN Mbindyo, TE Mallouk, M Cabassi, TS Mayer. *Mater Sci Eng C* 19:255, 2002.
12. HS Nalwa, Ed. *Handbook of Low and High Dielectric Constant Materials and Their Applications*, Vol. 1. Academic Press, New York, 1999.
13. VM Cepak, JC Hulteen, G Che, KB Jirage, BB Lakshmi, ER Fisher, CR Martin. *Chem Mater* 9:1065, 1997.
14. JS Yu, JY Kim, S Lee, JKN Mbindyo, BR Martin, TE Mallouk. *Chem Commun* 2445, 2000.
15. SO Obare, NR Jana, CJ Murphy. *Nano Lett* 1:601, 2001.
16. G Schnur, R Price, P Schoen, P Yager. *Thin Solid Films* 152:181, 1987.
17. E Braun, Y Eichen, U Sivan, G Ben-Yoseph. *Nature* 391:775, 1998.
18. R Yu, L Chen, Q Liu, J Lin, K Tan, S Ng, H Chan, G Xu, T Hor. *Chem Mater* 10:718, 1998.
19. Q Fu, C Lu, J Liu. *Nano Lett* 2:329, 2002.
20. X Shi, S Han, RJ Sanedrin, C Galvez, DJ Ho, B Hernandez, F Zhou, M Selke. *Nano Lett* 2:289, 2002.
21. BB Lakshmi, CJ Patrissi, CR Martin. *Chem Mater* 9:2544, 1997.
22. (a) D Al-Mawlawi, CZ Liu, M Moskovits. *J Mater Res* 9:1014, 1994. (b) M Nishizava, VP Menon, CR Martin. *Science* 268:700, 1995.
23. M Wirtz, M Parker, Y Kobayashi, CR Martin. *Chem Eur J* 8:3573, 2002.
24. NI Kovtyukhova, TE Mallouk, TS Mayer. *Adv Mater*, 15:780, 2003.
25. M Zhang, Y Bando, K Wada. *J Mater Res* 15:187, 2000.
26. BB Lakshmi, PK Dorhout, CR Martin. *Chem Mater* 9:857, 1997.
27. RK Iler. *J Colloid Interface Sci* 21:569, 1966.
28. G Decher. *Science* 277:1232, 1997.
29. VL Colvin, AN Golstein, AP Alivisatos. *J Am Chem Soc* 114:5221, 1992.
30. J Fendler. *Chem Mater* 8:1616, 1996.
31. TE Mallouk, H-N Kim, PJ Ollivier, SW Keller. In *Comprehensive Supramolecular Chemistry*, Vol. 7, G Alberti, T Bein, Eds. Elsevier Science, Oxford, 1996, pp. 189.
32. YF Nicolau, JC Menard. *J Crystal Growth* 92:128, 1988.
33. I Ichinose, H Senzu, T Kunitake. *Chem Mater* 9:1296, 1997.
34. M Fang, CH Kim, BR Martin, TE Mallouk. *J Nanoparticle Res* 1:43, 1999.
35. NI Kovtyukhova, EV Buzaneva, CC Waraksa, BR Martin, TE Mallouk. *Chem Mater* 12:383, 2000.
36. T Suntola. *Mater Sci Rep* 4:261, 1989.
37. NI Kovtyukhova, PJ Ollivier, S Chizhik, A Dubravin, EV Buzaneva, AD Gorchinskiy, A Marchenko, NP Smirnova. *Thin Solid Films* 337:166, 1999.
38. ER Kleinfeld, GS Ferguson. *Chem Mater* 8:1575, 1996.

39. GD Parfitt, CH Rochester, Eds. *Adsorption from Solution at the Solid/Liquid Interface.* Academic Press, London, 1983.
40. T Yokoyama, Y Matsukado, A Ichida, Y Motomura, K Watanabe, E Izawa. *J Colloid Interface Sci* 233:112, 2001.
41. VP Menon, CR Martin. *Anal Chem* 67:1920, 1995.
42. GO Mallory, JB Hajdu, Eds. *Electroless Plating: Fundamentals and Applications.* American Electroplaters and Surface Finishers Society, Orlando, FL, 1990.
43. PA Smith, CD Nordquist, TN Jackson, TS Mayer, BR Martin, JKN Mbindyo, TE Mallouk. *Appl Phys Lett* 77:1399, 2000.
44. C Jin, JD Luttmer, DM Smith, TA Ramos. *MRS Bull* 22:39, 1997.
45. W Mönch. *Surf Sci* 299/300:928, 1994.
46. NI Kovtyukhova, TE Mallouk. *Adv Mater* 17:187, 2005.

18 Coherent Plasmon Coupling and Cooperative Interactions in the Two-Dimensional Array of Silver Nanoparticles

George Chumanov and Serhiy Z. Malynych

CONTENTS

18.1 Introduction 488
18.2 Self-Assembly of Two-Dimensional Arrays of Ag Nanoparticles...........488
    18.2.1 Fabrication Methods for Metal Nanoparticle Arrays 488
    18.2.2 Self-Assembly of Ag Nanoparticles on a PVP-Modified
        Flat Surface 489
    18.2.3 Stabilization of Ag Nanoparticle Arrays by Embedding
        into the Polymer Matrix...492
18.3 Coherent Plasmon Coupling between Ag Nanoparticles 494
    18.3.1 Optical Properties of Interacting Nanoparticles 494
    18.3.2 Extinction Spectra of Ag Nanoparticles Arranged into
        Two-Dimensional Arrays 496
    18.3.3 Effect of Interparticle Spacing on the Coherent Plasmon
        Coupling 498
    18.3.4 Angular Dependence of the Coherent Plasmon Coupling.........500
    18.3.5 Effect of External Dielectric Medium 502
    18.3.6 Substrate Effects on Optical Properties of Ag Nanoparticle
        Arrays 503
18.4 Conclusions 506
Acknowledgments..507
References..507

18.1 INTRODUCTION

Continuously growing interest in noble metals' nanoparticles, which started in 1857 by M. Faraday,[1] is largely driven by unique optical properties associated with the excitation of the collective oscillations of the electron density, termed *plasmon resonances*, or *plasmons*. When confined to nanosize dimensions, plasmons can be excited by visible light and their frequency effectively tuned by varying the size and dielectric environment of the particles.[2–7] The excitation of plasmon resonances in nanoparticles induces a local electromagnetic field in a thin layer next to the surface. The intensity of this local field is enhanced by several orders of magnitude, compared to the incident light, resulting in the enhancement of Raman and elastic scattering,[8–11] one- and two-photon fluorescence,[12] visible and infrared absorption of light,[13,14] second harmonic generation and other nonlinear processes,[15,16] photoelectrochemistry, etc. for molecules placed in the close vicinity of nanoparticles.[17,18] The discovery of this enhancement, in particular the surface-enhanced Raman scattering in 1976, and the prospect associated with potential applications instigated a new wave of interest in synthesis, study, and utilization of metal nanoparticles.[19–22]

The excitation of the plasmon resonances in silver and gold nanoparticles represents the most efficient mechanism by which light interacts with matter. In fact, the cross section for plasmon excitation in 50-nm silver particles could be several orders of magnitude larger than the absorption cross section of the same-dimension particles composed of any other organic or inorganic chromophore. Three major factors contributing to such high efficiency for the interaction with light are the large density of conducting electrons, the unique frequency dependence of the real and imaginary parts of the dielectric function, and the size confinement to the dimension being smaller than the mean free path for electrons. High efficiency for the interaction with light and tunability of optical properties as well as robustness, in particular toward high-intensity radiation, open new perspectives for using metal nanoparticles as building blocks for various optical and opto-electronic devices.[23–26]

18.2 SELF-ASSEMBLY OF TWO-DIMENSIONAL ARRAYS OF Ag NANOPARTICLES

18.2.1 FABRICATION METHODS FOR METAL NANOPARTICLE ARRAYS

Ensembles of noble metal nanoparticles organized in two-dimensional and three-dimensional regular structures are especially interesting because their optical properties are determined by various interactions between nanoparticles, such as direct electron coupling, near-field interaction, and far-field diffraction of light. The understanding and controlling of these interactions provide extensive possibilities for designing new optical materials and photonic devices. Various methods for the fabrication of two-dimensional nanoparticle arrays have been developed in the last 20 years. Microlithography[27] or nanosphere lithography[28] combined with the vacuum deposition of metals is an inexpensive, high-throughput technique providing the control of the particles' size, shape, and interparticle spacing. The obtained particles have an initially pyramid-like shape but can be reshaped to spheroids by annealing.

Typically, defect-free domains are obtained in the 10- to 100-μm^2 range.[29] Nanometer metal particles can also be arranged on charged substrates by direct deposition from the gas phase.[30] Manipulation of nanoparticles with an atomic force microscope (AFM) allows their positioning in any arbitrary arrangement on a surface; however, this is a low-throughput technique that cannot be used for large-scale production.[31,32] Monodispersed two-dimensional superlattices of nanoparticles have been synthesized by the method of the diffusion of metal atoms deposited by pulsed-laser ablation into the first layer of a mesoporous silica film.[33] Large arrays of metal nanoparticles have also been synthesized within mesoporous silica[34,35] or in polymer matrices.[36]

Self-assembly is a widely used method for the fabrication of a two-dimensional array of nanoparticles due to its simplicity, low cost, high throughput, and the possibility to form arrays on nonflat surfaces.[37–40] The self-assembly process requires the nanoparticles to have affinity to a substrate as well as a repulsive or attractive interaction with each other. Whereas sizable electrostatic repulsion can be obtained between electrical double layers surrounding metal nanoparticles in low-ionic-strength media, surface modifications of substrates with functional groups such as thiol, pyridyl, amino and carboxyl, provide the required affinity. Thiol-containing silanes have been used to immobilize silver, gold, and other metal nanoparticles on glass, quartz, silicon, indium/tin oxide (ITO) glass, and other metal oxide surfaces.[41–44] The thiol group forms a covalent bond with metals, resulting in the attachment of nanoparticles to the surface. It is important to emphasize that even though the thiol and other groups form strong bonds with noble metals, nanoparticles are not firmly immobilized on the substrates. They can diffuse on the surface, forming either clusters (two-dimensional aggregates) or two-dimensional arrays with semi-regular interparticle spacing, depending on whether electrostatic repulsion is sufficient to overcome attractive Van der Waals and exchange forces between nanoparticles. This diffusion does not require additional energy into the system because the break of existing bonds and the formation of new ones with the surface take place simultaneously when nanoparticles roll on the substrate.

Linear polycations and polyanions have been used to form alternating multilayers of metal, semiconductor, and dielectric nanoparticles via electrostatic attraction.[45–50] Hydrogen bonding and charge transfer interactions have also been exploited for the formation of various nanostructure assemblies.[51,52] Polymers can be very efficient surface modifiers for the self-assembly of various nanoparticles because each molecule provides many binding sites that can simultaneously interact with particles and a substrate. We used the neutral form of poly(vinylpyridine) (PVP) to modify glass, quartz, ITO, and other metal oxide substrates to produce single-layered assemblies of Ag, Au, and other metal, metal oxide, and semiconductor nanoparticles.[53]

18.2.2 SELF-ASSEMBLY OF AG NANOPARTICLES ON A PVP-MODIFIED FLAT SURFACE

The pyridyl group in the neutral form develops a strong covalent bond to metals through the donation of the lone pair of electrons of the nitrogen atom. The formation of this σ-bonding is most efficient when the aromatic ring adopts the perpendicular orientation to the metal surface. For the parallel orientation of the aromatic ring, π-stacking interaction can take place. Dilute polymer solutions favor the flat orientation of the aromatic ring, whereas high concentrations yield predominantly σ-bonding and normal orientation, with the nitrogen atom down to the metal surface. Both normal and flat orientations have been confirmed by surface-enhanced Raman spectroscopy studies of PVP on silver.[54–56] Pyridyl groups can also interact with various nonmetallic polar surfaces terminated with amines, carboxyl, hydroxyl, and other groups capable of undergoing hydrogen bonding. The surface of glass or silica in dry nonaqueous solvents is terminated with protonated silanol groups that can form hydrogen bonds with the nitrogen atom on the pyridyl. Addition of water will cause deprotonation of silanol groups (pK = 3 to 3.4)[57] and possible protonation of the pyridyl groups (pK = 5 to 5.2), resulting in electrostatic or charge–dipole type interaction between the polymer and the surface. The same is true for other oxide and polar surfaces. The interaction on individual pyridyl groups with the surface does not have to be strong to provide overall strong adhesion of PVP, because the polymer molecules have many groups that simultaneously interact with the substrate, providing an entropic advantage for the quasi-irreversible adsorption of the polymer to the surface, even in the presence of strong solvents. Adsorbed PVP molecules retain numerous unbound pyridyl groups capable of binding nanoparticles. Essentially only one type of functionality (pyridyl group) is required for the adhesion to surfaces and binding nanoparticles. Because of the strong adsorption to many metal and nonmetal surfaces via the same group, the PVP can be used as a highly efficient monofunctional surface modifier for the assembling of different nanoparticles on various substrates.

Poly(2-vinylpyridine) (M_w = 159,000 and 37,500) and poly(4-vinylpyridine) (M_w = 160,000) were dissolved in reagent alcohol, and the solution was used for the preparation of polymer films on various substrates by adsorption, spin-coating, and dip-coating methods. Prior to the modification, glass and quartz substrates were cleaned in piranha solution or in oxygen (air) plasma. Both cleaning methods yielded adequate results.

The most uniform polymer films across large areas were obtained by the method of adsorption for 1 to 2 h from an ethanolic solution. The concentration of the polymer in the solution in the range of 0.01 to 3% does not have a profound effect on the quality of Ag nanoparticle arrays, despite the fact that UV-visible (UV-Vis) absorption measurements indicate different amounts of PVP adsorbed on substrates from solutions with various polymer concentrations (Figure 18.1). After the modification with PVP, the substrates should be thoroughly rinsed in alcohol to remove all weakly bound polymer molecules, leaving only a monolayer directly adhered to the surface. PVP-modified substrates can be annealed at temperatures slightly above the melting temperature of the polymer to improve surface reproducibility. Exposure of PVP-modified substrates to an aqueous suspension of silver nanoparticles results in the formation of two-dimensional nanoparticle arrays. The surface density of nanoparticles strongly depends on exposure time and the concentration of nanoparticles in the suspension until it reaches saturation. At saturation, the nanoparticles self-assemble into two-dimensional semiregular arrays, with the interparticle distance comparable to their diameter (Figure 18.2).

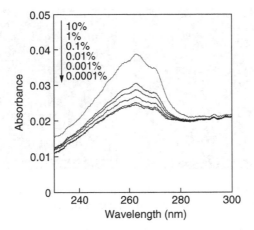

FIGURE 18.1 Absorption spectra of PVP films adsorbed on both sides of quartz substrates from alcohol solutions of different concentrations.

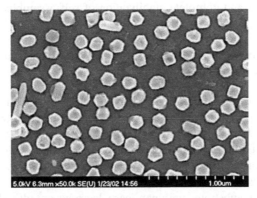

FIGURE 18.2 SEM image of silver nanoparticles immobilized on PVP-modified, ITO-coated glass.

As was mentioned above, the ionic strength of the suspension should be low for the long-range electrostatic repulsion that provides uniform spacing between nanoparticles. Drying substrates, placing them into less polar solvents, or increasing the ionic strength of an aqueous solution leads to collapsing of nanoparticle assembly and the formation of aggregates on the surface (Figure 18.3A). Presumably, large surface tension forces arising from the drying of the thin water layer drag particles together into aggregates. The stability of the arrays can be improved to a limited extent by modifying the surface of metal nanoparticles in the arrays with polymers (e.g., PVP) or molecules with long aliphatic chains. Drying of these modified arrays normally results in a smaller change in the UV-Vis spectra relative to the same unmodified arrays. More improvement of the stability was observed after coating the arrays with 5- to 20-nm-thick sol-gel silica layers. However, the sol-gel chemistry

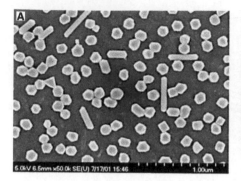
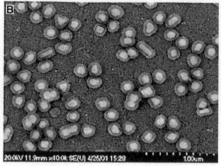

FIGURE 18.3 (A) SEM image of Ag nanoparticles immobilized on PVP-modified, ITO-coated glass. (B) The same for silica-coated Ag nanoparticles.

by itself can cause the aggregation of the particles on the surface (Figure 18.3B). In order to achieve the ultimate stability required for various device applications, the nanoparticles must be permanently immobilized on substrates. We developed a technique based on embedding of nanoparticles into a polymer matrix that provides chemical protection and complete stabilization of self-assembled two-dimensional arrays.[58]

18.2.3 STABILIZATION OF AG NANOPARTICLE ARRAYS BY EMBEDDING INTO THE POLYMER MATRIX

In this method, the two-dimensional arrays of self-assembled nanoparticles are embedded into a poly(dimethylsiloxane) (PDMS) (Sylgard® 184, Dow Corning Corp.) resin film. Unique properties of this elastomer, such as flexibility, large tunability range of Young's modulus, ability to be easily molded (soft lithography),[59,60] and conformability to complex shapes, make it attractive in many applications of nanomaterials.

To avoid drying, substrates with nanoparticle arrays are placed horizontally into a container with a water–alcohol mixture, and the resin mixed with the curing agent (10:1) is directly poured onto the surface of the substrate. By varying the concentration ratio of water and alcohol, the specific density of the mixture can be adjusted slightly below that of the polymer to provide "a soft landing" of the resin on top of the nanoparticles. The resin replaces the solvent layer from the surface; thus, drying of the substrate is completely avoided and nanoparticle assemblies remain preserved in their initial state. After the completion of the polymerization, the PDMS film containing embedded nanoparticles can be peeled off the substrate (Figure 18.4). These very stable and robust *immobilized nanoparticle assemblies* (INAs) can be easily handled without danger of aggregation and, after coating with another PDMS layer, safely exposed to different chemical environments. In addition to the complete stabilization of the nanoparticle assemblies, the PDMS resin provides an excellent matrix for various optical applications due to its outstanding transparency down to 300 nm.

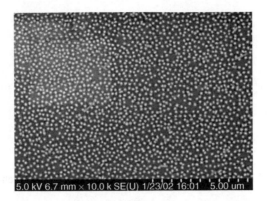

FIGURE 18.4 Scanning electron micrograph of coupled INA film. The mosaic structure is due to the thin platinum layer deposited on top of the film for conductivity.

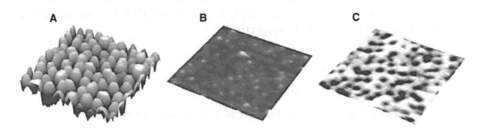

FIGURE 18.5 AFM images of Ag nanoparticles immobilized on a PVP-modified silicon wafer (A), coupled INA film (B), and nanocavities obtained after etching INA film (C). Scan range is 5 × 5 μm for all images. Vertical scale is the same for images (A) and (B).

Noncontact AFM measurements of INA films reveal that only a small part of each silver nanoparticle protrudes from the surface of the resin, implying that the particles are embedded into the resin and do not simply adhere to the surface of PDMS. An AFM image of a densely packed INA film and silver nanoparticles adsorbed on a PVP-modified silicon wafer prior to the immobilization into the resin are shown in Figure 18.5A. Protruded heights were consistently measured to be below 30 nm, which is smaller than the average radius of nanoparticles (55 nm). This result indicates that the nanoparticles are physically entrapped by PDMS resin covering more than half of their diameter. UV-Vis absorbance measurements of PVP-modified quartz substrates before the immobilization of nanoparticles and after peeling off of the resin film indicate that the PVP layer remains on the quartz substrates and does not adhere to PDMS resin, although it is not clear whether some PVP molecules stay adsorbed on embedded silver nanoparticles.

Because the nanoparticles in INA films are not completely covered with the resin, they can be subjected to different chemical reactions, such as surface modifications, etching, and redox reactions. Etching removes nanoparticles, leaving

FIGURE 18.6 AFM image of boundary region between original and etched INA films. Scan range, 10×10 μm.

behind an assembly of nanosize cavities in the PDMS film. The size, shape, and distribution of these cavities are the same as those for nanoparticles in the corresponding INA film. The nanocavities can be selectively modified and used as templates for synthesis of nanoparticles from other materials. AFM images of nanocavities, as well as the boundary between etched and original INA film, are presented in Figure 18.5C and Figure 18.6, respectively.

18.3 COHERENT PLASMON COUPLING BETWEEN AG NANOPARTICLES

18.3.1 OPTICAL PROPERTIES OF INTERACTING NANOPARTICLES

When irradiated with light, closely spaced silver and gold nanoparticles exhibit optical properties that are radically different from those of individual particles due to various light-induced interactions between nanoparticles. The large interest in studying these interactions is driven by the potential of discovering new physical phenomena and designing new materials and devices. Generally, three different types of interactions between nanoparticles can be distinguished. First, nanoparticles a few nanometers apart are electronically coupled due to the direct overlap of the electron wave functions.[61–63] This electron exchange interaction blurs the electronic boundaries between individual nanoparticles, resulting in the red shift and broadening of the plasmon resonance, compared to the resonance of noninteracting particles. For this reason, the extinction band of aggregated silver and gold colloids, in which nanoparticles are in intimate contact with each other, is broad and shifted to longer wavelengths.[64] The larger the number of electronically coupled particles, the larger is the red shift and broadening of the resonance; for an infinitely large number of such nanoparticles, the optical spectrum collapses to that of a continuous silver film. Various cluster shapes of electronically coupled nanoparticles exhibit complex frequency and spatial distributions of optical resonances. Theoretical modeling and the measurements using near-field optical microscopy have revealed that the distribution of optical modes in fractal-like clusters of silver nanoparticles is characterized by strong spatial localization of the electromagnetic field in certain regions of the cluster.[65–72] These so-called hot spots are believed to have a major contribution to a number of enhanced optical phenomena, such as surface-enhanced Raman scattering[73] and nonlinear wave mixing.[74]

A second type of interaction ($E \sim 1/r$, where r is the distance between nanoparticles and is larger than the wavelength of light) appears when the distance between nanoparticles is significantly larger than their diameter. This regime is designated as *far-field interaction*. Each particle possesses a characteristic plasmon resonance and, when irradiated with light, behaves as an independent oscillator. For the absorption component of the plasmon, all particles will absorb light independently and the total absorption is a simple summation of individual contributions. For the scattering component, individual particles also behave as independent emitters and the scattered light undergoes diffraction and interference in the far field. Because the density of scatterers is large, multiple diffraction plays an important role. The optical response of an ensemble of such nanoparticles is the interference of scattered light from individual particles. It is important to emphasize that in the far-field regime, polarizability of individual nanoparticles is the same as that of noninteracting particles; interaction between nanoparticles occurs only through the interference of radiative fields.

A particular interesting case of this interaction is when the metal nanoparticles are organized in regular two- and three-dimensional arrays. Because of the strong resonance scattering from individual nanoparticles, scattered fields will experience multiple diffraction. As a result, photonic-gap-type materials can be encountered with this approach. Strong resonance light scattering by individual metal nanoparticles plays the same role as high contrast of the refractive index, which is required for producing a photonic gap material through the regular packing of dielectric spheres. High dielectric contrast between the spheres and surrounding medium is required for the large intensity of scattered light by individual spheres. Large intensity of scattered light is required for multiple diffraction and, consequently, for the space and frequency localization of optical modes (photonic gap phenomenon).[75] In the case of the metal nanoparticles, the plasmon-assisted resonance scattering is already strong, and true photonic gap materials should be possible to achieve by simply organizing these particles in a regular fashion. Recent theoretical modeling supports this concept.[76] In addition to photonic gap properties, regular arrays of metal nanoparticles can be used for waveguiding purposes on plane surfaces.[77]

When the distance between nanoparticles is larger than that required for direct electron coupling but smaller or comparable to the particles' diameter, *near-field interactions* (Förster type, $E \sim 1/r^3$) play a key role in determining the optical properties of the system. Whereas each particle encounters only the incident field during the far-field regime, in the near-field regime each particle experiences, in addition to the incident field, the *local fields* from its neighbors. Moreover, not only do individual particles experience incident and local fields, but also electrostatic and plasmon-induced interactions between neighbors can potentially influence their polarizability. For closely spaced nanoparticles, especially those organized in regular structures, coherent interactions between particles can induce cooperative effects, resulting in the polarizability of individual particles being enhanced by their neighbors. Such cooperative interaction with the light of silver nanoparticles in two-dimensional semiregular arrays is described below.

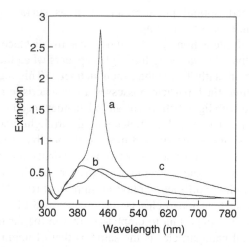

FIGURE 18.7 (a) UV-Vis absorption spectra of a silver colloidal suspension in water. (b) A closely spaced two-dimensional array of Ag nanoparticles assembled on a PVP-modified quartz surface from this suspension. (c) The same nanoparticle array after drying.

18.3.2 Extinction Spectra of Ag Nanoparticles Arranged into Two-Dimensional Arrays

The extinction spectrum of 100-nm Ag particles in aqueous suspension consists of two characteristic bands: dipole at 545-nm and quadrupole at 430-nm components of the plasmon resonance. A concentrated aqueous suspension of these nanoparticles appears greenish due to the strong resonance scattering of light in this spectral region by the dipole component of the plasmon. The spectral position of these bands is a function of particle size, shape, and the dielectric function of the external medium. When organized into a closely spaced two-dimensional array, the Ag particles exhibit drastically different optical spectra compared to those in the suspension.[78] Instead of two characteristic bands for dipole and quadrupole components, the spectrum degenerates into a single intense and sharp peak positioned at about 436 nm (Figure 18.7). The intensity of this peak is great, corresponding to the extinction of light about 2000 times at the maximum. Such a great extinction for a single layer of Ag nanoparticles in the array is indicative of the cooperative interaction in which light interacts not with individual particles but with an ensemble of particles as a whole.

The self-assembly of nanoparticles on PVP-modified surfaces results in formation of a semiregular array with average spacing between particles of approximately the particles' average diameter. Due to the electrostatic repulsion between particles, the final array has nearly hexagonal symmetry. This symmetry can be observed for arrays in water, in alcohol, or immobilized into PDMS resin. Drying of the samples instantly causes particle aggregation (Figure 18.8) and, consequently, damping of the sharp extinction peak (see Figure 18.7). By varying the substrates' exposure time to the nanoparticle suspension, the surface coverage (number of nanoparticles per unit area) can be controlled and, thus, the arrays can be obtained with different peak

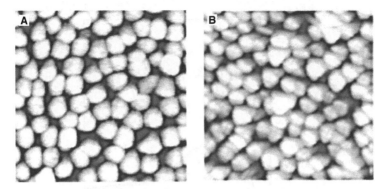

FIGURE 18.8 AFM topography images of silver nanoparticles immobilized on a PVP-modified silicon wafer: in water (A) and in the dry state (B). Scan range, 2×2 μm.

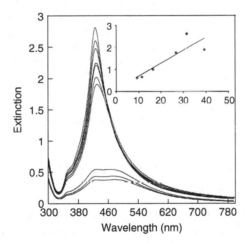

FIGURE 18.9 Changes in the extinction spectra during the formation of an Ag nanoparticle array. Insert: Maximal intensity of the plasmon resonance as a function of surface coverage (number of particles per square micron).

width and intensity (Figure 18.9). The sharpest peak is observed at the saturation coverage corresponding to about 25 μm^{-2}.

The sharp peak in the spectra of two-dimensional closely spaced arrays of 100-nm Ag particles arises from a coherent plasmon mode that results from the coupling of the electron density oscillations in neighboring particles. Because this peak is observed for the quite disordered array depicted in Figure 18.4, the long-range order in the arrays appears to not be a critical factor. This observation suggests that the coupling originates from light-induced, short-range (near-field) interactions between nanoparticles rather than from diffraction-type phenomena that require long-range order. The near-field interactions provide coherency (the phase correlation) for electron density oscillations in neighboring particles. It is not clear, however, how

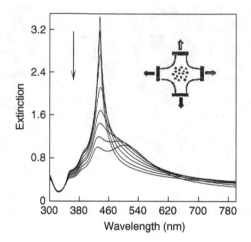

FIGURE 18.10 Extinction spectra of coupled INA film for different degrees of biaxial stretching. The second curve from the bottom corresponds to the film that is stretched 1.5 times from its original state. The sharp peak corresponds to the unstretched film. The arrow indicates the direction of spectra evolution with the degree of stretching. All curves are normalized to the same number of particles probed by the beam. (From Malynych, S. and Chumanov, G., *J. Am. Chem. Soc.*, 125, 2896–2898, 2003. With permission.)

great the coherent length is or what minimum number of coupled particles is required for establishing the new plasmon mode and the sharp peak in the extinction spectrum.

18.3.3 EFFECT OF INTERPARTICLE SPACING ON THE COHERENT PLASMON COUPLING

The intensity and width of the coherent plasmon mode are extremely sensitive to the interparticle distance in the array. This behavior has been studied by measuring the extinction spectra of coupled INA films as a function of biaxial stretching. As the film is stretched, the intensity of the sharp peak rapidly decreases and a new band appears growing in the 500-nm region (Figure 18.10). Surprisingly, after stretching by only 50%, the sharp peak nearly disappears and the extinction spectrum of the film becomes similar to that of noninteracting nanoparticles in suspension. Such strong dependence on the distance further emphasizes the near-field character of the coupling. It has been determined that the symmetry of the particle arrangement plays a major role in the coupling. Because several (more than two in-line) closely spaced neighbors surround each silver particle in the array, the symmetry requirement for such arrangement favors the coupling of the higher multipolar components of the plasmon resonance, in particular the quadrupole component, over the dipole one. Indeed, as the interparticle distance becomes smaller, the sharp peak gradually evolves from the quadrupole band in the extinction spectrum of noninteracting particles, whereas the dipole component of the plasmon resonance (band at 500 nm) progressively decreases (Figure 18.10).

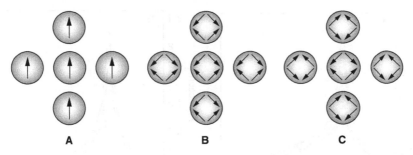

FIGURE 18.11 Model for dipolar (A) and quadrupolar (B), and couplings between (C), plasmon resonances in a two-dimensional array of closely spaced silver nanoparticles. Arrows indicate the direction of the electric vector induced by incident light. (B) illustrates antibonding and (C) bonding type quadrupolar couplings. (From Malynych, S. and Chumanov, G., *J. Am. Chem. Soc.*, 125, 2896–2898, 2003. With permission.)

The prevalence of quadrupolar coupling can be qualitatively justified by the following symmetry considerations. Each silver particle in the array is indistinguishable and should interact in the same way with all of its neighbors. Considering every particle surrounded by four neighbors, it becomes evident that for dipolar oscillations of the electron density, particles in the vertical direction interact with each other in a head-to-tail manner, whereas particles in the horizontal direction experience head-to-head and tail-to-tail interactions (Figure 18.11A). Because the requirement for the same interaction in all directions is not fulfilled in this case, the dipolar coupling is symmetry forbidden for the two-dimensional arrangement. At the same time, the quadrupolar coupling is allowed because it provides the same interaction in all directions (Figure 18.11B and C).[79] Two types of quadrupolar coupling can take place: repulsive (Figure 18.11B) and attractive (Figure 18.11C). The electron density oscillations in the adjacent particles are in-phase for repulsive and alternating in-phase/out-of-phase for attractive interactions, thereby constituting the coherent nature of the coupling. In two-dimensional arrays of closely spaced silver nanoparticles, the dipole coupling of plasmon resonances is symmetry forbidden for normal incidence of light (the electric vector is parallel to the plane of the array), whereas quadrupole coupling is allowed. This symmetry-determined selection rule enhances the quadrupole and suppresses the dipole coupling, even though the dipole interaction normally extends to longer distances than the quadrupole one.[80] The quadrupole coupling results in the electron density in particles oscillating in the plane of the substrate (perpendicular to the plane of incidence). This is different from noncoupled particles, in which quadrupole oscillations are in the plane of incidence.[79] However, for nearly spherical silver nanoparticles used in this study, these two modes are degenerate and have the same frequency. For this reason, the sharp peak in the coupled arrays appears as one gradually evolving from the quadrupole component of noninteracting particles (Figure 18.10). The described spectral changes are completely reversible during repeated stretching and relaxing of coupled INA films.

The effect of symmetry was further confirmed by studying the evolution of the extinction spectra during uniaxial stretching of coupled INA films. More pronounced

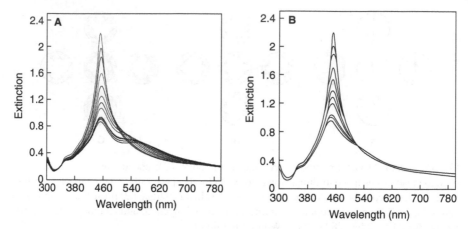

FIGURE 18.12 Extinction spectra of coupled INA films for different degrees of uniaxial stretching. The incident beam was polarized perpendicular (A) and parallel (B) to the stretching axis.

changes in the spectra occur for the polarization perpendicular to the direction of stretching than for the parallel polarization. The sharp peak decreases and a broad band at 580 nm, associated with the dipole component of plasmon resonance, grows up. In the end, the spectrum slightly resembles that of an aqueous suspension of Ag nanoparticles (Figure 18.12A). It should be mentioned that stretching of INA films along one direction is not truly uniaxial stretching because it is simultaneously accompanied with some shrinkage in the perpendicular direction. For the polarization parallel to the direction of stretching, only damping of the sharp peak is observed, with no pronounced appearance on the dipole peak in the red spectral range (Figure 18.12B). These experiments further emphasize the quadrupole nature of the plasmon coupling in closely spaced two-dimensional arrays of silver nanoparticles.

18.3.4 ANGULAR DEPENDENCE OF THE COHERENT PLASMON COUPLING

The coherent character of the near-field coupling is concluded from the extinction measurements as a function of the incidence angle for both *s*- and *p*-polarized light. As the incident angle increases from 0° (normal incidence) to 60°, the intensity of the sharp peak gradually decreases for both *s*- and *p*-polarizations (Figure 18.13). This decrease can be explained in terms of the damping of the coherent plasmon mode due to the dephasing of the electron density oscillations in neighboring particles. At oblique incident angles, the adjacent particles experience slightly different phases of incident light and, consequently, less efficient coupling. The dephasing increases with incident angle. The difference between *s*- and *p*-polarizations emerges at large incident angles, at which a new band is growing in the 460- to 470-nm spectral range for *p*-polarized light (Figure 18.13B). Whereas the electric vector for *s*-polarization at any incident angle has only a tangential component that excites the electron density oscillations in the plane of the array, an additional normal component arises at oblique angles for *p*-polarization. This normal component of the electric vector excites electron oscillations perpendicular to the plane of the array. Because of the proximity of nanoparticles to each other, the

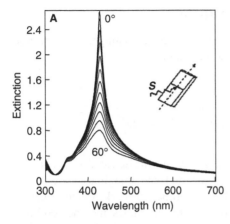
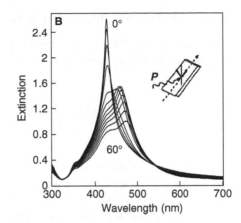

FIGURE 18.13 Extinction spectra of the silver nanoparticle array as a function of the incident angles for *s*- and *p*-polarized light. All curves are normalized to the same number of particles probed by the beam. (From Malynych, S. and Chumanov, G., *J. Am. Chem. Soc.*, 125, 2896–2898, 2003. With permission.)

normal oscillations are also coupled; however, the coupling is dipolar rather than quadrupolar. The dipole coupling overcomes the quadrupole one in this case because dipole–dipole is a stronger interaction and no symmetry restrictions are imposed on the dipole coupling. The growth of a new band in the 460- to 470-nm spectral region with the increase of incident angle is a manifestation of the dipole coupling of the normal component for electron density oscillations (Figure 18.13B). The intensity of this band is determined by the two competing factors. The efficiency for the excitation of the normal component and, consequently, the dipolar coupling between particles become larger as the incident angle increases. At the same time, simultaneously increasing dephasing reduces the efficiency of the coupling as well. This new band should reach maximum at a certain optimum incident angle. Indeed, as can be seen in Figure 18.13, the band at 460 to 470 nm is more intense at about a 35° angle than at any other incident angle.

The integrated area under the curves in Figure 18.13 (total oscillator strength of the system) decreases with an increase of the incident angle. At large incident angles, the silver particles in the array behave more like independent oscillators, and the observed decrease of the oscillator strength indicates the higher efficiency for the interaction with light of coupled particles, compared to the same number of non-coupled particles. The increase in the efficiency is due to the enhanced polarizability of individual particles that results from coherent, resonant, near-field interactions. The resonant character of the near-field interactions comes from the fact that all particles have the same plasmon (eigen) frequency. This near-field interaction takes place by virtue of the enhanced local electromagnetic field that is associated with the electron oscillations in individual silver particles. Because of the proximity to each other, every particle in the array experiences the enhanced local fields from its neighbors. These fields have the same phase and the same frequency, thereby resonantly enhancing the polarizability of nanoparticles.

18.3.5 EFFECT OF EXTERNAL DIELECTRIC MEDIUM

According to the Mie theory, optical properties of metal nanoparticles strongly depend on the dielectric function of the host medium as

$$\sigma_{ext}(\omega) = 9\frac{\omega}{c}\varepsilon_m^{3/2}V_0\frac{\varepsilon_2(\omega)}{\left[\varepsilon_1(\omega)+2\varepsilon_m\right]^2+\varepsilon_2(\omega)^2}$$

where $\sigma_{ext}(\omega)$ is the extinction coefficient, ε_m is the dielectric function of the surrounding medium, and $\varepsilon_1(\omega)$ and $\varepsilon_2(\omega)$ are real and imaginary parts of the dielectric function of the metal, respectively.[4] Experimental work on the effects of the external medium on the optical properties of nanoparticles includes colloidal suspensions,[81] polymer films,[82–85] partially oxidized Si matrix,[86] absorbed organic compounds,[87] and the interaction between a silver cluster with another metal, semiconductor, or dielectric cluster.[88] The effect of solvent on the extinction spectra of periodic arrays of silver nanoclusters fabricated by nanosphere lithography was studied in reference 89. An overview of nanoparticles–surrounding medium interaction is presented in reference 90.

The frequency of the coherent plasmon mode of a coupled two-dimensional array of Ag nanoparticles exhibits a strong dependence on the dielectric function of its host medium. Extinction spectra of the arrays prepared on PVP-modified substrates were first measured in a water–alcohol mixture with various concentrations of ethanol. A further increase of the dielectric function was reached by adding various amounts of m-Cresol (C_7H_8O, n = 1.54) to pure ethanol. In this way, the refractive index of the medium surrounding the two-dimensional array of nanoparticles was changed from water (n = 1.33) to pure Cresol (n = 1.54).

As the refractive index of the surrounding solvent increases, a steady shift of the coherent plasmon mode frequency is observed (Figure 18.14). The shift is linear and can be empirically expressed as (Figure 18.15)

$$\lambda = 167.24n_{med} + 213.66$$

The linear and reproducible behavior of the shift provides a basis for analytical techniques for the determination of the refractive indexes of liquids and gases. The shifts are fully reproducible when the refractive index of the solvent changes in the opposite direction (from m-Cresol to water), confirming the stability of the particle arrangement during the measurements. The sensitivity coefficient of 167.24 translates into a 1-nm wavelength shift for a 0.006-unit change of the refractive index. Similar spectral shifts have been reported for arrays of truncated-tetrahedron-shaped Ag nanoparticles fabricated by nanosphere lithography.[89] It has been shown that the truncated-tetrahedron-shaped particles are more sensitive to the refractive index of the external medium than spheroidal ones. An advantage of using the coherent plasmon mode of two-dimensional arrays of silver nanoparticles for measurements of refractive indexes comes from the extreme sharpness of the resonance, allowing

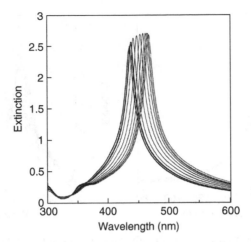

FIGURE 18.14 Extinction spectra of a two-dimensional silver nanoparticle array immersed into water, ethanol, m-Cresol, and their mixture. The peak suffers a shift as the value of the dielectric constant increases.

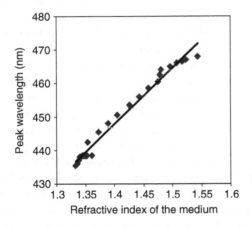

FIGURE 18.15 Shift in a plasmon peak position as a function of the refractive index of the external medium.

small changes in its position to be readily detected. The sharpness of the coherent plasmon peak permits an easy and direct measurement of a 0.1-nm spectral shift that corresponds to a 0.0006-unit change of the refractive index using a bench-type spectrophotometer.

18.3.6 SUBSTRATE EFFECTS ON OPTICAL PROPERTIES OF AG NANOPARTICLE ARRAYS

It is important to understand for practical purposes how the underlying substrate with complex dielectric function influences the coherent plasmon mode. The previously

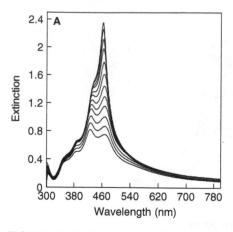
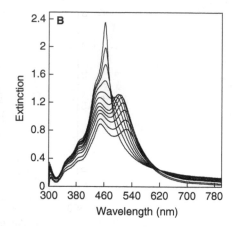

FIGURE 18.16 Extinction spectra of the silver nanoparticle array on ITO glass with 4-Ohm sheet resistance as a function of the incident angles for (A) *s*- and (B) *p*-polarized light. All curves are normalized to the same number of particles probed by the beam.

published work focused mainly on the optical properties of silver nanoparticle arrays fabricated on dielectric substrates.[91–94] Silver nanoparticles on conducting (metal) surfaces were considered mainly as substrates for surface-enhanced phenomena.[8–11,14,95] Interesting results were obtained for Ag nanoparticles positioned very close to an Al surface.[96] Light emission was observed from the surface plasmon caused by the presence of the Ag nanoparticles, and the intensity of emitted light rapidly decreased as the distance between the particles and the surface increased. We studied coupled two-dimensional arrays of Ag nanoparticles prepared on ITO glass and semitransparent chromium films. The ITO glass has a complex dielectric function and is well known as a material for transparent electrodes that are used in many practical applications. ITO and gold substrates were previously used for the deposition of gold oblate nanoparticles, and significant differences in the extinction spectra were noted for the two substrates.[97] A single plasmon band at 650 nm was observed for the particles deposited on ITO surfaces, whereas two bands at 525 and 650 nm appeared on the gold substrate. The short-wavelength band in the latter was assigned to the localized plasmon resonance of an isolated particle coupled to a propagating surface plasmon wave at the gold–glass interface.

We used ITO glass with a sheet resistance of 4, 20, and 100 Ohm·cm for the fabrication of coupled two-dimensional arrays of Ag nanoparticles by the same procedures as described above. The extinction spectra of these arrays reveal several prominent differences. A shoulder appears at about 450 nm, in addition to the sharp peak at 462 nm, that is due to the coherent plasmon mode. The frequency of the coherent plasmon mode on the ITO substrate is shifted by about 30 nm, compared to the same mode on other dielectric substrates (compare the most intense peaks in Figure 18.13 and Figure 18.16). There is a weakly pronounced trend that the shoulder becomes stronger with higher conductivity of the ITO substrate. Interestingly, the shoulder completely disappears after the arrays are lifted from ITO substrates with PDMS resin. The maximum of the coherent plasmon mode shifts back about 30 nm, and the extinction spectra in this case become practically indistinguishable from those of arrays lifted from glass, quartz, silicon, or any other dielectric substrate. This result indicates that the origin of the shoulder and the shift are from the conductive ITO substrates and not from a possibly different arrangement of the nanoparticles on ITO, compared to other dielectric substrates. The described changes are completely reversible: the shoulder appears again when the PDMS films with embedded nanoparticles are tightly pressed against ITO substrates.

Differences in the extinction spectra of coupled two-dimensional arrays on ITO substrates become even more pronounced when the measurements are taken with polarized light at various incident angles. As the incident angle for both polarizations increases, the shoulder gradually evolves into a separate band (Figure 18.16). For *s*-polarized light, the maximum of this band shifts slightly and somewhat increases in intensity, whereas for *p*-polarized light it shifts and also increases in intensity. The peak corresponding to the coherent plasmon mode evolves the same as that on other dielectric substrates described previously.

It is reasonable to assign the shoulder to two new plasmon modes that result from the direct electronic coupling of Ag nanoparticles and the underlying conducting ITO substrate. At normal incidence of light, these modes are degenerate and there is no difference in the spectra measured with *s*- and *p*-polarized light. However, at oblique angles, the normal component of the electric vector of *p*-polarized light excites a mode in which the electrons oscillate along the axis perpendicular to the substrate. This mode is also coupled, and the character of the coupling is dipolar. Because this is a light-induced coupling, the continuously increasing normal component of the electric vector induces more coupling with an increase of the incident angle (Figure 18.16A). That, in turn, lowers the frequency of this new plasmon mode, causing the shift in the position of the shoulder. This behavior is similar to that previously described for the dipolar coupling on dielectric substrates.[98]

S-polarized light excites a plasmon mode that corresponds to the oscillations of the electrons in the plane of the substrate. The efficiency of the excitation of this mode does not depend on the angle of incidence, so no change of the frequency and intensity of the correspondent band should be observed with the increase of the angle. The seeming increase in intensity and the blue shift in this case may be due to the phenomenon of a weak band overlapping on the left side with a very strong band. When the intensity of the latter (the coherent plasmon mode) goes down, the former (new band) visually appears as increasing in intensity and shifting to the left, even though in actuality it remains the same.

When Ag nanoparticles are self-assembled onto a PVP-modified, 20-nm-thick layer of chromium, a single broad featureless band at about 473 nm is observed in the extinction spectra, instead of a sharp peak, such as on dielectric substrates (Figure 18.17). Chromium forms an atomically smooth layer that can be made semitransparent. In addition, chromium provides a conducting substrate that is coated with a thin dielectric chromium oxide layer. The latter allows study of the plasmon coupling without direct electron exchange between Ag particles and the substrate. The presence of the thin dielectric layer and a much higher electron density in the underlying layer is a major difference between chromium and ITO substrates. The broad band on the chromium substrate represents the coherent plasmon mode, which is strongly

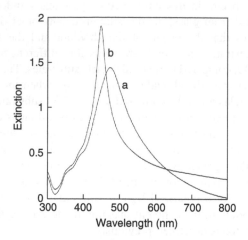

FIGURE 18.17 Extinction spectra of the silver nanoparticle array on a chromium substrate (a) and the correspondent INA film after peeling off (b).

damped by the high electron density of the underlying Cr layer. As in the case of ITO substrates, nanoparticle arrays lifted off with PDMS resin restore the sharp peak due to the coherent plasmon mode that is not damped by the substrate.

18.4 CONCLUSIONS

The physical nature of the coherent coupling of plasmon resonances in a two-dimensional array of closely spaced silver nanoparticles is different from that responsible for optical mode localization in photonic gap materials.[75] Whereas the former is due to the near-field interactions between plasmon resonances, the latter results from a Bragg diffraction of light from dielectric periodicities. The following arguments are used to support the different origins of these two phenomena. First, the coherent plasmon coupling has been observed for quite disordered two-dimensional arrays (Figure 18.4), whereas Bragg diffraction requires long-range order. Second, the spectral position of the transmission minimum in colloidal crystals depends on the distance between dielectric spheres in a characteristic way for diffraction phenomena.[99,100] With the increase of the distance between spheres, it shifts and retains essentially the same intensity. On the contrary, the sharp extinction peak of the coherent plasmon mode experiences hardly any spectral shift with increase of interparticle distance. Only a characteristic decrease of the peak intensity is observed due to the steep decline of the near-field interaction between plasmon resonances (Figure 18.11).

The observation of the coherent plasmon coupling in two-dimensional arrays of closely spaced silver nanoparticles exemplifies a new approach for designing optical materials and devices. Novel optical properties can be obtained by assembling metal nanoparticles in various regular and irregular one-, two-, and three-dimensional structures. Wide implementation of this approach will certainly result in unraveling of novel principles of both fundamental and practical importance. The extreme sensitivity of the coupled plasmon modes to the arrangement and spacing between particles and to the dielectric properties of the surrounding medium can be potentially applied in various sensors.

ACKNOWLEDGMENTS

Financial support was provided by the National Science Foundation and the U.S. Environmental Protection Agency, Grant R829603. The authors also thank Ames Laboratory (U.S. Department of Energy) for the equipment used for this work.

REFERENCES

1. Faraday, M., The Bakerian lecture: experimental relations of gold (and other metals) to light, *Philos. Trans.* 147, 145, 1857.
2. Mie, G., Contributions to the optics of turbid media, particularly of colloidal metal solutions, *Ann. Phys.* 25, 377, 1908.
3. Bohren, C.F., Huffman, D.R., *Absorption and Scattering of Light by Small Particles*, Wiley Interscience, New York, 1983.
4. Kreibig, U., Vollmer, M., *Optical Properties of Metal Clusters*, Springer-Verlag, Berlin, 1995.
5. Heath, J.R., Size-dependent surface-plasmon resonances of bare silver particles, *Phys. Rev. B* 40, 9982–9985, 1989.
6. Doyle, W.T., Optical properties of a suspension of metal spheres, *Phys. Rev. B* 39, 9852–9858, 1989.
7. Shahbazyan, T.V., Perakis, I.E., Surface collective excitations in ultrafast pump-probe spectroscopy of metal nanoparticles, *Chem. Phys.* 251, 37–49, 2000.
8. Vlckova, B., Gu, X.J., Moskovits, M., SERS excitation profiles of phthalazine adsorbed on single colloidal silver aggregates as a function of cluster size, *J. Phys. Chem.* 101, 1588–1593, 1997.
9. Keating, C.D., Kovaleski, K.K., Natan, M.J., Heightened electromagnetic fields between metal nanoparticles: surface enhanced Raman scattering from metal-cyto-chrome c-metal sandwiches, *J. Phys. Chem. B* 102, 9414–9425, 1998.
10. Doering, W.E., Nie, S.M., Single-molecule and single-nanoparticle SERS: examining the roles of surface active sites and chemical enhancement, *J. Phys. Chem. B* 106, 311–317, 2002.
11. Lee, C.R., Bae, S.J., Gong, M.S., et al., Surface-enhanced Raman scattering of 4,4'-dicyanobiphenyl on gold and silver nanoparticle surfaces, *J. Raman. Spectrosc.* 33, 429–433, 2002.
12. Maxwell, D.J., Taylor, J.R., Nie, S.M., Self-assembled nanoparticle probes for recognition and detection of biomolecules, *J. Am. Chem. Soc.* 124, 9606–9612, 2002.
13. Farbman, I., Levi, O., Efrima, S., Optical-response of concentrated colloids of coinage metals in the near-ultraviolet, visible, and infrared regions, *J. Chem. Phys.* 96, 6477–6485, 1992.
14. Jensen, T.R., Van Duyne, R.P., Jonson, S.A., Maroni, V.A., Surface-enhanced infrared spectroscopy: a comparison of metal island films with discrete and nondiscrete surface plasmons, *Appl. Spectrosc.* 54, 371–377, 2000.
15. Tuovinen, H., Kauranen, M., Jefimovs, K., et al., Linear and second-order nonlinear optical properties of arrays of noncentrosymmetric gold nanoparticles, *J. Nonlinear Opt. Phys.* 11, 421–432, 2002.

16. Perner, M., Bost, P., Lemmer, U., von Plessen, G., Feldmann, J. et al., Optically induced damping of the surface plasmon resonance in gold colloids, *Phys. Rev. Lett.* 78, 2192–2195, 1997.

17. Haes, A.J., Van Duyne, R.P., A nanoscale optical biosensor: sensitivity and selectivity of an approach based on the localized surface plasmon resonance spectroscopy of triangular silver nanoparticles, *J. Am. Chem. Soc.* 124, 10596–10604, 2002.

18. Xu, H.X., Kall, M., Modeling the optical response of nanoparticle-based surface plasmon resonance sensors, *Sensors Actuators B* 87, 244–249, 2002.

19. Andersen, P.C., Rowlen, K.L., Brilliant optical properties of nanometric noble metal spheres, rods, and aperture arrays, *Appl. Spectrosc.* 56, 124A–135A, 2002.

20. Shalaev, V.M., Botet, R., Butenko, A.V., Localization of collective dipole excitations on fractals, *Phys. Rev. B* 48, 6662–6664, 1993.

21. Stockman, M.I., Faleev, S.V., Bergman, D.J., Localization versus delocalization of surface plasmons in nanosystems: can one state have both characteristics? *Phys. Rev. Lett.* 87, 167401-1–167401-4, 2001.

22. Schatz, G.C., Electrodynamics of nonspherical noble metal nanoparticles and nanoparticle aggregates, *J. Mol. Struct. Theochem.* 573, 73–80, 2001.

23. Maier, S.A., Brongersma, M.L., Kik, P.G., Atwater, H.A., Observation of near-field coupling in metal nanoparticle chains using far-field polarization spectroscopy, *Phys. Rev. B* 65, 193408-1–193408-4, 2002.

24. Maier, S.A., Brongersma, M.L., Kik, P.G., Meltzer, S., Atwater, H.A., Plasmonics: a route to nanoscale optical devices, *Adv. Mater.* 13, 1501–1505, 2001.

25. Quinten, M., Leitner, A., Krenn, J.R., Aussenegg, F.R., Electromagnetic energy transport via linear chains of silver nanoparticles, *Optics Lett.* 23, 1331–1333, 1998.

26. Ditlbacher, H., Krenn, J.R., Schider, G., Leitner, A., Aussenegg, F.R., Two-dimensional optics with surface plasmon polaritons, *Appl. Phys. Lett.* 81, 1762–1764, 2002.

27. Russel, B.K., Mantovani, J.G., Anderson, V.E., Warmak, R.J., Ferrel, T.L., Experimental test of the Mie theory for microlithographically produced silver spheres, *Phys. Rev. B* 35, 2151–2154, 1987.

28. Hulteen, J.C., Van Duyne, R.P., Nanosphere lithography: a materials general fabrication process for periodic particle array surfaces, *J. Vac. Sci. Tech. A* 13, 1553–1558, 1995.

29. Haynes, C.L., Van Duyne, R.P., Nanosphere lithography: a versatile nanofabrication tool for studies of size-dependent nanoparticle optics, *J. Phys. Chem.* 105, 5599–5611, 2001.

30. Krinke, T.J., Fissan, H., Deppert, K., Magnusson, M.H., Samuelson, L., Positioning of nanometer-sized particles on flat surfaces by direct deposition from the gas phase, *Appl. Phys. Lett.* 78, 3708–3710, 2001.

31. Junno, T., Deppert, K., Montelius, L., Samuelson, L., Controlled manipulation of nanoparticles with an atomic-force microscope, *Appl. Phys. Lett.* 66, 3627–3629, 1995.

32. Martin, M., Roschier, L., Hakonen, P., Parts, U., Paalanen, M., Schleicher, B., Kauppinen, E.I., Manipulation of Ag nanoparticles utilizing noncontact atomic force microscopy, *Appl. Phys. Lett.* 73, 1505–1507, 1998.

33. Renard, C., Ricolleau, C., Fort, E., Besson, S., Gacoin, T., Boilot, J.-P., Coupled technique to produce two-dimensional superlattices of nanoparticles, *Appl. Phys. Lett.* 80, 300–302, 2002.

34. Cai, W., Hofmeister, H., Rainer, T., Chen, W., Optical properties of Ag and Au nanoparticles dispersed within the pores of monolithic mesoporous silica, *J. Nanoparticle Res.* 3, 443–453, 2001.

35. Fukuoka, A., Araki, H., Sakamoto, Y. et al., Template synthesis of nanoparticle arrays of gold and platinum in mesoporous silica films, *Nano Lett.* 2, 793–795, 2002.
36. Kunz, M.S., Shull, K.R., Kellock, A.J., Colloidal gold dispersions in polymeric matrices, *J. Colloid Interface Sci.* 156, 240–249, 1993.
37. Motte, L., Billoudet, F., Pileni, M.P., Self-assembled monolayer of nanosized particles differing by their sizes, *J Phys. Chem.* 99, 16425–16429, 1995.
38. Ohara, P.C., Heath, J.R., Gelbart, W.M., Self-assembly of submicrometer rings of particles from solutions of nanoparticles, *Angew. Chem. Int. Ed.* 36, 1078, 1997.
39. Pastoriza-Santos, I., Serra-Rodriguez, C., Liz-Marzan, L.M., Self-assembly of silver particle monolayers on glass from Ag+ solutions in DMF, *J. Colloid Interface Sci.* 221, 236–241, 2000.
40. Lopes, W.A., Jaeger, H.M., Hierarchical self-assembly of metal nanostructures on diblock copolymer scaffolds, *Nature* 414, 735–738, 2001.
41. Chumanov, G., Sokolov, K., Gregory, B., Cotton, T.M., Colloidal metal-films as a substrate for surface-enhanced spectroscopy, *J. Phys. Chem.* 99, 9466–9471, 1995.
42. Freeman, R.G., Grabar, K.C., Allison, K.J., Bright, R.M., Davis, J.A., Guthrie, A.P., Hommer, M.B., Jackson, M.A., Smith, P.C., Walter, D.G., Natan, M.J., Self-assembled metal colloid monolayers: an approach to SERS substrates, *Science* 267, 1629–1632, 1995.
43. Grabar, K.C., Freeman, R.G., Hommer, M.B., Natan, M.J., Preparation and characterization of Au colloid monolayers, *Anal. Chem.* 67, 735–743, 1995.
44. Fritzsche, W., Sokolov, K., Chumanov, G., Cotton, T.M., Henderson, E., Ultrastructural characterization of colloidal metal films for bioanalytical applications by scanning force microscopy, *J. Vac. Sci. Technol. A* 14, 1766–1769, 1996.
45. Decher, G., Fuzzy nanoassemblies: toward layered polymeric multicomposites, *Science* 277, 1232–1237, 1997.
46. Ostrander, J.W., Mamedov, A.A., Kotov, N.A., Two modes of linear layer-by-layer growth of nanoparticle–polyelectrolyte multilayers and different interactions in the layer-by-layer deposition, *J. Am. Chem. Soc.* 123, 1101–1110, 2001.
47. Kotov, N.A., Dekany, I., Fendler, J.H., Layer-by-layer self-assembly of polyelectrolyte–semiconductor nanoparticle composite films, *J. Phys. Chem.* 99, 13065–13069, 1995.
48. Kotov, N.A., Dekany, I., Fendler, J.H., Ultrathin graphite oxide–polyelectrolyte composites prepared by self-assembly: transition between conductive and non-conductive states, *Adv. Mater.* 8, 637–641, 1996.
49. Ariga, K., Lvov, Y., Onda, M., Ichinose, I., Kunitake, T., Alternately assembled ultrathin film of silica nanoparticles and linear polycations, *Chem. Lett.* 2, 125–126, 1997.
50. Liu, Y., Wang, A., Claus R.J., Molecular self-assembly of TiO_2/polymer nanocomposite films, *J. Phys. Chem. B* 101, 1385–1388, 1997.
51. Shimazaki, Y., Mitsuishi, M., Ito, S., Yamamoto, M., Preparation of the layer-by-layer deposited ultrathin film based on the charge-transfer interaction, *Langmuir* 13, 1385–1388, 1997.
52. Wang, L.Y., Cui, S.X., Wang, Z.Q., Zhang, X., Jiang, M., Chi, L.F., Fuchs, H., Multilayer assemblies of copolymer PSOH and PVP on the basis of hydrogen bonding, *Langmuir* 16, 10490–10494, 2000.
53. Malynych, S., Luzinov, I., Chumanov, G., Poly(Vinyl pyridine) as a surface modifier for immobilization of nanoparticles, *J. Phys. Chem. B.* 106, 1280–1285, 2002.
54. Lippert, J.L., Brandt, S.E., Surface-enhanced Raman-spectroscopy of poly(2-vinylpyridine) adsorbed on silver electrode surfaces, *Langmuir* 4, 127–132, 1988.

55. Garrel, R.L., Beer, K.D., Surface-enhanced Raman-scattering from 4-ethylpyridine and poly(4-vinylpyridine) on gold and silver electrodes, *Langmuir* 5, 452–458, 1989.

56. Lu, Y., Xue, G., Study of molecular-orientation and bonding at the polymer–metal interface by surface-enhanced Raman-scattering, *Polymer* 34, 3750–3752, 1993.

57. Iler, R.K., *Chemistry of Silica*, Wiley, New York, 1979.

58. Malynych, S., Robuck, H., Chumanov, G., Fabrication of two-dimensional assemblies of Ag nanoparticles and nanocavities in poly(dimethylsiloxane) resin, *Nano Lett.* 1, 647–649, 2001.

59. Quake, S.R., Scherer, A., From micro- to nanofabrication with soft materials, *Science* 290, 1536–1540, 2000.

60. Xia, Y., Whitesides, G.M., Soft lithography, *Angew. Chem. Int. Ed.* 37, 551–575, 1998.

61. Collier, C.P., Saykally, R.H., Shiang, J.J., Henrichs, S.E., Heath, J.R., Reversible tuning of silver quantum dots monolayer through the metal-insulator transition, *Science* 277, 1978–1981, 1997.

62. Shiang, J.J., Heath, J.R., Collier, C.P., Saykally, R.H., Cooperative phenomena in artificial solids made from silver quantum dots: the importance of classical coupling, *J. Phys. Chem. B* 102, 3425–3430, 1998.

63. Kim, S.-H., Medeiros-Ribiero, G., Ohlberg, D.A.A., Williams, R.S., Heath, J.R., Individual and collective electronic properties of Ag nanocrystals, *J. Phys. Chem. B* 103, 10341–10347, 1999.

64. Creighton, A., Metal colloids, in *Surface Enhanced Raman Scattering*, Chang, R.K., Furtak, T.E., Eds., Plenum Press, New York, 1983, pp. 315–337.

65. Weeber, J.-C., Krenn, J.R., Dereux, A., Lamprecht, B., Lacroute, Y., Goudonnet, J.P., Near-field observation of surface plasmon polariton propagation on thin metal stripes, *Phys. Rev. B* 64, 045411 (1–9), 2001.

66. Markel, V.A., Shalaev, V.M., Stechel, E.B., Kim, W., Armstrong, R.L., Small-particle composite. 1. Linear optical properties, *Phys. Rev. B* 53, 2425–2436, 1996.

67. Markel, V.A., Muratov, S.L., Stockman, M.I., George, T.F., Theory and numerical simulation of optical properties of fractal clusters, *Phys. Rev. B* 43, 8183–8195, 1991.

68. Shalaev, V.M., Electromagnetic properties of small-particle composites, *Phys. Rep.* 272, 61–137, 1996.

69. Shalaev, V.M., Poliakov, E.Y., Markel, V.A., Small-particle composite. 2. Nonlinear optical properties, *Phys. Rev. B* 53, 2437–2446, 1996.

70. Markel, V.A., Shalaev, V.M., Zhang, P., Huynh, W., Tay, L., Haslett, T.L., Moskovits, M., Near-field optical spectroscopy of individual surface-plasmon modes in colloid clusters, *Phys. Rev. B* 59, 10903–10909, 1999.

71. Stockman, M.I., Pandey, L.N., George, T.F., Inhomogeneous localization of polar eigenmodes in fractals, *Phys. Rev. B* 53, 2183–2186, 1996.

72. Zhang, P., Haslett, T.L., Douketis, C., Moskovits, M., Mode localization in self-affine fractal interfaces observed by near-field microscopy, *Phys. Rev. B* 57, 15513–15518, 1998.

73. Vlckova, B., Gu, X.J., Tsai, D.P., Moskovits, M., A microscopic surface-enhanced Raman study of a single adsorbate-covered colloidal silver aggregate, *J. Phys. Chem.* 100, 3169–3174, 1996.

74. Tsai, D.P., Kovacs, J., Wang, Z.H., Moskovits, M., Shalaev, V.M., Suh, J.S., Botet, R., Photon scanning–tunneling microscopy images of optical excitations of fractal metal colloidal clusters, *Phys. Rev. Lett.* 72, 4149–4152, 1994.

75. Yablonovitch, E., Inhibited spontaneous emission in solid-state physics and electronics, *Phys. Rev. Lett.* 58, 2059–2062, 1987.

76. Wang, Z., Chan, C.T., Zhang, W., Ming, N., Sheng, P., Three-dimensional self-assembly of metal nanoparticles: possible photonic crystal with complete gap below the plasma frequency, *Phys. Rev. B* 64, 113108 (1–4), 2001.

77. Bozhevolnyi, S.I., Erland, J., Leosson, K., Skovgaard, P.M.W., Hvam, J.M., Waveguiding in surface plasmon polariton band gap structures, *Phys. Rev. Lett.* 86, 3008–3011, 2001.

78. Chumanov, G., Sokolov, K., Cotton, T.M., Unusual extinction spectra of nanometer-sized silver particles arranged in two-dimensional arrays, *J. Phys. Chem.* 100, 5166–5168, 1996.

79. Born, M., Wolf, E., Optics of Metals in *Principles of Optics*, Pergamon Press, Oxford, 1993.

80. Jackson J.D., Multipoles, Electrostatics of Macroscopic Media, Dielectrics, in *Classical Electrodynamics*, John Wiley & Sons, New York, 1975.

81. Heilmann, A., Quinten, M., Werner, J., Optical response of thin plasma–polymer films with non-spherical silver nanoparticles, *Eur. Phys. J. B* 3, 455–461, 1998.

82. Heilmann, A., Kiesow, A., Gruner, M., Kreibig, U., Optical and electrical properties of embedded silver nanoparticles at low temperatures, *Thin Solid Films* 343/344, 175–178, 1999.

83. Quinten, M., Heilmann, A., Kiesow, A., Refined interpretation of optical extinction spectra of nanoparticles in plasma polymer films, *Appl. Phys. B* 68, 707–712, 1999.

84. Hilger, A., Cuppers, N., Tenfelde, M., Kreibig, U., Surface and interface effects in the optical properties of silver nanoparticles, *Eur. Phys. J. D* 10, 115–118, 2000.

85. Fritzsche, W., Porwol, H., Wiegand, A., Bormann, S., Kohler, J.M., *In-situ* formation of Ag-containing nanoparticles in thin polymer films, *Nanostruct. Mater.* 10, 89–97, 1998.

86. Yang, L., Li, G.H., Zhang, J.G., Zhang, L.D., Liu, Y.L., Wang, Q.M., Fine structure of the plasmon resonance absorption peak of Ag nanoparticles embedded in partially oxidized Si matrix, *Appl. Phys. Lett.* 78, 102–104, 2001.

87. Lebedev, A.N., Stenzel, O., Optical extinction of an assembly of spherical particles in an absorbing medium: application to silver clusters in absorbing organic materials, *Eur. Phys. J. D* 7, 83–88, 1999.

88. Quinten, M., Optical response of aggregates of clusters with different dielectric functions, *Appl. Phys. B* 67, 101–106, 1998.

89. Jensen, T.R., Duval, M.L., Kelly, K.L., Lazarides, A.A., Schatz, G.C., Van Duyne, R.P., Nanosphere lithography: effect of the external dielectric medium on the surface plasmon resonance spectrum of a periodic array of sliver nanoparticles, *J. Phys. Chem. B* 103, 9846–9853, 1999.

90. Kreibig, U., Gartz, M., Hilger, A., Neuendorf, R., Interfaces in nanostructures: optical investigations on cluster-matter, *Nanostruct. Mater.* 11, 1335–1342, 1999.

91. Hilger, A., Tenfelde, M., Kreibig, U., Silver nanoparticles deposited on dielectric surfaces, *Appl. Phys. B* 73, 361–372, 2001.

92. Malinsky, M.D., Kelly, K.L., Schatz, G.C., Van Duyne, R.P., Nanosphere lithography: effect of the substrate on the localized plasmon resonance spectrum of silver nano-particles, *J. Phys. Chem. B* 105, 2343–2350, 2001.

93. Stuart, H.R., Hall, D.G., Enhanced dipole–dipole interaction between elementary radiators near a surface, *Phys. Rev. Let.* 80, 5663–5666, 1998.

94. Schröter, U., Heitmann, D., Grating couplers for surface plasmons excited on thin metal films in the Kretschmann–Raether configuration, *Phys. Rev. B* 60, 4992–4999, 1999.

95. Chah, S., Hutter, E., Roy, D., Fendler, J.H., Yi, J., The effect of substrate metal on 2-aminoethanethiol and nanoparticle enhanced surface plasmon resonance imaging, *Chem. Phys.* 272, 127–136, 2001.

96. Kume, T., Hayashi, S., Yamamoto, K., Light emission from surface plasmon polaritons mediated by metallic fine particles, *Phys. Rev. B* 55, 4774–4782, 1997.

97. Felidj, N., Aubart, J., Levi, G. et al., Enhanced substrate-induced coupling in two-dimensional gold nanoparticle array, *Phys. Rev. B* 66, 245407-1–245407-7, 2002.

98. Malynych, S., Chumanov, G., Light-induced coherent interactions between silver nanoparticles in two-dimensional arrays, *J. Am. Chem. Soc.* 125, 2896–2898, 2003.

99. Asher, S.A., Holtz, J., Weissman, J., Pan, G., *MRS Bull.* 23, 44, 1998.

100. Foulger, S.H., Jiang, P., Ying, Y.R. et al., Photonic bandgap composites, *Adv. Mater.* 13, 1898–1901, 2001.

*Self-Assembly
of Nanoscale Colloids*

19 Self-Organization of Metallic Nanorods into Liquid Crystalline Arrays

Catherine J. Murphy, Nikhil R. Jana, Latha A. Gearheart, Sherine O. Obare, Stephen Mann, Christopher J. Johnson, and Karen J. Edler

CONTENTS

19.1 Experimental Details 516
19.2 Results and Discussion 517
19.3 Conclusions 523
Acknowledgments 523
References 523

The synthesis of inorganic nanomaterials is increasingly well developed; however, assembly of nanomaterials into well-organized arrays is challenging. Here we find that gold nanorods, prepared in aqueous solution by a seed-mediated growth approach in the presence of a cationic surfactant, assemble into liquid crystalline phases as a function of nanorod concentration. Liquid crystalline behavior is established by polarizing microscopy, small-angle x-ray scattering, and transmission electron microscopy.

The promises of nanotechnology require that scientists and engineers be able to manufacture and assembly nano-objects into working devices to be used in information technology, medicine, sensors, and a host of other applications.[1] In the last decade, the chemistry and materials science communities have developed numerous methods to synthesize nanometer-scale inorganic nanoparticles of various shapes and sizes.[2-4] However, linking nanoparticles via rational chemical approaches, or self-assembly approaches, is still challenging.

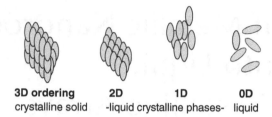

3D ordering | **2D** | **1D** | **0D**
crystalline solid | -liquid crystalline phases- | liquid

FIGURE 19.1 Far left: An array of rod-like particles ordered in three dimensions forms a crystalline solid. Middle: Ordering in one or two dimensions produces nematic and smectic liquid crystalline phases, respectively. Far right: A liquid has no long-range order.

In the late 1940s, Onsager[5] developed theories of the interaction of colloidal particles with each other, with an eye toward explaining phase separation in dilute solutions of highly anisotropic particles, such as the rod-shaped tobacco mosaic virus. Onsager predicted that solutions of hard rods would form liquid crystalline phases once some critical concentration had been reached; this concentration also depended on the aspect ratio of the rods.[5] Subsequent theory and experiment have extended and verified these ideas,[6-9] suggesting that anisotropic nanoparticles may self-organize into liquid crystalline arrays as a function of nanoparticle concentration. In general, one predicts that higher-aspect-ratio materials would require lower concentrations to self-order.[8] Figure 19.1 illustrates the differences between three-dimensional crystalline ordering, two- and one-dimensional liquid crystalline ordering, and zero-dimensional ordering in a liquid.

We have recently developed ways to make gold and silver nanorods of controllable size and aspect ratio.[10-13] The synthetic procedure leads to nanorods covered with a cationic surfactant; thus, the rods are not truly hard, and the ionic environment surrounding them, in terms of counterion association, is not well defined. Nonetheless, we have found that for gold nanorods of aspect ratios of 13 to 18, self-organization into liquid crystalline arrays does occur as the aqueous suspensions of the nanoparticles are concentrated to ~5 to 10% by weight.[14]

19.1 EXPERIMENTAL DETAILS

Gold nanorods were prepared as described previously.[11-13] All glassware must be cleaned rigorously with aqua regia and should be dedicated to the production of gold nanoparticles. Briefly, a 20-ml gold seed solution containing 2.5×10^{-4} M $HAuCl_4$ and 2.5×10^{-4} M trisodium citrate was prepared in a conical flask. Next, 0.6 ml of ice-cold aqueous 0.1 M $NaBH_4$ solution was added to the solution all at once while stirring. The solution turned pink immediately after adding $NaBH_4$, indicating nanoparticle formation. These nanoparticles were used as seeds within 2 to 5 h of preparation. The average particle diameter according to transmission electron microscopy (TEM) was 3.5 ± 0.7 nm.

To grow the gold nanorods from the seeds, 9 ml of 0.1 M aqueous cetyltrime-thylammonuim bromide (CTAB) was placed in three test tubes (labeled 1, 2, and 3, respectively). A 10^{-2} M HAuCl$_4$ stock solution (0.25 ml) and a 0.01 M ascorbic acid solution (0.05 ml) were added to each test tube. Ascorbic acid is a weak reducing agent that cannot reduce gold salt to elemental gold at room temperature in the absence of seeds. Next, to test tube 1, 1 ml of gold seed solution was added. The solution was rapidly stirred for about 30 sec, and then 1 ml of the contents of tube 1 was added to the contents of tube 2. The contents of tube 2 were stirred for 30 sec, and then 1 ml of the tube 2 solution was added to the contents of tube 3. The contents of tube 3 were stirred for 30 sec and allowed to sit overnight undisturbed, allowing the rods to form. Excess CTAB was removed from the nanoparticles by centrifuging tube 3 at 2000 rpm for 30 min at room temperature. The supernatant containing CTAB was removed, leaving behind gold rods that were redispersed in 1 ml of water. The aspect ratio of the resulting nanorods was 13 to 18.

Liquid crystalline assemblies were observed in the range of 1 to 100 mM CTAB. Samples were prepared for optical microscopy by concentrating 5 µl of concentrated Au nanorod solution on a glass slide by slow partial evaporation. A PZO Model ZM 100-T optical microscope was used to acquire optical microscopy images for both liquid crystalline Au nanorod assemblies and CTAB. TEM images were recorded from dilute and concentrated aqueous suspensions of gold nanorods that were dried by slow evaporation. A JEOL 1200EX analytical electron microscope operating at an 80-kV accelerating voltage was used.

Small-angle x-ray-scattering (SAXS) experiments were performed on concen-trated and diluted dispersions of gold nanorods in 1.5-mm-diameter x-ray glass capillary tubes using beam-line BM26B at the European Synchrotron Radiation Facility (ESRF), with a beam energy of 15 keV, as previously described.[14]

19.2 RESULTS AND DISCUSSION

The synthesis of gold and silver nanorods in solution has been developed by us using a seed-mediated growth approach.[10–13] The rationale is to separate nucleation of colloidal metal nanoparticles from growth into rods by performing the reduction reactions in two separate steps. In the first step, metal salt (HAuCl$_4$ or AgNO$_3$) is reduced in water with sodium borohydride, a strong agent, in the presence of sodium citrate to make about 4-nm metallic spheres, which are used as seeds in subsequent steps. Sodium citrate is a weak reducing agent, but here its function is to cap the growing nanoparticles to control their size. Sodium borohydride slowly reacts with water itself, so excess is used in this seed step; allowing the seed solution to sit for 2 to 5 h results in the reaction of the excess borohydride, so that in the subsequent steps only growth will take place.

For growth into nanorods, either single- or multistep growth procedures may be used.[10–13] In the single-step approach, one adds a portion of the seed solution to a fresh vessel containing more metal salt, the CTAB surfactant, and the weak reducing agent ascorbic acid. Ascorbic acid does not appreciably reduce the metal salt at room temperature alone, but in the presence of the seeds, the reaction is rapid. The aspect ratio of the resulting nanorods is controlled by the seed-to-salt ratio. Many spheres

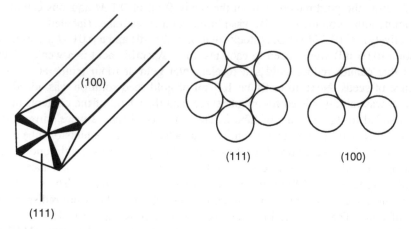

FIGURE 19.2 Left: Model of the crystal structure of gold nanorods, showing the faces of face-centered-cubic (fcc) gold that are on the short axis and long axis of the nanorods.[12] Right: Arrangement of Au atoms on the {111} and {100} faces of gold, respectively.

are present as well — in fact, the initial yield of nanorods is ~10 to 30%. Purification of the rods from the spheres is achieved by repeated washing and centrifugation.

The multistep method is one in which short rods are used as seeds to grow longer rods.[11] This is the method detailed in the experimental section, in which three reaction tubes are set up, and aliquots of solution from tube 1 (which produces short nanorods) are added to tube 2 to make longer nanorods, etc.

The gold nanorods originate from the surfactant-mediated growth of penta-twinned primary crystallites.[13] Originally, CTAB was chosen as a surfactant to direct the growth of inorganic nanorods because it formed a rod-like micelle structure on its own.[10–12] However, it is now thought that preferential adsorption of CTAB, or complex ions of it, to different faces of the growing gold nanorods is responsible for the shape anisotropy.[1,13]

The crystallography of the gold nanorods was determined by high-resolution transmission electron microscopy and single-area electron diffraction.[12] It was found that the gold nanorods (and presumably, by analogy, silver) are penta-tetrahedral twins. The ends of the nanorods consist of five tetrahedrons making a ring, with the {111} face of gold facing the solvent; the {100} face of gold is the face along the long axis of the nanorods (Figure 19.2).

With this crystallographic model, it becomes hard to reconcile how the rod-like micellar nature of CTAB functions as a soft template to promote the growth of gold nanoparticles into a rod-like shape. Instead, we now favor the hypothesis that pref-erential adsorption of CTAB, or complex ions of it, to different crystal faces controls the growth of the nanoparticles into rods. The {111} face of gold is the most stable, closely packed face of gold, whereas the {100} face has longer Au-Au average distances. It is believed that CTAB adsorbs with its cationic head group down to the gold surface and makes at least a bilayer on the surface, with a second layer of CTAB interdigitating its nonpolar tails together and the second layer of CTAB molecules aiming their polar heads at the solvent.[15] Modeling of the CTAB head group size compared to the Au-Au spacing of the two crystal faces of gold suggests that the {111} face is too tightly packed to facilitate epitaxy for CTAB and that the {100} face might allow more room for the surfactant to bind.[12] Additionally, the stabilization afforded by a CTAB bilayer, once started, may assist gold nanorod growth by zipping; this point will be discussed further below.

For liquid crystal array experiments, the gold nanorods were concentrated and separated from spherical nanoparticles by centrifugation. Thermogravimetric analysis showed that ~20% of the total mass of the nanorods burned off at ~200C and was presumably associated with the CTAB surfactant. This is significantly larger than that calculated for monolayer coverage (3 wt%) on the basis of particle surface area and indicates that the nanorods are covered with multiple CTAB layers, as shown also by previous data from Nikoobakht et al.[15] The presence of the surfactant coating was of key importance not only for the solubilization of the nanorods in water but also for controlling long-range self-assembly in concentrated dispersions. For example, we found that the optimum conditions required for *in situ* liquid crystalline ordering involved redispersing the nanorods in 1 to 100 mM CTAB after separating the rods from spheres by centrifugation. Above or below this surfactant concentration, the nanorods were unable to be redispersed in water.

The standard way to prove one has a liquid crystalline phase is to observe birefringence between crossed polarizers (e.g., in a polarizing optical microscope).[16] Because much of the work on inorganic colloidal liquid crystalline phases has been done with colorless materials such as aluminum hydroxide, this experiment is relatively easy to do if sufficient quantities of material are available. In our case, the 13- to 18-aspect-ratio nanorod solutions were dark brown in color (when concentrated) and had a weak absorbance maximum visible at ~530 nm, in addition to a near-infrared absorbance at ~1700 nm.[11] The fact that the concentrated solutions were so darkly colored and were small in volume (200 µl typically) made the polarizing microscopy experiments very difficult. Nonetheless, thin films of the concentrated dispersions supported on glass slides showed iridescent droplets of ~0.1 mm in diameter under polarizing light microscopy (Figure 19.3). The observed

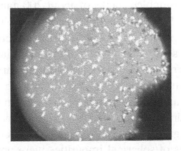

FIGURE 19.3 Polarizing microscopy image of ~5 to 10% dispersion of aspect ratio 13 to 18 gold nanorods in aqueous CTAB solution. The iridescent droplets indicate local regions of liquid crystalline ordering. (From Jana, N.R. et al., *J. Mater. Chem.*, 12, 2909, 2002. With permission of the Royal Society of Chemistry.)

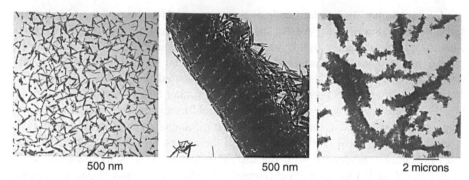

500 nm   500 nm   2 microns

FIGURE 19.4 Transmission electron micrographs showing concentration-dependent ordering of gold nanorods. Left: <1 wt% showing isolated nanorods; some spherical particles are still present even after centrifugation. Center: ~5 to 10 wt% showing liquid crystalline arrays near droplet edges. Right: ~5 to 10 wt% showing linear stacks of gold nanorods. (From Jana, N.R. et al., *J. Mater. Chem.*, 12, 2909, 2002. With permission of the Royal Society of Chemistry.)

textures were indicative of localized regions of liquid crystalline ordering and are similar to nematic droplets observed in boehmite nanoneedle solutions.[8] Significantly, the liquid crystalline droplets were stable up to 200°C in air, after which the surfactant began to degrade, although the nanorods remained unchanged in size and shape as judged by subsequent TEM experiments.

Another technique to estimate long-range ordering of species in solution is SAXS.[16] SAXS experiments were performed on concentrated (~5 to 10 wt% of solids) and diluted (by a factor of ~1000) dispersions of the gold nanorods. The data were fitted to a model consisting of core-shell cylinders stacked with a Gaussian distribution of interparticle distances using nonlinear least-squares fitting. The fitting parameters included the radius of the nanorods, number of particles in a stack, width of the Gaussian distribution of interparticle distances in the stack, surfactant layer thickness, and major radius of elliptical impurities (which were observed in transmission electron microscopy; see below). The fits suggest that the concentrated solutions contained self-assembled stacks of about 200 nanorods, each of which had a surfactant coating of 3.9 nm in thickness, consistent with a CTAB bilayer, as suggested by previous workers.[15] In contrast, smaller clusters of ~30 rods were present in the more dilute sample. The width of the Gaussian distribution of spacings between the particle centers was an order of magnitude narrower in the concentrated dispersions than in the dilute dispersions, consistent with more dense arrays in the concentrated dispersions and a high degree of disorder in the dilute dispersions.

TEM images of air-dried dispersions prepared at low nanorod concentrations (<1% by weight, including surfactant) showed mainly discrete nanoparticles (Figure 19.4). Significant amounts of spherical impurities were present; improved centrifugation and washing steps reduced the amount of spherical impurities. Some short-range ordering was observed, presumably due to capillary forces associated with the drying process.[15] At high nanorod concentrations (~5 to 10 wt%), in contrast, liquid crystalline arrays of closely packed nanorods were observed (Figure 19.4).

The arrays consisted of nanorods that were aligned parallel to each other in microme-ter-sized rows, which in turn were stacked side by side to produce the higher-order superstructure. Such structures were observed predominantly at the edges of dried droplets (apparent by visual inspection of the TEM grid as brown rings), suggesting that capillary forces were responsible for the organization. In contrast, other areas of the TEM grid showed a predominance of micrometer-long rows of ordered nanorods (Figure 19.4), which probably correspond more closely to the *in situ* organization of the nanorods within the concentrated dispersion. We estimate tens of rods assemble in vertical stacks, whereas ~100 stacks assemble into rows, in reasonable agreement with the SAXS data.

Nikoobakht et al.[15] have reported that gold nanorods of aspect ratio 4.6, coated with two different cationic surfactants, assemble into higher-order structures upon concentration from aqueous solution. Our results with higher-aspect-ratio nanorods are consistent with this and indicate that surfactant-mediated interactions between gold nanorods of uniform shape and size can give rise to ordered liquid crystalline arrays in concentrated suspensions. Multilayers of surface-adsorbed cationic surfac-tants such as CTAB can induce a remarkable degree of self-ordering of spherical gold nanoparticles due to a balance of short-range electrostatic repulsion and inter-chain attraction.[17] Moreover, interdigitation of surfactant chains on specific faces of prismatic nanocrystals can give rise to ordered single chains of other (e.g., $BaCrO_4$) nanoparticles.[18] Li et al.[19] has polarizing microscopy evidence for liquid crystalline phases in concentrated solutions of CdSe nanorods of aspect ratios 10 to 13. For the gold nanorods described here, liquid crystalline arrays were only routinely observed in aqueous solutions containing the proper concentration range of CTAB, suggesting that interactions between surfactant molecules in solution with surface adsorbates were important aspects of the assembly process. The hydrophilic nature of the gold nanoparticles prior to assembly indicates that the surfactant molecules in the outer layer of the surface coating are oriented with their cationic head groups exposed to the solvent. However, as the surfactant-coated nanorods approach each other in solution, expulsion of the outermost cationic head-out CTAB molecules and their associated counterions could result in the formation of hydrophobic nanorods in which the remaining CTAB hydrophobic tails face the solvent; thus, the resulting nanorods spontaneously self-assemble in a side-on fashion to minimize the unfavor-able hydrophilic–hydrophobic interactions with water and promote interdigitation of the surfactant tails (Figure 19.5).

To what degree does interdigitation of the hydrophobic chains of the absorbed CTAB surfactant influence the growth and assembly of gold nanorods? Approxi-mately 0.1 *M* CTAB is required for nanorod formation in the first place, and 1 to 100 m*M* CTAB is required for the assembly of the nanorods into liquid crystalline arrays. Assembly of CTAB, either by itself in the form of a micelle or on a gold surface in a bilayer, requires the interdigitation of hydrocarbon chains. The hydro-carbon tail contribution to the standard free energy of micellization for trimethy-lammonium bromides in water, following a mass-action model that uses empirical data, has been estimated to be[20]

$$\Delta G^\circ = 2.303 \, (2 - z/j) \, RT \, (0.1128 - 0.3074n) \qquad (19.1)$$

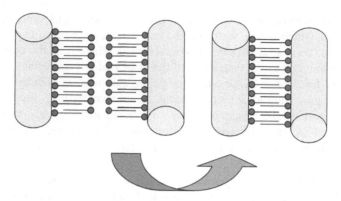

FIGURE 19.5 Cartoon showing gold nanorods (cylinders) covered with a bilayer of CTAB surfactant (balls with lines) being brought close together in a side-by-side alignment as CTABs are expelled from the interface (along with water and counterions, presumably). The arrow denotes the entropy-driven ion expulsion process.

where n is the number of carbon atoms in the surfactant tail (16 for CTAB), z is the charge on the micelle, j is the aggregation number of the micelle, z/j corresponds to the number of strongly bound anions to the micelle, R is the gas constant, and T is temperature. If we assume for the sake of estimation that z and j do not change upon bilayer formation on gold nanorods and that the z/j ratio is significantly smaller than 1 due to strongly bound counterions, then we estimate that $\Delta G^{\circ}_{n=16} = -48.1$ kJ/mol. Thus, formation of a CTAB bilayer (or multilayer) *in situ* as the gold nanorods grow would further assist in rod formation in a zipping mechanism. One would also predict, based on Equation 19.1, that shorter-chain surfactants would lead to less stabilization of the bilayer on the gold surface and perhaps influence the growth and stability of the nanorods. We have recently tested this idea with other trimethylammonium bromide surfactants for n = 10, 12, and 14.[21] These molecules are all surfactants, with the same head group as CTAB; all make micelles in water but have shorter tails than CTAB. We find that the longest gold nanorods possible using our standard procedure directly depend on the hydrocarbon chain length. For n = 10, no rods are formed (in spite of the supposed epitaxial nature of head group binding to the different crystal faces of gold); short rods are formed with n = 12; longer ones with n = 14; and the longest are formed with n = 16 (CTAB itself). These data also support the notion of a surfactant zipper assisting in nanorod growth, in addition to being critical for the formation of higher-ordered arrays of nanorods.[21]

Theory predicts that liquid crystal formation would occur in concentrated dispersions of anisotropic particles, with higher-aspect-ratio materials requiring lower concentrations to order, and also that the polydispersity of the colloidal solution influences liquid crystalline organization.[5,8,22,23] In the case of hard rods, two-dimensional smectic-like ordering is predicted to be observed for shape polydispersities less than 18%; for more polydisperse solutions, only nematic or columnar phases should be observed.[22] For nanorods of the aspect ratios examined here, Onsager[5] and Gabriel[8] predict approximately a minimum concentration of 20% to order. Our samples have (barely) reached these criteria, and we do indeed observe ordering in solution, as described above. More rigorous tests of Onsager's theories would require larger quantities of purified nanorods, with variable aspect ratios and polydispersities; indeed, in other inorganic colloidal systems, Onsager's theories are amply borne out.[6,7,16]

19.3 CONCLUSIONS

We have shown that gold nanorods of aspect ratio 13 to 18 self-organize into liquid crystalline arrays as a function of nanorod concentration, as evidenced by polarizing microscopy, small-angle x-ray scattering, and transmission electron microscopy. The assembly process is likely to be driven by entropy in the form of ion and water expulsion from the interface as the nanorods approach each other.

ACKNOWLEDGMENTS

We thank the EPSRC, U.K., for support of a postgraduate studentship to C.J., Dr. Igor Dolbyna and the other staff on BM26B at the ESRF for assistance in collecting the SAXS data, the Royal Society for a Dorothy Hodgkin Research Fellowship (K.J.E.), and the U.S. National Science Foundation (C.J.M.). We also thank N. Kotov for assembling a wonderful symposium on nanomaterials at the August 2002 American Chemical Society National Meeting in Boston, MA.

REFERENCES

1. Murphy, C.J. *Science* 2002, *289*, 2139.
2. Feldheim, D.L., Foss, C.A., Jr., Eds. *Metal Nanoparticles: Synthesis, Characterization, and Applications*. Marcel Dekker: New York, 2002.
3. Shipway, A.N., Katz, E., Willner, I. *ChemPhysChem* 2000, *1*, 18.
4. Hu, J., Odom, T.W., Lieber, C.M. *Acc. Chem. Res.* 1999, *32*, 435.
5. Onsager, L. *Ann. N.Y. Acad. Sci.* 1949, *51*, 627.
6. Frenkel, D., Lekkerkerker, H.N.W., Stroobants, A. *Nature* 1988, *332*, 822.
7. van der Kooij, F.M., Lekkerkerker, H.N.W. *Langmuir* 2000, *16*, 10144.
8. Gabriel, J.-C. P., Davidson, P. *Adv. Mater.* 2000, *12*, 9.
9. Kim, F., Kwan, S., Akana, J., Yang, P. *J. Am. Chem. Soc.* 2001, *123*, 4360.
10. Jana, N.R., Gearheart, L., Murphy, C.J. *Chem. Commun.* 2001, *7*, 617.
11. Jana, N.R., Gearheart, L., Murphy, C.J. *J. Phys. Chem. B* 2001, *105*, 4065.
12. Jana, N.R., Murphy, C.J. *Adv. Mater.* 2002, *14*, 80.
13. Johnson, C.J., Dujardin, E., Davis, S.A., Murphy, C.J., Mann, S. *J. Mater. Chem.* 2002, *12*, 1765.
14. Jana, N.R., Gearheart, L.A., Obare, S.O., Johnson, C.J., Edler, K.J., Mann, S., Murphy, C.J. *J. Mater. Chem.* 2002, *12*, 2909.
15. Nikoobakht, B., Wang, Z.L., El-Sayed, M.A. *J. Phys. Chem. B* 2000, *104*, 8635.
16. van der Kooij, F.M., Kassapidou, K., Lekkerkerker, H.N.W. *Nature* 2000, *406*, 868.
17. Fink, J., Kiely, C.J., Bethell, D., Schiffrin, D.J. *Chem. Mater.* 1998, *10*, 922.
18. Li, M., Schnablegger, H., Mann, S. *Nature* 1999, *402*, 393.
19. Li, L.-S., Walda, J., Manna, L., Alivisatos, A.P. *Nano Lett.* 2002, *2*, 557.

20. Anacker, E.W. In *Cationic Surfactants*, Jungermann, E., Ed. Marcel Dekker: New York, 1970, 203–288.
21. Gao, J., Bender, C.M., Murphy, C.J. *Langmuir* 2003, *19*(21): 9065–9070.
22. Bates, M.A., Frenkel, D. *J. Chem. Phys.* 1998, *109*, 6193.
23. Bates, M.A., Frenkel, D. *J. Chem. Phys.* 2000, *112*, 10034.

20 Tailoring the Morphology and Assembly of Silver Nanoparticles Formed in DMF

Isabel Pastoriza-Santos and Luis M. Liz-Marzán

CONTENTS

20.1 Introduction 525
    20.1.1 Metal Colloids...525
    20.1.2 Coated Nanoparticles 526
    20.1.3 Organic Solvents as Reducing Agents....................................527
20.2 Reduction of Silver with DMF...527
    20.2.1 DMF as a Reducing Agent 527
    20.2.2 Silver Deposition on Solid Substrates 528
    20.2.3 Polymer-Stabilized Silver Colloids..530
20.3 Synthesis of Coated Silver Nanoparticles in DMF.............................530
    20.3.1 Ag@SiO$_2$ Core-Shell Colloids..530
    20.3.2 Ag@TiO$_2$ Core-Shell Colloids..532
20.4 Complex Nanostructures from TiO2-Coated Nanoparticles 533
    20.4.1 TiO$_2$ Nanoshells...534
    20.4.2 Formation of AgI@TiO$_2$ 535
    20.4.3 Nanostructured Thin Films through Layer-by-Layer
            Assembly 535
20.5 Anisometric Nanoparticles...538
    20.5.1 Formation of Nanorods and Nanoprisms................................538
    20.5.2 Optical Spectra in Solution..539
    20.5.3 Layer-by-Layer Assembled Films..542
20.6 Summary and Outlook 543
References...544

20.1 INTRODUCTION

20.1.1 METAL COLLOIDS

Nanostructured materials with characteristic dimensions of their structural units of 1 to 100 nm represent one of the most dynamic areas of modern science.[1-3] The

525

interest of these systems is due to the remarkable and unusual properties when at least one dimension is in the nanometer range. Nanoparticles present a high fraction of surface atoms compared to conventional materials, which implies that most interparticle interactions are surface–interface interactions. In the case of nanosized metals, the unique optical and electronic properties stem from the presence of free electrons in the conduction band, which are responsible for the characteristic surface plasmon absorption band, due to coupling of the electron oscillation frequency and the alternating field of the incoming electromagnetic radiation. Such an absorption of light in the UV or visible region by metal particles is sensitive to particle size,[4,5] shape,[6–11] dispersion medium,[4,5,12] encapsulation layers,[13–16] electron density on the particles,[17,18] and temperature.[19–21] The origin and behavior of these optical effects have been explained by combining the classical Drude model for the dielectric properties of metals with the Mie theory for light absorption and scattering by spheres.[5]

One of the main applications of metal particles is found in catalysis.[22,23] The remarkable reactivity of nanoparticles is associated with their capability of storing excess electrons, which makes metallic particles act as catalysts in several photochemical reactions. In addition, the interest has been partly motivated by the potential for creating novel, nanostructured materials,[24] which enhances their possible applications to other fields such as biomedicine[25] and technology (solar absorbers, photovoltaic systems,[26,27] microelectronics,[28] or electrically conductive films[29]).

20.1.2 COATED NANOPARTICLES

Core-shell nanoparticles[30] are attracting more and more interest due to the combination of different properties in one particle based on different compositions of the core and the shell. The cores often (though not always, as will be shown below) carry the relevant property, e.g., semiconductors, metals, magnetic oxides, and encapsulated molecules, whereas the shells can stabilize the core, serve as a compatibilizer between the core and the environment, or change the charge, functionality, or reactivity of the surface. In recent years, especially for medical purposes, the shells have also become more and more important, e.g., for drug delivery applications.

A fundamental problem with conventional stabilization methods is that coalescence takes place when colloids are subject to severe conditions, such as high volume fractions, transfer into solvents dissimilar from that in which the particles were initially dispersed, or elevated temperature. These problems can be (partly) obviated by coating the particles with silica. Silica encapsulation provides high stability against coagulation under harsh conditions, including those stated above. Other requirements that cannot be achieved with conventional stabilization techniques, but which are satisfied by silica coating, include optical transparency, chemical inertness, thermal stability, control of interparticle spacing, low cost, and simple methodology for silica encapsulation of particles. These attributes of the silica shell as a stabilizer should provide a vast number of opportunities for the exploitation of properties of the core particles that could not be achieved with the bare particles, and numerous novel nanomaterials and nanostructures may be generated. A number of silica-coating procedures have been devised since the pioneering work by Iler,[31] based on the slow polymerization of soluble silicates, the hydrolysis of organosilanes,[32] both of them (without[33,34] or with preliminary surface priming[16,35–37]), or reactions in microemulsions.[38,39]

In a different direction, metal nanoparticles have also been coated with semi-conductor shells.[40–45] In this case, the coating is not directed in general to stability enhancement but, rather, to obtain special properties due to the combination of the two materials with different electronic structures. We shall deal later in this chapter with silver nanoparticles coated with TiO_2, showing not only how this affects the optical properties but also how, upon removal of the metal cores, the shells can be assembled and utilized for biomedical applications.

20.1.3 ORGANIC SOLVENTS AS REDUCING AGENTS

Several examples exist on the reduction of metallic salts by organic solvents. Prob-ably the most popular one is ethanol, which has been long used by Toshima and coworkers[22,46–48] for the preparation of metal nanoparticles such as Pt, Pd, Au, or Rh (suitable for catalytic applications) in the presence of a protecting polymer. Another interesting example is found in Figlarz's polyol method, which was initially devel-oped for the formation of larger colloid particles[49] though later was also adapted and improved for the production of silver nanoparticles,[50] and recently by Xia and coworkers[51,52] for the production of silver nanowires. Isopropanol was also used for the preparation of silver colloids in basic conditions.[53] Related to these processes, the reduction of noble metal salts has also been accomplished by nonionic surfactants and, more specifically, by those with a large number of ethoxy groups, to which the reducing ability has been assigned.[54] N,N-dimethylformamide (DMF) is also one of the usual organic solvents for numerous processes, including the preparation of colloids, which usually contain metals in their composition.[55,56] The oxidation of DMF has been studied with respect to the production of hydrogen from water–DMF mixtures[57] and even for the reduction of Ni(IV) to Ni(II) in alkaline medium.[58] These studies show that DMF can be an active reducing agent under suitable conditions. We have explored the possibilities of DMF for the synthesis of metal nanoparticles under various conditions,[25,42,59–63] and the results are summarized within this chapter. It has also been shown that formamide[64] can act as a powerful reductant for silver or gold salts in the absence of oxygen. More recently, Rodríguez-Gattorno and coworkers[65] reported the preparation of silver nanoparticles by the spontaneous reduction of silver 2-ethylhexanoate in dimethyl sulfoxide (DMSO) at room tem-perature.

20.2 REDUCTION OF SILVER WITH DMF

20.2.1 DMF AS A REDUCING AGENT

It has been demonstrated that DMF can reduce Ag^+ ions to the zero-valent metal, even at room temperature, in the dark, and in the absence of any additional reducing agent.[61] Based on the absence of gas evolution during the process, the following reaction was proposed among several possible routes reported in the literature:

$$HCONMe_2 + 2Ag^+ + H_2O \rightarrow 2Ag^0 + Me_2NCOOH + 2H^+ \qquad (20.1)$$

This mechanism is supported by a measured increase of conductivity as the reaction proceeds, which indicates that the larger Ag^+ ions are progressively exchanged for the more mobile H^+ ions.

Compared to the ethanol reduction method,[14] it is remarkable that the DMF reduction proceeds at a meaningful rate even when performed at room temperature and in the dark, which is readily observed as a yellow color showing up in the solution and deepening with time. It is also notable that the reaction is performed without degassing the solutions or taking any special care with regard to the presence of oxygen, which is in contrast with the observations of Han and coworkers,[64] who claimed that the reduction of $AuCl_4^-$ by formamide can only take place in an oxygen-free environment and is actually frozen when oxygen is allowed to come into contact with the system.

This reducing ability under mild conditions indicates a better efficiency of DMF for the production of Ag nanoparticles than ethanol or other organic solvents. Profit of this feature has been taken for the synthesis of various nanomaterials, as discussed in this chapter.

20.2.2 Silver Deposition on Solid Substrates

Because of the strong plasmon resonance of silver nanoparticles, the reduction of silver salts by DMF can be monitored from the color evolution of the solution. Visual observation shows that upon addition of $AgNO_3$ or $AgClO_4$ to DMF, the color shifts from light yellow to dark brown, through orange and olive green, and then starts to accumulate on the glass surfaces in contact with the solution, while the solution itself becomes increasingly clearer (the amount of stable Ag colloid in solution decreases) until it is completely colorless. Spectral monitoring of the solution indicates that the reduction is basically completed at this point, meaning that silver metal particles are formed in solution and then stick onto glass surfaces due to electrostatic attraction between the particles with excess positive charge (from adsorption of unreacted silver ions[66]) and the negatively charged SiO_2 surface. Shown in Figure 20.1A is an example of the spectral changes in solution for an initial silver salt concentration of 2.0 mM. Figure 20.1B shows the spectra of glass slides immersed in the solution for 24 h, for samples with different initial silver concentrations. In Figure 20.2, selected SEM images are presented, showing the morphology of the surface on which silver was deposited, for low and high concentrations. It can be observed that for low silver concentrations (0.5 mM, Figure 20.2A) the aggregates are constituted by particles that maintain their individuality, whereas for higher concentrations (2.0 mM, Figure 20.2B) the aggregate structure is more compact, being made up of particles that look sintered. The aggregate structure is also reflected in the shape of the spectra.

It is obvious from both SEM images and absorption spectra that regardless of concentration, silver nanoparticles are steadily deposited, initially with a uniform distribution but later leading to agglomeration, which is reflected in red shift and broadening of the plasmon absorption band,[67–70] initially centered around 400 nm. At high Ag concentrations, the high wavelength absorbance tail rises from zero, showing the transition from separated nanoparticles to bulk-like silver films.

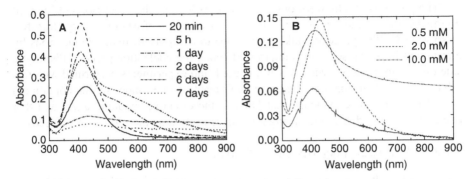

FIGURE 20.1 (A) Spectral UV-visible changes in solution for an initial silver salt concentration of 2.0 m*M*. (From Schlamp, M.C. et al., *J. Appl. Phys.*, 82, 5837–5842, 1997. With permission. Copyright © 2000 Academic Press.) (B) Spectra UV-visibility of films deposited from AgNO$_3$ solutions for a period of 24 h. (From Pastoriza-Santos, I. and Liz-Marzán, L.M., *Pure Appl. Chem.*, 72, 83–90, 2000. With permission. Copyright © 2000 IUPAC.)

FIGURE 20.2 Scanning electron micrographs showing the morphology of thin films deposited from DMF solutions of AgNO$_3$ with starting concentrations of 0.5 m*M* (A) and 2.0 m*M* (B). The insets show blow-ups from selected areas. (Adapted from Pastoriza-Santos, I. and Liz-Marzán, L.M., *Pure Appl. Chem.*, 72, 83–90, 2000. With permission. Copyright © 2000 IUPAC.)

The previous results were obtained using $AgNO_3$ as the precursor salt for sup-plying Ag^+ ions to the solution, which poses the question of whether charge transfer-to-solvent (CTTS) processes take place[71] due to a supply of hydrated electrons by nitrate ions in the presence of light, thus affecting the nucleation process. Photolytic side effects do not arise when $AgClO_4$ is used, and experiments performed with this salt (nanoparticle formation with comparable reaction rates under similar conditions) provided[59] conclusive evidence that DMF is the effective reducing agent and that nucleation occurs due to increasing Ag^0 concentration as the reduction proceeds.

20.2.3 POLYMER-STABILIZED SILVER COLLOIDS

The most widely used substances for the stabilization of metal nanoparticles are ligands and polymers, especially natural or synthetic polymers with a certain affinity toward metals, which are soluble in suitable solvents.[72,73] Such substances can also control the reduction rate of the metal ions and the aggregation process of zero-valent metal atoms. The preparation of polymer-stabilized nanoparticles (through chemical methods) basically involves two processes: reduction of metal ions into neutral atoms and coordination of the polymer to the metal nanoparticles.[74] During the last decade, Toshima's group[75-77] developed and optimized an effective technique for the preparation of colloidal dispersions through reduction with an alcohol at reflux, in the presence of a protecting polymer. Using poly(vinylpyrrolidone) (PVP) as a stabilizer, they managed to obtain stable dispersions of nanoparticles of gold, platinum, and other metals (not silver) in ethanol.

A systematic study has been recently performed[62] on the reduction of Ag^+ ions in and by DMF, in the presence of PVP. In that study, the influence of several parameters was monitored, such as the use of reflux conditions or microwave irra-diation to accelerate the reaction. Although under most of the experimental condi-tions stable silver colloids were obtained, the use of moderate concentrations ([Ag] = 0.76 mM) and microwave irradiation proved to be optimal for the production of monodisperse colloids with well-defined surface plasmon bands. The as-formed nanoparticles could also be transferred into other solvents, such as ethanol and water, with no need of intermediate surface modification.

20.3 SYNTHESIS OF COATED SILVER NANOPARTICLES IN DMF

In this section we provide a survey of various types of core-shell nanoparticles with silver cores synthesized by DMF reduction. The main interest of such coated particles relies on the possibility to create nanomaterials and nanostructures with unique, tailored properties.

20.3.1 AG@SiO$_2$ CORE-SHELL COLLOIDS

As mentioned in the introduction, the formation of inert silica shells on metallic cores has been accomplished through various chemical procedures. The advantages include an enhancement of colloidal and chemical stability, the possibility of surface

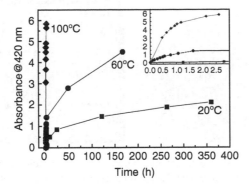

FIGURE 20.3 Kinetic traces for the formation of Ag particles in DMF solution with APS at different temperatures, with [Ag]/[APS] = 1 The inset shows the traces for the initial stage of the reaction. [Ag] = 0.5 m*M*. (From Pastoriza-Santos, I. and Liz-Marzán, L.M., *Langmuir*, 15, 948–951, 1999. With permission. Copyright © 1999 American Chemical Society.)

modification and solvent exchange, or tailored interparticle separation within assemblies, among others. In the context of the formation of silver nanoparticles in DMF, silica coating has been achieved by simple addition of a silica precursor to the Ag+ solution in DMF.[59] As a silica precursor, the silane coupling agent 3-aminopropyl-trimethoxysilane (APS) was used, because it can readily complex to silver atoms through its amine functionality, so that silane groups point toward solution, in a similar fashion to what was reported for the equivalent 3-mercaptopropyl-trimethoxysilane (MPS), during the synthesis of gold nanoparticles in ethanol.[78]

As mentioned above, the reduction of Ag+ by DMF can proceed at room temperature. However, it was observed that at higher temperatures the reaction rate is greatly enhanced. The effect of temperature was followed by UV-visible spectroscopy for three different values of the [Ag]/[APS] ratio *R* (0.38, 1.0, 1.2), because this ratio also affects the reaction rate. For each *R* value, three identical samples were prepared and the reaction was followed at 20, 60, and 100°C by measuring the UV-visible spectrum of aliquots extracted from the samples at selected time intervals. The reactions performed at reflux were so quick that the time evolution could not be followed. As can be observed in Figure 20.3 for *R* = 1.0, the reaction rate increases dramatically (in orders of magnitude) upon temperature increase.

The influence of temperature on the morphology and size of the resulting particles was studied by transmission electron microscopy (TEM). A different morphology was obtained for low- and high-temperature reactions with excess APS, which can be easily related to the respective reaction rates for silver reduction and APS condensation. At room temperature, silver reduction is so slow that silica nuclei can form and grow in solution, on which (slowly formed) silver particles attach (Figure 20.4A). Conversely, at high temperature silver formation is so quick that silica polymerizes directly onto the (APS-coated) silver particle surface (Figure 20.4B). Because the shells are formed by condensation of APS molecules, amine functionalities are present on the surface of the coated nanoparticles, which can be useful for further surface modification reactions, such as bioconjugation.

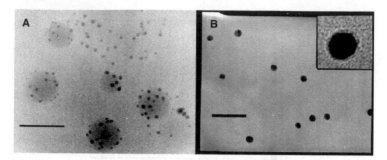

FIGURE 20.4 Transmission electron micrographs of Ag particles prepared in DMF. (A) Reduction at room temperature. (B) Reduction at reflux, where the inset is a close-up of one of the particles, showing more clearly the presence of a thin, homogeneous silica shell. [Ag] = 0.5 m*M*, [Ag]/[APS] = 1. Scale, 100 nm. (From Pastoriza-Santos, I. and Liz-Marzán, L.M., *Langmuir*, 15, 948–951, 1999. With permission. Copyright © 1999 American Chemical Society.)

20.3.2 Ag@TiO$_2$ Core-Shell Colloids

TiO$_2$-coated particles are exceptionally difficult to synthesize due to the high reactivity of most titania precursors, which makes it difficult to control their hydrolysis. However, the semiconducting properties of TiO$_2$ make these particles very useful as catalysts (other advantages include low cost and toxicity as well as convenient redox properties),[79–81] as pigments,[82,83] and as energy converters in solar cells.[84,85] The large dielectric constant of titania also makes it a material of choice for the fabrication of photonic crystals.[86–88]

In this section we show that silver nanoparticles coated with titania can be synthesized through a one-step process similar to that described in the previous section for Ag@SiO$_2$. The preparation of Ag@TiO$_2$ core-shell nanoparticles can be considered a combination of two processes that occur sequentially in a single reaction mixture: the formation of the silver core and the hydrolysis of titanium butoxide leading to TiO$_2$ formation. The first process comprises reduction of silver ions by DMF, which yields stable silver colloids in the presence of a stabilizer. The second process is the controlled hydrolysis and condensation of Ti(OC$_4$H$_9$)$_4$ (TOB) by refluxing in the presence of a chelating agent, such as acetylacetone, to slow down the hydrolysis.[89] The reactions involved in the hydrolysis can be represented as

$$Ti(OC_4H_9)_4 + 4H_2O \rightarrow Ti(OH)_4 + 4C_4H_9OH \qquad (20.2)$$

$$Ti(OH)_4 \rightarrow TiO_2 \cdot xH_2O + (x-2)H_2O \qquad (20.3)$$

Because of the surface plasmon band of silver cores, again the process can be easily monitored by UV-visible spectroscopy, as shown in Figure 20.5. The spectrum of the final aqueous sol shows a large red shift of the silver plasmon band from 400 nm, which is typical for uncoated silver nanoparticles, to 435 nm, as well as largely enhanced absorbance at lower wavelengths. Both of these spectral features are related to the high refractive index of TiO$_2$[5,16] being in contact with the silver surface.

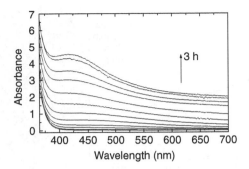

FIGURE 20.5 Time evolution of UV-visible absorption spectra during the formation of Ag@TiO₂ particles in a DMF:EtOH mixture (1:4).

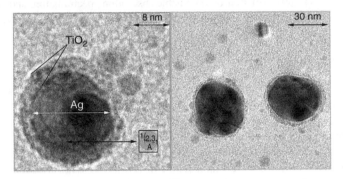

FIGURE 20.6 Transmission electron micrographs of Ag@TiO₂ particles. (From Pastoriza-Santos, I. et al., *Langmuir*, 16, 2731–2735, 2000. With permission. Copyright © 2000 American Chemical Society.)

The final nanoparticles were also characterized by high-resolution transmission electron microscopy (HRTEM), with the result that particles with two different average sizes were found on the carbon-coated copper grids (Figure 20.6). There is a population of larger silver particles with an average diameter of about 20 nm, which are homogeneously coated with a thin shell of amorphous TiO₂ as identified by energy dispersion analysis. Additionally, smaller silver particles can be seen on the same grid, with an average particle size of some 4 nm. For them, the low contrast of the image does not allow for the visualization of a shell. However, the stability of these dispersions at pH values at which standard naked silver colloids flocculate suggests that their surface is also modified by TOB.

20.4 COMPLEX NANOSTRUCTURES FROM TiO₂-COATED NANOPARTICLES

The porosity of the titania shells makes the titania-coated nanoparticles prone to undergoing chemical reactions *in situ*, thereby leading to new nanostructures with different properties. Such processes enhance the possibilities of these particles toward interesting applications involving such nanostructured materials.

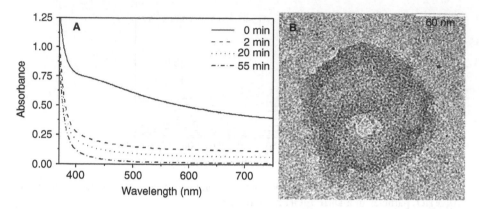

FIGURE 20.7 (A) Time evolution of UV-visible absorption spectra before and after adding CN^- to $Ag@TiO_2$ dispersion. (B) High-resolution transmission electron micrograph of a TiO_2 nanoshell.

20.4.1 TiO₂ NANOSHELLS

Because of the possible applications in fields such as catalysis, electronics, or biomedicine, the formation of titania nanoshells is one of the most interesting possibilities to achieve by *in situ* chemical reaction. Titanium dioxide nanoshells were obtained either by adding 30% solution of ammonium hydroxide or by adding potassium cyanide to the solution of the original dispersion of $Ag@TiO_2$ nanoparticles. The silver core dissolved as a result of oxidation with atmospheric oxygen and formation of a silver complex ion. The silver complex with NH_3 molecules is formed as follows[90]:

$$4Ag + 8NH_3 + O_2 + 2H_2O \rightarrow 4Ag(NH_3)_2^+ + 4OH^- \tag{20.4}$$

whereas the reaction with CN^- ions is[36]:

$$4Ag + 8CN^- + O_2 + 2H_2O \rightarrow 4Ag(CN)_2^- + 4OH^- \tag{20.5}$$

and a soluble, colorless complex ion is also formed. The formation of coordinate compounds reduces the redox potential of the Ag/Ag^+ pair and results in the oxidation of silver metal by oxygen from the environment. The oxidation products, i.e., silver ion complexes, diffuse from the interior of the $Ag@TiO_2$ nanoparticles to the solution through pores in the titania shell. The oxidation of silver was monitored by UV-visible spectroscopy through the gradual decrease of the silver surface plasmon band at 440 nm (Figure 20.7A). Upon complete extraction of the silver core, hollow porous TiO_2 spheres are obtained with a shell thickness of 3 to 5 nm, as evidenced by HRTEM (Figure 20.7B). The dissolution process was confirmed by x-ray photoelectron spectroscopy (XPS) measurements,[25] which provided cumulative information on the chemical composition of a large number of particles that might not have been surveyed by TEM.

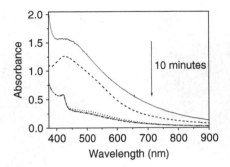

FIGURE 20.8 Time evolution of UV-visible absorption spectra before and after adding molecular I_2 to Ag@TiO$_2$ dispersion.

20.4.2 Formation of AgI@TiO$_2$

More complex nanoparticles can be created by following the same principle, i.e., chemical reactions on the metal core without affecting the oxide shell. One example is the oxidation by molecular iodine according to

$$2Ag + I_2 \rightarrow 2AgI \tag{20.6}$$

As previously reported by Giersig et al.[91] for Ag@SiO$_2$ particles, the surface plasmon band of the silver cores located at 435 nm quickly disappears upon iodine addition, and after 10 min the spectrum reveals the typical 420-nm peak of colloidal β-AgI, which confirms that the observed absorption changes are due to metallic silver oxidation and formation of colloidal AgI (Figure 20.8).

HRTEM images show that during the first stage, I_2 diffuses through the porous titania shells, where it starts oxidizing the silver surface. At the same time, an AgI nucleus is formed. Upon reaction, an incomplete titania shell is observed on the AgI surface, which indicates that the formation and growth of AgI nuclei take place inside the titania shell, but due to volume increase, the shell breaks since it is unable to accommodate the whole AgI particle formed. Figure 20.9 shows the initial shape of a Ag@TiO$_2$ nanoparticle and an intermediate stage of the reaction, where AgI still coexists with unreacted Ag.

20.4.3 Nanostructured Thin Films through Layer-by-Layer Assembly

In the most interesting and practically promising ideas for metallic and metal sulfide/oxide clusters, for instance, light-emitting diodes,[92–94] photovoltaic and energy conversion elements,[95–98] and memory devices,[99] the nanoparticles are utilized in the form of thin films deposited on suitable substrates.

Homogeneous, compact thin films made of core-shell nanoparticles can be constructed using the layer-by-layer (LBL) assembly technique.[100–102] LBL permits the preparation of materials that can hardly be obtained by other methods. In this method, a monolayer of polyelectrolyte is deposited on a hydrophilic substrate, which is subsequently exposed to a colloidal dispersion of oppositely charged particles. The

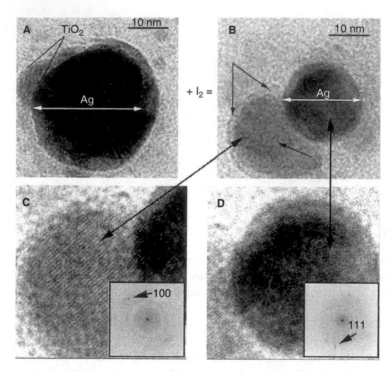

FIGURE 20.9 Electron micrographs of a single Ag@TiO$_2$ particle before (A) and after (B) partial reaction with I$_2$. (C, D) High-resolution images of the forming AgI@TiO$_2$ particle and the remaining silver particle.

combination of both electrostatic and hydrophobic interactions[103] results in a strong adsorption of the nanoparticles onto the polyelectrolyte layer, whereas a high surface charge impedes the formation of thick, rough films. The thickness and lateral particle density of the films can be varied by changing the strength of electrostatic and hydrophobic interactions.

In the case of Ag@TiO$_2$, the relatively hydrophobic environment in the DMF–ethanol mixture of the original dispersion and the low surface charge of the nanoparticles make LBL assembly problematic. Therefore, prior to film preparation the original dispersion is diluted (1:15) with water at pH <3.5. Under these conditions, the titania-coated particles become positively charged, whereas the large amount of dilution water activates the hydrophobic attraction to the polyelectrolyte strands. For positively charged Ag@TiO$_2$ nanoparticles, negatively charged poly(acrylic acid) (PAA) is used as a partner polyelectrolyte for LBL.

Atomic force microscopy (AFM) images reveal that the surface density of the nanoparticles can be varied by changing the pH of the nanoparticle dispersions.[42] The highest particle density is obtained at pH = 2.5, whereas the best ordering is observed at pH = 2.0, when the nanoparticles form closely packed films (Figure 20.10). A deviation in pH in both directions from the optimal pH range of 2.0 to 2.5 yields a decrease in particle density. This fact is also indicative of the changes in the nature of interparticle and particle–polyelectrolyte interactions.

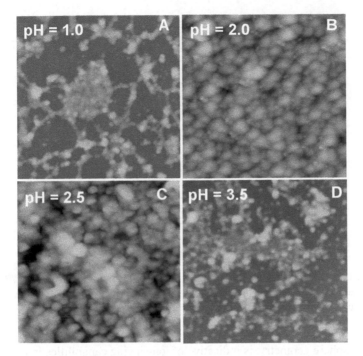

FIGURE 20.10 1×1 μm AFM images of monolayers of the nanoparticles shown in Figure 20.6 assembled from dispersions at pH = 1.0 (A), 2.0 (B), 2.5 (C), and 3.5 (D). In (B) the two populations of nanoparticles seen in Figure 20.6 can be identified. (From Pastoriza-Santos, I. et al., *Langmuir*, 16, 2731–2735, 2000. With permission. Copyright © 2000 American Chemical Society.)

The alternation of the surface charge when the layers of nanoparticles and polyelectrolyte are successfully deposited makes possible the cyclic repetition of the dipping procedure. This process of multilayer buildup on glass slides can be easily monitored by UV-visible spectroscopy (Figure 20.11). The consecutively acquired spectra display a virtually linear increase of absorbance in every deposition cycle (i.e., with each PAA/Ag@TiO$_2$ double layer), which is commonly observed for many LBL deposited systems.[102,104] Such linear behavior demonstrates a degree of vertical ordering of the multilayers imparted by the deposition procedure and high lateral density of polyelectrolyte and nanoparticle layers. The layer-by-layer character of the deposition procedure affords thickness control of the film with accuracy equal to the thickness of one particle layer. During deposition, the surface plasmon band remains at 434 nm, which indicates that the nanoparticles remain electrically insulated,[105,106] which can be attributed to the presence of an insulating (probably amorphous) shell of titania around the metal cores that impedes electron exchange between them.

After LBL assembly, the silver cores can be removed by adding a concentrated solution of ammonia, leading to TiO$_2$ nanoshell multilayer films. AFM and scanning tunneling microscopy (STM) studies[25] showed no difference in particle packing before and after removal of the silver cores. The existence of pores in the TiO$_2$ nanoshells, which are sufficient for Ag$^+$ ions to escape during oxidation, as well as

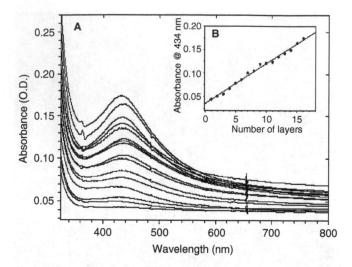

FIGURE 20.11 UV-visible absorption spectra of sequentially absorbed Ag@TiO$_2$/PAA bilayers assembled at pH 2.5. The inset shows the dependence at 434 nm vs. the number of deposition cycles. (From Pastoriza-Santos, I. et al., *Langmuir*, 16, 2731–2735, 2000. With permission. Copyright © 2000 American Chemical Society.)

the hollow sphere geometry, is indicative of ion-sieving capabilities in this nanomaterial, which were demonstrated using cyclic voltammetry with potassium ferricyanide Fe(CN)$_6^{3-}$ as a probe molecule. Ion permeability of the LBL nanoshell films strongly depends on pH and ionic strength, which permitted control of diffusion just by adjusting these parameters. This is related to charge reversal within the nanoshells when the pH changes, thus allowing or excluding the diffusion of ions of a certain sign.

The biocompatible nature of this material, together with the strong ion selectivity, makes it a suitable candidate for neurotransmitter detection. An example of this application was shown[25] in the selective enhancement of the permeation of positively charged dopamine, one of the most important neurotransmitters, and retardation of the transport of negatively charged ascorbic acid. Currently, the lack of dopamine/ascorbic acid selectivity for electrochemical *in vivo* detection of dopamine poses a significant problem for monitoring brain activity and for diagnostics of Parkinson's and similar neurological diseases.[107,108] The utilization of the nanoshell coatings enables a significant improvement in dopamine selectivity by suppressing the interference from oppositely charged electroactive ions.

20.5 ANISOMETRIC NANOPARTICLES

20.5.1 Formation of Nanorods and Nanoprisms

The shape (and size) of metal nanoparticles has a large influence on the optical properties,[6-11] and therefore control of these parameters is one of the most important targets in nanoparticle synthesis. We have recently demonstrated[62,63,109] that size and

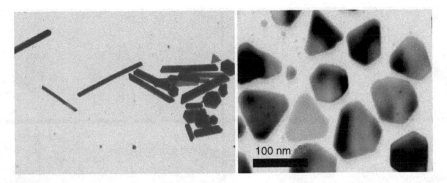

FIGURE 20.12 TEM micrographs showing Ag nanorods (A) and nanoprisms (B) obtained by reduction in DMF in the presence of PVP (0.06 and 0.4 mM, respectively), at high Ag$^+$ concentration (0.02 M).

shape of silver nanoparticles can be controlled through DMF reduction in the presence of a polymer (PVP). The relevant synthetic conditions are silver salt and PVP concentrations, temperature, and reaction time. Whereas mainly spherical and small nanoparticles are formed when the silver concentration is low (<1 mM),[62] increasing up to 0.02 M leads to the formation of anisometric particles. PVP concentration and reaction temperature strongly influence the shape of the final particles.

Figure 20.12 shows that both Ag nanorods and nanoprisms can be obtained for lower and higher PVP concentrations, respectively. TEM observation indicates that, initially, small spheres are formed, which then assemble in certain shapes and a melting-like process takes place, which leads to crystalline particles with well-defined shapes.[110] In the case of nanorods, this process is more obvious (Figure 20.13) and is probably related to an insufficient concentration of surfactant on the initial nanoparticles, which leads to a decrease in electrostatic repulsion and, in turn, an increase in attractive forces between nanoparticles, as previously observed by Tang et al.[111] for CdTe quantum dots. The facet at which the nanoparticles join together is determined by the energetically most favorable orientation, which seems to be the (111).

AFM measurements demonstrated the flat geometry of the nanoprisms. An example is shown in Figure 20.14, indicating an average thickness of about 35 nm for lateral dimensions on the order of 200 nm. It can also be observed in this image that most of the nanoprisms are truncated triangles, which sometimes leads to other polygonal shapes.

20.5.2 OPTICAL SPECTRA IN SOLUTION

The study of the optical properties of these colloids is very interesting, because very few reports have been published on the formation of metal nanoprisms,[112,113] which allow the experimental confirmation of theoretical calculations.[113,114] Such calculations are based on the discrete dipole approximation, so that the particle is divided into small elements that interact with each other through dipole–dipole interactions, and subsequent global evaluation of absorption and scattering. Using the discrete

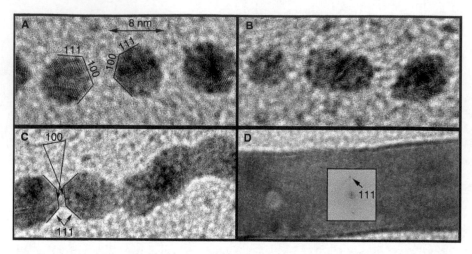

FIGURE 20.13 TEM micrographs showing separate Ag nanocrystals (A), two intermediate stages with partial nanocrystal aggregation (B, C), and one final single-crystalline Ag nanowire (D), formed in DMF (From Giersig, M. et al., *J. Mater. Chem.*, 14, 607–610, 2004. With permission by Royal Society of Chemistry.)

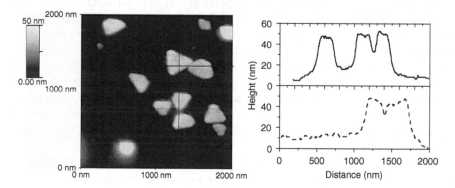

FIGURE 20.14 AFM image showing the flat nature of the Ag nanoprisms obtained. The plots are height profiles at the lines shown (solid, vertical; dashed, horizontal).

dipole approximation, Jin et al.[113] calculated the UV-visible spectrum of perfectly triangular nanoprisms (thickness, 16 nm; lateral dimension, 100 nm) and of triangular nanoprisms truncated at each tip. Their results indicate that for perfect triangles, one main peak should be observed at high wavelengths (770 nm), corresponding to the in-plane dipole resonance, and two peaks with much lower intensity at lower wavelengths, corresponding to the in-plane (470 nm) and out-of-plane (340 nm) quadrupole resonances, whereas the out-of-plane dipole resonance is observed only as a shoulder at 410 nm. In the case of truncated triangles, the main difference is a noticeable blue shift of the in-plane dipole resonance down to 670 nm.

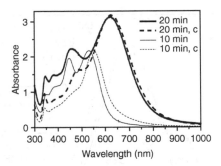

FIGURE 20.15 UV-visible spectra of Ag nanosprisms in DMF before (solid lines) and after (dashed lines) centrifugation and redispersion in ethanol. Boiling time is indicated. (From Pastoriza-Santos, I. and Liz-Marzán, L.M., *Nano Lett.*, 2, 903–905, 2002. With permission. Copyright © 2002 American Chemical Society.)

Figure 20.15 shows the experimental UV-visible spectra of two silver nanoprism dispersions prepared with different boiling times. Although the spectra of freshly made samples show some discrepancies with respect to those predicted by Jin et al., the agreement is very good after separation of small spheres present in the dispersion, which was achieved by centrifugation.[63] The presence of PVP on the surface prevents aggregation during the separation process. It should be mentioned that the in-plane resonance band is in general placed at lower wavelengths than expected, which can be explained on the basis of the deviations of the nanoparticles' shape with respect to perfect triangles, especially at short boiling times.

Once the main features of the spectra of silver nanoprisms in solution are characterized, it is interesting to study how other parameters affect such spectra. One of the main factors influencing the plasmon resonance is the refractive index of the surrounding medium.[12] An increase in solvent refractive index leads to a red shift of the plasmon band, as predicted by the Mie theory. We have studied this effect on colloids of PVP-protected silver spheres,[62] finding that when the solvent is changed from DMF (n = 1.426) to water (n = 1.333), the plasmon band is blue shifted by just 5 nm (see inset of Figure 20.16), which agrees with theoretical calculations. When a similar experiment is performed on PVP-protected Ag nanoprisms, a much stronger effect is observed, mainly on the in-plane dipole resonance, as shown in Figure 20.16. In the figure, spectra of two different nanoprism samples of different lateral dimensions (about 100 and 200 nm) were plotted, showing the increasingly larger shift in the in-plane dipole plasmon resonance wavelength as the size of the nanoprisms is increased.

For the larger particles, the red shift associated with solvent exchange from water to DMF is as large as 39 nm, whereas for the smaller nanoprisms, the shift is 18 nm. Interestingly, the peak at 340 nm, associated with the quadrupole out-of-plane resonance, remains almost constant (2-nm shift) during solvent exchange, which reflects the much higher sensitivity of the in-plane plasmon oscillation toward the nature of the surrounding medium. Such a sensitivity, which is expected to be of a similar magnitude toward adsorption of molecules to the nanoprism surface, can be exploited for the use of these systems as sensors.

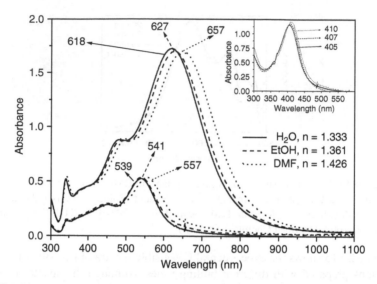

FIGURE 20.16 UV-visible spectra of Ag nanosprisms in DMF (solid lines), ethanol (dashed lines), and water (dotted lines). The upper curves correspond to average lateral dimensions of 200 nm, whereas the lower curves correspond to average lateral dimensions of about 100 nm. The inset shows the spectra for dispersions of Ag nanospheres.

20.5.3 Layer-by-Layer Assembled Films

The study of metal nanoparticles is interesting not only from the point of view of their single-particle behavior but also because of the complex collective properties when they are assembled close to each other. A number of studies have been made on assemblies of spherical nanoparticles,[24,115,116] but very few have been done on assemblies of anisometric nanoparticles.[112] The LBL technique has been previously used for the assembly of gold nanospheres[115,117] and nanoprisms,[112] and in both cases it was shown that the assembly of multilayers leads to a gradual red shift of the plasmon band(s), provided that the particles/layers are close enough to each other. Such a red shift is due to dipole–dipole interactions between neighboring nanoparticles. The optical properties of multilayer films prepared through LBL assembly of the silver nanoprisms described above are shown in Figure 20.17, where the inset shows the position of the in-plane dipole plasmon band vs. the silver monolayer number.

It is clear that in this case, opposite to what was found for gold, the plasmon band gradually blue shifts, with a change from 578 to 521 nm after eight dipping cycles. One could think that the reason for this is an opposite behavior for silver, compared to gold. However, similar experiments performed with 10-nm silver nanospheres yielded the expected trend of gradual red shift when increasing the number of monolayers. Thus, interactions of a different nature seem to occur in this system, which at this moment we do not understand and which need careful modeling.

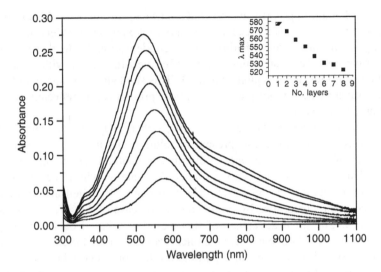

FIGURE 20.17 UV-visible spectra of layer-by-layer assembled films of Ag nanoprisms for the first eight monolayers. The inset shows the maximum position of the in-plane dipole as a function of layer number.

20.6 SUMMARY AND OUTLOOK

We have described in this chapter the preparation of silver nanoparticles through reduction with DMF, which is simultaneously used as a solvent. The reduction can be performed at room temperature and in the absence of light, though a temperature increase can affect the reaction rate and the morphology of the particles formed. In the absence of a stabilizing agent, the spontaneous reduction of Ag^+ ions into metallic Ag in DMF leads to nanoparticle deposition onto glass surfaces in contact with the system, so that homogeneous monolayers can be formed under suitable conditions.

Through the addition of stabilizing agents to the solution, the formation of stable silver nanoparticle colloids becomes possible. The use of poly(vinylpyrrolidone) provides steric stabilization through complexation onto the Ag particle surface. The reaction can be induced through refluxing conditions or through controlled microwave irradiation. We observed that the use of microwaves implies a higher level of control on the reduction process and permits the synthesis of smaller and more monodisperse particles.

The reduction of silver in DMF also allows for the preparation of metal nanoparticles coated with ceramic materials such as silica or titania, yielding nanocomposite materials with potential technological applications. The formation of ceramic shells takes place simultaneously to silver reduction, which means a clear advantage with respect to other existing methods for the synthesis of coated nanoparticles. The deposition of silica can only be produced at high temperature and is due to the hydrolysis of 3-aminopropil trimethoxysilane (APS). In the case of TiO_2, titanium tetrabutoxide was used as a precursor, but a complexing agent (acetylacetone) was necessary to control the rate of hydrolysis and condensation. Under these conditions, the silver particles get coated immediately after their formation, with a continuous, amorphous shell just a few nanometers thick. This shell alters the nature of the silver particles' surface, thus modifying their optical properties, as reflected in the UV-visible absorption spectra.

The porous nature of the ceramic shells, either TiO_2 or SiO_2, allows the Ag particles to react with chemicals present in solution and in this way to participate in a number of chemical reactions, so that their potential catalytic applications are greatly enhanced. As an example, we studied the reaction with molecular I_2, which leads to the formation of $AgI@TiO_2$. The porosity of the shell also allows for a dissolution of the cores, thereby producing, in the case of $Ag@TiO_2$, hollow TiO_2 nanocapsules. The great advantage of this material is based on its huge specific surface area, because both the outer and inner faces are accessible as reaction centers.

Further possibilities of this nanoparticle production process include the formation of anisometric silver nanoparticles (nanoprisms and nanorods) with very interesting optical properties, both in solution and upon layer-by-layer assembly.

REFERENCES

1. JH Fendler. *Nanoparticles and Nanostructured Films: Preparation, Characterisation and Applications*. New York: Wiley-VCH, 1998.
2. HS Nalwa. *Handbook of Advanced Electronic and Photonic Materials and Devices, Vol. 6, Nanostructured Materials*. New York: Academic Press, 2001.
3. LM Liz-Marzán, PV Kamat. *Nanoscale Materials*. Boston: Kluwer Academic Publishers, 2003.
4. U Kreibig, M Vollmer. *Optical Properties of Metal Clusters*. Berlin: Springer-Verlag, 1995.
5. P Mulvaney. Surface plasmon spectroscopy of nanosized metal particles. *Langmuir* 12:788–800, 1996.
6. I Lisiecki, F Billoudet, MP Pileni. Control of the shape and the size of copper metallic particles. *J Phys Chem* 100:4160–4166, 1996.
7. SS Chang, CW Shih, CD Chen, WC Lai, CRC Wang. The shape transition of gold nanorods. *Langmuir* 15:701–709, 1999.
8. JA Creighton, DG Eadon. Ultraviolet-visible absorption spectra of the colloidal metallic elements. *J Chem Soc Faraday Trans* 87:3881–3891, 1991.
9. BMI van der Zande, MR Böhmer, LG Fokkink, C Schöneberger. Colloidal dispersions of gold rods: synthesis and optical properties. *Langmuir* 16:451–458, 2000.
10. CA Foss Jr, GL Hornyak, JA Stockert, CR Martin. Template-synthesized nanoscopic gold particles: optical spectra and the effects of particle size and shape. *J Phys Chem* 98:2963–2971, 1994.
11. S Link, MB Mohamed, MA El-Sayed. Simulation of the optical absorption spectra of gold nanorods as a function of their aspect ratio and the effect of the medium dielectric constant. *J Phys Chem B* 103:3073–3077, 1999.
12. S Underwood, P Mulvaney. Effect of the solution refractive index on the color of gold colloids. *Langmuir* 10:3427–3430, 1994.

13. P Mulvaney, M Giersig, A Henglein. Electrochemistry of multilayer colloids: preparation and absorption spectrum of gold-coated silver particles. *J Phys Chem* 97:7061–7064, 1993.
14. N Toshima. Polymer-protected bimetallic clusters with various alloying structures. *Macromol Symp* 105:111–118, 1996.
15. M Michaelis, A Henglein, P Mulvaney. Composite Pd-Ag particles in aqueous solution. *J Phys Chem* 98:6212–6215, 1994.
16. LM Liz-Marzán, M Giersig, P Mulvaney. Synthesis of nanosized gold-silica core-shell particles. *Langmuir* 12:4329–4335, 1996.
17. T Ung, D Dunstan, M Giersig, P Mulvaney. Spectroelectrochemistry of colloidal silver. *Langmuir* 13:1773–1782, 1997.
18. T Ung, LM Liz-Marzán, P Mulvaney. Redox catalysis using Ag@SiO$_2$ colloids. *J Phys Chem B* 103:6770–6773, 1999.
19. LM Liz-Marzán, P Mulvaney. Au@SiO$_2$ colloids: temperature effect on the surface plasmon absorption. *New J Chem* 1285–1288, 1998.
20. RH Doremus. Optical properties of small gold particles. *J Chem Phys* 40:2389–2396, 1963.
21. S Link, MA El-Sayed. Size and temperature dependence of the plasmon absorption of colloidal gold nanoparticles. *J Phys Chem B* 103:4212–4217, 1999.
22. P Lu, T Teranishi, K Asakura, M Miyake, N Toshima. Polymer-protected Ni/Pd bimetallic nano-clusters: preparation, characterization and catalysis for hydrogenation of nitrobenzene. *J Phys Chem B* 103:9673–9682, 1999.
23. Y Nakao, K Kaeriyama. Adsorption of surfactant-stabilized colloidal noble metals by ion-exchange resins and their catalytic activity for hydrogenation. *J Colloid Interface Sci* 131:186–191, 1989.
24. S Henrichs, CP Collier, RJ Saykally, YR Shen, JR Heath. The dielectric function of silver nanoparticles Langmuir monolayers compressed through the metal insulator transition. *J Am Chem Soc* 122:4077–4083, 2000.
25. DS Koktysh, X Liang, B-G Yun, I Pastoriza-Santos, RL Matts, M Giersig, C Serra-Rodríguez, LM Liz-Marzán, NA Kotov. Biomaterials by design: layer-by-layer assembled ion-selective and biocompatible films of TiO$_2$ nanoshells for neurochemical monitoring. *Adv Funct Mater* 12:255–265, 2002.
26. R Nayak, J Galsworthy, P Dobson, J Hutchison. Synthesis of gold-cadmium selenide co-colloids. *J Mater Res* 13:905–908, 1998.
27. T Hirano, T Oku, K Suganuma. Fabrication and magnetic properties of boron nitride nanocapsules encaging iron oxide nanoparticles. *Diamond Relat Mater* 9:476–479, 2000.
28. HE Katz, ML Schilling. Electrical properties of multilayers based on zirconium phosphate/phosphonate bonds. *Chem Mater* 5:1162–1166, 1993.
29. NA Kotov, I Dèkány, JH Fendler. Ultrathin graphite oxide–polyelectrolyte composites prepared by self-assembly. Transition between conductive and non-conductive states. *Adv Mater* 8:637–641, 1996.
30. LM Liz-Marzán, MA Correa-Duarte, I Pastoriza-Santos, P Mulvaney, T Ung, M Giersig, NA Kotov. Core-shell nanoparticles and assemblies thereof. In *Handbook of Surfaces and Interfaces of Materials*, Vol. 3, HS Nalwa, Ed. New York: Academic Press, 2001, pp. 189–232.
31. RK Iler. Product comprising a skin of dense, hydrated amorphous silica bound upon a core of another solid material and process of making same. U.S. patent 2,885,366, 1959.
32. M Ohmori, E Matijevic. Preparation and properties of uniform coated colloidal particles. VII. Silica on hematite. *J Colloid Interface Sci* 150:594–598, 1992.

33. AP Philipse, MP van Bruggen, C Pathmamanoharan. Magnetic silica dispersions: preparation and stability of surface-modified silica particles with a magnetic core. *Langmuir* 10:92–99, 1994.

34. AP Philipse, AM Nechifor, C Pathmamanoharan. Isotropic and birefringent dispersions of surface modified silica rods with a boehmite-needle core. *Langmuir* 10:4451–4458, 1994.

35. LM Liz-Marzán, M Giersig, P Mulvaney. Homogeneous silica coating of vitreophobic colloids. *J Chem Soc Chem Commun* 731–732, 1996.

36. T Ung, LM Liz-Marzán, P Mulvaney. Controlled method for silica-coating of silver colloids. Influence of coating on the rate of chemical reactions. *Langmuir* 14:3740–3748, 1998.

37. MA Correa-Duarte, M Giersig, LM Liz-Marzán. Stabilization of CdS semiconductor nanoparticles against photodegradation by a silica coating procedure. *Chem Phys Lett* 286:497–501, 1998.

38. S Chang, L Liu, SA Asher. Preparation and properties of tailored morphology, monodisperse colloidal silica-cadmium sulfide nanocomposites. *J Am Chem Soc* 116:6739–6744, 1994.

39. T Li, J Moon, AA Morrone, JJ Mecholsky, DR Talhman, JH Adair. Preparation of Ag/SiO$_2$ nanosize composites by a reverse micelle and sol-gel technique. *Langmuir* 15, 4328–4334, 1999.

40. PV Kamat, B Shanghavi. Interparticle electron transfer in metal/semiconductor composites. Picosecond dynamics of CdS-capped gold nanoclusters. *J Phys Chem B* 101:7675–7679, 1997.

41. C Nayral, T Ould-Ely, A Maisonnat, B Chaudret, P Fau, L Lescouzères, A Peyre-Lavigne. A novel mechanism for the synthesis of tin/tin oxide nanoparticles of low-size dispersion and of nanostructured SnO$_2$ for the sensitive layers of gas sensors. *Adv Mater* 11:61–63, 1999.

42. I Pastoriza-Santos, D Koktysch, A Mamedov, NA Kotov, LM Liz-Marzán. One-pot synthesis of Ag@TiO$_2$ core-shell nanoparticles and their layer-by-layer assembly. *Langmuir* 16:2731–2735, 2000.

43. KS Mayya, DI Gittins, F Caruso. Gold-titania core-shell nanoparticles by polyelectrolyte complexation with a titania precursor. *Chem Mater* 13:3833–3836, 2001.

44. KS Mayya, DI Gittins, AM Dibaj, F Caruso. Nanotubes prepared by templating sacrificial nickel nanorods. *Nano Lett* 1:727–730, 2001.

45. G Oldfield, T Ung, P Mulvaney. Au@SnO$_2$ core-shell nanocapacitors. *Adv Mater* 12:1519–1522, 2000.

46. H Hirai, Y Nakao, N Toshima. Preparation of colloidal transition metals in polymers by reduction with for hydrogenation of cyclohexene at 30.0 deg under atm. H pressure *J Macromol Sci Chem A* 13:727–750, 1979.

47. N Toshima, T Yonezawa. Preparation of polymer-protected gold/platinum bimetallic clusters and their application to visible light-induced hydrogen evolution. *Makromol Chem Macromol Symp* 59:281–295, 1992.

48. Y Wang, N Toshima. Preparation of Pd-Pt bimetallic colloids with controllable core/shell structures. *J Phys Chem B* 101:5301–5306, 1997.

49. F Fievet, JP Lagier, M Figlarz. Preparing monodisperse metal powders in micrometer and submicrometer sizes by the polyol process. *MRS Bull* 14:29–34, 1989.

50. P-Y Silvert, R Herrera-Urbina, N Duvauchelle, V Vijayakrishnan, KJ Tekaia-Elhissen. Preparation of colloidal silver dispersions by the polyol process. Part 1. Synthesis and characterization. *J Mater Chem* 6:573–577, 1996.

51. Y Sun, B Gates, B Mayers, Y Xia. Crystalline silver nanowires by soft solution processing. *Nano Lett* 2:165–168, 2002.
52. Y Sun, Y Xia. Large-scale synthesis of uniform silver nanowires through a soft, self-seeding, polyol process. *Adv Mater* 14:833–837, 2002.
53. Z-Y Huang, G Mills, BJ Hajek. Spontaneous formation of silver particles in basic 2-propanol. *J Phys Chem* 97:11542–11550, 1993.
54. LM Liz-Marzán, I Lado-Touriño. Reduction and stabilization of silver nano-particles in ethanol by non-ionic surfactants. *Langmuir* 12:3585–3589, 1996.
55. K Osakada, A Taniguchi, E Kubota, S Dev, K Tanaka, K Kubota, T Yamamoto. New organosols of copper(II) sulfide, cadmium sulfide, zinc sulfide, mercury(II) sulfide, nickel(II) sulfide and mixed metal sulfides in N,N-dimethylformamide and dimethyl sulfoxide. Preparation, characterization, and physical properties. *Chem Mater* 4:562–570, 1992.
56. P Borowicz, J Hotta, K Sasaki, H Masuhara. Chemical and optical mechanism of microparticle formation of poly(N-vinylcarbazole) in N,N-dimethylformamide by photon pressure of a focused near-infrared laser beam. *J Phys Chem B* 102:1896–1901,1998.
57. JY Yu, S Schreiner, L Vaska. Homogeneous catalytic production of hydrogen and other molecules from water–DMF solutions. *Inorganica Chim Acta* 170:145–147, 1990.
58. GH Hugar, ST Nandibewoor. Kinetics of oxidation of dimethylformamide (DMF) by diperiodatonickelate(IV) (DPN) in aqueous alkaline medium. *Ind J Chem* 32A:1056–1059, 1993.
59. I Pastoriza-Santos, LM Liz-Marzán. Formation and stabilization of silver nanoparticles through reduction by N,N-dimethylformamide. *Langmuir* 15:948–951, 1999.
60. I Pastoriza-Santos, LM Liz-Marzán. Reduction of silver nanoparticles in DMF. Formation of monolayers and stable colloids. *Pure Appl Chem* 72:83–90, 2000.
61. I Pastoriza-Santos, C Serra-Rodríguez, LM Liz-Marzán. Self-assembly of silver particle monolayers on glass from Ag+ solutions in DMF. *J Colloid Interface Sci* 221:236–241, 2000.
62. I Pastoriza-Santos, LM Liz-Marzán. Preparation of PVP-protected metal nanoparticles in DMF. *Langmuir* 18:2888–2894, 2002.
63. I Pastoriza-Santos, LM Liz-Marzán. Synthesis of silver nanoprisms in DMF. *Nano Lett* 2:903–905, 2002.
64. MY Han, CH Quek, W Huang, CH Chew, LM Gan. A simple and effective chemical route for the preparation of uniform nonaqueous gold colloids. *Chem Mater* 11:1144–1147, 1999.
65. G Rodríguez-Gattorno, D Díaz, L Rendón, GO Hernández-Segura. Metallic nano-particles from spontaneous reduction of silver(I) in DMSO. Interaction between nitric oxide and silver nanoparticles. *J Phys Chem B* 106:2482–2487, 2002.
66. T Linnert, P Mulvaney, A Henglein, H Weller. Long-lived nonmetallic silver clusters in aqueous solution: preparation and photolysis. *J Am Chem Soc* 112:4567–4664, 1990.
67. J Rostalki, M Quinten. Effect of a surface charge on the halfwidth and peak position of cluster plasmons in colloidal metal particles. *Colloid Polym Sci* 274:648–653, 1996.
68. DV Leff, L Brandt, JR Heath. Synthesis and characterization of hydrophobic, organically soluble gold nanocrystals functionalized with primary amines. *Langmuir* 12:4723–4730, 1996.
69. RG Freeman, KC Grabar, KJ Allison, RM Bright, JA Davis, AP Guthrie, MB Hommer, MA Jackson, PC Smith, DG Walter, MJ Natan. Self-assembled metal colloid monolayers: an approach to SERS substrates. *Science* 267:1629–1631, 1995.

70. MJ Hostetler, SJ Green, JJ Stokes, RW Murray. Monolayers in three dimensions: synthesis and electrochemistry of w-functionalized alkanethiolate-stabilized gold cluster compounds. *J Am Chem Soc* 118:4212–4213, 1996.
71. N Takahashi, K Sakai, H Tanida, I Watanabe. Vertical ionization potentials and CTTS energies for anions in water and acetonitrile. *Chem Phys Lett* 246:183–186, 1995.
72. H Hirai. Formation and catalytic functionality of synthetic polymer–noble metal colloid. *J Macromol Sci Chem A* 13:633–649, 1979.
73. N Toshima, K Kushihashi, T Yonezawa, H Hirai. Colloidal dispersions of palladium–platinum bimetallic clusters protected by polymers. Preparation and application to catalysis. *Chem Lett* 1769–72, 1989.
74. N Toshima, T Yonezawa. Bimetallic nanoparticles: novel materials for chemical and physical applications. *New J Chem* 1179–1201, 1998.
75. T Yonezawa, M Sutoh, T Kunitake. Practical preparation of size-controlled gold nanoparticles in water. *Chem Lett* 619–620, 1997.
76. H Hirai, Y Nakao, N Toshima, K Adachi. Colloidal rhodium in poly(vinyl alcohol) as a hydrogenation catalyst of olefins. *Chem Lett* 905–910, 1976.
77. P Lu, T Teranishi, K Asakura, M Miyake, N Toshima. Polymer-protected Ni/Pd bimetallic nano-clusters: preparation, characterization and catalysis for hydrogenation of nitrobenzene. *J Phys Chem B* 103:9673–9682, 1999.
78. PA Buining, BM Humbel, AP Philipse, AJ Verkleij. Preparation of functional silane-stabilized gold colloids in the (sub)nanometer size range. *Langmuir* 13:3921–3926, 1997.
79. A Hanprasopwattana, S Srinivasan, AG Sault, AK Datye. Titania coatings on mono-disperse silica spheres (characterization using 2-propanol dehydration and TEM). *Langmuir* 12:3173–3179, 1996.
80. X-C Guo, P Dong. Multistep coating of thick titania layers on monodisperse silica nanospheres. *Langmuir* 15:5535–5540, 1999.
81. H Zhang, CC Wang, R Zakaria, JY Ying. Role of particle size in nanocrystalline TiO$_2$-based photocatalysts. *J Phys Chem B* 102:10871–10878, 1998.
82. M Ocana, WP Hsu, E Matijevic. Preparation and properties of uniform-coated colloidal particles. 6. Titania on zinc oxide. *Langmuir* 7:2911–2916, 1991.
83. WP Hsu, R Yu, E Matijevic. Paper whiteners. I. Titania coated silica. *J Colloid Interface Sci* 156:56–65, 1993.
84. JA Moss, JM Stipkala, JC Yang, CA Bignozzi, GJ Meyer, TJ Meyer, X Wen, RW Linton. Sensitization of nanocrystalline TiO$_2$ by electropolymerized thin films. *Chem Mater* 10:1748–1750, 1998.
85. NG Park, J van de Lagemaat, AJ Frank. Comparison of dye-sensitized rutile- and anatase-based TiO$_2$ solar cells. *J Phys Chem B* 104:8989–8994, 2000.
86. A Imhof, DJ Pine. Ordered macroporous materials by emulsion templating. *Nature* 389:948–951, 1997.
87. BT Holland, CF Blanford, A Stein. Synthesis of macroporous minerals with highly ordered three-dimensional arrays of spheroidal voids. *Science* 281:538–540, 1998.
88. J Wijnhoven, WL Vos. Preparation of photonic crystals made of air spheres in titania. *Science* 281:802–804, 1998.
89. E Scolan, C Sánchez. Synthesis and characterization of surface-protected nanocrystalline titania particles. *Chem Mater* 10:3217–3223, 1998.
90. P Papet, N Lebars, JF Baumard, A Lecomte, A Dauger. Transparent monolithic zirconia gels: effects of acetylacetone content on gelation. *J Mater Sci* 24:3850–3854, 1989.

91. M Giersig, T Ung, LM Liz-Marzan, P Mulvaney. Direct observation of chemical reactions in silica coated gold and silver nanoparticles. *Adv Mater* 9:570–575, 1997.

92. VL Colvin, MC Schlamp, AP Alivisatos. Light-emitting diodes made from cadmium selenide nanocrystals and a semiconducting polymer. *Nature* 370:354–357, 1994.

93. MC Schlamp, XG Peng, AP Alivisatos. Improved efficiencies in light emitting diodes made with CdSe(CdS) core/shell type nanocrystals and a semiconducting polymer. *J Appl Phys* 82:5837–5842, 1997.

94. H Mattoussi, LH Radzilowski, BO Dabbousi, EL Thomas, MG Bawendi, MF Rubner. Electroluminescence from heterostructures of poly(phenylene vinylene) and inorganic CdSe nanocrystals. *J Appl Phys* 83:7965–7974, 1998.

95. L Kronik, N Ashkenasy, M Leibovitch, E Fefer, Y Shapira, S Gorer, G Hodes. Surface states and photovoltaic effects in CdSe quantum dot films. *J Electrochem Soc* 145:1748–1755, 1998.

96. DE Fogg, LH Radzilowski, BO Dabbousi, RR Schrock, EL Thomas, MG Bawendi. Fabrication of quantum dot–polymer composites: semiconductor nanoclusters in dual-function polymer matrixes with electron-transporting and cluster-passivating properties. *Macromolecules* 30:8433–8439, 1997.

97. NP Gaponik, DV Sviridov. Synthesis and characterization of PbS quantum dots embedded in polyaniline film. *Ber Bunsen-Ges Phys Chem* 101:1657–1659, 1997.

98. NC Greenham, XG Peng, AP Alivisatos. Charge separation and transport in conjugated-polymer/semiconductor-nanocrystal composites studied by photoluminescence quenching and photoconductivity. *Phys Rev B* 54:17628–17637, 1996.

99. FG Aliev, MA Correa-Duarte, A Mamedov, JW Ostrander, M Giersig, LM Liz-Marzán, NA Kotov. Layer-by-layer assembly of core-shell magnetite nanoparticles: effect of silica coating on interparticle interactions and magnetic properties. *Adv Mater* 11:1006–1010, 1999.

100. MY Gao, B Richter, S Kirstein, H Möhwald. Electroluminescence studies on self-assembled films of PPV and CdSe nanoparticles. *J Phys Chem B* 102:4096–4103, 1998.

101. NA Kotov, I Dekany, JH Fendler. Layer-by-layer self-assembly of polyelectrolyte-semiconductor nanoparticle composite films. *J Phys Chem* 99:13065–13069, 1995.

102. NA Kotov, T Haraszti, L Turi, G Zavala, RE Geer, I Dekany, JH Fendler. Mechanism of and defect formation in the self-assembly of polymeric polycation-montmorillonite ultrathin films. *J Am Chem Soc* 119:6821–6832, 1997.

103. JW Ostrander, AA Mamedov, NA Kotov. Two modes of linear layer-by-layer growth of nanoparticle–polyelectrolyte multilayers and different interactions in the layer-by-layer deposition. *J Am Chem Soc* 123:1101–1110, 2001.

104. YJ Liu, AB Wang, R Claus. Molecular self-assembly of TiO_2/polymer nanocomposite films. *J Phys Chem B* 101:1385–1388, 1997.

105. W Schrof, S Rozouvan, E Vankeuren, D Horn, J Schmitt, G Decher. Nonlinear optical properties of polyelectrolyte thin films containing gold nanoparticles investigated by wavelength dispersive femtosecond degenerate four wave mixing (DFWM). *Adv Mater* 10:338–341, 1998.

106. P Mulvaney, LM Liz-Marzán, M Giersig, T Ung. Silica encapsulation of quantum dots and metal clusters. *J Mater Chem* 10:1259–1269, 2000.

107. JA Stamford, JB Justice. Probing brain chemistry. Voltammetry comes of age. *Anal Chem* 68:359A–363A, 1996.

108. J Smythies. Redox mechanisms at the glutamate synapse and their significance: a review. *Eur J Pharmacol* 370:1–7, 1999.

109. I Pastoriza-Santos, Y Hamanaka, K Fukuta, A Nakamura, LM Liz-Marzán. Anisotropic silver nanoparticles: formation and optical properties. In *Low-Dimensional Systems: Theory, Preparation and Some Applications*, LM Liz-Marzán, M Giersig, Eds. Dordrecht: Kluwer Academic Publishers, 2003, pp. 65–75.

110. M Giersig, I Pastoriza-Santos, LM Liz-Marzán, Evidence of aggregative mechanism during the formation of silver nanowires in N,N-dimethylformamide. *J. Mater. Chem.* 14:607–610, 2004.

111. Z Tang, NA Kotov, M Giersig. Spontaneous organization of single CdTe nanoparticles into luminescent nanowires. *Science* 297:237–240, 2002.

112. N Malikova, I Pastoriza-Santos, M Schierhorn, NA Kotov, LM Liz-Marzán. Layer-by-layer assembled mixed spherical and planar gold nanoparticles: control of interparticle interactions. *Langmuir* 18:3694–3697, 2002.

113. RC Jin, YW Cao, CA Mirkin, KL Kelly, GC Schatz, JG Zheng. Photoinduced conversion of silver nanospheres to nanoprisms. *Science* 294:1901–1903, 2001.

114. GC Schatz. Electrodynamics of nonspherical noble metal nanoparticles and nanoparticles aggregates. *J Mol Struct (Theochem)* 573:73–80, 2001.

115. T Ung, LM Liz-Marzán, P Mulvaney. Optical properties of thin films of Au@SiO₂ particles. *J Phys Chem B* 105:3441–3452, 2001.

116. G Chumanov, K Sokolov, TM Cotton. Unusual extinction spectra of nanometer-sized silver particles arranged in two-dimensional arrays. *J Phys Chem* 100:5166–5168, 1996.

117. T Ung, LM Liz-Marzán, P Mulvaney. Gold nanoparticle thin films. *Colloid Surf A* 202:119–126, 2002.

21 Interparticle Structural and Spatial Properties of Molecularly Mediated Assembly of Nanoparticles

Chuan-Jian Zhong, Li Han, Nancy Kariuki,
Mathew M. Maye, and Jin Luo

CONTENTS

21.1 Introduction 551
21.2 Nanoparticle Processing and Assembly.....................................554
    21.2.1 Thermally Activated Size Processing 554
    21.2.2 Hydrogen-Bonding-Mediated Assembly.......................554
    21.2.3 Thickness Control of the Assembly...............................555
    21.2.4 Reversibility of the Assembly..558
21.3 Interparticle Structural Properties...559
    21.3.1 Interparticle Hydrogen Bonding 559
    21.3.2 Interparticle Structural Evolution in Assembly Process............562
21.4 Interparticle Spatial Properties..563
    21.4.1 Ordered Array...564
    21.4.2 Branching Morphology 567
21.5 Conclusions and Implications..569
Acknowledgments...571
References..572

21.1 INTRODUCTION

The understanding of factors governing nanoparticle assembly is important for exploring nanostructured optical, electronic, magnetic, catalytic, and chemical or biological applications.[1,2] The preparation of a variety of nanoparticles and assemblies toward the fabrication of macroscopic materials of both fundamental and technological interests has been reported, including two-phase synthesis of monolayer-protected nanoparticles,[3-5] shell exchange reactivity,[6] stepwise layer-by-layer construction,[5,7-9] DNA linking,[10] molecular recognition,[11-14] and one-step exchange cross-linking precipitation.[15] Because many applications strongly depend on the precise control of nanoparticle ordering and morphological properties, it is becoming increasingly important to develop a full understanding of the interparticle structural and spatial properties.

Many studies on interparticle spatial properties of a variety of nanoparticle assemblies have documented the importance of either the Van der Waals (VW) interaction or covalent bonding in the assembly of nanoparticles. The VW interaction via interdigitation of alkyl chains on nanoparticles' shells is found to be one of the major driving forces for ordering nanocrystals. Table 21.1 lists some of the important findings in recent literature related to this subject.[16-26] This list is by no means exhaustive, but rather illustrates some examples.

TABLE 21.1
Interparticle Spatial Properties Reported for Several Nanoparticle Assemblies

Nanoparticle Assembly	Interparticle Distance (Edge to Edge)	Refs.
RS-capped Ag (core: 5.5 nm) Prep.: Casting	1.5–2.0 nm (R = n-dodecanethiol, C_{12})	16
RS-capped Au (core: 1.64 nm) Prep.: Casting	0.8 nm (C4), 1.0 nm (C6), 1.4 nm (C12), 2.2 nm (C18) (R = alkanethiols of C4, C6, C12, C18)	17
RS-capped Ag (core: 2–7.5 nm) Prep.: Casting	1–1.2 nm (R = dodecanthiol)	18
RS-capped Ag_2S (core: 5.8 nm) Prep.: Casting	2.0 nm (for 5.8-nm particles) RS = dodecanthiol	19
(1) RS-capped Au (core: 4.5 and 7.8 nm) (2) TOA$^+$Br$^-$-capped Au (core: 3 and 5 nm) Prep.: Casting	(1) ~1 nm, (2) 1.2 nm RS = decanethiol for (1) TOA$^+$Br$^-$ = tetraoctylammonium bromide	20
RS-capped Au (core: 3.7 nm) Prep.: (1) Casting (2) Dithiol linking RS (2)-capped Au	(1) 1.3 nm, (2) 1.7 nm RS = octyldecanethiol Dithiol = aryl dithiol (2-nm chain length)	21
R_1R_2-capped Fe_xPt_y Prep.: Casting (Core: 4.0 and 6.0 nm)	~4 nm R_1 = oleic acid, R_2 = oleic amine	22
R1- and R2-capped Au Prep.: PAMAN dentrimers linking (Core: ~1.5 nm)	~4 nm (for G4 denstrimer) R1 = octanethiol, R2 = MUA	23
DNA-capped Au (core: 13 nm) Capping agents: 3'-thiol-TACCGTTG-5' and 5'-AGTCGTTT-3'-thiol Prep.: DNA linking	~ 6.0 nm (9.5 nm expected from rigid DNA model) DNA liner: 5'-ATGGCAAC$IIII_{TCAGCAAA-5'}$[a]	10

[a] IIII stands for the hybridized 12-base-pair portion of the linking duplex.

It is clear that the interparticle distances are defined by the structure and interactions of the capping or linking molecules on the shell of the nanocrystals. In many cases, the interdigitation of alkyl chains plays a dominant role, as evidenced by the fact that the interparticle distance is often shorter than that based on a simple shell-to-shell interaction model for the core-shell-type nanoparticle assembly.

In this chapter, we will focus on the study of interparticle structural and spatial properties of a hydrogen-bonding-mediated assembly from core-shell nanoparticles consisting of a nanocrystal core and organic monolayer shell, e.g., alkanethiolate-capped gold nanoparticles. In comparison with covalent, electrostatic, or VW interactions, hydrogen bonding has an intermediate binding strength, which is usually reversible, and offers a flexible and controllable pathway for structural manipulation and correlation. Some of these attributes have been shown in several recent reports for supramolecular and nanoparticle assemblies.[14,27–31] Consider two extremes of a hypothesis of the nanoparticle assembly: (1) dense packing and (2) two-dimensional fractal growth of the linked particles. The interparticle properties are dependent on the core-shell size and the actual degree of packing or branching. Whereas the VW force is known to be responsible for the formation of ordered arrays in nanoparticles prepared by casting and evaporation,[5] the precise control of interparticle forces by such a nonspecific force has not been fully realized on a macroscopic materials scale. In contrast, the chemically specific binding (e.g., covalent or H-bonding) could overcome the weakness, though observations of large domains of highly ordered organization are very limited.[21] It has been recently rationalized that it might be possible to use a combination of these two forces for constructing nanoparticle assembly by manipulating the chemically specific constituents in the core-shell structure. For example, if the relative molar ratio of the hydrogen-bonding ligand vs. methyl-terminated alkyl-capping molecules (the latter is proportional to the concentration of the R_1-capped nanoparticles (NPs)) is manipulated, one would expect that the interparticle interaction could consist of a balance of hydrogen bonding and VW forces, leading to the regulation of the interparticle packing or spatial properties.

The outline of this chapter is as follows. We first describe the preparation of highly monodispersed nanoparticles and their assemblies based on hydrogen-bonding mediation. This description is followed by discussions of the interparticle structures based on spectroscopic characterizations. In the third section, we discuss the interparticle spatial properties based on transmission electron microscopic characterization of highly ordered arrays for the nanoparticle assembly. In the last section, implications of the understanding of the interparticle structural and spatial properties to potential technological applications are highlighted by examples exploring the nanostructured materials for chemical and electrochemical sensing. In view of the diversity of place-exchange reactivities of monolayer-capped nanoparticles to tailor the shell functional properties,[6] and the importance of hydrogen bonding in materials and biological or biomimetic systems,[15b] an in-depth understanding of the interparticle spatial properties could also facilitate the design of nanostructures for mimicking biological or synthetic receptors[14,32] and a wide range of applications such as filtration, sensors, and catalysis.[2]

21.2 NANOPARTICLE PROCESSING AND ASSEMBLY

Although a variety of metal and semiconductor nanoparticles has been studied, gold nanoparticles and assemblies provide an ideal model system because of the well-established surface chemistry in interfacial reactivities. This section first outlines the size processing of gold nanoparticles and then describes the hydrogen-bonding-mediated assemblies and their basic chemical or physical properties.

21.2.1 THERMALLY ACTIVATED SIZE PROCESSING

A thermally activated processing strategy for processing nanoparticles in terms of size and shape has been developed in our laboratory, which uses gold nanoparticles of 1- to 2-nm core size encapsulated with alkanethiolate monolayer shells as precursors.[33] The precursors were synthesized according to Schiffrin's two-phase method.[3,4] For example, starting from 2-nm gold nanoparticles (Au_{2-nm}, 1.9 ± 0.7 nm), the preparation of gold nanoparticles of 3- to 7-nm sizes with size monodispersity down to ± 0.4 nm has been demonstrated.[33] In this route, the size evolution was controlled by controlling the heating temperature (130–150°C), annealing temperature and time, and nanoparticle concentration. Mechanistically, the core-shell structure undergoes desorption/redeposition and coalescence/growth and evolves toward a larger core size in this process. Many research laboratories have used this thermal processing protocol to successfully prepare gold and alloy nanoparticles in the 1- to 10-nm range.[33–35] In this chapter, we largely focus on gold nanoparticles of 5- to 6-nm core sizes. Figure 21.1 shows a representative transmission electron microscopy (TEM) micrograph for the processed gold nanoparticles capped with a decanethiolate (DT) monolayer. These particles are highly crystalline, as evidenced by the visualization of the crystal facets in the high-resolution TEM (HRTEM). The particles are also highly monodispersed, as shown by the size distribution histogram, which shows an average diameter of 6.2 nm with a standard deviation of ± 0.5 nm.[35,36] Au nanoparticles are abbreviated as Au_{nm} in the rest of the chapter (e.g., Au_{6-nm}).

21.2.2 HYDROGEN-BONDING-MEDIATED ASSEMBLY

Based on the exchange reactivity demonstrated first by Murray's group,[6] we developed an effective exchange cross-linking precipitation route for assembling nanoparticles on almost any substrate (Scheme 21.1),[15] where the linker molecules serve as mediators. The nanoparticle assembly process involves immersing an appropriate substrate in a hexane solution of the gold nanoparticles and the mediation agent. Two mediation agents described in this chapter are 11-mercaptoundecanoic acid (MUA) and 15-mercaptohexadecanoic acid (MHA). The relative molar ratio of the carboxylic acid ligand (L) vs. the methyl-capped (NPs), i.e., N_L/N_{NP}, is controlled in the preparation procedure (typically 0.2 ~ 1.0 μM Au_{6-nm} and 0.3 ~ 1.5 μM MUA or MHA; higher concentrations of MUA or MHA were also used). The nanoparticle concentrations were determined by spectrophotometric calibration, and a truncated octahedron model[4a] was used for the particle shape. The nanoparticle thin-film formation immediately follows the exchange reaction.[15] The substrates were immersed vertically to ensure that the film was free of powder deposition. After a

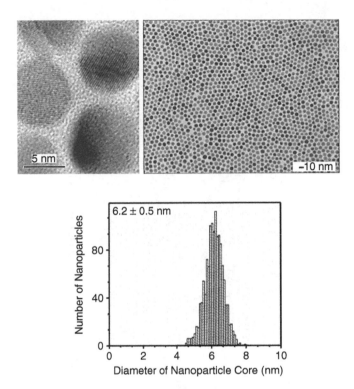

FIGURE 21.1 TEM micrograph and size distribution histogram (based on 1300 counts) of DT-capped Au nanoparticles produced by the thermally activated processing route. (From L Han et al., *Chem Mater* 15:29–37, 2003. With permission.)

controlled immersion time, the film-deposited substrates were emersed and immediately rinsed thoroughly with the solvent. The thin-film assembly was dried under nitrogen or air before characterizations.

This assembly route is entirely general, as demonstrated for nanoparticle assemblies on other substrates such as glass, mica, and graphite. The adhesion, however, depends on the nature of the substrate. Both the VW interactions between hydrophobic alkyl components on the nanoparticle shells and hydrophobic component on the substrate, and hydrogen-bonding interactions between carboxylic acid groups on the nanoparticle shells and the hydrophilic component on the substrate, play important roles in the adhesion. These interactions involve multiple binding sites because of the network structure of the linked nanoparticles. In the case of a thin film of MUA-mediated assembly of Au nanoparticles on a Au/mica substrate, the combination of thiolate–Au bonding and hydrogen-bonding linkages functions as the bridging force for the adhesion on the substrate.

21.2.3 THICKNESS CONTROL OF THE ASSEMBLY

One of the unique properties of nanoparticles is the display of a surface plasmon (SP) resonance band. The characteristic of the band depends on the particle size, interparticle

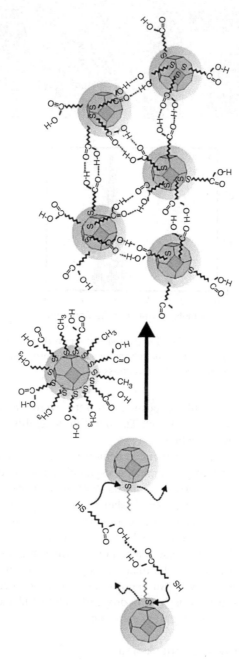

SCHEME 21.1 A schematic illustration of the hydrogen-bonding-mediated assembly of gold nanoparticles in an exchange cross-linking precipitation route.

distance, and dielectric medium properties. In general, the SP resonance band for both MUA-Au$_{5\text{-nm}}$ films appears at ~590 nm in a UV-visible (UV-Vis) spectrum, which shows a small shift in wavelength and broadening in bandwidth in comparison with its solution counterpart (520 nm).[15] Thin films of MUA-Au$_{2\text{-nm}}$ display a weak band superimposed in a rising spectral background, resembling its solution counterpart.[15] Figure 21.2A presents a set of UV-Vis spectra for the formation of MUA-Au$_{5\text{-nm}}$ film from a hexane solution of 0.05 mM MUA and 1.0 μM Au$_{5\text{-nm}}$. The time interval between emersion and reimmersion, i.e., the time spent for the optical measurement, was about 2 ~ 5 min. The spectra show two major features. First, the SP band characteristics showed evolutions in both wavelength and absorbance with a subtle shift (from 575 to 590 nm) from its solution counterpart, reflecting a reduction of the interparticle distances to less than the particle sizes by the MUA-based linkages. The bandwidth is also shown to be greater for the film than for the solution. These subtle changes can be qualitatively related to reduction in intercore distance and in effective medium effect.[37] The bandwidth broadening relates to a range of interband transitions due to a range of interparticle distance distributions for the closely spaced particles in the film, in comparison with those in the solution counterpart. Secondly, the intensity of the SP band increases with increasing immersion time and levels off at ~200 min (Figure 21.2B, curve a). This spectral evolution is accompanied by thickness growth of the film, as evidenced by darkening of the bluish color of the film.

Quartz crystal microbalance (QCM) is also used to monitor the mass loading for the thin-film formation. The frequency change (Δf) determined by QCM measurement is proportional to the mass loading of the deposited nanoparticle film, which is determined as a function of immersion time (Figure 21.2B, curve b). Δf increases with the immersion time, indicating an increase of the thin-film thickness. Qualitatively, the trend of the QCM result is comparable with the UV-Vis data. The rate of frequency change increases with increasing nanoparticle concentration. The coverage of the deposited nanoparticles, Γ$_{\text{NP film}}$, is proportional to concentration change in the film, i.e., (C$_{\text{NP}}^0$ − C$_{\text{NP}}$),[15c]

$$\Gamma_{\text{NP film}} = -K(C_{\text{NP}}^0 - C_{\text{NP}}) = K'\left(1 - e^{-kt}\right) \qquad (21.1)$$

where C$_{\text{NP}}^0$ is the initial concentration of the nanoparticles and K′ is a constant. Equation 21.1 predicts a simple exponential-rising character for the thin-film growth. By assuming that both the SP absorbance and the QCM frequency change are proportional to the nanoparticle surface coverage, i.e., A$_{590\,\text{nm}}$ or Δm ∝ Γ$_{\text{NP film}}$, it was found that the first-order kinetics could fit reasonably well with the exponential-rising trend. The fitted k parameters for the UV-Vis absorbance data and the QCM data differ by a factor of ~4. While the two sets of data (Figure 21.2B) are comparable for thinner films (i.e., immersion time < 200 min), the time length for leveling off of the mass change (curve a, >1500 min) is greater than that for the absorbance data (curve b, >500 min). Further considerations of both diffusion and surface crystallization effects are needed for a more general assessment of the assembly reactivity and kinetics.

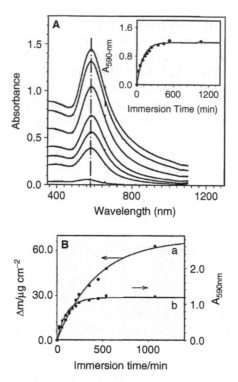

FIGURE 21.2 (A) UV-Vis spectra as a function of immersion time for the formation of MUA-Au$_{5\text{-nm}}$ film from a hexane solution containing 1.0 μM Au$_{5\text{-nm}}$ and 0.05 mM MUA. Immersion time increases from the bottom to the top. The immersion times are (from bottom to top curves) 22, 53, 94, 131, 237, 447, and 537 min. (B) A comparison of the UV-Vis absorbance (a) and QCM mass change (b) data for MUA-Au$_{5\text{-nm}}$ films, where the solid lines represent theoretically fitted curves based on Equation 21.4. (From L Han et al., *J Mater Chem* 11:1259–1264, 2001. With permission.)

21.2.4 Reversibility of the Assembly

The MUA-linked Au$_{5\text{-nm}}$ or Au$_{6\text{-nm}}$ thin film was not soluble in either nonpolar solvent (e.g., hexane) or polar solvent (e.g., ethanol). The blue color of the thin film remained intact for years. Whereas the insolubility in nonpolar solvent was expected, the insolubility in polar ethanol solvent was due to a relatively low extent of exchange of the capping DT by MUA, as discussed later. However, when a large excess MUA (~90 mM) was added to the ethanol solvent, the thin-film assembly became completely soluble in the solvent. Evidently, a further exchange between the excess MUA in the solution and the remaining DT molecules on the nanocrystals occurred, leading to the eventual dissolution of the assembled gold nanoparticles in solution and the reappearance of the original color.

As a further confirmation of the individual nanoparticle integrity, the nanoparticles dissolved in the presence of excess MUA were characterized using UV-Vis spectrophotometry. Figure 21.3 compares the SP band characteristics for the starting DT-capped Au$_{5\text{-nm}}$ nanoparticles and the MUA-linked Au$_{5\text{-nm}}$ film before and after

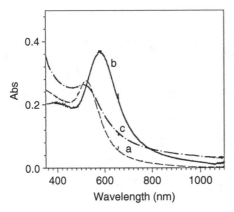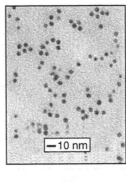

FIGURE 21.3 Left: UV-Vis spectra: DT/Au$_{6\text{-nm}}$ in hexane (a), MUA-Au$_{5\text{-nm}}$ film on glass substrate (b), and the MUA-Au$_{5\text{-nm}}$ film dissolved in an ethanolic solution of 90 mM MUA (c). Right: TEM image of the nanoparticles sampled from the dissolved solution. (From L Han et al., *Chem Mater* 15:29–37, 2003.

dissolution. It is evident that the wavelength of the SP band for the dissolved nanoparticles basically reversed from 580 nm for the assembled film to 520 nm, characteristic of Au$_{5\text{-nm}}$ in the solution. A similar result was also obtained for Au$_{6\text{-nm}}$. The shift of the SP band from solution to film reflects changes in the interparticle distance and the dielectric medium.[15c,38] The subtle difference in peak shape between the initial and redissolved nanoparticles is due to differences in solvent, capping agent, and concentration.[38] A TEM examination of the nanoparticles sampled from the dissolved nanoparticle solution (Figure 21.3, right) clearly supports the UV-Vis data, confirming that the redissolved nanoparticle cores remain the same as the starting nanoparticles in terms of size. The assembling and disassembling processes do not change the individual character of the original nanocrystals.

The solubility of the MUA-linked films is also dependent on the nanocrystal core size and the surface concentration of –CO$_2$H groups in the shell. In contrast to the poor solubility for MUA-linked Au$_{5\text{-nm}}$ or Au$_{6\text{-nm}}$ film, MUA-linked Au$_{2\text{-nm}}$ film was found to be completely soluble in an ethanol solvent. In this case, the solubility was due to a combination of factors, including low-molecular-mass effect and the more extensive exchange of MUA for the small-sized particles. A nuclear magnetic resonance (NMR) study of the extent of exchange revealed ~8% for Au$_{5\text{-nm}}$ particles prepared from a relatively high MUA-to-Au$_{5\text{-nm}}$ ratio (500:1), and ~46% for Au$_{2\text{-nm}}$ particles prepared from a relatively high MUA-to-Au$_{2\text{-nm}}$ ratio (200:1).[39] The latter is in close agreement with a recent report for similar systems under a different preparative condition.[40] The MUA-linked Au$_{6\text{-nm}}$ film from the dilute MUA concentrations should then have a much lower extent of exchange, which is addressed next by IRS characterization.

21.3 INTERPARTICLE STRUCTURAL PROPERTIES

21.3.1 INTERPARTICLE HYDROGEN BONDING

The infrared reflectance spectroscopic (IRS) technique is used to characterize the detailed hydrogen-bonding structural properties of the hydrogen-bonding-mediated

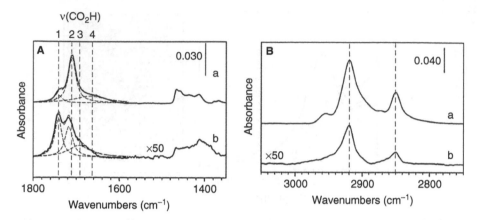

FIGURE 21.4 IRS spectra in low- (A) and high- (B) energy regions for an MUA-Au$_{5\text{-nm}}$ film (a) and an MUA monolayer (b) on Au/glass substrates. The dashed lines represent spectral deconvolution based on Lorentzian-type peak profiles. Each of the $v_{C=O}(CO_2H)$ spectral components is numerically labeled as 1, 2, 3, 4. (From WX Zheng et al., *Anal Chem* 72:2190–2199, 2000. With permission.)

assembly.[15b] Figure 21.4 presents a set of IRS spectra in the low- and high-energy regions for MUA-Au$_{5\text{-nm}}$ and MUA monolayer films. The diagnostic bands of the COOH ($v(COOH)$) in the low-energy region (Figure 21.4A) provide important structural information for assessing the hydrogen-bonding properties. The spectral components under the $v(COOH)$ vibration envelope in the 1740- to 1650-cm^{-1} region differ significantly between these films, whereas those in the 1500- to 1400-cm^{-1} region (mainly C–H bending modes) show little variations. Note that both spectra have an overall absorbance of ~50 times larger than the MUA monolayer on planar substrate, presumably due to a combination of nanoparticle multilayer (10 ~ 20 equivalent) and surface enhancement[41] effects. The multiple-band components in the spectral envelope are represented by the dashed curves based on spectral deconvolution (numerically numbered as 1, 2, 3, 4). The assignments of bands diagnostic of –CO$_2$H, –CO$_2^-$, and C–H stretching vibrations are listed in Table 21.2.

For the self-assembled MUA monolayer on the planar Au substrate, three $v(C=O)$ bands are observed, which can be assigned[42] to –COOH groups of the free or non-hydrogen-bonded (1741 cm^{-1}), side-by-side dimeric hydrogen-bonded (1718 cm^{-1}), and polymeric hydrogen-bonded (~1690 cm^{-1}) modes, respectively. In comparison, the three bands identified for the MUA-Au$_{5\text{-nm}}$ film are displayed at 1740 cm^{-1} (1), 1710 cm^{-1} (2) and ~1670 cm^{-1} (4). Whereas the free acid band (1) remains basically unchanged, the shifts in the hydrogen-bonded modes (2, 4) are remarkable. We attribute these bands to the formation of head-to-head hydrogen bonding of –COOH groups in the nanostructure. In fact, the wave number of band 2 is identical to the band position observed for the head-to-head hydrogen-bonding dimer with cis-configuration in condensed phases of alkanoic acids.[43,44] Such dimer structures are depicted in Scheme 21.2.

The trans-configuration is basically absent in the MUA-Au$_{5\text{-nm}}$ film. The broad band 4 is due mostly to polymeric hydrogen bonding. The spectral difference from that for

TABLE 21.2
Mode Assignments for the Diagnostic IRS Bands (cm⁻¹)
in the High- and Low-Energy Regions for MUA-Au$_{5\text{-nm}}$
Film, MUA-Au$_{2\text{-nm}}$ Film, and MUA Monolayer

Mode Assignment	MUA-Au$_{5\text{-nm}}$	MUA-Au$_{2\text{-nm}}$	MUA
$v_a(CH_3)$	~2956[b]	—	—
$v_s(CH_3)$	~2872[b]	—	—
$v_a(CH_2)$	2920	2920	2920
$v_s(CH_2)$	2850	2850	2849
$v_{C=O}$ (CO$_2$H)[a]			
v_1	1739	1738	1741
v_2	1710	1709	1717
v_3	—	1693	—
v_4	1668	1665	1695
$v_a(CO_2^-)$	1578	1569	1556
$v_s(CO_2^-)$	~1435	~1432	1431

[a] See Figure 21.5 and the text for assignments of the v_1 to v_4 components of the v(COOH) bands.
[b] Weak shoulder peaks.

Source: Han, L. et al., *Chem. Mater.*, 15, 29–37, 2003. With permission.

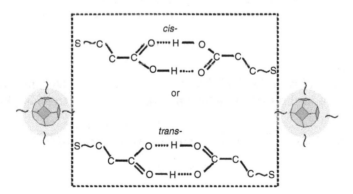

SCHEME 21.2 Schematic illustration of the possible cis- and trans-configurations of the head-to-head hydrogen bonding at carboxylic acid groups. The C$_\alpha$–C$_\beta$ bond is on the same side of the C=O group for cis-configuration, whereas the C$_\alpha$–C$_\beta$ bond is on the opposite side of the C=O group for trans-configuration.

the MUA monolayer in band position reflects the difference of the hydrogen-bonding properties between head-to-head and side-by-side configurations. Moreover, whereas the two-dimensional monolayer exhibits an ~40% free acid component, the MUA-Au$_{5\text{-nm}}$ film shows only a level of <10%. The results therefore provide evidence for the predominant head-to-head hydrogen-bonding linkages in the MUA-Au$_{5\text{-nm}}$ network. Interestingly, MUA-Au$_{2\text{-nm}}$ film exhibits a striking distinction from MUA-Au$_{5\text{-nm}}$ film by displaying a new peak component at 1694 cm^{-1} (3). The relative absorbance of the free acid band (~1738 cm^{-1} (1)) remains a small fraction, supporting the nature of predominant hydrogen bonding. We attribute the bands at 1708 cm^{-1} (2) and ~1694 cm^{-1} (3) to cis- and trans-modes of the head-to-head hydrogen-bonded dimers,[43] in sharp contrast to the MUA-Au$_{5-nm}$ film. It is known that the trans-configuration of alkanoic acid dimer in solids is more stable than cis-configuration.[43,44] Although the exact origin is not yet identified, the two different hydrogen-bonding configurations are intriguing because they demonstrate that fine structures of the hydrogen-bonding linkages are strongly dependent on particle sizes.

In the C-H region (Figure 21.4B), the predominant methylene stretching bands at 2918 and 2848 cm^{-1} for the two films further support that the network structure is primarily composed of MUA chains. The slight difference from those for the MUA monolayer (2920 and 2850 cm^{-1}) is indicative of comparable chain–chain packing properties. The methyl stretching bands are nearly absent for the Au$_{2-nm}$ film but identifiable at 2955 and 2870 cm^{-1} for the Au$_{5-nm}$ film. The result is consistent with a significant replacement of the original capping thiolates by MUA.

The interparticle linking carboxylic acid groups have been further shown to be reversibly tunable by pH. The protonation–deprotonation reactivity and the associated pK$_a$ have been characterized.[15b]

21.3.2 INTERPARTICLE STRUCTURAL EVOLUTION IN ASSEMBLY PROCESS

The IRS technique was used to further examine the thin-film assembly as a function of assembling time. Table 21.3 displays the IRS data for samples prepared at a relatively low MUA concentration (3.0 μM). The MUA-to-Au$_{6-nm}$ molar ratio is ~6.

TABLE 21.3
The Relative Ratios of the Peak Areas Based on Spectral Deconvolution of the Spectra in the C-H Stretching and the Carbonyl Stretching Regions

	IRS Peak Area Ratio	
Assembling Time (min)	A[v$_a$(CH$_3$)]/A[v$_a$(CH$_2$)]	A[v(free CO$_2$H)]/A[v(H-bonded CO$_2$H)]
	[MUA] = 3.0 μM	
0	0.26	—
15	0.16	0.31
30	0.13	0.17
120	0.15	0.27
	[MUA] = 0.4 μM	
0	0.26	—
5	0.21	1.02
15	0.20	1.05
45	0.20	1.23
90	0.20	1.11

Source: Han, L. et al., *Chem. Mater.*, 15, 29–37, 2003. With permission.

In comparison with the relative absorbance ratio of the methyl band ($v_a(CH_3)$) vs. methylene band ($v_a(CH_2)$), i.e., $A(v_a(CH_3))/A(v_a(CH_2))$, for the cast DT-Au nanoparticles, a reduction in absorbance is evident for the MUA-linked nanoparticle films, indicating a partial replacement of the capping DT by MUA molecules. The fact that there is little shift in both $v_a(CH_2)$ and $v_s(CH_2)$ bands as a function of immersion time is in contrast to those observed for self-assembled monolayers with a similar chain length,[45] where $v_a(CH_2)$ shifted from 2926 to 2918 cm^{-1} due to an increased packing of the alkyl chains. This observation demonstrates that the packing structure of the shell in the thin-film assembly remains crystalline-like.[46–48] The relatively small change in the ratio of $A(v_a(CH_3))/A(v_a(CH_2))$ (Table 21.3) after the initial layer formation indicates the structural similarity of the linked nanoparticles in the different growth layers. The relative ratio of free (1740 cm^{-1}) vs. hydrogen-bonded (1710 cm^{-1}) CO_2H groups, i.e., $A[v(free)]/A[v(H\text{-bonded})]$, is also dependent on the immersion time. The relatively small change after the initial layer formation indicates that the structure is similar between the initial and subsequent film growth. In general, the free acid component is less than 20% of the total $-CO_2H$ groups, again suggesting a predominant head-to-head hydrogen bonding.

The spectral evolution was dependent on the concentration of MUA. For example, the thin films were prepared at an eightfold lower concentration of MUA (0.4 μM), in which the MUA-to-$Au_{6\text{-nm}}$ molar ratio 1.6:1 was reduced by a factor of ~4. It was found that whereas the C–H stretching region shows a very similar evolution, the carbonyl region reveals two major distinctions. First, the absorbance is much weaker for the low concentration case, suggesting a relatively smaller extent of exchange. The relatively small change in absorbance is again indicative of the structural similarity in layer-by-layer growth. Secondly, the spectral deconvolution shows that the relative absorbance ratio of the hydrogen-bonded (1710 cm^{-1}) vs. free (1740 cm^{-1}) $-CO_2H$ groups remains constant during the film growth, each of which is ~40% of the total (in addition to ~20% of the polymeric hydrogen-bonding component (~1640 cm^{-1})). This finding is in sharp contrast to the fact that less than 20% of free acid was observed for the films from a higher MUA concentration, indicating a relatively lower extent of hydrogen bonding. The dependence of the nanoparticle assembly on the concentration of MUA shows the effective mediation of the interparticle hydrogen bonding. The MUA-to-nanoparticle molar ratio of ~1 translates to approximately one MUA per ~500 capping DT molecules on the nanocrystal ($Au_{6\text{-nm}}$) in the solution. The extent of exchange clearly decreases when the MUA concentration is decreased. These insights are important for assessing the interparticle morphological properties, as described next.

21.4 INTERPARTICLE SPATIAL PROPERTIES

The interparticle spatial properties of the hydrogen-bonding-mediated assembly of nanoparticles are dependent on the size and monodispersity of the nanocrystals. Whereas smaller size and less uniform nanoparticles do not form large domains of ordered structure, large size and high monodispersity show a propensity for forming relatively large domains of ordered arrays. The results of the ordered assemblies are discussed in the following subsections.

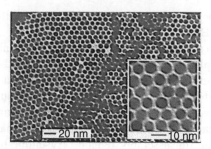

FIGURE 21.5 TEM micrographs of MHA-linked Au_{6-nm} assembly from a hexane solution of $[Au_{6-nm}] = 0.24 \ \mu M$ and $[MHA] = 0.39 \ \mu M$ ($t_{assembly}$ = 15 min). (From L Han et al., *Chem Mater* 15:29–37, 2003. With permission.)

21.4.1 ORDERED ARRAY

We examined several carboxylic acid–functionalized alkyl thiols in the hydrogen-bonding-mediated assembly of nanoparticles; TEM was used to characterize the interparticle spatial properties.[35] The thin-film assembly of MUA-Au_{6-nm} displays large domains of an ordered array with hexagonal packing. Such morphology was not observed in a control experiment. For a longer-chain-length mediator, i.e., MHA, the assembly of Au_{6-nm} particles (Figure 21.5) also displays large domains of ordered arrays with hexagonal packing. The apparent patches are due to either the formation of a second layer of nanoparticles on the first layer or nanoparticles assembled on the other side of the grid. A comparison with the evaporated DT/Au_{6-nm} particles reveals subtle differences in the periodicity or interparticle spacing and in the solubility. It is important to emphasize that whereas the evaporated DT/Au_{6-nm} film can be redissolved in hexane, the MUA-linked film is soluble neither in hexane nor in ethanol solvent. The driving force for the interparticle packing in the DT/Au_{6-nm} case involves only VW interaction through interdigitation of alkyl chains.[15] For the MUA- or MHA-linked nanoparticles, a combination of hydrogen-bonding and VW interactions must be involved, as revealed by IRS data.

The analysis of the average center-to-center or edge-to-edge distances from the TEM data provide important insights into the interparticle spatial property. Because the edge-to-edge distances were derived from the center-to-center distances (Table 21.4), they should be within the same statistical uncertainty. The DT/Au_{6-nm} array formed by evaporation has an average edge-to-edge distance of 1.5 nm, which is quite consistent with the expectation based on a model of interdigitation of alkyl thiolates on the shells between neighboring nanoparticles.[1,5,49] Interestingly, the average edge-to-edge distance is slightly smaller or comparable for both MUA-linked Au_{6-nm} (1.2 nm) and MHA-linked Au_{6-nm} (1.4 nm). The edge-to-edge distance for the MHA-linked Au_{6-nm} assembly appears comparable to that for the MUA-linked Au_{6-nm} in view of the standard deviation. A key observation here is that the average edge-to-edge distance is much smaller than the expectation, based on a model of 100% head-to-head hydrogen bonding between –CO_2H terminal groups of the trans alkyl chain shell structure. The analysis of a hydrogen-bonding dimer model would yield ~3.6 nm for $(MUA)_2$ and ~4.8 nm for $(MHA)_2$.

TABLE 21.4
The Average Interparticle Distances Derived from TEM Image Analysis of the Data in Figures 21.1 and 21.5

Assembly	Interparticle Distance	
	d_1 (Center to Center) (nm)[a]	d_2 (Edge to Edge) (nm)[b]
(a) DT-Au$_{6\text{-nm}}$	7.7 (± 0.5)	1.5
(b) MUA-linked Au$_{6\text{-nm}}$	7.4 (± 0.5)	1.2
(c) MHA-linked Au$_{6\text{-nm}}$	7.6 (± 0.4)	1.4

[a] Based on a computer analysis of domains with 1300 counts of particles for (a) and 250 counts for (b) and (c).
[b] Based on d_1 and a spherical model for the particle (average diameter, 6.2 ± 0.5 nm).

Source: Han, L. et al., *Chem. Mater.*, 15, 29–37, 2003. With permission.

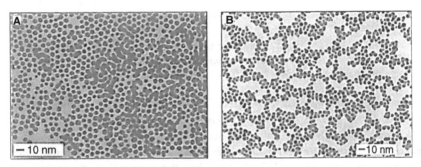

FIGURE 21.6 TEM images. MUA-linked Au$_{5\text{-nm}}$ films assembled from a hexane solution of [Au$_{5\text{-nm}}$] = 0.90 μM and [MUA] = 1.5 μM ($t_{assembly}$ = 14 min) (A) and [Au$_{5\text{-nm}}$] = 0.90 μM and [MUA] = 45.0 μM ($t_{assembly}$ = 5 min) (B). Au$_{5\text{-nm}}$: 5.6 ± 1.1 nm.

Using 2- ~ 5-nm core-sized particles, which are relatively less monodispersed (≥0.6), we did not detect significant domains of ordered arrays (Figure 21.6). The capability of forming large and monodispersed nanoparticles to maximize the inter-shell cohesive interaction is consistent with recent findings from a study of resorcinarene-encapsulated gold nanoparticles (16 to 170 nm),[14e] which showed that the interparticle distance decreases with core size and is much smaller than the expected size of the capping molecule.

To account for the interparticle distance being smaller than the expected lengths, we have considered several interparticle interaction forces. Two of the most important forces are VW and hydrogen-bonding interactions for the examined systems. The former occurs via interpenetration of the alkyl chains,[49] whereas the latter occurs at the terminal functional groups. We believe that a combination of the two interactions played a significant role in the final interparticle packing after the initial mediation by exchange and hydrogen bonding. This assessment is supported by the analysis of the shell composition. In comparison with the extent of exchange (~8%

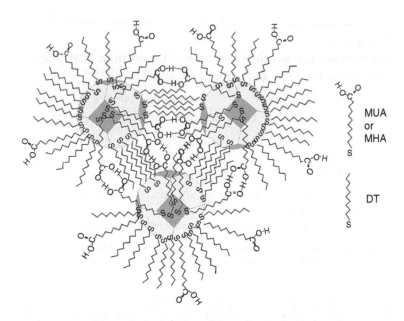

SCHEME 21.3 Schematic illustration of the squeezed packing model involving a combination of hydrogen bonding at the carboxylic acid groups and the cohesive interaction through interdigitation of the alkyl chains.

DT) for samples prepared at high N_L/N_{NP} ratios, the shell for thin films prepared at very small N_L/N_{NP} ratios (e.g., 1 ~ 2) must consist of a larger fraction of DT. This is evidenced by the IRS and solubility data. For the mixed $-CO_2H$ and $-CH_3$ shell, whereas the relatively limited hydrogen-bonding sites mediate the interparticle linking, the interpenetration of DT thiolates has the propensity of maximizing the interdigitative cohesive interaction. The combination of the two forces leads to a squeezed interparticle spatial organization, as illustrated in Scheme 21.3, which is neither purely VW interdigitation, as for cast alkanethiolate-capped nanoparticles,[5,49] nor purely interdigitated hydrogen bonding, as proposed for nanoparticles capped with amide-functionalized thiolates.[27] We note that smaller interparticle distances have been reported for gold or silver nanoparticle assemblies[1,5,38,49-52] that do not involve interparticle hydrogen-bonding interactions.

Evidence for the squeezed interparticle spatial model has been provided by grazing angle x-ray diffraction (GXRD) characterization and a comparison of the nanoparticle assembly thickness data from two different techniques, atomic force microscopy (AFM) and QCM. In the GXRD result, a d-spacing value of 5.1 nm was obtained from the data for a MUA-linked $Au_{5\text{-}nm}$ film deposited on a glass substrate. Two peaks were observed at small angles. Whereas the origin for peak 2 is not completely clear, the d-spacing value for peak 1 (d_1) calculated from $\lambda = 2d \sin \theta$ yields 5.04 nm, which is largely consistent with the TEM result.

The determination of the total mass of assembled nanoparticles also provided an estimate of the thickness based on the particle size and interparticle edge-to-edge distance. The 6.2-nm nanoparticle is modeled as a truncoctahedron,[4a] which gives

TABLE 21.5
Comparison of Thin-Film Thickness Determined from QCM and AFM
for the Mediator-Linked Gold Nanoparticle Films

	Film Thickness[a,d] (nm)			
	By AFM	By QCM		
Assembly	d	Δf(Hz) [b]	d[c]	d[d]
(a) MUA-linked $Au_{5\text{-nm}}$ film	8.0 ± 0.3	2523 ± 10	19.0	8.2
(b) NDT-linked $Au_{5\text{-nm}}$ film[e]	17.2 ± 0.4	5426 ± 10	20.9	20.1

[a] The nanoparticle films were assembled on a 10-MHz polished quartz crystal coated with a gold film.

[b] Δf is translated to mass loading (Δm) based on Sauerbrey's equation, $\Delta m = -C_f \Delta f$, where the mass sensitivity $C_f = 5.5$ ng/cm^2/Hz.

[c] Calculated from Δf, using a theoretical edge-to-edge distance of 3.6 nm based on an idealized model of the head-to-head hydrogen bonding by Au-S(CH$_2$)$_{10}$CO$_2$H ... HO$_2$C(CH$_2$)$_{10}$S-Au.

[d] Calculated from Δf, using the experimental edge-to-edge distance of 1.2 nm, which is determined from the ordered array (see Table 21.3) and is hypothesized as hydrogen-bonding-mediated cohesive interparticle linkage (see Scheme 21.2).

[e] Based on very small domains (average domain size ~15 particles) of ordered arrays in a TEM micrograph for NDT-Au$_{6\text{-nm}}$ film. The determined average edge-to-edge distance was 1.4 ± 0.5 nm.

Source: Han, L. et al., *Chem. Mater.*, 15, 29–37, 2003. With permission.

8307 gold atoms and 713 thiolates. Table 21.5 shows the data determined for a MUA-linked Au$_{6\text{-nm}}$ film. The data for an NDT (1,9-nonanedithiol)-linked Au$_{6\text{-nm}}$ film are also included for comparison. The results reveal a very good agreement between the AFM-determined (8.0 nm) and QCM-determined thickness (8.2 nm) when the edge-to-edge distance of 1.2 nm is used. The data therefore support the squeezed interparticle spatial model in Scheme 21.2. In the case of NDT-mediated assembly, whereas only very small domains of ordered arrays (~15 particles) could be identified, the estimated average edge-to-edge distance (1.4 nm) was only slightly smaller than the theoretical distance (1.5 nm). The close agreement between the AFM- (17.2 nm) and QCM- (20.1 nm) determined thickness further supports the above assessment.

21.4.2 BRANCHING MORPHOLOGY

By manipulating particle size and mediator concentrations, branching morphologies have also been observed for the nanoparticle assembly. Figure 21.7A shows a TEM micrograph for an MUA-linked Au$_{6\text{-nm}}$ thin film prepared from a dilute solution, which is characteristic of the two-dimensional branching, as depicted in Figure 21.7B(b). The film displays a nanoporous network created by the branched architecture with an average pore size of ~80 nm. A certain fraction of the nanoparticle array, similar to the ordered arrays described in the previous subsection, is evidently embedded in the two-dimensional branching architecture. Under a given concentration, the relative coverage of the densely packed array area vs. the two-dimensional

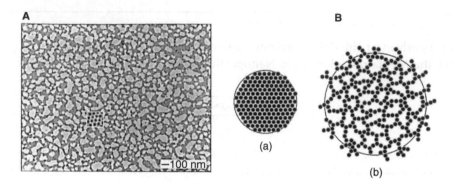

FIGURE 21.7 (A) TEM micrograph of the branching morphology for an MUA-linked Au_{6-nm} film assembly from a hexane solution of $[Au_{6-nm}] = 0.24 \mu M$ and $[MUA] = 0.39 \mu M$ ($t_{assembly}$ =15 min). Inset: Enlarged view. (B) Schematic illustration of two extremes for 2D interparticle assembly of nanoparticles toward a domain size of radius R: dense packing (a) and branching growth (b). (From L Han, et al., *Chem Mater* 15:29–37, 2003. With permission.)

branched-pore area is highly dependent on the monodispersity of the nanoparticles. By increasing monodispersity, the average size of the pores decreases. The two-dimensional branching morphology is also dependent on the N_L/N_{NP} ratio. For example, pore sizes of an average diameter of 30 to 40 nm were observed using a much smaller ratio (e.g., 0.3). In contrast, much smaller pore sizes were observed when using a much larger ratio (≥ 50). In the latter case, the particles were densely packed in a three-dimensional architecture.

Similar branching or fractal growth phenomena have also been observed for nanoparticle assembly on other substrates. For example, AFM data revealed domains of dense packing and branching morphology as large as a few microns for a thin film assembled on an Au/mica substrate. The cross-section data also reveal a height variation corresponding to the size of the assembled nanoparticles.

Theoretically, there are two extremes of two-dimensional interparticle assembly of nanoparticles toward a domain size of radius R: dense packing and branching growth, as schematically illustrated in Figure 21.7B. The mechanistic details for the control factors are a subject of our ongoing investigations. To a certain degree, the branched structure resembles those of fractal aggregation for colloids[53] known in sol-gel technology. The fractal growth can be described by mass fractal dimension, d, i.e., $M \propto R^d$,[54] where M represents the mass of the branched structure and R is the corresponding radius. The fractal component d equals 2 for two-dimensional growth on a planar substrate and 3 for growth in three-dimensional space. Importantly, based on our earlier kinetic studies of the mass changes in the nanoparticle assembling process, to which the fractal component can be related,[15c] we found an exponential growth that is characteristic of a reaction-limited aggregation process mediated by the linker molecules for the formation of both compact and branching morphologies.[55] Under a very low concentration of mediator molecules, it is possible that some of the exchanged $-CO_2H$ ligands are not symmetrically distributed in the shell. We further note that the use of simple alkanoic acid did not produce any of the phenomena observed, thus ruling out possible surfactant effects.

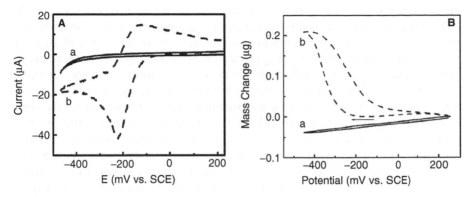

FIGURE 21.8 Cyclic voltammetric curves (A) and EQCN response curves (B) for MUA-$Au_{2\text{-nm}}$ film-coated electrode in 1.0 mM $Ru(NH_3)_6^{3+}$/0.5 M KCl solutions at pH = 2.7 (a) and 10.7 (b). Scan rate, 50 mV/sec. (The electrode substrate is GC (0.28 cm^2) for (A) and Au/quartz (0.2 cm^2) for (B).) (From WX Zheng et al., *Anal Chem* 72:2190–2199, 2000, and from J Luo, *J Phys Chem* 106:9313–9321, 2002. Both with permission.)

21.5 CONCLUSIONS AND IMPLICATIONS

It is clear that the interparticle spatial properties are dependent on not only the physical dimension of the capping or mediator molecules but also the nature of the molecular interactions. The particle size and monodispersity play important roles as well. The use of carboxylic acid–functionalized alkyl thiol as the mediation agent in the assembly of decanethiolate-capped gold nanoparticles leads to the formation of controllable interparticle packing. The interparticle structural and spatial properties can be characterized by a squeezed interparticle interaction model involving both hydrogen bonding at the carboxylic acid terminal groups and cohesive VW interdigitation of the capping decanethiolate molecules. This model implies that the $-CO_2H$ groups are mostly located at the interparticle void spaces that are surrounded by hydrophobic alkyl chains, which are interdigitated.

As a result of the spatial packing of nanocrystals, the nanostructured microenvironment has a framework-type architecture. It is envisioned that this type of architecture could be exploited as membrane-like channels in which the channeling properties can be tuned chemically or electrochemically. It is also envisioned that the structure can be exploited as a ligand framework for chemical sensing. A specific interfacial partition or flux induces subtle changes in interparticle spacing or dielectric medium properties, which can be detected by electronic transducers. Two examples are described next to illustrate the above implications.

The first example concerns ion-channeling properties of the nanostructured thin films.[15b,56] Figure 21.8 shows the proof of concept demonstrating the exploitation of interparticle spatial properties for ion-gating application.[15b,56] Data from both cyclic voltammetry and EQCN (electrochemical quartz crystal nanobalance) experiments are included for assessing ion passages through the network regulated by the nanostructured ligand framework binding specificity. The redox reaction of $[Ru(NH_3)_6]^{3+/2+}$ was examined for a piezoelectrode coated with an MUA-$Au_{2\text{-nm}}$ film.

At low pH, the redox currents are basically shut off by the film (close state), with only small capacitive currents being detectable. In contrast, the response at the high pH is large (open state) and comparable with that at the bare GC. The electrostatic effect is operative for the ion-gating process. When pH is high, the network carries negative charges ($-CO_2^-$) and admits the positively charged redox species via an attractive force. When pH is low, the neutral network does not admit the probe at all, suggesting that the channels in the neutral network are too small for the probe to penetrate. The EQCN data provided further information to assess the ion-channeling property. At pH 2, the mass change is insignificant, in excellent agreement with the lack of redox current in the voltammetric curve. In contrast, a distinctive mass change profile is detected at pH 11. The mass increases in the negative potential sweep and returns in the positive potential sweep. Qualitatively, this observation is suggestive of incorporation of electrolyte species during the reduction of $Ru(NH_3)_6^{3+}$, which are electrostatically associated with the negatively fixed charges ($-CO_2^-$ group) in the film. For reasons of electroneutrality, the reduction process requires the supply of positive charges to compensate the fixed negative $-CO_2^-$ groups. Therefore, cations (K^+) must flux into the film during the reduction and flux out upon reoxidation. To confirm the electrostatic effect, negatively charged probes, $[Fe(CN)]_6^{3-}$s, were examined. The redox responses at both pH values were found to be effectively suppressed, which is in sharp contrast to the result observed for the cationic redox probe. The effective blockage at the high pH is indeed consistent with the repulsive effect between the probe and the negative charges residing at the rim of the channels. A similar and elegant study has recently been reported,[58] which involved the use of negatively charged poly(acrylic acid) (PAA) to assemble titania nanoshells via a layer-by-layer fashion to produce a network of voids and channels for ion separation of dopamine and ascorbic acid.

The second example deals with the vapor phase–sensing properties of the nanostructured thin films.[57] Using interdigitated microelectrodes and piezoelectric oscillators, we examined responses of MUA-linked nanoparticle films to a series of alcohols such as methanol, ethanol, propanol, and butanol. Although the data for water and methanol sorption display a negative response, the sorption of ethanol displays a positive change in resistance (Figure 21.9I). Evidently, the increased hydrophobic character for the vapor molecules reversed the response to the usual positive response profile. Data for propanol and butanol showed the same positive profile and increased sensitivity. The $\Delta R/R_i - C_{ppm}$ and $\Delta f - C_{ppm}$ responses to water, methanol, and ethanol exhibit approximately linear relationships for the resistance and frequency changes, as a function of the vapor concentration (Figure 21.9II). The response sensitivity shows negative signs for both water and methanol vapors and positive signs for ethanol vapor. The positive response sensitivity was found to further increase with increasing alkyl chain length of the vapor, as observed for propanol and butanol (not shown). The striking difference of the response sensitivity to vapor sorption demonstrates the high selectivity of the sensor in response to interfacial vapor sorption to the molecular interaction and the detailed nanostructure. Several factors could be responsible for this difference, including the formation of hydrogen bonding between the polar vapor molecule and the head-to-head hydrogen-bonded MUA linkage, the dielectric medium effect in the nanostructured film partitioned

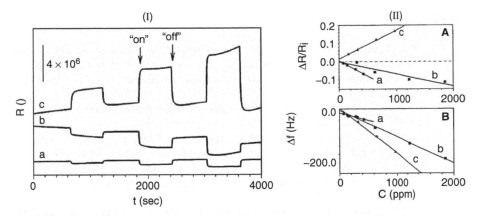

FIGURE 21.9 (I) Response curves of resistance change (R) for vapor sorption of methanol and water on MUA-Au$_{2\text{-nm}}$ film (N $\approx$ 31). (II) The baselines are offset for comparison. Vapor concentrations: 142, 284, and 425 ppm for water (A, a); 620, 1239, and 1859 ppm for methanol (A, b); and 303, 606, and 909 ppm for ethanol (A, c). (II) Plots of $\Delta R/R_i$ (A) and Δf (B) vs. C_{ppm} for vapor sorption of water (a), methanol (b), and ethanol (c) at MUA-Au$_{2\text{-nm}}$ film. (From L Han et al., *Anal Chem* 73:4441–4449, 2001. With permission.)

with the vapor molecules, and the solubility effect of the nanostructured film in alcoholic solvents. We note that using different methods to assemble nanostructured interfaces, several other studies[51,59,60] have also demonstrated vapor-responsive properties.

In summary, the interparticle structure and spatial properties for the assembly of monolayer-capped nanoparticles can be controlled by manipulating the interparticle mediation agent toward constructing chemically responsive architectures. This ability is significant because the controllable three-dimensional morphology with a regulation in interparticle properties and mass transport may find applications in nanostructured sensing or catalysis, where the precise control of the interparticle distance is essential. The interparticle phenomena may also be generalized to many other nanoparticle assemblies using different mediators, including organic, inorganic, polymeric, and biological molecules, such as vesicle aggregation with a controlled ligand-receptor stoichiometry.[61] An in-depth delineation of the detailed interfacial nanochemistry for the assembly of nanoparticles in terms of size, shape, space, and surface properties should aid greatly our exploration of nanoscience and nanotechnology.

ACKNOWLEDGMENTS

We gratefully thank many of our collaborators for their help or contributions to different parts of the work described in this chapter, including Drs. M. Hepel, Y. Lin, and V.W. Jones. Financial support of this work, in part from the National Science Foundation (CHE0316322, CHE0349040), the Petroleum Research Fund administered by the American Chemical Society (40253-AC5M), and 3M Corporation, is gratefully acknowledged. L.H. acknowledges the ACS Division of Analytical Chemistry Graduate Fellowship sponsored by Merck. M.M.M. thanks the Department of Defense (Army Research Office) for support via a National Defense Science and Engineering Graduate Fellowship.

REFERENCES

1. (a) AC Templeton, WP Wuelfing, RW Murray, Monolayer protected cluster molecules, *Acc Chem Res* 33: 27–36, 2000. (b) AN Shipway, E Katz, I Willner, Nanoparticle arrays on surfaces for electronic, optical, and sensor applications, *ChemPhysChem* 1: 18–52, 2000.
2. CJ Zhong, MM Maye, Core-shell assembled nanoparticles as catalysts, *Adv Mater* 13: 1507–1511, 2001.
3. (a) M Brust, M Walker, D Bethell, DJ Schiffrin, R Whyman, Synthesis of thiool-derivatized gold nanoparticles in a 2-phase liquid–liquid system, *J Chem Soc Chem Commun* 801–802, 1994. (b) DI Gittins, D Bethell, DJ Schiffrin, RJ Nichols, A nanometre-scale electronic switch consisting of a metal cluster and redox-addressable groups, *Nature* 408: 67–69, 2000.
4. (a) MJ Hostetler, JE Wingate, CJ Zhong, JE Harris, RW Vachet, MR Clark, JD Londono, SJ Green, JJ Stokes, GD Wignall, GL Glish, MD Porter, ND Evans, RW Murray, Alkanethiolate gold cluster molecules with core diameters from 1.5 to 5.2 nm: core and monolayer properties as a function of core size, *Langmuir* 14: 17–30,1998. (b) FP Zamborini, SM Gross, RW Murray, Synthesis, characterization, reactivity, and electrochemistry of palladium monolayer protected clusters, *Langmuir* 17: 481–488, 2001.
5. RL Whetten, JT Khoury, MM Alvarez, S Murthy, I Vezmar, ZL Wang, PW Stephens, CL Cleveland, WD Luedtke, U Landman, Nanocrystal gold molecules, *Adv Mater* 8: 428–434, 1996.
6. (a) AC Templeton, MJ Hostetler, CT Kraft, RW Murray, Reactivity of monolayer-protected gold cluster molecules: steric effects, *J Am Chem Soc* 120: 1906–1911, 1998. (b) MJ Hostetler, SJ Green, JJ Stokes, RW Murray, Monolayers in three dimensions: synthesis and electrochemistry of omega-functionalized alkanethiolate-stabilized gold cluster compounds, *J Am Chem Soc* 118: 4212–4213, 1996.
7. (a) M Brust, D Bethell, CJ Kiely, DJ Schiffrin, Self-assembled gold nanoparticle thin films with nonmetallic optical and electronic properties, *Langmuir* 14: 5425–5429, 1998. (b) D Bethell, M Brust, DJ Schiffrin, C Kiely, From monolayers to nanostructured materials: an organic chemist's view of self-assembly, *J Electroanal Chem* 409: 137–143, 1996. (c) M Brust, CJ Kiely, D Bethell, DJ Schiffrin, C-60 mediated aggregation of gold nanoparticles, *J Am Chem Soc* 120: 12367–12368, 1998. (d) T Baum, D Bethell, M Brust, DJ Schiffrin, Electrochemical charge injection into immobilized nanosized gold particle ensembles: potential modulated transmission and reflectance spectroscopy, *Langmuir* 15: 866–871, 1999.
8. MD Musick, DJ Pena, SL Botsko, TM McEvoy, JN Richardson, MJ Natan, Electrochemical properties of colloidal Au-based surfaces: multilayer assemblies and seeded colloid films, *Langmuir*, 15: 844–850, 1999.
9. (a) FP Zamborini, JF Hicks, RW Murray, Quantized double layer charging of nanoparticle films assembled using carboxylate/(Cu2+ or Zn2+)/carboxylate bridges, *J Am Chem Soc* 122: 4514–4515, 2000. (b) AC Templeton, FP Zamborini, WP Wuelfing, RW Murray, Controlled and reversible formation of nanoparticle aggregates and films using Cu2+-carboxylate chemistry, *Langmuir* 16: 6682–6688, 2000. (c) WP Wuelfing, FP Zamborini, AC Templeton, XG Wen, H Yoon, RW Murray, Monolayer-protected clusters: molecular precursors to metal films, *Chem Mater* 13: 87–95, 2001.
10. CA Mirkin, RL Letsinger, RC Mucic, JJ Storhoff, A DNA-based method for rationally assembling nanoparticles into macroscopic materials, *Nature* 382: 607–609, 1996.

11. (a) R Elghanian, JJ Storhoff, RC Mucic, RL Letsinger, CA Mirkin, Selective colorimetric detection of polynucleotides based on the distance-dependent optical properties of gold nanoparticles, *Science* 277: 1078–1081, 1997. (b) TA Taton, RC Mucic, CA Mirkin, RL Letsinger, The DNA-mediated formation of supramolecular mono- and multilayered nanoparticle structures, *J Am Chem Soc* 122: 6305–6306, 2000.

12. (a) M Bruchez, M Moronne, P Gin, S Weiss, AP Alivisatos, Semiconductor nanocrystals as fluorescent biological labels, *Science* 281: 2013–2016, 1998. (b) AP Alivisatos, KP Johnsson, XG Peng, TE Wilson, CJ Loweth, MP Bruchez, PG Schultz, Organization of "nanocrystal molecules" using DNA, *Nature* 382: 609–611, 1996.

13. JKN Mbindyo, BD Reiss, BR Martin, CD Keating, MJ Natan, TE Mallouk, DNA-directed assembly of gold nanowires on complementary surfaces, *Adv Mater* 13: 249–254, 2001.

14. (a) AK Boal, VM Rotello, Redox-modulated recognition of flavin by functionalized gold nanoparticles, *J Am Chem Soc* 121: 4914–4915, 1999. (b) J Liu, S Mendoza, E Roman, MJ Lynn, RL Xu, AE Kaifer, Cyclodextrin-modified gold nanospheres. Host-guest interactions at work to control colloidal properties, *J Am Chem Soc* 121: 4304–4305, 1999. (c) M Lahav, AN Shipway, I Willner, Au-nanoparticle-bis-bipyridinium cyclophane superstructures: assembly, characterization and sensoric applications, *J Chem Soc Perkin Trans* 2: 1925–1931, 1999. (d) SW Chen, RJ Pei, Ion-induced rectification of nanoparticle quantized capacitance charging in aqueous solutions, *J Am Chem Soc* 123: 10607–10615, 2001. (e) B Kim, SL Tripp, A Wei, Self-organization of large gold nanoparticle arrays, *J Am Chem Soc* 123: 7955–7956, 2001.

15. (a) FL Leibowitz, WX Zheng, MM Maye, CJ Zhong, Structures and properties of nanoparticle thin films formed via a one-step exchange–crosslinking–precipitation route, *Anal Chem* 71: 5076–5083, 1999. (b) WX Zheng, MM Maye, FL Leibowitz, CJ Zhong, Imparting biomimetic ion-gating recognition properties to electrodes with hydrogen-bonding structured core-shell nanoparticle network, *Anal Chem* 72: 2190–2199, 2000. (c) L Han, MM Maye, FL Leibowitz, NK Ly, CJ Zhong, Quartz-crystal microbalance and spectrophotometric assessments of inter-core and inter-shell reactivities in nanoparticle film formation and growth, *J Mater Chem* 11: 1259–1264, 2001.

16. ZL Wang, SA Harfenist, RL Whetten, J Bentley, ND Evans, Bundling and interdigitation of adsorbed thiolate groups in self-assembled nanocrystal superlattices, *J Phys Chem B* 102: 3068–3072, 1998.

17. TG Schaaff, MN Shafigullin, JT Khoury, I Vezmar, RL Whetten, Properties of a ubiquitous 29 kDa Au:SR cluster compound, *J Phys Chem B* 105: 8785–8796, 2001.

18. JR Heath, CM Knobler, DV Leff, Pressure/temperature phase diagrams and superlattices of organically functionalized metal nanocrystal monolayers: the influence of particle size, size distribution, and surface passivant, *J Phys Chem B* 101: 189–197, 1997.

19. MP Pileni, Nanocrystal self-assemblies: fabrication and collective properties, *J Phys Chem B* 105: 3358–3371, 2001.

20. J Fink, CJ Kiely, D Bethell, DJ Schiffrin, Self-organization of nanosized gold particles, *Chem Mater* 10: 922–926, 1998.

21. RP Andres, JD Bielefeld, JI Henderson, DB Janes, VR Kolagunta, CP Kubiak, WJ Mahoney, RG Osifchin, Self-assembly of a two-dimensional superlattice of molecularly linked metal clusters, *Science* 273: 1690–1693, 1996.

22. SH Sun, CB Murray, D Weller, L Folks, A Moser, Monodisperse FePt nanoparticles and ferromagnetic FePt nanocrystal superlattices, *Science* 287: 1989–1992, 2000.

23. BL Frankamp, AK Boal, VM Rotello, Controlled interparticle spacing through self-assembly of Au nanoparticles and poly(amidoamine) dendrimers, *J Am Chem Soc* 124: 15146–15147, 2002.

24. SR Nicewarner-Pena, RG Freeman, BD Reiss, L He, DJ Pena, ID Walton, R Cromer, CD Keating, MJ Natan, Submicrometer metallic barcodes, *Science* 294: 137–141, 2001.

25. P Mulvaney, LM Liz-Marzan, M Giersig, T Ung, Silica encapsulation of quantum dots and metal clusters, *J Mater Chem* 10:1259–1270, 2000.

26. AP Alivisatos, KP Johnsson, XG Peng, TE Wilson, CJ Loweth, MP Bruchez, PG Schultz, Organization of "nanocrystal molecules" using DNA, *Nature* 382: 609–611, 1996.

27. AK Boal, VM Rotello, Intra- and intermonolayer hydrogen bonding in amide-functionalized alkanethiol self-assembled monolayers on gold nanoparticles, *Langmuir* 16: 9527–9532. 2000.

28. YJ Kim, RC Johnson, JT Hupp, Gold nanoparticle-based sensing of "spectroscopically silent" heavy metal ions, *Nano Lett* 1: 165–167, 2001.

29. CS Weisbecker, MV Merritt, GM Whitesides, Molecular self-assembly of aliphatic thiols on gold colloids, *Langmuir* 12: 3763–3772, 1996.

30. SR Johnson, SD Evans, R Brydson, Influence of a terminal functionality on the physical properties of surfactant-stabilized gold nanoparticles, *Langmuir* 14: 6639–6647, 1998.

31. KS Mayya, V Patil, M Sastry, On the stability of carboxylic acid derivatized gold colloidal particles: the role of colloidal solution pH studied by optical absorption spectroscopy, *Langmuir* 13: 3944–3947, 1997.

32. (a) TE Mallouk, JA Gavin, Molecular recognition in lamellar solids and thin films, *Acc Chem Res* 31: 209–217, 1998. (b) J-H Fuhrhop, J Köning, *Membrane and Molecular Assemblies: The Synkinetic Approach*, London, RSC Publications, 1994.

33. (a) MM Maye, WX Zheng, FL Leibowitz, NK Ly, CJ Zhong, Heating-induced evolution of thiolate-encapsulated gold nanoparticles: a strategy for size and shape manipulations, *Langmuir* 16: 490–497, 2000. (b) MM Maye, CJ Zhong, Manipulating core-shell reactivities for processing nanoparticle sizes and shapes, *J Mater Chem* 10: 1895–1901, 2000.

34. (a) SW Chen, Langmuir–Blodgett fabrication of two-dimensional robust cross-linked nanoparticle assemblies, *Langmuir* 17: 2878–2884, 2001. (b) NZ Clarke, C Waters, KA Johnson, J Satherley, DJ. Schiffrin, Size-dependent solubility of thiol-derivatized gold nanoparticles in supercritical ethane, *Langmuir* 17:6048–6050, 2001. (c) T Teranishi, S Hasegawa, T Shimizu, M Miyake, Heat-induced size evolution of gold nanoparticles in the solid state, *Adv Mater* 13: 1699–1701, 2001.

35. L Han, J Luo, NN Kariuki, MM Maye, VW Jones, CJ Zhong, Novel interparticle spatial properties of hydrogen-bonding mediated nanoparticle assembly, *Chem Mater* 15: 29–37, 2003.

36. L Han, MM Maye, CJ Zhong, Nanoparticle assembly via hydrogen-bonding: IRS, TEM and AFM characterizations, *Mater Res Soc Symp Proc* 635: 4.5.1–4.5.6, 2000.

37. MM Alvarez, JT Khoury, TG Schaaff, MN Shafigullin, I Vezmar, RL Whetten, Optical absorption spectra of nanocrystal gold molecules, *J Phys Chem B* 101: 3706–3712, 1997.

38. S Link, MA El-Sayed, Shape and size dependence of radiative, non-radiative and photothermal properties of gold nanocrystals, *Int Rev Phys Chem* 19: 409–453, 2000.

39. N Kariuki, L Han, NK Ly, MJ Peterson, MM Maye, G Liu, CJ Zhong, Preparation and characterization of gold nanoparticles dispersed in poly(2-hydroxyethyl methacrylate), *Langmuir* 18: 8255–8259, 2002.

40. JF Hicks, Y Seok-Shon, RW Murray, Layer-by-layer growth of polymer/nanoparticle films containing monolayer-protected gold clusters, *Langmuir* 18: 2288–2294, 2002.

41. CW Brown, Y Li, JA Seelenbinder, P Pivarnik, A Rand, SV Letcher, OJ Gregory, MJ Platek, Immunoassays based on surface enhanced infrared absorption spectroscopy, *Anal Chem* 70: 2991–2996, 1998.

42. (a) RG Nuzzo, LH Dubois, DL Allara, Fundamental studies of microscopic wetting on organic surfaces. 1. Formation and structural characterization of a self-consistent series of polyfunctional organic monolayers, *J Am Chem Soc* 112: 558–569, 1990. (b) EL Smith, CA Alves, JW Anderegg, LM Siperko, MD Porter, Deposition of metal overlayers at end-group-functionalized thiolate monolayers adsorbed at Au 111 surface and interfactial chemical characterization of deposited Cu-overlayers at carboxylic acid-terminated structures, *Langmuir* 8: 2707–2714, 1992. (c) EL Smith, MD Porter, Structure of monolayers of short-chain n-alkanoic acids ($CH_3(CH_2)NCOOH$, N = 0–9) spontaneously adsorbed from the gas-phase at silver as probed by infrared reflection spectroscopy, *J Phys Chem* 97: 8032–8038, 1993. (d) LJ Kepley, RM Crooks, Selective surface acoustic wave–based organophosphonate chemical sensor employing a self-assembled composite monolayer: a new paradigm for sensor design, *Anal Chem* 64: 3191–3193, 1992. (e) Y-T Tao, W-L Lin, GD Hietpas, DL Allara, Infrared spectroscopic study of chemically induced dewetting in liquid crystalline types of self-assembled monolayers, *J Phys Chem B* 101: 9732–9740, 1997.

43. S Hayashi, J Umemura, Infrared spectroscopic evidence for the coexistence of two molecular configurations in crystalline fatty acid, *J Chem Phys* 63: 1732–1740, 1975.

44. J Umemura, Change of molecular configuration in crystalline decanoic acid as studied by infrared spectra and normal coordinate analysis, *J Chem Phys* 1: 42–48, 1978.

45. F Bensebaa, R Voicu, L Huron, TH Ellis, E Kruus, Kinetics of formation of long-chain n-alkanethiolate monolayers on polycrystalline gold, *Langmuir* 13: 5335–5340, 1997.

46. MD Porter, TB Bright, DL Allara, CED Chidsey, Spontaneously organized molecular assemblies. 4. Structural characterization of n-alkyl thiol monolayers on gold by optical ellipsometry, infrared spectroscopy, and electrochemistry, *J Am Chem Soc* 109: 3559–3568, 1987.

47. MJ Hostetler, JJ Stokes, RW Murray, Infrared spectroscopy of three-dimensional self-assembled monolayers: N-alkanethiolate monolayers on gold cluster compounds, *Langmuir* 12: 3604–3612, 1996.

48. MM Maye, J Luo, L Han, CJ Zhong, Probing pH-tuned morphological changes in core-shell nanoparticle assembly using AFM, *Nano Lett* 1: 575–580, 2001.

49. (a) ZL Wang, SA Harfenist, RL Whetten, J Bentley, ND Evans, Bundling and interdigitation of adsorbed thiolate groups in self-assembled nanocrystal superlattices, *J Phys Chem B* 102: 3068–3072, 1998. (b) JM Petroski, TC Green, MA El-Sayed, Self-assembly of platinum nanoparticles of various size and shape, *J Phys Chem A* 105: 5542–5547, 2001.

50. L Motte, E Lacaze, M Maillard, MP Pileni, Self-assemblies of silver sulfide nanocrystals on various substrates, *Langmuir* 16: 3803–3812, 2000.

51. FP Zamborini, MC Leopold, JF Hicks, PJ Kulesza, MA Malik, RW Murray, Electron hopping conductivity and vapor sensing properties of flexible network polymer films of metal nanoparticles, *J Am Chem Soc* 124: 8958–8964, 2002.

52. (a) PC Ohara, WM Gelbart, Interplay between hole instability and nanoparticle array formation in ultrathin liquid films, *Langmuir* 14: 3418–3424, 1998. (b) PC Ohara, JR Heath, WM Gelbart, Self-assembly of sub-micro-micrometer rings of particles from solutions of nanoparticles, *Angew Chem Int Ed Engl* 36: 1077–1080, 1997. (c) S Maenosono, CD Dushkin, S Saita, Y Yamaguchi, Growth of a semiconductor nanoparticle ring during the drying of a suspension droplet, *Langmuir* 15: 957–965, 1985.

53. J Liu, Y Shin, ZM Nie, JH Chang, LQ Wang, GE Fryxell, WD Samuels, GJ Exarhos, Molecular assembly in ordered mesoporosity: a new class of highly functional nanoscale materials, *J Phys Chem A* 104: 8328–8339, 2000.

54. T Fujii, T Yano, K Nakamura, O Miyawaki, The sol-gel preparation and characterization of nanoporous silica membrane with controlled pore size, *J Membrane Sci* 187: 171–180, 2001.

55. (a) GL Gulley, JE Martin, Stabilization of colloidal silica using polyols, *J Colloid Interface Sci* 241: 340–345, 2001. (b) AL Hiddessen, SD Rodgers, DA Weitz, DA Hammer, Assembly of binary colloidal structures via specific biological adhesion, *Langmuir* 16: 9744–9753, 2000. (c) LH Hanus, K Sooklal, CJ Murphy, HJ Ploehn, Aggregation kinetics of dendrimer-stabilized CdS nanoclusters, *Langmuir* 16: 2621–2626, 2000.

56. J Luo, N Kariuki, L Han, MM Maye, LW Moussa, SR Kowaleski, FL Kirk, M Hepel, CJ Zhong, Interfacial mass flux at 11-mercaptoundecanoic acid linked nanoparticle assembly on electrodes, *J Phys Chem* 106: 9313–9321, 2002.

57. L Han, DR Daniel, MM Maye, CJ Zhong, Core-shell nanostructured nanoparticle films as chemically-sensitive interfaces, *Anal Chem* 73: 4441–4449, 2001.

58. DS Koktysh, XR Liang, BG Yun, I Pastoriza-Santos, RL Matts, M Giersig, C Serra-Rodriguez, LM Liz-Marzan, NA Kotov, Biomaterials by design: layer-by-layer assembled ion-selective and biocompatible films of TiO_2 nanoshells for neurochemical monitoring, *Adv Funct Mater* 12: 255–265, 2002.

59. H Wohltjen, AW Snow, Colloidal metal–insulator–metal ensemble chemiresistor sensor, *Anal Chem* 70: 2856–2859, 1998.

60. GM Gross, DA Nelson, RE Synovek, Monolayer-protected gold nanoparticles as a stationary phase for open tubular gas chromatography, *Anal Chem* 75: 4558–4564, 2003.

61. ET Kisak, MT Kennedy, D Trommeshauser, JA Zasadzinski, Self-limiting aggregation by controlled ligand-receptor stoichiometry, *Langmuir* 16: 2825–2831, 2000.

22 Langmuir–Blodgett Thin Films of Gold Nanoparticle Molecules: Fabrication of Cross-Linked Networks and Interfacial Dynamics

Shaowei Chen

CONTENTS

22.1 Introduction 577
22.2 Two-Dimensional Nanoparticle Cross-Linked Superlattices 581
    22.2.1 Networking 581
    22.2.2 Effects of Protecting Monolayers on Particle Cross-Linking 584
    22.2.3 Nanoparticle Cross-Linked Networks for Surface
             Nanofabrications..587
22.3 Interfacial Dynamics of Nanoparticles' LB Thin Films............................589
    22.3.1 Langmuir Thin Films of Electroactive Nanoparticles 589
    22.3.2 Interfacial Dynamics of Nanoparticle LB Thin Films 590
    22.3.3 Multilayer Structures of Electroactive Nanoparticles................594
22.4 Conclusion..595
Acknowledgments...596
References and Notes 596

22.1 INTRODUCTION

The recent intense research interest in nanosized particle materials has been largely driven by the technological implication of these novel structural elements in the nano-aturizations of optical, electronic, and magnetic devices.[1,2] One of the technical challenges in the bottom-up approach[3] in the construction of these nanoscale entities is to develop efficient methods to organize nanoparticles into ordered and robust arrays in a controllable manner, where the collective properties of the particle assemblies can be easily manipulated by the nature and composition of the particles as well as the physical distribution of the particles in the ensembles. In particular, much research effort has been focused on the construction of two-dimensional organized assemblies that can serve as the structural basis for more complicated nanostructures. Several approaches have been reported. For instance, nanoparticles of narrow dispersity can self-organize into ordered structures by simply dropcasting the particle solution onto a substrate.[4] However, the particle assemblies generally lack long-range ordering and good reproducibility. Another approach relies on bifunctional ligands that immobilize the particles to a substrate surface by a sequential anchoring mechanism.[5,6] Further improvement is made by first incorporating the bifunctional linkers into the nanoparticle surface, rendering the particles surface active with peripheral anchoring groups, which can then be self-assembled onto substrate surfaces, akin to monomeric alkanethiol molecules.[7] This self-assembling approach is very effective in achieving particle assemblies of high surface coverage and, thanks to the strong chemical interactions between the anchor sites and the substrate (and, hence, a vertical approach), leads to the fabrications of mechanically robust particle structures. Apparently, the latter is the limiting factor in the application of this route. In case the interactions between the particles and the substrate are weak, equally stable particle network structures can be constructed by using the Langmuir–Blodgett (LB) technique (a horizontal approach), where neighboring particles are cross-linked by, again, bifunctional linkers at the air–water interface.[8]

It is a well-known procedure to use the LB technique to fabricate and manipulate the molecular ordering of nanoparticle thin films.[9–14] One well-known example is the insulator–metal transition of silver nanoparticles at the air–water interface when sufficient electronic coupling between neighboring particles is allowed at very small interparticle separation.[9a] Nonlinear current potential (I-V) profiles have also been observed for gold nanoparticle LB thin films at an interdigitated array (IDA) electrode, which was interpreted on the basis of the Coulomb blockade theory.[13b] More interestingly, when incorporated into a metal–insulator–semiconductor (MIS) junction structure, organically capped gold nanoparticles exhibited a hysteresis in its capacitance vs. voltage characteristics, the magnitude of which is dependent upon the voltage sweep conditions.[13c] This was ascribed to the electronic memory effects (i.e., charge storage). These earlier studies demonstrate strong application potentialities of nanoparticles as the building blocks for the development of novel electronic nanodevices and nanocircuits.

However, it should be noted that the resulting particle assemblies by LB deposition typically lack long-term stability, in particular when deposited at relatively low surface pressures, mainly due to the relatively weak interactions between neighboring particles as well as between particles and substrates. Consequently, this will seriously affect the durability and stability of their device application. Thus, by cross-linking the neighboring particles with strong chemical bridges, we aim to develop an effective method to construct particle ensembles with enhanced mechanical stability, which might ultimately lead to the fabrication of stand-alone nanoparticle thin films. Recently, we showed that nanoparticle cross-linking could be initiated at high surface pressures by confining the particles and the cross-linking reagents on the water surface, where robust and compact particle networks were readily constructed.[8a,b] In addition, the networking between neighboring particles at the air–water interface has been demonstrated by, for instance, the cross-linking of aminoethanethiol-modified CdS semiconductor nanoparticles by glutaraldehyde in the *subphase*, where the resulting particle networks exhibited enhanced n-type photosensitivity.[15]

One of the intriguing potential applications of these robust nanoparticle assemblies lies in the construction of surface nanostructures, where long-range (metal) quantum dot arrays can be readily fabricated in a nonlithographic manner. Here the organic components can be removed photochemically, leaving the quantum dots deposited onto the substrate surfaces. As the nanodot dimensions are mainly determined by the original particle sizes, this nanoparticle-based approach might serve as a complementary route to other surface fabrication techniques where nanometer-sized patterning remains a challenge. For instance, in nanosphere lithography,[16] closely packed micrometer-sized colloidal particles are used as the masking template and metal islands are grown in the particle interstices, with a dimension as big as a few tenths of the colloid radius. In addition, it is anticipated that the surface quantum dot arrays will provide a well-defined structural basis for surface nanoscale chemical functionalization that might not be readily accessible by other methods (e.g., micro-contact printing[17] or scanning probe microscopy-based surface nanografting[18]).

Here it is worthwhile to briefly review the recent progress of using the LB technique as a powerful tool to create nanoscale surface-ordered structures. For instance, Iakovenko and coworkers[19] showed that one- and two-dimensional arrays of magnetic (Fe_3O_4 and Fe_2O_3) nanoparticles could be fabricated by the LB technique in which the particles were adsorbed from the subphase onto a stearic acid monolayer at the air–water interface. This was ascribed to the electrostatic attraction between the magnetic particles and the surfactant head groups. In addition, in combination with microcontact printing (μCP), μ-dot arrays of γ-Fe_2O_3 nanoparticles were created by first depositing the particles onto a patterned poly(dimethylsiloxane) (PDMS) stamp using the LB technique; the particles were then transferred onto a substrate surface (e.g., Si wafer).[20] Patterned structures of nanoparticles have also been developed by electron beam irradiation of an LB thin film of particle precursors.[21] More interestingly, nanometer-scale wire structures have also been created by electron beam lithography on LB films of alkanethiolate-protected gold nanoparticles.[22]

Using polymer thin films (again, by the LB technique) as the supporting matrices,[23] nanoparticle networks have also been created. One of the examples includes the two-dimensional networks of gold nanoclusters by using a poly(vinyl-pyrrolidone) (PVP) template to (quasi-) one-dimensionally organize the $Au_{55}(PPh_3)_{12}Cl_6$ clusters.[23a] In another study, gold nanoparticles were anchored onto ultrathin films of poly(N-dodecyl acrylamide-co-4-vinyl pyridine) prepared by the LB technique, where patterned arrays of nanoparticles can then be constructed by further photo-lithographic treatment.[23b]

Ordered arrays of nanowires have also been fabricated by using the LB technique.[24,25] For instance, Liu et al.[24] used cationic poly(N-vinylcarbazole) as templates to control the composite assembly of thallic oxide (Tl_2O_3) nanowires. Tao and coworkers[25] fabricated LB monolayers of silver nanowires for molecular-sensing applications using surface-enhanced Ramam spectroscopy (SERS).

These earlier studies demonstrate that by using nanoparticles as the building blocks, a wide variety of unique and functional nanostructures can be easily accessible by the LB technique. In particular, in combination with other surface nanofabrication tools (e.g., electron beam lithography and photolithography), more complicated nanoscale architectures can be created for more diverse applications.

In our previous report,[8a] we demonstrated that alkanethiolate-protected gold nanoparticles (AuMPCs) could be networked by rigid aryldithiol linkers (e.g., 4,4′-thiobisbenzenethiol (TBBT)) at relatively high surface pressures, as evidenced by optical and electron microscopic measurements. These particle superlattices can then be exploited as an effective alternative for surface nanofabrication. Dithiols have been used previously to cross-link gold particles, forming three-dimensional aggregates in which the dithiol ligands were introduced into the particle-protecting monolayers, and upon the removal of solvents, this assembling process occurred rather quickly, efficiently, and yet uncontrollably.[26] To use dithiols as the linkers to achieve two-dimensional cross-linked nanoparticle networks, we brought the particles and the rigid dithiol linkers into contact at the air–water interface, where the cross-linking process was confined to the interfacial region and initiated at relatively high surface pressures, presumably due to ligand intercalation and surface place–exchange reactions.[27c] However, the resulting particle networks[8a] did not show ordered superlattice structure, which was ascribed to the modest dispersivity of the particles. In contrast, when highly monodisperse nanoparticles were used, long-range ordered nanoparticle superlattices were readily achieved.[8b]

As mentioned above, classical LB thin films of nanoparticles (without lateral cross-linking) lack long-term mechanical stability. This should be manifested easily by, for instance, using nanoparticles with electroactive functional groups that can then be probed by electrochemical measurements. However, this subject has not received extensive attention so far. More importantly, part of the major motivation arises from the more complicated chemical functionalization of the nanoparticle assemblies. For instance, ferrocene has been used quite extensively as a model system in the investigation of interfacial electron transfer kinetics. One of the typical approaches is based on immobilizing the ferrocene moieties onto electrode surfaces through an aliphatic chain where the separation of reaction products and the detection of the electron-mediation effect might be facilitated. Two-dimensional self-assembled monolayers of monomeric alkanethiols with ω-ferrocene moieties are among the most extensively studied examples.[28,29] Nanosized particle–supported ferrocene functional groups represent a novel and yet more complicated three-dimensional system, where, in contrast to their two-dimensional counterparts,[30] the ferrocene moieties are concentrated on the nanoscale particle surface. Previous electrochemical studies[30] of ω-ferrocenated nanoparticles in solutions have shown that the particle-bound ferrocene moieties undergo (independent) successive 1– e redox reactions due to the rapid spinning of the nanoparticles in solutions.

However, when the particles are immobilized onto a substrate surface, due to the three-dimensional nature of the particle core, the associated interfacial electron transfer between the ferrocene moieties and the electrode might be complicated by the interactions with electrolyte ions (charge compensation, for instance) as well as double-layer effects.[31] The spatial distribution of the ferrocene moieties on the particle surface is anticipated to lead to the variation of the energetic states, demonstrated in different redox (over)potentials. In addition, in multilayer thin films, the electrochemical responses of the electroactive moieties buried within the thin films might be further complicated by ion penetration/compensation as well as charge propagation.

Thus, in this chapter, we will first focus on our recent studies of the two-dimensional cross-linking of monodispersive gold nanoparticles at the air–water interface, where long-range ordered and robust particle superlattice assemblies are obtained. The effect of the particle-protecting monolayers will be investigated by using particles of varied monolayer thickness. In addition, we will present some preliminary studies of using these particle networks as the structural basis for surface nonlithographic nanofabrication. We will then demonstrate that without chemical cross-linkers, the particles in the deposited thin films exhibit substantial mobility on substrate (electrode) surfaces. Here we carry out a series of electrochemical studies of Langmuir–Blodgett thin films of ferrocenated gold nanoparticles, where effects of the particle film thickness on the ferrocene electrochemistry will be investigated. In particular, the electroactive ferrocene moieties will serve as the molecular probe to electrochemically examine the interfacial dynamics of the MPC surface thin films.

22.2 TWO-DIMENSIONAL NANOPARTICLE CROSS-LINKED SUPERLATTICES

22.2.1 NETWORKING

The formation of nanoparticle networks is achieved by using the Langmuir–Blodgett method, in which interparticle cross-linking is initiated on the water surface at relatively high surface pressures in the presence of rigid bifunctional ligands (e.g., aryl dithiols).[8a,b] This is ascribed to ligand intercalation and, hence, surface exchange reactions between the linking reagents and the particle-bound thiolates. The networking process can be manifested by monitoring the variation of the surface isotherms. Figure 22.1 shows the isotherms of n-octanethiolate-protected gold (C_8Au) particles before and after the addition of 4,4′-thiobisbenzenethiol (TBBT). One can see that both isotherms show rather similar rising slopes, indicating a similarity of the compressibility in these two cases. However, by holding the barrier at the CLOSED position for about 15 h, the subsequent compression shows an isotherm that is vastly different from the previous two: (1) the take-off area is even much smaller than that prior to the addition of TBBT and (2) the rising slope is much steeper. Both observations indicate a more rigid and compact nature of the particle thin films on the water surface and strongly suggest the formation of nanoparticle cross-linked networks (Figure 22.1, inset). In fact, one can see that on the water surface, macroscale purple particle patches are very visible, whereas the original particle thin films without the addition of TBBT look brown/pink (*vide infra*). These particle patches were very stable, virtually unchanged for days, even with the barrier moved back to the OPEN position.

The formation of nanoparticle networks is further supported by more direct and visual characterizations, such as transmission electron microscopic (TEM) measurements. Figure 22.2 shows the TEM micrographs of the C_8Au particles deposited

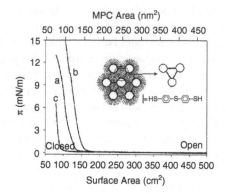

FIGURE 22.1 LB isotherms of C_8Au particles before and after TBBT cross-linking. Initially, 150 µl of the C_8Au particle solution in hexane (1 mg/ml) were spread onto the water surface. The TBBT concentration was 4 mM in CH_2Cl_2, and 75 µl were added to the water surface. (a) Isotherm of the C_8Au particles prior to the addition of TBBT. (b) First compression after the introduction of TBBT. (c) Isotherm of the resulting particle networks after holding the barrier at CLOSED for 15 h. Inset: Schematic of the particle cross-linked networks. (From Chen, S., *Langmuir*, 17, 2878–2884, 2001. With permission. Copyright © 2001 American Chemical Society.)

onto SiO_x-coated Cu grids at varied surface pressures prior to the addition of TBBT as well as after holding the barrier at the CLOSED position in the presence of TBBT. One can see that in the original particle thin films without TBBT, only rather random distribution of the particles is observed; in addition, there appear to be some particle coalescence–forming twins or higher-order oligomers of the particles. In sharp contrast, after compressing the particles with added TBBT molecules, the resulting particle assemblies show striking long-range ordered structures, where it can be clearly seen that rather closely packed superlattices are formed without the appearance of particle aggregation (defects are relatively rare and generally observed near particles that are not uniform in size, e.g., Figure 22.2C).

It should also be noted that the deposition pressure appears not to be an important factor in controlling the interparticle spacing once they are cross-linked by TBBT ligands. For instance, in Figure 22.2C and D, where the particles were deposited at 4 and 7 mN/m, respectively, the interparticle spacing is found at about 1.6 nm. By taking into account the Au-S bond length, this spacing is very close to a single molecular length of TBBT (the molecular length of TBBT is 1.2 nm, as calculated by Hyperchem®). In contrast, without the bridging ligands, the particle spacing (and hence particle density) is predictably varied with surface pressure, as shown in Figure 22.2A and B.

Optical characterization provides further evidence of the formation of particle cross-linked networks. Figure 22.3 (top panel) shows the UV-visible (UV-Vis) spectra of the C_8Au particles that were deposited onto clean glass slide surfaces at varied surface pressures before and after TBBT cross-linking. Also shown is the absorption spectrum of the particles dissolved in hexane. It is well known that nanosized gold particles exhibit a unique surface plasmon (SP) absorption band, which is superimposed

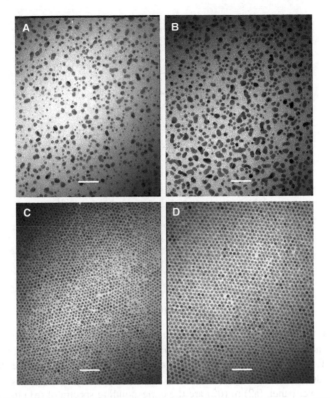

FIGURE 22.2 TEM micrographs of the C_8Au particles deposited onto an SiO_x-coated Cu grid at varied surface pressures before and after TBBT cross-linking. (A) No TBBT, 4 mN/m. (B) No TBBT, 7 mN/m. (C) After TBBT networking, 4 mN/m. (D) After TBBT networking, 7 mN/m. Scale bars, 33 nm. (From Chen, S., *Langmuir*, 17, 2878–2884, 2001. With permission. Copyright © 2001 American Chemical Society.)

onto the exponential-decay Mie scattering profiles.[27,32] The specific position of the SP band is very sensitive to the size, shape, and chemical environments of the particles (e.g., interparticle spacing),[27c,32] a property that has been exploited as the sensing responses for chemical/biological identifications, for instance.[33] Here, as anticipated, for the C_8Au particles dissolved in solution, the SP band is found at about 510 nm (Figure 22.3e). However, when the C_8Au particle monolayers were deposited at 4 or 7 mN/m onto clean glass substrates (even without the cross-linking of TBBT), it can be seen that the corresponding absorption profiles were broadened and the SP band red shifted to about 550 nm. When TBBT cross-linking is initiated, the red shifting of the SP energy is even more pronounced, to 580 nm, where the deposition pressures are 4 and 7 mN/m, respectively. These observations are consistent with the colorimetric change of the particle monolayers mentioned above as well as the interparticle spacing manifested by the TEM measurements (Figure 22.2). The red shifting is clearly due to the electronic coupling interactions between adjacent particles and increases with decreasing interparticle separation. It should be pointed out that within the experimental context here, there is no significant

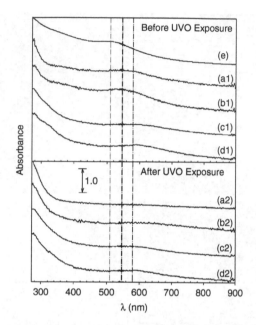

FIGURE 22.3 UV-Vis spectra of the C_8Au particle monolayers deposited onto cleaned glass slide surfaces at varied surface pressures before or after TBBT cross-linking (upper panel). (a1) No TBBT, $\pi = 4$ mN/m. (a2) No TBBT, $\pi = 7$ mN/m. (c1) TBBT added, $\pi = 4$ mN/m. (d1) TBBT added, $\pi = 7$ mN/m. Effects of UV–ozone exposure to the particle thin films are shown in the lower panel. (a2) to (d2) are the corresponding spectra of (a1) to (d1) after UV exposure, respectively. Also shown is the spectrum (e) of the C_8Au particles dissolved in hexane (concentration, about 0.1 mg/ml). All spectra were normalized to their respective absorbance at 272 nm and offset for the sake of clarity. Other experimental conditions are the same as in Figure 22.2. (From Chen, S., *Langmuir*, 17, 2878–2884, 2001. With permission. Copyright © 2001 American Chemical Society.)

variation of the SP energy with deposition pressures either before or after TBBT cross-linking.

These studies clearly demonstrate that the interparticle cross-linking can be effectively achieved at relatively high surface pressures by using bifunctional chemical bridges, leading to the formation of mechanically robust and long-range ordered nanoparticle superlattices.

22.2.2 EFFECTS OF PROTECTING MONOLAYERS ON PARTICLE CROSS-LINKING

As speculated above, the interparticle cross-linking is effected by the intercalation and exchange reactions of bifunctional TBBT molecules with particle-protecting (alkanethiolate) monolayers at high surface pressures. The above LB protocol was then used to fabricate robust cross-linked networks of gold nanoparticles with varied protecting monolayers. The LB isotherms observed in the formation of these varied particle networks are very similar to each other and to those shown in Figure 22.1

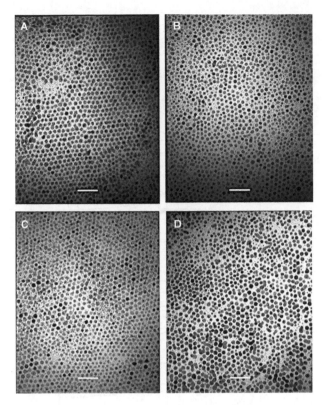

FIGURE 22.4 TEM micrographs of varied gold particle networks cross-linked by TBBT. (A) C_6Au MPC, 200-ml spread, and 50-ml TBBT added. Particle film was deposited at 6 mN/m. (B) $C_{10}Au$ MPC, 100-ml spread, and 50-ml TBBT added. Particle film was deposited at 4 mN/m. (C) $C_{12}Au$ MPC, 100 ml spread, and 50-ml TBBT added. Particle film was deposited at 6 mN/m. (D) Same particle film as in (C), but treated with UV–ozone exposure for 15 min. All particle solutions were 1 mg/ml in hexane. TBBT concentration, 4 mM. Scale bars, 20 nm in (A), 25 nm in (B), and 33 nm in (C) and (D). (From Chen, S., *Langmuir*, 17, 2878–2884, 2001. With permission. Copyright © 2001 American Chemical Society.)

for C_8Au particles (hence, not shown). Again, large-scale purple-colored particle patches were very visible on the water surface. Figure 22.4 shows the representative TEM micrographs of these particle assemblies that were originally protected by varied alkanethiolate monolayers: (A) 1-hexanethiolate (C_6); (B) 1-decanethiolate (C_{10}); and (C) 1-dodencanethiolate (C_{12}). Despite the variation of the thickness of particle-protecting monolayers (C_6, 0.78 nm; C_8, 1.04 nm; C_{10}, 1.29 nm; C_{12}, 1.55 nm — all calculated by Hyperchem®), the closest interparticle spacing is found to be about 1.6 nm in all of these TBBT-bridged particle networks, which equals a single TBBT molecular length, as described earlier. Certainly, at the moment, it is hard to tell if all neighboring particles are linked to each other. However, it is believed that interparticle networking is rampant enough to sustain the entire particle structures, with some disruptions at the defect sites that are largely ascribed to particle size dispersity. Although it might be somewhat surprising that for particles with

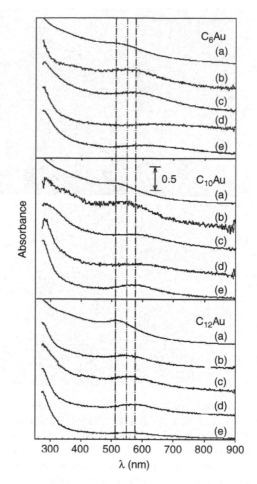

FIGURE 22.5 UV-Vis spectra of varied gold nanoparticles deposited onto glass slide surfaces. (a) Particles in solution. (b) Particle monolayers without TBBT linkers and subsequent UVO exposure. (c) Particle monolayers with TBBT linkers but without subsequent UVO exposure. (d) Particle monolayers without TBBT linkers and after subsequent UVO exposure. (e) Particle monolayers with TBBT linkers and after subsequent UVO exposure. The experimental conditions were the same as those described in Figure 22.4. All spectra were normalized to their respective absorbance at 272 nm and offset for the sake of clarity. (From Chen, S., *Langmuir*, 17, 2878–2884, 2001. With permission. Copyright © 2001 American Chemical Society.)

protecting ligands longer than the chemical bridges (TBBT) the cross-linking is still achievable, it has been demonstrated previously that the organic shells of nanosized particles could be readily compressed.[9a]

Additionally, the optical properties of these particle cross-linked networks exhibit similar responses in the UV-Vis spectroscopic measurements. Figure 22.5 shows the UV-Vis spectra of the series of particle assemblies before and after TBBT cross-linking, along with the corresponding solution phase spectra. The red shifting of the SP band positions, relative to that of the particles dissolved in solution (all near 510 nm), is again very obvious and similar to those observed earlier (Figure 22.3): for the particle thin films without the addition of TBBT, the SP bands are typically found at 550 nm, whereas after TBBT cross-linking, the resulting particle assemblies exhibit the SP bands at an even longer wavelength position (about 580 nm).

The above studies clearly demonstrate that by strengthening the interparticle interactions through strong chemical linkages, the resulting particle assemblies exhibit significantly enhanced mechanical stability, compared to the LB thin films of nanoparticles, without further chemical modifications (*vide infra*). The interparticle spacing is found to be mainly determined by the molecular length of the rigid chemical bridges, regardless of the original thickness of the particle-protecting layers. The virtue of the approach described here is partly attributed to the diversity of surface decoration of the nanoparticle core surface, which might then be exploited in the fabrication of robust multifunctional stand-alone quantum dot thin films.

22.2.3 NANOPARTICLE CROSS-LINKED NETWORKS FOR SURFACE NANOFABRICATIONS

The above-obtained robust particle superlattice networks might be useful for surface nanofabrication without involving sophisticated instrumentation. This nonlithographic approach is taking advantage of the easy fabrication of nanoparticle monolayers by LB deposition and the ready removal of simple organic molecules by photooxidation. The photosensitized cleaning protocol has been a common practice in surface science to remove organic contaminant molecules by the absorption of short-wavelength UV radiation. Here we deposited the particle patches onto an optically stable substrate (e.g., glass slides and SiO_x-coated TEM grids), which was then subject to ultraviolet and ozone (UVO) molecular cleaning to remove the organic components (mainly the alkanethiolates and the TBBT linkers), and thus have the quantum arrays deposited onto the substrate surface.

Here we take the $C_{12}Au$ particle networks as the illustrating example. Figure 22.4D shows the TEM micrograph of the same $C_{12}Au$ assemblies (Figure 22.4C) after the treatment of UVO cleaning for 15 min. One can see that although the superlattice ordering was disrupted somewhat, only a small number of particles were found coalesced, and a rather large fraction of the particles were found to even maintain the hexagonal ordered structures. In addition, the average particle (cross-sectional) diameter was found to increase somewhat from the original 5.1 to 6.1 nm after UVO exposure, suggesting that the particle shape might change from the original spherical morphology to a disc-like island structure (Figure 22.6, inset). Alternatively, it might be due to the formation of gold oxide with larger crystal lattice.

Energy-dispersive x-ray (EDX) analysis was carried out to examine the removal of the organic components. Figure 22.6 shows the spectra before and after the UVO treatment of the $C_{12}Au$ particle ensembles. One can see that the sulfur signal is virtually absent, indicating the efficient desorption of the thiolates from the particle core surface; however, the carbon peak is still quite strong, most likely due to the deposition of carbon onto the metal quantum dot surfaces (the Cu, Si, and O peaks

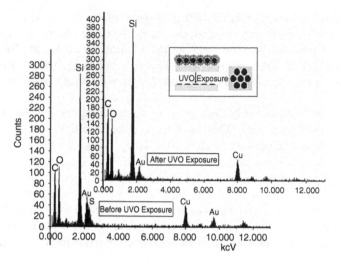

FIGURE 22.6 Energy-dispersive x-ray spectra of $C_{12}Au$ particle networks deposited on SiO_x-coated Cu grids before (Figure 22.4C) and after (Figure 22.4D) UVO exposure. Inset shows the schematic of the resulting gold quantum dot arrays after UVO treatment. (From Chen, S., *Langmuir*, 17, 2878–2884, 2001. With permission. Copyright © 2001 American Chemical Society.)

are from the TEM grids). A similar observation of carbon deposition was found with organic-capped PtFe alloy particles that were thermally annealed at elevated temperatures.[4a]

As the majority of the particles within the cross-linked networks are still well separated in the superlattice matrix (Figure 22.4D), one would anticipate only very subtle variation of the optical responses of the resulting quantum dot arrays, as compared to those prior to the photo treatment. Figure 22.3 (bottom panel) and Figure 22.5d and e show the UV-Vis spectra of a series of gold particle networks after UVO exposure. One can see that after UV exposure, the TBBT-linked particle patches essentially demonstrate invariant absorption profiles (in particular, the SP energy; Figure 22.3(c2) and (d2), and Figure 22.5e), consistent with the above TEM measurements (Figure 22.4C and D). In contrast, for the particle assemblies fabricated without the cross-linking of TBBT, after UVO exposure, the absorption profiles (Figure 22.3(a2) and (b2), and Figure 22.5d) are much broader where the SP band positions are mostly hard to locate. It is most likely that the latter is attributed to the coalescence of the particles upon UV and ozone excitation, considering the observed aggregation of particles even prior to the photo treatment (e.g., Figure 22.2A and B).

The resulting metal quantum dot arrays exhibit even stronger mechanical stability than the original cross-linked particle networks. For instance, it was found that the particle patches deposited onto a glass surface could be easily peeled off with tape, whereas the quantum dot arrays survived the tape test and stayed on the substrate surface without any visible damage.

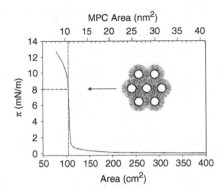

FIGURE 22.7 Langmuir isotherm of C_8FcAu particles on the water surface. Seventy microliters of 1.75 mg/ml particle solution were spread onto the water surface. Compression rate, 20 cm²/min. Dashed lines depict the deposition pressure and surface area. Inset: Schematic of the structure of the nanoparticle assemblies on the water surface. (From Chen, S., *Langmuir*, 17, 6664–6668, 2001. With permission. Copyright © 2001 American Chemical Society.)

In addition, by using other metal nanoparticles,[27c,34] one will anticipate that diverse surface nanostructures will be readily constructed. In particular, in combination with other surface decoration methods (e.g., surface nanografting[18] and lithography[11–14]), more complicated nanoscale surface chemical patterning can be achieved.

22.3 INTERFACIAL DYNAMICS OF NANOPARTICLES' LB THIN FILMS

22.3.1 LANGMUIR THIN FILMS OF ELECTROACTIVE NANOPARTICLES

As mentioned above, conventional LB thin films lack long-term mechanical stability. For nanoparticle thin films, this can be manifested by using an electrochemically active species as the molecular probe.[8c] Figure 22.7 shows a representative surface pressure (π)-area isotherm of n-octanethiolate-protected gold nanoparticles functionalized with ω-ferrocene moieties (C_8FcAu, particle core diameter of 2 nm, with approximately 4.6 Fc per particle), which is quite similar to those observed with simple alkanethiolate-protected gold nanoparticles (Figure 22.1).[8a,b] One can see that at surface areas greater than 12 nm²/MPC, the surface pressure is virtually zero, indicating a two-dimensional gaseous phase of the particles on the water surface. However, at areas smaller than 12 nm²/MPC, the surface pressure starts to increase drastically, where the steep rising slope suggests the rigid nature of particle closely packed thin films. By assuming a hexagonal distribution of the particles in the thin films (Figure 22.7, inset), this take-off area (12 nm²/MPC) corresponds to a center-to-center distance of about 3.72 nm. As the fully extended chain length of an octanethiolate is 1.01 nm (calculated by Hyperchem®), this distance is somewhat smaller than the nanoparticle molecular diameter (core + two monolayer ligands),

Nanoparticle Assemblies and Superstructures

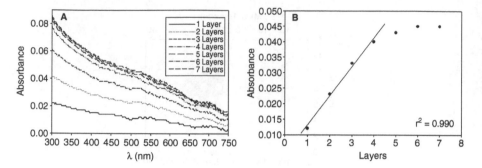

FIGURE 22.8 (A) UV-Vis spectra of varied layers of C_8FcAu particles deposited at 8 mN/m onto a microscope glass slide. A blank glass slide was used as the reference. (B) Variation of absorbance (at 520 nm) with the number of particle layers. (From Chen, S., *Langmuir*, 17, 6664–6668, 2001. With permission. Copyright © 2001 American Chemical Society.)

indicating the initiation of ligand intercalation at this surface area, leading to the observed sharp increase in surface pressure with further compression.

A second transition starts at 10.5 nm^2/MPC, with a somewhat smaller rising slope. This might be ascribed to the further ligand intercalation between neighboring particles, leading to the formation of solid-like closely packed MPC thin films on the water surface. In fact, on the water surface, brown-colored MPC thin films are very visible.

The particle monolayers were then deposited onto a glass slide using the Langmuir–Blodgett (LB) technique (vertical deposition) at $\pi = 8$ mN/m, with a lifting speed of 1 mm/min. Figure 22.8A shows the optical responses of the resulting particle LB assemblies. Although the surface plasmon band of nanosized gold particles is not very well defined here, partly due to the small particle core size (about 2 nm in diameter), the absorption profiles do exhibit a Mie character, where the absorption intensity decreases exponentially with decreasing photon energy. In addition, the absorption intensity at 520 nm (the surface plasmon band position of alkanethiolate-protected gold nanoparticles[27]) shows a linear increase with the thickness of nanoparticle layers up to four monolayers (Figure 22.8B), indicating a rather repeatable and efficient transfer of the nanoparticle thin films. This is similar to the observation with self-assembled nanoparticle thin films fabricated by alternating layers of nanoparticles and dithiol linkers.[35] However, further deposition appears to be less efficient, and the corresponding optical absorbance of the particle assemblies starts to level off.

22.3.2 INTERFACIAL DYNAMICS OF NANOPARTICLE LB THIN FILMS

As the MPC molecules in these LB thin films are not chemically cross-linked to each other, the particles might exhibit some degrees of mobility when deposited onto the substrate surface, though not as free as in solution. This could lead to particle reorganization (i.e., annealing) and be monitored by electrochemical measurements using the electroactive ferrocene moieties as the molecular probes. Figure 22.9 shows the cyclic voltammogram of a C_8FcAu particle monolayer freshly

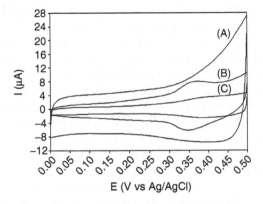

FIGURE 22.9 Cyclic voltammograms of a naked gold thin film electrode (0.48 cm^2) (A), with a C_8FcAu particle monolayer freshly deposited at 8 mN/m (B), and with the same MPC monolayer after about 20 min in 0.10 M KNO$_3$ (C). Potential sweep rate, 100 mV/sec. (From Chen, S., *Langmuir*, 17, 6664–6668, 2001. With permission. Copyright © 2001 American Chemical Society.)

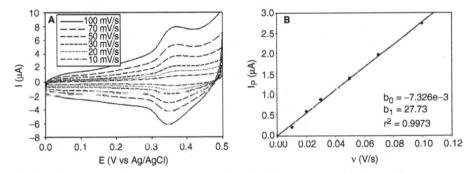

FIGURE 22.10 (A) Cyclic voltammograms of the freshly deposited C_8FcAu MPC monolayer (as in Figure 22.9B) in 0.10 M KNO$_3$ at various potential sweep rates. (B) Variation of the voltammetric peak currents with potential scan rate. The line is the linear regression. (From Chen, S., *Langmuir*, 17, 6664–6668, 2001. With permission. Copyright © 2001 American Chemical Society.)

deposited onto a gold thin film electrode in 0.10 M KNO$_3$. One can see that compared to that at the same bare electrode, the charging current within the potential range of 0 to +0.5 V is suppressed substantially, indicating the coating of organically functionalized gold MPCs onto the electrode surface. In addition, a pair of voltammetric peaks are observed at +0.36 V, with a peak splitting of only 20 mV, suggesting a very facile electron transfer reaction. These peaks are ascribed to the oxidation–reduction of the particle-bound ferrocene moieties, where the peak currents are found to increase linearly with the potential sweep rate (Figure 22.10), consistent with a surface-confined system. From the slope (Figure 22.10B), one can estimate the effective surface coverage of ferrocene moieties as $\Gamma_{Fc} = 3.87 \times 10^{13}$ molecules/cm^2, and hence the MPC surface coverage (Γ_{MPC}) is 8.41×10^{12} particles/cm^2, corresponding to an effective particle area of 11.9 nm^2 — somewhat larger than the original MPC area during LB deposition (10.7 nm^2; Figure 22.7). This might be accounted for by at least two possibilities; e.g., the MPC molecular area was expanded due to interfacial dynamics (relaxation), or part of the MPC-supported ferrocene moieties were not accessible and active (vide infra).

Additionally, this ferrocene surface concentration corresponds to 20% of the approximate maximum coverage of α-mercaptoalkyl ferrocene derivatives self-assembled onto flat gold surfaces.[28,29] The appearance of a single pair of voltammetric waves for the ferrocene sites in the LB monolayers suggests a similarity of the energetic states of these surface electroactive groups. In our earlier (unpublished) studies, in which a compact monolayer of ω-ferrocenated particles was formed by chemisorptive linkages to the electrode surface (particles were first rendered surface active with peripheral thiol groups by incorporating multiple copies of alkanedithiols into the particle-protecting monolayer through exchange reactions[7]), the resulting electrochemical measurements revealed two pairs of voltammetric waves, which were ascribed to the variation of ferrocene energetic states due to spatial distribution on the particle surface.[36] Similar phenomena were also reported with two-dimensional mixed monolayers of ω-ferrocenated alkanethiols and n-alkanethiol derivatives at a relatively high ferrocene surface coverage.[28,29] There appear to be at least two possible sources to account for the discrepancy observed here. First, it might be attributed to the fact that the LB-deposited particle molecules still possess some degree of conformational mobility, in part because the interactions between the particles and between the particles and the electrode surface mainly involve weak Van der Waals forces,[8b] compared to the chemisorptive interactions by thiols on gold in the self-assembled systems.[28,29] Despite the fact that the particle layers were deposited at relatively high surface pressures (e.g., 8 mN/m), not all particles appeared to be in the crystalline configuration, i.e., intercalated to each other, due to particle size dispersity.[8a] Thus, in this case, part of the particle assemblies might be free to rotate on the electrode surface (though unlikely to desorb from the electrode surface). This then helps initiate the interfacial dynamic transitions of the MPC monolayers, rendering the ferrocene sites accessible to the electrode electron (Scheme 22.1). A second possibility is that the initial LB thin film might possess some defects that facilitate ion transport for charge compensation during the electrooxidation/reduction of the ferrocene groups (more discussion below).

The dynamic characters of the electrode-supported particle assemblies were further observed voltammetrically. Figure 22.9C shows the cyclic voltammogram of the above same particle monolayer assemblies after about 20 min of potential sweeps. Compared to that observed with a freshly deposited MPC monolayer (Figure 22.9B), there are two aspects that warrant attention here. First, the interfacial charging current (within 0 to +0.3 V) actually decreases upon the aging of the monolayer, suggesting some surface rearrangement of the deposited particle molecules. This is akin to potential-induced annealing. The driving force of these particle rearrangements (Scheme 22.1) seems likely to originate from the bulky ferrocene sites, which, when oxidized (to positively charged ferrocenium), became energetically less favorable to stay in a hydrophobic environment and, hence, less likely to have an interpenetration conformation with the monolayer ligands of neighboring particles. This deintercalation

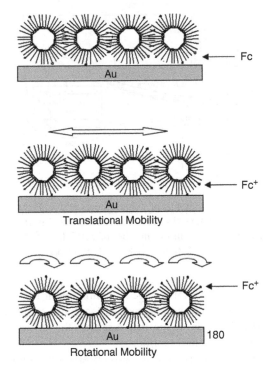

SCHEME 22.1 Interfacial dynamics of nanoparticle thin films upon the (electro)oxidation of the ω-ferrocene moieties (Fc $\Leftrightarrow$ Fc$^+$ + e).

leads to more spread out MPC molecules and enhances hydrophobic interactions between the surface particle molecules, consequently decreasing the double-layer charging current (by reducing the electrode surface defect sites, for instance). This is also consistent with the above observation of expanded MPC molecular areas.

Second, the voltammetric peaks of the ferrocene groups became much broader and the formal potential shifted anodically ($E^{\circ\prime}$ = +0.40 V). Considering the suppression of the double-layer current envelope discussed above, it is rather unlikely that the decrease in the ferrocene faradaic current is due to particle loss from the electrode surface. Additionally, due to the hydrophobic nature of the nanoparticles, it is energetically unfavorable that the particles would desorb from the electrode surface into the aqueous solution. Also, the peak potential splitting appears to be virtually unchanged, suggesting a thermodynamic rather than a kinetic effect. In earlier studies involving self-assembled monolayers of ferrocene-terminated alkanethiols with coadsorbed electroinactive ω-functionalized n-alkanethiol derivatives,[28,29] it was found that in aqueous electrolyte solutions, the formal potential of the ferrocene moiety also shifted anodically when the ω-terminal groups of the coadsorbates became less polar. This observation was interpreted on the basis of the combined effects of local ion solvation and double-layer potential distribution.[28,29,31] These effects might also account for the present observations with the LB particle layers, in which the enhanced hydrophobic interaction resulting from the structural rearrangements leads to more difficult penetration for electrolyte ions for charge

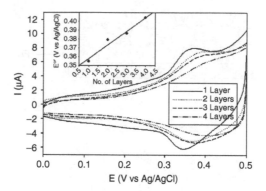

FIGURE 22.11 Cyclic voltammograms of freshly deposited C_8FcAu MPC thin films of varied layers in $0.10\,M$ KNO_3. Potential sweep rate, 100 mV/sec. Inset: Variation of the formal potential of the ferrocene moiety with the number of MPC layers. Line is for eye-guiding only. (From Chen, S., *Langmuir*, 17, 6664–6668, 2001. With permission. Copyright © 2001 American Chemical Society.)

compensation. The broad voltammetric peak widths might reflect the wide spatial distribution of ferrocene sites on the electrode surface[31] due to the three-dimensional nature of the nanoparticle cores, suggesting that after the surface structural reorganization, the particles became less mobile.

Similar dynamic reorganizations were also observed with multilayer structures (not shown); however, the ferrocene voltammetric currents decreased to a lesser extent with more MPC layers. This might be understood in terms of the decrease in MPC molecular mobility, especially for those buried underneath the topmost layer, based on the above speculation of interfacial structural dynamics.[36]

22.3.3 MULTILAYER STRUCTURES OF ELECTROACTIVE NANOPARTICLES

Electrochemistry of the C_8FcAu MPC multilayers shows additional unusual characters. Figure 22.11 depicts the cyclic voltammograms of varied layers of C_8FcAu MPCs freshly deposited onto a gold thin film electrode. First, one can see that the double-layer charging current within the potential range of $0 \sim +0.5$ V decreases with the deposition of additional MPC layers, consistent with the presence of more organically functionalized MPC molecules on the electrode surface (*vide ante*). However, surprisingly, the voltammetric current due to the ferrocene moieties does not increase proportionally with the number of MPC layers, which is rather counterintuitive. Whereas the ferrocene faradaic currents increase linearly with potential sweep rates with varied layers of C_8FcAu MPCs (Figure 22.12), again, consistent with surface-confined systems, the slopes (S) normalized to that with an MPC monolayer (Figure 22.10) are $S_1:S_2:S_3:S_4 = 1:1.87:1.59:1.52$, with the subscripts referring to the number of MPC layers. This is drastically smaller than the theoretical value of $S_1:S_2:S_3:S_4 = 1:2:3:4$, which is anticipated from the aforementioned optical measurements (Figure 22.8) by assuming that all ferrocene sites are active and accessible. This is in great contrast to colloidal multilayers fabricated by the alternated-dipping self-assembling method, with electroactive species incorporated

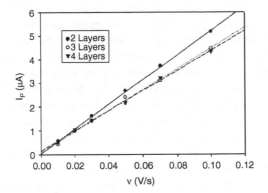

FIGURE 22.12 Variation of ferrocene faradaic currents with potential sweep rates for the varied layers of C_8FcAu MPC LB thin films. Experimental conditions are the same as in Figure 22.11. Symbols are experimental data and lines are the corresponding linear regressions. (From Chen, S., *Langmuir*, 17, 6664–6668, 2001. With permission. Copyright © 2001 American Chemical Society.)

between the particle layers,[37,38] where the voltammetric currents due to the faradaic processes increase linearly with the number of particle layers, indicating that all sites are accessible in the electron transfer reactions and the porous nature of the resulting colloidal thin films. The discrepancy observed here suggests that in the LB thin films of nanoparticles, due to ligand intercalation, some ferrocene sites are buried within a hydrophobic environment, which creates a large energetic barrier for charge transfer as well as charge compensation.[31] In other words, the film behaves in a less conductive manner with more MPC layers, resulting in the decrease of accessible ferrocene sites and, hence, smaller voltammetric currents (Figure 22.11 and Figure 22.12). This might also account, at least in part, for the anodic shift of the formal potential of the ferrocene moieties with more MPC layers (Figure 22.11, inset), where an overpotential of 15 mV is found with each additional layer of MPC molecules.[39]

It should be noted that of these active ferrocene sites, their peak splittings remain very small (<40 mV), again indicating that the effect of the local environmental variation is reflected thermodynamically rather than kinetically, as mentioned earlier.

22.4 CONCLUSION

There have been numerous studies using the LB technique to create varied ordered structures of nanoparticles (metals as well as semiconductors).[9–15] In this chapter, we focus on the investigations of the interfacial dynamics of nanoparticle LB thin films and the fabrication of long-range ordered and robust nanoparticle assemblies through interparticle cross-linking with bifunctional linkers, which have not been explored previously. The cross-linking protocol exploits the unique properties of nanoparticle surface functionalization by initiating surface exchange reactions between dithiol chemical bridges and nanoparticle-bound thiolate ligands at high surface pressures where ligand intercalation is possible. The obtained nanoparticle networks might be used in surface nanofabrication in a nonlithographic manner, leading to the construction of ordered arrays of metal quantum dots. It is anticipated that the protocols described here can be extended to other nanoparticle materials (e.g., other metal or semiconductor nanoparticles) and chemical bridges where diverse nanoparticle superlattice assemblies can be achieved.

Without cross-linkers, nanoparticle LB thin films exhibit drastic interfacial dynamics. This can be probed by electrochemical measurements of the LB thin films of ferrocene-functionalized gold nanoparticles (using electroactive ferrocene moieties as the molecular probes). It was found that in the nanoparticle monolayers, the molecular mobility due to relatively weak interparticle interactions led to the potential-induced reorganization of the particle molecules, reflected in a decrease in the electrode double-layer charging current and the surprising suppression of the faradaic currents of the ferrocene moieties. With the deposition of additional nanoparticle layers, the particle thin films became less conductive, leading to the anodic shift of the ferrocene formal potential as well as the drastic decrease in ferrocene faradaic currents.

ACKNOWLEDGMENTS

The author gratefully acknowledges the following agencies for their generous financial support of the work described herein: the National Science Foundation (CAREER Award CHE — 0092760), the Office of Naval Research, the Petroleum Research Fund administered by the American Chemical Society, the Illinois Department of Commerce and Community Affairs, and SIU Materials Technology Center. S.C. is a Cottrell Scholar of Research Corporation.

REFERENCES AND NOTES

1. (a) G Schmid. *Clusters and Colloids: From Theory to Applications*. New York: VCH, 1994. (b) R Turton. *The Quantum Dot: A Journey into the Future of Microelectronics*. New York: Oxford University Press, 1995.
2. See also all of the review articles in the February 16, 1996, and November 24, 2000, issues of *Science*.
3. See, for instance, the Special Reports in the October 16, 2000, issue of *Chemical and Engineering News*.
4. (a) S Sun, CB Murray, D Weller, L Folks, A Moser. Monodisperse FePt nanoparticles and ferromagnetic FePt nanocrystal superlattices. *Science* 287:1989–1992, 2000. (b) CJ Kiely, J Fink, M Brust, D Bethell, DJ Schiffrin. Spontaneous ordering of bimodal ensembles of nanoscopic gold clusters. *Nature* 396:444–446, 1998. (c) ZL Wang. Structural analysis of self-assembling nanocrystal superlattices. *Adv Mater* 10:13–30, 1998. (d) M-P Pileni. Optical properties of nanosized particles dispersed in colloidal solutions or arranged in 2D and 3D superlattices. *New J Chem* 693–702, 1998. (e) RL Whetten, JT Khoury, MM Alvarez, S Murthy, I Vezmar, ZL Wang, PW Stephens, CL Cleveland, WD Luetke, U Landman. Nanocrystal gold molecules. *Adv Mater* 8:428–433, 1996. (f) CB Murray, CR Kagan, MG Bawendi. Self-organization of CdSe

nanocrystallites into three-dimensional quantum dot superlattices. *Science* 270: 1335–1338, 1995. (g) RG Osifchin, RP Andres, JI Henderson, CP Kubiak, RN Dominey. Synthesis of nanoscale arrays of coupled metal dots. *Nanotechnology* 7: 412–416, 1996. (h) Z Tang, NA Kotov, M Giersig. Spontaneous organization of single CdTe nanoparticles into luminescent nanowires. *Science* 297:237–240, 2002.

5. (a) G Schmid. The role of big metal clusters in nanoscience. *J Chem Soc Dalton Trans* 1077–1082, 1998. (b) RP Andres, JD Bielefeld, JI Henderson, DB Janes, VR Kolagunta, CP Kubiak, WJ Mahoney, RG Osifchin. Self-assembly of a two-dimensional superlattice of molecularly linked metal clusters. *Science* 273:1690–1693, 1996. (c) DL Feldheim, KC Grabar, MJ Natan, TE Mallouk. Electron transfer in self-assembled inorganic polyelectrolyte/metal nanoparticle heterostructures. *J Am Chem Soc* 118:7640–7641, 1996. (d) TA Taton, RC Mucic, CA Mirkin, RL Letsinger. The DNA-mediated formation of supramolecular mono- and multilayered nanoparticle structures. *J Am Chem Soc* 122:6305–6306, 2000. (e) S Mann, W Shenton, M Li, S Connolly, D Fitzmaurice. Biologically programmed nanoparticle assembly. *Adv Mater* 12:147–150, 2000.

6. (a) DL Feldheim, CD Keating. Self-assembly of single electron transistors and related devices. *Chem Soc Rev* 27:1–12, 1998. (b) CP Collier, T Vossmeyer, JR Heath. Nanocrystal superlattices. *Ann Rev Phys Chem* 49:371–404, 1998.

7. (a) S Chen. Self-assembling of monolayer-protected gold nanoparticles. *J Phys Chem B* 104:663–667, 2000. (b) S Chen. Nanoparticle assemblies: "rectified" quantized charging in aqueous media. *J Am Chem Soc* 122:7420–7421, 2000.

8. (a) S Chen. Two-dimensional crosslinked nanoparticle networks. *Adv Mater* 12:186–189, 2000. (b) S Chen. Langmuir–Blodgett fabrication of two-dimensional robust crosslinked nanoparticle networks. *Langmuir* 17:2878–2884, 2001. (c) S Chen. Electrochemical studies of Langmuir–Blodgett thin films of electroactive nanoparticles. *Langmuir* 17:6664–6668, 2001.

9. (a) CP Collier, RJ Saykally, JJ Shiang, SE Henrichs, JR Heath. Reversible tunneling of silver quantum dot monolayers through the metal-insulator transition. *Science* 277:1978–1981, 1997. (b) W-Y Lee, MJ Hostetler, RW Murray, M Majda. Electron hopping and electronic conductivity in monolayers of alkanethiol-stabilized gold nano-clusters at the air–water interface. *Isr J Chem* 37:213–223, 1997.

10. (a) KC Yi, Z Hórvölgyi, JH Fendler. Chemical formation of silver particulates films under monolayers. *J Phys Chem* 98:3872–3881, 1994. (b) JH Fendler, FC Meldrum. The colloid chemical approach to nanostructured materials. *Adv Mater* 7:607–631, 1995. (c) Y Tian, JH Fendler. Langmuir–Blodgett film formation from fluorescence-activated, surfactant-capped, size-selected CdS nanoparticles spread on water surface. *Chem Mater* 8:969–974, 1996.

11. (a) C Gutiérrez-Wing, JA Ascencio, M Pérez-Alvarez, M Marín-Almazo, M José-Yacamán. On the structure and formation of self-assembled lattices of gold nanoparticles. *J Cluster Sci* 9:529–545, 1998. (b) M Sastry, A Gole, V Patil. Lamellar Langmuir–Blodgett films of hydrophobic colloidal gold nanoparticles by organization at the air–water interface. *Thin Solid Films* 384:125–131, 2001. (c) S Huang, G Tsutsui, H Sakaue, S Shingubara, T Takahagi. Experimental conditions for a highly ordered monolayer of gold nanoparticles fabricated by the Langmuir–Blodgett method. *J Vac Sci Technol B* 19:2045–2049, 2001. (d) M Brust, N Stuhr-Hansen, K Nørgaard, JB Christensen, LK Nielsen, T Bjørnholm. Langmuir–Blodgett films of alkane chalcogenide (S, Se, Te) stabilized gold nanoparticles. *Nano Lett* 1:189–191, 2001.

12. (a) YH Sun, JR Li, BF Li, L Jiang. Study of the improvement of photoelectric response and stability on a 9-cis-retinal Langmuir–Blodgett film containing ultrafine gold particles. *Langmuir* 13:5799–5801, 1997. (b) C Fan, T Lu, B Li, L Jiang. A new method for preparing nanometer-scale particles onto LB films. *Thin Solid Films* 286: 37–39, 1996.

13. (a) H Perez, RM Lisboa de Sousa, J-P Pradeau, P-A Albouy. Elaboration and electrical characterization of Langmuir–Blodgett films of 4-mercaptoaniline functionalized platinum nanoparticles. *Chem Mater* 13:1512–1517, 2001. (b) J-P Bourgoin, C Kergueris, E Lefèvre, S Palacin. Langmuir-Blodgett films of thiol-capped gold nanoclusters fabrication and electrical properties. *Thin Solid Films* 327–329:515–519, 1998. (c) S Paul, C Pearson, A Molloy, MA Cousins, M Green, S Kollipoulou, P Dimitrakis, P Normand, D Tsoukalas, MC Petty. Langmuir–Blodgett film deposition of metallic nanoparticles and their application to electronic memory structures. *Nano Lett* 3:533–536, 2003.

14. P Poddar, T Telem-Shafir, T Fried, G Markovich. Dipolar interactions in two- and three-dimensional magnetic nanoparticle arrays. *Phys Rev B* 66:060403-1–060403-4, 2002.

15. (a) T Torimoto, N Tsumura, M Miyake, M Nishizawa, T Sakata, H Mori, H Yoneyama. Preparation and photoelectrochemical properties of two-dimensionally organized CdS nanoparticle thin films. *Langmuir* 15:1853–1858, 1999. (b) T Torimoto, N Tsumura, H Nakamura, S Kuwabata, T Sakata, H Mori, H Yoneyama. Photoelectrochemical properties of size-quantized semiconductor photoelectrodes prepared by two-dimensional cross-linking of monodisperse CdS nanoparticles. *Electrochim Acta* 45:3269–3276, 2000.

16. (a) J Boneberg, F Burmeister, C Schäfle, P Leiderer, D Reim, A Fery, S Herminghaus. The formation of nano-dot and nano-ring structures in colloidal monolayer lithography. *Langmuir* 13:7080–7084, 1997. (b) TR Jensen, ML Duval, KL Kelly, AA Lazarides, GC Schatz, RP van Duyne. Nanosphere lithography: effect of the external dielectric medium on the surface plasmon resonance spectrum of a periodic array of silver nanoparticles. *J Phys Chem B* 103:9846–9853, 1999.

17. Y Xia, M Mrksich, E Kim, GM Whitesides. Microcontact printing of octadecylsiloxane on the surface of silicon dioxide and its application in microfabrication. *J Am Chem Soc* 117:9576–9577, 1995.

18. (a) G-Y Liu, S Xu, Y Qian. Nanofabrication of self-assembled monolayers using scanning probe lithography. *Acc Chem Res* 33:457–466, 2000. (b) JC Garno, Y Yang, NA Amro, S Cruchon-Dupeyrat, S Chen, G-Y Liu. Precise positioning of nanoparticles on surfaces using scanning probe lithography. *Nano Lett* 3:389–395, 2003.

19. SA Iakovenko, AS Trifonov, M Giesig, A Mamedov, DK Nagesha, VV Hanin, EC Solatov, NA Kotov. One- and two-dimensional arrays of magnetic nanoparticles by the Langmuir–Blodgett technique. *Adv Mater* 11:388–392, 1999.

20. Q Guo, X Teng, S Rahmanm, H Yang. Patterned Langmuir–Blodgett films of monodisperse nanoparticles of iron oxide using soft lithography. *J Am Chem Soc* 125:630–631, 2003.

21. V Erokhin, V Troitsky, S Erokhina, G Mascetti, C Nicolini. In-plane patterning of aggregated nanoparticle layers. *Langmuir* 18:3185–3190, 2002.

22. (a) MHV Werts, M Lambert, J-P Bourgoin, M Brust. Nanometer scale patterning of Langmuir–Blodgett films of gold nanoparticles by electron beam lithography. *Nano Lett* 2:43–47, 2002. (b) T Hassenkam, K Nørgaard, L Iversen, CJ Kiely, M Brust, T Bjørnholm. Fabrication of 2D gold nanowires by self-assembly of gold nanoparticles on water surfaces in the presence of surfactants. *Adv Mater* 14:1126–1130, 2002.

23. (a) T Reuter, O Vidoni, V Torma, G Schmid, L Nan, M Gleiche, L Chi, H Fuchs. Two-dimensional networks via quasi one-dimensional arrangements of gold clusters. *Nano Lett* 2:709–711, 2002. (b) H Tanaka, M Mitsuishi, T Miyashita. Tailor-control of gold nanoparticle adsorption onto polymer nanosheets. *Langmuir* 19:3103–3105, 2003.

24. JF Liu, KZ Yang, ZH Lu. Controlled assembly of regular composite nanowire arrays and their multilayers using electropolymerized polymers as templates. *J Am Chem Soc* 119:11061–11065, 1997.

25. A Tao, F Kim, C Hess, J Goldberger, R He, Y Sun, Y Xia, P Yang. Langmuir–Blodgett silver nanowire monolayers for molecular sensing using surface enhanced Raman spectroscopy. *Nano Lett* 3:1229–1233, 2003.

26. M Brust, D Bethell, DJ Schiffrin, CJ Kiely. Novel gold-dithiol nano-networks with non-metallic electronic properties. *Adv Mater* 7:795–797, 1995.

27. (a) M Brust, M Walker, D Bethell, DJ Schiffrin, R Kiely. Synthesis of thiol-derivatised gold nanoparticles in a two-phase liquid–liquid system. *J Chem Soc Chem Commun* 801–802, 1994. (b) RL Whetten, MN Shafigullin, JT Khoury, TG Schaaff, I Vezmar, MM Alvarez, A Wilkinson. Crystal structures of molecular gold nanocrystal arrays. *Acc Chem Res* 32:397–406, 1999. (c) AC Templeton, WP Wuelfing, RW Murray. Monolayer-protected cluster molecules. *Acc Chem Res* 33:27–36, 2000.

28. For instance, (a) HD Sikes, JF Smalley, SP Dudek, AR Cook, MD Newton, CED Chidsey, SW Feldberg. Rapid electron tunneling through oligophenylenevinylene bridges. *Science* 291:1519–1523, 2001. (b) CED Chidsey, CR Bertozzi, TM Putvinski, AM Mujsce. Coadsorption of ferrocene-terminated and unsubstituted alkanethiols on gold: electroactive self-assembled monolayers. *J Am Chem Soc* 112:4301–4306, 1990. (c) L Tender, MT Carter, RW Murray. Cyclic voltammetric analysis of ferrocene alkanethiol monolayer electrode kinetics based on Marcus theory. *Anal Chem* 66:3173–3181, 1994.

29. (a) JJ Sumner, SE Creager. Topological effects in bridge-mediated electron transfer between redox molecules and metal electrodes. *J Am Chem Soc* 122:11914–11920, 2000. (b) SE Creager, GK Rowe. Interfacial salvation and double-layer effects on redox reactions in organized assemblies. *J Phys Chem* 98:5500–5507, 1994.

30. (a) MJ Hostetler, SJ Green, JJ Stokes, RW Murray. Monolayers in three dimensions: synthesis and electrochemistry of ω-functionalized alkanethiolate-stabilized gold cluster compounds. *J Am Chem Soc* 118:4212–4213, 1998. (b) SJ Green, JJ Stokes, MJ Hostetler, JJ Pietron, RW Murray. Three-dimensional monolayers: nanometer-sized electrodes of alkanethiolate-stabilized gold cluster molecules. *J Phys Chem B* 101:2663–2668, 1997.

31. CP Smith, HS White. Theory of the interfacial potential distribution and reversible voltammetric response of electrodes coated with electroactive molecular films. *Anal Chem* 64:2398–2405, 1992.

32. S Underwood, P Mulvaney. Effect of the solution refractive index on the color of gold colloids. *Langmuir* 10:3427–3430, 1994.

33. CA Mirkin, RL Letsinger, RC Mucic, JJ Storhoff. A DNA-based method for rationally assembling nanoparticles into macroscopic materials. *Nature* 382:607–610, 1996.

34. S Chen, K Huang, JA Stearns. Alkanethiolate-protected palladium nanoparticles. *Chem Mater* 12:540–547, 2000.

35. M Brust, D Bethell, CJ Kiely, DJ Schiffrin. Self-assembled gold nanoparticle thin films with nonmetallic optical and electronic properties. *Langmuir* 14:5425–5429, 1998.

36. In another study, in which the ω-ferrocenated particles were cross-linked at the air–water interface by rigid dithiol linkers prior to deposition onto the electrode surface, an additional pair of voltammetric peaks was found at a more positive potential position; however, the voltammetric responses were very stable. These observations are consistent with the argument of molecular mobility and consequential interfacial structural dynamics.

37. AN Shipway, M Lahav, I Willner. Nanostructured gold colloid electrodes. *Adv Mater* 12:993–998, 2000.

38. DI Gittins, D Bethell, RJ Nichols, DJ Schiffrin. Redox-connected multilayers of discrete gold particles: a novel electroactive nanomaterial. *Adv Mater* 11:737–740, 1999.

39. In the case of four layers of gold nanoclusters, the ferrocene peak potential was determined by differential pulse voltammetry (DPV). Also, in a dropcast thick film (~μm) of the same ω-ferrocenated nanoparticles, the formal potential of the ferrocene moieties was found at an even more positive position, +0.45 V.

23 Self-Assembling of Gold Nanoparticles at an Air–Water Interface

Hiroshi Yao, Seiichi Sato, and Keisaku Kimura

CONTENTS

23.1 Introduction 601
23.2 Why Carboxylate-Modified Water-Soluble Nanoparticles for
Crystallization? 602
23.3 Synthesis and Characterization of Carboxylate-Modified Gold
Nanoparticles 603
23.4 Gold Nanoparticle Crystals Grown at an Air–Water Interface 604
23.5 Optical Properties of Gold Nanoparticle Crystals 609
23.6 Controls of Interparticle Spacing in the Two-Dimensional Ordered
Arrays of Nanoparticles 611
23.7 Conclusion 613
Acknowledgment 613
References 613

23.1 INTRODUCTION

Metal nanoparticles with well-defined surfaces have attracted much attention due to their potential use as nanoelectronic/optical device elements.[1] So far, there are many reports on the preparation and surface modifications of metal nanoparticles using thiols or phosphines. Recent investigations have been focused on the syntheses of ordered nanoparticle assembly into two- or three-dimensional superlattice structures, because it will provide a new device performance based on collective physical behaviors resulting from electronic interactions between neighboring particles.[2] In most cases, ordered nanoparticle arrays are spontaneously prepared by evaporating the organic solvent on a substrate-like amorphous carbon, highly oriented pyrolytic graphite (HOPG), or mica, because these nanoparticles are soluble in organic liquids owing to the modification with long alkanethiols or alkylphosphines.[3–5] These techniques have utilized one-way formation of particle arrays and no regulation has been performed.

On the other hand, we have developed the large-scale synthesis of carboxylate-modified, water-soluble, and isolable metal nanoparticles using mercaptosuccinic acid as a surface modifier.[6-9] Application of water-soluble nanoparticles is a new field for crystallization of these particles into organized two- and three-dimensional systems, not by the weak van der Waals interaction but by ionic or hydrogen-bonding interactions. In this review chapter, we will show our recent development on the structural formation and characterization of ordered assemblies of water-soluble gold nanoparticles at an air–water interface.

23.2 WHY CARBOXYLATE-MODIFIED WATER-SOLUBLE NANOPARTICLES FOR CRYSTALLIZATION?

If one wants to prepare well-defined small particles, (1) the size of the particles should be unique and (2) the surface must be identified clearly. If one can treat these small particles as a molecule, it might be possible to crystallize the particles into a three-dimensional superlattice (particle crystal) as if it were a molecular crystal. This is a driving force for the synthesis of uniform-sized and surface-modified nanoparticles not only from a view of nanotechnology but also from a physical or crystallographic standpoint. Moreover, using carboxylic acid as a surface modifier is one of the good candidates for the formation of a hydrogen-bonding system, which is a potential technology for hydrogen-bonding network device systems.

Typically, a particle is composed of the central substances (core), including pure metals or compounds, and the surrounding organic films (shell) with different thicknesses. Thiols are widely used as a shell material. By an appropriate selection of the size or functional groups of shell organic molecules, we can control the distance and the relevant interaction forces between particles and, hence, can regulate the physical properties of particle crystals, such as optical and electronic characteristics based on a band structure of three-dimensional systems. The cohesive energy and structure of this crystal are then functions of these controllable parameters. This crystal has a superstructure with two kinds of periodicity: angstrom (atomic unit) and nanometer (particle regularity). The combination of double periodicity enables the inducing of peculiar electronic properties in a particle crystal.[2]

The surface modifier constituting the shell will be classified roughly into two categories: (1) oil soluble (hydrophobic) or (2) water soluble (hydrophilic). Therefore, prepared nanoparticles have a nonpolar or polar nature, depending on the property of the surface modifiers. Carboxylic acids or carboxylates provide useful advantages of versatile and flexible functionalization based on specific interactions, such as electrostatic or hydrogen-bonding interactions. Unlike nonpolar alkanethiol-modified nanoparticles, the successful synthesis of uniform-sized, water-soluble, carboxylic acid (or carboxylate)–modified nanoparticles implies the possibility to assemble the nanoparticles into large artificially constructed nanostructures via electrostatic or hydrogen-bonding interactions. Since molecules possessing carboxylic acid functional groups have been widely applied to form large self-assembled architectures in supramolecular chemistry, it is reasonable to suppose that the stable nanoparticles covered with carboxylic acid (or carboxylate) groups can also be used as building blocks for generating two- or three-dimensional nanoparticle superlattices or for the hybridized self-assembling between supramolecular architectures and nanoparticles.

23.3 SYNTHESIS AND CHARACTERIZATION OF CARBOXYLATE-MODIFIED GOLD NANOPARTICLES

We selected mercaptosuccinic acid (MSA), $HOOCCH_2CH(SH)COOH$, as a surface modifier that is soluble in water in a wide pH range. The synthesis of gold nanoparticles was conducted by direct chemical reduction of chloroauric acid ($HAuCl_4$) in the presence of MSA in methanol. The initial molar ratio between MSA and $HAuCl_4$, denoted as S/Au, was adjusted to control the composition of intermediates, the Au-MSA complex, and the relative rates of particle nucleation and growth. This parameter finally determined the particle size. For a typical preparation of gold nanoparticles at S/Au = 3, 0.5 mmol $HAuCl_4$ aqueous solution (5 w/v%) was mixed with 1.5 mmol MSA in 100 ml of methanol to give a transparent solution. A freshly prepared 0.2 M ice-cold aqueous $NaBH_4$ solution (25 ml) was then added under vigorous stirring. The color of the solution changed to dark brown immediately, and a flocculent dark brown precipitate was yielded. After further stirring for 1 h, the precipitate was gathered by decantation and centrifugation. The precipitate was washed twice with water/methanol (20 v/v%) solution through ultrasonic redispersion/centrifugation processes and then with pure methanol and ethanol. Continuous-flow dialysis was applied against distilled water for 1 day. Finally, the sample was obtained as a solid powder by freeze-drying. The mean particle diameter, determined by transmission electron microscopy (TEM), was 3.5 nm (full width at half maximum (fwhm), 0.4 nm).

To clearly learn the structures of organics on the particle surface, elemental analysis of the sample was conducted by using energy-dispersive x-ray (EDX) spectroscopy attached to a scanning electron microscope (SEM). Although we could not know the content of the hydrogen atom, the apparent molecular formula of $C_{4.3}O_{4.1}SAu_{1.9}Na_{0.99}$ was determined. Compared with the chemical formula of MSA, $C_4O_4SH_6$, the MSA moiety is attached on the gold surface through the thiol group and exists apparently in the form of monosodium carboxylate. Note that MSA moiety adsorbed on the gold nanoparticle surface in the form of a disodium carboxylate without applying dialysis,[6] indicating that dialysis influenced the formation of the carboxylic acid from carboxylate.

The infrared (IR) spectrum reveals the existence of carboxylate (–COO), together with carboxylic acid (–COOH), in the solid powder. Figure 23.1 shows the Fourier transform (FT)-IR spectrum of the sample powder measured at an ambient condition. The strong bands at ~1580 and ~1400 cm^{-1} are attributed to the asymmetric and symmetric stretching vibration of a carboxylate group, respectively. The band observed at ~1695 cm^{-1} is ascribed to the carbonyl (C=O) stretching vibration in the carboxylic acid. Note that this C=O stretching band cannot be observed without

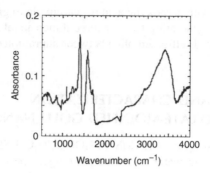

FIGURE 23.1 IR absorption spectrum of MSA-modified gold nanoparticle powder.

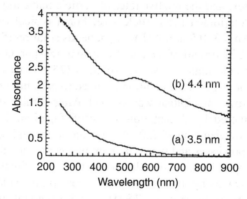

FIGURE 23.2 Optical absorption spectra of MSA-modified gold nanoparticles with different sizes dispersed in water: S/Au = 3.0 (diameter of 3.5 nm) (a) and S/Au = 1.0 (4.4 nm) (b). The sizes were determined by the size histogram from TEM observations. Note that a characteristic absorption band at about 520 nm is due to the surface plasmon for a particle with a 4.4-nm diameter.

conducting dialysis. The broad peak at about 3400 cm⁻¹ belongs to the O-H stretching vibration, indicating the existence of water molecules in the sample. The disappearance of the band at 2548 cm⁻¹, owing to the S-H stretching vibration mode, confirmed the formation of S-Au bonding, namely, MSA anchors on the gold surface through the S atom in the mercapto group. Figure 23.2 shows the optical absorption spectra of the samples with S/Au = 3.0 and 1.0, respectively. The larger the value of the ratio, the smaller the size of particles becomes. Hence, the characteristic absorption band due to surface plasmon locating at about 520 nm smears out for a small particle.

23.4 GOLD NANOPARTICLE CRYSTALS GROWN AT AN AIR–WATER INTERFACE

For nanoparticle crystallization, we used the MSA-modified gold nanoparticles prepared at the condition of S/Au = 3.0, as mentioned in the previous section. The

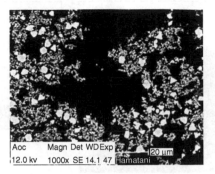

FIGURE 23.3 An SEM micrograph of nanoparticle crystals on an HOPG substrate mixed with amorphous particle assemblies deposited from suspensions.

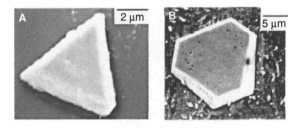

FIGURE 23.4 Close-up images of SEM micrographs of a nanoparticle crystal with well-resolved facets: triangular (A) and hexagonal-shaped (B) crystal.

crystallization took place at an air–water interface in 4 to 10 days in the presence of an appropriate amount of hydrochloric acid (~0.4 M) in a sealed vial, giving numerous micrometer-sized nanoparticle crystals with clear crystal habit.[10] Figure 23.3 shows an SEM image of micrometer-sized nanoparticle crystals (bright images in the picture) mixed with amorphous powders deposited from a suspension on a conductive substrate. To date, this is the first example of gold nanoparticle crystals grown in a bulk aqueous phase. Figure 23.4 shows SEM micrographs of crystals with well-resolved facets. In most cases, we could obtain triangular (Figure 23.4A) or hexagonal-shaped (Figure 23.4B) particle crystals. The results imply that the crystals grew under equilibrium conditions[11] largely different from those of alkanethiol-coated nanoparticles, which grew under the condition of a drying process, namely, an irreversible process, of organic solvents on the solid substrate. Furthermore, clear SEM images of these particle crystals indicate that these crystals are more or less conductive. Figure 23.5 shows a typical TEM image of the crystal. According to a magnified image of the edge of the particle crystal (Figure 23.5B), we can observe a periodic lattice of nanoparticles in hexagonal arrangements, which suggests an ordered structure made from nanoparticles. Figure 23.6 shows a TEM image of a fractured crystal with the cleavage face clearly showing arrays of nanoparticles.

A scanning laser microscopic measurement revealed that the thickness of the crystals ranged from 100 ~ 1000 nm, showing that the crystals were plate-like. In

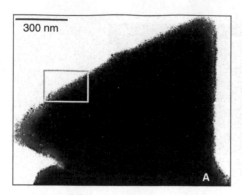
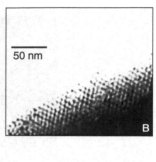

FIGURE 23.5 (A) TEM image of a triangular crystal. (B) Magnified TEM image of the edge part of a crystal in which the rectangular region is indicated.

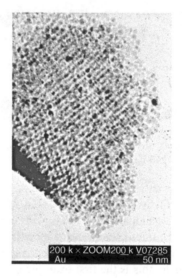

FIGURE 23.6 A TEM micrograph of a nanoparticle crystal with a fractured surface.

Figure 23.7, cross-sectional heights of three crystals are 450, 1000, and 450 nm, respectively. We assume that the crystal growth is most probably by the layer-by-layer step at an air–water interface, suggesting the step structure at the surfaces. The existence of nanometer-sized steps on the surface of a particle crystal is indicated by an atomic force microscopic measurement on a large crystal, as shown in Figure 23.8. The step height at point a is 3.5 nm and at point b is 4.3 nm, and these values are almost consistent with the size of particles, 3.5 nm.

Initial crystal growth stages of gold nanoparticles with various layers were also observed by occasionally sampling the sample suspension at different time intervals. Figure 23.9A to C shows the TEM images of monolayer, bilayer, and multilayer ordered arrays of nanoparticles. Figure 23.9A indicates that the perfectly arranged monolayer domain size is less than ~30 nm. Such an imperfection might be inappropriate for the

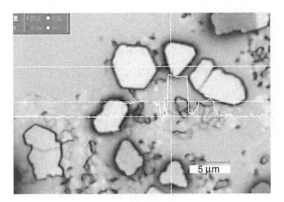

FIGURE 23.7 A scanning laser micrograph of nanoparticle crystals. The thickness is 450 nm for a hexagonal-shaped crystal and 1000 nm for a triangular crystal.

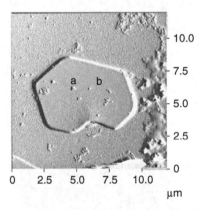

FIGURE 23.8 AFM micrographs of a nanoparticle crystal with well-resolved surfaces: (a) and (b) are positions of a step measurement.

construction of a high-quality micrometer-sized particle crystal. However, Figure 23.9B or C shows that the perfectly arranged domain size is about 100 nm, larger than that in the monolayer film. The results indicate that the quality of the closely packed structure in the superlattice was improved during the growth process, probably by the self-correcting.[11] This is consistent with a scanning laser microscopic observation, but it is not yet clear when the growth mechanism switches in the bulk growth mode.

Stacking of the closely packed particles shows either a face-centered-cubic (fcc) or hexagonal closely packed (hcp) structure. These closely packed structures can be distinguished by observing the sequence of the stacking layer. In the fcc structure, each interstitial site caused by the first layer packing (denoted as *A*) is covered by the second (denoted as *B*) and third (denoted as *C*) layers, namely, *ABCABC...* stacking. In the hcp structure, half of the sites are not covered due to the absence of site *C*, namely, *ABABAB...* stacking. Hexagonally arranged unoccupied sites in

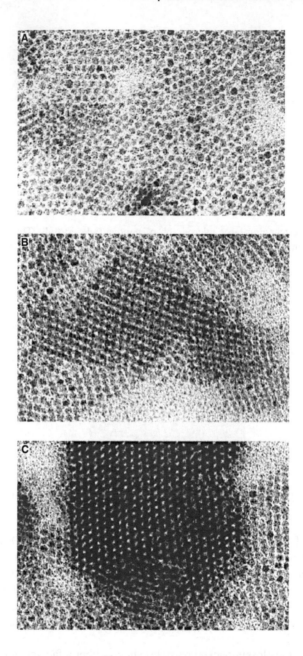

FIGURE 23.9 TEM images of monolayer (A), bilayer (B), and multilayer (C) ordered nanoparticle arrays. Image (C) verifies the hcp nanoparticle arrangements.

the (0001) plane can be seen for the hcp structure, whereas those in the (111) plane cannot be seen for the fcc structure. According to Figure 23.9c, we could detect the hexagonally arranged unoccupied gaps within the multilayer arrays, indicating that the particle crystals possess an hcp structure. Note that the distance between these gaps was obtained to be ~4.9 nm from Figure 23.9C. Because the core diameter and the shell MSA thickness of these nanoparticles are about 3.5 and 0.6 nm, respectively, the distance between the gaps is calculated to be 4.7 nm, quite similar to the obtained value. Therefore, we conclude that the nanoparticle crystals grown at an air–water interface are classified as an hcp structure. The hcp structure formed at an air–water interface was recently confirmed by the method of low-angle x-ray diffraction, together with the transmission electron diffraction technique.[12]

23.5 OPTICAL PROPERTIES OF GOLD NANOPARTICLE CRYSTALS

The color of the particle crystals seemed to be metallic gold, indicating that they have high reflectivity. The inset in Figure 23.10A shows an example of an optical

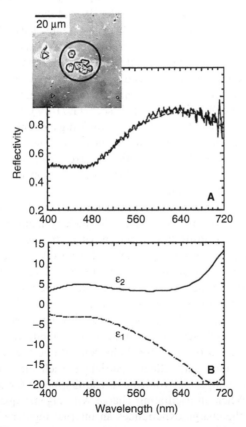

FIGURE 23.10 (A) Reflectance spectrum of gold nanoparticle crystals. The dashed curve shows the simulation. The sample area measured is shown in the optical micrograph in the inset. (B) Simulated dielectric functions (ε_1, real part; ε_2, imaginary part) of particle crystals obtained by analyzing the spectrum (A).

microscope image of particle crystals on a glass substrate. We measured a reflectance spectrum of these crystals by using a polychromator–multichannel photodetector set on the optical microscope. Figure 23.10A (solid curve) shows the reflectance spectrum of the crystals. The probe beam diameter was ~30 μm, as shown in the circle in the inset in Figure 23.10A. The overall features of the reflectance spectrum are similar to that of bulk gold.[13] However, we note that there is a significant difference in reflectivity at a long-wavelength region between bulk gold and nanoparticle crystals, suggesting an absorption band not existing in bulk gold.

We can simulate the observed reflectance spectrum by assuming two contributions: Drude-type free electrons and interband dipole transitions.[13,14] Since a decrease in the reflectivity at the long-wavelength region was observed for the nanoparticle crystals, we have considered two kinds of dipole transitions in the present model. The complex dielectric constant, $\varepsilon(\omega)$, is then a sum of these contributions, which is characterized by three parameters from the Drude model, together with six parameters from two interband transitions. For the Drude model, the three parameters are as follows: free electron relaxation time ($\tau = 9.3 \times 10^{-15}$ sec, bulk value[14]), plasmon frequency ($\omega_p = 1.35 \times 10^{16}$ sec^{-1}, bulk value[13]), and dielectric constant of gold at an infinite frequency (ε_∞). Note that ε_∞ influences the intensity of reflectivity. Therefore, we regard ε_∞ as a determinable parameter. The other six parameters are related to two interband dipole transitions as follows: eigenfrequencies of transition dipoles (ω_1 and ω_2), oscillator strengths (S_1 and S_2), and damping constants (Γ_1 and Γ_2). The suffix denotes each dipole transition. The best fit obtained was $\varepsilon_\infty = 7.9$, $\omega_1 = 4.2 \times 10^{15}$ sec^{-1} (2.76 eV), $\omega_2 = 2.6 \times 10^{15}$ sec^{-1} (1.71 eV), $S_1 = 1.6$, $S_2 = 1.3$, $\Gamma_1 = 1.55 \times 10^{15}$ sec^{-1}, and $\Gamma_2 = 3.0 \times 10^{14}$ sec^{-1}, providing a quantitative agreement with the observed spectrum shown by the dashed curve in Figure 23.10A. By using these obtained parameters, the dielectric functions ε_1 (real part) and ε_2 (imaginary part) are calculated as shown in Figure 23.10B. The spectrum of ε_2 suggests a strong absorption band at about 730 nm.

In order to clarify further the optical properties, we conducted a direct absorption measurement for very thin polycrystalline films made of nanoparticle crystals. Figure 23.11 shows the optical absorption spectrum of particle crystals, together with a deposited film having no periodicity, measured from UV to near-IR regions. Obviously, there is a distinct peak at about 670 nm in the polycrystalline film. This band corresponds to the dip in the reflectance spectrum or the rise in the imaginary part of the dielectric function observed at about 700 nm. Note that dispersed nanoparticles in aqueous solution (Figure 23.2b), as well as in a deposited film (Figure 23.11, amorphous), show a small surface plasmon peak at about 520 nm. Therefore, the red-shifted peak observed is a new band characteristic to the nanoparticle crystals. This low-energy transition may reflect a particle–particle electronic interaction and can be explained by a simple electronic polarizability model for an oscillating charge in the assembly.[15] When an oscillating charge with eigenfrequency ω_0 is arranged like a solid phase, the charge experiences an internal regular electric field caused by surrounding other charges together with an external field. This internal field essentially can give rise to a low-energy shift of the eigenfrequency.[16] We think this interaction can be directly tuned by further modifying the surface ligands with chemical bonding, such as ester linkage or hydrogen bonding.

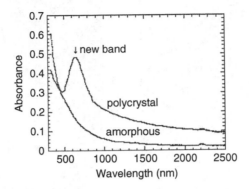

FIGURE 23.11 Absorption spectrum of a particle crystal film (polycrystalline) and a deposited film (amorphous).

23.6 CONTROLS OF INTERPARTICLE SPACING IN THE TWO-DIMENSIONAL ORDERED ARRAYS OF NANOPARTICLES

Metal-to-insulator transition can be designed with the superlattices by fine-tuning of the spacing between adjacent nanoparticles.[2] The interparticle separation might be controllable if an incorporation of rigid molecules between adjacent gold nanoparticles is successful. For water-soluble nanoparticle assemblies, the spacing control by hydrogen bonding is one of the potential candidates. A hydrogen bond is based on an attractive, directional, and selective interaction[17] and expected to work quite effectively at an air–water (hydrophobic) interface.[18] In this section, we demonstrate control of interparticle separation distances by incorporating a rigid hydrogen-bonding molecular unit among MSA-modified gold nanoparticles.

Gold nanoparticle self-assembling occurred at an air–water interface within several days, in the presence of a small amount of hydrochloric acid together with 4-pyridinecarboxylic acid (PyC) in a sealed vial. Because carboxylic acid and pyridine are good hydrogen-bonding partners,[19] we selected PyC as a rigid molecular unit that would be incorporated between MSA moieties on gold nanoparticles. Under the present synthesis condition, the concentration of HCl is much lower (~1/10) than that in the absence of PyC. Therefore, both PyC and HCl are indispensable for the ordered array formation, indicating that the suppression of the dissociation of carboxylic acid is essential.[20]

Figure 23.12A shows a typical TEM image of gold nanoparticle ordered arrays representing a long-range ordered structure, where it can be clearly seen that hexagonal closely packed structures are formed. The mean particle core diameter (d_{core}) and the standard deviation are 5.7 and 0.81 nm, respectively. Because no arrays were generated in the absence of PyC under the condition as a reference experiment, the formation of two-dimensional ordered arrays at an air–water interface suggests that hydrogen-bonding interactions between PyC and MSA moieties would drive the particle assembly.[18] Figure 23.12B shows the FT pattern of Figure 23.12A, which demonstrates the long-range nanoparticle ordering and its hexagonal symmetry. The center-to-center distance between nearest-neighbor particles (D_{cc}) was calculated to

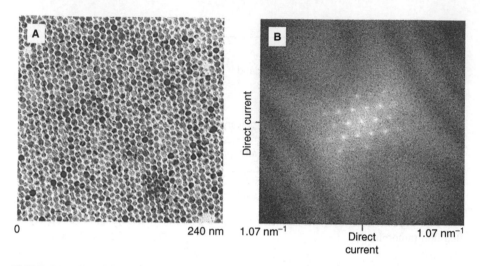

FIGURE 23.12 (A) TEM image of a two-dimensional ordered nanoparticle array prepared in the presence of 4-pyridinecarboxylic acid (PyC). (B) Fourier transform pattern of image (A).

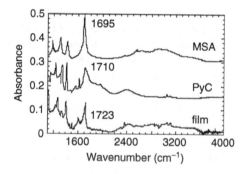

FIGURE 23.13 IR spectra of a two-dimensional ordered array film, PyC, and MSA.

be 7.6 nm by analyzing Figure 23.12B. Considering that the two-dimensional array was composed of 5.7-nm nanoparticles in the core diameter, we could estimate the interparticle separation gap (D_{gap}: $D_{cc} - d_{core}$) of 1.9 nm. Since the obtained D_{gap} is about three times larger than the thickness of the MSA modifier (~0.6 nm), it is expected that PyC (molecular length, ~0.6 nm) is incorporated between adjacent MSA-modified nanoparticles. We believe that the controlling of the interparticle separation gap in nanoparticle ordered arrays is possible by use of such an intercalating molecule.

Hydrogen bond formation between MSA and PyC moieties in the array film was identified by attenuated total reflectance (ATR)-IR spectroscopy. Figure 23.13 shows the IR absorption spectra of an array film, PyC, and MSA. The spectrum of the film has characteristic features of both PyC and MSA. A broad band around 2800 to 3500 cm^{-1} is assigned to the carboxylic acid O-H stretching owing to the MSA molecule.[6] The peaks at 1603 and 1406 cm^{-1} are assigned to ring vibrations of pyridine groups. The absorption bands around 1900 to 2000 and 2400 to 2500 cm^{-1} are attributed to the O-H stretching bands based on hydrogen bonding between pyridine and carboxylic acid groups in a splitting pattern.[20,21] The peak shift (1723 cm^{-1}) assigned to the C=O stretching vibration is the crucial evidence for hydrogen-bonding interactions between MSA and PyC moieties; the peak position was different from that for pure MSA (1695 cm^{-1}) or PyC (1710 cm^{-1}). We concluded that strong hydrogen bonding between MSA on the surface of gold nanoparticles and PyC was present in the array films. Future research in our group includes the arbitrary controls of interparticle separation gaps inside nanoparticle two- and three-dimensional ordered arrays by incorporating various-sized hydrogen-bonding molecular units.

23.7 CONCLUSION

By choosing MSA as an adequate ligand attachable to the surface of gold nanoparticles, it is possible to prepare unique surface-modified particles that are soluble, isolable, and redispersible in water. With an acid treatment, this particle easily crystallizes at an air–water interface to make micrometer-sized, plate-like crystals with well-resolved facets, together with two-dimensional ordered arrays. The crystals possessed an hcp structure of nanoparticles as elucidated by TEM. Optical properties of the crystals showed a characteristic reflectance and a new absorption band around near-IR regions. Tentative control of interparticle spacing in the two-dimensional nanoparticle arrays through hydrogen-bonding networks by intercalating molecules was also demonstrated. Further research includes the synthesis and evaluation of physical properties of crystals composed of nanoparticles with different diameters, which can be accomplished by changing the preparation parameters of S/Au.

ACKNOWLEDGMENT

This work is supported in part by a grant-in-aid for scientific research (B: 13440212) from the Ministry of Education, Culture, Sports, Science and Technology; the Japan Space Utilization Promotion Center (JSUP); the Mitsubishi Research Institute (MRI); and the Hyogo Science and Technology Association (HSTA). AFM imaging by Drs. M. Ishikawa and K. Honda of MRI is acknowledged.

REFERENCES

1. RP Andres, T Bein, M Dorogi, S Feng, JI Henderson, CP Kubiak, W Mahoney, RG Osifchin, R Reifenberger. "Coulomb staircase" at room temperature in a self-assembled molecular nanostructure. *Science* 272: 1323–1325, 1996.
2. CP Collier, RJ Saykally, JJ Shiang, SE Henrichs, JR Heath. Reversible tuning of silver quantum dot monolayers through the metal-insulator transition. *Science* 277: 1978–1981, 1997.
3. CB Murray, CR Kagan, MG Bawendi. Self-organization of CdSe nanocrystallites into 3-dimensional quantum dot superlattices. *Science* 270: 1335–1338, 1995.

4. L Motte, F Billoudet, E Lacaze, J Douin, MP Pileni. Self-organization into 2D and 3D superlattices of nanosized particles differing by their size. *J Phys Chem B* 101: 138–144, 1997.

5. C J Kiely, J Fink, M Brust, D Bethell, DJ Schiffrin. Spontaneous ordering of bimodal ensembles of nanoscopic gold clusters. *Nature* 396: 444–446, 1998.

6. S Chen, K Kimura. Synthesis and characterization of carboxylate-modified gold nanoparticle powders dispersible in water. *Langmuir* 15: 1075–1082, 1999.

7. S Chen, K Kimura. Water soluble silver nanoparticles functionalized with thiolate. *Chem Lett* 1169–1170, 1999.

8. H Yao, O Momozawa, T Hamatani, K Kimura. Phase transfer of gold nanoparticles across a water/oil interface by stoichiometric ion-pair formation on particle surfaces. *Bull Chem Soc Jpn* 73: 2675–2678, 2000.

9. H Yao, O Momozawa, T Hamatani, K Kimura. Stepwise size-selective extraction of carboxylate-modified gold nanoparticles from an aqueous suspension into toluene with tetraoctylammonium cations. *Chem Mater* 13: 4692–4697, 2001.

10. K Kimura, S Sato, H Yao. Particle crystals of surface modified gold nanoparticles grown from water. *Chem Lett* 372–373, 2001.

11. S Sato, H Yao, K Kimura. Equilibrium growth of three-dimensional gold nanoparticle superlattices. *Physica E*, 17: 521–522, 2003.

12. S Wang, S Sato, K Kimura. Self-assembly of hydrophilic gold nanocrystals into nanoparticle crystals. *Trans Mater Res Soc Jpn* 28: 903–906, 2003.

13. MM Alvarez, JK Khoury, TG Schaaff, MN Shafigullin, I Vezmar, RL Whetten. Optical absorption spectra of nanocrystal gold molecules. *J Phys Chem B* 101: 3706–3712, 1997.

14. U Kreibig, M Vollmer. *Optical Properties of Metal Clusters*. Berlin: Springer, 1995, Chapter 2.

15. S Sato, H Yao, K Kimura. Optical absorption properties of three-dimensional Au nanoparticle crystals. *Chem Lett* 526–527, 2002.

16. PA Cox. *The Electronic Structure and Chemistry of Solid*. Oxford: Oxford University Press, 1987, Chapter 2 (Spectroscopic Methods).

17. GA Jeffrey. *An Introduction to Hydrogen Bonding*. New York: Oxford University Press, 1997, Chapter 2 (Nature and Properties).

18. H Hoshi, M Sakurai, Y Inoue, R Chujo. Medium effects on the molecular electronic structure. I. The formulation of a theory for the estimation of a molecular electronic structure surrounded by an anisotropic medium. *J Chem Phys* 87: 1107–1115, 1987.

19. E Hao, T Lian. Layer-by-layer assembly of CdSe nanoparticles based on hydrogen bonding. *Langmuir* 16: 7879–7881, 2000.

20. H Yao, H Kojima, S Sato, K Kimura. Construction of 2-D superlattices of gold nanoparticles at an air/water interface based on hydrogen-bonding networks. *Chem Lett* 32: 698–699, 2003.

21. JY Lee, PC Painter, MM Coleman. Hydrogen bonding in polymer blends. 4. Blends involving polymers containing methacrylic acid and vinylpyridine groups. *Macromolecules* 21: 954–960, 1988.

Index

A

Adsorption kinetics, 442–443
  interrupted-deposition protocol, 445–446
  mechanism, 444, *445*
  salt content effect on, 446–449
  via two kinetics regimes, 443–444
Aggregates, metal nanoparticle, 200–203
  DNA-directed assembly of, 234–235, *236*
Air-water interface, gold nanoparticle crystals
      grown at, 604–609
Alignment methods, nanowire
  biological, 52
  chemical, 50–51
  physical, 48–50
Angular dependence of coherent plasmon
      coupling, 500–501
Anisometric nanoparticles, 538–542
Annealing, postpreparative, 352–354
Anodic and cathodic photocurrents, 451–452
Arrays
  FePtAg and FePtAu nanoparticle, 378–381
  interparticle spacing in two-dimensional
      ordered, 611–613
  liquid crystalline, 515–523
  metal nanoparticle, 488–489
  metallic quantum dot
    computed current-voltage curves, 172–174
    Fermi window function, 169–172
    Green's function and, 159–160
    Hamiltonian of, 157–158
    measurement of electronic structures of,
      153–157
    mechanisms for conduction in, 160–161
    quantum phase transitions of, 153
    rate of charge transport across, 169–172
    transmission function, 155–169
    variations in sizes of individual dots in,
      154–155
  ordered, 564–567, 611–613
  silver nanoparticle, 488, 496–498
Artificial organic templates, supramolecular self-
      assembly from, 288–291
Assembly. *See also* Bioinspired assembly
  liquid crystalline arrays, 515–523
  magnetic nanocrystal
    capping ligands effects on, 348–349
    general considerations, 343–346
    growth via additional injections of
      precursors, 349–350

    heating time effect on, 346
    improvement of particle crystallinity in,
      352–354
    molar ratio between components in,
      350–351
    postpreparative annealing, 352–355
    postpreparative size-selective fractionation,
      350
    reaction temperature effect on, 346–348
    shape control, 351–352
    size tuning of, 343–350
    superstructures of, 356–363
    surface ligand exchange and phase transfer,
      355
  magnetic particles
    colloidal, 387–391
    ferrofluid syntheses, 391–394
    methods, 391–398
  metal nanowire, 414–415
  metallic nanorods, 517–523
  nanobuilding block, 96
  nanocrystal
    from artificial organic templates,
      288–291
    inorganic species effect on, 271–273
    metal, 283–284
    micelles in, 280–284
    microemulsions in, 284–285
    by molecular template mechanism,
      209–212
    monolayers in, 285–288
    organic species effect on, 273–275
    shape control in, 282–284
    small-molecule additive-mediated shape
      control in, 271–280
    solvent effect on, 275–280
    stability of, 219–220
    strategies, 270–271
    surfactant-mediated templating and
      crystallization, 280–288
    with TOPO/TBP surfactant, 212–221
    vesicles in, 288
  nanoparticle
    biopolymers in, 292–297
    dendrimers in, 298
    DHBCs in, 299–320
    double hydrophilic block copolymers in,
      299–320
    hydrogen-bonding-mediated, 554–555

interparticle structural evolution in process
of, 562–563
layer-by-layer, 44–47, 439–456
one-dimensional, 31–37
oriented attachment growth and
spontaneous organization, 320–324
polyamides in, 298–299
polymer-controlled crystallization and
morphosynthesis, 291–292
reversibility of, 558–559
silver, 488–494
single nanowire, 47–52
small-molecule additive-mediated shape
control in, 271–280
supramolecular polymers used in,
288–289
surfactant-mediated templating and
crystallization, 280–288
thermally activated size processing, 554
thickness control of, 555–557
three-dimensional, 41–47
transcription from organic matrix template,
289–291
two-dimensional, 37–41
nanospheres and nanoprisms, 538–539, 542
nanotube
-encapsulated nanowire devices, 473–475
nanoparticle/polymer, 465–469
SiO_2, 469–473
nanowire
combined in-membrane and on-wire,
482–483
layer-by-layer, 476–478

B

$BaCO_3$, 312–313
$BaCrO_4$, 302–310
$BaSO_4$, 302–310
Bioinspired assembly
from artificial organic templates
organic matrix templates transcription,
289–291
supramolecular polymers in, 288–289
crystallization and morphosynthesis
biopolymers in, 292–297
dendrimers in, 298
double hydrophilic block copolymers,
299–320
polyamides in, 298–299
oriented attachment growth and spontaneous
organization via, 320–324
small-molecule additive-mediated shape
control in
inorganic species, 271–273

organic species, 273–275
solvent effects on, 275–280
strategies, 270–271
surfactant-mediated templating and
crystallization
micelles in, 280–284
microemulsions in, 284–285
monolayers in, 285–288
vesicles in, 288
Biolabels, 248–251
Biomolecules. *See also* DNA
applications of nanoparticles assembled with,
31–32, 248–258
directed nanoparticle organization, 230–231
protein-based recognition systems in,
231–233
lithography and, 243
recognition in nanowire assembly, 428–430
scale length of nanoparticles and, 228–229
Biopolymers, 292–297
Biotemplates, 31–33
Branching morphology, 567–568
Building blocks, nanoscale
anisotropic basic elements
one-dimensional rods and wires, 81–90
two-dimensional discs and plates, 90–91
aspects of, 75–77
assemblies of, 96
construction of DNA-protein conjugates for
biotemplating, 240–246, *247*
DNA-directed oligomerization of, 234–235,
236
geometrical control of, 92–94
high symmetry, 91–95
isotropic basic elements
cubical, 79–81
spherical, 77–79

C

$CaCO_3$
formation mechanism of calcite hollow
spheres, 313–314, *315*
other complex structures by modified
crystallization of, 315
polymeric templates and, 316–320
sphere-to-rod-to-dumbbell-to-sphere
transformation mechanism,
310–313
Calcite hollow spheres, 313–314, *315*
Calcium phosphates, 301, *302*
Capping ligands, 348–349
Carbon nanotubes (CNTs), 4–5, 10–11, 414–415
growth, 400–401

Carboxylate-modified water-soluble nanoparticles, 602–603
   gold, 603–604
Catalysis, 526
$CdWO_4$, 315–316
Charge coupled devices (CCD), 106–107
Chemical method of nanowire alignment, 50–52
Chemical modification of nanowire surfaces, 417–418
Coated nanoparticles, 526–527
Coherent plasmon coupling between silver nanoparticles
   angular dependence of, 500–501
   effect of external dielectric medium on, 502–503
   effect of interparticle spacing on, 498–500
   extinction spectra and, 496–498
   optical properties, 494–495
   substrate effects on optical properties, 503–506
Colloidal crystals, 41–42, 274. *See also* Magnetic nanocrystals; PbSe nanocrystals
   oriented attachment growth, 320
Colloidal particles
   formation in DMF, 530–533, 538–542
   magnetism, 389–391
   metal, 525–526
   syntheses and assembly, 387–389
Conduction mechanisms, elastic transmission, 160–161
Conductivity modulation, 29
Confocal microscopy, 107
Copolymers, double hydrophilic block. *See also* Polymers
   $BaCrO_4$ and $BaSO_4$ from, 302–310
   $CaCO_3$ and $BaCO_3$ from, 312–313
   calcium phosphates from, 301, *302*
   fine-tuning of nanostructure of inorganic nanocrystals, 315–316
   morphosynthesis
      and hierarchical structures, 301–310
      of metal carbonate minerals using, 310–315
   polymeric templates, 316–320
   selective adsorption and crystal growth using, 299–301
$CoPt_3$ nanocrystals, 350–351
Core-shell nanoparticles, 526–527
Cross-linking, nanoparticle
   effects of protecting monolayers on, 584–587
   networking and, 581–584
   for surface nanofabrications, 587–589
Crystallinity, nanoparticle, 5–6, 186–188, 352–354
Cubes, nanocrystal, 79–81

Currents
   -voltage curves, 172–174
   expressed as Fermi window function, 169–172

D

Data storage devices, 398–400
Decay rates, fluorescence, 117–118
Dendrimers, 298
Density, surface, 446–449
Diffusion, surface, 428–430
Dip-pen lithography, 37, 40–41
Dipole-dipole interaction
   in nanocrystal assemblies, 220
   in nanoparticle layers, 183–185
Discs and plates, two-dimensional, 90–91
DMF (N,N-dimethylformamide)
   formation of nanorods and nanoprisms using, 538–539
   as a reducing agent, 527–528
   synthesis of coated silver nanoparticles in, 530–533
DNA, 31–33, 201–202, 229
   in individual nanoscaled particle assemblies, 236–238
   and nucleic acid-coated particles, 251–257
   protein conjugates
      construction of, 240–246, *247*
      as model systems, 238–239
   therapeutic applications of nanoparticles and, 257
   used in assembly of nanoparticles, 232, 233 235, *236*

E

Electrical properties of nanowires, 8–9, 418–425, 478–482
Electrochromic devices, 18–21
Electrodeposition of nanowires, 415–416
Electroluminescence devices
   nanowires in, 27–28
   thin film nanoparticles in, 13–15
Electron relaxation dynamics
   in gold nanoparticle superlattices and aggregates, 200–203
   spherical gold nanoparticles, 195–198
   spheroidal, rod-shaped, and gold nanoshell particles, 199–200
Electronic devices
   nanoscience research and, 4–5
   physical dimensions of, 4
   structural design of, 4–5
Electronic interaction between nanoparticles in closely packed layers, 188–190

Electroluminescence devices, 13–15
Erbium-doped silicon
  dots
    photophysical measurements of, 145
    structural characterization of, 141–144
  nanowires
    fabrication of, 145–149
    photophysical measurements of, 149–150
Extended x-ray absorption fine-structure
    spectroscopy (EXAFS), 144
Extinction spectra of silver nanoparticles arranged
    into two-dimensional arrays,
    496–498

F

Far-field microscopy, 106–107
FePt nanocrystals
  molar ratio and, 350–351
  superstructures of, 357–360
  transition from fcc to $L1_0$ phase in, 355
FePt nanoparticles
  FePtAg and FePtAu ternary alloy nanoparticles
    from, 375–378
  FePtCu ternary alloy nanoparticles from,
    371–375
  for granular thin-film magnetic media,
    369–371
  magnetic properties of FePtAg and FePtAu
    nanoparticle arrays from, 378–381
  significance of work in, 381–382
Fermi window function, 169–172
Ferrofluid syntheses, 391–394
Field sensors, 29–30
Fine-tuning of nanostructure of inorganic
    nanocrystals, 315–316
Fluid dynamics, 38–39
Fluorescence *in situ* hybridization (FISH), 250
Fluorescence microscopy
  blinking in C_DS_E nanocrystals and, 107–108,
    110–115
  experimental, 105–110
  influence of alternating argon and oxygen in,
    112, 113–115
  influence of gas environments on, 113–115
  nanocrystal properties and, 103–104
Fluorescence spectroscopy, 115–119, 430–431
Fourier transform infrared (FTIR) spectrum,
    467–468
Fractionation, size-selective, 350

G

Gas sensors, 22
Gelatin gels, 318

Gold. *See also* Metal nanoparticles; Metal
    nanowires
  added to FePt nanoparticles, 376–378
  crystals, nanoparticle
    grown at air-water interface, 604–609
    optical properties of, 609–610
  nanoparticles
    as biolabels, 248–251
    carboxylate-modified, 603–604
    DNA-linked, 234–235
    grown at air-water interface, 604–609
    hot electron relaxation dynamics in,
     195–198
    nucleic acid-coated, 251–257
    optical properties, 609–610
    properties of, 193–194
    protein-coated, 248–251
    spherical, 195–198
    spheroidal, rod-shaped, and nanoshell,
     199–200
    surface plasma resonance (SPR), 194–195
    therapeutic applications of, 257
  nanorods
    beat signals of, 127
    experiments, 516–517
    growth procedures, 517–523
    hole-burning experiments in, 133–135
    laser-induced heating and coherent
     excitation of vibrational modes of,
     128–130
    optical properties of, 125–126, 198–199
    transmission electron microscopy (TEM)
     of, 127–128
    vibrational period versus probe laser
     wavelength and rod length in,
     130–133
Granular thin-film magnetic media, 369–371
Green's function and arrays, 159–160
Group II-VI semiconductors, 81–83
Group III-V semiconductors, 84–85
Group IV-semiconductors, 85–87

H

Hamiltonian of metallic quantum dots arrays,
    157–158
Heating time effect on size of magnetic
    nanocrystals, 346
High symmetry nanoscale building blocks, 91–95
Hole-burning experiments in nanorods, 133–135
Hollow-shaped nanocrystals, 95
Hydrogen bonding
  -mediated nanoparticle assembly, 554–555
  interparticle, 559–562

I

In-membrane and on-wire assembly of nanowire devices, 482–483
Infrared reflectance spectroscopic (IRS) technique, 559–562
Inorganic crystals shape control, 271–273, 315–316
Interfaces
air-water, 604–609
controlling particle-surface, 426–427
Interparticle hydrogen bonding, 559–562
Interrupted-deposition protocol, 445–446
Inverse opals, template synthesis of, 361–363, *364*
Iron oxide nanoparticles, 255–257

K

Kinetics of adsorption, 442–443
interrupted-deposition protocol, 445–446
mechanism, 444, *445*
salt content effect on, 446–449
via two kinetics regimes, 443–444

L

Langmuir-Blodgett films, 44
Langmuir-Blodgett (LB) technique, 39, 450, 577–581
cross-linked networks for surface nanofabrications, 587–589
effects of protecting monolayers on particle cross-linking in, 584–587
nanoparticle networks achieved using, 581–584
for thin films, 589–595
Laser
assisted catalytic growth (LCG), 270
induced heating and coherent excitation of vibrational modes in gold nanorods, 128–133
wavelength, rod length and vibrational period, 130–133
Layer-by-layer (LBL) assembly technique, 44–47, 51, 463–465
light conversion efficiency and, 454–456
linear, 439–440
matrix effect on optical properties, 440–442
of nanoparticle/polymer nanotubes, 465–469
of nanospheres and nanoprisms, 542
nanostructured thin films through, 535–538
photogenerated charge transfer and transport at, 449–456

of polyacrylate-modified Q-CdS and Nano-Pt, 439–440
of SiO$_2$ nanotubes from molecular precursors, 469–473
on top of nanowire templates, 476–478
Layers, nanoparticle
absorption spectra of, 181–183
building up of, 181
closely packed, 180–183, 185–186
dipole-dipole interaction in, 183–185
electroactive, 594–595
electronic interaction in, 188–190
homogeneity of, 181–182
and three-dimensional assemblies of nanoparticles, 44–47
Ligands
capping, 348–349
exchange and phase transfer, surface, 355
Light conversion efficiency and multilayered assembly, 454–456
Light-emitting diodes (LEDs), 13–15
Linear LBL assembly by polyacrylate-modified Q-CdS and Nano-Pt, 439–440
Liquid crystalline arrays, assembly of, 515–523
Liquid evaporation, 38–39
Lithographically defined wells, 427–428
Lithography in nanoparticle assembly, 37–38, 40–41
biomolecular, 243
Luminescence
metal nanoparticles, 198–199
nanowire devices, 27–28
thin film devices, 13–15

M

Magnetic nanocrystals
colloidal crystals of, 361
colloidal synthesis of
annealing, postpreparative, 352–355
capping ligands effects on, 348–349
growth via additional injections of precursors, 349–350
heating time effect on, 346
molar ratio between components of, 350–351
postpreparative size-selective fractionation, 350
reaction temperature effect on, 346–348
shape control, 351–352
size tuning of, 343–350
surface ligand exchange and phase transfer, 355

postpreparative annealing
  improvement of particle crystallinity in,
    352–354
  transition from fcc to $L1_0$ phase in FePt,
    355
potential applications of, 342
superstructures of
  colloidal crystals of, 361
  template synthesis of inverse opals,
    361–363, *364*
  two- and three-dimensional ordered,
    356–360
Magnetic particles
  applications
    carbon nanotube growth, 400–401
    data storage, 398–400
    photonic crystals, 401–402
  assembly methods, 394–398
  colloidal, 387–389
  ferrofluid syntheses, 391–394
  magnetism of, 389–391
  nanostructured devices using, 385–387
Magnetic properties of FePtAg and FePtAu
    nanoparticle arrays, 378–381
Magnetism of colloidal particles, 389–391
Metal carbonate minerals
  morphosynthesis of, 310–315
  sphere-to-rod-to-dumbbell-to-sphere
    transformation mechanism,
    310–313
Metal nanoparticles. *See also* Gold; Nanoparticles;
    Silver
  applications of, 601–602
  as biolabels, 248–251
  hot electron relaxation dynamics in, 195–198,
    199–203
  as model systems, 257–258
  and nanoclusters, 198–199
  nucleic acid-coated, 251–257
  optical emission of, 125–126, 198–199
  properties of, 193–194
  protein-coated, 248–251
  spherical, 195–198
  spheroidal, rod-shaped, and nanoshell,
    199–200
  superlattices and aggregates, 200–203
  surface plasma resonance (SPR) of, 194–195
  therapeutic applications of, 257
Metal nanowires
  assembly, 414–415
    biomolecular recognition in, 428–430
    in lithographically defined wells, 427–428
  chemical modification of surfaces of, 417–418
  controlling particle-surface interactions,
    426–427

diffusion on surfaces, 428–430
electrical measurements of, 418–421
electrodeposition of, 415–416
electronic properties with functional
    components, 421–425
at interfaces, 426–431
synthesis, 414
template growth of, 415–418
Metallic nanorods
  applications of, 515–516
  experiments, 516–517
  growth procedures, 517–523
  nanocrystal, 87–89
Micelles
  mineralization of inorganic minerals and
    mesoscale self-assembly with,
    281–282
  reverse, 280–281, 283–284
  shape control
    of low-dimensional semiconductor
      nanocrystals, 282
    of metal nanocrystals, 283–284
Microemulsions, 284–285
Molar ratio, 350–351
Molecular beacons, 256
Molecular template mechanism for growth for
    wire-like PbSe nanocrystals,
    209–212
Monolayers, 285–288, 584–587
Morphology, branching, 567–568
Morphosynthesis
  biopolymers in, 292–297
  controlled crystallization and, 291–292
  of metal carbonate minerals, 310–315
  using double hydrophilic block copolymers,
    301–320

N

Nanocircuits, 24–26
Nanoclusters, optical emission of, 198–199
Nanocolloids, thin films of nanoparticles in, 11–23
Nanocrystals. *See also* Magnetic nanocrystals
  anisotropic basic elements of, 81–90
  assembly
    from artificial organic templates, 288–291
    microemulsions in, 284–285
    monolayers in, 285–288
    ordered or disordered, 218
    oriented attachment growth, 320–323
    stability, 219–220
    vesicles in, 288
  colloidal crystals of, 361
  cubical, 79–81
  electrical properties of, 220–221

erbium-doped silicon, 139–151
fine-tuning of nanostructure of inorganic,
    315–316
fluorescence microscopy of
    blinking in C_DS_E, 107–108, 110–115
    experimental, 105–110
    physical properties and, 103–104
fluorescence spectroscopy of, 115–119
hollow-shaped, 95
isotropic basic elements, 77–81
magnetic alloy, 350–351
metal, 87–89, 283–284
one-dimensional, 81–90
PbSe
    applications of, 207–209
    wire-like assemblies grown with
        TOPO/TBP surfactant, 212–221
    wire-like assembly grown by molecular
        template mechanism, 209–212
polymer-controlled crystallization and
        morphosynthesis of, 291–292
    biopolymers in, 292–297
semiconductor, 81–87, 207–209, 282
shape control
    double hydrophilic block copolymers in,
        299–301
    inorganic species influence on,
        271–273
    of low-dimensional semiconductor, 282
    magnetic, 351–352
    metal, 283–284
    organic species influence on, 273–275
    solvent effects on, 275–280
spherical, 77–79
surfactant-mediated templating and
        crystallization
    micelles in, 280–284
    microemulsions in, 284–285
    monolayers in, 285–288
    vesicles in, 288
transition metal oxide, 89–90
two-dimensional discs and plates, 90–91
Nanoparticles. See also FePt nanoparticles; Gold;
        Metal nanoparticles; Silver
anisometric, 538–542
from artificial organic templates, 288–291
assembly in polyelectrolytes by electrostatic
        interactions, 439–442
biomolecule-directed organization of
    applications of, 248–258
    into biolabels, 248–251
    DNA-based, 233–234
    DNA-directed oligomerization of,
        234–235, 236
    into molecular beacon conjugates, 256

and nucleic acid-coated particles, 251–257
protein-based recognition systems in,
        231–233
as building blocks for higher-order
        architectures, 438
carboxylate-modified water-soluble, 602–603
closely packed layers of, 180–183, 185–186
coated, 526–527
colloidal synthesis of metal or semiconductor,
        5
compared to biomolecules length, 228–229
complex nanostructures from TiO$_2$-coated,
        533–538
core-shell, 526–527
crystal structure of, 5–6, 186–188, 352–354
DNA
    -protein conjugates for biotemplating,
        240–246, 247
    and assembly of, 233–235, 236
electroactive, 589–590, 594–595
fluid dynamics and, 38–39
future prospects, 53–54
hydrogen-bonding-mediated assembly,
        554–555
interparticle spatial properties, 551–553, 563
    branching morphology, 567–568
    ordered array, 564–567
interparticle structural properties
    evolution in assembly process, 562–563
    hydrogen bonding, 559–562
iron oxide, 255–257
kinetics of adsorption and factors governing
        self-assembly of, 442–449
Langmuir-Blodgett (LB) technique and, 39,
        450, 577–581
    cross-linked networks for surface
        fabrications, 587–589
    effects of protecting monolayers, 584–587
    networking, 581–584
    thin film, 589–595
layer-by-layer assembly, 44–47, 439–456,
        535–538
    of nanoparticle/polymer nanotubes,
        465–469
as model systems, 257–258
multilayer structures of electroactive, 594–595
nanodevices based on, 5–7
networks, 581–584
one-dimensional, 31–37
optical properties of, 125–126, 198–199,
        440–442
oriented attachment and spontaneous
        organization, 320–324
photocurrents in, 451–454

polymer-controlled crystallization and
morphosynthesis of, 291–320
protein-based, 40–41, 231–233
reversibility of assembly of, 558–559
ringed, 34–35
salt content effect on self-assembly of,
446–449
semiconductor, 35–36
single-electron tunneling (SET) effect in, 5–6
small-molecule additive-mediated shape
control in, 271–280
stability of, 589–590
superstructures of, 52–53, 186–188
surface density, 446–449
surfactant-mediated templating and
crystallization of, 280–288
templated, 41–42
thermally activated size processing, 554
thickness control of, 555–557
thin film, 535–538
electroluminescence devices, 13–15
electrochromic devices, 18–21
FePt, 369–371
Langmuir-Blodgett, 589–595
photoelectronic devices, 12–13
photovoltaic devices, 15–18
sensors and biosensors, 21–23
three-dimensional, 41–47, 356–360
two-dimensional, 37–41, 581–587, 611–613
water-soluble carboxylate-modified, 602–603
Nanoprisms
formation in DMF, 538–539
optical properties of, 539–541, *542*
Nanorods
beat signals of, 127
formation in DMF, 538–539
gold, 125–133, 133–135, 198–199, 515–523
hold-burning experiments in, 133–135
laser-induced heating and coherent excitation
of vibrational modes of, 128–130
metallic
applications of, 515–516
experiments, 516–517
growth procedures, 517–523
one-dimensional, 81–90
optical properties of, 125–126, 539–541, *542*
oriented attachment growth of, 321–322
spontaneous organization of nanoparticles into,
323
transmission electron microscopy (TEM)
experiments on, 127–128
vibrational period versus probe laser
wavelength and rod length in,
130–133
Nanoshells, TiO$_2$, 534

Nanotubes, 290–291
-encapsulated nanowire devices, 473–475
carbon, 4–5, 10–11, 400–401, 414–415
layer-by-layer
self-assembly of nanoparticle/polymer,
465–469
synthesis of SiO$_2$, 469–473
Nanowires
alignment
biological method of, 52
chemical method of, 50–51
physical method of, 48–50
assemblies of single, 47–52
biosensors, 29–30
chemical properties of, 7–8
combined in-membrane and on-wire assembly
of devices of, 482–483
devices, nanotube-encapsulated, 473–475
electrical characteristics of, 8–9, 418–425,
478–482
electroluminescence devices, 27–28
erbium-doped silicon, 145–150
future prospects, 53–54
insulation of, 9
layer-by-layer assembly
on top of nanowire templates, 476–478
metal
assembly, 414–415, 427–428, 430–431,
428–430
biomolecular recognition in, 428–430
chemical modification of surfaces of,
417–418
diffusion on surfaces, 428–430
electrical measurements of, 418–421
electrodeposition of, 415–416
electronic properties with functional
components, 421–425
at interfaces, 426–431
in lithographically defined wells, 427–428
synthesis, 414
template growth of, 415–418
as molecular electronic devices, 23–30
nanocircuit, 24–26
nanodevices based on semiconductor, 7–11
sensors, 28–30
single, 47–52
spontaneous organization of nanoparticles into,
323–324
strengths of devices made from, 10–11
superstructures of, 52–53
synthesis of nanotube-encapsulated devices of,
473–475
Near-field scanning optical microscopy (NSOM),
105–106
Nucleic acid-coated particles, 251–257

O

Oligomerization of nanoparticles, DNA-directed, 234–235, *236*
One-dimensional assemblies of nanoparticles, 31–37
One-dimensional rods and wires, 81–90
Optical properties
    gold nanoparticle crystals, 609–610
    interacting nanoparticles, 494–495
    and layer-by-layer assembly of nanoparticles, 440–442
    metal nanoparticles and nanoclusters, 125–126, 198–199
    nanorods and nanoprisms, 539–541, *542*
    PbSe nanocrystals, 216–217
    silver nanoparticle arrays, 503–506
Ordered arrays, 564–567, 611–613
Organic solvents as reducing agents, 527
Oriented attachment growth, 320–323
Ostwald ripening, 345–346

P

PbSe nanocrystals
    applications of, 207–209
    conductivity measurements, 217–218
    optical properties, 216–217
    ordered or disordered assembly, 218
    spherical and wire-like assemblies grown with TOPO/TBP surfactant, 212–221
    stability, 219 220
    wire-like assembly grown by molecular template mechanism, 209–212
Phase transfer and surface ligand exchange, 355
Photocurrents
    anodic and cathodic, 451–452
    generation and ensembles of Q-particles, 452–454
Photoelectronic devices from NP thin films, 12–13
Photogenerated charge transfer and transport at LBL assemblies, 449–456
Photonic band gaps, 41–42
Photonic crystals, 401–402
Photovoltaic devices, 15–18
Physical method of aligning nanowires, 48–50
Polyamides, 298–299
Polymeric templates, 316–320
Polymers. *See also* Copolymers, double hydrophilic block
    controlled crystallization and morphosynthesis, 291–292
    biopolymers in, 292–297
    dendrimers in, 298

    double hydrophilic block copolymers in, 299–320
    polyamides in, 298–299
    induced liquid precursor (PILP) process, 296–297
    stabilization of silver nanoparticle arrays by embedding into matrix of, 492–494
    supramolecular, 288–289
Polyoxometalates (POMs), 18
Proteins
    based recognition systems in biomolecule-directed nanoparticle organization, 231–233
    coated nanoparticles as biolabels, 248–251
    DNA conjugates
        construction of, 240–246, *247*
        as model systems, 238–239
    used in assembly of nanoparticles, 40–41
Prototype devices based on single nanoparticles and nanowires, 5–11
Prussian Blue, 20–21
PVP-modified flat surface, self-assembly of silver nanoparticles on, 489–492

Q

Q-particles, photocurrent generation at ensembles of, 452–454
Q-state, 451–452
Quartz crystal microbalance (QCM), 557

R

Rate of charge transport across arrays, 169–172
Reaction temperature effect on size of magnetic nanocrystals, 346–348
Recognition systems, protein-based, 231–233
Redox transmetallation, 78–79
Reducing agents, 527–528
Reversibility of nanoparticle assembly, 558–559
Rings, nanoparticle, 34–35
Rinsing-drying dependence, 445–446
Rods. *See* Nanorods

S

Salt content effect on nanoparticles, 446–449
Scanning electron microscopy (SEM), 147, *149*, 303, 319
    high resolution, 214–216, 361
Scanning-probe microscopy (SPM), 37–38
Semiconductors
    attached to surfaces of nanowires, 52–53
    electrical properties of nanowire, 422

group II-VI, 81–83
group III-V, 84–85
group IV, 85–87
nanocrystals, 207–221, 282
nanoparticles as fluorescent probes,
    250
one-dimensional assembly of, 35–36
photogenerated charge transfer and transport
    in, 450
Sensors and biosensors
    field, 29–30
    nanowires in, 28–30
    thin film nanoparticles in, 21–23
Shape control
    double hydrophilic block copolymers in,
        299–301
    inorganic nanocrystals, 271–273,
        315–316
    low-dimensional semiconductor nanocrystal,
        282
    magnetic nanocrystals, 351–352
    of metal nanocrystals, 283–284
    small-molecule additive-mediated
        inorganic species, 271–273
        organic species, 273–275
        solvent effect on, 275–280
Silicon dots, erbium-doped
    energy transfer in, 140–141
    fabrication of, 141
    photophysical measurements of, 145
    structural characterization of, 141–144
Silicon nanowires, erbium-doped
    fabrication of, 145–149
    photophysical measurements of, 149–150
    scanning electron microscopy (SEM) of, 147,
        149
Silver
    AgI@TiO$_2$ formation from, 535
    deposition on solid substrates, 528–530
    nanoparticles
        coated, 526–527
        coherent plasmon coupling
            angular dependence of, 500–501
            effect of interparticle spacing on,
                498–500
            external dielectric medium effect on,
                502–503
            and extinction spectra of, 496–498
            and optical properties of interacting,
                494–495
            substrate effects on optical properties,
                503–506
        stabilization through embedding into
            polymer matrix, 492–494
    synthesis in DMF, 530–533

two-dimensional arrays of
    fabrication methods, 488–489
    on PVP-modified flat surface, 489–492
and nanostructured thin films through layer-by-
        layer assembly, 535–538
organic solvents as reducing agents for, 527
reduction with DMF, 527–530
@SiO$_2$ core-shell colloids, 530–531
@TiO$_2$ core-shell colloids, 532–533
Single-electron transistors, 6, 7
Single-electron tunneling (SET) effect, 5–6, 12
SiO$_2$ nanotubes, 469–473
Size tuning of magnetic nanocrystals, 343–350
Small-molecule additive-mediated shape control
    inorganic species, 271–273
    organic species, 273–275
    solvent effects on, 275–280
Solvent effects on crystal shape, 275–280
Spatial properties, interparticle, 551–553
    branching morphology, 567–568
    implications of, 569–571
    ordered array, 564–567
    in two-dimensional ordered arrays of
            nanoparticles, 611–613
Spectroscopy
    extended x-ray absorption fine-structure
        (EXAFS), 144
    of gold nanorods, 126–127
    of nanowires, 430–431
    time-resolved, 115–119
Sphere-to-rod-to-dumbbell-to-sphere
            transformation mechanism,
            310–313
Spheres
    calcite hollow, 313–314, 315
    gold nanoparticle, 195–198
    magnetic nanoparticle, 351–352
    nanocrystal, 77–79
Spontaneous organization, 323–324
Streptavidin (STV) protein, 231–232, 237,
            238–240, 244–246, 247
Strongly interacting aggregates of gold (SIAs),
            200–203
Structural properties, interparticle, 559–563
Superlattices
    magnetic nanocrystal, 356–360
    metal nanoparticle, 200–203
    two-dimensional nanoparticle cross-linked
        networking, 581–584
Superstructures
    BaCrO$_4$, 302–310
    built with nanobuilding blocks, 96
    crystalline, 186–188
    double hydrophilic block copolymers in
        assembly of, 301–310

of magnetic nanocrystals, 356–364
of nanowires and nanoparticles, 52–53
self-assembly of, 38–41, 42–43
Supramolecular chemistry, 227–228
Supramolecular polymers, 288–289
Supramolecular self-assembly
supramolecular polymers used in,
288–289
transcription from organic matrix template,
289–291
Surface(s)
-particle interactions, 426–427
chemical modification of nanowire,
417–418
density, salt content effect on, 446–449
diffusion of nanowires, 428–430
ligand exchange and phase transfer, 355
nanofabrications, nanoparticle cross-linked
networks for, 587–589
plasma resonance (SPR) and hot electron
relaxation dynamics, 194–198
plasmon (SP) resonance band, 555–557
PVP-modified flat surface, 489–492
Surfactant-mediated templating and crystallization
micelles in, 280–284
microemulsions in, 284–285
monolayers in, 285–288
vesicles in, 288
Synthetic opals, 41–42

T

Template
growth of nanowires, 415–418
synthesis of inverse opals, 361–363, *364*
three-dimensional assemblies of nanoparticles,
41–42
Thermally activated size processing of
nanoparticles, 554
Thickness control of nanoparticle assembly,
555–557
Thin films, nanoparticle
electroactive, 589–590
of electroactive nanoparticles, 589–590
electrochromic devices, 18–21
electroluminescence devices, 13–15
FePt, 369–371
interfacial dynamics of, 590–594
Langmuir-Blodgett, 577–581
layer-by-layer assembly of nanostructured,
535–538
photoelectronic devices, 12–13
photovoltaic devices, 15–18
sensors and biosensors, 21–23
single-electron charging, 12

Three-dimensional assemblies of nanoparticles,
41–47
ordered superlattices, 356–360
Time-resolved spectroscopy
of gold nanorods, 126–127
of nanocrystals, 115–119
TiO_2
-coated nanoparticles, complex nanostructures
from, 533–538
core-shell colloids, Ag@, 532–533
formation of AgI@, 535
nanoshells, 534
TOPO/TBP surfactant used in PbSe nanocrystal
growth, 212–221, 352
Transcription from organic matrix template,
289–291
Transient differential absorption (TDA), 105
Transient hold burning, 105
Transition dipole moment, 183–185
Transition from fcc to $L1_0$ phase in FePt
nanocrystals, 355
Transition metal oxides, 89–90
Transmission function of arrays
computational results for, 164–169
matrix elements working expression, 161–164
measurement of, 155–157, 158–160
mechanisms for conduction, 160–161
Two-dimensional assemblies of nanoparticles
cross-linked superlattices
controls of, 611–613
networking, 581–584
discs and plates, 90–91
ordered arrays, 37–41, 356–360
silver arrays, 488–494
Two-dimensional nanoparticles
cross-linked superlattices
effects of protecting monolayers on,
584–587

V

Vapor-liquid-solid (VLS) method, 51
Vesicles, 288
Vibrational modes, gold nanorods
hole-burning experiments and, 133–135
laser-induced heating and coherent excitation
of, 128–130
period versus probe laser wavelength and rod
length, 130–133

W

Water-soluble nanoparticles, carboxylate-
modified, 602–603

Weakly interacting aggregates of gold (WIAs),
     200–203
Wells, lithographically defined, 427–428
Wire-like assembly of PbSe nanocrystals

grown with molecular template mechanism,
     209–212
grown with TOPO/TBP surfactant,
     212–221

Milton Keynes UK
Ingram Content Group UK Ltd.
UKHW021933071024
449327UK00022B/1791

9 780367 392284